Study and Solutions Guide

CALCULUS

SEVENTH EDITION

Larson/Hostetler/Edwards

Volume I
Chapters P-10

Bruce H. Edwards
University of Florida

HOUGHTON MIFFLIN COMPANY   Boston   New York

Printed in the United States of America

ISBN 0-618-14922-8

56789-CRS-06 05 04

CONTENTS

Chapter P Preparation for Calculus 1

Chapter 1 Limits and Their Properties 26

Chapter 2 Differentiation 52

Chapter 3 Applications of Differentiation...102

Chapter 4 Integration...176

Chapter 5 Logarithmic, Exponential, and Other Transcendental Functions 217

Chapter 6 Applications of Integration...281

Chapter 7 Integration Techniques, L'Hôpital's Rule, and Improper Integrals.......................329

Chapter 8 Infinite Series...400

Chapter 9 Conics, Parametric Equations, and Polar Coordinates 456

Chapter 10 Vectors and the Geometry of Space 506

PREFACE

This *Study and Solutions Guide* is designed as a supplement to *Calculus*, Seventh Edition, by Ron Larson, Robert P. Hostetler, and Bruce H. Edwards. All references to chapters, theorems, and exercises relate to the main text. Solutions to every odd-numbered exercise in the text are given with all essential algebraic steps included. Although this supplement is not a substitute for good study habits, it can be valuable when incorporated into a well-planned course of study. For suggestions that may assist you in the use of this text, your lecture notes, and this *Guide*, please refer to the student web site for your text at *college.hmco.com*.

I have made every effort to see that the solutions are correct. However, I would appreciate hearing about any errors or other suggestions for improvement. Good luck with your study of calculus.

Bruce H. Edwards
University of Florida
Gainesville, Florida 32611
(be@math.ufl.edu)

CHAPTER P
Preparation for Calculus

Section P.1   Graphs and Models...2

Section P.2   Linear Models and Rates of Change 7

Section P.3   Functions and Their Graphs 14

Section P.4   Fitting Models to Data....................................18

Review Exercises 19

Problem Solving...23

C H A P T E R P
Preparation for Calculus

Section P.1   Graphs and Models

Solutions to Odd-Numbered Exercises

1. $y = -\frac{1}{2}x + 2$

  x-intercept: $(4, 0)$

  y-intercept: $(0, 2)$

  Matches graph (b)

3. $y = 4 - x^2$

  x-intercepts: $(2, 0), (-2, 0)$

  y-intercept: $(0, 4)$

  Matches graph (a)

5. $y = \frac{3}{2}x + 1$

x	-4	-2	0	2	4
y	-5	-2	1	4	7

7. $y = 4 - x^2$

x	-3	-2	0	2	3
y	-5	0	4	0	-5

9. $y = |x + 2|$

x	-5	-4	-3	-2	-1	0	1
y	3	2	1	0	1	2	3

11. $y = \sqrt{x} - 4$

x	0	1	4	9	16
y	-4	-3	-2	-1	0

13.

Xmin = -3
Xmax = 5
Xscl = 1
Ymin = -3
Ymax = 5
Yscl = 1

Note that $y = 4$ when $x = 0$.

15.

(a) $(2, y) = (2, 1.73)$ $\left(y = \sqrt{5 - 2} = \sqrt{3} \approx 1.73\right)$

(b) $(x, 3) = (-4, 3)$ $\left(3 = \sqrt{5 - (-4)}\right)$

17. $y = x^2 + x - 2$

y-intercept: $y = 0^2 + 0 - 2$

$\quad\quad\quad\quad\quad y = -2;\ (0, -2)$

x-intercepts: $0 = x^2 + x - 2$

$\quad\quad\quad\quad\quad 0 = (x + 2)(x - 1)$

$\quad\quad\quad\quad\quad x = -2, 1;\ (-2, 0),\ (1, 0)$

19. $y = x^2\sqrt{25 - x^2}$

y-intercept: $y = 0^2\sqrt{25 - 0^2}$

$\quad\quad\quad\quad\quad y = 0;\ (0, 0)$

x-intercepts: $0 = x^2\sqrt{25 - x^2}$

$\quad\quad\quad\quad\quad 0 = x^2\sqrt{(5 - x)(5 + x)}$

$\quad\quad\quad\quad\quad x = 0, \pm 5;\ (0, 0);\ (\pm 5, 0)$

21. $y = \dfrac{3\left(2 - \sqrt{x}\right)}{x}$

y-intercept: None. x cannot equal 0.

x-intercepts: $0 = \dfrac{3\left(2 - \sqrt{x}\right)}{x}$

$\quad\quad\quad\quad\quad 0 = 2 - \sqrt{x}$

$\quad\quad\quad\quad\quad x = 4;\ (4, 0)$

23. $x^2y - x^2 + 4y = 0$

y-intercept:

$\quad\quad 0^2(y) - 0^2 + 4y = 0$

$\quad\quad\quad\quad\quad\quad y = 0;\ (0, 0)$

x-intercept:

$\quad\quad x^2(0) - x^2 + 4(0) = 0$

$\quad\quad\quad\quad\quad\quad x = 0;\ (0, 0)$

25. Symmetric with respect to the y-axis since

$\quad y = (-x)^2 - 2 = x^2 - 2.$

27. Symmetric with respect to the x-axis since

$\quad (-y)^2 = y^2 = x^3 - 4x.$

29. Symmetric with respect to the origin since

$\quad (-x)(-y) = xy = 4.$

31. $y = 4 - \sqrt{x + 3}$

No symmetry with respect to either axis or the origin.

33. Symmetric with respect to the origin since

$\quad -y = \dfrac{-x}{(-x)^2 + 1}$

$\quad\quad y = \dfrac{x}{x^2 + 1}.$

35. $y = |x^3 + x|$ is symmetric with respect to the y-axis

since $y = |(-x)^3 + (-x)| = |-(x^3 + x)| = |x^3 + x|.$

37. $y = -3x + 2$

Intercepts:

$\left(\tfrac{2}{3}, 0\right),\ (0, 2)$

Symmetry: none

39. $y = \dfrac{x}{2} - 4$

Intercepts:

$(8, 0), (0, -4)$

Symmetry: none

41. $y = 1 - x^2$

Intercepts:

$(1, 0), (-1, 0), (0, 1)$

Symmetry: y-axis

43. $y = (x + 3)^2$

Intercepts:

$(-3, 0), (0, 9)$

Symmetry: none

45. $y = x^3 + 2$

Intercepts:

$\left(- \sqrt[3]{2}, 0\right), (0, 2)$

Symmetry: none

47. $y = x\sqrt{x + 2}$

Intercepts:

$(0, 0), (-2, 0)$

Symmetry: none

Domain: $x \geq -2$

49. $x = y^3$

Intercepts: $(0, 0)$

Symmetry: origin

51. $y = \dfrac{1}{x}$

Intercepts: none

Symmetry: origin

53. $y = 6 - |x|$

Intercepts:

$(0, 6), (-6, 0), (6, 0)$

Symmetry: y-axis

55. $y^2 - x = 9$

$\quad y^2 = x + 9$

$\quad y = \pm\sqrt{x + 9}$

Intercepts:

$(0, 3), (0, -3), (-9, 0)$

Symmetry: x-axis

57. $x + 3y^2 = 6$

$\quad 3y^2 = 6 - x$

$\quad y = \pm\sqrt{2 - \dfrac{x}{3}}$

Intercepts:

$(6, 0), \left(0, \sqrt{2}\right), \left(0, -\sqrt{2}\right)$

Symmetry: x-axis

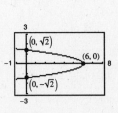

59. $y = (x + 2)(x - 4)(x - 6)$ (other answers possible)

61. Some possible equations:

$$y = x$$
$$y = x^3$$
$$y = 3x^3 - x$$
$$y = \sqrt[3]{x}$$

63. $x + y = 2 \Longrightarrow y = 2 - x$

$2x - y = 1 \Longrightarrow y = 2x - 1$

$2 - x = 2x - 1$

$3 = 3x$

$1 = x$

The corresponding y-value is $y = 1$.

Point of intersection: $(1, 1)$

65. $x + y = 7 \Longrightarrow y = 7 - x$

$3x - 2y = 11 \Longrightarrow y = \dfrac{3x - 11}{2}$

$7 - x = \dfrac{3x - 11}{2}$

$14 - 2x = 3x - 11$

$-5x = -25$

$x = 5$

The corresponding y-value is $y = 2$.

Point of intersection: $(5, 2)$

67. $x^2 + y = 6 \Longrightarrow y = 6 - x^2$

$x + y = 4 \Longrightarrow y = 4 - x$

$6 - x^2 = 4 - x$

$0 = x^2 - x - 2$

$0 = (x - 2)(x + 1)$

$x = 2, -1$

The corresponding y-values are $y = 2$ (for $x = 2$) and $y = 5$ (for $x = -1$).

Points of intersection: $(2, 2), (-1, 5)$

69. $x^2 + y^2 = 5 \Longrightarrow y^2 = 5 - x^2$

$x - y = 1 \Longrightarrow y = x - 1$

$5 - x^2 = (x - 1)^2$

$5 - x^2 = x^2 - 2x + 1$

$0 = 2x^2 - 2x - 4 = 2(x + 1)(x - 2)$

$x = -1 \text{ or } x = 2$

The corresponding y-values are $y = -2$ and $y = 1$.

Points of intersection: $(-1, -2), (2, 1)$

71. $\qquad y = x^3$

$\qquad y = x$

$\qquad x^3 = x$

$x^3 - x = 0$

$\qquad x(x + 1)(x - 1) = 0$

$x = 0, x = -1, \text{ or } x = 1$

The corresponding y-values are $y = 0$, $y = -1$, and $y = 1$.

Points of intersection: $(0, 0), (-1, -1), (1, 1)$

73. $\qquad y = x^3 - 2x^2 + x - 1$

$\qquad y = -x^2 + 3x - 1$

$x^3 - 2x^2 + x - 1 = -x^2 + 3x - 1$

$\qquad x^3 - x^2 - 2x = 0$

$x(x - 2)(x + 1) = 0$

$\qquad x = -1, 0, 2$

$(-1, -5), (0, -1), (2, 1)$

75. $C = R$

$$5.5\sqrt{x} + 10,000 = 3.29x$$

$$\left(5.5\sqrt{x}\right)^2 = (3.29x - 10,000)^2$$

$$30.25x = 10.8241x^2 - 65,800x + 100,000,000$$

$$0 = 10.8241x^2 - 65,830.25x + 100,000,000 \qquad \text{Use the Quadratic Formula.}$$

$$x \approx 3133 \text{ units}$$

The other root, $x \approx 2949$, does not satisfy the equation $R = C$.

This problem can also be solved by using a graphing utility and finding the intersection of the graphs of C and R.

77. (a) Using a graphing utility, you obtain

$$y = -0.0153t^2 + 4.9971t + 34.9405.$$

(b)

(c) For the year 2004, $t = 34$ and
$y \approx 187.2$ CPI.

79. $y = \dfrac{10,770}{x^2} - 0.37$

If the diameter is doubled, the resistance is changed by approximately a factor of $\frac{1}{4}$. For instance, $y(20) \approx 26.555$ and $y(40) \approx 6.36125$.

81. False; x-axis symmetry means that if $(1, -2)$ is on the graph, then $(1, 2)$ is also on the graph.

83. True; the x-intercepts are

$$\left(\frac{-b \pm \sqrt{b^2 - 4ac}}{2a}, 0\right).$$

85.   Distance from the origin $= K \times$ Distance from $(2, 0)$

$$\sqrt{x^2 + y^2} = K\sqrt{(x - 2)^2 + y^2}, K \neq 1$$

$$x^2 + y^2 = K^2(x^2 - 4x + 4 + y^2)$$

$$(1 - K^2)x^2 + (1 - K^2)y^2 + 4K^2x - 4K^2 = 0$$

Note: This is the equation of a circle!

Section P.2   Linear Models and Rates of Change

1. $m = 1$

3. $m = 0$

5. $m = -12$

7.

9. $m = \dfrac{2 - (-4)}{5 - 3}$

$= \dfrac{6}{2} = 3$

11. $m = \dfrac{5 - 1}{2 - 2}$

$= \dfrac{4}{0}$

undefined

13. $m = \dfrac{2/3 - 1/6}{-1/2 - (-3/4)}$

$= \dfrac{1/2}{1/4} = 2$

15. Since the slope is 0, the line is horizontal and its equation is $y = 1$. Therefore, three additional points are $(0, 1)$, $(1, 1)$, and $(3, 1)$.

17. The equation of this line is

$$y - 7 = -3(x - 1)$$

$$y = -3x + 10.$$

Therefore, three additional points are $(0, 10)$, $(2, 4)$, and $(3, 1)$.

19. Given a line L, you can use any two distinct points to calculate its slope. Since a line is straight, the ratio of the change in y-values to the change in x-values will always be the same. See Section P.2 Exercise 93 for a proof.

21. (a)

(b) The slopes of the line segments are

$$\frac{255.0 - 252.1}{2 - 1} = 2.9$$

$$\frac{257.7 - 255.0}{3 - 2} = 2.7$$

$$\frac{260.3 - 257.7}{4 - 3} = 2.6$$

$$\frac{262.8 - 260.3}{5 - 4} = 2.5$$

$$\frac{265.2 - 262.8}{6 - 5} = 2.4$$

$$\frac{267.7 - 265.2}{7 - 6} = 2.5$$

$$\frac{270.3 - 267.7}{8 - 7} = 2.6$$

The population increased most rapidly from 1991 to 1992.

$(m = 2.9)$

23. $x + 5y = 20$

$$y = -\frac{1}{5}x + 4$$

Therefore, the slope is $m = -\frac{1}{5}$ and the y-intercept is $(0, 4)$.

25. $x = 4$

The line is vertical. Therefore, the slope is undefined and there is no y-intercept.

27. $y = \frac{3}{4}x + 3$

$4y = 3x + 12$

$0 = 3x - 4y + 12$

29. $y = \frac{2}{3}x$

$3y = 2x$

$2x - 3y = 0$

31. $y + 2 = 3(x - 3)$

$y + 2 = 3x - 9$

$y = 3x - 11$

$y - 3x + 11 = 0$

33. $m = \dfrac{6 - 0}{2 - 0} = 3$

$y - 0 = 3(x - 0)$

$y = 3x$

35. $m = \dfrac{1 - (-3)}{2 - 0} = 2$

$y - 1 = 2(x - 2)$

$y - 1 = 2x - 4$

$0 = 2x - y - 3$

37. $m = \dfrac{8 - 0}{2 - 5} = -\dfrac{8}{3}$

$$y - 0 = -\frac{8}{3}(x - 5)$$

$$y = -\frac{8}{3}x + \frac{40}{3}$$

$3y + 8x - 40 = 0$

39. $m = \dfrac{8 - 1}{5 - 5}$ Undefined.

Vertical line $x = 5$

41. $m = \dfrac{7/2 - 3/4}{1/2 - 0} = \dfrac{11/4}{1/2} = \dfrac{11}{2}$

$$y - \frac{3}{4} = \frac{11}{2}(x - 0)$$

$$y = \frac{11}{2}x + \frac{3}{4}$$

$$22x - 4y + 3 = 0$$

43. $x = 3$

$x - 3 = 0$

45. $\dfrac{x}{2} + \dfrac{y}{3} = 1$

$3x + 2y - 6 = 0$

47. $\dfrac{x}{a} + \dfrac{y}{a} = 1$

$$\frac{1}{a} + \frac{2}{a} = 1$$

$$\frac{3}{a} = 1$$

$$a = 3 \Rightarrow x + y = 3$$

$$x + y - 3 = 0$$

49. $y = -3$

$y + 3 = 0$

51. $y = -2x + 1$

53. $y - 2 = \frac{3}{2}(x - 1)$

$y = \frac{3}{2}x + \frac{1}{2}$

$2y - 3x - 1 = 0$

55. $2x - y - 3 = 0$

$y = 2x - 3$

57. (a)

The lines do not appear perpendicular.

(b)

The lines appear perpendicular.

The lines are perpendicular because their slopes 1 and -1 are negative reciprocals of each other.
You must use a square setting in order for perpendicular lines to appear perpendicular.

59. $4x - 2y = 3$

$\qquad y = 2x - \frac{3}{2}$

$\qquad m = 2$

(a) $\qquad y - 1 = 2(x - 2)$

$\qquad\quad y - 1 = 2x - 4$

$\quad 2x - y - 3 = 0$

(b) $\qquad y - 1 = -\frac{1}{2}(x - 2)$

$\qquad 2y - 2 = -x + 2$

$\quad x + 2y - 4 = 0$

61. $5x - 3y = 0$

$\qquad y = \frac{5}{3}x$

$\qquad m = \frac{5}{3}$

(a) $\qquad y - \frac{7}{8} = \frac{5}{3}\left(x - \frac{3}{4}\right)$

$\qquad 24y - 21 = 40x - 30$

$\quad 24y - 40x + 9 = 0$

(b) $\qquad y - \frac{7}{8} = -\frac{3}{5}\left(x - \frac{3}{4}\right)$

$\qquad 40y - 35 = -24x + 18$

$\quad 40y + 24x - 53 = 0$

63. The given line is vertical.

(a) $x = 2 \Longrightarrow x - 2 = 0$

(b) $y = 5 \Longrightarrow y - 5 = 0$

65. The slope is 125. Hence, $V = 125(t - 1) + 2540$

$\qquad\qquad\qquad\qquad = 125t + 2415$

67. The slope is -2000. Hence, $V = -2000(t - 1) + 20,400$

$\qquad\qquad\qquad\qquad\qquad = -2000t + 22,400$

69.

You can use the graphing utility to determine that the points of intersection are $(0, 0)$ and $(2, 4)$. Analytically,

$\qquad\quad x^2 = 4x - x^2$

$\quad 2x^2 - 4x = 0$

$\quad 2x(x - 2) = 0$

$\qquad x = 0 \Longrightarrow y = 0 \Longrightarrow (0, 0)$

$\qquad x = 2 \Longrightarrow y = 4 \Longrightarrow (2, 4).$

The slope of the line joining $(0, 0)$ and $(2, 4)$ is $m = (4 - 0)/(2 - 0) = 2$. Hence, an equation of the line is

$\quad y - 0 = 2(x - 0)$

$\qquad\quad y = 2x.$

71. $m_1 = \dfrac{1 - 0}{-2 - (-1)} = -1$

$m_2 = \dfrac{-2 - 0}{2 - (-1)} = -\dfrac{2}{3}$

$m_1 \neq m_2$

The points are not collinear.

73. Equations of perpendicular bisectors:

$y - \dfrac{c}{2} = \dfrac{a - b}{c}\left(x - \dfrac{a + b}{2}\right)$

$y - \dfrac{c}{2} = \dfrac{a + b}{-c}\left(x - \dfrac{b - a}{2}\right)$

Setting the right-hand sides of the two equations equal and solving for x yields $x = 0$.

Letting $x = 0$ in either equation gives the point of intersection:

$\left(0, \dfrac{-a^2 + b^2 + c^2}{2c}\right)$.

This point lies on the third perpendicular bisector, $x = 0$.

75. Equations of altitudes:

$y = \dfrac{a - b}{c}(x + a)$

$x = b$

$y = -\dfrac{a + b}{c}(x - a)$

Solving simultaneously, the point of intersection is

$\left(b, \dfrac{a^2 - b^2}{c}\right)$.

77. Find the equation of the line through the points $(0, 32)$ and $(100, 212)$.

$m = \dfrac{180}{100} = \dfrac{9}{5}$

$F - 32 = \dfrac{9}{5}(C - 0)$

$F = \dfrac{9}{5}C + 32$

or

$C = \dfrac{1}{9}(5F - 160)$

$5F - 9C - 160 = 0$

For $F = 72°$, $C \approx 22.2°$.

79. (a) $W_1 = 0.75x + 12.50$

$W_2 = 1.30x + 9.20$

(b)

Using a graphing utility, the point of intersection is $(6, 17)$. Analytically,

$0.75x + 12.50 = 1.30x + 9.20$

$3.3 = 0.55x \Rightarrow x = 6$

$y = 0.75(6) + 12.50 = 17$.

(c) Both jobs pay \$17 per hour if 6 units are produced. For someone who can produce more than 6 units per hour, the second offer would pay more. For a worker who produces less than 6 units per hour, the first offer pays more.

81. (a) Two points are (50, 580) and (47, 625). The slope is

$$m = \frac{625 - 580}{47 - 50} = -15.$$

$$p - 580 = -15(x - 50)$$

$$p = -15x + 750 + 580 = -15x + 1330$$

$$\text{or } x = \tfrac{1}{15}(1330 - p)$$

(b)

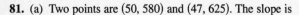

If $p = 655$, $x = \tfrac{1}{15}(1330 - 655) = 45$ units.

(c) If $p = 595$, $x = \tfrac{1}{15}(1330 - 595) = 49$ units.

83. $4x + 3y - 10 = 0 \Rightarrow d = \dfrac{|4(0) + 3(0) - 10|}{\sqrt{4^2 + 3^2}} = \dfrac{10}{5} = 2$

85. $x - y - 2 = 0 \Rightarrow d = \dfrac{|1(-2) + (-1)(1) - 2|}{\sqrt{1^2 + 1^2}} = \dfrac{5}{\sqrt{2}} = \dfrac{5\sqrt{2}}{2}$

87. A point on the line $x + y = 1$ is $(0, 1)$. The distance from the point $(0, 1)$ to $x + y - 5 = 0$ is

$$d = \frac{|1(0) + 1(1) - 5|}{\sqrt{1^2 + 1^2}} = \frac{|1 - 5|}{\sqrt{2}} = \frac{4}{\sqrt{2}} = 2\sqrt{2}.$$

89. If $A = 0$, then $By + C = 0$ is the horizontal line $y = -C/B$. The distance to (x_1, y_1) is

$$d = \left| y_1 - \left(\frac{-C}{B} \right) \right| = \frac{|By_1 + C|}{|B|} = \frac{|Ax_1 + By_1 + C|}{\sqrt{A^2 + B^2}}.$$

If $B = 0$, then $Ax + C = 0$ is the vertical line $x = -C/A$. The distance to (x_1, y_1) is

$$d = \left| x_1 - \left(\frac{-C}{A} \right) \right| = \frac{|Ax_1 + C|}{|A|} = \frac{|Ax_1 + By_1 + C|}{\sqrt{A^2 + B^2}}.$$

(Note that A and B cannot both be zero.)

The slope of the line $Ax + By + C = 0$ is $-A/B$. The equation of the line through (x_1, y_1) perpendicular to $Ax + By + C = 0$ is:

$$y - y_1 = \frac{B}{A}(x - x_1)$$

$$Ay - Ay_1 = Bx - Bx_1$$

$$Bx_1 - Ay_1 = Bx - Ay$$

The point of intersection of these two lines is:

$$Ax + By = -C \quad\quad \Rightarrow \quad A^2x + ABy = -AC \quad\quad\quad (1)$$

$$Bx - Ay = Bx_1 - Ay_1 \Rightarrow \quad \underline{B^2x - ABy = B^2x_1 - ABy_1} \; (2)$$

$$(A^2 + B^2)x = -AC + B^2x_1 - ABy_1 \;\; \text{(By adding equations (1) and (2))}$$

$$x = \frac{-AC + B^2x_1 - ABy_1}{A^2 + B^2}$$

$$Ax + By = -C \quad\quad \Rightarrow \quad ABx + B^2y = -BC \quad\quad\quad (3)$$

$$Bx - Ay = Bx_1 - Ay_1 \Rightarrow \quad \underline{-ABx + A^2y = -ABx_1 + A^2y_1} \; (4)$$

$$(A^2 + B^2)y = -BC - ABx_1 + A^2y_1 \;\; \text{(By adding equations (3) and (4))}$$

$$y = \frac{-BC - ABx_1 + A^2y_1}{A^2 + B^2}$$

—CONTINUED—

89. **—CONTINUED—**

$$\left(\frac{-AC + B^2x_1 - ABy_1}{A^2 + B^2}, \frac{-BC - ABx_1 + A^2y_1}{A^2 + B^2}\right) \text{ point of intersection}$$

The distance between (x_1, y_1) and this point gives us the distance between (x_1, y_1) and the line $Ax + By + C = 0$.

$$
\begin{aligned}
d &= \sqrt{\left[\frac{-AC + B^2x_1 - ABy_1}{A^2 + B^2} - x_1\right]^2 + \left[\frac{-BC - ABx_1 + A^2y_1}{A^2 + B^2} - y_1\right]^2} \\
&= \sqrt{\left[\frac{-AC - ABy_1 - A^2x_1}{A^2 + B^2}\right]^2 + \left[\frac{-BC - ABx_1 - B^2y_1}{A^2 + B^2}\right]^2} \\
&= \sqrt{\left[\frac{-A(C + By_1 + Ax_1)}{A^2 + B^2}\right]^2 + \left[\frac{-B(C + Ax_1 + By_1)}{A^2 + B^2}\right]^2} \\
&= \sqrt{\frac{(A^2 + B^2)(C + Ax_1 + By_1)^2}{(A^2 + B^2)^2}} \\
&= \frac{|Ax_1 + By_1 + C|}{\sqrt{A^2 + B^2}}
\end{aligned}
$$

91. For simplicity, let the vertices of the rhombus be $(0, 0)$, $(a, 0)$, (b, c), and $(a + b, c)$, as shown in the figure. The slopes of the diagonals are then

$$m_1 = \frac{c}{a + b} \text{ and } m_2 = \frac{c}{b - a}.$$

Since the sides of the Rhombus are equal, $a^2 = b^2 + c^2$, and we have

$$m_1 m_2 = \frac{c}{a + b} \cdot \frac{c}{b - a} = \frac{c^2}{b^2 - a^2} = \frac{c^2}{-c^2} = -1.$$

Therefore, the diagonals are perpendicular.

93. Consider the figure below in which the four points are collinear. Since the triangles are similar, the result immediately follows.

$$\frac{y_2{}^* - y_1{}^*}{x_2{}^* - x_1{}^*} = \frac{y_2 - y_1}{x_2 - x_1}$$

95. True.

$$ax + by = c_1 \Rightarrow y = -\frac{a}{b}x + \frac{c_1}{b} \Rightarrow m_1 = -\frac{a}{b}$$

$$bx - ay = c_2 \Rightarrow y = \frac{b}{a}x - \frac{c_2}{a} \Rightarrow m_2 = \frac{b}{a}$$

$$m_2 = -\frac{1}{m_1}$$

Section P.3   Functions and Their Graphs

1. (a) $f(0) = 2(0) - 3 = -3$

  (b) $f(-3) = 2(-3) - 3 = -9$

  (c) $f(b) = 2b - 3$

  (d) $f(x - 1) = 2(x - 1) - 3 = 2x - 5$

3. (a) $g(0) = 3 - 0^2 = 3$

  (b) $g(\sqrt{3}) = 3 - (\sqrt{3})^2 = 3 - 3 = 0$

  (c) $g(-2) = 3 - (-2)^2 = 3 - 4 = -1$

  (d) $g(t - 1) = 3 - (t - 1)^2 = -t^2 + 2t + 2$

5. (a) $f(0) = \cos(2(0)) = \cos 0 = 1$

  (c) $f\left(\dfrac{\pi}{3}\right) = \cos\left(2\left(\dfrac{\pi}{3}\right)\right) = \cos\dfrac{2\pi}{3} = -\dfrac{1}{2}$

  (b) $f\left(-\dfrac{\pi}{4}\right) = \cos\left(2\left(-\dfrac{\pi}{4}\right)\right) = \cos\left(-\dfrac{\pi}{2}\right) = 0$

7. $\dfrac{f(x + \Delta x) - f(x)}{\Delta x} = \dfrac{(x + \Delta x)^3 - x^3}{\Delta x} = \dfrac{x^3 + 3x^2\Delta x + 3x(\Delta x)^2 + (\Delta x)^3 - x^3}{\Delta x} = 3x^2 + 3x\Delta x + (\Delta x)^2, \Delta x \neq 0$

9. $\dfrac{f(x) - f(2)}{x - 2} = \dfrac{(1/\sqrt{x - 1} - 1)}{x - 2}$

$= \dfrac{1 - \sqrt{x - 1}}{(x - 2)\sqrt{x - 1}} \cdot \dfrac{1 + \sqrt{x - 1}}{1 + \sqrt{x - 1}} = \dfrac{2 - x}{(x - 2)\sqrt{x - 1}(1 + \sqrt{x - 1})} = \dfrac{-1}{\sqrt{x - 1}(1 + \sqrt{x - 1})}, x \neq 2$

11. $h(x) = -\sqrt{x + 3}$

Domain: $x + 3 \geq 0 \Rightarrow [-3, \infty)$

Range: $(-\infty, 0]$

13. $f(t) = \sec\dfrac{\pi t}{4}$

$\dfrac{\pi t}{4} \neq \dfrac{(2k + 1)\pi}{2} \Rightarrow t \neq 4k + 2$

Domain: all $t \neq 4k + 2$, k an integer

Range: $(-\infty, -1], [1, \infty)$

15. $f(x) = \dfrac{1}{x}$

Domain: $(-\infty, 0), (0, \infty)$

Range: $(-\infty, 0), (0, \infty)$

17. $f(x) = \begin{cases} 2x + 1, x < 0 \\ 2x + 2, x \geq 0 \end{cases}$

  (a) $f(-1) = 2(-1) + 1 = -1$

  (b) $f(0) = 2(0) + 2 = 2$

  (c) $f(2) = 2(2) + 2 = 6$

  (d) $f(t^2 + 1) = 2(t^2 + 1) + 2 = 2t^2 + 4$

  (Note: $t^2 + 1 \geq 0$ for all t)

  Domain: $(-\infty, \infty)$

  Range: $(-\infty, 1), [2, \infty)$

19. $f(x) = \begin{cases} |x| + 1, x < 1 \\ -x + 1, x \geq 1 \end{cases}$

  (a) $f(-3) = |-3| + 1 = 4$

  (b) $f(1) = -1 + 1 = 0$

  (c) $f(3) = -3 + 1 = -2$

  (d) $f(b^2 + 1) = -(b^2 + 1) + 1 = -b^2$

  Domain: $(-\infty, \infty)$

  Range: $(-\infty, 0] \cup [1, \infty)$

21. $f(x) = 4 - x$

Domain: $(-\infty, \infty)$

Range: $(-\infty, \infty)$

23. $h(x) = \sqrt{x - 1}$

Domain: $[1, \infty)$

Range: $[0, \infty)$

25. $f(x) = \sqrt{9 - x^2}$

Domain: $[-3, 3]$

Range: $[0, 3]$

27. $g(t) = 2 \sin \pi t$

Domain: $(-\infty, \infty)$

Range: $[-2, 2]$

29. $x - y^2 = 0 \Longrightarrow y = \pm\sqrt{x}$

y is not a function of x. Some vertical lines intersect the graph twice.

31. y is a function of x. Vertical lines intersect the graph at most once.

33. $x^2 + y^2 = 4 \Longrightarrow y = \pm\sqrt{4 - x^2}$

y is not a function of x since there are two values of y for some x.

35. $y^2 = x^2 - 1 \Longrightarrow y = \pm\sqrt{x^2 - 1}$

y is not a function of x since there are two values of y for some x.

37. $f(x) = |x| + |x - 2|$

If $x < 0$, then $f(x) = -x - (x - 2) = -2x + 2 = 2(1 - x)$.

If $0 \le x < 2$, then $f(x) = x - (x - 2) = 2$.

If $x \ge 2$, then $f(x) = x + (x - 2) = 2x - 2 = 2(x - 1)$.

Thus,

$$f(x) = \begin{cases} 2(1 - x), & x < 0 \\ 2, & 0 \le x < 2 \\ 2(x - 1), & x \ge 2 \end{cases}.$$

39. The function is $g(x) = cx^2$. Since $(1, -2)$ satisfies the equation, $c = -2$. Thus, $g(x) = -2x^2$.

41. The function is $r(x) = c/x$, since it must be undefined at $x = 0$. Since $(1, 32)$ satisfies the equation, $c = 32$. Thus, $r(x) = 32/x$.

43. (a) For each time t, there corresponds a depth d.

(b) Domain: $0 \le t \le 5$

Range: $0 \le d \le 30$

(c)

45.

47. (a) The graph is shifted
3 units to the left.

(b) The graph is shifted
1 unit to the right.

(c) The graph is shifted
2 units upward.

(d) The graph is shifted
4 units downward.

(e) The graph is stretched
vertically by a factor of 3.

(f) The graph is stretched
vertically by a factor
of $\frac{1}{4}$.

49. (a) $y = \sqrt{x} + 2$

Vertical shift 2 units upward

(b) $y = -\sqrt{x}$

Reflection about the x-axis

(c) $y = \sqrt{x - 2}$

Horizontal shift 2 units to the
right

51. (a) $T(4) = 16°$, $T(15) \approx 23°$

(b) If $H(t) = T(t - 1)$, then the program would turn on (and off) one hour later.

(c) If $H(t) = T(t) - 1$, then the overall temperature would be reduced 1 degree.

53. $f(x) = x^2, g(x) = \sqrt{x}$

$(f \circ g)(x) = f(g(x)) = f(\sqrt{x}) = (\sqrt{x})^2 = x, \quad x \geq 0$

Domain: $[0, \infty)$

$(g \circ f)(x) = g(f(x)) = g(x^2) = \sqrt{x^2} = |x|$

Domain: $(-\infty, \infty)$

No. Their domains are different. $(f \circ g) = (g \circ f)$ for $x \geq 0$.

55. $f(x) = \frac{3}{x}, g(x) = x^2 - 1$

$(f \circ g)(x) = f(g(x)) = f(x^2 - 1) = \frac{3}{x^2 - 1}$

Domain: all $x \neq \pm 1$

$(g \circ f)(x) = g(f(x)) = g\left(\frac{3}{x}\right) = \left(\frac{3}{x}\right)^2 - 1 = \frac{9}{x^2} - 1 = \frac{9 - x^2}{x^2}$

Domain: all $x \neq 0$

No, $f \circ g \neq g \circ f$.

57. $(A \circ r)(t) = A(r(t)) = A(0.6t) = \pi(0.6t)^2 = 0.36\pi t^2$

$(A \circ r)(t)$ represents the area of the circle at time t.

59. $f(-x) = (-x)^2(4 - (-x)^2) = x^2(4 - x^2) = f(x)$

Even

61. $f(-x) = (-x)\cos(-x) = -x\cos x = -f(x)$

Odd

63. (a) If f is even, then $\left(\frac{3}{2}, 4\right)$ is on the graph.

(b) If f is odd, then $\left(\frac{3}{2}, -4\right)$ is on the graph.

65. $f(-x) = a_{2n+1}(-x)^{2n+1} + \cdots + a_3(-x)^3 + a_1(-x)$

$\qquad = -[a_{2n+1}x^{2n+1} + \cdots + a_3x^3 + a_1x]$

$\qquad = -f(x)$

Odd

67. Let $F(x) = f(x)g(x)$ where f and g are even. Then

$\qquad F(-x) = f(-x)g(-x) = f(x)g(x) = F(x).$

Thus, $F(x)$ is even. Let $F(x) = f(x)g(x)$ where f and g are odd. Then

$\qquad F(-x) = f(-x)g(-x) = [-f(x)][-g(x)] = f(x)g(x) = F(x).$

Thus, $F(x)$ is even.

69. $f(x) = x^2 + 1$ and $g(x) = x^4$ are even.

$f(x)g(x) = (x^2 + 1)(x^4) = x^6 + x^4$ is even.

$f(x) = x^3 - x$ is odd and $g(x) = x^2$ is even.

$f(x)g(x) = (x^3 - x)(x^2) = x^5 - x^3$ is odd.

71. (a)

x	length and width	volume V
1	$24 - 2(1)$	484
2	$24 - 2(2)$	800
3	$24 - 2(3)$	972
4	$24 - 2(4)$	1024
5	$24 - 2(5)$	980
6	$24 - 2(6)$	864

The maximum volume appears to be 1024 cm³.

(c) $V = x(24 - 2x)^2 = 4x(12 - x)^2$

Domain: $0 < x < 12$

(b)

Yes, V is a function of x.

(d)

Maximum volume is $V = 1024$ cm³ for box having dimensions $4 \times 16 \times 16$ cm.

73. False; let $f(x) = x^2$.

Then $f(-3) = f(3) = 9$, but $-3 \neq 3$.

75. True, the function is even.

Section P.4   Fitting Models to Data

1. Quadratic function

3. Linear function

5. (a), (b)

Yes. The cancer mortality increases linearly with increased exposure to the carcinogenic substance.

(c) If $x = 3$, then $y \approx 136$.

7. (a) $d = 0.066F$ or $F = 15.1d + 0.1$

(b)

The model fits well.

(c) If $F = 55$, then $d \approx 0.066(55) = 3.63$ cm.

9. (a) Let x = per capita energy usage (in millions of Btu)

$\qquad\qquad$ y = per capita gross national product (in thousands)

$\qquad$ $y = 0.0764x + 4.9985 \approx 0.08x + 5.0$

$\qquad$ $r = 0.7052$

(b)

(c) Denmark, Japan, and Canada

(d) Deleting the data for the three countries above,

$\qquad$ $y = 0.0959x + 1.0539$

$\qquad$ ($r = 0.9202$ is much closer to 1.)

11. (a) $y_1 = 0.0343t^3 - 0.3451t^2 + 0.8837t + 5.6061$

$\qquad$ $y_2 = 0.1095t + 2.0667$

$\qquad$ $y_3 = 0.0917t + 0.7917$

(b)

For 2002, $t = 12$ and $y_1 + y_2 + y_3 \approx 31.06$ cents/mile

13. (a) $y_1 = 4.0367t + 28.9644$

$\qquad$ $y_2 = -0.0099t^3 + 0.5488t^2 + 0.2399t + 33.1414$

(b)

(c) The cubic model is better.

(d) $y_3 = 0.4297t^2 + 0.5994t + 32.9745$

(e) The slope represents the average increase per year in the number of people (in millions) in HMOs.

(f) For 2000, $t = 10$, and $y_1 \approx 69.3$ million. (linear)

$\qquad$ $y_2 \approx 80.5$ million (cubic)

15. (a) $y = -1.81x^3 + 14.58x^2 + 16.39x + 10$

(b)

(c) If $x = 4.5$, $y \approx 214$ horsepower.

17. (a) Yes, y is a function of t. At each time t, there is one and only one displacement y.

(b) The amplitude is approximately

$$(2.35 - 1.65)/2 = 0.35.$$

The period is approximately

$$2(0.375 - 0.125) = 0.5.$$

(c) One model is $y = 0.35 \sin(4\pi t) + 2$.

(d)

19. Answers will vary.

Review Exercises for Chapter P

1. $y = 2x - 3$

$x = 0 \Rightarrow y = 2(0) - 3 = -3 \Rightarrow (0, -3)$ y-intercept

$y = 0 \Rightarrow 0 = 2x - 3 \Rightarrow x = \frac{3}{2} \Rightarrow \left(\frac{3}{2}, 0\right)$ x-intercept

3. $y = \dfrac{x - 1}{x - 2}$

$x = 0 \Rightarrow y = \dfrac{0 - 1}{0 - 2} = \dfrac{1}{2} \Rightarrow \left(0, \dfrac{1}{2}\right)$ y-intercept

$y = 0 \Rightarrow 0 = \dfrac{x - 1}{x - 2} \Rightarrow x = 1 \Rightarrow (1, 0)$ x-intercept

5. Symmetric with respect to y-axis since

$$(-x)^2 y - (-x)^2 + 4y = 0$$
$$x^2 y - x^2 + 4y = 0.$$

7. $y = -\frac{1}{2}x + \frac{3}{2}$

9. $-\frac{1}{3}x + \frac{5}{6}y = 1$

$-\frac{2}{5}x + y = \frac{6}{5}$

$y = \frac{2}{5}x + \frac{6}{5}$

Slope: $\frac{2}{5}$

y-intercept: $\frac{6}{5}$

11. $y = 7 - 6x - x^2$

13. $y = \sqrt{5 - x}$

Domain: $(-\infty, 5]$

15. $y = 4x^2 - 25$

Xmin = -5
Xmax = 5
Xscl = 1
Ymin = -30
Ymax = 10
Yscl = 5

17. $3x - 4y = 8$

$\dfrac{4x + 4y = 20}{7x \qquad = 28}$

$x = 4$

$y = 1$

Point: $(4, 1)$

19. You need factors $(x + 2)$ and $(x - 2)$. Multiply by x to obtain origin symmetry

$y = x(x + 2)(x - 2)$

$= x^3 - 4x.$

21.

Slope $= \dfrac{(5/2) - 1}{5 - (3/2)} = \dfrac{3/2}{7/2} = \dfrac{3}{7}$

23. $\dfrac{1 - t}{1 - 0} = \dfrac{1 - 5}{1 - (-2)}$

$1 - t = -\dfrac{4}{3}$

$t = \dfrac{7}{3}$

25. $\quad y - (-5) = \tfrac{3}{2}(x - 0)$

$\quad\quad\quad y = \tfrac{3}{2}x - 5$

$2y - 3x + 10 = 0$

27. $\quad\quad y - 0 = -\tfrac{2}{3}(x - (-3))$

$\quad\quad\quad\quad y = -\tfrac{2}{3}x - 2$

$3y + 2x + 6 = 0$

29. (a) $y - 4 = \dfrac{7}{16}(x + 2)$

$16y - 64 = 7x + 14$

$0 = 7x - 16y + 78$

(c) $m = \dfrac{4 - 0}{-2 - 0} = -2$

$y = -2x$

$2x + y = 0$

(b) Slope of line is $\dfrac{5}{3}$.

$y - 4 = \dfrac{5}{3}(x + 2)$

$3y - 12 = 5x + 10$

$0 = 5x - 3y + 22$

(d) $x = -2$

$x + 2 = 0$

31. The slope is -850. $V = -850t + 12,500$.

$V(3) = -850(3) + 12,500 = \$9950$

33. $x - y^2 = 0$

$y = \pm\sqrt{x}$

Not a function of x since there are two values of y for some x.

35. $y = x^2 - 2x$

Function of x since there is one value of y for each x.

37. $f(x) = \dfrac{1}{x}$

(a) $f(0)$ does not exist.

(b) $\dfrac{f(1 + \Delta x) - f(1)}{\Delta x} = \dfrac{\dfrac{1}{1 + \Delta x} - \dfrac{1}{1}}{\Delta x} = \dfrac{1 - 1 - \Delta x}{(1 + \Delta x)\Delta x}$

$= \dfrac{-1}{1 + \Delta x}, \; \Delta x \neq -1, 0$

39. (a) Domain: $36 - x^2 \geq 0 \Longrightarrow -6 \leq x \leq 6$ or $[-6, 6]$

Range: $[0, 6]$

(b) Domain: all $x \neq 5$ or $(-\infty, 5), (5, \infty)$

Range: all $y \neq 0$ or $(-\infty, 0), (0, \infty)$

(c) Domain: all x or $(-\infty, \infty)$

Range: all y or $(-\infty, \infty)$

41. (a) $f(x) = x^3 + c, c = -2, 0, 2$

(b) $f(x) = (x - c)^3, c = -2, 0, 2$

41. —CONTINUED—

(c) $f(x) = (x - 2)^3 + c, c = -2, 0, 2$

(d) $f(x) = cx^3, c = -2, 0, 2$

43. (a) Odd powers: $f(x) = x, g(x) = x^3, h(x) = x^5$

Even powers: $f(x) = x^2, g(x) = x^4, h(x) = x^6$

The graphs of f, g, and h all rise to the right and fall to the left. As the degree increases, the graph rises and falls more steeply. All three graphs pass through the points $(0, 0)$, $(1, 1)$, and $(-1, -1)$.

The graphs of f, g, and h all rise to the left and to the right. As the degree increases, the graph rises more steeply. All three graphs pass through the points $(0, 0)$, $(1, 1)$, and $(-1, 1)$.

(b) $y = x^7$ will look like $h(x) = x^5$, but rise and fall even more steeply.

$y = x^8$ will look like $h(x) = x^6$, but rise even more steeply.

45. (a)

$2x + 2y = 24$

$y = 12 - x$

$A = xy = x(12 - x) = 12x - x^2$

(b) Domain: $0 < x < 12$

(c) Maximum area is $A = 36$. In general, the maximum area is attained when the rectangle is a square. In this case, $x = 6$.

47. (a) 3 (cubic), negative leading coefficient

(b) 4 (quartic), positive leading coefficient

(c) 2 (quadratic), negative leading coefficient

(d) 5, positive leading coefficient

49. (a) Yes, y is a function of t. At each time t, there is one and only one displacement y.

(b) The amplitude is approximately

$$(0.25 - (-0.25))/2 = 0.25.$$

The period is approximately 1.1.

(c) One model is $y = \dfrac{1}{4} \cos\left(\dfrac{2\pi}{1.1}t\right) \approx \dfrac{1}{4} \cos(5.7t)$

(d)

Problem Solving for Chapter P

1. (a)
$$x^2 - 6x + y^2 - 8y = 0$$
$$(x^2 - 6x + 9) + (y^2 - 8y + 16) = 9 + 16$$
$$(x - 3)^2 + (y - 4)^2 = 25$$

Center: $(3, 4)$ Radius: 5

(c) Slope of line from $(6, 0)$ to $(3, 4)$ is $\dfrac{4 - 0}{3 - 6} = -\dfrac{4}{3}$.

Slope of tangent line is $\dfrac{3}{4}$. Hence,

$$y - 0 = \frac{3}{4}(x - 6) \implies y = \frac{3}{4}x - \frac{9}{2} \quad \text{Tangent line}$$

(b) Slope of line from $(0, 0)$ to $(3, 4)$ is $\dfrac{4}{3}$ Slope of tangent line is $-\dfrac{3}{4}$ Hence,

$$y - 0 = -\frac{3}{4}(x - 0) \implies y = -\frac{3}{4}x \quad \text{Tangent line}$$

(d) $-\dfrac{3}{4}x = \dfrac{3}{4}x - \dfrac{9}{2}$

$\dfrac{3}{2}x = \dfrac{9}{2}$

$x = 3$

Intersection: $\left(3, -\dfrac{9}{4}\right)$

3. $H(x) = \begin{cases} 1 & x \geq 0 \\ 0 & x < 0 \end{cases}$

(a) $H(x) - 2$

(b) $H(x - 2)$

(c) $-H(x)$

(d) $H(-x)$

(e) $\frac{1}{2}H(x)$

(f) $-H(x - 2) + 2$

5. (a) $x + 2y = 100 \implies y = \dfrac{100 - x}{2}$

$$A(x) = xy = x\left(\dfrac{100 - x}{2}\right) = -\dfrac{x^2}{2} + 50x$$

Domain: $0 < x < 100$

(b)

Maximum of 1250 m^2 at $x = 50$ m, $y = 25$ m.

(c) $A(x) = -\dfrac{1}{2}(x^2 - 100x)$

$\qquad = -\dfrac{1}{2}(x^2 - 100x + 2500) + 1250$

$\qquad = -\dfrac{1}{2}(x - 50)^2 + 1250$

$A(50) = 1250$ m^2 is the maximum. $x = 50$ m, $y = 25$ m.

9. (a) Slope $= \dfrac{9 - 4}{3 - 2} = 5$. Slope of tangent line is less than 5.

(b) Slope $= \dfrac{4 - 1}{2 - 1} = 3$. Slope of tangent line is greater than 3.

(c) Slope $= \dfrac{4.41 - 4}{2.1 - 2} = 4.1$. Slope of tangent line is less than 4.1.

(d) Slope $= \dfrac{f(2 + h) - f(2)}{(2 + h) - 2}$

$\qquad = \dfrac{(2 + h)^2 - 4}{h}$

$\qquad = \dfrac{4h + h^2}{h}$

$\qquad = 4 + h,\ h \neq 0$

(e) Letting h get closer and closer to 0, the slope approaches 4. Hence, the slope at $(2, 4)$ is 4.

11. (a) $\qquad \dfrac{I}{x^2} = \dfrac{2I}{(x - 3)^2}$

$x^2 - 6x + 9 = 2x^2$

$x^2 + 6x - 9 = 0$

$\qquad x = \dfrac{-6 \pm \sqrt{36 + 36}}{2} = -3 \pm \sqrt{18} \approx 1.2426, -7.2426$

(b) $\qquad \dfrac{I}{x^2 + y^2} = \dfrac{2I}{(x - 3)^2 + y^2}$

$\qquad (x - 3)^2 + y^2 = 2(x^2 + y^2)$

$x^2 - 6x + 9 + y^2 = 2x^2 + 2y^2$

$x^2 + y^2 + 6x - 9 = 0$

$\qquad (x + 3)^2 + y^2 = 18$

Circle of radius $\sqrt{18}$ and center $(-3, 0)$.

7. The length of the trip in the water is $\sqrt{2^2 + x^2}$, and the length of the trip over land is $\sqrt{1 + (3 - x)^2}$. Hence, the total time is

$$T = \dfrac{\sqrt{4 + x^2}}{2} + \dfrac{\sqrt{1 + (3 - x)^2}}{4} \text{ hours.}$$

13.

$$d_1 d_2 = 1$$

$$[(x + 1)^2 + y^2][(x - 1)^2 + y^2] = 1$$

$$(x + 1)^2(x - 1)^2 + y^2[(x + 1)^2 + (x - 1)^2] + y^4 = 1$$

$$(x^2 - 1)^2 + y^2[2x^2 + 2] + y^4 = 1$$

$$x^4 - 2x^2 + 1 + 2x^2y^2 + 2y^2 + y^4 = 1$$

$$(x^4 + 2x^2y^2 + y^4) - 2x^2 + 2y^2 = 0$$

$$(x^2 + y^2)^2 = 2(x^2 - y^2)$$

Let $y = 0$. Then $x^4 = 2x^2 \implies x = 0$ or $x^2 = 2$.

Thus, $(0, 0), \left(\sqrt{2}, 0\right)$ and $\left(-\sqrt{2}, 0\right)$ are on the curve.

C H A P T E R 1
Limits and Their Properties

Section 1.1   A Preview of Calculus 27

Section 1.2   Finding Limits Graphically and Numerically 27

Section 1.3   Evaluating Limits Analytically 31

Section 1.4   Continuity and One-Sided Limits 37

Section 1.5   Infinite Limits 42

Review Exercises 47

Problem Solving 49

C H A P T E R 1
Limits and Their Properties

Section 1.1   A Preview of Calculus

Solutions to Odd-Numbered Exercises

1. Precalculus: $(20 \text{ ft/sec})(15 \text{ seconds}) = 300$ feet

3. Calculus required: slope of tangent line at $x = 2$ is rate of change, and equals about 0.16.

5. Precalculus: Area $= \frac{1}{2}bh = \frac{1}{2}(5)(3) = \frac{15}{2}$ sq. units

7. Precalculus: Volume $= (2)(4)(3) = 24$ cubic units

9. (a)

(b) The graphs of y_2 are approximations to the tangent line to y_1 at $x = 1$.

(c) The slope is approximately 2. For a better approximation make the list numbers smaller:
$\{0.2, 0.1, 0.01, 0.001\}$

11. (a) $D_1 = \sqrt{(5 - 1)^2 + (1 - 5)^2} = \sqrt{16 + 16} \approx 5.66$

(b) $D_2 = \sqrt{1 + \left(\frac{5}{2}\right)^2} + \sqrt{1 + \left(\frac{5}{2} - \frac{5}{3}\right)^2} + \sqrt{1 + \left(\frac{5}{3} - \frac{5}{4}\right)^2} + \sqrt{1 + \left(\frac{5}{4} - 1\right)^2}$

$\approx 2.693 + 1.302 + 1.083 + 1.031 \approx 6.11$

(c) Increase the number of line segments.

Section 1.2   Finding Limits Graphically and Numerically

1.

x	1.9	1.99	1.999	2.001	2.01	2.1
$f(x)$	0.3448	0.3344	0.3334	0.3332	0.3322	0.3226

$$\lim_{x \to 2} \frac{x - 2}{x^2 - x - 2} \approx 0.3333 \quad \left(\text{Actual limit is } \tfrac{1}{3}.\right)$$

3.

x	-0.1	-0.01	-0.001	0.001	0.01	0.1
$f(x)$	0.2911	0.2889	0.2887	0.2887	0.2884	0.2863

$$\lim_{x \to 0} \frac{\sqrt{x + 3} - \sqrt{3}}{x} \approx 0.2887 \quad \left(\text{Actual limit is } 1/(2\sqrt{3}).\right)$$

27

5.

x	2.9	2.99	2.999	3.001	3.01	3.1
f(x)	−0.0641	−0.0627	−0.0625	−0.0625	−0.0623	−0.0610

$$\lim_{x \to 3} \frac{[1/(x+1)] - (1/4)}{x - 3} \approx -0.0625 \quad \left(\text{Actual limit is } -\tfrac{1}{16}.\right)$$

7.

x	−0.1	−0.01	−0.001	0.001	0.01	0.1
f(x)	0.9983	0.99998	1.0000	1.0000	0.99998	0.9983

$$\lim_{x \to 0} \frac{\sin x}{x} \approx 1.0000 \quad \text{(Actual limit is 1.) (Make sure you use radian mode.)}$$

9. $\lim_{x \to 3} (4 - x) = 1$

11. $\lim_{x \to 2} f(x) = \lim_{x \to 2} (4 - x) = 2$

13. $\lim_{x \to 5} \dfrac{|x - 5|}{x - 5}$ does not exist. For values of x to the left of 5, $|x - 5|/(x - 5)$ equals -1,

whereas for values of x to the right of 5, $|x - 5|/(x - 5)$ equals 1.

15. $\lim_{x \to \pi/2} \tan x$ does not exist since the function increases and decreases without bound as x approaches $\pi/2$.

17. $\lim_{x \to 0} \cos(1/x)$ does not exist since the function oscillates between -1 and 1 as x approaches 0.

19. $C(t) = 0.75 - 0.50[\![-(t - 1)]\!]$

(a)

(b)

t	3	3.3	3.4	3.5	3.6	3.7	4
C	1.75	2.25	2.25	2.25	2.25	2.25	2.25

$$\lim_{t \to 3.5} C(t) = 2.25$$

(c)

t	2	2.5	2.9	3	3.1	3.5	4
C	1.25	1.75	1.75	1.75	2.25	2.25	2.25

$\lim_{t \to 3} C(t)$ does not exist. The values of C jump from 1.75 to 2.25 at $t = 3$.

21. You need to find δ such that $0 < |x - 1| < \delta$ implies

$$|f(x) - 1| = \left|\frac{1}{x} - 1\right| < 0.1. \text{ That is,}$$

$$-0.1 < \frac{1}{x} - 1 < 0.1$$

$$1 - 0.1 < \frac{1}{x} < 1 + 0.1$$

$$\frac{9}{10} < \frac{1}{x} < \frac{11}{10}$$

$$\frac{10}{9} > x > \frac{10}{11}$$

$$\frac{10}{9} - 1 > x - 1 > \frac{10}{11} - 1$$

$$\frac{1}{9} > x - 1 > -\frac{1}{11}.$$

So take $\delta = \dfrac{1}{11}$. Then $0 < |x - 1| < \delta$ implies

$$-\frac{1}{11} < x - 1 < \frac{1}{11}$$

$$-\frac{1}{11} < x - 1 < \frac{1}{9}.$$

Using the first series of equivalent inequalities, you obtain

$$|f(x) - 1| = \left|\frac{1}{x} - 1\right| < 0.1.$$

23. $\lim_{x \to 2} (3x + 2) = 8 = L$

$|(3x + 2) - 8| < 0.01$

$|3x - 6| < 0.01$

$3|x - 2| < 0.01$

$0 < |x - 2| < \dfrac{0.01}{3} \approx 0.0033 = \delta$

Hence, if $0 < |x - 2| < \delta = \dfrac{0.01}{3}$, you have

$3|x - 2| < 0.01$

$|3x - 6| < 0.01$

$|(3x + 2) - 8| < 0.01$

$|f(x) - L| < 0.01$

25. $\lim_{x \to 2} (x^2 - 3) = 1 = L$

$|(x^2 - 3) - 1| < 0.01$

$|x^2 - 4| < 0.01$

$|(x + 2)(x - 2)| < 0.01$

$|x + 2|\,|x - 2| < 0.01$

$|x - 2| < \dfrac{0.01}{|x + 2|}$

If we assume $1 < x < 3$, then $\delta = 0.01/5 = 0.002$.

Hence, if $0 < |x - 2| < \delta = 0.002$, you have

$|x - 2| < 0.002 = \dfrac{1}{5}(0.01) < \dfrac{1}{|x + 2|}(0.01)$

$|x + 2|\,|x - 2| < 0.01$

$|x^2 - 4| < 0.01$

$|(x^2 - 3) - 1| < 0.01$

$|f(x) - L| < 0.01$

27. $\lim_{x \to 2} (x + 3) = 5$

Given $\epsilon > 0$:

$|(x + 3) - 5| < \epsilon$

$|x - 2| < \epsilon = \delta$

Hence, let $\delta = \epsilon$.

Hence, if $0 < |x - 2| < \delta = \epsilon$, you have

$|x - 2| < \epsilon$

$|(x + 3) - 5| < \epsilon$

$|f(x) - L| < \epsilon$

29. $\lim_{x \to -4} \left(\frac{1}{2}x - 1\right) = \frac{1}{2}(-4) - 1 = -3$

Given $\epsilon > 0$:

$\left|\left(\frac{1}{2}x - 1\right) - (-3)\right| < \epsilon$

$\left|\frac{1}{2}x + 2\right| < \epsilon$

$\frac{1}{2}|x - (-4)| < \epsilon$

$|x - (-4)| < 2\epsilon$

Hence, let $\delta = 2\epsilon$.

Hence, if $0 < |x - (-4)| < \delta = 2\epsilon$, you have

$|x - (-4)| < 2\epsilon$

$\left|\frac{1}{2}x + 2\right| < \epsilon$

$\left|\left(\frac{1}{2}x - 1\right) + 3\right| < \epsilon$

$|f(x) - L| < \epsilon$

31. $\lim_{x \to 6} 3 = 3$

Given $\epsilon > 0$:

$|3 - 3| < \epsilon$

$0 < \epsilon$

Hence, any $\delta > 0$ will work.

Hence, for any $\delta > 0$, you have

$|3 - 3| < \epsilon$

$|f(x) - L| < \epsilon$

33. $\lim_{x \to 0} \sqrt[3]{x} = 0$

Given $\epsilon > 0$: $\left|\sqrt[3]{x} - 0\right| < \epsilon$

$\left|\sqrt[3]{x}\right| < \epsilon$

$|x| < \epsilon^3 = \delta$

Hence, let $\delta = \epsilon^3$.

Hence for $0 < |x - 0| < \delta = \epsilon^3$, you have

$|x| < \epsilon^3$

$\left|\sqrt[3]{x}\right| < \epsilon$

$\left|\sqrt[3]{x} - 0\right| < \epsilon$

$|f(x) - L| < \epsilon$

35. $\lim\limits_{x \to -2} |x - 2| = |(-2) - 2| = 4$

Given $\epsilon > 0$:

$$||x - 2| - 4| < \epsilon$$

$$|-(x - 2) - 4| < \epsilon \quad (x - 2 < 0)$$

$$|-x - 2| = |x + 2| = |x - (-2)| < \epsilon$$

Hence, $\delta = \epsilon$.

Hence for $0 < |x - (-2)| < \delta = \epsilon$, you have

$$|x + 2| < \epsilon$$

$$|-(x + 2)| < \epsilon$$

$$|-(x - 2) - 4| < \epsilon$$

$$||x - 2| - 4| < \epsilon \quad \text{(because } x - 2 < 0)$$

$$|f(x) - L| < \epsilon$$

37. $\lim\limits_{x \to 1} (x^2 + 1) = 2$

Given $\epsilon > 0$:

$$|(x^2 + 1) - 2| < \epsilon$$

$$|x^2 - 1| < \epsilon$$

$$|(x + 1)(x - 1)| < \epsilon$$

$$|x - 1| < \frac{\epsilon}{|x + 1|}$$

If we assume $0 < x < 2$, then $\delta = \epsilon/3$.

Hence for $0 < |x - 1| < \delta = \dfrac{\epsilon}{3}$, you have

$$|x - 1| < \frac{1}{3}\epsilon < \frac{1}{|x + 1|}\epsilon$$

$$|x^2 - 1| < \epsilon$$

$$|(x^2 + 1) - 2| < \epsilon$$

$$|f(x) - 2| < \epsilon$$

39. $f(x) = \dfrac{\sqrt{x + 5} - 3}{x - 4}$

$\lim\limits_{x \to 4} f(x) = \dfrac{1}{6}$

The domain is $[-5, 4) \cup (4, \infty)$.
The graphing utility does not show the hole at $\left(4, \frac{1}{6}\right)$.

41. $f(x) = \dfrac{x - 9}{\sqrt{x} - 3}$

$\lim\limits_{x \to 9} f(x) = 6$

The domain is all $x \geq 0$ except $x = 9$. The graphing utility does not show the hole at $(9, 6)$.

43. $\lim\limits_{x \to 8} f(x) = 25$ means that the values of f approach 25 as x gets closer and closer to 8.

45. (i) The values of f approach different numbers as x approaches c from different sides of c:

(ii) The values of f increase without bound as x approaches c:

(iii) The values of f oscillate between two fixed numbers as x approaches c:

47. $f(x) = (1 + x)^{1/x}$

$\lim\limits_{x \to 0} (1 + x)^{1/x} = e \approx 2.71828$

x	$f(x)$	x	$f(x)$
-0.1	2.867972	0.1	2.593742
-0.01	2.731999	0.01	2.704814
-0.001	2.719642	0.001	2.716942
-0.0001	2.718418	0.0001	2.718146
-0.00001	2.718295	0.00001	2.718268
-0.000001	2.718283	0.000001	2.718280

49. False; $f(x) = (\sin x)/x$ is undefined when $x = 0$. From Exercise 7, we have

$$\lim_{x \to 0} \frac{\sin x}{x} = 1.$$

51. False; let

$$f(x) = \begin{cases} x^2 - 4x, & x \neq 4 \\ 10, & x = 4 \end{cases}.$$

$$f(4) = 10$$

$$\lim_{x \to 4} f(x) = \lim_{x \to 4} (x^2 - 4x) = 0 \neq 10$$

53. Answers will vary.

55. If $\lim_{x \to c} f(x) = L_1$ and $\lim_{x \to c} f(x) = L_2$, then for every $\epsilon > 0$, there exists $\delta_1 > 0$ and $\delta_2 > 0$ such that $|x - c| < \delta_1 \Rightarrow |f(x) - L_1| < \epsilon$ and $|x - c| < \delta_2 \Rightarrow |f(x) - L_2| < \epsilon$. Let δ equal the smaller of δ_1 and δ_2. Then for $|x - c| < \delta$, we have

$$|L_1 - L_2| = |L_1 - f(x) + f(x) - L_2| \leq |L_1 - f(x)| + |f(x) - L_2| < \epsilon + \epsilon.$$

Therefore, $|L_1 - L_2| < 2\epsilon$. Since $\epsilon > 0$ is arbitrary, it follows that $L_1 = L_2$.

57. $\lim_{x \to c} [f(x) - L] = 0$ means that for every $\epsilon > 0$ there exists $\delta > 0$ such that if

$$0 < |x - c| < \delta,$$

then

$$|(f(x) - L) - 0| < \epsilon.$$

This means the same as $|f(x) - L| < \epsilon$ when

$$0 < |x - c| < \delta.$$

Thus, $\lim_{x \to c} f(x) = L$.

Section 1.3   Evaluating Limits Analytically

1.

$h(x) = x^2 - 5x$

(a) $\lim_{x \to 5} h(x) = 0$

(b) $\lim_{x \to -1} h(x) = 6$

3.

$f(x) = x \cos x$

(a) $\lim_{x \to 0} f(x) = 0$

(b) $\lim_{x \to \pi/3} f(x) \approx 0.524$

$$\left(= \frac{\pi}{6} \right)$$

5. $\lim_{x \to 2} x^4 = 2^4 = 16$

7. $\lim_{x \to 0} (2x - 1) = 2(0) - 1 = -1$

9. $\lim_{x \to -3} (x^2 + 3x) = (-3)^2 + 3(-3) = 9 - 9 = 0$

11. $\lim_{x \to -3} (2x^2 + 4x + 1) = 2(-3)^2 + 4(-3) + 1 = 18 - 12 + 1 = 7$

13. $\lim_{x \to 2} \dfrac{1}{x} = \dfrac{1}{2}$

15. $\lim_{x \to 1} \dfrac{x - 3}{x^2 + 4} = \dfrac{1 - 3}{1^2 + 4} = \dfrac{-2}{5} = -\dfrac{2}{5}$

17. $\lim_{x \to 7} \dfrac{5x}{\sqrt{x + 2}} = \dfrac{5(7)}{\sqrt{7 + 2}} = \dfrac{35}{\sqrt{9}} = \dfrac{35}{3}$

19. $\lim_{x \to 3} \sqrt{x + 1} = \sqrt{3 + 1} = 2$

21. $\lim\limits_{x \to -4} (x + 3)^2 = (-4 + 3)^2 = 1$

23. (a) $\lim\limits_{x \to 1} f(x) = 5 - 1 = 4$

(b) $\lim\limits_{x \to 4} g(x) = 4^3 = 64$

(c) $\lim\limits_{x \to 1} g(f(x)) = g(f(1)) = g(4) = 64$

25. (a) $\lim\limits_{x \to 1} f(x) = 4 - 1 = 3$

(b) $\lim\limits_{x \to 3} g(x) = \sqrt{3 + 1} = 2$

(c) $\lim\limits_{x \to 1} g(f(x)) = g(3) = 2$

27. $\lim\limits_{x \to \pi/2} \sin x = \sin \dfrac{\pi}{2} = 1$

29. $\lim\limits_{x \to 2} \cos \dfrac{\pi x}{3} = \cos \dfrac{\pi 2}{3} = -\dfrac{1}{2}$

31. $\lim\limits_{x \to 0} \sec 2x = \sec 0 = 1$

33. $\lim\limits_{x \to 5\pi/6} \sin x = \sin \dfrac{5\pi}{6} = \dfrac{1}{2}$

35. $\lim\limits_{x \to 3} \tan\left(\dfrac{\pi x}{4}\right) = \tan \dfrac{3\pi}{4} = -1$

37. (a) $\lim\limits_{x \to c} [5g(x)] = 5 \lim\limits_{x \to c} g(x) = 5(3) = 15$

(b) $\lim\limits_{x \to c} [f(x) + g(x)] = \lim\limits_{x \to c} f(x) + \lim\limits_{x \to c} g(x) = 2 + 3 = 5$

(c) $\lim\limits_{x \to c} [f(x)g(x)] = \left[\lim\limits_{x \to c} f(x)\right]\left[\lim\limits_{x \to c} g(x)\right] = (2)(3) = 6$

(d) $\lim\limits_{x \to c} \dfrac{f(x)}{g(x)} = \dfrac{\lim\limits_{x \to c} f(x)}{\lim\limits_{x \to c} g(x)} = \dfrac{2}{3}$

39. (a) $\lim\limits_{x \to c} [f(x)]^3 = \left[\lim\limits_{x \to c} f(x)\right]^3 = (4)^3 = 64$

(b) $\lim\limits_{x \to c} \sqrt{f(x)} = \sqrt{\lim\limits_{x \to c} f(x)} = \sqrt{4} = 2$

(c) $\lim\limits_{x \to c} [3 f(x)] = 3 \lim\limits_{x \to c} f(x) = 3(4) = 12$

(d) $\lim\limits_{x \to c} [f(x)]^{3/2} = \left[\lim\limits_{x \to c} f(x)\right]^{3/2} = (4)^{3/2} = 8$

41. $f(x) = -2x + 1$ and $g(x) = \dfrac{-2x^2 + x}{x}$ agree except at $x = 0$.

(a) $\lim\limits_{x \to 0} g(x) = \lim\limits_{x \to 0} f(x) = 1$

(b) $\lim\limits_{x \to -1} g(x) = \lim\limits_{x \to -1} f(x) = 3$

43. $f(x) = x(x + 1)$ and $g(x) = \dfrac{x^3 - x}{x - 1}$ agree except at $x = 1$.

(a) $\lim\limits_{x \to 1} g(x) = \lim\limits_{x \to 1} f(x) = 2$

(b) $\lim\limits_{x \to -1} g(x) = \lim\limits_{x \to -1} f(x) = 0$

45. $f(x) = \dfrac{x^2 - 1}{x + 1}$ and $g(x) = x - 1$ agree except at $x = -1$.

$$\lim\limits_{x \to -1} f(x) = \lim\limits_{x \to -1} g(x) = -2$$

47. $f(x) = \dfrac{x^3 - 8}{x - 2}$ and $g(x) = x^2 + 2x + 4$ agree except at $x = 2$.

$$\lim\limits_{x \to 2} f(x) = \lim\limits_{x \to 2} g(x) = 12$$

49. $\lim\limits_{x \to 5} \dfrac{x - 5}{x^2 - 25} = \lim\limits_{x \to 5} \dfrac{x - 5}{(x + 5)(x - 5)}$

$$= \lim\limits_{x \to 5} \dfrac{1}{x + 5} = \dfrac{1}{10}$$

51. $\lim\limits_{x \to -3} \dfrac{x^2 + x - 6}{x^2 - 9} = \lim\limits_{x \to -3} \dfrac{(x + 3)(x - 2)}{(x + 3)(x - 3)}$

$$= \lim\limits_{x \to -3} \dfrac{x - 2}{x - 3} = \dfrac{-5}{-6} = \dfrac{5}{6}$$

53. $\displaystyle\lim_{x\to0}\frac{\sqrt{x+5}-\sqrt{5}}{x}=\lim_{x\to0}\frac{\sqrt{x+5}-\sqrt{5}}{x}\cdot\frac{\sqrt{x+5}+\sqrt{5}}{\sqrt{x+5}+\sqrt{5}}$

$$=\lim_{x\to0}\frac{(x+5)-5}{x\left(\sqrt{x+5}+\sqrt{5}\right)}=\lim_{x\to0}\frac{1}{\sqrt{x+5}+\sqrt{5}}=\frac{1}{2\sqrt{5}}=\frac{\sqrt{5}}{10}$$

55. $\displaystyle\lim_{x\to4}\frac{\sqrt{x+5}-3}{x-4}=\lim_{x\to4}\frac{\sqrt{x+5}-3}{x-4}\cdot\frac{\sqrt{x+5}+3}{\sqrt{x+5}+3}$

$$=\lim_{x\to4}\frac{(x+5)-9}{(x-4)\left(\sqrt{x+5}+3\right)}=\lim_{x\to4}\frac{1}{\sqrt{x+5}+3}=\frac{1}{\sqrt{9}+3}=\frac{1}{6}$$

57. $\displaystyle\lim_{x\to0}\frac{\dfrac{1}{3+x}-\dfrac{1}{3}}{x}=\lim_{x\to0}\frac{\dfrac{3-(3+x)}{3(3+x)}}{x}=\lim_{x\to0}\frac{-1}{3(3+x)}=-\frac{1}{9}$

59. $\displaystyle\lim_{\Delta x\to0}\frac{2(x+\Delta x)-2x}{\Delta x}=\lim_{\Delta x\to0}\frac{2x+2\Delta x-2x}{\Delta x}=\lim_{\Delta x\to0}2=2$

61. $\displaystyle\lim_{\Delta x\to0}\frac{(x+\Delta x)^2-2(x+\Delta x)+1-(x^2-2x+1)}{\Delta x}=\lim_{\Delta x\to0}\frac{x^2+2x\Delta x+(\Delta x)^2-2x-2\Delta x+1-x^2+2x-1}{\Delta x}$

$$=\lim_{\Delta x\to0}(2x+\Delta x-2)=2x-2$$

63. $\displaystyle\lim_{x\to0}\frac{\sqrt{x+2}-\sqrt{2}}{x}\approx0.354$

x	-0.1	-0.01	-0.001	0	0.001	0.01	0.1
$f(x)$	0.358	0.354	0.345	?	0.354	0.353	0.349

Analytically, $\displaystyle\lim_{x\to0}\frac{\sqrt{x+2}-\sqrt{2}}{x}=\lim_{x\to0}\frac{\sqrt{x+2}-\sqrt{2}}{x}\cdot\frac{\sqrt{x+2}+\sqrt{2}}{\sqrt{x+2}+\sqrt{2}}$

$$=\lim_{x\to0}\frac{x+2-2}{x\left(\sqrt{x+2}+\sqrt{2}\right)}=\lim_{x\to0}\frac{1}{\sqrt{x+2}+\sqrt{2}}=\frac{1}{2\sqrt{2}}=\frac{\sqrt{2}}{4}\approx0.354$$

65. $\displaystyle\lim_{x\to0}\frac{\dfrac{1}{2+x}-\dfrac{1}{2}}{x}=-\frac{1}{4}$

x	-0.1	-0.01	-0.001	0	0.001	0.01	0.1
$f(x)$	-0.263	-0.251	-0.250	?	-0.250	-0.249	-0.238

Analytically, $\displaystyle\lim_{x\to0}\frac{\dfrac{1}{2+x}-\dfrac{1}{2}}{x}=\lim_{x\to0}\frac{2-(2+x)}{2(2+x)}\cdot\frac{1}{x}=\lim_{x\to0}\frac{-x}{2(2+x)}\cdot\frac{1}{x}=\lim_{x\to0}\frac{-1}{2(2+x)}=-\frac{1}{4}.$

67. $\lim\limits_{x \to 0} \dfrac{\sin x}{5x} = \lim\limits_{x \to 0} \left[\left(\dfrac{\sin x}{x} \right) \left(\dfrac{1}{5} \right) \right] = (1) \left(\dfrac{1}{5} \right) = \dfrac{1}{5}$

69. $\lim\limits_{x \to 0} \dfrac{\sin x (1 - \cos x)}{2x^2} = \lim\limits_{x \to 0} \left[\dfrac{1}{2} \cdot \dfrac{\sin x}{x} \cdot \dfrac{1 - \cos x}{x} \right]$

$$= \dfrac{1}{2}(1)(0) = 0$$

71. $\lim\limits_{x \to 0} \dfrac{\sin^2 x}{x} = \lim\limits_{x \to 0} \left[\dfrac{\sin x}{x} \sin x \right] = (1) \sin 0 = 0$

73. $\lim\limits_{h \to 0} \dfrac{(1 - \cos h)^2}{h} = \lim\limits_{h \to 0} \left[\dfrac{1 - \cos h}{h}(1 - \cos h) \right]$

$$= (0)(0) = 0$$

75. $\lim\limits_{x \to \pi/2} \dfrac{\cos x}{\cot x} = \lim\limits_{x \to \pi/2} \sin x = 1$

77. $\lim\limits_{t \to 0} \dfrac{\sin 3t}{2t} = \lim\limits_{t \to 0} \left(\dfrac{\sin 3t}{3t} \right) \left(\dfrac{3}{2} \right) = (1) \left(\dfrac{3}{2} \right) = \dfrac{3}{2}$

79. $f(t) = \dfrac{\sin 3t}{t}$

t	-0.1	-0.01	-0.001	0	0.001	0.01	0.1
$f(t)$	2.96	2.9996	3	?	3	2.9996	2.96

Analytically, $\lim\limits_{t \to 0} \dfrac{\sin 3t}{t} = \lim\limits_{t \to 0} 3 \left(\dfrac{\sin 3t}{3t} \right) = 3(1) = 3.$

The limit appears to equal 3.

81. $f(x) = \dfrac{\sin x^2}{x}$

x	-0.1	-0.01	-0.001	0	0.001	0.01	0.1
$f(x)$	-0.099998	-0.01	-0.001	?	0.001	0.01	0.099998

Analytically, $\lim\limits_{x \to 0} \dfrac{\sin x^2}{x} = \lim\limits_{x \to 0} x \left(\dfrac{\sin x^2}{x^2} \right) = 0(1) = 0.$

83. $\lim\limits_{\Delta x \to 0} \dfrac{f(x + \Delta x) - f(x)}{\Delta x} = \lim\limits_{\Delta x \to 0} \dfrac{2(x + \Delta x) + 3 - (2x + 3)}{\Delta x} = \lim\limits_{\Delta x \to 0} \dfrac{2x + 2\Delta x + 3 - 2x - 3}{\Delta x} = \lim\limits_{\Delta x \to 0} \dfrac{2\Delta x}{\Delta x} = 2$

85. $\lim\limits_{\Delta x \to 0} \dfrac{f(x + \Delta x) - f(x)}{\Delta x} = \lim\limits_{\Delta x \to 0} \dfrac{\dfrac{4}{x + \Delta x} - \dfrac{4}{x}}{\Delta x} = \lim\limits_{\Delta x \to 0} \dfrac{4x - 4(x + \Delta x)}{(x + \Delta x)x \Delta x} = \lim\limits_{\Delta x \to 0} \dfrac{-4}{(x + \Delta x)x} = \dfrac{-4}{x^2}$

87. $\lim\limits_{x \to 0} (4 - x^2) \leq \lim\limits_{x \to 0} f(x) \leq \lim\limits_{x \to 0} (4 + x^2)$

$$4 \leq \lim\limits_{x \to 0} f(x) \leq 4$$

Therefore, $\lim\limits_{x \to 0} f(x) = 4.$

89. $f(x) = x \cos x$

$$\lim\limits_{x \to 0} (x \cos x) = 0$$

91. $f(x) = |x| \sin x$

$$\lim_{x \to 0} |x| \sin x = 0$$

93. $f(x) = x \sin \dfrac{1}{x}$

$$\lim_{x \to 0} \left(x \sin \frac{1}{x} \right) = 0$$

95. We say that two functions f and g agree at all but one point (on an open interval) if $f(x) = g(x)$ for all x in the interval except for $x = c$, where c is in the interval.

97. An indeterminant form is obtained when evaluating a limit using direct substitution produces a meaningless fractional expression such as $0/0$. That is,

$$\lim_{x \to c} \frac{f(x)}{g(x)}$$

for which $\lim_{x \to c} f(x) = \lim_{x \to c} g(x) = 0$

99. $f(x) = x$, $g(x) = \sin x$, $h(x) = \dfrac{\sin x}{x}$

When you are "close to" 0 the magnitude of f is approximately equal to the magnitude of g. Thus, $|g|/|f| \approx 1$ when x is "close to" 0.

101. $s(t) = -16t^2 + 1000$

$$\lim_{t \to 5} \frac{s(5) - s(t)}{5 - t} = \lim_{t \to 5} \frac{600 - (-16t^2 + 1000)}{5 - t} = \lim_{t \to 5} \frac{16(t + 5)(t - 5)}{-(t - 5)} = \lim_{t \to 5} -16(t + 5) = -160 \text{ ft/sec.}$$

Speed = 160 ft/sec

103. $s(t) = -4.9t^2 + 150$

$$\lim_{t \to 3} \frac{s(3) - s(t)}{3 - t} = \lim_{t \to 3} \frac{-4.9(3^2) + 150 - (-4.9t^2 + 150)}{3 - t} = \lim_{t \to 3} \frac{-4.9(9 - t^2)}{3 - t}$$

$$= \lim_{x \to 3} \frac{-4.9(3 - t)(3 + t)}{3 - t} = \lim_{x \to 3} -4.9(3 + t) = -29.4 \text{ m/sec}$$

105. Let $f(x) = 1/x$ and $g(x) = -1/x$. $\lim_{x \to 0} f(x)$ and $\lim_{x \to 0} g(x)$ do not exist.

$$\lim_{x \to 0} [f(x) + g(x)] = \lim_{x \to 0} \left[\frac{1}{x} + \left(-\frac{1}{x} \right) \right] = \lim_{x \to 0} [0] = 0$$

107. Given $f(x) = b$, show that for every $\epsilon > 0$ there exists a $\delta > 0$ such that $|f(x) - b| < \epsilon$ whenever $|x - c| < \delta$. Since $|f(x) - b| = |b - b| = 0 < \epsilon$ for any $\epsilon > 0$, then any value of $\delta > 0$ will work.

109. If $b = 0$, then the property is true because both sides are equal to 0. If $b \neq 0$, let $\epsilon > 0$ be given. Since $\lim_{x \to c} f(x) = L$, there exists $\delta > 0$ such that $|f(x) - L| < \epsilon/|b|$ whenever $0 < |x - c| < \delta$. Hence, wherever $0 < |x - c| < \delta$, we have

$$|b||f(x) - L| < \epsilon \quad \text{or} \quad |bf(x) - bL| < \epsilon$$

which implies that $\lim_{x \to c} [bf(x)] = bL$.

111.
$$-M|f(x)| \le f(x)g(x) \le M|f(x)|$$
$$\lim_{x \to c}(-M|f(x)|) \le \lim_{x \to c}f(x)g(x) \le \lim_{x \to c}(M|f(x)|)$$
$$-M(0) \le \lim_{x \to c}f(x)g(x) \le M(0)$$
$$0 \le \lim_{x \to c}f(x)g(x) \le 0$$

Therefore, $\lim_{x \to c}f(x)g(x) = 0$.

113. False. As x approaches 0 from the left, $\dfrac{|x|}{x} = -1$.

115. True.

117. False. The limit does not exist.

119. Let
$$f(x) = \begin{cases} 4, & \text{if } x \ge 0 \\ -4, & \text{if } x < 0 \end{cases}$$
$$\lim_{x \to 0}|f(x)| = \lim_{x \to 0}4 = 4.$$

$\lim_{x \to 0}f(x)$ does not exist since for $x < 0, f(x) = -4$ and for $x \ge 0, f(x) = 4$.

121. $f(x) = \begin{cases} 0, & \text{if } x \text{ is rational} \\ 1, & \text{if } x \text{ is irrational} \end{cases}$

$g(x) = \begin{cases} 0, & \text{if } x \text{ is rational} \\ x, & \text{if } x \text{ is irrational} \end{cases}$

$\lim_{x \to 0}f(x)$ does not exist.

No matter how "close to" 0 x is, there are still an infinite number of rational and irrational numbers so that $\lim_{x \to 0}f(x)$ does not exist.

$\lim_{x \to 0}g(x) = 0$.

When x is "close to" 0, both parts of the function are "close to" 0.

123. (a) $\displaystyle\lim_{x \to 0}\frac{1 - \cos x}{x^2} = \lim_{x \to 0}\frac{1 - \cos x}{x^2} \cdot \frac{1 + \cos x}{1 + \cos x}$

$= \displaystyle\lim_{x \to 0}\frac{1 - \cos^2 x}{x^2(1 + \cos x)}$

$= \displaystyle\lim_{x \to 0}\frac{\sin^2 x}{x^2} \cdot \frac{1}{1 + \cos x}$

$= (1)\left(\dfrac{1}{2}\right) = \dfrac{1}{2}$

(b) Thus, $\dfrac{1 - \cos x}{x^2} \approx \dfrac{1}{2} \Rightarrow 1 - \cos x \approx \dfrac{1}{2}x^2$

$\Rightarrow \cos x \approx 1 - \dfrac{1}{2}x^2$ for $x \approx 0$.

(c) $\cos(0.1) \approx 1 - \dfrac{1}{2}(0.1)^2 = 0.995$

(d) $\cos(0.1) \approx 0.9950$, which agrees with part (c).

Section 1.4   Continuity and One-Sided Limits

1. (a) $\lim\limits_{x \to 3^+} f(x) = 1$

(b) $\lim\limits_{x \to 3^-} f(x) = 1$

(c) $\lim\limits_{x \to 3} f(x) = 1$

The function is continuous at $x = 3$.

3. (a) $\lim\limits_{x \to 3^+} f(x) = 0$

(b) $\lim\limits_{x \to 3^-} f(x) = 0$

(c) $\lim\limits_{x \to 3} f(x) = 0$

The function is NOT continuous at $x = 3$.

5. (a) $\lim\limits_{x \to 4^+} f(x) = 2$

(b) $\lim\limits_{x \to 4^-} f(x) = -2$

(c) $\lim\limits_{x \to 4} f(x)$ does not exist

The function is NOT continuous at $x = 4$.

7. $\lim\limits_{x \to 5^+} \dfrac{x - 5}{x^2 - 25} = \lim\limits_{x \to 5^+} \dfrac{1}{x + 5} = \dfrac{1}{10}$

9. $\lim\limits_{x \to -3^-} \dfrac{x}{\sqrt{x^2 - 9}}$ does not exist because $\dfrac{x}{\sqrt{x^2 - 9}}$ grows without bound as $x \to -3^-$.

11. $\lim\limits_{x \to 0^-} \dfrac{|x|}{x} = \lim\limits_{x \to 0^-} \dfrac{-x}{x} = -1$.

13. $\lim\limits_{\Delta x \to 0^-} \dfrac{\dfrac{1}{x + \Delta x} - \dfrac{1}{x}}{\Delta x} = \lim\limits_{\Delta x \to 0^-} \dfrac{x - (x + \Delta x)}{x(x + \Delta x)} \cdot \dfrac{1}{\Delta x} = \lim\limits_{\Delta x \to 0^-} \dfrac{-\Delta x}{x(x + \Delta x)} \cdot \dfrac{1}{\Delta x}$

$$= \lim\limits_{\Delta x \to 0^-} \dfrac{-1}{x(x + \Delta x)}$$

$$= \dfrac{-1}{x(x + 0)} = -\dfrac{1}{x^2}$$

15. $\lim\limits_{x \to 3^-} f(x) = \lim\limits_{x \to 3^-} \dfrac{x + 2}{2} = \dfrac{5}{2}$

17. $\lim\limits_{x \to 1^+} f(x) = \lim\limits_{x \to 1^+} (x + 1) = 2$

$\lim\limits_{x \to 1^-} f(x) = \lim\limits_{x \to 1^-} (x^3 + 1) = 2$

$\lim\limits_{x \to 1} f(x) = 2$

19. $\lim\limits_{x \to \pi} \cot x$ does not exist since

$\lim\limits_{x \to \pi^+} \cot x$ and $\lim\limits_{x \to \pi^-} \cot x$ do not exist.

21. $\lim\limits_{x \to 4^-} (3[\![x]\!] - 5) = 3(3) - 5 = 4$

$([\![x]\!] = 3 \text{ for } 3 < x < 4)$

23. $\lim\limits_{x \to 3} (2 - [\![-x]\!])$ does not exist

because

$\lim\limits_{x \to 3^-} (2 - [\![-x]\!]) = 2 - (-3) = 5$

and

$\lim\limits_{x \to 3^+} (2 - [\![-x]\!]) = 2 - (-4) = 6$.

25. $f(x) = \dfrac{1}{x^2 - 4}$

has discontinuities at $x = -2$ and $x = 2$ since $f(-2)$ and $f(2)$ are not defined.

27. $f(x) = \dfrac{[\![x]\!]}{2} + x$

has discontinuities at each integer k since $\lim\limits_{x \to k^-} f(x) \neq \lim\limits_{x \to k^+} f(x)$.

29. $g(x) = \sqrt{25 - x^2}$ is continuous on $[-5, 5]$.

31. $\lim\limits_{x \to 0^-} f(x) = 3 = \lim\limits_{x \to 0^+} f(x)$.

f is continuous on $[-1, 4]$.

33. $f(x) = x^2 - 2x + 1$ is continuous for all real x.

35. $f(x) = 3x - \cos x$ is continuous for all real x.

37. $f(x) = \dfrac{x}{x^2 - x}$ is not continuous at $x = 0, 1$. Since

$$\frac{x}{x^2 - x} = \frac{1}{x - 1} \text{ for } x \neq 0, \, x = 0 \text{ is a removable}$$

discontinuity, whereas $x = 1$ is a nonremovable discontinuity.

39. $f(x) = \dfrac{x}{x^2 + 1}$ is continuous for all real x.

41. $f(x) = \dfrac{x + 2}{(x + 2)(x - 5)}$

has a nonremovable discontinuity at $x = 5$ since $\lim\limits_{x \to 5} f(x)$ does not exist, and has a removable discontinuity at $x = -2$ since

$$\lim_{x \to -2} f(x) = \lim_{x \to -2} \frac{1}{x - 5} = -\frac{1}{7}.$$

43. $f(x) = \dfrac{|x + 2|}{x + 2}$ has a nonremovable discontinuity at $x = -2$ since $\lim\limits_{x \to -2} f(x)$ does not exist.

45. $f(x) = \begin{cases} x, & x \leq 1 \\ x^2, & x > 1 \end{cases}$

has a **possible** discontinuity at $x = 1$.

1. $f(1) = 1$

2. $\left. \begin{array}{l} \lim\limits_{x \to 1^-} f(x) = \lim\limits_{x \to 1^-} x = 1 \\ \lim\limits_{x \to 1^+} f(x) = \lim\limits_{x \to 1^+} x^2 = 1 \end{array} \right\} \lim\limits_{x \to 1} f(x) = 1$

3. $f(1) = \lim\limits_{x \to 1} f(x)$

f is continuous at $x = 1$, therefore, f is continuous for all real x.

47. $f(x) = \begin{cases} \dfrac{x}{2} + 1, & x \leq 2 \\ 3 - x, & x > 2 \end{cases}$ has a **possible** discontinuity at $x = 2$.

1. $f(2) = \dfrac{2}{2} + 1 = 2$

2. $\left. \begin{array}{l} \lim\limits_{x \to 2^-} f(x) = \lim\limits_{x \to 2^-} \left(\dfrac{x}{2} + 1 \right) = 2 \\ \lim\limits_{x \to 2^+} f(x) = \lim\limits_{x \to 2^+} (3 - x) = 1 \end{array} \right\} \lim\limits_{x \to 2} f(x)$ does not exist.

Therefore, f has a nonremovable discontinuity at $x = 2$.

49. $f(x) = \begin{cases} \tan \dfrac{\pi x}{4}, & |x| < 1 \\ x, & |x| \geq 1 \end{cases} = \begin{cases} \tan \dfrac{\pi x}{4}, & -1 < x < 1 \\ x, & x \leq -1 \text{ or } x \geq 1 \end{cases}$ has **possible** discontinuities at $x = -1, x = 1$.

1. $f(-1) = -1$ $f(1) = 1$

2. $\lim\limits_{x \to -1} f(x) = -1$ $\lim\limits_{x \to 1} f(x) = 1$

3. $f(-1) = \lim\limits_{x \to -1} f(x)$ $f(1) = \lim\limits_{x \to 1} f(x)$

f is continuous at $x = \pm 1$, therefore, f is continuous for all real x.

51. $f(x) = \csc 2x$ has nonremovable discontinuities at integer multiples of $\pi/2$.

53. $f(x) = [\![x - 1]\!]$ has nonremovable discontinuities at each integer k.

55. $\lim\limits_{x \to 0^+} f(x) = 0$

$\lim\limits_{x \to 0^-} f(x) = 0$

f is not continuous at $x = -2$.

57. $f(2) = 8$

Find a so that $\lim\limits_{x \to 2^+} ax^2 = 8 \implies a = \dfrac{8}{2^2} = 2$.

59. Find a and b such that $\lim\limits_{x \to -1^+} (ax + b) = -a + b = 2$ and $\lim\limits_{x \to 3^-} (ax + b) = 3a + b = -2$.

$$a - b = -2$$
$$\underline{(+)\; 3a + b = -2}$$
$$4a \qquad = -4$$
$$a = -1$$
$$b = \;\; 2 + (-1) = 1$$

$$f(x) = \begin{cases} 2, & x \le -1 \\ -x + 1, & -1 < x < 3 \\ -2, & x \ge 3 \end{cases}$$

61. $f(g(x)) = (x - 1)^2$

Continuous for all real x.

63. $f(g(x)) = \dfrac{1}{(x^2 + 5) - 6} = \dfrac{1}{x^2 - 1}$

Nonremovable discontinuities at $x = \pm 1$

65. $y = [\![x]\!] - x$

Nonremovable discontinuity at each integer

67. $f(x) = \begin{cases} 2x - 4, & x \le 3 \\ x^2 - 2x, & x > 3 \end{cases}$

Nonremovable discontinuity at $x = 3$

69. $f(x) = \dfrac{x}{x^2 + 1}$

Continuous on $(-\infty, \infty)$

71. $f(x) = \sec \dfrac{\pi x}{4}$

Continuous on:
$$\ldots, (-6, -2), (-2, 2), (2, 6), (6, 10), \ldots$$

73. $f(x) = \dfrac{\sin x}{x}$

75. $f(x) = \frac{1}{16}x^4 - x^3 + 3$ is continuous on $[1, 2]$.

$f(1) = \frac{33}{16}$ and $f(2) = -4$. By the Intermediate Value Theorem, $f(c) = 0$ for at least one value of c between 1 and 2.

The graph **appears** to be continuous on the interval $[-4, 4]$. Since $f(0)$ is not defined, we know that f has a discontinuity at $x = 0$. This discontinuity is removable so it does not show up on the graph.

77. $f(x) = x^2 - 2 - \cos x$ is continuous on $[0, \pi]$.

$f(0) = -3$ and $f(\pi) = \pi^2 - 1 > 0$. By the Intermediate Value Theorem, $f(c) = 0$ for the least one value of c between 0 and π.

79. $f(x) = x^3 + x - 1$

$f(x)$ is continuous on $[0, 1]$.

$f(0) = -1$ and $f(1) = 1$

By the Intermediate Value Theorem, $f(x) = 0$ for at least one value of c between 0 and 1. Using a graphing utility, we find that $x \approx 0.6823$.

81. $g(t) = 2 \cos t - 3t$

g is continuous on $[0, 1]$.

$g(0) = 2 > 0$ and $g(1) \approx -1.9 < 0$.

By the Intermediate Value Theorem, $g(t) = 0$ for at least one value c between 0 and 1. Using a graphing utility, we find that $t \approx 0.5636$.

83. $f(x) = x^2 + x - 1$

f is continuous on $[0, 5]$.

$f(0) = -1$ and $f(5) = 29$

$$-1 < 11 < 29$$

The Intermediate Value Theorem applies.

$$x^2 + x - 1 = 11$$
$$x^2 + x - 12 = 0$$
$$(x + 4)(x - 3) = 0$$
$$x = -4 \text{ or } x = 3$$
$$c = 3 \ (x = -4 \text{ is not in the interval.})$$

Thus, $f(3) = 11$.

85. $f(x) = x^3 - x^2 + x - 2$

f is continuous on $[0, 3]$.

$f(0) = -2$ and $f(3) = 19$

$$-2 < 4 < 19$$

The Intermediate Value Theorem applies.

$$x^3 - x^2 + x - 2 = 4$$
$$x^3 - x^2 + x - 6 = 0$$
$$(x - 2)(x^2 + x + 3) = 0$$
$$x = 2$$
$$(x^2 + x + 3 \text{ has no real solution.})$$
$$c = 2$$

Thus, $f(2) = 4$.

87. (a) The limit does not exist at $x = c$.

(b) The function is not defined at $x = c$.

(c) The limit exists at $x = c$, but it is not equal to the value of the function at $x = c$.

(d) The limit does not exist at $x = c$.

89.

The function is not continuous at $x = 3$ because $\lim\limits_{x \to 3^+} f(x) = 1 \neq 0 = \lim\limits_{x \to 3^-} f(x)$.

91. The functions agree for integer values of x:

$$\left. \begin{array}{l} g(x) = 3 - [\![-x]\!] = 3 - (-x) = 3 + x \\ f(x) = 3 + [\![x]\!] = 3 + x \end{array} \right\} \text{ for } x \text{ an integer}$$

However, for non-integer values of x, the functions differ by 1.

$$f(x) = 3 + [\![x]\!] = g(x) - 1 = 2 - [\![-x]\!].$$

For example, $f\left(\frac{1}{2}\right) = 3 + 0 = 3$, $g\left(\frac{1}{2}\right) = 3 - (-1) = 4$.

93. $N(t) = 25\left(2\left[\!\left[\dfrac{t+2}{2}\right]\!\right] - t\right)$

t	0	1	1.8	2	3	3.8
$N(t)$	50	25	5	50	25	5

Discontinuous at every positive even integer. The company replenishes its inventory every two months.

95. Let $V = \frac{4}{3}\pi r^3$ be the volume of a sphere of radius r. V is continuous on $[1, 5]$.

$V(1) = \frac{4}{3}\pi \approx 4.19$

$V(5) = \frac{4}{3}\pi(5^3) \approx 523.6$

Since $4.19 < 275 < 523.6$, the Intermediate Value Theorem implies that there is at least one value r between 1 and 5 such that $V(r) = 275$. (In fact, $r \approx 4.0341$.)

97. Let c be any real number. Then $\lim\limits_{x \to c} f(x)$ does not exist since there are both rational and irrational numbers arbitrarily close to c. Therefore, f is not continuous at c.

99. $\text{sgn}(x) = \begin{cases} -1, & \text{if } x < 0 \\ 0, & \text{if } x = 0 \\ 1, & \text{if } x > 0 \end{cases}$

(a) $\lim\limits_{x \to 0^-} \text{sgn}(x) = -1$

(b) $\lim\limits_{x \to 0^+} \text{sgn}(x) = 1$

(c) $\lim\limits_{x \to 0} \text{sgn}(x)$ does not exist.

101. True; if $f(x) = g(x)$, $x \neq c$, then $\lim\limits_{x \to c} f(x) = \lim\limits_{x \to c} g(x)$ (if they exist) and at least one of these limits then does not equal the corresponding function value at $x = c$.

103. False; $f(1)$ is not defined and $\lim\limits_{x \to 1} f(x)$ does not exist.

105. (a) $f(x) = \begin{cases} 0 & 0 \le x < b \\ b & b < x \le 2b \end{cases}$

NOT continuous at $x = b$.

(b) $g(x) = \begin{cases} \dfrac{x}{2} & 0 \le x \le b \\ b - \dfrac{x}{2} & b < x \le 2b \end{cases}$

Continuous on $[0, 2b]$.

107. $f(x) = \dfrac{\sqrt{x + c^2} - c}{x}, \ c > 0$

Domain: $x + c^2 \geq 0 \implies x \geq -c^2$ and $x \neq 0, \ [-c^2, 0) \cup (0, \infty)$

$$\lim_{x \to 0} \frac{\sqrt{x + c^2} - c}{x} = \lim_{x \to 0} \frac{\sqrt{x + c^2} - c}{x} \cdot \frac{\sqrt{x + c^2} + c}{\sqrt{x + c^2} + c}$$

$$= \lim_{x \to 0} \frac{(x + c^2) - c^2}{x\left[\sqrt{x + c^2} + c\right]} = \lim_{x \to 0} \frac{1}{\sqrt{x + c^2} + c} = \frac{1}{2c}$$

Define $f(0) = 1/(2c)$ to make f continuous at $x = 0$.

109. $h(x) = x[\![x]\!]$

h has nonremovable discontinuities at

$x = \pm 1, \pm 2, \pm 3, \ldots$

Section 1.5   Infinite Limits

1. $\displaystyle \lim_{x \to -2^+} 2\left|\frac{x}{x^2 - 4}\right| = \infty$

$\displaystyle \lim_{x \to -2^-} 2\left|\frac{x}{x^2 - 4}\right| = \infty$

3. $\displaystyle \lim_{x \to -2^+} \tan \frac{\pi x}{4} = -\infty$

$\displaystyle \lim_{x \to -2^-} \tan \frac{\pi x}{4} = \infty$

5. $f(x) = \dfrac{1}{x^2 - 9}$

x	-3.5	-3.1	-3.01	-3.001	-2.999	-2.99	-2.9	-2.5
$f(x)$	0.308	1.639	16.64	166.6	-166.7	-16.69	-1.695	-0.364

$\displaystyle \lim_{x \to -3^-} f(x) = \infty$

$\displaystyle \lim_{x \to -3^+} f(x) = -\infty$

7. $f(x) = \dfrac{x^2}{x^2 - 9}$

x	-3.5	-3.1	-3.01	-3.001	-2.999	-2.99	-2.9	-2.5
$f(x)$	3.769	15.75	150.8	1501	-1499	-149.3	-14.25	-2.273

$\displaystyle \lim_{x \to -3^-} f(x) = \infty$

$\displaystyle \lim_{x \to -3^+} f(x) = -\infty$

9. $\lim\limits_{x \to 0^+} \dfrac{1}{x^2} = \infty = \lim\limits_{x \to 0^-} \dfrac{1}{x^2}$

Therefore, $x = 0$ is a vertical asymptote.

11. $\lim\limits_{x \to 2^+} \dfrac{x^2 - 2}{(x - 2)(x + 1)} = \infty$

$\lim\limits_{x \to 2^-} \dfrac{x^2 - 2}{(x - 2)(x + 1)} = -\infty$

Therefore, $x = 2$ is a vertical asymptote.

$\lim\limits_{x \to -1^+} \dfrac{x^2 - 2}{(x - 2)(x + 1)} = \infty$

$\lim\limits_{x \to -1^-} \dfrac{x^2 - 2}{(x - 2)(x + 1)} = -\infty$

Therefore, $x = -1$ is a vertical asymptote.

13. $\lim\limits_{x \to -2^-} \dfrac{x^2}{x^2 - 4} = \infty$ and $\lim\limits_{x \to -2^+} \dfrac{x^2}{x^2 - 4} = -\infty$

Therefore, $x = -2$ is a vertical asymptote.

$\lim\limits_{x \to 2^-} \dfrac{x^2}{x^2 - 4} = -\infty$ and $\lim\limits_{x \to 2^+} \dfrac{x^2}{x^2 - 4} = \infty$

Therefore, $x = 2$ is a vertical asymptote.

15. No vertical asymptote since the denominator is never zero.

17. $f(x) = \tan 2x = \dfrac{\sin 2x}{\cos 2x}$ has vertical asymptotes at

$x = \dfrac{(2n + 1)\pi}{4} = \dfrac{\pi}{4} + \dfrac{n\pi}{2}$, n any integer.

19. $\lim\limits_{t \to 0^+} \left(1 - \dfrac{4}{t^2} \right) = -\infty = \lim\limits_{t \to 0^-} \left(1 - \dfrac{4}{t^2} \right)$

Therefore, $t = 0$ is a vertical asymptote.

21. $\lim\limits_{x \to -2^+} \dfrac{x}{(x + 2)(x - 1)} = \infty$

$\lim\limits_{x \to -2^-} \dfrac{x}{(x + 2)(x - 1)} = -\infty$

Therefore, $x = -2$ is a vertical asymptote.

$\lim\limits_{x \to 1^+} \dfrac{x}{(x + 2)(x - 1)} = \infty$

$\lim\limits_{x \to 1^-} \dfrac{x}{(x + 2)(x - 1)} = -\infty$

Therefore, $x = 1$ is a vertical asymptote.

23. $f(x) = \dfrac{x^3 + 1}{x + 1} = \dfrac{(x + 1)(x^2 - x + 1)}{x + 1}$

has no vertical asymptote since

$\lim\limits_{x \to -1} f(x) = \lim\limits_{x \to -1} (x^2 - x + 1) = 3$

25. $f(x) = \dfrac{(x - 5)(x + 3)}{(x - 5)(x^2 + 1)} = \dfrac{x + 3}{x^2 + 1}, x \neq 5$

No vertical asymptotes. The graph has a hole at $x = 5$.

27. $s(t) = \dfrac{t}{\sin t}$ has vertical asymptotes at $t = n\pi$, n

a nonzero integer. There is no vertical asymptote at $t = 0$ since

$\lim\limits_{t \to 0} \dfrac{t}{\sin t} = 1.$

29. $\lim\limits_{x\to -1} \dfrac{x^2-1}{x+1} = \lim\limits_{x\to -1} (x-1) = -2$

Removable discontinuity at $x = -1$

31. $\lim\limits_{x\to -1^+} \dfrac{x^2+1}{x+1} = \infty$

$\lim\limits_{x\to -1^-} \dfrac{x^2+1}{x+1} = -\infty$

Vertical asymptote at
$x = -1$

33. $\lim\limits_{x\to 2^+} \dfrac{x-3}{x-2} = -\infty$

35. $\lim\limits_{x\to 3^+} \dfrac{x^2}{(x-3)(x+3)} = \infty$

37. $\lim\limits_{x\to -3^-} \dfrac{x^2+2x-3}{x^2+x-6} = \lim\limits_{x\to -3^-} \dfrac{(x-1)(x+3)}{(x-2)(x+3)} = \lim\limits_{x\to -3^-} \dfrac{x-1}{x-2} = \dfrac{4}{5}$

39. $\lim\limits_{x\to 1} \dfrac{x^2-x}{(x^2+1)(x-1)} = \lim\limits_{x\to 1} \dfrac{x}{x^2+1} = \dfrac{1}{2}$

41. $\lim\limits_{x\to 0^-} \left(1 + \dfrac{1}{x}\right) = -\infty$

43. $\lim\limits_{x\to 0^+} \dfrac{2}{\sin x} = \infty$

45. $\lim\limits_{x\to \pi} \dfrac{\sqrt{x}}{\csc x} = \lim\limits_{x\to \pi} \left(\sqrt{x}\,\sin x\right) = 0$

47. $\lim\limits_{x\to (1/2)^-} x\sec(\pi x) = \infty$ and $\lim\limits_{x\to (1/2)^+} x\sec(\pi x) = -\infty.$
Therefore, $\lim\limits_{x\to (1/2)} x\sec(\pi x)$ does not exist.

49. $f(x) = \dfrac{x^2+x+1}{x^3-1}$

$\lim\limits_{x\to 1^+} f(x) = \lim\limits_{x\to 1^+} \dfrac{1}{x-1} = \infty$

51. $f(x) = \dfrac{1}{x^2-25}$

$\lim\limits_{x\to 5^-} f(x) = -\infty$

53. A limit in which $f(x)$ increases or decreases without bound as x approaches c is called an infinite limit. ∞ is not a number. Rather, the symbol

$\lim\limits_{x\to c} f(x) = \infty$

says how the limit fails to exist.

55. One answer is $f(x) = \dfrac{x-3}{(x-6)(x+2)} = \dfrac{x-3}{x^2-4x-12}.$

57.

59. $S = \dfrac{k}{1-r}, \ 0 < |r| < 1.$ Assume $k \neq 0.$

$\lim\limits_{r\to 1^-} S = \lim\limits_{r\to 1^-} \dfrac{k}{1-r} = \infty$ (or $-\infty$ if $k < 0$)

61. $C = \dfrac{528x}{100 - x}, 0 \leq x < 100$

    (a) $C(25) = \$176$ million

    (b) $C(50) = \$528$ million

    (c) $C(75) = \$1584$ million

    (d) $\displaystyle\lim_{x \to 100^-} \dfrac{528}{100 - x} = \infty$ Thus, it is not possible.

63. (a) $r = \dfrac{2(7)}{\sqrt{625 - 49}} = \dfrac{7}{12}$ ft/sec

    (b) $r = \dfrac{2(15)}{\sqrt{625 - 225}} = \dfrac{3}{2}$ ft/sec

    (c) $\displaystyle\lim_{x \to 25^-} \dfrac{2x}{\sqrt{625 - x^2}} = \infty$

65. (a)

x	1	0.5	0.2	0.1	0.01	0.001	0.0001
$f(x)$	0.1585	0.0411	0.0067	0.0017	≈ 0	≈ 0	≈ 0

$\displaystyle\lim_{x \to 0^+} \dfrac{x - \sin x}{x} = 0$

(b)

x	1	0.5	0.2	0.1	0.01	0.001	0.0001
$f(x)$	0.1585	0.0823	0.0333	0.0167	0.0017	≈ 0	≈ 0

$\displaystyle\lim_{x \to 0^+} \dfrac{x - \sin x}{x^2} = 0$

(c)

x	1	0.5	0.2	0.1	0.01	0.001	0.0001
$f(x)$	0.1585	0.1646	0.1663	0.1666	0.1667	0.1667	0.1667

$\displaystyle\lim_{x \to 0^+} \dfrac{x - \sin x}{x^3} = 0.1167 \; (1/6)$

(d)

x	1	0.5	0.2	0.1	0.01	0.001	0.0001
$f(x)$	0.1585	0.3292	0.8317	1.6658	16.67	166.7	1667.0

$\displaystyle\lim_{x \to 0^+} \dfrac{x - \sin x}{x^4} = \infty$

For $n \geq 3$, $\displaystyle\lim_{x \to 0^+} \dfrac{x - \sin x}{x^n} = \infty$.

67. (a) Because the circumference of the motor is half that of the saw arbor, the saw makes $1700/2 = 850$ revolutions per minute.

(c) $2(20 \cot \phi) + 2(10 \cot \phi)$: straight sections. The angle subtended in each circle is

$$2\pi - \left(2\left(\frac{\pi}{2} - \phi\right)\right) = \pi + 2\phi.$$

Thus, the length of the belt around the pulleys is

$$20(\pi + 2\phi) + 10(\pi + 2\phi) = 30(\pi + 2\phi).$$

Total length $= 60 \cot \phi + 30(\pi + 2\phi)$

Domain: $\left(0, \dfrac{\pi}{2}\right)$

(b) The direction of rotation is reversed.

(d)

ϕ	0.3	0.6	0.9	1.2	1.5
L	306.2	217.9	195.9	189.6	188.5

(e)

(f) $\lim\limits_{\phi \to (\pi/2)^-} L = 60\pi \approx 188.5$

(All the belts are around pulleys.)

(g) $\lim\limits_{\phi \to 0^+} L = \infty$

69. False; for instance, let

$$f(x) = \frac{x^2 - 1}{x - 1} \text{ or}$$

$$g(x) = \frac{x}{x^2 + 1}.$$

71. False; let

$$f(x) = \begin{cases} \dfrac{1}{x}, & x \neq 0 \\ 3, & x = 0. \end{cases}$$

The graph of f has a vertical asymptote at $x = 0$, but $f(0) = 3$.

73. Given $\lim\limits_{x \to c} f(x) = \infty$ and $\lim\limits_{x \to c} g(x) = L$:

(2) Product:

If $L > 0$, then for $\epsilon = L/2 > 0$ there exists $\delta_1 > 0$ such that $|g(x) - L| < L/2$ whenever $0 < |x - c| < \delta_1$. Thus, $L/2 < g(x) < 3L/2$. Since $\lim\limits_{x \to c} f(x) = \infty$ then for $M > 0$, there exists $\delta_2 > 0$ such that $f(x) > M(2/L)$ whenever $|x - c| < \delta_2$. Let δ be the smaller of δ_1 and δ_2. Then for $0 < |x - c| < \delta$, we have $f(x)g(x) > M(2/L)(L/2) = M$. Therefore $\lim\limits_{x \to c} f(x)g(x) = \infty$. The proof is similar for $L < 0$.

(3) Quotient: Let $\epsilon > 0$ be given.

There exists $\delta_1 > 0$ such that $f(x) > 3L/2\epsilon$ whenever $0 < |x - c| < \delta_1$ and there exists $\delta_2 > 0$ such that $|g(x) - L| < L/2$ whenever $0 < |x - c| < \delta_2$. This inequality gives us $L/2 < g(x) < 3L/2$. Let δ be the smaller of δ_1 and δ_2. Then for $0 < |x - c| < \delta$, we have

$$\left|\frac{g(x)}{f(x)}\right| < \frac{3L/2}{3L/2\epsilon} = \epsilon.$$

Therefore, $\lim\limits_{x \to c} \dfrac{g(x)}{f(x)} = 0$.

75. Given $\lim\limits_{x \to c} \dfrac{1}{f(x)} = 0$.

Suppose $\lim\limits_{x \to c} f(x)$ exists and equals L. Then,

$$\lim_{x \to c} \frac{1}{f(x)} = \frac{\lim\limits_{x \to c} 1}{\lim\limits_{x \to c} f(x)} = \frac{1}{L} = 0.$$

This is not possible. Thus, $\lim\limits_{x \to c} f(x)$ does not exist.

Review Exercises for Chapter 1

1. Calculus required. Using a graphing utility, you can estimate the length to be 8.3.
Or, the length is slightly longer than the distance between the two points, 8.25.

3.

x	-0.1	-0.01	-0.001	0.001	0.01	0.1
$f(x)$	-1.05	-1.01	-1.00	-0.9995	-0.995	-0.952

$\lim\limits_{x \to 0} f(x) \approx -1$

5. $h(x) = \dfrac{x^2 - 2x}{x}$

    (a) $\lim\limits_{x \to 0} h(x) = -2$

    (b) $\lim\limits_{x \to -1} h(x) = -3$

7. $\lim\limits_{x \to 1} (3 - x) = 3 - 1 = 2$

Let $\epsilon > 0$ be given. Choose $\delta = \epsilon$. Then for

$$0 < |x - 1| < \delta = \epsilon, \text{ you have}$$

$$|x - 1| < \epsilon$$
$$|1 - x| < \epsilon$$
$$|(3 - x) - 2| < \epsilon$$
$$|f(x) - L| < \epsilon$$

9. $\lim\limits_{x \to 2} (x^2 - 3) = 1$

Let $\epsilon > 0$ be given. We need $|x^2 - 3 - 1| < \epsilon \Rightarrow |x^2 - 4| = |(x - 2)(x + 2)| < \epsilon \Rightarrow |x - 2| < \dfrac{1}{|x + 2|}\epsilon.$

Assuming, $1 < x < 3$, you can choose $\delta = \epsilon/5$. Hence, for $0 < |x - 2| < \delta = \epsilon/5$ you have

$$|x - 2| < \frac{\epsilon}{5} < \frac{1}{|x + 2|}\epsilon$$
$$|x - 2||x + 2| < \epsilon$$
$$|x^2 - 4| < \epsilon$$
$$|(x^2 - 3) - 1| < \epsilon$$
$$|f(x) - L| < \epsilon$$

11. $\lim\limits_{t \to 4} \sqrt{t + 2} = \sqrt{4 + 2} = \sqrt{6} \approx 2.45$

13. $\lim\limits_{t \to -2} \dfrac{t + 2}{t^2 - 4} = \lim\limits_{t \to -2} \dfrac{1}{t - 2} = -\dfrac{1}{4}$

15. $\lim\limits_{x \to 4} \dfrac{\sqrt{x} - 2}{x - 4} = \lim\limits_{x \to 4} \dfrac{\sqrt{x} - 2}{(\sqrt{x} - 2)(\sqrt{x} + 2)}$

$$= \lim\limits_{x \to 4} \frac{1}{\sqrt{x} + 2} = \frac{1}{\sqrt{4} + 2} = \frac{1}{4}$$

17. $\lim\limits_{x \to 0} \dfrac{[1/(x + 1)] - 1}{x} = \lim\limits_{x \to 0} \dfrac{1 - (x + 1)}{x(x + 1)}$

$$= \lim\limits_{x \to 0} \frac{-1}{x + 1} = -1$$

19. $\lim\limits_{x \to -5} \dfrac{x^3 + 125}{x + 5} = \lim\limits_{x \to -5} \dfrac{(x + 5)(x^2 - 5x + 25)}{x + 5}$

$$= \lim\limits_{x \to -5} (x^2 - 5x + 25)$$
$$= 75$$

21. $\lim\limits_{x \to 0} \dfrac{1 - \cos x}{\sin x} = \lim\limits_{x \to 0} \left(\dfrac{x}{\sin x}\right)\left(\dfrac{1 - \cos x}{x}\right) = (1)(0) = 0$

23. $\displaystyle\lim_{\Delta x \to 0} \frac{\sin[(\pi/6) + \Delta x] - (1/2)}{\Delta x} = \lim_{\Delta x \to 0} \frac{\sin(\pi/6)\cos \Delta x + \cos(\pi/6)\sin \Delta x - (1/2)}{\Delta x}$

$$= \lim_{\Delta x \to 0} \frac{1}{2} \cdot \frac{(\cos \Delta x - 1)}{\Delta x} + \lim_{\Delta x \to 0} \frac{\sqrt{3}}{2} \cdot \frac{\sin \Delta x}{\Delta x}$$

$$= 0 + \frac{\sqrt{3}}{2}(1) = \frac{\sqrt{3}}{2}$$

25. $\displaystyle\lim_{x \to c}\,[f(x) \cdot g(x)] = \left(-\tfrac{3}{4}\right)\left(\tfrac{2}{3}\right) = -\tfrac{1}{2}$

27. $f(x) = \dfrac{\sqrt{2x + 1} - \sqrt{3}}{x - 1}$

(a)

x	1.1	1.01	1.001	1.0001
$f(x)$	0.5680	0.5764	0.5773	0.5773

$\displaystyle\lim_{x \to 1^+} \frac{\sqrt{2x + 1} - \sqrt{3}}{x - 1} \approx 0.577$ (Actual limit is $\sqrt{3}/3$.)

(b)

(c) $\displaystyle\lim_{x \to 1^+} \frac{\sqrt{2x + 1} - \sqrt{3}}{x - 1} = \lim_{x \to 1^+} \frac{\sqrt{2x + 1} - \sqrt{3}}{x - 1} \cdot \frac{\sqrt{2x + 1} + \sqrt{3}}{\sqrt{2x + 1} + \sqrt{3}}$

$$= \lim_{x \to 1^+} \frac{(2x + 1) - 3}{(x - 1)\left(\sqrt{2x + 1} + \sqrt{3}\right)}$$

$$= \lim_{x \to 1^+} \frac{2}{\sqrt{2x + 1} + \sqrt{3}}$$

$$= \frac{2}{2\sqrt{3}} = \frac{1}{\sqrt{3}} = \frac{\sqrt{3}}{3}$$

29. $\displaystyle\lim_{t \to a} \frac{s(a) - s(t)}{a - t} = \lim_{t \to 4} \frac{(-4.9(4)^2 + 200) - (-4.9t^2 + 200)}{4 - t}$

$$= \lim_{t \to 4} \frac{4.9(t - 4)(t + 4)}{4 - t}$$

$$= \lim_{t \to 4} -4.9(t + 4) = -39.2 \text{ m/sec}$$

31. $\displaystyle\lim_{x \to 3^-} \frac{|x - 3|}{x - 3} = \lim_{x \to 3^-} \frac{-(x - 3)}{x - 3} = -1$

33. $\displaystyle\lim_{x \to 2^-} f(x) = 0$

35. $\displaystyle\lim_{t \to 1} h(t)$ does not exist because $\displaystyle\lim_{t \to 1^-} h(t) = 1 + 1 = 2$ and $\displaystyle\lim_{t \to 1^+} h(t) = \frac{1}{2}(1 + 1) = 1$.

37. $f(x) = [\![x + 3]\!]$

$\displaystyle\lim_{x \to k^+} [\![x + 3]\!] = k + 3$ where k is an integer.

$\displaystyle\lim_{x \to k^-} [\![x + 3]\!] = k + 2$ where k is an integer.

Nonremovable discontinuity at each integer k

Continuous on $(k, k + 1)$ for all integers k

39. $f(x) = \dfrac{3x^2 - x - 2}{x - 1} = \dfrac{(3x + 2)(x - 1)}{x - 1}$

$\displaystyle\lim_{x \to 1} f(x) = \lim_{x \to 1} (3x + 2) = 5$

Removable discontinuity at $x = 1$

Continuous on $(-\infty, 1) \cup (1, \infty)$

41. $f(x) = \dfrac{1}{(x - 2)^2}$

$\displaystyle\lim_{x \to 2} \frac{1}{(x - 2)^2} = \infty$

Nonremovable discontinuity at $x = 2$

Continuous on $(-\infty, 2) \cup (2, \infty)$

43. $f(x) = \dfrac{3}{x + 1}$

$\displaystyle\lim_{x \to 1^-} f(x) = -\infty$

$\displaystyle\lim_{x \to 1^+} f(x) = \infty$

Nonremovable discontinuity at $x = -1$

Continuous on $(-\infty, -1) \cup (-1, \infty)$

45. $f(x) = \csc \dfrac{\pi x}{2}$

Nonremovable discontinuities at each even integer. Continuous on

$$(2k, 2k + 2)$$

for all integers k.

47. $f(2) = 5$

Find c so that $\lim\limits_{x \to 2^+} (cx + 6) = 5$.

$$c(2) + 6 = 5$$
$$2c = -1$$
$$c = -\dfrac{1}{2}$$

49. f is continuous on $[1, 2]$. $f(1) = -1 < 0$ and $f(2) = 13 > 0$. Therefore by the Intermediate Value Theorem, there is at least one value c in $(1, 2)$ such that $2c^3 - 3 = 0$.

51. $f(x) = \dfrac{x^2 - 4}{|x - 2|} = (x + 2)\left[\dfrac{x - 2}{|x - 2|}\right]$

(a) $\lim\limits_{x \to 2^-} f(x) = -4$

(b) $\lim\limits_{x \to 2^+} f(x) = 4$

(c) $\lim\limits_{x \to 2} f(x)$ does not exist.

53. $g(x) = 1 + \dfrac{2}{x}$

Vertical asymptote at $x = 0$

55. $f(x) = \dfrac{8}{(x - 10)^2}$

Vertical asymptote at $x = 10$

57. $\lim\limits_{x \to -2^-} \dfrac{2x^2 + x + 1}{x + 2} = -\infty$

59. $\lim\limits_{x \to -1^+} \dfrac{x + 1}{x^3 + 1} = \lim\limits_{x \to -1^+} \dfrac{1}{x^2 - x + 1} = \dfrac{1}{3}$

61. $\lim\limits_{x \to 1^-} \dfrac{x^2 + 2x + 1}{x - 1} = -\infty$

63. $\lim\limits_{x \to 0^+} \left(x - \dfrac{1}{x^3}\right) = -\infty$

65. $\lim\limits_{x \to 0^+} \dfrac{\sin 4x}{5x} = \lim\limits_{x \to 0^+} \left[\dfrac{4}{5}\left(\dfrac{\sin 4x}{4x}\right)\right] = \dfrac{4}{5}$

67. $\lim\limits_{x \to 0^+} \dfrac{\csc 2x}{x} = \lim\limits_{x \to 0^+} \dfrac{1}{x \sin 2x} = \infty$

69. $C = \dfrac{80,000p}{100 - p}, \ 0 \le 0 < 100$

(a) $C(15) \approx \$14,117.65$ 　(b) $C(50) = \$80.000$

(c) $C(90) = \$720,000$ 　(d) $\lim\limits_{p \to 100^-} \dfrac{80,000p}{100 - p} = \infty$

Problem Solving for Chapter 1

1. (a) Perimeter $\triangle PAO = \sqrt{x^2 + (y - 1)^2} + \sqrt{x^2 + y^2} + 1$
$= \sqrt{x^2 + (x^2 - 1)^2} + \sqrt{x^2 + x^4} + 1$

Perimeter $\triangle PBO = \sqrt{(x - 1)^2 + y^2} + \sqrt{x^2 + y^2} + 1$
$= \sqrt{(x - 1)^2 + x^4} + \sqrt{x^2 + x^4} + 1$

(b) $r(x) = \dfrac{\sqrt{x^2 + (x^2 - 1)^2} + \sqrt{x^2 + x^4} + 1}{\sqrt{(x - 1)^2 + x^4} + \sqrt{x^2 + x^4} + 1}$

x	4	2	1	0.1	0.01
Perimeter $\triangle PAO$	33.02	9.08	3.41	2.10	2.01
Perimeter $\triangle PBO$	33.77	9.60	3.41	2.00	2.00
$r(x)$	0.98	0.95	1	1.05	1.005

(c) $\lim\limits_{x \to 0^+} r(x) = \dfrac{1 + 0 + 1}{1 + 0 + 1} = \dfrac{2}{2} = 1$

3. (a) There are 6 triangles, each with a central angle of $60° = \pi/3$. Hence,

$$\text{Area hexagon} = 6\left[\frac{1}{2}bh\right] = 6\left[\frac{1}{2}(1)\sin\frac{\pi}{3}\right]$$

$$= \frac{3\sqrt{3}}{2} \approx 2.598.$$

Error: $\pi - \dfrac{3\sqrt{3}}{2} \approx 0.5435.$

(b) There are n triangles, each with central angle of $\theta = 2\pi/n$. Hence,

$$A_n = n\left[\frac{1}{2}bh\right] = n\left[\frac{1}{2}(1)\sin\frac{2\pi}{n}\right] = \frac{n\sin(2\pi/n)}{2}.$$

(c)

n	6	12	24	48	96
A_n	2.598	3	3.106	3.133	3.139

(d) As n gets larger and larger, $2\pi/n$ approaches 0.

Letting $x = 2\pi/n$,

$$A_n = \frac{\sin(2\pi/n)}{2/n} = \frac{\sin(2\pi/n)}{(2\pi/n)}\pi = \frac{\sin x}{x}\pi$$

which approaches $(1)\pi = \pi$.

5. (a) Slope $= -\dfrac{12}{5}$

(b) Slope of tangent line is $\dfrac{5}{12}$.

$$y + 12 = \frac{5}{12}(x - 5)$$

$$y = \frac{5}{12}x - \frac{169}{12} \quad \text{Tangent line}$$

(c) $Q = (x, y) = \left(x, -\sqrt{169 - x^2}\right)$

$$m_x = \frac{-\sqrt{169 - x^2} + 12}{x - 5}$$

(d) $\displaystyle\lim_{x\to 5} m_x = \lim_{x\to 5}\frac{12 - \sqrt{169 - x^2}}{x - 5} \cdot \frac{12 + \sqrt{169 - x^2}}{12 + \sqrt{169 - x^2}}$

$$= \lim_{x\to 5}\frac{144 - (169 - x^2)}{(x - 5)\left(12 + \sqrt{169 - x^2}\right)}$$

$$= \lim_{x\to 5}\frac{x^2 - 25}{(x - 5)\left(12 + \sqrt{169 - x^2}\right)}$$

$$= \lim_{x\to 5}\frac{(x + 5)}{12 + \sqrt{169 - x^2}}$$

$$= \frac{10}{12 + 12} = \frac{5}{12}$$

This is the same slope as part (b).

7. (a) $3 + x^{1/3} \geq 0$

$$x^{1/3} \geq -3$$

$$x \geq -27$$

Domain: $x \geq -27, x \neq 1$

(b)

(c) $\displaystyle\lim_{x\to -27^+} f(x) = \frac{\sqrt{3 + (-27)^{1/3}} - 2}{-27 - 1}$

$$= \frac{-2}{-28} = \frac{1}{14} \approx 0.0714$$

(d) $\displaystyle\lim_{x\to 1} f(x) = \lim_{x\to 1}\frac{\sqrt{3 + x^{1/3}} - 2}{x - 1} \cdot \frac{\sqrt{3 + x^{1/3}} + 2}{\sqrt{3 + x^{1/3}} + 2}$

$$= \lim_{x\to 1}\frac{3 + x^{1/3} - 4}{(x - 1)\left(\sqrt{3 + x^{1/3}} + 2\right)}$$

$$= \lim_{x\to 1}\frac{x^{1/3} - 1}{(x^{1/3} - 1)(x^{2/3} + x^{1/3} + 1)\left(\sqrt{3 + x^{1/3}} + 2\right)}$$

$$= \lim_{x\to 1}\frac{1}{(x^{2/3} + x^{1/3} + 1)\left(\sqrt{3 + x^{1/3}} + 2\right)}$$

$$= \frac{1}{(1 + 1 + 1)(2 + 2)} = \frac{1}{12}$$

9. (a) $\displaystyle\lim_{x\to 2} f(x) = 3: g_1, g_4$

(b) f continuous at 2: g_1

(c) $\displaystyle\lim_{x\to 2^-} f(x) = 3: g_1, g_3, g_4$

11.

(a) $f(1) = [\![1]\!] + [\![-1]\!] = 1 + (-1) = 0$

   $f(0) = 0$

   $f\left(\frac{1}{2}\right) = 0 + (-1) = -1$

   $f(-2.7) = -3 + 2 = -1$

(b) $\lim_{x \to 1^-} f(x) = -1$

   $\lim_{x \to 1^+} f(x) = -1$

   $\lim_{x \to 1/2} f(x) = -1$

(c) f is continuous for all real numbers except

   $x = 0, \pm 1, \pm 2, \pm 3, \ldots$

13. (a)

(b) (i) $\lim_{x \to a^+} P_{a,\,b}(x) = 1$

   (ii) $\lim_{x \to a^-} P_{a,\,b}(x) = 0$

   (iii) $\lim_{x \to b^+} P_{a,\,b}(x) = 0$

   (iv) $\lim_{x \to b^-} P_{a,\,b}(x) = 1$

(c) $P_{a,\,b}$ is continuous for all positive real numbers except $x = a, b$.

(d) The area under the graph of u, and above the x-axis, is 1.

CHAPTER 2
Differentiation

Section 2.1   The Derivative and the Tangent Line Problem . . . 53

Section 2.2   Basic Differentiation Rules and Rates of Change . 60

Section 2.3   The Product and Quotient Rules and
Higher-Order Derivatives 67

Section 2.4   The Chain Rule 73

Section 2.5   Implicit Differentiation 79

Section 2.6   Related Rates 85

Review Exercises   92

Problem Solving 98

CHAPTER 2
Differentiation

Section 2.1   The Derivative and the Tangent Line Problem

Solutions to Odd-Numbered Exercises

1. (a) $m = 0$

(b) $m = -3$

3. (a), (b)

$$y = \frac{f(4) - f(1)}{4 - 1}(x - 1) + f(1) = x + 1$$

(c) $y = \dfrac{f(4) - f(1)}{4 - 1}(x - 1) + f(1)$

$$= \frac{3}{3}(x - 1) + 2$$

$$= 1(x - 1) + 2$$

$$= x + 1$$

5. $f(x) = 3 - 2x$ is a line. Slope $= -2$

7. Slope at $(1, -3) = \lim\limits_{\Delta x \to 0} \dfrac{g(1 + \Delta x) - g(1)}{\Delta x}$

$$= \lim\limits_{\Delta x \to 0} \frac{(1 + \Delta x)^2 - 4 - (-3)}{\Delta x}$$

$$= \lim\limits_{\Delta x \to 0} \frac{1 + 2(\Delta x) + (\Delta x)^2 - 1}{\Delta x}$$

$$= \lim\limits_{\Delta x \to 0} [2 + 2(\Delta x)] = 2$$

9. Slope at $(0, 0) = \lim\limits_{\Delta t \to 0} \dfrac{f(0 + \Delta t) - f(0)}{\Delta t}$

$$= \lim\limits_{\Delta t \to 0} \frac{3(\Delta t) - (\Delta t)^2 - 0}{\Delta t}$$

$$= \lim\limits_{\Delta t \to 0} (3 - \Delta t) = 3$$

11. $f(x) = 3$

$$f'(x) = \lim\limits_{\Delta x \to 0} \frac{f(x + \Delta x) - f(x)}{\Delta x}$$

$$= \lim\limits_{\Delta x \to 0} \frac{3 - 3}{\Delta x}$$

$$= \lim\limits_{\Delta x \to 0} 0 = 0$$

13. $f(x) = -5x$

$$f'(x) = \lim\limits_{\Delta x \to 0} \frac{f(x + \Delta x) - f(x)}{\Delta x}$$

$$= \lim\limits_{\Delta x \to 0} \frac{-5(x + \Delta x) - (-5x)}{\Delta x}$$

$$= \lim\limits_{\Delta x \to 0} -5 = -5$$

15. $h(s) = 3 + \dfrac{2}{3}s$

$$h'(s) = \lim\limits_{\Delta s \to 0} \frac{h(s + \Delta s) - h(s)}{\Delta s}$$

$$= \lim\limits_{\Delta s \to 0} \frac{3 + \dfrac{2}{3}(s + \Delta s) - \left(3 + \dfrac{2}{3}s\right)}{\Delta s}$$

$$= \lim\limits_{\Delta s \to 0} \frac{\dfrac{2}{3}\Delta s}{\Delta s} = \frac{2}{3}$$

17. $f(x) = 2x^2 + x - 1$

$$f'(x) = \lim_{\Delta x \to 0} \frac{f(x + \Delta x) - f(x)}{\Delta x}$$

$$= \lim_{\Delta x \to 0} \frac{[2(x + \Delta x)^2 + (x + \Delta x) - 1] - [2x^2 + x - 1]}{\Delta x}$$

$$= \lim_{\Delta x \to 0} \frac{(2x^2 + 4x\Delta x + 2(\Delta x)^2 + x + \Delta x - 1) - (2x^2 + x - 1)}{\Delta x}$$

$$= \lim_{\Delta x \to 0} \frac{4x\Delta x + 2(\Delta x)^2 + \Delta x}{\Delta x} = \lim_{\Delta x \to 0} (4x + 2\Delta x + 1) = 4x + 1$$

19. $f(x) = x^3 - 12x$

$$f'(x) = \lim_{\Delta x \to 0} \frac{f(x + \Delta x) - f(x)}{\Delta x}$$

$$= \lim_{\Delta x \to 0} \frac{[(x + \Delta x)^3 - 12(x + \Delta x)] - [x^3 - 12x]}{\Delta x}$$

$$= \lim_{\Delta x \to 0} \frac{x^3 + 3x^2\Delta x + 3x(\Delta x)^2 + (\Delta x)^3 - 12x - 12\Delta x - x^3 + 12x}{\Delta x}$$

$$= \lim_{\Delta x \to 0} \frac{3x^2\Delta x + 3x(\Delta x)^2 + (\Delta x)^3 - 12\Delta x}{\Delta x}$$

$$= \lim_{\Delta x \to 0} (3x^2 + 3x\Delta x + (\Delta x)^2 - 12) = 3x^2 - 12$$

21. $f(x) = \dfrac{1}{x - 1}$

$$f'(x) = \lim_{\Delta x \to 0} \frac{f(x + \Delta x) - f(x)}{\Delta x}$$

$$= \lim_{\Delta x \to 0} \frac{\dfrac{1}{x + \Delta x - 1} - \dfrac{1}{x - 1}}{\Delta x}$$

$$= \lim_{\Delta x \to 0} \frac{(x - 1) - (x + \Delta x - 1)}{\Delta x(x + \Delta x - 1)(x - 1)}$$

$$= \lim_{\Delta x \to 0} \frac{-\Delta x}{\Delta x(x + \Delta x - 1)(x - 1)}$$

$$= \lim_{\Delta x \to 0} \frac{-1}{(x + \Delta x - 1)(x - 1)}$$

$$= -\frac{1}{(x - 1)^2}$$

23. $f(x) = \sqrt{x + 1}$

$$f'(x) = \lim_{\Delta x \to 0} \frac{f(x + \Delta x) - f(x)}{\Delta x}$$

$$= \lim_{\Delta x \to 0} \frac{\sqrt{x + \Delta x + 1} - \sqrt{x + 1}}{\Delta x} \cdot \left(\frac{\sqrt{x + \Delta x + 1} + \sqrt{x + 1}}{\sqrt{x + \Delta x + 1} + \sqrt{x + 1}} \right)$$

$$= \lim_{\Delta x \to 0} \frac{(x + \Delta x + 1) - (x + 1)}{\Delta x\left[\sqrt{x + \Delta x + 1} + \sqrt{x + 1} \right]}$$

$$= \lim_{\Delta x \to 0} \frac{1}{\sqrt{x + \Delta x + 1} + \sqrt{x + 1}}$$

$$= \frac{1}{\sqrt{x + 1} + \sqrt{x + 1}} = \frac{1}{2\sqrt{x + 1}}$$

25. (a) $f(x) = x^2 + 1$

$$f'(x) = \lim_{\Delta x \to 0} \frac{f(x + \Delta x) - f(x)}{\Delta x}$$

$$= \lim_{\Delta x \to 0} \frac{[(x + \Delta x)^2 + 1] - [x^2 + 1]}{\Delta x}$$

$$= \lim_{\Delta x \to 0} \frac{2x\Delta x + (\Delta x)^2}{\Delta x}$$

$$= \lim_{\Delta x \to 0} (2x + \Delta x) = 2x$$

At $(2, 5)$, the slope of the tangent line is
$m = 2(2) = 4$. The equation of the tangent line is

$$y - 5 = 4(x - 2)$$

$$y - 5 = 4x - 8$$

$$y = 4x - 3.$$

(b)

27. (a) $f(x) = x^3$

$$f'(x) = \lim_{\Delta x \to 0} \frac{f(x + \Delta x) - f(x)}{\Delta x}$$

$$= \lim_{\Delta x \to 0} \frac{(x + \Delta x)^3 - x^3}{\Delta x}$$

$$= \lim_{\Delta x \to 0} \frac{3x^2\Delta x + 3x(\Delta x)^2 + (\Delta x)^3}{\Delta x}$$

$$= \lim_{\Delta x \to 0} (3x^2 + 3x\Delta x + (\Delta x)^2) = 3x^2$$

At $(2, 8)$, the slope of the tangent is $m = 3(2)^2 = 12$.
The equation of the tangent line is

$$y - 8 = 12(x - 2)$$

$$y = 12x - 16.$$

(b)

29. (a) $f(x) = \sqrt{x}$

$$f'(x) = \lim_{\Delta x \to 0} \frac{f(x + \Delta x) - f(x)}{\Delta x}$$

$$= \lim_{\Delta x \to 0} \frac{\sqrt{x + \Delta x} - \sqrt{x}}{\Delta x} \cdot \frac{\sqrt{x + \Delta x} + \sqrt{x}}{\sqrt{x + \Delta x} + \sqrt{x}}$$

$$= \lim_{\Delta x \to 0} \frac{(x + \Delta x) - x}{\Delta x(\sqrt{x + \Delta x} + \sqrt{x})}$$

$$= \lim_{\Delta x \to 0} \frac{1}{\sqrt{x + \Delta x} + \sqrt{x}} = \frac{1}{2\sqrt{x}}$$

At $(1, 1)$, the slope of the tangent line is

$$m = \frac{1}{2\sqrt{1}} = \frac{1}{2}.$$

The equation of the tangent line is

$$y - 1 = \frac{1}{2}(x - 1)$$

$$y = \frac{1}{2}x + \frac{1}{2}.$$

(b)

31. (a) $f(x) = x + \dfrac{4}{x}$

$$f'(x) = \lim_{\Delta x \to 0} \frac{f(x + \Delta x) - f(x)}{\Delta x}$$

$$= \lim_{\Delta x \to 0} \frac{(x + \Delta x) + \dfrac{4}{x + \Delta x} - \left(x + \dfrac{4}{x}\right)}{\Delta x}$$

$$= \lim_{\Delta x \to 0} \frac{x(x + \Delta x)(x + \Delta x) + 4x - x^2(x + \Delta x) - 4(x + \Delta x)}{x(\Delta x)(x + \Delta x)}$$

$$= \lim_{\Delta x \to 0} \frac{x^3 + 2x^2(\Delta x) + x(\Delta x)^2 - x^3 - x^2(\Delta x) - 4(\Delta x)}{x(\Delta x)(x + \Delta x)}$$

$$= \lim_{\Delta x \to 0} \frac{x^2(\Delta x) + x(\Delta x)^2 - 4(\Delta x)}{x(\Delta x)(x + \Delta x)}$$

$$= \lim_{\Delta x \to 0} \frac{x^2 + x(\Delta x) - 4}{x(x + \Delta x)}$$

$$= \frac{x^2 - 4}{x^2} = 1 - \frac{4}{x^2}$$

At $(4, 5)$, the slope of the tangent line is

$$m = 1 - \frac{4}{16} = \frac{3}{4}.$$

The equation of the tangent line is

$$y - 5 = \frac{3}{4}(x - 4)$$

$$y = \frac{3}{4}x + 2.$$

(b)

33. From Exercise 27 we know that $f'(x) = 3x^2$. Since the slope of the given line is 3, we have

$$3x^2 = 3$$

$$x = \pm 1.$$

Therefore, at the points $(1, 1)$ and $(-1, -1)$ the tangent lines are parallel to $3x - y + 1 = 0$. These lines have equations

$$y - 1 = 3(x - 1) \qquad \text{and} \qquad y + 1 = 3(x + 1)$$

$$y = 3x - 2 \qquad\qquad\qquad\qquad y = 3x + 2.$$

35. Using the limit definition of derivative,

$$f'(x) = \frac{-1}{2x\sqrt{x}}.$$

Since the slope of the given line is $-\frac{1}{2}$, we have

$$-\frac{1}{2x\sqrt{x}} = -\frac{1}{2}$$

$$x = 1.$$

Therefore, at the point $(1, 1)$ the tangent line is parallel to $x + 2y - 6 = 0$. The equation of this line is

$$y - 1 = -\frac{1}{2}(x - 1)$$

$$y - 1 = -\frac{1}{2}x + \frac{1}{2}$$

$$y = -\frac{1}{2}x + \frac{3}{2}.$$

37. $g(5) = 2$ because the tangent line passes through $(5, 2)$.

$$g'(5) = \frac{2 - 0}{5 - 9} = \frac{2}{-4} = -\frac{1}{2}$$

39. $f(x) = x \Rightarrow f'(x) = 1$ matches (b)

41. $f(x) = \sqrt{x} \Rightarrow f'(x)$ matches (a)

(decreasing slope as $x \to \infty$)

43.

Answers will vary.

Sample answer: $y = -x$

45. (a) If $f'(c) = 3$ and f is odd, then $f'(-c) = f'(c) = 3$

(b) If $f'(c) = 3$ and f is even, then $f'(-c) = -f'(c) = -3$

47. Let (x_0, y_0) be a point of tangency on the graph of f. By the limit definition for the derivative, $f'(x) = 4 - 2x$. The slope of the line through $(2, 5)$ and (x_0, y_0) equals the derivative of f at x_0:

$$\frac{5 - y_0}{2 - x_0} = 4 - 2x_0$$

$$5 - y_0 = (2 - x_0)(4 - 2x_0)$$

$$5 - (4x_0 - x_0^2) = 8 - 8x_0 + 2x_0^2$$

$$0 = x_0^2 - 4x_0 + 3$$

$$0 = (x_0 - 1)(x_0 - 3) \Rightarrow x_0 = 1, 3$$

Therefore, the points of tangency are $(1, 3)$ and $(3, 3)$, and the corresponding slopes are 2 and -2. The equations of the tangent lines are

$$y - 5 = 2(x - 2) \qquad y - 5 = -2(x - 2)$$

$$y = 2x + 1 \qquad\qquad y = -2x + 9$$

49. (a) $g'(0) = -3$

(b) $g'(3) = 0$

(c) Because $g'(1) = -\frac{8}{3}$, g is decreasing (falling) at $x = 1$.

(d) Because $g'(-4) = \frac{7}{3}$, g is increasing (rising) at $x = -4$.

(e) Because $g'(4)$ and $g'(6)$ are both positive, $g(6)$ is greater than $g(4)$, and $g(6) - g(4) > 0$.

(f) No, it is not possible. All you can say is that g is decreasing (falling) at $x = 2$.

51. $f(x) = \frac{1}{4}x^3$

By the limit definition of the derivative we have $f'(x) = \frac{3}{4}x^2$.

x	-2	-1.5	-1	-0.5	0	0.5	1	1.5	2
$f(x)$	-2	$-\frac{27}{32}$	$-\frac{1}{4}$	$-\frac{1}{32}$	0	$\frac{1}{32}$	$\frac{1}{4}$	$\frac{27}{32}$	2
$f'(x)$	3	$\frac{27}{16}$	$\frac{3}{4}$	$\frac{3}{16}$	0	$\frac{3}{16}$	$\frac{3}{4}$	$\frac{27}{16}$	3

53. $g(x) = \dfrac{f(x + 0.01) - f(x)}{0.01}$

$\qquad = [(2(x + 0.01) - (x + 0.01)^2 - 2x + x^2)] \cdot 100$

$\qquad = 2 - 2x - 0.01$

The graph of $g(x)$ is approximately the graph of $f'(x)$.

55. $f(2) = 2(4 - 2) = 4$, $f(2.1) = 2.1(4 - 2.1) = 3.99$

$\quad f'(2) \approx \dfrac{3.99 - 4}{2.1 - 2} = -0.1$ [Exact: $f'(2) = 0$]

57. $f(x) = \dfrac{1}{\sqrt{x}}$ and $f'(x) = \dfrac{-1}{2x^{3/2}}$.

As $x \to \infty$, f is nearly horizontal and thus $f' \approx 0$.

59. $f(x) = 4 - (x - 3)^2$

$\quad S_{\Delta x}(x) = \dfrac{f(2 + \Delta x) - f(2)}{\Delta x}(x - 2) + f(2)$

$\qquad = \dfrac{4 - (2 + \Delta x - 3)^2 - 3}{\Delta x}(x - 2) + 3 = \dfrac{1 - (\Delta x - 1)^2}{\Delta x}(x - 2) + 3 = (-\Delta x + 2)(x - 2) + 3$

(a) $\Delta x = \quad 1$: $S_{\Delta x} = (x - 2) + 3 = x + 1$

$\quad \Delta x = 0.5$: $S_{\Delta x} = \left(\dfrac{3}{2}\right)(x - 2) + 3 = \dfrac{3}{2}x$

$\quad \Delta x = 0.1$: $S_{\Delta x} = \left(\dfrac{19}{10}\right)(x - 2) + 3 = \dfrac{19}{10}x - \dfrac{4}{5}$

(b) As $\Delta x \to 0$, the line approaches the tangent line to f at $(2, 3)$.

61. $f(x) = x^2 - 1$, $c = 2$

$\quad f'(2) = \lim\limits_{x \to 2} \dfrac{f(x) - f(2)}{x - 2} = \lim\limits_{x \to 2} \dfrac{(x^2 - 1) - 3}{x - 2} = \lim\limits_{x \to 2} \dfrac{(x - 2)(x + 2)}{x - 2} = \lim\limits_{x \to 2}(x + 2) = 4$

63. $f(x) = x^3 + 2x^2 + 1$, $c = -2$

$\quad f'(-2) = \lim\limits_{x \to -2} \dfrac{f(x) - f(-2)}{x + 2} = \lim\limits_{x \to -2} \dfrac{(x^3 + 2x^2 + 1) - 1}{x + 2} = \lim\limits_{x \to -2} \dfrac{x^2(x + 2)}{x + 2} = \lim\limits_{x \to -2} x^2 = 4$

65. $g(x) = \sqrt{|x|}$, $c = 0$

$\quad g'(0) = \lim\limits_{x \to 0} \dfrac{g(x) - g(0)}{x - 0} = \lim\limits_{x \to 0} \dfrac{\sqrt{|x|}}{x}$. Does not exist.

$\quad$ As $x \to 0^-$, $\dfrac{\sqrt{|x|}}{x} = \dfrac{-1}{\sqrt{|x|}} \to -\infty$

$\quad$ As $x \to 0^+$, $\dfrac{\sqrt{|x|}}{x} = \dfrac{1}{\sqrt{x}} \to \infty$

67. $f(x) = (x - 6)^{2/3}$, $c = 6$

$\quad f'(6) = \lim\limits_{x \to 6} \dfrac{f(x) - f(6)}{x - 6}$

$\qquad = \lim\limits_{x \to 6} \dfrac{(x - 6)^{2/3} - 0}{x - 6}$

$\qquad = \lim\limits_{x \to 6} \dfrac{1}{(x - 6)^{1/3}}$

Does not exist.

69. $h(x) = |x + 5|$, $c = -5$

$$h'(-5) = \lim_{x \to -5} \frac{h(x) - h(-5)}{x - (-5)}$$

$$= \lim_{x \to -5} \frac{|x + 5| - 0}{x + 5}$$

$$= \lim_{x \to -5} \frac{|x + 5|}{x + 5}$$

Does not exist.

71. $f(x)$ is differentiable everywhere except at $x = -3$. (Sharp turn in the graph.)

73. $f(x)$ is differentiable everywhere except at $x = -1$. (Discontinuity)

75. $f(x)$ is differentiable everywhere except at $x = 3$. (Sharp turn in the graph)

77. $f(x)$ is differentiable on the interval $(1, \infty)$. (At $x = 1$ the tangent line is vertical)

79. $f(x)$ is differentiable everywhere except at $x = 0$. (Discontinuity)

81. $f(x) = |x - 1|$

The derivative from the left is

$$\lim_{x \to 1^-} \frac{f(x) - f(1)}{x - 1} = \lim_{x \to 1^-} \frac{|x - 1| - 0}{x - 1} = -1.$$

The derivative from the right is

$$\lim_{x \to 1^+} \frac{f(x) - f(1)}{x - 1} = \lim_{x \to 1^+} \frac{|x - 1| - 0}{x - 1} = 1.$$

The one-sided limits are not equal. Therefore, f is not differentiable at $x = 1$.

83. $f(x) = \begin{cases} (x - 1)^3, & x \le 1 \\ (x - 1)^2, & x > 1 \end{cases}$

The derivative from the left is

$$\lim_{x \to 1^-} \frac{f(x) - f(1)}{x - 1} = \lim_{x \to 1^-} \frac{(x - 1)^3 - 0}{x - 1}$$

$$= \lim_{x \to 1^-} (x - 1)^2 = 0.$$

The derivative from the right is

$$\lim_{x \to 1^+} \frac{f(x) - f(1)}{x - 1} = \lim_{x \to 1^+} \frac{(x - 1)^2 - 0}{x - 1}$$

$$= \lim_{x \to 1^+} (x - 1) = 0.$$

These one-sided limits are equal. Therefore, f is differentiable at $x = 1$. $(f'(1) = 0)$

85. Note that f is continuous at $x = 2$. $f(x) = \begin{cases} x^2 + 1, & x \le 2 \\ 4x - 3, & x > 2 \end{cases}$

The derivative from the left is $\lim_{x \to 2^-} \dfrac{f(x) - f(2)}{x - 2} = \lim_{x \to 2^-} \dfrac{(x^2 + 1) - 5}{x - 2} = \lim_{x \to 2^-} (x + 2) = 4.$

The derivative from the right is $\lim_{x \to 2^+} \dfrac{f(x) - f(2)}{x - 2} = \lim_{x \to 2^+} \dfrac{(4x - 3) - 5}{x - 2} = \lim_{x \to 2^+} 4 = 4.$

The one-sided limits are equal. Therefore, f is differentiable at $x = 2$. $(f'(2) = 4)$

87. (a) The distance from $(3, 1)$ to the line $mx - y + 4 = 0$ is

$$d = \frac{|Ax_1 + By_1 + C|}{\sqrt{A^2 + B^2}}$$

$$= \frac{|m(3) - 1(1) + 4|}{\sqrt{m^2 + 1}} = \frac{|3m + 3|}{\sqrt{m^2 + 1}}.$$

(b)

The function d is not differentiable at $m = -1$. This corresponds to the line $y = -x + 4$, which passes through the point $(3, 1)$.

89. False. the slope is $\lim\limits_{\triangle x \to 0} \dfrac{f(2 + \triangle x) - f(2)}{\triangle x}$.

91. False. If the derivative from the left of a point does not equal the derivative from the right of a point, then the derivative does not exist at that point. For example, if $f(x) = |x|$, then the derivative from the left at $x = 0$ is -1 and the derivative from the right at $x = 0$ is 1. At $x = 0$, the derivative does not exist.

93. $f(x) = \begin{cases} x \sin(1/x), & x \neq 0 \\ 0, & x = 0 \end{cases}$

Using the Squeeze Theorem, we have $-|x| \leq x \sin(1/x) \leq |x|$, $x \neq 0$. Thus, $\lim\limits_{x \to 0} x \sin(1/x) = 0 = f(0)$ and f is continuous at $x = 0$. Using the alternative form of the derivative we have

$$\lim_{x \to 0} \frac{f(x) - f(0)}{x - 0} = \lim_{x \to 0} \frac{x \sin(1/x) - 0}{x - 0} = \lim_{x \to 0} \left(\sin\frac{1}{x} \right).$$

Since this limit does not exist (it oscillates between -1 and 1), the function is not differentiable at $x = 0$.

$$g(x) = \begin{cases} x^2 \sin(1/x), & x \neq 0 \\ 0, & x = 0 \end{cases}$$

Using the Squeeze Theorem again we have $-x^2 \leq x^2 \sin(1/x) \leq x^2$, $x \neq 0$. Thus, $\lim\limits_{x \to 0} x^2 \sin(1/x) = 0 = f(0)$ and f is continuous at $x = 0$. Using the alternative form of the derivative again we have

$$\lim_{x \to 0} \frac{f(x) - f(0)}{x - 0} = \lim_{x \to 0} \frac{x^2 \sin(1/x) - 0}{x - 0} = \lim_{x \to 0} x \sin\frac{1}{x} = 0.$$

Therefore, g is differentiable at $x = 0$, $g'(0) = 0$.

Section 2.2   Basic Differentiation Rules and Rates of Change

1. (a) $y = x^{1/2}$

$y' = \frac{1}{2}x^{-1/2}$

$y'(1) = \frac{1}{2}$

(b) $y = x^{3/2}$

$y' = \frac{3}{2}x^{1/2}$

$y'(1) = \frac{3}{2}$

(c) $y = x^2$

$y' = 2x$

$y'(1) = 2$

(d) $y = x^3$

$y' = 3x^2$

$y'(1) = 3$

3. $y = 8$

$y' = 0$

5. $y = x^6$

$y' = 6x^5$

7. $y = \dfrac{1}{x^7} = x^{-7}$

$y' = -7x^{-8} = \dfrac{-7}{x^8}$

9. $y = \sqrt[5]{x} = x^{1/5}$

$y' = \dfrac{1}{5}x^{-4/5} = \dfrac{1}{5x^{4/5}}$

11. $f(x) = x + 1$

$f'(x) = 1$

13. $f(t) = -2t^2 + 3t - 6$

$f'(x) = -4t + 3$

15. $g(x) = x^2 + 4x^3$

$g'(x) = 2x + 12x^2$

17. $s(t) = t^3 - 2t + 4$

$s'(t) = 3t^2 - 2$

19. $y = \dfrac{\pi}{2} \sin \theta - \cos \theta$

$y' = \dfrac{\pi}{2} \cos \theta + \sin \theta$

21. $y = x^2 - \dfrac{1}{2} \cos x$

$y' = 2x + \dfrac{1}{2} \sin x$

23. $y = \dfrac{1}{x} - 3 \sin x$

$y' = -\dfrac{1}{x^2} - 3 \cos x$

Function	*Rewrite*	*Derivative*	*Simplify*
25. $y = \dfrac{5}{2x^2}$	$y = \dfrac{5}{2}x^{-2}$	$y' = -5x^{-3}$	$y' = \dfrac{-5}{x^3}$
27. $y = \dfrac{3}{(2x)^3}$	$y = \dfrac{3}{8}x^{-3}$	$y' = \dfrac{-9}{8}x^{-4}$	$y' = \dfrac{-9}{8x^4}$
29. $y = \dfrac{\sqrt{x}}{x}$	$y = x^{-1/2}$	$y' = -\dfrac{1}{2}x^{-3/2}$	$y' = -\dfrac{1}{2x^{3/2}}$

31. $f(x) = \dfrac{3}{x^2} = 3x^{-2}, (1, 3)$

$\qquad f'(x) = -6x^{-3} = \dfrac{-6}{x^3}$

$\qquad f'(1) = -6$

33. $f(x) = -\dfrac{1}{2} + \dfrac{7}{5}x^3, \left(0, -\dfrac{1}{2}\right)$

$\qquad f'(x) = \dfrac{21}{5}x^2$

$\qquad f'(0) = 0$

35. $\qquad y = (2x + 1)^2, (0, 1)$

$\qquad\qquad = 4x^2 + 4x + 1$

$\qquad\quad y' = 8x + 4$

$\qquad\quad y'(0) = 4$

37. $f(\theta) = 4 \sin \theta - \theta, (0, 0)$

$\qquad f'(\theta) = 4 \cos \theta - 1$

$\qquad f'(0) = 4(1) - 1 = 3$

39. $f(x) = x^2 + 5 - 3x^{-2}$

$\qquad f'(x) = 2x + 6x^{-3} = 2x + \dfrac{6}{x^3}$

41. $g(t) = t^2 - \dfrac{4}{t^3} = t^2 - 4t^{-3}$

$\qquad g'(t) = 2t + 12t^{-4} = 2t + \dfrac{12}{t^4}$

43. $f(x) = \dfrac{x^3 - 3x^2 + 4}{x^2} = x - 3 + 4x^{-2}$

$\qquad f'(x) = 1 - \dfrac{8}{x^3} = \dfrac{x^3 - 8}{x^3}$

45. $y = x(x^2 + 1) = x^3 + x$

$\qquad y' = 3x^2 + 1$

47. $f(x) = \sqrt{x} - 6 \sqrt[3]{x} = x^{1/2} - 6x^{1/3}$

$\qquad f'(x) = \dfrac{1}{2}x^{-1/2} - 2x^{-2/3} = \dfrac{1}{2\sqrt{x}} - \dfrac{2}{x^{2/3}}$

49. $h(s) = s^{4/5} - s^{2/3}$

$\qquad h'(s) = \dfrac{4}{5}s^{-1/5} - \dfrac{2}{3}s^{-1/3} = \dfrac{4}{5s^{1/5}} - \dfrac{2}{3s^{1/3}}$

51. $f(x) = 6\sqrt{x} + 5 \cos x = 6x^{1/2} + 5 \cos x$

$\qquad f'(x) = 3x^{-1/2} - 5 \sin x = \dfrac{3}{\sqrt{x}} - 5 \sin x$

53. (a) $y = x^4 - 3x^2 + 2$

$\qquad\quad y' = 4x^3 - 6x$

$\qquad\quad$ At $(1, 0)$: $y' = 4(1)^3 - 6(1) = -2$.

$\qquad\quad$ Tangent line: $\qquad y - 0 = -2(x - 1)$

$\qquad\qquad\qquad\qquad\quad 2x + y - 2 = 0$

(b)

55. (a) $f(x) = \dfrac{2}{\sqrt[4]{x^3}} = 2x^{-3/4}$

$\qquad\quad f'(x) = \dfrac{-3}{2}x^{-7/4} = \dfrac{-3}{2x^{7/4}}$

$\qquad\quad$ At $(1, 2)$, $f'(1) = \dfrac{-3}{2}$

$\qquad\quad$ Tangent line: $\qquad y - 2 = -\dfrac{3}{2}(x - 1)$

$\qquad\qquad\qquad\qquad\qquad y = -\dfrac{3}{2}x + \dfrac{7}{2}$

$\qquad\qquad\qquad\qquad\quad 3x + 2y - 7 = 0$

(b)

57. $y = x^4 - 8x^2 + 2$

$y' = 4x^3 - 16x$

$\quad = 4x(x^2 - 4)$

$\quad = 4x(x - 2)(x + 2)$

$y' = 0 \implies x = 0, \pm 2$

Horizontal tangents: $(0, 2), (2, -14), (-2, -14)$

59. $y = \dfrac{1}{x^2} = x^{-2}$

$y' = -2x^{-3} = \dfrac{-2}{x^3}$ cannot equal zero.

Therefore, there are no horizontal tangents.

61. $y = x + \sin x, 0 \le x < 2\pi$

$y' = 1 + \cos x = 0$

$\cos x = -1 \implies x = \pi$

At $x = \pi, y = \pi$.

Horizontal tangent: (π, π)

63. $x^2 - kx = 4x - 9$ Equate functions

$\quad 2x - k = 4$ Equate derivatives

Hence, $k = 2x - 4$ and

$x^2 - (2x - 4)x = 4x - 9 \implies -x^2 = -9 \implies x = \pm 3.$

For $x = 3, k = 2$ and for $x = -3, k = -10.$

65. $\dfrac{k}{x} = -\dfrac{3}{4}x + 3$ Equate functions

$-\dfrac{k}{x^2} = -\dfrac{3}{4}$ Equate derivatives

Hence, $k = \dfrac{3}{4}x^2$ and $\dfrac{\frac{3}{4}x^2}{x} = \dfrac{-3}{4}x + 3 \implies \dfrac{3}{4}x = -\dfrac{3}{4}x + 3 \implies \dfrac{3}{2}x = 3 \implies x = 2 \implies k = 3.$

67. (a) The slope appears to be steepest between A and B.

(b) The average rate of change between A and B is **greater** than the instantaneous rate of change at B.

(c)

69. $g(x) = f(x) + 6 \implies g'(x) = f'(x)$

71.

If f is linear then its derivative is a constant function.

$\qquad f(x) = ax + b$

$\qquad f'(x) = a$

73. Let (x_1, y_1) and (x_2, y_2) be the points of tangency on $y = x^2$ and $y = -x^2 + 6x - 5$, respectively. The derivatives of these functions are

$$y' = 2x \implies m = 2x_1 \quad \text{and} \quad y' = -2x + 6 \implies m = -2x_2 + 6.$$

$$m = 2x_1 = -2x_2 + 6$$

$$x_1 = -x_2 + 3$$

Since $y_1 = x_1^2$ and $y_2 = -x_2^2 + 6x_2 - 5$,

$$m = \frac{y_2 - y_1}{x_2 - x_1} = \frac{(-x_2^2 + 6x_2 - 5) - (x_1^2)}{x_2 - x_1} = -2x_2 + 6.$$

$$\frac{(-x_2^2 + 6x_2 - 5) - (-x_2 + 3)^2}{x_2 - (-x_2 + 3)} = -2x_2 + 6$$

$$(-x_2^2 + 6x_2 - 5) - (x_2^2 - 6x_2 + 9) = (-2x_2 + 6)(2x_2 - 3)$$

$$-2x_2^2 + 12x_2 - 14 = -4x_2^2 + 18x_2 - 18$$

$$2x_2^2 - 6x_2 + 4 = 0$$

$$2(x_2 - 2)(x_2 - 1) = 0$$

$$x_2 = 1 \text{ or } 2$$

$x_2 = 1 \implies y_2 = 0, x_1 = 2$ and $y_1 = 4$

Thus, the tangent line through $(1, 0)$ and $(2, 4)$ is

$$y - 0 = \left(\frac{4 - 0}{2 - 1}\right)(x - 1) \implies y = 4x - 4.$$

$x_2 = 2 \implies y_2 = 3, x_1 = 1$ and $y_1 = 1$

Thus, the tangent line through $(2, 3)$ and $(1, 1)$ is

$$y - 1 = \left(\frac{3 - 1}{2 - 1}\right)(x - 1) \implies y = 2x - 1.$$

75. $f(x) = \sqrt{x}, \ (-4, 0)$

$$f'(x) = \frac{1}{2}x^{-1/2} = \frac{1}{2\sqrt{x}}$$

$$\frac{1}{2\sqrt{x}} = \frac{0 - y}{-4 - x}$$

$$4 + x = 2\sqrt{x}y$$

$$4 + x = 2\sqrt{x}\sqrt{x}$$

$$4 + x = 2x$$

$$x = 4, y = 2$$

The point $(4, 2)$ is on the graph of f.

Tangent line: $y - 2 = \dfrac{0 - 2}{-4 - 4}(x - 4)$

$$4y - 8 = x - 4$$

$$0 = x - 4y + 4$$

77. $f'(1) = -1$

79. (a) One possible secant is between (3.9, 7.7019) and (4, 8):

$$y - 8 = \frac{8 - 7.7019}{4 - 3.9}(x - 4)$$

$$y - 8 = 2.981(x - 4)$$

$$y = S(x) = 2.981x - 3.924$$

(b) $f'(x) = \frac{3}{2}x^{1/2} \Rightarrow f'(4) = \frac{3}{2}(2) = 3$

$$T(x) = 3(x - 4) + 8 = 3x - 4$$

$S(x)$ is an approximation of the tangent line $T(x)$.

(c) As you move further away from (4, 8), the accuracy of the approximation T gets worse.

(d)

Δx	-3	-2	-1	-0.5	-0.1	0	0.1	0.5	1	2	3
$f(4 + \Delta x)$	1	2.828	5.196	6.548	7.702	8	8.302	9.546	11.180	14.697	18.520
$T(4 + \Delta x)$	-1	2	5	6.5	7.7	8	8.3	9.5	11	14	17

81. False. Let $f(x) = x^2$ and $g(x) = x^2 + 4$. Then $f'(x) = g'(x) = 2x$, but $f(x) \neq g(x)$.

83. False. If $y = \pi^2$, then $dy/dx = 0$. (π^2 is a constant.)

85. True. If $g(x) = 3f(x)$, then $g'(x) = 3f'(x)$.

87. $f(t) = 2t + 7, [1, 2]$

$f'(t) = 2$

Instantaneous rate of change is the constant 2.
Average rate of change:

$$\frac{f(2) - f(1)}{2 - 1} = \frac{[2(2) + 7] - [2(1) + 7]}{1} = 2$$

(These are the same because f is a line of slope 2.)

89. $f(x) = -\frac{1}{x}, [1, 2]$

$$f'(x) = \frac{1}{x^2}$$

Instantaneous rate of change:

$$(1, -1) \Rightarrow f'(1) = 1$$

$$\left(2, -\frac{1}{2}\right) \Rightarrow f'(2) = \frac{1}{4}$$

Average rate of change:

$$\frac{f(2) - f(1)}{2 - 1} = \frac{(-1/2) - (-1)}{2 - 1} = \frac{1}{2}$$

91. (a) $s(t) = -16t^2 + 1362$

$v(t) = -32t$

(b) $\dfrac{s(2) - s(1)}{2 - 1} = 1298 - 1346 = -48$ ft/sec

(c) $v(t) = s'(t) = -32t$

When $t = 1$: $v(1) = -32$ ft/sec.

When $t = 2$: $v(2) = -64$ ft/sec.

(d) $-16t^2 + 1362 = 0$

$$t^2 = \frac{1362}{16} \Rightarrow t = \frac{\sqrt{1362}}{4} \approx 9.226 \text{ sec}$$

(e) $v\left(\dfrac{\sqrt{1362}}{4}\right) = -32\left(\dfrac{\sqrt{1362}}{4}\right)$

$\qquad = -8\sqrt{1362} \approx -295.242$ ft/sec

93. $s(t) = -4.9t^2 + v_0 t + s_0$

$\qquad = -4.9t^2 + 120t$

$v(t) = -9.8t + 120$

$v(5) = -9.8(5) + 120 = 71$ m/sec

$v(10) = -9.8(10) + 120 = 22$ m/sec

95.

(The velocity has been converted to miles per hour)

97. $v = 40$ mph $= \frac{2}{3}$ mi/min

$\left(\frac{2}{3} \text{ mi/min}\right)(6 \text{ min}) = 4$ mi

$v = 0$ mph $= 0$ mi/min

$(0 \text{ mi/min})(2 \text{ min}) = 0$ mi

$v = 60$ mph $= 1$ mi/min

$(1 \text{ mi/min})(2 \text{ min}) = 2$ mi

99. (a) Using a graphing utility, you obtain

$R = 0.167v - 0.02.$

(b) Using a graphing utility, you obtain

$B = 0.00586v^2 - 0.0239v + 0.46.$

(c) $T = R + B = 0.00586v^2 + 0.1431v + 0.44$

(d)

(e) $\dfrac{dT}{dv} = 0.01172v + 0.1431$

For $v = 40$, $T'(40) \approx 0.612$.

For $v = 80$, $T'(80) \approx 1.081$.

For $v = 100$, $T'(100) \approx 1.315$.

(f) For increasing speeds, the total stopping distance increases.

101. $A = s^2, \dfrac{dA}{ds} = 2s$

When $s = 4$ m,

$\dfrac{dA}{ds} = 8$ square meters per meter change in s.

103. $\qquad C = \dfrac{1{,}008{,}000}{Q} + 6.3Q$

$\qquad \dfrac{dC}{dQ} = -\dfrac{1{,}008{,}000}{Q^2} + 6.3$

$C(351) - C(350) \approx 5083.095 - 5085 \approx -\$1.91$

When $Q = 350$, $\dfrac{dC}{dQ} \approx -\$1.93$.

105. (a) $f'(1.47)$ is the rate of change of the amount of gasoline sold when the price is \$1.47 per gallon.

(b) $f'(1.47)$ is usually negative. As prices go up, sales go down.

107. $y = ax^2 + bx + c$

Since the parabola passes through $(0, 1)$ and $(1, 0)$, we have

$(0, 1)$: $1 = a(0)^2 + b(0) + c \implies c = 1$

$(1, 0)$: $0 = a(1)^2 + b(1) + 1 \implies b = -a - 1$.

Thus, $y = ax^2 + (-a - 1)x + 1$. From the tangent line $y = x - 1$, we know that the derivative is 1 at the point $(1, 0)$.

$y' = 2ax + (-a - 1)$

$1 = 2a(1) + (-a - 1)$

$1 = a - 1$

$a = 2$

$b = -a - 1 = -3$

Therefore, $y = 2x^2 - 3x + 1$.

109. $y = x^3 - 9x$

$y' = 3x^2 - 9$

Tangent lines through $(1, -9)$:

$y + 9 = (3x^2 - 9)(x - 1)$

$(x^3 - 9x) + 9 = 3x^3 - 3x^2 - 9x + 9$

$0 = 2x^3 - 3x^2 = x^2(2x - 3)$

$x = 0 \quad \text{or} \quad x = \frac{3}{2}$

The points of tangency are $(0, 0)$ and $\left(\frac{3}{2}, -\frac{81}{8}\right)$. At $(0, 0)$ the slope is $y'(0) = -9$. At $\left(\frac{3}{2}, -\frac{81}{8}\right)$ the slope is $y'\left(\frac{3}{2}\right) = -\frac{9}{4}$.

Tangent lines:

$y - 0 = -9(x - 0) \quad \text{and} \quad y + \frac{81}{8} = -\frac{9}{4}\left(x - \frac{3}{2}\right)$

$y = -9x \qquad\qquad\qquad\qquad y = -\frac{9}{4}x - \frac{27}{4}$

$9x + y = 0 \qquad\qquad\qquad 9x + 4y + 27 = 0$

111. $f(x) = \begin{cases} ax^3, & x \le 2 \\ x^2 + b, & x > 2 \end{cases}$

f must be continuous at $x = 2$ to be differentiable at $x = 2$.

$\left.\begin{array}{l} \displaystyle\lim_{x \to 2^-} f(x) = \lim_{x \to 2^-} ax^3 = 8a \\[2mm] \displaystyle\lim_{x \to 2^+} f(x) = \lim_{x \to 2^+} (x^2 + b) = 4 + b \end{array}\right\} \quad \begin{array}{l} 8a = 4 + b \\[2mm] 8a - 4 = b \end{array}$

$f'(x) = \begin{cases} 3ax^2, & x < 2 \\ 2x, & x > 2 \end{cases}$

For f to be differentiable at $x = 2$, the left derivative must equal the right derivative.

$3a(2)^2 = 2(2)$

$12a = 4$

$a = \frac{1}{3}$

$b = 8a - 4 = -\frac{4}{3}$

113. Let $f(x) = \cos x$.

$$f'(x) = \lim_{\Delta x \to 0} \frac{f(x + \Delta x) - f(x)}{\Delta x}$$

$$= \lim_{\Delta x \to 0} \frac{\cos x \cos \Delta x - \sin x \sin \Delta x - \cos x}{\Delta x}$$

$$= \lim_{\Delta x \to 0} \frac{\cos x(\cos \Delta x - 1)}{\Delta x} - \lim_{\Delta x \to 0} \sin x\left(\frac{\sin \Delta x}{\Delta x}\right)$$

$$= 0 - \sin x(1) = -\sin x$$

Section 2.3   The Product and Quotient Rules and Higher-Order Derivatives

1. $g(x) = (x^2 + 1)(x^2 - 2x)$

$g'(x) = (x^2 + 1)(2x - 2) + (x^2 - 2x)(2x)$

$\quad = 2x^3 - 2x^2 + 2x - 2 + 2x^3 - 4x^2$

$\quad = 4x^3 - 6x^2 + 2x - 2$

3. $h(t) = \sqrt[3]{t}(t^2 + 4) = t^{1/3}(t^2 + 4)$

$h'(t) = t^{1/3}(2t) + (t^2 + 4)\frac{1}{3}t^{-2/3}$

$\quad = 2t^{4/3} + \frac{t^2 + 4}{3t^{2/3}}$

$\quad = \frac{7t^2 + 4}{3t^{2/3}}$

5. $f(x) = x^3 \cos x$

$f'(x) = x^3(-\sin x) + \cos x(3x^2)$

$\quad = 3x^2 \cos x - x^3 \sin x$

7. $f(x) = \dfrac{x}{x^2 + 1}$

$f'(x) = \dfrac{(x^2 + 1)(1) - x(2x)}{(x^2 + 1)^2} = \dfrac{1 - x^2}{(x^2 + 1)^2}$

9. $h(x) = \dfrac{\sqrt[3]{x}}{x^3 + 1} = \dfrac{x^{1/3}}{x^3 + 1}$

$h'(x) = \dfrac{(x^3 + 1)\frac{1}{3}x^{-2/3} - x^{1/3}(3x^2)}{(x^3 + 1)^2}$

$\quad = \dfrac{(x^3 + 1) - x(9x^2)}{3x^{2/3}(x^3 + 1)^2}$

$\quad = \dfrac{1 - 8x^3}{3x^{2/3}(x^3 + 1)^2}$

11. $g(x) = \dfrac{\sin x}{x^2}$

$g'(x) = \dfrac{x^2(\cos x) - \sin x(2x)}{(x^2)^2} = \dfrac{x \cos x - 2 \sin x}{x^3}$

13. $f(x) = (x^3 - 3x)(2x^2 + 3x + 5)$

$f'(x) = (x^3 - 3x)(4x + 3) + (2x^2 + 3x + 5)(3x^2 - 3)$

$\quad = 10x^4 + 12x^3 - 3x^2 - 18x - 15$

$f'(0) = -15$

15. $f(x) = \dfrac{x^2 - 4}{x - 3}$

$f'(x) = \dfrac{(x - 3)(2x) - (x^2 - 4)(1)}{(x - 3)^2} = \dfrac{2x^2 - 6x - x^2 + 4}{(x - 3)^2}$

$\quad = \dfrac{x^2 - 6x + 4}{(x - 3)^2}$

$f'(1) = \dfrac{1 - 6 + 4}{(1 - 3)^2} = -\dfrac{1}{4}$

17. $f(x) = x \cos x$

$f'(x) = (x)(-\sin x) + (\cos x)(1) = \cos x - x \sin x$

$f'\left(\dfrac{\pi}{4}\right) = \dfrac{\sqrt{2}}{2} - \dfrac{\pi}{4}\left(\dfrac{\sqrt{2}}{2}\right) = \dfrac{\sqrt{2}}{8}(4 - \pi)$

| *Function* | *Rewrite* | *Derivative* | *Simplify* |

19. $y = \dfrac{x^2 + 2x}{3}$ $\qquad$ $y = \dfrac{1}{3}x^2 + \dfrac{2}{3}x$ $\qquad$ $y' = \dfrac{2}{3}x + \dfrac{2}{3}$ $\qquad$ $y' = \dfrac{2x + 2}{3}$

21. $y = \dfrac{7}{3x^3}$ $\qquad$ $y = \dfrac{7}{3}x^{-3}$ $\qquad$ $y' = -7x^{-4}$ $\qquad$ $y' = -\dfrac{7}{x^4}$

23. $y = \dfrac{4x^{3/2}}{x}$ $\qquad$ $y = 4\sqrt{x}, x > 0$ $\qquad$ $y' = 2x^{-1/2}$ $\qquad$ $y' = \dfrac{2}{\sqrt{x}}$

25. $f(x) = \dfrac{3 - 2x - x^2}{x^2 - 1}$

$\qquad f'(x) = \dfrac{(x^2 - 1)(-2 - 2x) - (3 - 2x - x^2)(2x)}{(x^2 - 1)^2}$

$\qquad\qquad = \dfrac{2x^2 - 4x + 2}{(x^2 - 1)^2} = \dfrac{2(x - 1)^2}{(x^2 - 1)^2}$

$\qquad\qquad = \dfrac{2}{(x + 1)^2}, x \neq 1$

27. $f(x) = x\left(1 - \dfrac{4}{x + 3}\right) = x - \dfrac{4x}{x + 3}$

$\qquad f'(x) = 1 - \dfrac{(x + 3)4 - 4x(1)}{(x + 3)^2} = \dfrac{(x^2 + 6x + 9) - 12}{(x + 3)^2}$

$\qquad\qquad = \dfrac{x^2 + 6x - 3}{(x + 3)^2}$

29. $f(x) = \dfrac{2x + 5}{\sqrt{x}} = 2x^{1/2} + 5x^{-1/2}$

$\qquad f'(x) = x^{-1/2} - \dfrac{5}{2}x^{-3/2} = x^{-3/2}\left[x - \dfrac{5}{2}\right]$

$\qquad\qquad = \dfrac{2x - 5}{2x\sqrt{x}} = \dfrac{2x - 5}{2x^{3/2}}$

31. $h(s) = (s^3 - 2)^2 = s^6 - 4s^3 + 4$

$\qquad h'(s) = 6s^5 - 12s^2 = 6s^2(s^3 - 2)$

33. $f(x) = \dfrac{2 - \dfrac{1}{x}}{x - 3} = \dfrac{2x - 1}{x(x - 3)} = \dfrac{2x - 1}{x^2 - 3x}$

$\qquad f'(x) = \dfrac{(x^2 - 3x)2 - (2x - 1)(2x - 3)}{(x^2 - 3x)^2} = \dfrac{2x^2 - 6x - 4x^2 + 8x - 3}{(x^2 - 3x)^2}$

$\qquad\qquad = \dfrac{-2x^2 + 2x - 3}{(x^2 - 3x)^2} = -\dfrac{2x^2 - 2x + 3}{x^2(x - 3)^2}$

35. $f(x) = (3x^3 + 4x)(x - 5)(x + 1)$

$\qquad f'(x) = (9x^2 + 4)(x - 5)(x + 1) + (3x^3 + 4x)(1)(x + 1) + (3x^3 + 4x)(x - 5)(1)$

$\qquad\qquad = (9x^2 + 4)(x^2 - 4x - 5) + 3x^4 + 3x^3 + 4x^2 + 4x + 3x^4 - 15x^3 + 4x^2 - 20x$

$\qquad\qquad = 9x^4 - 36x^3 - 41x^2 - 16x - 20 + 6x^4 - 12x^3 + 8x^2 - 16x$

$\qquad\qquad = 15x^4 - 48x^3 - 33x^2 - 32x - 20$

37. $f(x) = \dfrac{x^2 + c^2}{x^2 - c^2}$

$\qquad f'(x) = \dfrac{(x^2 - c^2)(2x) - (x^2 + c^2)(2x)}{(x^2 - c^2)^2}$

$\qquad\qquad = \dfrac{-4xc^2}{(x^2 - c^2)^2}$

39. $f(t) = t^2 \sin t$

$\qquad f'(t) = t^2 \cos t + 2t \sin t$

$\qquad\qquad = t(t \cos t + 2 \sin t)$

41. $f(t) = \dfrac{\cos t}{t}$

$f'(t) = \dfrac{-t \sin t - \cos t}{t^2} = -\dfrac{t \sin t + \cos t}{t^2}$

43. $f(x) = -x + \tan x$

$f'(x) = -1 + \sec^2 x = \tan^2 x$

45. $g(t) = \sqrt[4]{t} + 8 \sec t = t^{1/4} + 8 \sec t$

$g'(t) = \dfrac{1}{4} t^{-3/4} + 8 \sec t \tan t = \dfrac{1}{4t^{3/4}} + 8 \sec t \tan t$

47. $y = \dfrac{3(1 - \sin x)}{2 \cos x} = \dfrac{3}{2}(\sec x - \tan x)$

$y' = \dfrac{3}{2}(\sec x \tan x - \sec^2 x) = \dfrac{3}{2} \sec x(\tan x - \sec x)$

$\qquad = \dfrac{3}{2}(\sec x \tan x - \tan^2 x - 1)$

49. $y = -\csc x - \sin x$

$y' = \csc x \cot x - \cos x$

$\quad = \dfrac{\cos x}{\sin^2 x} - \cos x$

$\quad = \cos x(\csc^2 x - 1)$

$\quad = \cos x \cot^2 x$

51. $f(x) = x^2 \tan x$

$f'(x) = x^2 \sec^2 x + 2x \tan x$

$\qquad = x(x \sec^2 x + 2 \tan x)$

53. $y = 2x \sin x + x^2 \cos x$

$y' = 2x \cos x + 2 \sin x + x^2(-\sin x) + 2x \cos x$

$\quad = 4x \cos x + 2 \sin x - x^2 \sin x$

55. $g(x) = \left(\dfrac{x + 1}{x + 2}\right)(2x - 5)$

$g'(x) = \dfrac{2x^2 + 8x - 1}{(x + 2)^2}$ \quad (form of answer may vary)

57. $g(\theta) = \dfrac{\theta}{1 - \sin \theta}$

$g'(\theta) = \dfrac{1 - \sin \theta + \theta \cos \theta}{(\sin \theta - 1)^2}$ \quad (form of answer may vary)

59. $y = \dfrac{1 + \csc x}{1 - \csc x}$

$y' = \dfrac{(1 - \csc x)(-\csc x \cot x) - (1 + \csc x)(\csc x \cot x)}{(1 - \csc x)^2} = \dfrac{-2 \csc x \cot x}{(1 - \csc x)^2}$

$y\left(\dfrac{\pi}{6}\right) = \dfrac{-2(2)\left(\sqrt{3}\right)}{(1 - 2)^2} = -4\sqrt{3}$

61. $h(t) = \dfrac{\sec t}{t}$

$h'(t) = \dfrac{t(\sec t \tan t) - (\sec t)(1)}{t^2}$

$\quad = \dfrac{\sec t(t \tan t - 1)}{t^2}$

$h'(\pi) = \dfrac{\sec \pi(\pi \tan \pi - 1)}{\pi^2} = \dfrac{1}{\pi^2}$

63. (a) $f(x) = (x^3 - 3x + 1)(x + 2),\ (1, -3)$

$f'(x) = (x^3 - 3x + 1)(1) + (x + 2)(3x^2 - 3)$

$\qquad = 4x^3 + 6x^2 - 6x - 5$

$f'(1) = -1 =$ slope at $(1, -3)$.

Tangent line: $y + 3 = -1(x - 1) \Longrightarrow y = -x - 2$

(b)

65. (a) $f(x) = \dfrac{x}{x - 1},\ (2, 2)$

$f'(x) = \dfrac{(x - 1)(1) - x(1)}{(x - 1)^2} = \dfrac{-1}{(x - 1)^2}$

$f'(2) = \dfrac{-1}{(2 - 1)^2} = -1 =$ slope at $(2, 2)$.

Tangent line: $y - 2 = -1(x - 2) \Longrightarrow y = -x + 4$

(b)

67. (a) $f(x) = \tan x,\ \left(\dfrac{\pi}{4}, 1\right)$

$f'(x) = \sec^2 x$

$f'\!\left(\dfrac{\pi}{4}\right) = 2 =$ slope at $\left(\dfrac{\pi}{4}, 1\right)$.

Tangent line:

$$y - 1 = 2\left(x - \dfrac{\pi}{4}\right)$$

$$y - 1 = 2x - \dfrac{\pi}{2}$$

$$4x - 2y - \pi + 2 = 0$$

(b)

69. $f(x) = \dfrac{x^2}{x - 1}$

$f'(x) = \dfrac{(x - 1)(2x) - x^2(1)}{(x - 1)^2}$

$\qquad = \dfrac{x^2 - 2x}{(x - 1)^2} = \dfrac{x(x - 2)}{(x - 1)^2}$

$f'(x) = 0$ when $x = 0$ or $x = 2$.

Horizontal tangents are at $(0, 0)$ and $(2, 4)$.

71. $f'(x) = \dfrac{(x + 2)3 - 3x(1)}{(x + 2)^2} = \dfrac{6}{(x + 2)^2}$

$g'(x) = \dfrac{(x + 2)5 - (5x + 4)(1)}{(x + 2)^2} = \dfrac{6}{(x + 2)^2}$

$g(x) = \dfrac{5x + 4}{(x + 2)} = \dfrac{3x}{(x + 2)} + \dfrac{2x + 4}{(x + 2)} = f(x) + 2$

f and g differ by a constant.

73. $f(x) = x^n \sin x$

$f'(x) = x^n \cos x + nx^{n-1} \sin x$

$\qquad = x^{n-1}(x \cos x + n \sin x)$

When $n = 1$: $f'(x) = x \cos x + \sin x$.

When $n = 2$: $f'(x) = x(x \cos x + 2 \sin x)$.

When $n = 3$: $f'(x) = x^2(x \cos x + 3 \sin x)$.

When $n = 4$: $f'(x) = x^3(x \cos x + 4 \sin x)$.

For general n, $f'(x) = x^{n-1}(x \cos x + n \sin x)$.

75. Area $= A(t) = (2t + 1)\sqrt{t} = 2t^{3/2} + t^{1/2}$

$A'(t) = 2\left(\dfrac{3}{2}t^{1/2}\right) + \dfrac{1}{2}t^{-1/2}$

$\qquad = 3t^{1/2} + \dfrac{1}{2}t^{-1/2}$

$\qquad = \dfrac{6t + 1}{2\sqrt{t}}$ cm²/sec

77. $C = 100\left(\dfrac{200}{x^2} + \dfrac{x}{x + 30}\right), \quad 1 \le x$

$\dfrac{dC}{dx} = 100\left(-\dfrac{400}{x^3} + \dfrac{30}{(x + 30)^2}\right)$

(a) When $x = 10$: $\dfrac{dC}{dx} = -\$38.13$.

(b) When $x = 15$: $\dfrac{dC}{dx} = -\$10.37$.

(c) When $x = 20$: $\dfrac{dC}{dx} = -\$3.80$.

As the order size increases, the cost per item decreases.

79. $P(t) = 500\left[1 + \dfrac{4t}{50 + t^2}\right]$

$P'(t) = 500\left[\dfrac{(50 + t^2)(4) - (4t)(2t)}{(50 + t^2)^2}\right]$

$= 500\left[\dfrac{200 - 4t^2}{(50 + t^2)^2}\right]$

$= 2000\left[\dfrac{50 = t^2}{(50 + t^2)^2}\right]$

$P'(2) \approx 31.55$ bacteria per hour

81. (a) $\sec x = \dfrac{1}{\cos x}$

$\dfrac{d}{dx}[\sec x] = \dfrac{d}{dx}\left[\dfrac{1}{\cos x}\right] = \dfrac{(\cos x)(0) - (1)(-\sin x)}{(\cos x)^2} = \dfrac{\sin x}{\cos x \cos x} = \dfrac{1}{\cos x} \cdot \dfrac{\sin x}{\cos x} = \sec x \tan x$

(b) $\csc x = \dfrac{1}{\sin x}$

$\dfrac{d}{dx}[\csc x] = \dfrac{d}{dx}\left[\dfrac{1}{\sin x}\right] = \dfrac{(\sin x)(0) - (1)(\cos x)}{(\sin x)^2} = -\dfrac{\cos x}{\sin x \sin x} = -\dfrac{1}{\sin x} \cdot \dfrac{\cos x}{\sin x} = -\csc x \cot x$

(c) $\cot x = \dfrac{\cos x}{\sin x}$

$\dfrac{d}{dx}[\cot x] = \dfrac{d}{dx}\left[\dfrac{\cos x}{\sin x}\right] = \dfrac{\sin x(-\sin x) - (\cos x)(\cos x)}{(\sin x)^2} = -\dfrac{\sin^2 x + \cos^2 x}{\sin^2 x} = -\dfrac{1}{\sin^2 x} = -\csc^2 x$

83. $f(x) = 4x^{3/2}$

$f'(x) = 6x^{1/2}$

$f''(x) = 3x^{-1/2} = \dfrac{3}{\sqrt{x}}$

85. $f(x) = \dfrac{x}{x - 1}$

$f'(x) = \dfrac{(x - 1)(1) - x(1)}{(x - 1)^2} = \dfrac{-1}{(x - 1)^2}$

$f''(x) = \dfrac{2}{(x - 1)^3}$

87. $f(x) = 3 \sin x$

$f'(x) = 3 \cos x$

$f''(x) = -3 \sin x$

89. $f'(x) = x^2$

$f''(x) = 2x$

91. $f'''(x) = 2\sqrt{x}$

$f^{(4)}(x) = \dfrac{1}{2}(2)x^{-1/2} = \dfrac{1}{\sqrt{x}}$

93.

$f(2) = 0$

One such function is $f(x) = (x - 2)^2$.

95. $f(x) = 2g(x) + h(x)$

$f'(x) = 2g'(x) + h'(x)$

$f'(2) = 2g'(2) + h'(2)$

$= 2(-2) + 4$

$= 0$

97. $f(x) = \dfrac{g(x)}{h(x)}$

$f'(x) = \dfrac{h(x)g'(x) - g(x)h'(x)}{[h(x)]^2}$

$f'(2) = \dfrac{h(2)g'(2) - g(2)h'(2)}{[h(2)]^2}$

$= \dfrac{(-1)(-2) - (3)(4)}{(-1)^2}$

$= -10$

99.

It appears that f is cubic; so f' would be quadratic and f'' would be linear.

101. $v(t) = 36 - t^2, \; 0 \le t \le 6$

$a(t) = -2t$

$v(3) = 27 \text{ m/sec}$

$a(3) = -6 \text{ m/sec}^2$

The speed of the object is decreasing.

103. $v(t) = \dfrac{100t}{2t + 15}$

$a(t) = \dfrac{(2t + 15)(100) - (100t)(2)}{(2t + 15)^2}$

$= \dfrac{1500}{(2t + 15)^2}$

(a) $a(5) = \dfrac{1500}{[2(5) + 15]^2} = 2.4 \text{ ft/sec}^2$

(b) $a(10) = \dfrac{1500}{[2(10) + 15]^2} \approx 1.2 \text{ ft/sec}^2$

(c) $a(20) = \dfrac{1500}{[2(20) + 15]^2} \approx 0.5 \text{ ft/sec}^2$

105. $f(x) = g(x)h(x)$

(a) $\quad f'(x) = g(x)h'(x) + h(x)g'(x)$

$f''(x) = g(x)h''(x) + g'(x)h'(x) + h(x)g''(x) + h'(x)g'(x)$

$= g(x)h''(x) + 2g'(x)h'(x) + h(x)g''(x)$

$f'''(x) = g(x)h'''(x) + g'(x)h''(x) + 2g'(x)h''(x) + 2g''(x)h'(x) + h(x)g'''(x) + h'(x)g''(x)$

$= g(x)h'''(x) + 3g'(x)h''(x) + 3g''(x)h'(x) + g'''(x)h(x)$

$f^{(4)}(x) = g(x)h^{(4)}(x) + g'(x)h'''(x) + 3g'(x)h'''(x) + 3g''(x)h''(x) + 3g''(x)h''(x) + 3g'''(x)h'(x)$

$\quad + g'''(x)h'(x) + g^{(4)}(x)h(x)$

$= g(x)h^{(4)}(x) + 4g'(x)h'''(x) + 6g''(x)h''(x) + 4g'''(x)h'(x) + g^{(4)}(x)h(x)$

(b) $f^{(n)}(x) = g(x)h^{(n)}(x) + \dfrac{n(n-1)(n-2)\cdots(2)(1)}{1[(n-1)(n-2)\cdots(2)(1)]}g'(x)h^{(n-1)}(x) + \dfrac{n(n-1)(n-2)\cdots(2)(1)}{(2)(1)[(n-2)(n-3)\cdots(2)(1)]}g''(x)h^{(n-2)}(x)$

$\quad + \dfrac{n(n-1)(n-2)\cdots(2)(1)}{(3)(2)(1)[(n-3)(n-4)\cdots(2)(1)]}g'''(x)h^{(n-3)}(x) + \cdots$

$\quad + \dfrac{n(n-1)(n-2)\cdots(2)(1)}{[(n-1)(n-2)\cdots(2)(1)](1)}g^{(n-1)}(x)h'(x) + g^{(n)}(x)h(x)$

$= g(x)h^{(n)}(x) + \dfrac{n!}{1!(n-1)!}g'(x)h^{(n-1)}(x) + \dfrac{n!}{2!(n-2)!}g''(x)h^{(n-2)}(x) + \cdots$

$\quad + \dfrac{n!}{(n-1)!1!}g^{(n-1)}(x)h'(x) + g^{(n)}(x)h(x)$

Note: $n! = n(n-1)\ldots 3 \cdot 2 \cdot 1$ (read "n factorial.")

107. $f(x) = \cos x$ $\qquad f\left(\dfrac{\pi}{3}\right) = \cos\dfrac{\pi}{3} = \dfrac{1}{2}$

$\quad f'(x) = -\sin x$ $\qquad f'\left(\dfrac{\pi}{3}\right) = -\sin\dfrac{\pi}{3} = -\dfrac{\sqrt{3}}{2}$

$\quad f''(x) = -\cos x$ $\qquad f''\left(\dfrac{\pi}{3}\right) = -\cos\dfrac{\pi}{3} = -\dfrac{1}{2}$

(a) $P_1(x) = f'(a)(x - a) + f(a) = -\dfrac{\sqrt{3}}{2}\left(x - \dfrac{\pi}{3}\right) + \dfrac{1}{2}$

$\quad P_2(x) = \dfrac{1}{2}f''(a)(x - a)^2 + f'(a)(x - a) + f(a)$

$\qquad = -\dfrac{1}{4}\left(x - \dfrac{\pi}{3}\right)^2 - \dfrac{\sqrt{3}}{2}\left(x - \dfrac{\pi}{3}\right) + \dfrac{1}{2}$

(b)

(c) P_2 is a better approximation.

(d) The accuracy worsens as you move farther away from $x = a = (\pi/3)$.

109. False. If $y = f(x)g(x)$, then

$\quad \dfrac{dy}{dx} = f(x)g'(x) + g(x)f'(x).$

111. True

$\quad h'(c) = f(c)g'(c) + g(c)f'(c)$

$\qquad = f(c)(0) + g(c)(0)$

$\qquad = 0$

113. True

115. $f(x) = x|x| = \begin{cases} x^2, & \text{if } x \geq 0 \\ -x^2, & \text{if } x < 0 \end{cases}$

$\quad f'(x) = \begin{cases} 2x, & \text{if } x \geq 0 \\ -2x, & \text{if } x < 0 \end{cases} = 2|x|$

$\quad f''(x) = \begin{cases} 2, & \text{if } x > 0 \\ -2, & \text{if } x < 0 \end{cases}$

$f''(0)$ does not exist since the left and right derivatives are not equal.

Section 2.4   The Chain Rule

$y = f(g(x))$	$u = g(x)$	$y = f(u)$
1. $y = (6x - 5)^4$	$u = 6x - 5$	$y = u^4$
3. $y = \sqrt{x^2 - 1}$	$u = x^2 - 1$	$y = \sqrt{u}$
5. $y = \csc^3 x$	$u = \csc x$	$y = u^3$

7. $y = (2x - 7)^3$

$\quad y' = 3(2x - 7)^2(2) = 6(2x - 7)^2$

9. $g(x) = 3(4 - 9x)^4$

$\quad g'(x) = 12(4 - 9x)^3(-9) = -108(4 - 9x)^3$

11. $f(x) = (9 - x^2)^{2/3}$

$\quad f'(x) = \dfrac{2}{3}(9 - x^2)^{-1/3}(-2x) = -\dfrac{4x}{3(9 - x^2)^{1/3}}$

13. $f(t) = (1 - t)^{1/2}$

$\quad f'(t) = \dfrac{1}{2}(1 - t)^{-1/2}(-1) = -\dfrac{1}{2\sqrt{1 - t}}$

15. $y = (9x^2 + 4)^{1/3}$

$$y' = \frac{1}{3}(9x^2 + 4)^{-2/3}(18x) = \frac{6x}{(9x^2 + 4)^{2/3}}$$

17. $y = 2(4 - x^2)^{1/4}$

$$y' = 2\left(\frac{1}{4}\right)(4 - x^2)^{-3/4}(-2x)$$

$$= \frac{-x}{\sqrt[4]{(4 - x^2)^3}}$$

19. $y = (x - 2)^{-1}$

$$y' = -1(x - 2)^{-2}(1) = \frac{-1}{(x - 2)^2}$$

21. $f(t) = (t - 3)^{-2}$

$$f'(t) = -2(t - 3)^{-3} = \frac{-2}{(t - 3)^3}$$

23. $y = (x + 2)^{-1/2}$

$$\frac{dy}{dx} = -\frac{1}{2}(x + 2)^{-3/2} = -\frac{1}{2(x + 2)^{3/2}}$$

25. $f(x) = x^2(x - 2)^4$

$$f'(x) = x^2[4(x - 2)^3(1)] + (x - 2)^4(2x)$$

$$= 2x(x - 2)^3[2x + (x - 2)]$$

$$= 2x(x - 2)^3(3x - 2)$$

27. $y = x\sqrt{1 - x^2} = x(1 - x^2)^{1/2}$

$$y' = x\left[\frac{1}{2}(1 - x^2)^{-1/2}(-2x)\right] + (1 - x^2)^{1/2}(1)$$

$$= -x^2(1 - x^2)^{-1/2} + (1 - x^2)^{1/2}$$

$$= (1 - x^2)^{-1/2}[-x^2 + (1 - x^2)]$$

$$= \frac{1 - 2x^2}{\sqrt{1 - x^2}}$$

29. $y = \dfrac{x}{\sqrt{x^2 + 1}} = x(x^2 + 1)^{-1/2}$

$$y' = x\left[-\frac{1}{2}(x^2 + 1)^{-3/2}(2x)\right] + (x^2 + 1)^{-1/2}(1)$$

$$= -x^2(x^2 + 1)^{-3/2} + (x^2 + 1)^{-1/2}$$

$$= (x^2 + 1)^{-3/2}[-x^2 + (x^2 + 1)]$$

$$= \frac{1}{(x^2 + 1)^{3/2}}$$

31. $g(x) = \left(\dfrac{x + 5}{x^2 + 2}\right)^2$

$$g'(x) = 2\left(\frac{x + 5}{x^2 + 2}\right)\left(\frac{(x^2 + 2) - (x + 5)(2x)}{(x^2 + 2)^2}\right)$$

$$= \frac{2(x + 5)(2 - 10x - x^2)}{(x^2 + 2)^3}$$

33. $f(v) = \left(\dfrac{1 - 2v}{1 + v}\right)^3$

$$f'(v) = 3\left(\frac{1 - 2v}{1 + v}\right)^2\left(\frac{(1 + v)(-2) - (1 - 2v)}{(1 + v)^2}\right)$$

$$= \frac{-9(1 - 2v)^2}{(1 + v)^4}$$

35. $y = \dfrac{\sqrt{x} + 1}{x^2 + 1}$

$$y' = \frac{1 - 3x^2 - 4x^{3/2}}{2\sqrt{x}(x^2 + 1)^2}$$

The zero of y' corresponds to the point on the graph of y where the tangent line is horizontal.

37. $g(t) = \dfrac{3t^2}{\sqrt{t^2 + 2t - 1}}$

$$g'(t) = \frac{3t(t^2 + 3t - 2)}{(t^2 + 2t - 1)^{3/2}}$$

The zeros of g' correspond to the points on the graph of g where the tangent lines are horizontal.

39. $y = \sqrt{\dfrac{x + 1}{x}}$

$y' = -\dfrac{\sqrt{(x + 1)/x}}{2x(x + 1)}$

y' has no zeros.

41. $s(t) = \dfrac{-2(2 - t)\sqrt{1 + t}}{3}$

$s'(t) = \dfrac{t}{\sqrt{1 + t}}$

The zero of $s'(t)$ corresponds to the point on the graph of $s(t)$ where the tangent line is horizontal.

43. $y = \dfrac{\cos \pi x + 1}{x}$

$\dfrac{dy}{dx} = \dfrac{-\pi x \sin \pi x - \cos \pi x - 1}{x^2}$

$\quad = -\dfrac{\pi x \sin \pi x + \cos \pi x + 1}{x^2}$

The zeros of y' correspond to the points on the graph of y where the tangent lines are horizontal.

45. (a) $\quad y = \sin x$

$\quad\quad y' = \cos x$

$\quad y'(0) = 1$

$\quad 1$ cycle in $[0, 2\pi]$

(b) $y = \sin 2x$

$\quad y' = 2 \cos 2x$

$\quad y'(0) = 2$

$\quad 2$ cycles in $[0, 2\pi]$

The slope of $\sin ax$ at the origin is a.

47. $y = \cos 3x$

$\dfrac{dy}{dx} = -3 \sin 3x$

49. $g(x) = 3 \tan 4x$

$g'(x) = 12 \sec^2 4x$

51. $y = \sin (\pi x)^2 = \sin (\pi^2 x^2)$

$y' = \cos (\pi^2 x^2)[2\pi^2 x] = 2\pi^2 x \cos(\pi^2 x^2)$

53. $h(x) = \sin 2x \cos 2x$

$h'(x) = \sin 2x(-2 \sin 2x) + \cos 2x(2 \cos 2x)$

$\quad = 2 \cos^2 2x - 2 \sin^2 2x$

$\quad = 2 \cos 4x.$

Alternate solution: $h(x) = \dfrac{1}{2} \sin 4x$

$\quad\quad\quad\quad h'(x) = \dfrac{1}{2} \cos 4x(4) = 2 \cos 4x$

55. $f(x) = \dfrac{\cot x}{\sin x} = \dfrac{\cos x}{\sin^2 x}$

$f'(x) = \dfrac{\sin^2 x(-\sin x) - \cos x(2 \sin x \cos x)}{\sin^4 x}$

$\quad = \dfrac{-\sin^2 x - 2 \cos^2 x}{\sin^3 x} = \dfrac{-1 - \cos^2 x}{\sin^3 x}$

57. $y = 4 \sec^2 x$

$y' = 8 \sec x \cdot \sec x \tan x = 8 \sec^2 x \tan x$

59. $f(\theta) = \frac{1}{4} \sin^2 2\theta = \frac{1}{4}(\sin 2\theta)^2$

$f'(\theta) = 2(\frac{1}{4})(\sin 2\theta)(\cos 2\theta)(2)$

$= \sin 2\theta \cos 2\theta = \frac{1}{2} \sin 4\theta$

61. $f(x) = 3 \sec^2(\pi t - 1)$

$f'(t) = 6 \sec(\pi t - 1) \sec(\pi t - 1) \tan(\pi t - 1)(\pi)$

$= 6\pi \sec^2(\pi t - 1) \tan(\pi t - 1) = \dfrac{6\pi \sin(\pi t - 1)}{\cos^3(\pi t - 1)}$

63. $y = \sqrt{x} + \frac{1}{4} \sin(2x)^2$

$= \sqrt{x} + \frac{1}{4} \sin(4x^2)$

$\dfrac{dy}{dx} = \frac{1}{2}x^{-1/2} + \frac{1}{4} \cos(4x^2)(8x)$

$= \dfrac{1}{2\sqrt{x}} + 2x \cos(2x)^2$

65. $y = \sin(\cos x)$

$\dfrac{dy}{dx} = \cos(\cos x) \cdot (-\sin x)$

$= -\sin x \cos(\cos x)$

67. $s(t) = (t^2 + 2t + 8)^{1/2}, \quad (2, 4)$

$s'(t) = \frac{1}{2}(t^2 + 2t + 8)^{-1/2}(2t + 2)$

$= \dfrac{t + 1}{\sqrt{t^2 + 2t + 8}}$

$s'(2) = \dfrac{3}{4}$

69. $f(x) = \dfrac{3}{x^3 - 4} = 3(x^3 - 4)^{-1}, \quad \left(-1, -\dfrac{3}{5}\right)$

$f'(x) = -3(x^3 - 4)^{-2}(3x^2) = -\dfrac{9x^2}{(x^3 - 4)^2}$

$f'(-1) = -\dfrac{9}{25}$

71. $f(t) = \dfrac{3t + 2}{t - 1}, \quad (0, -2)$

$f'(t) = \dfrac{(t - 1)(3) - (3t + 2)(1)}{(t - 1)^2} = \dfrac{-5}{(t - 1)^2}$

$f'(0) = -5$

73. $y = 37 - \sec^3(2x), \quad (0, 36)$

$y' = -3 \sec^2(2x)[2 \sec(2x) \tan(2x)]$

$= -6 \sec^3(2x) \tan(2x)$

$y'(0) = 0$

75. (a) $f(x) = \sqrt{3x^2 - 2}, \quad (3, 5)$

$f'(x) = \frac{1}{2}(3x^2 - 2)^{-1/2}(6x)$

$= \dfrac{3x}{\sqrt{3x^2 - 2}}$

$f'(3) = \dfrac{9}{5}$

Tangent line:

$y - 5 = \dfrac{9}{5}(x - 3) \Longrightarrow 9x - 5y - 2 = 0$

(b)

77. (a) $f(x) = \sin 2x, \quad (\pi, 0)$

$f'(x) = 2 \cos 2x$

$f'(\pi) = 2$

Tangent line:

$y = 2(x - \pi) \Longrightarrow 2x - y - 2\pi = 0$

(b)

79. $f(x) = 2(x^2 - 1)^3$

$f'(x) = 6(x^2 - 1)^2(2x)$

$\quad = 12x(x^4 - 2x^2 + 1)$

$\quad = 12x^5 - 24x^3 + 12x$

$f''(x) = 60x^4 - 72x^2 + 12$

$\quad = 12(5x^2 - 1)(x^2 - 1)$

81. $f(x) = \sin x^2$

$f'(x) = 2x \cos x^2$

$f''(x) = 2x[2x(-\sin x^2)] + 2 \cos x^2$

$\quad = 2[\cos x^2 - 2x^2 \sin x^2]$

83.

The zeros of f' correspond to the points where the graph of f has horizontal tangents.

85.

The zeros of f' correspond to the points where the graph of f has horizontal tangents.

87. $g(x) = f(3x)$

$g'(x) = f'(3x)(3) \Rightarrow g'(x) = 3f'(3x)$

89. (a) $f(x) = g(x)h(x)$

$f'(x) = g(x)h'(x) + g'(x)h(x)$

$f'(5) = (-3)(-2) + (6)(3) = 24$

(b) $f(x) = g(h(x))$

$f'(x) = g'(h(x))h'(x)$

$f'(5) = g'(3)(-2) = -2g'(3)$

Need $g'(3)$ to find $f'(5)$.

(c) $f(x) = \dfrac{g(x)}{h(x)}$

$f'(x) = \dfrac{h(x)g'(x) - g(x)h'(x)}{[h(x)]^2}$

$f'(5) = \dfrac{(3)(6) - (-3)(-2)}{(3)^2} = \dfrac{12}{9} = \dfrac{4}{3}$

(d) $f(x) = [g(x)]^3$

$f'(x) = 3[g(x)]^2 g'(x)$

$f'(5) = 3(-3)^2(6) = 162$

91. (a) $F = 132,400(331 - v)^{-1}$

$F' = (-1)(132,400)(331 - v)^{-2}(-1)$

$\quad = \dfrac{132,400}{(331 - v)^2}$

When $v = 30$, $F' \approx 1.461$.

(b) $F = 132,400(331 + v)^{-1}$

$F' = (-1)(132,400)(331 + v)^{-2}(1)$

$\quad = \dfrac{-132,400}{(331 + v)^2}$

When $v = 30$, $F' \approx -1.016$.

93. $\theta = 0.2 \cos 8t$

The maximum angular displacement is $\theta = 0.2$ (since $-1 \le \cos 8t \le 1$).

$\dfrac{d\theta}{dt} = 0.2[-8 \sin 8t] = -1.6 \sin 8t$

When $t = 3$, $d\theta/dt = -1.6 \sin 24 \approx 1.4489$ radians per second.

95. $S = C(R^2 - r^2)$

$\dfrac{dS}{dt} = C\left(2R\dfrac{dR}{dt} - 2r\dfrac{dr}{dt}\right)$

Since r is constant, we have $dr/dt = 0$ and

$\dfrac{dS}{dt} = (1.76 \times 10^5)(2)(1.2 \times 10^{-2})(10^{-5})$

$\quad = 4.224 \times 10^{-2} = 0.04224.$

97. (a) $x = -1.6372t^3 + 19.3120t^2 - 0.5082t - 0.6162$

(b) $C = 60x + 1350$

$$= 60(-1.6372t^3 + 19.3120t^2 - 0.5082t - 0.6162) + 1350$$

$$\frac{dC}{dt} = 60(-4.9116t^2 + 38.624t - 0.5082)$$

$$= -294.696t^2 + 2317.44t - 30.492$$

The function $\dfrac{dC}{dt}$ is quadratic, not linear. The cost function levels off at the end of the day, perhaps due to fatigue.

99. $f(x) = \sin \beta x$

(a) $f'(x) = \beta \cos \beta x$

$f''(x) = -\beta^2 \sin \beta x$

$f'''(x) = -\beta^3 \cos \beta x$

$f^{(4)} = \beta^4 \sin \beta x$

(b) $f''(x) + \beta^2 f(x) = -\beta^2 \sin \beta x + \beta^2(\sin \beta x) = 0$

(c) $f^{(2k)}(x) = (-1)^k \beta^{2k} \sin \beta x$

$f^{(2k-1)}(x) = (-1)^{k+1} \beta^{2k-1} \cos \beta x$

101. (a) $r'(x) = f'(g(x))g'(x)$

$r'(1) = f'(g(1))g'(1)$

Note that $g(1) = 4$ and $f'(4) = \dfrac{5 - 0}{6 - 2} = \dfrac{5}{4}$

Also, $g'(1) = 0$. Thus, $r'(1) = 0$

(b) $s'(x) = g'(f(x))f'(x)$

$s'(4) = g'(f(4))f'(4)$

Note that $f(4) = \dfrac{5}{2}$, $g'\left(\dfrac{5}{2}\right) = \dfrac{6 - 4}{6 - 2} = \dfrac{1}{2}$ and

$f'(4) = \dfrac{5}{4}$.

Thus, $s'(4) = \dfrac{1}{2}\left(\dfrac{5}{4}\right) = \dfrac{5}{8}$.

103. $g = \sqrt{x(x + n)}$

$= \sqrt{x^2 + nx}$

$\dfrac{dg}{dx} = \dfrac{1}{2}(x^2 + nx)^{-1/2}(2x + n)$

$= \dfrac{2x + n}{2\sqrt{x^2 + nx}}$

$= \dfrac{(2x + n)/2}{\sqrt{x(x + n)}}$

$= \dfrac{[x + (x + n)]/2}{\sqrt{x(x + n)}}$

$= \dfrac{a}{g}$

105. $g(x) = |2x - 3|$

$g'(x) = 2\left(\dfrac{2x - 3}{|2x - 3|}\right), \quad x \neq \dfrac{3}{2}$

107. $h(x) = |x|\cos x$

$h'(x) = -|x| \sin x + \dfrac{x}{|x|} \cos x, \quad x \neq 0$

109. (a) $f(x) = \tan\dfrac{\pi x}{4}$ $f(1) = 1$

$$f'(x) = \frac{\pi}{4}\sec^2\frac{\pi x}{4} \qquad\qquad f'(1) = \frac{\pi}{4}(2) = \frac{\pi}{2}$$

$$f''(x) = \frac{\pi}{2}\sec^2\frac{\pi x}{4}\cdot\tan\frac{\pi x}{4}\left(\frac{\pi}{4}\right) \qquad f''(1) = \frac{\pi^2}{8}(2)(1) = \frac{\pi^2}{4}$$

$$P_1(x) = f'(1)(x-1) + f(1) = \frac{\pi}{2}(x-1) + 1.$$

$$P_2(x) = \frac{1}{2}\left(\frac{\pi^2}{4}\right)(x-1)^2 + f'(1)(x-1) + f(1) = \frac{\pi^2}{8}(x-1)^2 + \frac{\pi}{2}(x-1) + 1$$

(b)

(c) P_2 is a better approximation than P_1.

(d) The accuracy worsens as you move away from $x = c = 1$.

111. False. If $y = (1-x)^{1/2}$, then $y' = \frac{1}{2}(1-x)^{-1/2}(-1)$.   **113.** True

Section 2.5   Implicit Differentiation

1. $x^2 + y^2 = 36$

     $2x + 2yy' = 0$

$$y' = \frac{-x}{y}$$

3. $x^{1/2} + y^{1/2} = 9$

$$\frac{1}{2}x^{-1/2} + \frac{1}{2}y^{-1/2}y' = 0$$

$$y' = -\frac{x^{-1/2}}{y^{-1/2}} = -\sqrt{\frac{y}{x}}$$

5. $x^3 - xy + y^2 = 4$

   $3x^2 - xy' - y + 2yy' = 0$

       $(2y - x)y' = y - 3x^2$

$$y' = \frac{y - 3x^2}{2y - x}$$

7. $x^3y^3 - y - x = 0$

  $3x^3y^2y' + 3x^2y^3 - y' - 1 = 0$

     $(3x^3y^2 - 1)y' = 1 - 3x^2y^3$

$$y' = \frac{1 - 3x^2y^3}{3x^3y^2 - 1}$$

9. $x^3 - 3x^2y + 2xy^2 = 12$

  $3x^2 - 3x^2y' - 6xy + 4xyy' + 2y^2 = 0$

     $(4xy - 3x^2)y' = 6xy - 3x^2 - 2y^2$

$$y' = \frac{6xy - 3x^2 - 2y^2}{4xy - 3x^2}$$

11. $\sin x + 2\cos 2y = 1$

    $\cos x - 4(\sin 2y)y' = 0$

$$y' = \frac{\cos x}{4\sin 2y}$$

13. $\sin x = x(1 + \tan y)$

$\cos x = x(\sec^2 y)y' + (1 + \tan y)(1)$

$$y' = \frac{\cos x - \tan y - 1}{x \sec^2 y}$$

15. $\qquad y = \sin(xy)$

$y' = [xy' + y]\cos(xy)$

$y' - x\cos(xy)y' = y\cos(xy)$

$$y' = \frac{y\cos(xy)}{1 - x\cos(xy)}$$

17. (a) $x^2 + y^2 = 16$

$y^2 = 16 - x^2$

$y = \pm\sqrt{16 - x^2}$

(b)

(c) Explicitly:

$$\frac{dy}{dx} = \pm\frac{1}{2}(16 - x^2)^{-1/2}(-2x)$$

$$= \frac{\mp x}{\sqrt{16 - x^2}} = \frac{-x}{\pm\sqrt{16 - x^2}} = \frac{-x}{y}$$

(d) Implicitly:

$2x + 2yy' = 0$

$$y' = -\frac{x}{y}$$

19. (a) $16y^2 = 144 - 9x^2$

$y^2 = \frac{1}{16}(144 - 9x^2) = \frac{9}{16}(16 - x^2)$

$y = \pm\frac{3}{4}\sqrt{16 - x^2}$

(b)

(c) Explicitly:

$$\frac{dy}{dx} = \pm\frac{3}{8}(16 - x^2)^{-1/2}(-2x)$$

$$= \mp\frac{3x}{4\sqrt{16 - x^2}} = \frac{-3x}{4(4/3)y} = \frac{-9x}{16y}$$

(d) Implicitly:

$18x + 32yy' = 0$

$$y' = \frac{-9x}{16y}$$

21. $\qquad xy = 4$

$xy' + y(1) = 0$

$xy' = -y$

$$y' = \frac{-y}{x}$$

At $(-4, -1)$: $y' = -\frac{1}{4}$

23. $y^2 = \dfrac{x^2 - 4}{x^2 + 4}$

$2yy' = \dfrac{(x^2 + 4)(2x) - (x^2 - 4)(2x)}{(x^2 + 4)^2}$

$2yy' = \dfrac{16x}{(x^2 + 4)^2}$

$y' = \dfrac{8x}{y(x^2 + 4)^2}$

At $(2, 0)$, y' is undefined.

25. $x^{2/3} + y^{2/3} = 5$

$\dfrac{2}{3}x^{-1/3} + \dfrac{2}{3}y^{-1/3}y' = 0$

$$y' = \frac{-x^{-1/3}}{y^{-1/3}} = -\sqrt[3]{\frac{y}{x}}$$

At $(8, 1)$: $y' = -\dfrac{1}{2}$.

27. $\tan(x + y) = x$

$(1 + y')\sec^2(x + y) = 1$

$$y' = \frac{1 - \sec^2(x + y)}{\sec^2(x + y)}$$

$$= \frac{-\tan^2(x + y)}{\tan^2(x + y) + 1} = -\sin^2(x + y)$$

$$= -\frac{x^2}{x^2 + 1}$$

At $(0, 0)$: $y' = 0$.

29. $(x^2 + 4)y = 8$

$(x^2 + 4)y' + y(2x) = 0$

$$y' = \frac{-2xy}{x^2 + 4}$$

$$= \frac{-2x[8/(x^2 + 4)]}{x^2 + 4}$$

$$= \frac{-16x}{(x^2 + 4)^2}$$

At $(2, 1)$: $y' = \dfrac{-32}{64} = -\dfrac{1}{2}$

$\left(\text{Or, you could just solve for } y: \ y = \dfrac{8}{x^2 + 4}\right)$

31. $(x^2 + y^2)^2 = 4x^2y$

$2(x^2 + y^2)(2x + 2yy') = 4x^2y' + y(8x)$

$4x^3 + 4x^2yy' + 4xy^2 + 4y^3y' = 4x^2y' + 8xy$

$4x^2yy' + 4y^3y' - 4x^2y' = 8xy - 4x^3 - 4xy^2$

$4y'(x^2y + y^3 - x^2) = 4(2xy - x^3 - xy^2)$

$$y' = \frac{2xy - x^3 - xy^2}{x^2y + y^3 - x^2}$$

At $(1, 1)$: $y' = 0$.

33. $\tan y = x$

$y'\sec^2 y = 1$

$$y' = \frac{1}{\sec^2 y} = \cos^2 y, \ -\frac{\pi}{2} < y < \frac{\pi}{2}$$

$\sec^2 y = 1 + \tan^2 y = 1 + x^2$

$$y' = \frac{1}{1 + x^2}$$

35. $x^2 + y^2 = 36$

$2x + 2yy' = 0$

$$y' = \frac{-x}{y}$$

$$y'' = \frac{y(-1) + xy'}{y^2} = \frac{-y + x\left(-\dfrac{x}{y}\right)}{y^2} = \frac{-y^2 - x^2}{y^3} = \frac{-36}{y^3}$$

37. $x^2 - y^2 = 16$

$2x - 2yy' = 0$

$$y' = \frac{x}{y}$$

$x - yy' = 0$

$1 - yy'' - (y')^2 = 0$

$1 - yy'' - \left(\dfrac{x}{y}\right)^2 = 0$

$y^2 - y^3y'' = x^2$

$$y'' = \frac{y^2 - x^2}{y^3} = \frac{-16}{y^3}$$

39. $y^2 = x^3$

$2yy' = 3x^2$

$$y' = \frac{3x^2}{2y} = \frac{3x^2}{2y} \cdot \frac{xy}{xy} = \frac{3y}{2x} \cdot \frac{x^3}{y^2} = \frac{3y}{2x}$$

$$y'' = \frac{2x(3y') - 3y(2)}{4x^2}$$

$$= \frac{2x[3 \cdot (3y/2x)] - 6y}{4x^2}$$

$$= \frac{3y}{4x^2} = \frac{3x}{4y}$$

41. $\sqrt{x} + \sqrt{y} = 4$

$\frac{1}{2}x^{-1/2} + \frac{1}{2}y^{-1/2}y' = 0$

$y' = \frac{-\sqrt{y}}{\sqrt{x}}$

At $(9, 1), y' = -\frac{1}{3}$

Tangent line: $y - 1 = -\frac{1}{3}(x - 9)$

$y = -\frac{1}{3}x + 4$

$x + 3y - 12 = 0$

43. $x^2 + y^2 = 25$

$2x + 2yy' = 0$

$y' = \frac{-x}{y}$

At $(4, 3)$:

Tangent line: $y - 3 = \frac{-4}{3}(x - 4) \Longrightarrow 4x + 3y - 25 = 0$

Normal line: $y - 3 = \frac{3}{4}(x - 4) \Longrightarrow 3x - 4y = 0$.

At $(-3, 4)$:

Tangent line: $y - 4 = \frac{3}{4}(x + 3) \Longrightarrow 3x - 4y + 25 = 0$

Normal line: $y - 4 = \frac{-4}{3}(x + 3) \Longrightarrow 4x + 3y = 0$.

45. $x^2 + y^2 = r^2$

$2x + 2yy' = 0$

$y' = \frac{-x}{y} = $ slope of tangent line

$\frac{y}{x} = $ slope of normal line

Let (x_0, y_0) be a point on the circle. If $x_0 = 0$, then the tangent line is horizontal, the normal line is vertical and, hence, passes through the origin. If $x_0 \neq 0$, then the equation of the normal line is

$y - y_0 = \frac{y_0}{x_0}(x - x_0)$

$y = \frac{y_0}{x_0}x$

which passes through the origin.

47. $25x^2 + 16y^2 + 200x - 160y + 400 = 0$

$$50x + 32yy' + 200 - 160y' = 0$$

$$y' = \frac{200 + 50x}{160 - 32y}$$

Horizontal tangents occur when $x = -4$:

$$25(16) + 16y^2 + 200(-4) - 160y + 400 = 0$$

$$y(y - 10) = 0 \Rightarrow y = 0, 10$$

Horizontal tangents: $(-4, 0), (-4, 10)$.

Vertical tangents occur when $y = 5$:

$$25x^2 + 400 + 200x - 800 + 400 = 0$$

$$25x(x + 8) = 0 \Rightarrow x = 0, -8$$

Vertical tangents: $(0, 5), (-8, 5)$.

49. Find the points of intersection by letting $y^2 = 4x$ in the equation $2x^2 + y^2 = 6$.

$$2x^2 + 4x = 6 \quad \text{and} \quad (x + 3)(x - 1) = 0$$

The curves intersect at $(1, \pm 2)$.

Ellipse:	_Parabola:_
$4x + 2yy' = 0$	$2yy' = 4$
$y' = -\dfrac{2x}{y}$	$y' = \dfrac{2}{y}$

At $(1, 2)$, the slopes are:

$y' = -1$ $\qquad\qquad$ $y' = 1.$

At $(1, -2)$, the slopes are:

$y' = 1$ $\qquad\qquad$ $y' = -1.$

Tangents are perpendicular.

51. $y = -x$ and $x = \sin y$

Point of intersection: $(0, 0)$

$y = -x$:	$x = \sin y$:
$y' = -1$	$1 = y'\cos y$
	$y' = \sec y$

At $(0, 0)$, the slopes are:

$y' = -1$ $\qquad$ $y' = 1.$

Tangents are perpendicular.

53. $\qquad xy = C \qquad x^2 - y^2 = K$

$xy' + y = 0 \qquad 2x - 2yy' = 0$

$y' = -\dfrac{y}{x} \qquad\qquad y' = \dfrac{x}{y}$

At any point of intersection (x, y) the product of the slopes is $(-y/x)(x/y) = -1$. The curves are orthogonal.

55. $2y^2 - 3x^4 = 0$

(a) $4yy' - 12x^3 = 0$

$\quad\quad 4yy' = 12x^3$

$\quad\quad y' = \dfrac{12x^3}{4y} = \dfrac{3x^3}{y}$

(b) $4y\dfrac{dy}{dt} - 12x^3\dfrac{dx}{dt} = 0$

$\quad\quad y\dfrac{dy}{dt} = 3x^3\dfrac{dx}{dt}$

57. $\cos \pi y - 3 \sin \pi x = 1$

(a) $-\pi \sin(\pi y)y' - 3\pi \cos \pi x = 0$

$\quad\quad y' = \dfrac{-3\cos \pi x}{\sin \pi y}$

(b) $-\pi \sin(\pi y)\dfrac{dy}{dt} - 3\pi \cos(\pi x)\dfrac{dx}{dt} = 0$

$\quad\quad -\sin(\pi y)\dfrac{dy}{dt} = 3\cos(\pi x)\dfrac{dx}{dt}$

59. A function is in explicit form if y is written as a function of x: $y = f(x)$. For example, $y = x^3$. An implicit equation is not in the form $y = f(x)$. For example, $x^2 + y^2 = 5$.

61. (a) $x^4 = 4(4x^2 - y^2)$

$\quad\quad 4y^2 = 16x^2 - x^4$

$\quad\quad y^2 = 4x^2 - \dfrac{1}{4}x^4$

$\quad\quad y = \pm\sqrt{4x^2 - \dfrac{1}{4}x^4}$

(b) $\quad y = 3 \Longrightarrow 9 = 4x^2 - \dfrac{1}{4}x^4$

$\quad\quad\quad 36 = 16x^2 - x^4$

$\quad x^4 - 16x^2 + 36 = 0$

$\quad\quad\quad x^2 = \dfrac{16 \pm \sqrt{256 - 144}}{2} = 8 \pm \sqrt{28}$

Note that $x^2 = 8 \pm \sqrt{28} = 8 \pm 2\sqrt{7} = \left(1 \pm \sqrt{7}\right)^2$.

Hence, there are four values of x:

$\quad -1 - \sqrt{7},\ 1 - \sqrt{7},\ -1 + \sqrt{7},\ 1 + \sqrt{7}$

To find the slope, $2yy' = 8x - x^3 \Longrightarrow y' = \dfrac{x(8 - x^2)}{2(3)}$.

For $x = -1 - \sqrt{7}$, $y' = \frac{1}{3}\left(\sqrt{7} + 7\right)$, and the line is

$\quad y_1 = \frac{1}{3}\left(\sqrt{7} + 7\right)\left(x + 1 + \sqrt{7}\right) + 3 = \frac{1}{3}\left[\left(\sqrt{7} + 7\right)x + 8\sqrt{7} + 23\right]$.

For $x = 1 - \sqrt{7}$, $y' = \frac{1}{3}\left(\sqrt{7} - 7\right)$, and the line is

$\quad y_2 = \frac{1}{3}\left(\sqrt{7} - 7\right)\left(x - 1 + \sqrt{7}\right) + 3 = \frac{1}{3}\left[\left(\sqrt{7} - 7\right)x + 23 - 8\sqrt{7}\right]$.

For $x = -1 + \sqrt{7}$, $y' = -\frac{1}{3}\left(\sqrt{7} - 7\right)$, and the line is

$\quad y_3 = -\frac{1}{3}\left(\sqrt{7} - 7\right)\left(x + 1 - \sqrt{7}\right) + 3 = -\frac{1}{3}\left[\left(\sqrt{7} - 7\right)x - \left(23 - 8\sqrt{7}\right)\right]$.

For $x = 1 + \sqrt{7}$, $y' = -\frac{1}{3}\left(\sqrt{7} + 7\right)$, and the line is

$\quad y_4 = -\frac{1}{3}\left(\sqrt{7} + 7\right)\left(x - 1 - \sqrt{7}\right) + 3 = -\frac{1}{3}\left[\left(\sqrt{7} + 7\right)x - \left(8\sqrt{7} + 23\right)\right]$.

—CONTINUED—

61. —CONTINUED—

(c) Equating y_3 and y_4:

$$-\frac{1}{3}\left(\sqrt{7} - 7\right)\left(x + 1 - \sqrt{7}\right) + 3 = -\frac{1}{3}\left(\sqrt{7} + 7\right)\left(x - 1 - \sqrt{7}\right) + 3$$

$$\left(\sqrt{7} - 7\right)\left(x + 1 - \sqrt{7}\right) = \left(\sqrt{7} + 7\right)\left(x - 1 - \sqrt{7}\right)$$

$$\sqrt{7}x + \sqrt{7} - 7 - 7x - 7 + 7\sqrt{7} = \sqrt{7}x - \sqrt{7} - 7 + 7x - 7 - 7\sqrt{7}$$

$$16\sqrt{7} = 14x$$

$$x = \frac{8\sqrt{7}}{7}$$

If $x = \dfrac{8\sqrt{7}}{7}$, then $y = 5$ and the lines intersect at $\left(\dfrac{8\sqrt{7}}{7}, 5\right)$.

63.

$$y = x^{p/q}$$

$$y^q = x^p$$

$$qy^{q-1}y' = px^{p-1}$$

$$y' = \frac{p}{q}\frac{x^{p-1}}{y^{q-1}} = \frac{p}{q} \cdot \frac{x^{p-1}y}{x^p} = \frac{p}{q}x^{p/q-1}$$

Thus, if $y = x^n = x^{p/q}, y' = nx^{n-1}$.

Section 2.6   Related Rates

1. $y = \sqrt{x}$

$$\frac{dy}{dt} = \left(\frac{1}{2\sqrt{x}}\right)\frac{dx}{dt}$$

$$\frac{dx}{dt} = 2\sqrt{x}\,\frac{dy}{dt}$$

(a) When $x = 4$ and $dx/dt = 3$,

$$\frac{dy}{dt} = \frac{1}{2\sqrt{4}}(3) = \frac{3}{4}.$$

(b) When $x = 25$ and $dy/dt = 2$,

$$\frac{dx}{dt} = 2\sqrt{25}\,(2) = 20.$$

3. $xy = 4$

$$x\frac{dy}{dt} + y\frac{dx}{dt} = 0$$

$$\frac{dy}{dt} = \left(-\frac{y}{x}\right)\frac{dx}{dt}$$

$$\frac{dx}{dt} = \left(-\frac{x}{y}\right)\frac{dy}{dt}$$

(a) When $x = 8$, $y = 1/2$, and $dx/dt = 10$,

$$\frac{dy}{dt} = -\frac{1/2}{8}(10) = -\frac{5}{8}.$$

(b) When $x = 1$, $y = 4$, and $dy/dt = -6$,

$$\frac{dx}{dt} = -\frac{1}{4}(-6) = \frac{3}{2}.$$

5. $y = x^2 + 1$

$\dfrac{dx}{dt} = 2$

$\dfrac{dy}{dt} = 2x \dfrac{dx}{dt}$

(a) When $x = -1$,

$\dfrac{dy}{dt} = 2(-1)(2) = -4 \text{ cm/sec.}$

(b) When $x = 0$,

$\dfrac{dy}{dt} = 2(0)(2) = 0 \text{ cm/sec.}$

(c) When $x = 1$,

$\dfrac{dy}{dt} = 2(1)(2) = 4 \text{ cm/sec.}$

9. (a) $\dfrac{dx}{dt}$ negative $\Rightarrow \dfrac{dy}{dt}$ positive

(b) $\dfrac{dy}{dt}$ positive $\Rightarrow \dfrac{dx}{dt}$ negative

13. $D = \sqrt{x^2 + y^2} = \sqrt{x^2 + (x^2 + 1)^2} = \sqrt{x^4 + 3x^2 + 1}$

$\dfrac{dx}{dt} = 2$

$\dfrac{dD}{dt} = \dfrac{1}{2}(x^4 + 3x^2 + 1)^{-1/2}(4x^3 + 6x)\dfrac{dx}{dt} = \dfrac{2x^3 + 3x}{\sqrt{x^4 + 3x^2 + 1}} \dfrac{dx}{dt} = \dfrac{4x^3 + 6x}{\sqrt{x^4 + 3x^2 + 1}}$

15. $A = \pi r^2$

$\dfrac{dr}{dt} = 3$

$\dfrac{dA}{dt} = 2\pi r \dfrac{dr}{dt}$

(a) When $r = 6$,

$\dfrac{dA}{dt} = 2\pi(6)(3) = 36\pi \text{ cm}^2/\text{min.}$

(b) When $r = 24$,

$\dfrac{dA}{dt} = 2\pi(24)(3) = 144\pi \text{ cm}^2/\text{min.}$

7. $y = \tan x$

$\dfrac{dx}{dt} = 2$

$\dfrac{dy}{dt} = \sec^2 x \dfrac{dx}{dt}$

(a) When $x = -\pi/3$,

$\dfrac{dy}{dt} = (2)^2(2) = 8 \text{ cm/sec.}$

(b) When $x = -\pi/4$,

$\dfrac{dy}{dt} = \left(\sqrt{2}\right)^2(2) = 4 \text{ cm/sec.}$

(c) When $x = 0$,

$\dfrac{dy}{dt} = (1)^2(2) = 2 \text{ cm/sec.}$

11. Yes, y changes at a constant rate: $\dfrac{dy}{dt} = a \cdot \dfrac{dx}{dt}$.

No, the rate $\dfrac{dy}{dt}$ is a multiple of $\dfrac{dx}{dt}$.

17. (a) $\sin \dfrac{\theta}{2} = \dfrac{(1/2)b}{s} \Rightarrow b = 2s \sin \dfrac{\theta}{2}$

$\cos \dfrac{\theta}{2} = \dfrac{h}{s} \Rightarrow h = s \cos \dfrac{\theta}{2}$

$A = \dfrac{1}{2}bh = \dfrac{1}{2}\left(2s \sin \dfrac{\theta}{2}\right)\left(s \cos \dfrac{\theta}{2}\right)$

$\quad = \dfrac{s^2}{2}\left(2 \sin \dfrac{\theta}{2} \cos \dfrac{\theta}{2}\right) = \dfrac{s^2}{2} \sin \theta$

(b) $\dfrac{dA}{dt} = \dfrac{s^2}{2} \cos \theta \dfrac{d\theta}{dt}$ where $\dfrac{d\theta}{dt} = \dfrac{1}{2} \text{ rad/min.}$

When $\theta = \dfrac{\pi}{6}, \dfrac{dA}{dt} = \dfrac{s^2}{2}\left(\dfrac{\sqrt{3}}{2}\right)\left(\dfrac{1}{2}\right) = \dfrac{\sqrt{3}s^2}{8}$

When $\theta = \dfrac{\pi}{3}, \dfrac{dA}{dt} = \dfrac{s^2}{2}\left(\dfrac{1}{2}\right)\left(\dfrac{1}{2}\right) = \dfrac{s^2}{8}$

(c) If $d\theta/dt$ is constant, dA/dt is proportional to $\cos\theta$.

19. $V = \frac{4}{3}\pi r^3, \frac{dV}{dt} = 800$

$$\frac{dV}{dt} = 4\pi r^2 \frac{dr}{dt}$$

$$\frac{dr}{dt} = \frac{1}{4\pi r^2}\left(\frac{dV}{dt}\right) = \frac{1}{4\pi r^2}(800)$$

(a) When $r = 30, \frac{dr}{dt} = \frac{1}{4\pi(30)^2}(800) = \frac{2}{9\pi}$ cm/min.

(b) When $r = 60, \frac{dr}{dt} = \frac{1}{4\pi(60)^2}(800) = \frac{1}{18\pi}$ cm/min.

23. $V = \frac{1}{3}\pi r^2 h = \frac{1}{3}\pi\left(\frac{9}{4}h^2\right)h$ [since $2r = 3h$]

$$= \frac{3\pi}{4}h^3$$

$$\frac{dV}{dt} = 10$$

$$\frac{dV}{dt} = \frac{9\pi}{4}h^2 \frac{dh}{dt} \Rightarrow \frac{dh}{dt} = \frac{4(dV/dt)}{9\pi h^2}$$

When $h = 15, \frac{dh}{dt} = \frac{4(10)}{9\pi(15)^2} = \frac{8}{405\pi}$ ft/min.

25.

(a) Total volume of pool $= \frac{1}{2}(2)(12)(6) + (1)(6)(12) = 144$ m³

Volume of 1m. of water $= \frac{1}{2}(1)(6)(6) = 18$ m³

(see similar triangle diagram)

% pool filled $= \frac{18}{144}(100\%) = 12.5\%$

(b) Since for $0 \le h \le 2, b = 6h$, you have

$$V = \frac{1}{2}bh(6) = 3bh = 3(6h)h = 18h^2$$

$$\frac{dV}{dt} = 36h \frac{dh}{dt} = \frac{1}{4} \Rightarrow \frac{dh}{dt} = \frac{1}{144h} = \frac{1}{144(1)} = \frac{1}{144}$$ m/min.

21. $s = 6x^2$

$$\frac{dx}{dt} = 3$$

$$\frac{ds}{dt} = 12x \frac{dx}{dt}$$

(a) When $x = 1,$

$$\frac{ds}{dt} = 12(1)(3) = 36 \text{ cm}^2/\text{sec.}$$

(b) When $x = 10,$

$$\frac{ds}{dt} = 12(10)(3) = 360 \text{ cm}^2/\text{sec.}$$

27. $x^2 + y^2 = 25^2$

$2x\dfrac{dx}{dt} + 2y\dfrac{dy}{dt} = 0$

$\quad\quad\dfrac{dy}{dt} = \dfrac{-x}{y}\cdot\dfrac{dx}{dt} = \dfrac{-2x}{y}$ since $\dfrac{dx}{dt} = 2.$

(a) When $x = 7, y = \sqrt{576} = 24, \dfrac{dy}{dt} = \dfrac{-2(7)}{24} = \dfrac{-7}{12}$ ft/sec.   (b) $A = \dfrac{1}{2}xy$

When $x = 15, y = \sqrt{400} = 20, \dfrac{dy}{dt} = \dfrac{-2(15)}{20} = \dfrac{-3}{2}$ ft/sec. $\dfrac{dA}{dt} = \dfrac{1}{2}\left(x\dfrac{dy}{dt} + y\dfrac{dx}{dt}\right)$

When $x = 24, y = 7, \dfrac{dy}{dt} = \dfrac{-2(24)}{7} = \dfrac{-48}{7}$ ft/sec.   From part (a) we have $x = 7, y = 24, \dfrac{dx}{dt} = 2,$

and $\dfrac{dy}{dt} = -\dfrac{7}{12}.$

Thus, $\dfrac{dA}{dt} = \dfrac{1}{2}\left[7\left(-\dfrac{7}{12}\right) + 24(2)\right]$

$\quad\quad\quad = \dfrac{527}{24} \approx 21.96$ ft²/sec.

(c) $\tan\theta = \dfrac{x}{y}$

$\sec^2\theta\,\dfrac{d\theta}{dt} = \dfrac{1}{y}\cdot\dfrac{dx}{dt} - \dfrac{x}{y^2}\cdot\dfrac{dy}{dt}$

$\quad\quad\dfrac{d\theta}{dt} = \cos^2\theta\left[\dfrac{1}{y}\cdot\dfrac{dx}{dt} - \dfrac{x}{y^2}\cdot\dfrac{dy}{dt}\right]$

Using $x = 7, y = 24, \dfrac{dx}{dt} = 2, \dfrac{dy}{dt} = -\dfrac{7}{12}$ and $\cos\theta = \dfrac{24}{25}$, we have $\dfrac{d\theta}{dt} = \left(\dfrac{24}{25}\right)^2\left[\dfrac{1}{24}(2) - \dfrac{7}{(24)^2}\left(-\dfrac{7}{12}\right)\right] = \dfrac{1}{12}$ rad/sec.

29. When $y = 6, x = \sqrt{12^2 - 6^2} = 6\sqrt{3}$, and

$s = \sqrt{x^2 + (12 - y)^2}$

$\quad = \sqrt{108 + 36} = 12.$

$\quad\quad x^2 + (12 - y)^2 = s^2$

$2x\dfrac{dx}{dt} + 2(12 - y)(-1)\dfrac{dy}{dt} = 2s\dfrac{ds}{dt}$

$\quad x\dfrac{dx}{dt} + (y - 12)\dfrac{dy}{dt} = s\dfrac{ds}{dt}$

Also, $x^2 + y^2 = 12^2$

$\quad 2x\dfrac{dx}{dt} + 2y\dfrac{dy}{dt} = 0 \Rightarrow \dfrac{dy}{dt} = \dfrac{-x}{y}\dfrac{dx}{dt}.$

Thus, $x\dfrac{dx}{dt} + (y - 12)\left(\dfrac{-x}{y}\dfrac{dx}{dt}\right) = s\dfrac{ds}{dt}$

$\dfrac{dx}{dt}\left[x - x + \dfrac{12x}{y}\right] = s\dfrac{ds}{dt} \Rightarrow \dfrac{dx}{dt} = \dfrac{sy}{12x}\cdot\dfrac{ds}{dt} = \dfrac{(12)(6)}{(12)\left(6\sqrt{3}\right)}(-0.2) = \dfrac{-1}{5\sqrt{3}} = \dfrac{-\sqrt{3}}{15}$ m/sec (horizontal)

$\dfrac{dy}{dt} = \dfrac{-x}{y}\dfrac{dx}{dt} = \dfrac{-6\sqrt{3}}{6}\cdot\dfrac{\left(-\sqrt{3}\right)}{15} = \dfrac{1}{5}$ m/sec (vertical).

31. (a) $s^2 = x^2 + y^2$

$$\frac{dx}{dt} = -450$$

$$\frac{dy}{dt} = -600$$

$$2s\frac{ds}{dt} = 2x\frac{dx}{dt} + 2y\frac{dy}{dt}$$

$$\frac{ds}{dt} = \frac{x(dx/dt) + y(dy/dt)}{s}$$

When $x = 150$ and $y = 200$, $s = 250$ and

$$\frac{ds}{dt} = \frac{150(-450) + 200(-600)}{250} = -750 \text{ mph.}$$

(b) $t = \dfrac{250}{750} = \dfrac{1}{3} \text{ hr} = 20 \text{ min}$

33. $s^2 = 90^2 + x^2$

$x = 30$

$$\frac{dx}{dt} = -28$$

$$2s\frac{ds}{dt} = 2x\frac{dx}{dt} \Longrightarrow \frac{ds}{dt} = \frac{x}{s} \cdot \frac{dx}{dt}$$

When $x = 30$,

$$s = \sqrt{90^2 + 30^2} = 30\sqrt{10}$$

$$\frac{ds}{dt} = \frac{30}{30\sqrt{10}}(-28) = \frac{-28}{\sqrt{10}} \approx -8.85 \text{ ft/sec.}$$

35. (a) $\dfrac{15}{6} = \dfrac{y}{y-x} \Longrightarrow 15y - 15x = 6y$

$$y = \frac{5}{3}x$$

$$\frac{dx}{dt} = 5$$

$$\frac{dy}{dt} = \frac{5}{3} \cdot \frac{dx}{dt} = \frac{5}{3}(5) = \frac{25}{3} \text{ ft/sec}$$

(b) $\dfrac{d(y-x)}{dt} = \dfrac{dy}{dt} - \dfrac{dx}{dt} = \dfrac{25}{3} - 5 = \dfrac{10}{3} \text{ ft/sec}$

37. $x(t) = \dfrac{1}{2} \sin \dfrac{\pi t}{6}, \, x^2 + y^2 = 1$

(a) Period: $\dfrac{2\pi}{\pi/6} = 12$ seconds

(b) When $x = \dfrac{1}{2}, \, y = \sqrt{1^2 - \left(\dfrac{1}{2}\right)^2} = \dfrac{\sqrt{3}}{2}$ m.

Lowest point: $\left(0, \dfrac{\sqrt{3}}{2}\right)$

(c) When $x = \dfrac{1}{4}, \, y = \sqrt{1 - \left(\dfrac{1}{4}\right)^2} = \dfrac{\sqrt{15}}{4}$ and $t = 1$

$\dfrac{dx}{dt} = \dfrac{1}{2}\left(\dfrac{\pi}{6}\right) \cos \dfrac{\pi t}{6} = \dfrac{\pi}{12} \cos \dfrac{\pi t}{6}$

$x^2 + y^2 = 1$

$2x \dfrac{dx}{dt} + 2y \dfrac{dy}{dt} = 0 \Longrightarrow \dfrac{dy}{dt} = \dfrac{-x}{y} \dfrac{dx}{dt}.$

Thus,

$\dfrac{dy}{dt} = -\dfrac{1/4}{\sqrt{15}/4} \cdot \dfrac{\pi}{12} \cos\left(\dfrac{\pi}{6}\right)$

$\quad = \dfrac{-\pi}{\sqrt{15}}\left(\dfrac{1}{12}\right)\dfrac{\sqrt{3}}{2} = \dfrac{-\pi}{24}\dfrac{1}{\sqrt{5}} = \dfrac{-\sqrt{5}\pi}{120}.$

Speed $= \left|\dfrac{-\sqrt{5}\pi}{120}\right| = \dfrac{\sqrt{5}\pi}{120}$ m/sec

41. $\qquad\qquad pV^{1.3} = k$

$1.3\, pV^{0.3}\dfrac{dV}{dt} + V^{1.3}\dfrac{dp}{dt} = 0$

$V^{0.3}\left(1.3p\dfrac{dV}{dt} + V\dfrac{dp}{dt}\right) = 0$

$\qquad\qquad 1.3p\dfrac{dV}{dt} = -V\dfrac{dp}{dt}$

43. $\qquad \tan\theta = \dfrac{y}{30}$

$\dfrac{dy}{dt} = 3$ m/sec.

$\sec^2\theta \cdot \dfrac{d\theta}{dt} = \dfrac{1}{30}\dfrac{dy}{dt}$

$\dfrac{d\theta}{dt} = \dfrac{1}{30}\cos^2\theta \cdot \dfrac{dy}{dt}$

When $y = 30, \, \theta = \pi/4$ and $\cos\theta = \sqrt{2}/2$. Thus,

$\dfrac{d\theta}{dt} = \dfrac{1}{30}\left(\dfrac{1}{2}\right)(3) = \dfrac{1}{20}$ rad/sec.

39. Since the evaporation rate is proportional to the surface area, $dV/dt = k(4\pi r^2)$. However, since $V = (4/3)\pi r^3$, we have

$\dfrac{dV}{dt} = 4\pi r^2 \dfrac{dr}{dt}.$

Therefore,

$k(4\pi r^2) = 4\pi r^2 \dfrac{dr}{dt} \Longrightarrow k = \dfrac{dr}{dt}.$

45. $\tan \theta = \dfrac{y}{x}, y = 5$

$$\dfrac{dx}{dt} = -600 \text{ mi/hr}$$

$$(\sec^2 \theta) \dfrac{d\theta}{dt} = -\dfrac{5}{x^2} \cdot \dfrac{dx}{dt}$$

$$\dfrac{d\theta}{dt} = \cos^2 \theta \left(-\dfrac{5}{x^2}\right) \dfrac{dx}{dt} = \dfrac{x^2}{L^2}\left(-\dfrac{5}{x^2}\right) \dfrac{dx}{dt}$$

$$= \left(-\dfrac{5^2}{L^2}\right)\left(\dfrac{1}{5}\right) \dfrac{dx}{dt} = (-\sin^2 \theta)\left(\dfrac{1}{5}\right)(-600) = 120 \sin^2 \theta$$

(a) When $\theta = 30°$, $\dfrac{d\theta}{dt} = \dfrac{120}{4} = 30 \text{ rad/hr} = \dfrac{1}{2} \text{ rad/min.}$

(b) When $\theta = 60°$, $\dfrac{d\theta}{dt} = 120\left(\dfrac{3}{4}\right) = 90 \text{ rad/hr} = \dfrac{3}{2} \text{ rad/min.}$

(c) When $\theta = 75°$, $\dfrac{d\theta}{dt} = 120 \sin^2 75° \approx 111.96 \text{ rad/hr} \approx 1.87 \text{ rad/min.}$

47. $\dfrac{d\theta}{dt} = (10 \text{ rev/sec})(2\pi \text{ rad/rev}) = 20\pi \text{ rad/sec}$

(a) $\cos \theta = \dfrac{x}{30}$

$$-\sin \theta \dfrac{d\theta}{dt} = \dfrac{1}{30} \dfrac{dx}{dt}$$

$$\dfrac{dx}{dt} = -30 \sin \theta \dfrac{d\theta}{dt} = -30 \sin \theta(20\pi) = -600\pi \sin \theta$$

(b)

(c) $|dx/dt| = |-600\pi \sin \theta|$ is greatest when $|\sin \theta| = 1 \implies \theta = (\pi/2) + n\pi$ (or $90° + n \cdot 180°$).

$|dx/dt|$ is least when $\theta = n\pi$ (or $n \cdot 180°$).

(d) For $\theta = 30°$, $\dfrac{dx}{dt} = -600\pi \sin(30°) = -600\pi \dfrac{1}{2} = -300\pi \text{ cm/sec.}$

For $\theta = 60°$, $\dfrac{dx}{dt} = -600\pi \sin(60°) = -600\pi \dfrac{\sqrt{3}}{2} = -300\sqrt{3}\pi \text{ cm/sec.}$

49. $\tan \theta = \dfrac{x}{50} \implies x = 50 \tan \theta$

$$\dfrac{dx}{dt} = 50 \sec^2 \theta \dfrac{d\theta}{dt}$$

$$2 = 50 \sec^2 \theta \dfrac{d\theta}{dt}$$

$$\dfrac{d\theta}{dt} = \dfrac{1}{25} \cos^2 \theta, \quad -\dfrac{\pi}{4} \le \theta \le \dfrac{\pi}{4}$$

51. $x^2 + y^2 = 25$; acceleration of the top of the ladder $= \dfrac{d^2y}{dt^2}$

First derivative: $2x \dfrac{dx}{dt} + 2y \dfrac{dy}{dt} = 0$

$$x \dfrac{dx}{dt} + y \dfrac{dy}{dt} = 0$$

Second derivative: $x \dfrac{d^2x}{dt^2} + \dfrac{dx}{dt} \cdot \dfrac{dx}{dt} + y \dfrac{d^2y}{dt^2} + \dfrac{dy}{dt} \cdot \dfrac{dy}{dt} = 0$

$$\dfrac{d^2y}{dt^2} = \left(\dfrac{1}{y} \right) \left[-x \dfrac{d^2x}{dt^2} - \left(\dfrac{dx}{dt} \right)^2 - \left(\dfrac{dy}{dt} \right)^2 \right]$$

When $x = 7$, $y = 24$, $\dfrac{dy}{dt} = -\dfrac{7}{12}$, and $\dfrac{dx}{dt} = 2$ (see Exercise 27). Since $\dfrac{dx}{dt}$ is constant, $\dfrac{d^2x}{dt^2} = 0$.

$$\dfrac{d^2y}{dt^2} = \dfrac{1}{24} \left[-7(0) - (2)^2 - \left(-\dfrac{7}{12} \right)^2 \right] = \dfrac{1}{24} \left[-4 - \dfrac{49}{144} \right] = \dfrac{1}{24} \left[-\dfrac{625}{144} \right] \approx -0.1808 \text{ ft/sec}^2$$

53. (a) Using a graphing utility, you obtain $m(s) = -0.881s^2 + 29.10s - 206.2$

(b) $\dfrac{dm}{dt} = \dfrac{dm}{ds} \dfrac{ds}{dt} = (-1.762s + 29.10) \dfrac{ds}{dt}$

(c) If $t = 5$ (1995), then $s = 15.5$ and $\dfrac{ds}{dt} = 1.2$.

Thus, $\dfrac{dm}{dt} = (-1.762(15.5) + 29.10)(1.2) \approx 2.15$ million.

Review Exercises for Chapter 2

1. $f(x) = x^2 - 2x + 3$

$f'(x) = \lim\limits_{\Delta x \to 0} \dfrac{f(x + \Delta x) - f(x)}{\Delta x}$

$\quad = \lim\limits_{\Delta x \to 0} \dfrac{[(x + \Delta x)^2 - 2(x + \Delta x) + 3] - [x^2 - 2x + 3]}{\Delta x}$

$\quad = \lim\limits_{\Delta x \to 0} \dfrac{(x^2 + 2x(\Delta x) + (\Delta x)^2 - 2x - 2(\Delta x) + 3) - (x^2 - 2x + 3)}{\Delta x}$

$\quad = \lim\limits_{\Delta x \to 0} \dfrac{2x(\Delta x) + (\Delta x)^2 - 2(\Delta x)}{\Delta x} = \lim\limits_{\Delta x \to 0} (2x + \Delta x - 2) = 2x - 2$

3. $f(x) = \sqrt{x} + 1$

$f'(x) = \lim\limits_{\Delta x \to 0} \dfrac{f(x + \Delta x) - f(x)}{\Delta x}$

$\quad = \lim\limits_{\Delta x \to 0} \dfrac{(\sqrt{x + \Delta x} + 1) - (\sqrt{x} + 1)}{\Delta x}$

$\quad = \lim\limits_{\Delta x \to 0} \dfrac{\sqrt{x + \Delta x} - \sqrt{x}}{\Delta x} \cdot \dfrac{\sqrt{x + \Delta x} + \sqrt{x}}{\sqrt{x + \Delta x} + \sqrt{x}}$

$\quad = \lim\limits_{\Delta x \to 0} \dfrac{(x + \Delta x) - x}{\Delta x(\sqrt{x + \Delta x} + \sqrt{x})}$

$\quad = \lim\limits_{\Delta x \to 0} \dfrac{1}{\sqrt{x + \Delta x} + \sqrt{x}} = \dfrac{1}{2\sqrt{x}}$

5. f is differentiable for all $x \neq -1$.

7. $f(x) = 4 - |x - 2|$

   (a) Continuous at $x = 2$.

   (b) Not differentiable at $x = 2$ because of the sharp turn in the graph.

9. Using the limit definition, you obtain $g'(x) = \dfrac{4}{3}x - \dfrac{1}{6}$.

   At $x = -1$, $g'(-1) = -\dfrac{4}{3} - \dfrac{1}{6} = \dfrac{-3}{2}$.

11. (a) Using the limit definition, $f'(x) = 3x^2$.

     At $x = -1$, $f'(-1) = 3$. The tangent line is

$$y - (-2) = 3(x - (-1))$$

$$y = 3x + 1$$

   (b)

13. $g'(2) = \lim\limits_{x \to 2} \dfrac{g(x) - g(2)}{x - 2}$

$$= \lim_{x \to 2} \frac{x^2(x - 1) - 4}{x - 2}$$

$$= \lim_{x \to 2} \frac{x^3 - x^2 - 4}{x - 2}$$

$$= \lim_{x \to 2} \frac{(x - 2)(x^2 + x + 2)}{x - 2}$$

$$= \lim_{x \to 2} (x^2 + x + 2) = 8$$

15.

17. $y = 25$

    $y' = 0$

19. $f(x) = x^8$

    $f'(x) = 8x^7$

21. $h(t) = 3t^4$

    $h'(t) = 12t^3$

23. $f(x) = x^3 - 3x^2$

    $f'(x) = 3x^2 - 6x = 3x(x - 2)$

25. $h(x) = 6\sqrt{x} + 3\sqrt[3]{x} = 6x^{1/2} + 3x^{1/3}$

    $h'(x) = 3x^{-1/2} + x^{-2/3} = \dfrac{3}{\sqrt{x}} + \dfrac{1}{\sqrt[3]{x^2}}$

27. $g(t) = \dfrac{2}{3}t^{-2}$

    $g'(t) = \dfrac{-4}{3}t^{-3} = \dfrac{-4}{3t^3}$

29. $f(\theta) = 2\theta - 3\sin\theta$

    $f'(\theta) = 2 - 3\cos\theta$

31. $f(\theta) = 3\cos\theta - \dfrac{\sin\theta}{4}$

    $f'(\theta) = -3\sin\theta - \dfrac{\cos\theta}{4}$

33. $F = 200\sqrt{T}$

$F'(t) = \dfrac{100}{\sqrt{T}}$

(a) When $T = 4$, $F'(4) = 50$ vibrations/sec/lb.

(b) When $T = 9$, $F'(9) = 33\frac{1}{3}$ vibrations/sec/lb.

35. $s(t) = -16t^2 + s_0$

$s(9.2) = -16(9.2)^2 + s_0 = 0$

$s_0 = 1354.24$

The building is approximately 1354 feet high (or 415 m).

37. (a)

Total horizontal distance: 50

(b) $0 = x - 0.02x^2$

$0 = x\left(1 - \dfrac{x}{50}\right)$ implies $x = 50$.

(c) Ball reaches maximum height when $x = 25$.

(d) $y = x - 0.02x^2$

$y' = 1 - 0.04x$

$y'(0) = 1$

$y'(10) = 0.6$

$y'(25) = 0$

$y'(30) = -0.2$

$y'(50) = -1$

(e) $y'(25) = 0$

39. $x(t) = t^2 - 3t + 2 = (t - 2)(t - 1)$

(a) $v(t) = x'(t) = 2t - 3$

(c) $v(t) = 0$ for $t = \frac{3}{2}$.

$x = \left(\frac{3}{2} - 2\right)\left(\frac{3}{2} - 1\right) = \left(-\frac{1}{2}\right)\left(\frac{1}{2}\right) = -\frac{1}{4}$

(b) $v(t) < 0$ for $t < \frac{3}{2}$.

(d) $x(t) = 0$ for $t = 1, 2$.

$|v(1)| = |2(1) - 3| = 1$

$|v(2)| = |2(2) - 3| = 1$

The speed is 1 when the position is 0.

41. $f(x) = (3x^2 + 7)(x^2 - 2x + 3)$

$f'(x) = (3x^2 + 7)(2x - 2) + (x^2 - 2x + 3)(6x)$

$= 2(6x^3 - 9x^2 + 16x - 7)$

43. $h(x) = \sqrt{x} \sin x = x^{1/2} \sin x$

$h'(x) = \dfrac{1}{2\sqrt{x}} \sin x + \sqrt{x} \cos x$

45. $f(x) = 2x - x^{-2}$

$f'(x) = 2 + 2x^{-3} = 2\left(1 + \dfrac{1}{x^3}\right)$

$= \dfrac{2(x^3 + 1)}{x^3}$

47. $f(x) = \dfrac{x^2 + x - 1}{x^2 - 1}$

$f'(x) = \dfrac{(x^2 - 1)(2x + 1) - (x^2 + x - 1)(2x)}{(x^2 - 1)^2}$

$= \dfrac{-(x^2 + 1)}{(x^2 - 1)^2}$

49. $f(x) = (4 - 3x^2)^{-1}$

$f'(x) = -(4 - 3x^2)^{-2}(-6x) = \dfrac{6x}{(4 - 3x^2)^2}$

51. $y = \dfrac{x^2}{\cos x}$

$y' = \dfrac{\cos x\,(2x) - x^2(-\sin x)}{\cos^2 x} = \dfrac{2x \cos x + x^2 \sin x}{\cos^2 x}$

53. $y = 3x^2 \sec x$

$y' = 3x^2 \sec x \tan x + 6x \sec x$

55. $y = -x \tan x$

$y' = -x \sec^2 x - \tan x$

57. $y = x \cos x - \sin x$

$y' = -x \sin x + \cos x - \cos x = -x \sin x$

59. $g(t) = t^3 - 3t + 2$

$g'(t) = 3t^2 - 3$

$g''(t) = 6t$

61. $f(\theta) = 3 \tan \theta$

$f'(\theta) = 3 \sec^2 \theta$

$f''(\theta) = 6 \sec \theta (\sec \theta \tan \theta) = 6 \sec^2 \theta \tan \theta$

63. $\quad y = 2 \sin x + 3 \cos x$

$y' = 2 \cos x - 3 \sin x$

$y'' = -2 \sin x - 3 \cos x$

$y'' + y = -(2 \sin x + 3 \cos x) + (2 \sin x + 3 \cos x)$

$\quad = 0$

65. $f(x) = (1 - x^3)^{1/2}$

$f'(x) = \dfrac{1}{2}(1 - x^3)^{-1/2}(-3x^2)$

$\quad = -\dfrac{3x^2}{2\sqrt{1 - x^3}}$

67. $h(x) = \left(\dfrac{x - 3}{x^2 + 1}\right)^2$

$h'(x) = 2\left(\dfrac{x - 3}{x^2 + 1}\right)\left(\dfrac{(x^2 + 1)(1) - (x - 3)(2x)}{(x^2 + 1)^2}\right)$

$\quad = \dfrac{2(x - 3)(-x^2 + 6x + 1)}{(x^2 + 1)^3}$

69. $f(s) = (s^2 - 1)^{5/2}(s^3 + 5)$

$f'(s) = (s^2 - 1)^{5/2}(3s^2) + (s^3 + 5)\left(\dfrac{5}{2}\right)(s^2 - 1)^{3/2}(2s)$

$\quad = s(s^2 - 1)^{3/2}[3s(s^2 - 1) + 5(s^3 + 5)]$

$\quad = s(s^2 - 1)^{3/2}(8s^3 - 3s + 25)$

71. $y = 3 \cos(3x + 1)$

$y' = -9 \sin(3x + 1)$

73. $y = \dfrac{1}{2} \csc 2x$

$y' = \dfrac{1}{2}(-\csc 2x \cot 2x)(2)$

$\quad = -\csc 2x \cot 2x$

75. $y = \dfrac{x}{2} - \dfrac{\sin 2x}{4}$

$y' = \dfrac{1}{2} - \dfrac{1}{4} \cos 2x(2)$

$\quad = \dfrac{1}{2}(1 - \cos 2x) = \sin^2 x$

77. $y = \dfrac{2}{3} \sin^{3/2} x - \dfrac{2}{7} \sin^{7/2} x$

$y' = \sin^{1/2} x \cos x - \sin^{5/2} x \cos x$

$\quad = (\cos x)\sqrt{\sin x}(1 - \sin^2 x)$

$\quad = (\cos^3 x)\sqrt{\sin x}$

79. $y = \dfrac{\sin \pi x}{x + 2}$

$y' = \dfrac{(x + 2)\pi \cos \pi x - \sin \pi x}{(x + 2)^2}$

81. $f(t) = t^2(t - 1)^5$

$f'(t) = t(t - 1)^4(7t - 2)$

The zeros of f' correspond to the points on the graph of f where the tangent line is horizontal.

83. $g(x) = 2x(x + 1)^{-1/2}$

$g'(x) = \dfrac{x + 2}{(x + 1)^{3/2}}$

g' does not equal zero for any value of x in the domain. The graph of g has no horizontal tangent lines.

85. $f(t) = \sqrt{t+1}\,\sqrt[3]{t+1}$

$f(t) = (t+1)^{1/2}(t+1)^{1/3} = (t+1)^{5/6}$

$f'(t) = \dfrac{5}{6(t+1)^{1/6}}$

f' does not equal zero for any x in the domain. The graph of f has no horizontal tangent lines.

87. $y = \tan\sqrt{1-x}$

$y' = -\dfrac{\sec^2\sqrt{1-x}}{2\sqrt{1-x}}$

y' does not equal zero for any x in the domain. The graph has no horizontal tangent lines.

89. $y = 2x^2 + \sin 2x$

$y' = 4x + 2\cos 2x$

$y'' = 4 - 4\sin 2x$

91. $f(x) = \cot x$

$f'(x) = -\csc^2 x$

$f''(x) = -2\csc x(-\csc x \cdot \cot x)$

$\qquad = 2\csc^2 x \cot x$

93. $f(t) = \dfrac{t}{(1-t)^2}$

$f'(t) = \dfrac{t+1}{(1-t)^3}$

$f''(t) = \dfrac{2(t+2)}{(1-t)^4}$

95. $g(\theta) = \tan 3\theta - \sin(\theta - 1)$

$g'(\theta) = 3\sec^2 3\theta - \cos(\theta - 1)$

$g''(\theta) = 18\sec^2 3\theta \tan 3\theta + \sin(\theta - 1)$

97. $T = \dfrac{700}{t^2 + 4t + 10}$

$T = 700(t^2 + 4t + 10)^{-1}$

$T' = \dfrac{-1400(t+2)}{(t^2 + 4t + 10)^2}$

(a) When $t = 1$,

$T' = \dfrac{-1400(1+2)}{(1+4+10)^2} \approx -18.667$ deg/hr.

(b) When $t = 3$,

$T' = \dfrac{-1400(3+2)}{(9+12+10)^2} \approx -7.284$ deg/hr.

(c) When $t = 5$,

$T' = \dfrac{-1400(5+2)}{(25+20+10)^2} \approx -3.240$ deg/hr.

(d) When $t = 10$,

$T' = \dfrac{-1400(10+2)}{(100+40+10)^2} \approx -0.747$ deg/hr.

99. $\qquad x^2 + 3xy + y^3 = 10$

$2x + 3xy' + 3y + 3y^2 y' = 0$

$3(x + y^2)y' = -(2x + 3y)$

$y' = \dfrac{-(2x + 3y)}{3(x + y^2)}$

103.
$$x \sin y = y \cos x$$
$$(x \cos y)y' + \sin y = -y \sin x + y' \cos x$$
$$y'(x \cos y - \cos x) = -y \sin x - \sin y$$
$$y' = \frac{y \sin x + \sin y}{\cos x - x \cos y}$$

105.
$$x^2 + y^2 = 20$$
$$2x + 2yy' = 0$$
$$y' = -\frac{x}{y}$$

At $(2, 4)$: $y' = -\frac{1}{2}$

Tangent line: $y - 4 = -\frac{1}{2}(x - 2)$
$$x + 2y - 10 = 0$$

Normal line: $y - 4 = 2(x - 2)$
$$2x - y = 0$$

107. $y = \sqrt{x}$

$$\frac{dy}{dt} = 2 \text{ units/sec}$$

$$\frac{dy}{dt} = \frac{1}{2\sqrt{x}} \frac{dx}{dt} \implies \frac{dx}{dt} = 2\sqrt{x}\frac{dy}{dt} = 4\sqrt{x}$$

(a) When $x = \frac{1}{2}$, $\frac{dx}{dt} = 2\sqrt{2}$ units/sec.

(b) When $x = 1$, $\frac{dx}{dt} = 4$ units/sec.

(c) When $x = 4$, $\frac{dx}{dt} = 8$ units/sec.

109. $\dfrac{s}{h} = \dfrac{1/2}{2}$

$$s = \frac{1}{4}h$$

$$\frac{dV}{dt} = 1$$

Width of water at depth h:

$$w = 2 + 2s = 2 + 2\left(\frac{1}{4}h\right) = \frac{4 + h}{2}$$

$$V = \frac{5}{2}\left(2 + \frac{4 + h}{2}\right)h = \frac{5}{4}(8 + h)h$$

$$\frac{dV}{dt} = \frac{5}{2}(4 + h)\frac{dh}{dt}$$

$$\frac{dh}{dt} = \frac{2(dV/dt)}{5(4 + h)}$$

When $h = 1$, $\dfrac{dh}{dt} = \dfrac{2}{25}$ m/min.

111. $s(t) = 60 - 4.9t^2$
$$s'(t) = -9.8t$$
$$s = 35 = 60 - 4.9t^2$$
$$4.9t^2 = 25$$
$$t = \frac{5}{\sqrt{4.9}}$$

$$\tan 30° = \frac{1}{\sqrt{3}} = \frac{s(t)}{x(t)}$$
$$x(t) = \sqrt{3}\,s(t)$$
$$\frac{dx}{dt} = \sqrt{3}\frac{ds}{dt} = \sqrt{3}(-9.8)\frac{5}{\sqrt{4.9}}$$
$$\approx -38.34 \text{ m/sec}$$

Problem Solving for Chapter 2

1. (a) $x^2 + (y - r)^2 = r^2$ Circle

$x^2 = y$ Parabola

Substituting,

$$(y - r)^2 = r^2 - y$$

$$y^2 - 2ry + r^2 = r^2 - y$$

$$y^2 - 2ry + y = 0$$

$$y(y - 2r + 1) = 0$$

Since you want only one solution, let $1 - 2r = 0 \Longrightarrow r = \frac{1}{2}$

Graph $y = x^2$ and $x^2 + \left(y - \frac{1}{2}\right)^2 = \frac{1}{4}$

(b) Let (x, y) be a point of tangency: $x^2 + (y - b)^2 = 1 \Longrightarrow 2x + 2(y - b)y' = 0 \Longrightarrow y' = \dfrac{x}{b - y}$ (circle).

$y = x^2 \Longrightarrow y' = 2x$ (parabola). Equating,

$$2x = \frac{x}{b - y}$$

$$2(b - y) = 1$$

$$b - y = \frac{1}{2} \Longrightarrow b = y + \frac{1}{2}$$

Also, $x^2 + (y - b)^2 = 1$ and $y = x^2$ imply

$$y + (y - b)^2 = 1 \Longrightarrow y + \left[y - \left(y + \frac{1}{2}\right)\right]^2 = 1 \Longrightarrow y + \frac{1}{4} = 1 \Longrightarrow y = \frac{3}{4} \text{ and } b = \frac{5}{4}.$$

Center: $\left(0, \dfrac{5}{4}\right)$

Graph $y = x^2$ and $x^2 + \left(y - \dfrac{5}{4}\right)^2 = 1$

3. (a) $f(x) = \cos x \qquad P_1(x) = a_0 + a_1 x$

$\quad f(0) = 1 \qquad P_1(0) = a_0 \Longrightarrow a_0 = 1$

$\quad f'(0) = 0 \qquad P'_1(0) = a_1 \Longrightarrow a_1 = 0$

$\quad P_1(x) = 1$

(b) $f(x) = \cos x \qquad P_2(x) = a_0 + a_1 x + a_2 x^2$

$\quad f(0) = 1 \qquad P_2(0) = a_0 \Longrightarrow a_0 = 1$

$\quad f'(0) = 0 \qquad P'_2(0) = a_1 \Longrightarrow a_1 = 0$

$\quad f''(0) = -1 \qquad P''_2(0) = 2a_2 \Longrightarrow a_2 = -\frac{1}{2}$

$\quad P_2(x) = 1 - \frac{1}{2}x^2$

(c)

x	-1.0	-0.1	-0.001	0	0.001	0.1	1.0
$\cos x$	0.5403	0.9950	≈ 1	1	≈ 1	0.9950	0.5403
$P_2(x)$	0.5	0.9950	≈ 1	1	≈ 1	0.9950	0.5

$P_2(x)$ is a good approximation of $f(x) = \cos x$ when x is near 0.

(d) $f(x) = \sin x \qquad P_3(x) = a_0 + a_1 x + a_2 x^2 + a_3 x^3$

$\quad f(0) = 0 \qquad P_3(0) = a_0 \Longrightarrow a_0 = 0$

$\quad f'(0) = 1 \qquad P'_3(0) = a_1 \Longrightarrow a_1 = 1$

$\quad f''(0) = 0 \qquad P''_3(0) = 2a_2 \Longrightarrow a_2 = 0$

$\quad f'''(0) = -1 \qquad P'''_3(0) = 6a_3 \Longrightarrow a_3 = -\frac{1}{6}$

$\quad P_3(x) = x - \frac{1}{6}x^3$

5. Let $p(x) = Ax^3 + Bx^2 + Cx + D$

$p'(x) = 3Ax^2 + 2Bx + C$

At $(1, 1)$: $\quad A + \quad B + C + D = \quad 1 \qquad$ Equation 1

$\qquad\qquad 3A + 2B + C \qquad = 14 \qquad$ Equation 2

At $(-1, -3)$: $\quad -A + \quad B - C + D = -3 \qquad$ Equation 3

$\qquad\qquad\qquad 3A - 2B + C \qquad = -2 \qquad$ Equation 4

Adding Equations 1 and 3: $2B + 2D = -2$

Subtracting Equations 1 and 3: $2A + 2C = 4$

Adding Equations 2 and 4: $6A + 2C = 12$

Subtracting Equations 2 and 4: $4B = 16$

Hence, $B = 4$ and $D = \frac{1}{2}(-2 - 2B) = -5$

Subtracting $2A + 2C = 4$ and $6A + 2C = 12$, you obtain $4A = 8 \Rightarrow A = 2$. Finally, $C = \frac{1}{2}(4 - 2A) = 0$

Thus, $p(x) = 2x^3 + 4x^2 - 5$.

7. (a) $\quad x^4 = a^2x^2 - a^2y^2$

$a^2y^2 = a^2x^2 - x^4$

$y = \dfrac{\pm\sqrt{a^2x^2 - x^4}}{a}$

Graph: $y_1 = \dfrac{\sqrt{a^2x^2 - x^4}}{a}$ and $y_2 = -\dfrac{\sqrt{a^2x^2 - x^4}}{a}$

(b)

$(\pm a, 0)$ are the x-intercepts, along with $(0, 0)$.

(c) Differentiating implicitly,

$4x^3 = 2a^2x - 2a^2yy'$

$y' = \dfrac{2a^2x - 4x^3}{2a^2y} = \dfrac{x(a^2 - 2x^2)}{a^2y} = 0 \Rightarrow 2x^2 = a^2 \Rightarrow x = \dfrac{\pm a}{\sqrt{2}}.$

$\left(\dfrac{a^2}{2}\right)^2 = a^2\left(\dfrac{a^2}{2}\right) - a^2y^2$

$\dfrac{a^4}{4} = \dfrac{a^4}{2} - a^2y^2$

$a^2y^2 = \dfrac{a^4}{4}$

$y^2 = \dfrac{a^2}{4}$

$y = \pm\dfrac{a}{2}$

Four points: $\left(\dfrac{a}{\sqrt{2}}, \dfrac{a}{2}\right), \left(\dfrac{a}{\sqrt{2}}, -\dfrac{a}{2}\right), \left(-\dfrac{a}{\sqrt{2}}, \dfrac{a}{2}\right), \left(\dfrac{-a}{\sqrt{2}}, -\dfrac{a}{2}\right)$

9. (a)

Not drawn to scale

Line determined by $(0, 30)$ and $(90, 6)$:

$$y - 30 = \frac{30 - 6}{0 - 90}(x - 0) = -\frac{24}{90}x = -\frac{4}{15}x \Rightarrow y = -\frac{4}{15}x + 30$$

When $x = 100$, $y = \frac{-4}{15}(100) + 30 = \frac{10}{3} > 3 \Rightarrow$ Shadow determined by man.

(b)

Not drawn to scale

Line determined by $(0, 30)$ and $(60, 6)$:

$$y - 30 = \frac{30 - 6}{0 - 60}(x - 0) = -\frac{2}{5}x \Rightarrow y = -\frac{2}{5}x + 30$$

When $x = 70$, $y = \frac{-2}{5}(70) + 30 = 2 < 3 \Rightarrow$ Shadow determined by child.

(c) Need $(0, 30)$, $(d, 6)$, $(d + 10, 3)$ collinear.

$$\frac{30 - 6}{0 - d} = \frac{6 - 3}{d - (d + 10)} \Rightarrow \frac{24}{d} = \frac{3}{10} \Rightarrow d = 80 \text{ feet}$$

(d) Let y be the distance from the base of the street light to the tip of the shadow. We know that $\frac{dx}{dt} = -5$.

For $x > 80$, the shadow is determined by the man.

$$\frac{y}{30} = \frac{y - x}{6} \Rightarrow y = \frac{5}{4}x \text{ and } \frac{dy}{dt} = \frac{5}{4}\frac{dx}{dt} = \frac{-25}{4}.$$

For $x < 80$, the shadow is determined by the child.

$$\frac{y}{30} = \frac{y - x - 10}{3} \Rightarrow y = \frac{10}{9}x + \frac{100}{9} \text{ and } \frac{dy}{dt} = \frac{10}{9}\frac{dx}{dt} = \frac{-50}{9}.$$

Therefore,

$$\frac{dy}{dt} = \begin{cases} \dfrac{-25}{4} & x > 80 \\ \dfrac{-50}{9} & 0 < x < 80 \end{cases}$$

$\frac{dy}{dt}$ is not continuous at $x = 80$.

11. $L'(x) = \lim_{\Delta x \to 0} \dfrac{L(x + \Delta x) - L(x)}{\Delta x}$

$ = \lim_{\Delta x \to 0} \dfrac{L(x) + L(\Delta x) - L(x)}{\Delta x}$

$ = \lim_{\Delta x \to 0} \dfrac{L(\Delta x)}{\Delta x}$

Also, $L'(0) = \lim_{\Delta x \to 0} \dfrac{L(\Delta x) - L(0)}{\Delta x}$

But, $L(0) = 0$ because $L(0) = L(0 + 0) = L(0) + L(0) \Rightarrow L(0) = 0$.

Thus, $L'(x) = L'(0)$, for all x.

The graph of L is a line through the origin of slope $L'(0)$.

13. (a)

z (degrees)	0.1	0.01	0.0001
$\dfrac{\sin z}{z}$	0.0174524	0.0174533	0.0174533

(b) $\displaystyle\lim_{z\to 0}\frac{\sin z}{z}\approx 0.0174533$

In fact, $\displaystyle\lim_{z\to 0}\frac{\sin z}{z}=\frac{\pi}{180}$.

(c) $\displaystyle\frac{d}{dz}(\sin z)=\lim_{\Delta z\to 0}\frac{\sin(z+\Delta z)-\sin z}{\Delta z}$

$\displaystyle\qquad=\lim_{\Delta z\to 0}\frac{\sin z\cdot\cos\Delta z+\sin\Delta z\cdot\cos z-\sin z}{\Delta z}$

$\displaystyle\qquad=\lim_{\Delta z\to 0}\left[\sin z\left(\frac{\cos\Delta z-1}{\Delta z}\right)\right]+\lim_{\Delta z\to 0}\left[\cos z\left(\frac{\sin\Delta z}{\Delta z}\right)\right]$

$\displaystyle\qquad=(\sin z)(0)+(\cos z)\left(\frac{\pi}{180}\right)=\frac{\pi}{180}\cos z$

(d) $S(90)=\sin\left(\dfrac{\pi}{180}90\right)=\sin\dfrac{\pi}{2}=1;\ C(180)=\cos\left(\dfrac{\pi}{180}180\right)=-1$

$\dfrac{d}{dz}S(z)=\dfrac{d}{dz}\sin(cz)=c\cdot\cos(cz)=\dfrac{\pi}{180}C(z)$

(e) The formulas for the derivatives are more complicated in degrees.

15. $j(t)=a'(t)$

(a) $j(t)$ is the rate of change of the acceleration.

(b) From Exercise 102 in Section 2.3:

$s(t)=-8.25t^2+66t$

$v(t)=-16.5t+66$

$a(t)=-16.5$

$a'(t)=j(t)=0$

The acceleration is constant.

C H A P T E R   3
Applications of Differentiation

Section 3.1 Extrema on an Interval 103

Section 3.2 Rolle's Theorem and the Mean Value Theorem . 107

Section 3.3 Increasing and Decreasing Functions and
the First Derivative Test 113

Section 3.4 Concavity and the Second Derivative Test 121

Section 3.5 Limits at Infinity 129

Section 3.6 A Summary of Curve Sketching 136

Section 3.7 Optimization Problems 145

Section 3.8 Newton's Method 155

Section 3.9 Differentials 160

Review Exercises 163

Problem Solving 172

C H A P T E R   3
Applications of Differentiation

Section 3.1   Extrema on an Interval
Solutions to Odd-Numbered Exercises

1. $f(x) = \dfrac{x^2}{x^2 + 4}$

$f'(x) = \dfrac{(x^2 + 4)(2x) - (x^2)(2x)}{(x^2 + 4)^2} = \dfrac{8x}{(x^2 + 4)^2}$

$f'(0) = 0$

3. $f(x) = x + \dfrac{27}{2x^2} = x + \dfrac{27}{2}x^{-2}$

$f'(x) = 1 - 27x^{-3} = 1 - \dfrac{27}{x^3}$

$f'(3) = 1 - \dfrac{27}{3^3} = 1 - 1 = 0$

5. $f(x) = (x + 2)^{2/3}$

$f'(x) = \dfrac{2}{3}(x + 2)^{-1/3}$

$f'(-2)$ is undefined.

7. Critical numbers: $x = 2$

$x = 2$: absolute maximum

9. Critical numbers: $x = 1, 2, 3$

$x = 1, 3$: absolute maximum

$x = 2$: absolute minimum

11. $f(x) = x^2(x - 3) = x^3 - 3x^2$

$f'(x) = 3x^2 - 6x = 3x(x - 2)$

Critical numbers: $x = 0, x = 2$

13. $g(t) = t\sqrt{4 - t}, \ t < 3$

$g'(t) = t\left[\dfrac{1}{2}(4 - t)^{-1/2}(-1)\right] + (4 - t)^{1/2}$

$= \dfrac{1}{2}(4 - t)^{-1/2}[-t + 2(4 - t)]$

$= \dfrac{8 - 3t}{2\sqrt{4 - t}}$

Critical number is $t = \dfrac{8}{3}$.

15. $h(x) = \sin^2 x + \cos x, \ 0 < x < 2\pi$

$h'(x) = 2 \sin x \cos x - \sin x = \sin x(2 \cos x - 1)$

On $(0, 2\pi)$, critical numbers: $x = \dfrac{\pi}{3}, x = \pi, x = \dfrac{5\pi}{3}$

17. $f(x) = 2(3 - x), [-1, 2]$

$f'(x) = -2 \Rightarrow$ No critical numbers

Left endpoint: $(-1, 8)$ Maximum

Right endpoint: $(2, 2)$ Minimum

19. $f(x) = -x^2 + 3x, [0, 3]$

$f'(x) = -2x + 3$

Left endpoint: $(0, 0)$ Minimum

Critical number: $\left(\frac{3}{2}, \frac{9}{4}\right)$ Maximum

Right endpoint: $(3, 0)$ Minimum

21. $f(x) = x^3 - \frac{3}{2}x^2, \ [-1, 2]$

$f'(x) = 3x^2 - 3x = 3x(x - 1)$

Left endpoint: $\left(-1, -\frac{5}{2}\right)$ Minimum

Right endpoint: $(2, 2)$ Maximum

Critical number: $(0, 0)$

Critical number: $\left(1, -\frac{1}{2}\right)$

23. $f(x) = 3x^{2/3} - 2x, [-1, 1]$

$f'(x) = 2x^{-1/3} - 2 = \dfrac{2\left(1 - \sqrt[3]{x}\right)}{\sqrt[3]{x}}$

Left endpoint: $(-1, 5)$ Maximum

Critical number: $(0, 0)$ Minimum

Right endpoint: $(1, 1)$

25. $g(t) = \dfrac{t^2}{t^2 + 3}, [-1, 1]$

$g'(t) = \dfrac{6t}{(t^2 + 3)^2}$

Left endpoint: $\left(-1, \frac{1}{4}\right)$ Maximum

Critical number: $(0, 0)$ Minimum

Right endpoint: $\left(1, \frac{1}{4}\right)$ Maximum

27. $h(s) = \dfrac{1}{s - 2}, [0, 1]$

$h'(s) = \dfrac{-1}{(s - 2)^2}$

Left endpoint: $\left(0, -\frac{1}{2}\right)$ Maximum

Right endpoint: $(1, -1)$ Minimum

29. $f(x) = \cos \pi x, \left[0, \frac{1}{6}\right]$

$f'(x) = -\pi \sin \pi x$

Left endpoint: $(0, 1)$ Maximum

Right endpoint: $\left(\frac{1}{6}, \frac{\sqrt{3}}{2}\right)$ Minimum

31. $y = \dfrac{4}{x} + \tan\dfrac{\pi x}{8}, \ [1, 2]$

$y' = \dfrac{-4}{x^2} + \dfrac{\pi}{8} \sec^2 \dfrac{\pi x}{8} = 0$

$$\frac{\pi}{8} \sec^2 \frac{\pi x}{8} = \frac{4}{x^2}$$

On the interval $[1, 2]$, this equation has no solutions. Thus, there are no critical numbers.

Left endpoint: $\left(1, \sqrt{2} + 3\right) \approx (1, 4.4142)$ Maximum

Right endpoint: $(2, 3)$ Minimum

33. (a) Minimum: $(0, -3)$

Maximum: $(2, 1)$

(b) Minimum: $(0, -3)$

(c) Maximum: $(2, 1)$

(d) No extrema

35. $f(x) = x^2 - 2x$

(a) Minimum: $(1, -1)$

Maximum: $(-1, 3)$

(b) Maximum: $(3, 3)$

(c) Minimum: $(1, -1)$

(d) Minimum: $(1, -1)$

37. $f(x) = \begin{cases} 2x + 2, & 0 \le x \le 1 \\ 4x^2, & 1 < x \le 3 \end{cases}$

Left endpoint: $(0, 2)$ Minimum

Right endpoint: $(3, 36)$ Maximum

39. $f(x) = \dfrac{3}{x - 1}, (1, 4]$

Right endpoint: $(4, 1)$ Minimum

41. (a)

Maximum: $(1, 4.7)$ (endpoint)

Minimum: $(0.4398, -1.0613)$

(b)
$$f(x) = 3.2x^5 + 5x^3 - 3.5x, [0, 1]$$
$$f'(x) = 16x^4 + 15x^2 - 3.5$$
$$16x^4 + 15x^2 - 3.5 = 0$$
$$x^2 = \frac{-15 \pm \sqrt{(15)^2 - 4(16)(-3.5)}}{2(16)}$$
$$= \frac{-15 \pm \sqrt{449}}{32}$$
$$x = \sqrt{\frac{-15 + \sqrt{449}}{32}} \approx 0.4398$$
$$f(0) = 0$$
$$f(1) = 4.7 \text{ Maximum (endpoint)}$$
$$f\left(\sqrt{\frac{-15 + \sqrt{449}}{32}}\right) \approx -1.0613$$

Minimum: $(0.4398, -1.0613)$

43. $f(x) = (1 + x^3)^{1/2}, [0, 2]$

$f'(x) = \dfrac{3}{2}x^2(1 + x^3)^{-1/2}$

$f''(x) = \dfrac{3}{4}(x^4 + 4x)(1 + x^3)^{-3/2}$

$f'''(x) = -\dfrac{3}{8}(x^6 + 20x^3 - 8)(1 + x^3)^{-5/2}$

Setting $f''' = 0$, we have $x^6 + 20x^3 - 8 = 0$.

$x^3 = \dfrac{-20 \pm \sqrt{400 - 4(1)(-8)}}{2}$

$x = \sqrt[3]{-10 \pm \sqrt{108}} = \sqrt{3} - 1$

In the interval $[0, 2]$, choose

$x = \sqrt[3]{-10 + \sqrt{108}} = \sqrt{3} - 1 \approx 0.732.$

$\left| f''\left(\sqrt[3]{-10 + \sqrt{108}}\right) \right| \approx 1.47$ is the maximum value.

45. $f(x) = (x + 1)^{2/3}, [0, 2]$

$f'(x) = \dfrac{2}{3}(x + 1)^{-1/3}$

$f''(x) = -\dfrac{2}{9}(x + 1)^{-4/3}$

$f'''(x) = \dfrac{8}{27}(x + 1)^{-7/3}$

$f^{(4)}(x) = -\dfrac{56}{81}(x + 1)^{-10/3}$

$f^{(5)}(x) = \dfrac{560}{243}(x + 1)^{-13/3}$

$\left| f^{(4)}(0) \right| = \dfrac{56}{81}$ is the maximum value.

47. $f(x) = \tan x$

f is continuous on $[0, \pi/4]$ but not on $[0, \pi]$. $\displaystyle\lim_{x \to \pi/2^-} \tan x = \infty$.

49.

51. (a) Yes

    (b) No

53. (a) No

    (b) Yes

55. $P = VI - RI^2 = 12I - 0.5I^2, 0 \le I \le 15$

$P = 0$ when $I = 0$.

$P = 67.5$ when $I = 15$.

$P' = 12 - I = 0$

Critical number: $I = 12$ amps

When $I = 12$ amps, $P = 72$, the maximum output.

No, a 20-amp fuse would not increase the power output.
P is decreasing for $I > 12$.

57.

$$S = 6hs + \frac{3s^2}{2}\left(\frac{\sqrt{3} - \cos\theta}{\sin\theta}\right), \frac{\pi}{6} \le \theta \le \frac{\pi}{2}$$

$$\frac{dS}{d\theta} = \frac{3s^2}{2}\left(-\sqrt{3}\csc\theta\cot\theta + \csc^2\theta\right)$$

$$= \frac{3s^2}{2}\csc\theta\left(-\sqrt{3}\cot\theta + \csc\theta\right) = 0$$

$\csc\theta = \sqrt{3}\cot\theta$

$\sec\theta = \sqrt{3}$

$\theta = \operatorname{arcsec}\sqrt{3} \approx 0.9553$ radians

$$S\left(\frac{\pi}{6}\right) = 6hs + \frac{3s^2}{2}\left(\sqrt{3}\right)$$

$$S\left(\frac{\pi}{2}\right) = 6hs + \frac{3s^2}{2}\left(\sqrt{3}\right)$$

$$S\left(\operatorname{arcsec}\sqrt{3}\right) = 6hs + \frac{3s^2}{2}\left(\sqrt{2}\right)$$

S is minimum when $\theta = \operatorname{arcsec}\sqrt{3} \approx 0.9553$ radians.

59. (a) $y = ax^2 + bx + c$

$y' = 2ax + b$

The coordinates of B are $(500, 30)$, and those of A are $(-500, 45)$.
From the slopes at A and B,

$-1000a + b = -0.09$

$1000a + b = 0.06$.

Solving these two equations, you obtain $a = 3/40000$ and $b = -3/200$. From the points $(500, 30)$ and $(-500, 45)$,
you obtain

$$30 = \frac{3}{40000}500^2 + 500\left(\frac{-3}{200}\right) + c$$

$$45 = \frac{3}{40000}500^2 - 500\left(\frac{-3}{200}\right) + c.$$

In both cases, $c = 18.75 = \dfrac{75}{4}$. Thus,

$$y = \frac{3}{40000}x^2 - \frac{3}{200}x + \frac{75}{4}.$$

—CONTINUED—

59. —CONTINUED—

(b)

x	-500	-400	-300	-200	-100	0	100	200	300	400	500
d	0	$.75$	3	6.75	12	18.75	12	6.75	3	$.75$	0

For $-500 \leq x \leq 0$, $d = (ax^2 + bx + c) - (-0.09x)$.

For $0 \leq x \leq 500$, $d = (ax^2 + bx + c) - (0.06x)$.

(c) The lowest point on the highway is $(100, 18)$, which is not directly over the point where the two hillsides come together.

61. True. See Exercise 25.   **63.** True.

Section 3.2   Rolle's Theorem and the Mean Value Theorem

1. Rolle's Theorem does not apply to $f(x) = 1 - |x - 1|$ over $[0, 2]$ since f is not differentiable at $x = 1$.

3. $f(x) = x^2 - x - 2 = (x - 2)(x + 1)$

x-intercepts: $(-1, 0), (2, 0)$

$f'(x) = 2x - 1 = 0$ at $x = \dfrac{1}{2}$.

5. $f(x) = x\sqrt{x + 4}$

x-intercepts: $(-4, 0), (0, 0)$

$f'(x) = x\dfrac{1}{2}(x + 4)^{-1/2} + (x + 4)^{1/2}$

$\quad = (x + 4)^{-1/2}\left(\dfrac{x}{2} + (x + 4)\right)$

$f'(x) = \left(\dfrac{3}{2}x + 4\right)(x + 4)^{-1/2} = 0$ at $x = -\dfrac{8}{3}$

7. $f(x) = x^2 - 2x, [0, 2]$

$f(0) = f(2) = 0$

f is continuous on $[0, 2]$. f is differentiable on $(0, 2)$. Rolle's Theorem applies.

$\quad f'(x) = 2x - 2$

$\quad 2x - 2 = 0 \Rightarrow x = 1$

c value: 1

9. $f(x) = (x - 1)(x - 2)(x - 3), [1, 3]$

$f(1) = f(3) = 0$

f is continuous on $[1, 3]$. f is differentiable on $(1, 3)$. Rolle's Theorem applies.

$\quad f(x) = x^3 - 6x^2 + 11x - 6$

$\quad f'(x) = 3x^2 - 12x + 11$

$3x^2 - 12x + 11 = 0 \Rightarrow x = \dfrac{6 \pm \sqrt{3}}{3}$

$c = \dfrac{6 - \sqrt{3}}{3}, c = \dfrac{6 + \sqrt{3}}{3}$

11. $f(x) = x^{2/3} - 1, [-8, 8]$

$f(-8) = f(8) = 3$

f is continuous on $[-8, 8]$. f is not differentiable on $(-8, 8)$ since $f'(0)$ does not exist. Rolle's Theorem does not apply.

13. $f(x) = \dfrac{x^2 - 2x - 3}{x + 2},\ [-1, 3]$

$f(-1) = f(3) = 0$

f is continuous on $[-1, 3]$. (**Note:** The discontinuity, $x = -2$, is not in the interval.) f is differentiable on $(-1, 3)$. Rolle's Theorem applies.

$$f'(x) = \frac{(x + 2)(2x - 2) - (x^2 - 2x - 3)(1)}{(x + 2)^2} = 0$$

$$\frac{x^2 + 4x - 1}{(x + 2)^2} = 0$$

$$x = \frac{-4 \pm 2\sqrt{5}}{2} = -2 \pm \sqrt{5}$$

c value: $-2 + \sqrt{5}$

15. $f(x) = \sin x,\ [0, 2\pi]$

$f(0) = f(2\pi) = 0$

f is continuous on $[0, 2\pi]$. f is differentiable on $(0, 2\pi)$. Rolle's Theorem applies.

$$f'(x) = \cos x$$

c values: $\dfrac{\pi}{2}, \dfrac{3\pi}{2}$

17. $f(x) = \dfrac{6x}{\pi} - 4\sin^2 x,\ \left[0, \dfrac{\pi}{6}\right]$

$$f(0) = f\left(\frac{\pi}{6}\right) = 0$$

f is continuous on $[0, \pi/6]$. f is differentiable on $(0, \pi/6)$. Rolle's Theorem applies.

$$f'(x) = \frac{6}{\pi} - 8\sin x \cos x = 0$$

$$\frac{6}{\pi} = 8\sin x \cos x$$

$$\frac{3}{4\pi} = \frac{1}{2}\sin 2x$$

$$\frac{3}{2\pi} = \sin 2x$$

$$\frac{1}{2}\arcsin\left(\frac{3}{2\pi}\right) = x$$

$$x \approx 0.2489$$

c value: 0.2489

19. $f(x) = \tan x,\ [0, \pi]$

$f(0) = f(\pi) = 0$

f is not continuous on $[0, \pi]$ since $f(\pi/2)$ does not exist. Rolle's Theorem does not apply.

21. $f(x) = |x| - 1,\ [-1, 1]$

$f(-1) = f(1) = 0$

f is continuous on $[-1, 1]$. f is not differentiable on $(-1, 1)$ since $f'(0)$ does not exist. Rolle's Theorem does not apply.

23. $f(x) = 4x - \tan \pi x, \quad \left[-\frac{1}{4}, \frac{1}{4} \right]$

$f\left(-\frac{1}{4} \right) = f\left(\frac{1}{4} \right) = 0$

f is continuous on $[-1/4, 1/4]$. f is differentiable on $(-1/4, 1/4)$. Rolle's Theorem applies.

$$f'(x) = 4 - \pi \sec^2 \pi x = 0$$

$$\sec^2 \pi x = \frac{4}{\pi}$$

$$\sec \pi x = \pm \frac{2}{\sqrt{\pi}}$$

$$x = \pm \frac{1}{\pi} \operatorname{arcsec} \frac{2}{\sqrt{\pi}} = \pm \frac{1}{\pi} \arccos \frac{\sqrt{\pi}}{2}$$

$$\approx \pm 0.1533 \text{ radian}$$

c values: ± 0.1533 radian

25. $f(t) = -16t^2 + 48t + 32$

(a) $f(1) = f(2) = 64$

(b) $v = f'(t)$ must be 0 at some time in $(1, 2)$.

$$f'(t) = -32t + 48 = 0$$

$$t = \frac{3}{2} \text{ seconds}$$

27.

29. $f(x) = \dfrac{1}{x - 3}, \quad [0, 6]$

f has a discontinuity at $x = 3$.

31. $f(x) = x^2$ is continuous on $[-2, 1]$ and differentiable on $(-2, 1)$.

$$\frac{f(1) - f(-2)}{1 - (-2)} = \frac{1 - 4}{3} = -1$$

$f'(x) = 2x = -1$ when $x = -\dfrac{1}{2}$. Therefore,

$$c = -\frac{1}{2}.$$

33. $f(x) = x^{2/3}$ is continuous on $[0, 1]$ and differentiable on $(0, 1)$.

$$\frac{f(1) - f(0)}{1 - 0} = 1$$

$$f'(x) = \frac{2}{3} x^{-1/3} = 1$$

$$x = \left(\frac{2}{3} \right)^3 = \frac{8}{27}$$

$$c = \frac{8}{27}$$

35. $f(x) = \sqrt{2 - x}$ is continuous on $[-7, 2]$ and differentiable on $(-7, 2)$.

$$\frac{f(2) - f(-7)}{2 - (-7)} = \frac{0 - 3}{9} = -\frac{1}{3}$$

$$f'(x) = \frac{-1}{2\sqrt{2 - x}} = -\frac{1}{3}$$

$$2\sqrt{2 - x} = 3$$

$$\sqrt{2 - x} = \frac{3}{2}$$

$$2 - x = \frac{9}{4}$$

$$x = -\frac{1}{4}$$

$$c = -\frac{1}{4}$$

37. $f(x) = \sin x$ is continuous on $[0, \pi]$ and differentiable on $(0, \pi)$.

$$\frac{f(\pi) - f(0)}{\pi - 0} = \frac{0 - 0}{\pi} = 0$$

$$f'(x) = \cos x = 0$$

$$c = \frac{\pi}{2}$$

39. $f(x) = \dfrac{x}{x + 1}$ on $\left[-\dfrac{1}{2}, 2\right]$.

(a)

(b) Secant line:

$$\text{slope} = \frac{f(2) - f(-1/2)}{2 - (-1/2)} = \frac{2/3 - (-1)}{5/2} = \frac{2}{3}$$

$$y - \frac{2}{3} = \frac{2}{3}(x - 2)$$

$$3y - 2 = 2x - 4$$

$$3y - 2x + 2 = 0$$

(c) $f'(x) = \dfrac{1}{(x + 1)^2} = \dfrac{2}{3}$

$$(x + 1)^2 = \frac{3}{2}$$

$$x = -1 \pm \sqrt{\frac{3}{2}} = -1 \pm \frac{\sqrt{6}}{2}$$

In the interval $[-1/2, 2]$, $c = -1 + \left(\sqrt{6}/2\right)$.

$$f(c) = \frac{-1 + \left(\sqrt{6}/2\right)}{\left[-1 + \left(\sqrt{6}/2\right)\right] + 1} = \frac{-2 + \sqrt{6}}{\sqrt{6}} = \frac{-2}{\sqrt{6}} + 1$$

Tangent line: $y - 1 + \dfrac{2}{\sqrt{6}} = \dfrac{2}{3}\left(x - \dfrac{\sqrt{6}}{2} + 1\right)$

$$y - 1 + \frac{\sqrt{6}}{3} = \frac{2}{3}x - \frac{\sqrt{6}}{3} + \frac{2}{3}$$

$$3y - 2x - 5 + 2\sqrt{6} = 0$$

41. $f(x) = \sqrt{x},\ [1, 9]$

$(1, 1), (9, 3)$

$$m = \frac{3 - 1}{9 - 1} = \frac{1}{4}$$

(a)

(b) Secant line: $y - 1 = \frac{1}{4}(x - 1)$

$$y = \frac{1}{4}x + \frac{3}{4}$$

$$0 = x - 4y + 3$$

(c)

$$f'(x) = \frac{1}{2\sqrt{x}}$$

$$\frac{f(9) - f(1)}{9 - 1} = \frac{1}{4}$$

$$\frac{1}{2\sqrt{c}} = \frac{1}{4}$$

$$\sqrt{c} = 2$$

$$c = 4$$

$$(c, f(c)) = (4, 2)$$

$$m = f'(4) = \frac{1}{4}$$

Tangent line: $y - 2 = \frac{1}{4}(x - 4)$

$$y = \frac{1}{4}x + 1$$

$$0 = x - 4y + 4$$

43. $s(t) = -4.9t^2 + 500$

(a) $V_{avg} = \dfrac{s(3) - s(0)}{3 - 0} = \dfrac{455.9 - 500}{3} = -14.7\ \text{m/sec}$

(b) $s(t)$ is continuous on $[0, 3]$ and differentiable on $(0, 3)$. Therefore, the Mean Value Theorem applies.

$$v(t) = s'(t) = -9.8t = -14.7\ \text{m/sec}$$

$$t = \frac{-14.7}{-9.8} = 1.5\ \text{seconds}$$

45. No. Let $f(x) = x^2$ on $[-1, 2]$.

$$f'(x) = 2x$$

$f'(0) = 0$ and zero is in the interval $(-1, 2)$ but $f(-1) \neq f(2)$.

47. Let $S(t)$ be the position function of the plane. If $t = 0$ corresponds to 2 P.M., $S(0) = 0$, $S(5.5) = 2500$ and the Mean Value Theorem says that there exists a time t_0, $0 < t_0 < 5.5$, such that

$$S'(t_0) = v(t_0) = \frac{2500 - 0}{5.5 - 0} \approx 454.54.$$

Applying the Intermediate Value Theorem to the velocity function on the intervals $[0, t_0]$ and $[t_0, 5.5]$, you see that there are at least two times during the flight when the speed was 400 miles per hour. $(0 < 400 < 454.54)$

49. (a) f is continuous on $[-10, 4]$ and changes sign, $(f(-8) > 0, f(3) < 0)$. By the Intermediate Value Theorem, there exists at least one value of x in $[-10, 4]$ satisfying $f(x) = 0$.

(b) There exist real numbers a and b such that $-10 < a < b < 4$ and $f(a) = f(b) = 2$. Therefore, by Rolle's Theorem there exists at least one number c in $(-10, 4)$ such that $f'(c) = 0$. This is called a critical number.

(c)

(d)

(e) No, f' did not have to be continuous on $[-10, 4]$.

51. f is continuous on $[-5, 5]$ and does not satisfy the conditions of the Mean Value Theorem.
$\Rightarrow f$ is not differentiable on $(-5, 5)$.
Example: $f(x) = |x|$

53. False. $f(x) = 1/x$ has a discontinuity at $x = 0$.

55. True. A polynomial is continuous and differentiable everywhere.

57. Suppose that $p(x) = x^{2n+1} + ax + b$ has two real roots x_1 and x_2. Then by Rolle's Theorem, since $p(x_1) = p(x_2) = 0$, there exists c in (x_1, x_2) such that $p'(c) = 0$. But $p'(x) = (2n + 1)x^{2n} + a \neq 0$, since $n > 0, a > 0$. Therefore, $p(x)$ cannot have two real roots.

59. If $p(x) = Ax^2 + Bx + C$, then

$$p'(x) = 2Ax + B = \frac{f(b) - f(a)}{b - a} = \frac{(Ab^2 + Bb + C) - (Aa^2 + Ba + C)}{b - a}$$

$$= \frac{A(b^2 - a^2) + B(b - a)}{b - a}$$

$$= \frac{(b - a)[A(b + a) + B]}{b - a}$$

$$= A(b + a) + B.$$

Thus, $2Ax = A(b + a)$ and $x = (b + a)/2$ which is the midpoint of $[a, b]$.

61. $f(x) = \frac{1}{2} \cos x$ differentiable on $(-\infty, \infty)$.

$$f'(x) = -\frac{1}{2} \sin x$$

$$-\frac{1}{2} \leq f'(x) \leq \frac{1}{2} \Rightarrow f'(x) < 1 \text{ for all real numbers.}$$

Thus, from Exercise 60, f has, at most, one fixed point. $(x \approx 0.4502)$

Section 3.3   Increasing and Decreasing Functions and the First Derivative Test

1. $f(x) = x^2 - 6x + 8$

Increasing on: $(3, \infty)$

Decreasing on: $(-\infty, 3)$

3. $y = \dfrac{x^3}{4} - 3x$

Increasing on: $(-\infty, -2), (2, \infty)$

Decreasing on: $(-2, 2)$

5. $f(x) = \dfrac{1}{x^2} = x^{-2}$

$f'(x) = \dfrac{-2}{x^3}$

Discontinuity: $x = 0$

Test intervals:	$-\infty < x < 0$	$0 < x < \infty$
Sign of $f'(x)$:	$f' > 0$	$f' < 0$
Conclusion:	Increasing	Decreasing

Increasing on $(-\infty, 0)$

Decreasing on $(0, \infty)$

7. $g(x) = x^2 - 2x - 8$

$g'(x) = 2x - 2$

Critical number: $x = 1$

Test intervals:	$-\infty < x < 1$	$1 < x < \infty$
Sign of $g'(x)$:	$g' < 0$	$g' > 0$
Conclusion:	Decreasing	Increasing

Increasing on: $(1, \infty)$

Decreasing on: $(-\infty, 1)$

9. $y = x\sqrt{16 - x^2}$ Domain: $[-4, 4]$

$y' = \dfrac{-2(x^2 - 8)}{\sqrt{16 - x^2}} = \dfrac{-2}{\sqrt{16 - x^2}}(x - 2\sqrt{2})(x + 2\sqrt{2})$

Critical numbers: $x = \pm 2\sqrt{2}$

Test intervals:	$-4 < x < -2\sqrt{2}$	$-2\sqrt{2} < x < 2\sqrt{2}$	$2\sqrt{2} < x < 4$
Sign of y':	$y' < 0$	$y' > 0$	$y' < 0$
Conclusion:	Decreasing	Increasing	Decreasing

Increasing on $\left(-2\sqrt{2}, 2\sqrt{2}\right)$

Decreasing on $\left(-4, -2\sqrt{2}\right), \left(2\sqrt{2}, 4\right)$

11. $f(x) = x^2 - 6x$

$f'(x) = 2x - 6 = 0$

Critical number: $x = 3$

Test intervals:	$-\infty < x < 3$	$3 < x < \infty$
Sign of $f'(x)$:	$f' < 0$	$f' > 0$
Conclusion:	Decreasing	Increasing

Increasing on: $(3, \infty)$

Decreasing on: $(-\infty, 3)$

Relative minimum: $(3, -9)$

13. $f(x) = -2x^2 + 4x + 3$

$f'(x) = -4x + 4 = 0$

Critical number: $x = 1$

Test intervals:	$-\infty < x < 1$	$1 < x < \infty$
Sign of $f'(x)$:	$f' > 0$	$f' < 0$
Conclusion:	Increasing	Decreasing

Increasing on: $(-\infty, 1)$

Decreasing on: $(1, \infty)$

Relative maximum: $(1, 5)$

15. $f(x) = 2x^3 + 3x^2 - 12x$

$f'(x) = 6x^2 + 6x - 12 = 6(x + 2)(x - 1) = 0$

Critical numbers: $x = -2, 1$

Test intervals:	$-\infty < x < -2$	$-2 < x < 1$	$1 < x < \infty$
Sign of $f'(x)$:	$f' > 0$	$f' < 0$	$f' > 0$
Conclusion:	Increasing	Decreasing	Increasing

Increasing on: $(-\infty, -2), (1, \infty)$

Decreasing on: $(-2, 1)$

Relative maximum: $(-2, 20)$

Relative minimum: $(1, -7)$

17. $f(x) = x^2(3 - x) = 3x^2 - x^3$

$f'(x) = 6x - 3x^2 = 3x(2 - x)$

Critical numbers: $x = 0, 2$

Test intervals:	$-\infty < x < 0$	$0 < x < 2$	$2 < x < \infty$
Sign of $f'(x)$:	$f' < 0$	$f' > 0$	$f' < 0$
Conclusion:	Decreasing	Increasing	Decreasing

Increasing on: $(0, 2)$

Decreasing on: $(-\infty, 0), (2, \infty)$

Relative maximum: $(2, 4)$

Relative minimum: $(0, 0)$

19. $f(x) = \dfrac{x^5 - 5x}{5}$

$f'(x) = x^4 - 1$

Critical numbers: $x = -1, 1$

Test intervals:	$-\infty < x < -1$	$-1 < x < 1$	$1 < x < \infty$
Sign of $f'(x)$:	$f' > 0$	$f' < 0$	$f' > 0$
Conclusion:	Increasing	Decreasing	Increasing

Increasing on: $(-\infty, -1), (1, \infty)$

Decreasing on: $(-1, 1)$

Relative maximum: $\left(-1, \frac{4}{5}\right)$

Relative minimum: $\left(1, -\frac{4}{5}\right)$

21. $f(x) = x^{1/3} + 1$

$f'(x) = \frac{1}{3}x^{-2/3} = \frac{1}{3x^{2/3}}$

Critical number: $x = 0$

Test intervals:	$-\infty < x < 0$	$0 < x < \infty$
Sign of $f'(x)$:	$f' > 0$	$f' > 0$
Conclusion:	Increasing	Increasing

Increasing on: $(-\infty, \infty)$

No relative extrema

23. $f(x) = (x - 1)^{2/3}$

$f'(x) = \frac{2}{3(x - 1)^{1/3}}$

Critical number: $x = 1$

Test intervals:	$-\infty < x < 1$	$1 < x < \infty$
Sign of $f'(x)$:	$f' < 0$	$f' > 0$
Conclusion:	Decreasing	Increasing

Increasing on: $(1, \infty)$

Decreasing on: $(-\infty, 1)$

Relative minimum: $(1, 0)$

25. $f(x) = 5 - |x - 5|$

$f'(x) = -\frac{x - 5}{|x - 5|} = \begin{cases} 1, & x < 5 \\ -1, & x > 5 \end{cases}$

Critical number: $x = 5$

Test intervals:	$-\infty < x < 5$	$5 < x < \infty$
Sign of $f'(x)$:	$f' > 0$	$f' < 0$
Conclusion:	Increasing	Decreasing

Increasing on: $(-\infty, 5)$

Decreasing on: $(5, \infty)$

Relative maximum: $(5, 5)$

27. $f(x) = x + \frac{1}{x}$

$f'(x) = 1 - \frac{1}{x^2} = \frac{x^2 - 1}{x^2}$

Critical numbers: $x = -1, 1$

Discontinuity: $x = 0$

Test intervals:	$-\infty < x < -1$	$-1 < x < 0$	$0 < x < 1$	$1 < x < \infty$
Sign of $f'(x)$:	$f' > 0$	$f' < 0$	$f' < 0$	$f' > 0$
Conclusion:	Increasing	Decreasing	Decreasing	Increasing

Increasing on: $(-\infty, -1), (1, \infty)$

Decreasing on: $(-1, 0), (0, 1)$

Relative maximum: $(-1, -2)$

Relative minimum: $(1, 2)$

29. $f(x) = \dfrac{x^2}{x^2 - 9}$

$f'(x) = \dfrac{(x^2 - 9)(2x) - (x^2)(2x)}{(x^2 - 9)^2} = \dfrac{-18x}{(x^2 - 9)^2}$

Critical number: $x = 0$

Discontinuities: $x = -3, 3$

Test intervals:	$-\infty < x < -3$	$-3 < x < 0$	$0 < x < 3$	$3 < x < \infty$
Sign of $f'(x)$:	$f' > 0$	$f' > 0$	$f' < 0$	$f' < 0$
Conclusion:	Increasing	Increasing	Decreasing	Decreasing

Increasing on: $(-\infty, -3), (-3, 0)$

Decreasing on: $(0, 3), (3, \infty)$

Relative maximum: $(0, 0)$

31. $f(x) = \dfrac{x^2 - 2x + 1}{x + 1}$

$f'(x) = \dfrac{(x + 1)(2x - 2) - (x^2 - 2x + 1)(1)}{(x + 1)^2} = \dfrac{x^2 + 2x - 3}{(x + 1)^2} = \dfrac{(x + 3)(x - 1)}{(x + 1)^2}$

Critical numbers: $x = -3, 1$

Discontinuity: $x = -1$

Test intervals:	$-\infty < x < -3$	$-3 < x < -1$	$-1 < x < 1$	$1 < x < \infty$
Sign of $f'(x)$:	$f' > 0$	$f' < 0$	$f' < 0$	$f' > 0$
Conclusion:	Increasing	Decreasing	Decreasing	Increasing

Increasing on: $(-\infty, -3), (1, \infty)$

Decreasing on: $(-3, -1), (-1, 1)$

Relative maximum: $(-3, -8)$

Relative minimum: $(1, 0)$

33. $f(x) = \dfrac{x}{2} + \cos x, \, 0 < x < 2\pi$

$f'(x) = \dfrac{1}{2} - \sin x = 0$

Critical numbers: $x = \dfrac{\pi}{6}, \dfrac{5\pi}{6}$

Test intervals:	$0 < x < \dfrac{\pi}{6}$	$\dfrac{\pi}{6} < x < \dfrac{5\pi}{6}$	$\dfrac{5\pi}{6} < x < 2\pi$
Sign of $f'(x)$:	$f' > 0$	$f' < 0$	$f' > 0$
Conclusion:	Increasing	Decreasing	Increasing

Increasing on: $\left(0, \dfrac{\pi}{6}\right), \left(\dfrac{5\pi}{6}, 2\pi\right)$

Decreasing on: $\left(\dfrac{\pi}{6}, \dfrac{5\pi}{6}\right)$

Relative maximum: $\left(\dfrac{\pi}{6}, \dfrac{\pi + 6\sqrt{3}}{12}\right)$

Relative minimum: $\left(\dfrac{5\pi}{6}, \dfrac{5\pi - 6\sqrt{3}}{12}\right)$

35. $f(x) = \sin^2 x + \sin x, \ 0 < x < 2\pi$

$f'(x) = 2 \sin x \cos x + \cos x = \cos x(2 \sin x + 1) = 0$

Critical numbers: $x = \dfrac{\pi}{2}, \dfrac{7\pi}{6}, \dfrac{3\pi}{2}, \dfrac{11\pi}{6}$

Test intervals:	$0 < x < \dfrac{\pi}{2}$	$\dfrac{\pi}{2} < x < \dfrac{7\pi}{6}$	$\dfrac{7\pi}{6} < x < \dfrac{3\pi}{2}$	$\dfrac{3\pi}{2} < x < \dfrac{11\pi}{6}$	$\dfrac{11\pi}{6} < x < 2\pi$
Sign of $f'(x)$:	$f' > 0$	$f' < 0$	$f' > 0$	$f' < 0$	$f' > 0$
Conclusion:	Increasing	Decreasing	Increasing	Decreasing	Increasing

Increasing on: $\left(0, \dfrac{\pi}{2}\right), \left(\dfrac{7\pi}{6}, \dfrac{3\pi}{2}\right), \left(\dfrac{11\pi}{6}, 2\pi\right)$

Decreasing on: $\left(\dfrac{\pi}{2}, \dfrac{7\pi}{6}\right), \left(\dfrac{3\pi}{2}, \dfrac{11\pi}{6}\right)$

Relative minima: $\left(\dfrac{7\pi}{6}, -\dfrac{1}{4}\right), \left(\dfrac{11\pi}{6}, -\dfrac{1}{4}\right)$

Relative maxima: $\left(\dfrac{\pi}{2}, 2\right), \left(\dfrac{3\pi}{2}, 0\right)$

37. $f(x) = 2x\sqrt{9 - x^2}, \ [-3, 3]$

(a) $f'(x) = \dfrac{2(9 - 2x^2)}{\sqrt{9 - x^2}}$

(b)

(c) $\dfrac{2(9 - 2x^2)}{\sqrt{9 - x^2}} = 0$

Critical numbers: $x = \pm\dfrac{3}{\sqrt{2}} = \pm\dfrac{3\sqrt{2}}{2}$

(d) Intervals:

$$\left(-3, -\dfrac{3\sqrt{2}}{2}\right) \quad \left(-\dfrac{3\sqrt{2}}{2}, \dfrac{3\sqrt{2}}{2}\right) \quad \left(\dfrac{3\sqrt{2}}{2}, 3\right)$$

$$f'(x) < 0 \qquad\qquad f'(x) > 0 \qquad\qquad f'(x) < 0$$

Decreasing $\qquad\quad$ Increasing $\qquad\quad$ Decreasing

f is increasing when f' is positive and decreasing when f' is negative.

39. $f(t) = t^2 \sin t, \ [0, 2\pi]$

(a) $f'(t) = t^2 \cos t + 2t \sin t$

$\qquad = t(t \cos t + 2 \sin t)$

(b)

(c) $t(t \cos t + 2 \sin t) = 0$

$t = 0$ or $t = -2 \tan t$

$t \cot t = -2$

$t \approx 2.2889, \ 5.0870 \ \text{(graphing utility)}$

Critical numbers: $t = 2.2889, \ t = 5.0870$

(d) Intervals:

$$(0, 2.2889) \qquad (2.2889, 5.0870) \qquad (5.0870, 2\pi)$$

$$f'(t) > 0 \qquad\quad f'(t) < 0 \qquad\quad f'(t) > 0$$

Increasing $\qquad\quad$ Decreasing $\qquad\quad$ Increasing

f is increasing when f' is positive and decreasing when f' is negative.

41. $f(x) = \dfrac{x^5 - 4x^3 + 3x}{x^2 - 1} = \dfrac{(x^2 - 1)(x^3 - 3x)}{x^2 - 1} = x^3 - 3x, \; x \neq \pm 1$

$f(x) = g(x) = x^3 - 3x$ for all $x \neq \pm 1$.

$f'(x) = 3x^2 - 3 = 3(x^2 - 1), \; x \neq \pm 1 \quad f'(x) \neq 0$

f symmetric about origin

zeros of f: $(0, 0), \left(\pm \sqrt{3}, 0\right)$

No relative extrema

Holes at $(-1, 2)$ and $(1, -2)$

43. $f(x) = c$ is constant $\Rightarrow f'(x) = 0$

45. f is quadratic $\Rightarrow f'$ is a line.

47. f has positive, but decreasing slope

In Exercises 49–53, $f'(x) > 0$ **on** $(-\infty, -4), f'(x) < 0$ **on** $(-4, 6)$ **and** $f'(x) > 0$ **on** $(6, \infty)$.

49. $g(x) = f(x) + 5$

$g'(x) = f'(x)$

$g'(0) = f'(0) < 0$

51. $g(x) = -f(x)$

$g'(x) = -f'(x)$

$g'(-6) = -f'(-6) < 0$

53. $g(x) = f(x - 10)$

$g'(x) = f'(x - 10)$

$g'(0) = f'(-10) > 0$

55. $f'(x) = \begin{cases} > 0, & x < 4 \Rightarrow f \text{ is increasing on } (-\infty, 4). \\ \text{undefined}, & x = 4 \\ < 0, & x > 4 \Rightarrow f \text{ is decreasing on } (4, \infty). \end{cases}$

Two possibilities for $f(x)$ are given below.

(a)

(b)

57. The critical numbers are in intervals $(-0.50, -0.25)$ and $(0.25, 0.50)$ since the sign of f' changes in these intervals. f is decreasing on approximately $(-1, -0.40)$, $(0.48, 1)$, and increasing on $(-0.40, 0.48)$.

Relative minimum when $x \approx -0.40$.

Relative maximum when $x \approx 0.48$.

59. $f(x) = x, g(x) = \sin x, 0 < x < \pi$

(a)

x	0.5	1	1.5	2	2.5	3
$f(x)$	0.5	1	1.5	2	2.5	3
$g(x)$	0.479	0.841	0.997	0.909	0.598	0.141

$f(x)$ seems greater than $g(x)$ on $(0, \pi)$.

(b)

$x > \sin x$ on $(0, \pi)$

(c) Let $h(x) = f(x) - g(x) = x - \sin x$

$$h'(x) = 1 - \cos x > 0 \text{ on } (0, \pi).$$

Therefore, $h(x)$ is increasing on $(0, \pi)$. Since $h(0) = 0$, $h(x) > 0$ on $(0, \pi)$. Thus,

$$x - \sin x > 0$$
$$x > \sin x$$
$$f(x) > g(x) \text{ on } (0, \pi).$$

61. $v = k(R - r)r^2 = k(Rr^2 - r^3)$

$v' = k(2Rr - 3r^2)$

$\quad = kr(2R - 3r) = 0$

$r = 0 \text{ or } \frac{2}{3}R$

Maximum when $r = \frac{2}{3}R$.

63. $P = \dfrac{vR_1R_2}{(R_1 + R_2)^2}$, v and R_1 are constant

$\dfrac{dP}{dR_2} = \dfrac{(R_1 + R_2)^2(vR_1) - vR_1R_2[2(R_1 + R_2)(1)]}{(R_1 + R_2)^4}$

$\quad = \dfrac{vR_1(R_1 - R_2)}{(R_1 + R_2)^3} = 0 \Rightarrow R_2 = R_1$

Maximum when $R_1 = R_2$.

65. (a) $B = 0.1198t^4 - 4.4879t^3 + 56.9909t^2 - 223.0222t + 579.9541$

(b)

(c) $B' = 0$ for $t \approx 2.78$, or 1983, (311.1 thousand bankruptcies)

Actual minimum: 1984 (344.3 thousand bankruptcies)

67. (a) Use a cubic polynomial
$f(x) = a_3x^3 + a_2x^2 + a_1x + a_0$.

(b) $f'(x) = 3a_3x^2 + 2a_2x + a_1$.

$(0, 0)$:	$0 = a_0$	$(f(0) = 0)$
	$0 = a_1$	$(f'(0) = 0)$
$(2, 2)$:	$2 = 8a_3 + 4a_2$	$(f(2) = 2)$
	$0 = 12a_3 + 4a_2$	$(f'(2) = 0)$

(c) The solution is $a_0 = a_1 = 0$, $a_2 = \dfrac{3}{2}$, $a_3 = -\dfrac{1}{2}$:

$$f(x) = -\frac{1}{2}x^3 + \frac{3}{2}x^2.$$

(d)

69. (a) Use a fourth degree polynomial $f(x) = a_4x^4 + a_3x^3 + a_2x^2 + a_1x + a_0$.

(b) $f'(x) = 4a_4x^3 + 3a_3x^2 + 2a_2x + a_1$

$(0, 0)$: $0 = a_0$ $(f(0) = 0)$

           $0 = a_1$ $(f'(0) = 0)$

$(4, 0)$: $0 = 256a_4 + 64a_3 + 16a_2$ $(f(4) = 0)$

           $0 = 256a_4 + 48a_3 + 8a_2$ $(f'(4) = 0)$

$(2, 4)$: $4 = 16a_4 + 8a_3 + 4a_2$ $(f(2) = 4)$

           $0 = 32a_4 + 12a_3 + 4a_2$ $(f'(2) = 0)$

(c) The solution is $a_0 = a_1 = 0$, $a_2 = 4$, $a_3 = -2$, $a_4 = \dfrac{1}{4}$.

$$f(x) = \frac{1}{4}x^4 - 2x^3 + 4x^2$$

(d)

71. True

Let $h(x) = f(x) + g(x)$ where f and g are increasing. Then $h'(x) = f'(x) + g'(x) > 0$ since $f'(x) > 0$ and $g'(x) > 0$.

73. False

Let $f(x) = x^3$, then $f'(x) = 3x^2$ and f only has one critical number. Or, let $f(x) = x^3 + 3x + 1$, then $f'(x) = 3(x^2 + 1)$ has no critical numbers.

75. False. For example, $f(x) = x^3$ does not have a relative extrema at the critical number $x = 0$.

77. Assume that $f'(x) < 0$ for all x in the interval (a, b) and let $x_1 < x_2$ be any two points in the interval. By the Mean Value Theorem, we know there exists a number c such that $x_1 < c < x_2$, and

$$f'(c) = \frac{f(x_2) - f(x_1)}{x_2 - x_1}.$$

Since $f'(c) < 0$ and $x_2 - x_1 > 0$, then $f(x_2) - f(x_1) < 0$, which implies that $f(x_2) < f(x_1)$. Thus, f is decreasing on the interval.

79. Let $f(x) = (1 + x)^n - nx - 1$. Then

$f'(x) = n(1 + x)^{n-1} - n$

       $= n[(1 + x)^{n-1} - 1] > 0$ since $x > 0$ and $n > 1$.

Thus, $f(x)$ is increasing on $(0, \infty)$. Since $f(0) = 0 \Rightarrow f(x) > 0$ on $(0, \infty)$

$$(1 + x)^n - nx - 1 > 0 \Rightarrow (1 + x)^n > 1 + nx.$$

Section 3.4   Concavity and the Second Derivative Test

1. $y = x^2 - x - 2, y'' = 2$

Concave upward: $(-\infty, \infty)$

3. $f(x) = \dfrac{24}{x^2 + 12}, y'' = \dfrac{-144(4 - x^2)}{(x^2 + 12)^3}$

Concave upward: $(-\infty, -2), (2, \infty)$

Concave downward: $(-2, 2)$

5. $f(x) = \dfrac{x^2 + 1}{x^2 - 1}, \ y'' = \dfrac{4(3x^2 + 1)}{(x^2 - 1)^3}$

Concave upward: $(-\infty, -1), (1, \infty)$

Concave downward: $(-1, 1)$

7. $f(x) = 3x^2 - x^3$

$f'(x) = 6x - 3x^2$

$f''(x) = 6 - 6x$

Concave upward: $(-\infty, 1)$

Concave downward: $(1, \infty)$

9. $y = 2x - \tan x, \ \left(-\dfrac{\pi}{2}, \dfrac{\pi}{2}\right)$

$y' = 2 - \sec^2 x$

$y'' = -2 \sec^2 x \tan x$

Concave upward: $\left(-\dfrac{\pi}{2}, 0\right)$

Concave downward: $\left(0, \dfrac{\pi}{2}\right)$

11. $f(x) = x^3 - 6x^2 + 12x$

$f'(x) = 3x^2 - 12x + 12$

$f''(x) = 6(x - 2) = 0$ when $x = 2$.

The concavity changes at $x = 2$. $(2, 8)$ is a point of inflection.

Concave upward: $(2, \infty)$

Concave downward: $(-\infty, 2)$

13. $f(x) = \dfrac{1}{4}x^4 - 2x^2$

$f'(x) = x^3 - 4x$

$f''(x) = 3x^2 - 4$

$f''(x) = 3x^2 - 4 = 0$ when $x = \pm\dfrac{2}{\sqrt{3}}$.

Test interval:	$-\infty < x < -\dfrac{2}{\sqrt{3}}$	$-\dfrac{2}{\sqrt{3}} < x < \dfrac{2}{\sqrt{3}}$	$\dfrac{2}{\sqrt{3}} < x < \infty$
Sign of $f''(x)$:	$f''(x) > 0$	$f''(x) < 0$	$f''(x) > 0$
Conclusion:	Concave upward	Concave downward	Concave upward

Points of inflection: $\left(\pm\dfrac{2}{\sqrt{3}}, -\dfrac{20}{9}\right)$

15. $f(x) = x(x - 4)^3$

$f'(x) = x[3(x - 4)^2] + (x - 4)^3$

$\quad = (x - 4)^2(4x - 4)$

$f''(x) = 4(x - 1)[2(x - 4)] + 4(x - 4)^2$

$\quad = 4(x - 4)[2(x - 1) + (x - 4)]$

$\quad = 4(x - 4)(3x - 6) = 12(x - 4)(x - 2)$

$f''(x) = 12(x - 4)(x - 2) = 0$ when $x = 2, 4$.

Test interval:	$-\infty < x < 2$	$2 < x < 4$	$4 < x < \infty$
Sign of $f''(x)$:	$f''(x) > 0$	$f''(x) < 0$	$f''(x) > 0$
Conclusion:	Concave upward	Concave downward	Concave upward

Points of inflection: $(2, -16), (4, 0)$

17. $f(x) = x\sqrt{x + 3}$, Domain: $[-3, \infty)$

$f'(x) = x\left(\dfrac{1}{2}\right)(x + 3)^{-1/2} + \sqrt{x + 3} = \dfrac{3(x + 2)}{2\sqrt{x + 3}}$

$f''(x) = \dfrac{6\sqrt{x + 3} - 3(x + 2)(x + 3)^{-1/2}}{4(x + 3)} = \dfrac{3(x + 4)}{4(x + 3)^{3/2}}$

$f''(x) > 0$ on the entire domain of f (except for $x = -3$, for which $f''(x)$ is undefined). There are no points of inflection.
Concave upward on $(-3, \infty)$

19. $f(x) = \dfrac{x}{x^2 + 1}$

$f'(x) = \dfrac{1 - x^2}{(x^2 + 1)^2}$

$f''(x) = \dfrac{2x(x^2 - 3)}{(x^2 + 1)^3} = 0$ when $x = 0, \pm\sqrt{3}$

Test intervals:	$-\infty < x < -\sqrt{3}$	$-\sqrt{3} < x < 0$	$0 < x < \sqrt{3}$	$\sqrt{3} < x < \infty$
Sign of $f'(x)$:	$f'' < 0$	$f'' > 0$	$f'' < 0$	$f'' > 0$
Conclusion:	Concave downward	Concave upward	Concave downward	Concave upward

Points of inflection: $\left(-\sqrt{3}, -\dfrac{\sqrt{3}}{4}\right), (0, 0), \left(\sqrt{3}, \dfrac{\sqrt{3}}{4}\right)$

21. $f(x) = \sin\left(\dfrac{x}{2}\right), 0 \le x \le 4\pi$

$f'(x) = \dfrac{1}{2}\cos\left(\dfrac{x}{2}\right)$

$f''(x) = -\dfrac{1}{4}\sin\left(\dfrac{x}{2}\right)$

$f''(x) = 0$ when $x = 0, 2\pi, 4\pi$.

Point of inflection: $(2\pi, 0)$

Test interval:	$0 < x < 2\pi$	$2\pi < x < 4\pi$
Sign of $f''(x)$:	$f'' < 0$	$f'' > 0$
Conclusion:	Concave downward	Concave upward

23. $f(x) = \sec\left(x - \dfrac{\pi}{2}\right), 0 < x < 4\pi$

$f'(x) = \sec\left(x - \dfrac{\pi}{2}\right)\tan\left(x - \dfrac{\pi}{2}\right)$

$f''(x) = \sec^3\left(x - \dfrac{\pi}{2}\right) + \sec\left(x - \dfrac{\pi}{2}\right)\tan^2\left(x - \dfrac{\pi}{2}\right) \neq 0$ for any x in the domain of f.

Concave upward: $(0, \pi), (2\pi, 3\pi)$

Concave downward: $(\pi, 2\pi), (3\pi, 4\pi)$

No points of inflection

25. $f(x) = 2\sin x + \sin 2x, 0 \leq x \leq 2\pi$

$f'(x) = 2\cos x + 2\cos 2x$

$f''(x) = -2\sin x - 4\sin 2x = -2\sin x(1 + 4\cos x)$

$f''(x) = 0$ when $x = 0, 1.823, \pi, 4.460$.

Test interval:	$0 < x < 1.823$	$1.823 < x < \pi$	$\pi < x < 4.460$	$4.460 < x < 2\pi$
Sign of $f''(x)$:	$f'' < 0$	$f'' > 0$	$f'' < 0$	$f'' > 0$
Conclusion:	Concave downward	Concave upward	Concave downward	Concave upward

Points of inflection: $(1.823, 1.452), (\pi, 0), (4.46, -1.452)$

27. $f(x) = x^4 - 4x^3 + 2$

$f'(x) = 4x^3 - 12x^2 = 4x^2(x - 3)$

$f''(x) = 12x^2 - 24x = 12x(x - 2)$

Critical numbers: $x = 0, x = 3$

However, $f''(0) = 0$, so we must use the First Derivative Test. $f'(x) < 0$ on the intervals $(-\infty, 0)$ and $(0, 3)$; hence, $(0, 2)$ is not an extremum. $f''(3) > 0$ so $(3, -25)$ is a relative minimum.

29. $f(x) = (x - 5)^2$

$f'(x) = 2(x - 5)$

$f''(x) = 2$

Critical number: $x = 5$

$f''(5) > 0$

Therefore, $(5, 0)$ is a relative minimum.

31. $f(x) = x^3 - 3x^2 + 3$

$f'(x) = 3x^2 - 6x = 3x(x - 2)$

$f''(x) = 6x - 6 = 6(x - 1)$

Critical numbers: $x = 0, x = 2$

$f''(0) = -6 < 0$

Therefore, $(0, 3)$ is a relative maximum.

$f''(2) = 6 > 0$

Therefore, $(2, -1)$ is a relative minimum.

33. $g(x) = x^2(6 - x)^3$

$g'(x) = x(x - 6)^2(12 - 5x)$

$g''(x) = 4(6 - x)(5x^2 - 24x + 18)$

Critical numbers: $x = 0, \dfrac{12}{5}, 6$

$g''(0) = 432 > 0$

Therefore, $(0, 0)$ is a relative minimum.

$g''\left(\dfrac{12}{5}\right) = -155.52 < 0$

Therefore, $\left(\dfrac{12}{5}, 268.7\right)$ is a relative minimum.

$g''(6) = 0$

Test fails by the First Derivative Test, $(6, 0)$ is not an extremum.

35. $f(x) = x^{2/3} - 3$

$f'(x) = \dfrac{2}{3x^{1/3}}$

$f''(x) = \dfrac{-2}{9x^{4/3}}$

Critical number: $x = 0$

However, $f''(0)$ is undefined, so we must use the First Derivative Test. Since $f'(x) < 0$ on $(-\infty, 0)$ and $f'(x) > 0$ on $(0, \infty)$, $(0, -3)$ is a relative minimum.

37. $f(x) = x + \dfrac{4}{x}$

$f'(x) = 1 - \dfrac{4}{x^2} = \dfrac{x^2 - 4}{x^2}$

$f''(x) = \dfrac{8}{x^3}$

Critical numbers: $x = \pm 2$

$f''(-2) < 0$

Therefore, $(-2, -4)$ is a relative maximum.

$f''(2) > 0$

Therefore, $(2, 4)$ is a relative minimum.

39. $f(x) = \cos x - x$, $0 \le x \le 4\pi$

$f'(x) = -\sin x - 1 \le 0$

Therefore, f is non-increasing and there are no relative extrema.

41. $f(x) = 0.2x^2(x - 3)^3$, $[-1, 4]$

(a) $f'(x) = 0.2x(5x - 6)(x - 3)^2$

$f''(x) = (x - 3)(4x^2 - 9.6x + 3.6)$

$ = 0.4(x - 3)(10x^2 - 24x + 9)$

(b) $f''(0) < 0 \Longrightarrow (0, 0)$ is a relative maximum.

$f''\!\left(\dfrac{6}{5}\right) > 0 \Longrightarrow (1.2, -1.6796)$ is a relative minimum.

Points of inflection:

$(3, 0)$, $(0.4652, -0.7049)$, $(1.9348, -0.9049)$

(c)

f is increasing when $f' > 0$ and decreasing when $f' < 0$. f is concave upward when $f'' > 0$ and concave downward when $f'' < 0$.

43. $f(x) = \sin x - \dfrac{1}{3}\sin 3x + \dfrac{1}{5}\sin 5x$, $[0, \pi]$

(a) $f'(x) = \cos x - \cos 3x + \cos 5x$

$f'(x) = 0$ when $x = \dfrac{\pi}{6}$, $x = \dfrac{\pi}{2}$, $x = \dfrac{5\pi}{6}$.

$f''(x) = -\sin x + 3\sin 3x - 5\sin 5x$

$f''(x) = 0$ when $x = \dfrac{\pi}{6}$, $x = \dfrac{5\pi}{6}$, $x \approx 1.1731$, $x \approx 1.9685$

(b) $f''\!\left(\dfrac{\pi}{2}\right) < 0 \Longrightarrow \left(\dfrac{\pi}{2}, 1.53333\right)$ is a relative maximum.

Points of inflection: $\left(\dfrac{\pi}{6}, 0.2667\right)$, $(1.1731, 0.9638)$,

$(1.9685, 0.9637)$, $\left(\dfrac{5\pi}{6}, 0.2667\right)$

Note: $(0, 0)$ and $(\pi, 0)$ are not points of inflection since they are endpoints.

(c)

The graph of f is increasing when $f' > 0$ and decreasing when $f' < 0$. f is concave upward when $f'' > 0$ and concave downward when $f'' < 0$.

45. (a)

$f' < 0$ means f decreasing

f' increasing means
concave upward

(b)

$f' > 0$ means f increasing

f' increasing means
concave upward

47. Let $f(x) = x^4$.

$f''(x) = 12x^2$

$f''(0) = 0$, but $(0, 0)$ is not a point of inflection.

49.

51.

53.

55.

57.

f'' is linear.

f' is quadratic.

f is cubic.

f concave upwards on $(-\infty, 3)$, downward on $(3, \infty)$.

59. (a) $n = 1$:

$$f(x) = x - 2$$
$$f'(x) = 1$$
$$f''(x) = 0$$

No inflection points

$n = 2$:

$$f(x) = (x - 2)^2$$
$$f'(x) = 2(x - 2)$$
$$f''(x) = 2$$

No inflection points

Relative minimum:
(2, 0)

$n = 3$:

$$f(x) = (x - 2)^3$$
$$f'(x) = 3(x - 2)^2$$
$$f''(x) = 6(x - 2)$$

Inflection point: (2, 0)

$n = 4$:

$$f(x) = (x - 2)^4$$
$$f'(x) = 4(x - 2)^3$$
$$f''(x) = 12(x - 2)^2$$

No inflection points:

Relative minimum:
(2, 0)

Conclusion: If $n \geq 3$ and n is odd, then (2, 0) is an inflection point. If $n \geq 2$ and n is even, then (2, 0) is a relative minimum.

(b) Let $f(x) = (x - 2)^n$, $f'(x) = n(x - 2)^{n-1}$, $f''(x) = n(n - 1)(x - 2)^{n-2}$.

For $n \geq 3$ and odd, $n - 2$ is also odd and the concavity changes at $x = 2$.

For $n \geq 4$ and even, $n - 2$ is also even and the concavity does not change at $x = 2$.

Thus, $x = 2$ is an inflection point if and only if $n \geq 3$ is odd.

61. $f(x) = ax^3 + bx^2 + cx + d$

Relative maximum: (3, 3)

Relative minimum: (5, 1)

Point of inflection: (4, 2)

$f'(x) = 3ax^2 + 2bx + c, f''(x) = 6ax + 2b$

$\left. \begin{array}{l} f(3) = 27a + 9b + 3c + d = 3 \\ f(5) = 125a + 25b + 5c + d = 1 \end{array} \right\} 98a + 16b + 2c = -2 \Rightarrow 49a + 8b + c = -1$

$f'(3) = 27a + 6b + c = 0, f''(4) = 24a + 2b = 0$

$$
\begin{array}{ll}
49a + 8b + c = -1 & 24a + 2b = 0 \\
\underline{27a + 6b + c = 0} & \underline{22a + 2b = -1} \\
22a + 2b = -1 & 2a = 1
\end{array}
$$

$a = \frac{1}{2}, b = -6, c = \frac{45}{2}, d = -24$

$f(x) = \frac{1}{2}x^3 - 6x^2 + \frac{45}{2}x - 24$

63. $f(x) = ax^3 + bx^2 + cx + d$

Maximum: $(-4, 1)$

Minimum: $(0, 0)$

(a) $f'(x) = 3ax^2 + 2bx + c$, $f''(x) = 6ax + 2b$

$\quad f(0) = 0 \Rightarrow d = 0$

$\quad f(-4) = 1 \Rightarrow -64a + 16b - 4c = 1$

$\quad f'(-4) = 0 \Rightarrow \quad 48a - 8b + c = 0$

$\quad f'(0) = 0 \Rightarrow \quad\quad\quad\quad c = 0$

Solving this system yields $a = \frac{1}{32}$ and $b = 6a = \frac{3}{16}$.

$\quad f(x) = \frac{1}{32}x^3 + \frac{3}{16}x^2$

(b) The plane would be descending at the greatest rate at the point of inflection.

$$f''(x) = 6ax + 2b = \frac{3}{16}x + \frac{3}{8} = 0 \Rightarrow x = -2.$$

Two miles from touchdown.

65. $D = 2x^4 - 5Lx^3 + 3L^2x^2$

$D' = 8x^3 - 15Lx^2 + 6L^2x = x(8x^2 - 15Lx + 6L^2) = 0$

$x = 0$ or $x = \dfrac{15L \pm \sqrt{33}L}{16} = \left(\dfrac{15 \pm \sqrt{33}}{16}\right)L$

By the Second Derivative Test, the deflection is maximum when

$x = \left(\dfrac{15 - \sqrt{33}}{16}\right)L \approx 0.578L.$

67. $C = 0.5x^2 + 15x + 5000$

$\overline{C} = \dfrac{C}{x} = 0.5x + 15 + \dfrac{5000}{x}$

$\overline{C} =$ average cost per unit

$\dfrac{d\overline{C}}{dx} = 0.5 - \dfrac{5000}{x^2} = 0$ when $x = 100$

By the First Derivative Test, $\overline{C}$ is minimized when $x = 100$ units.

69. $S = \dfrac{5000t^2}{8 + t^2}$

$S'(t) = \dfrac{80,000t}{(8 + t^2)^2}$

$S''(t) = \dfrac{80,000(8 - 3t^2)}{(8 + t^2)^3}$

$S''(t) = 0$ for $t = \sqrt{8/3} \approx 1.633.$

Sales are increasing at the greatest rate at $t = 1.633$ years.

71. $f(x) = 2(\sin x + \cos x)$, $f\left(\dfrac{\pi}{4}\right) = 2\sqrt{2}$

$\quad f'(x) = 2(\cos x - \sin x)$, $f'\left(\dfrac{\pi}{4}\right) = 0$

$\quad f''(x) = 2(-\sin x - \cos x)$, $f''\left(\dfrac{\pi}{4}\right) = -2\sqrt{2}$

$P_1(x) = 2\sqrt{2} + 0\left(x - \dfrac{\pi}{4}\right) = 2\sqrt{2}$

$P_1'(x) = 0$

$P_2(x) = 2\sqrt{2} + 0\left(x - \dfrac{\pi}{4}\right) + \dfrac{1}{2}(-2\sqrt{2})\left(x - \dfrac{\pi}{4}\right)^2 = 2\sqrt{2} - \sqrt{2}\left(x - \dfrac{\pi}{4}\right)^2$

$P_2'(x) = -2\sqrt{2}\left(x - \dfrac{\pi}{4}\right)$

$P_2''(x) = -2\sqrt{2}$

The values of f, P_1, P_2, and their first derivatives are equal at $x = \pi/4$. The values of the second derivatives of f and P_2 are equal at $x = \pi/4$. The approximations worsen as you move away from $x = \pi/4$.

73. $f(x) = \sqrt{1-x}, \qquad f(0) = 1$

$f'(x) = -\dfrac{1}{2\sqrt{1-x}}, \qquad f'(0) = -\dfrac{1}{2}$

$f''(x) = -\dfrac{1}{4(1-x)^{3/2}}, \qquad f''(0) = -\dfrac{1}{4}$

$P_1(x) = 1 + \left(-\dfrac{1}{2}\right)(x-0) = 1 - \dfrac{x}{2}$

$P_1{}'(x) = -\dfrac{1}{2}$

$P_2(x) = 1 + \left(-\dfrac{1}{2}\right)(x-0) + \dfrac{1}{2}\left(-\dfrac{1}{4}\right)(x-0)^2 = 1 - \dfrac{x}{2} - \dfrac{x^2}{8}$

$P_2{}'(x) = -\dfrac{1}{2} - \dfrac{x}{4}$

$P_2{}''(x) = -\dfrac{1}{4}$

The values of f, P_1, P_2, and their first derivatives are equal at $x = 0$. The values of the second derivatives of f and P_2 are equal at $x = 0$. The approximations worsen as you move away from $x = 0$.

75. $f(x) = x\sin\left(\dfrac{1}{x}\right)$

$f'(x) = x\left[-\dfrac{1}{x^2}\cos\left(\dfrac{1}{x}\right)\right] + \sin\left(\dfrac{1}{x}\right) = -\dfrac{1}{x}\cos\left(\dfrac{1}{x}\right) + \sin\left(\dfrac{1}{x}\right)$

$f''(x) = -\dfrac{1}{x}\left[\dfrac{1}{x^2}\sin\left(\dfrac{1}{x}\right)\right] + \dfrac{1}{x^2}\cos\left(\dfrac{1}{x}\right) - \dfrac{1}{x^2}\cos\left(\dfrac{1}{x}\right) = -\dfrac{1}{x^3}\sin\left(\dfrac{1}{x}\right) = 0$

$x = \dfrac{1}{\pi}$

Point of inflection: $\left(\dfrac{1}{\pi}, 0\right)$

When $x > 1/\pi, f'' < 0$, so the graph is concave downward.

77. Assume the zeros of f are all real. Then express the function as $f(x) = a(x - r_1)(x - r_2)(x - r_3)$ where r_1, r_2, and r_3 are the distinct zeros of f. From the Product Rule for a function involving three factors, we have

$f'(x) = a[(x - r_1)(x - r_2) + (x - r_1)(x - r_3) + (x - r_2)(x - r_3)]$

$f''(x) = a[(x - r_1) + (x - r_2) + (x - r_1) + (x - r_3) + (x - r_2) + (x - r_3)]$

$\qquad = a[6x - 2(r_1 + r_2 + r_3)].$

Consequently, $f''(x) = 0$ if

$x = \dfrac{2(r_1 + r_2 + r_3)}{6} = \dfrac{r_1 + r_2 + r_3}{3} = $ (Average of r_1, r_2, and r_3).

79. True. Let $y = ax^3 + bx^2 + cx + d, a \neq 0$. Then $y'' = 6ax + 2b = 0$ when $x = -(b/3a)$, and the concavity changes at this point.

81. False.

$$f(x) = 3 \sin x + 2 \cos x$$

$$f'(x) = 3 \cos x - 2 \sin x$$

$$3 \cos x - 2 \sin x = 0$$

$$3 \cos x = 2 \sin x$$

$$\tfrac{3}{2} = \tan x$$

Critical number: $x = \tan^{-1}\left(\tfrac{3}{2}\right)$

$f\left(\tan^{-1} \tfrac{3}{2}\right) \approx 3.60555$ is the maximum value of y.

83. False. Concavity is determined by f''.

For example, let $f(x) = x$ and $C = 2$. Then
$f'(C) = f'(2) > 0$ but f is not concave upward at $x = 2$.

Section 3.5   Limits at Infinity

1. $f(x) = \dfrac{3x^2}{x^2 + 2}$

No vertical asymptotes

Horizontal asymptote: $y = 3$

Matches (f)

3. $f(x) = \dfrac{x}{x^2 + 2}$

No vertical asymptotes

Horizontal asymptote: $y = 0$

$f(1) < 1$

Matches (d)

5. $f(x) = \dfrac{4 \sin x}{x^2 + 1}$

No vertical asymptotes

Horizontal asymptote: $y = 0$

$f(1) > 1$

Matches (b)

7. $f(x) = \dfrac{4x + 3}{2x - 1}$

x	10^0	10^1	10^2	10^3	10^4	10^5	10^6
$f(x)$	7	2.26	2.025	2.0025	2.0003	2	2

$\lim\limits_{x \to \infty} f(x) = 2$

9. $f(x) = \dfrac{-6x}{\sqrt{4x^2 + 5}}$

x	10^0	10^1	10^2	10^3	10^4	10^5	10^6
$f(x)$	-2	-2.98	-2.9998	-3	-3	-3	-3

$\lim\limits_{x \to \infty} f(x) = -3$

11. $f(x) = 5 - \dfrac{1}{x^2 + 1}$

x	10^0	10^1	10^2	10^3	10^4	10^5	10^6
$f(x)$	4.5	4.99	4.9999	4.999999	5	5	5

$\lim\limits_{x \to \infty} f(x) = 5$

13. (a) $h(x) = \dfrac{f(x)}{x^2} = \dfrac{5x^3 - 3x^2 + 10}{x^2} = 5x - 3 + \dfrac{10}{x^2}$

$\displaystyle\lim_{x \to \infty} h(x) = \infty$ (Limit does not exist)

(b) $h(x) = \dfrac{f(x)}{x^3} = \dfrac{5x^3 - 3x^2 + 10}{x^3} = 5 - \dfrac{3}{x} + \dfrac{10}{x^3}$

$\displaystyle\lim_{x \to \infty} h(x) = 5$

(c) $h(x) = \dfrac{f(x)}{x^4} = \dfrac{5x^3 - 3x^2 + 10}{x^4} = \dfrac{5}{x} - \dfrac{3}{x^2} + \dfrac{10}{x^4}$

$\displaystyle\lim_{x \to \infty} h(x) = 0$

15. (a) $\displaystyle\lim_{x \to \infty} \dfrac{x^2 + 2}{x^3 - 1} = 0$

(b) $\displaystyle\lim_{x \to \infty} \dfrac{x^2 + 2}{x^2 - 1} = 1$

(c) $\displaystyle\lim_{x \to \infty} \dfrac{x^2 + 2}{x - 1} = \infty$ (Limit does not exist)

17. (a) $\displaystyle\lim_{x \to \infty} \dfrac{5 - 2x^{3/2}}{3x^2 - 4} = 0$

(b) $\displaystyle\lim_{x \to \infty} \dfrac{5 - 2x^{3/2}}{3x^{3/2} - 4} = -\dfrac{2}{3}$

(c) $\displaystyle\lim_{x \to \infty} \dfrac{5 - 2x^{3/2}}{3x - 4} = -\infty$ (Limit does not exist)

19. $\displaystyle\lim_{x \to \infty} \dfrac{2x - 1}{3x + 2} = \lim_{x \to \infty} \dfrac{2 - (1/x)}{3 + (2/x)} = \dfrac{2 - 0}{3 + 0} = \dfrac{2}{3}$

21. $\displaystyle\lim_{x \to \infty} \dfrac{x}{x^2 - 1} = \lim_{x \to \infty} \dfrac{1/x}{1 - (1/x^2)} = \dfrac{0}{1} = 0$

23. $\displaystyle\lim_{x \to -\infty} \dfrac{5x^2}{x + 3} = \lim_{x \to -\infty} \dfrac{5x}{1 + (3/x)} = -\infty$

Limit does not exist.

25. $\displaystyle\lim_{x \to -\infty} \dfrac{x}{\sqrt{x^2 - x}} = \lim_{x \to -\infty} \dfrac{1}{\dfrac{\sqrt{x^2 - x}}{-\sqrt{x^2}}}, \left(\text{for } x < 0 \text{ we have } x = -\sqrt{x^2} \right)$

$= \displaystyle\lim_{x \to -\infty} \dfrac{-1}{\sqrt{1 - (1/x)}} = -1$

27. $\displaystyle\lim_{x \to -\infty} \dfrac{2x + 1}{\sqrt{x^2 - x}} = \lim_{x \to -\infty} \dfrac{2 + \dfrac{1}{x}}{\left(\dfrac{\sqrt{x^2 - x}}{-\sqrt{x^2}} \right)}$ $\left(\text{for } x < 0, x = -\sqrt{x^2} \right)$

$= \displaystyle\lim_{x \to -\infty} \dfrac{-2 - (1/x)}{\sqrt{1 - (1/x)}} = -2$

29. Since $(-1/x) \le (\sin(2x))/x \le (1/x)$ for all $x \neq 0$, we have by the Squeeze Theorem,

$\displaystyle\lim_{x \to \infty} -\dfrac{1}{x} \le \lim_{x \to \infty} \dfrac{\sin(2x)}{x} \le \lim_{x \to \infty} \dfrac{1}{x}$

$0 \le \displaystyle\lim_{x \to \infty} \dfrac{\sin(2x)}{x} \le 0.$

Therefore, $\displaystyle\lim_{x \to \infty} \dfrac{\sin(2x)}{x} = 0.$

31. $\displaystyle\lim_{x \to \infty} \dfrac{1}{2x + \sin x} = 0$

33. (a) $f(x) = \dfrac{|x|}{x+1}$

$$\lim_{x \to \infty} \frac{|x|}{x+1} = 1$$

$$\lim_{x \to -\infty} \frac{|x|}{x+1} = -1$$

Therefore, $y = 1$ and $y = -1$ are both horizontal asymptotes.

35. $\displaystyle\lim_{x \to \infty} x \sin \frac{1}{x} = \lim_{t \to 0^+} \frac{\sin t}{t} = 1$

(Let $x = 1/t$.)

37. $\displaystyle\lim_{x \to -\infty} \left(x + \sqrt{x^2+3} \right) = \lim_{x \to -\infty} \left[\left(x + \sqrt{x^2+3} \right) \cdot \frac{x - \sqrt{x^2+3}}{x - \sqrt{x^2+3}} \right] = \lim_{x \to -\infty} \frac{-3}{x - \sqrt{x^2+3}} = 0$

39. $\displaystyle\lim_{x \to \infty} \left(x - \sqrt{x^2+x} \right) = \lim_{x \to \infty} \left[\left(x - \sqrt{x^2+x} \right) \cdot \frac{x + \sqrt{x^2+x}}{x + \sqrt{x^2+x}} \right]$

$$= \lim_{x \to \infty} \frac{-x}{x + \sqrt{x^2+x}} = \lim_{x \to \infty} \frac{-1}{1 + \sqrt{1 + (1/x)}} = -\frac{1}{2}$$

41.

x	10^0	10^1	10^2	10^3	10^4	10^5	10^6
$f(x)$	1	0.513	0.501	0.500	0.500	0.500	0.500

$$\lim_{x \to \infty} \left(x - \sqrt{x(x-1)} \right) = \lim_{x \to \infty} \frac{x - \sqrt{x^2 - x}}{1} \cdot \frac{x + \sqrt{x^2 - x}}{x + \sqrt{x^2 - x}}$$

$$= \lim_{x \to \infty} \frac{x}{x + \sqrt{x^2 - x}}$$

$$= \lim_{x \to \infty} \frac{1}{1 + \sqrt{1 - (1/x)}}$$

$$= \frac{1}{2}$$

43.

x	10^0	10^1	10^2	10^3	10^4	10^5	10^6
$f(x)$	0.479	0.500	0.500	0.500	0.500	0.500	0.500

Let $x = 1/t$.

$$\lim_{x \to \infty} x \sin\left(\frac{1}{2x} \right) = \lim_{t \to 0^+} \frac{\sin(t/2)}{t} = \lim_{t \to 0^+} \frac{1}{2} \frac{\sin(t/2)}{t/2} = \frac{1}{2}$$

45. (a)

(b) $\lim_{x \to \infty} f(x) = 3 \qquad \lim_{x \to \infty} f'(x) = 0$

(c) Since $\lim_{x \to \infty} f(x) = 3$, the graph approaches that of a

horizontal line, $\lim_{x \to \infty} f'(x) = 0$.

47. Yes. For example, let $f(x) = \dfrac{6|x - 2|}{\sqrt{(x - 2)^2 + 1}}$.

49. $y = \dfrac{2 + x}{1 - x}$

Intercepts: $(-2, 0), (0, 2)$

Symmetry: none

Horizontal asymptote: $y = -1$ since

$$\lim_{x \to -\infty} \frac{2 + x}{1 - x} = -1 = \lim_{x \to \infty} \frac{2 + x}{1 - x}.$$

Discontinuity: $x = 1$ (Vertical asymptote)

51. $y = \dfrac{x}{x^2 - 4}$

Intercept: $(0, 0)$

Symmetry: origin

Horizontal asymptote: $y = 0$

Vertical asymptote: $x = \pm 2$

53. $y = \dfrac{x^2}{x^2 + 9}$

Intercept: $(0, 0)$

Symmetry: y-axis

Horizontal asymptote: $y = 1$ since

$$\lim_{x \to -\infty} \frac{x^2}{x^2 + 9} = 1 = \lim_{x \to \infty} \frac{x^2}{x^2 + 9}.$$

Relative minimum: $(0, 0)$

55. $y = \dfrac{2x^2}{x^2 - 4}$

Intercept: $(0, 0)$

Symmetry: y-axis

Horizontal asymptote: $y = 2$

Vertical asymptotes: $x = \pm 2$

Relative maximum: $(0, 0)$

57. $xy^2 = 4$

Domain: $x > 0$

Intercepts: none

Symmetry: x-axis

Horizontal asymptote: $y = 0$ since

$$\lim_{x \to \infty} \frac{2}{\sqrt{x}} = 0 = \lim_{x \to \infty} -\frac{2}{\sqrt{x}}.$$

Discontinuity: $x = 0$ (Vertical asymptote)

59. $y = \dfrac{2x}{1 - x}$

Intercept: $(0, 0)$

Symmetry: none

Horizontal asymptote: $y = -2$ since

$$\lim_{x \to -\infty} \frac{2x}{1 - x} = -2 = \lim_{x \to \infty} \frac{2x}{1 - x}.$$

Discontinuity: $x = 1$ (Vertical asymptote)

61. $y = 2 - \dfrac{3}{x^2}$

Intercepts: $\left(\pm\sqrt{3/2}, 0 \right)$

Symmetry: y-axis

Horizontal asymptote: $y = 2$ since

$$\lim_{x \to -\infty} \left(2 - \frac{3}{x^2} \right) = 2 = \lim_{x \to \infty} \left(2 - \frac{3}{x^2} \right).$$

Discontinuity: $x = 0$ (Vertical asymptote)

63. $y = 3 + \dfrac{2}{x}$

Intercept: $y = 0 = 3 + \dfrac{2}{x} \Longrightarrow \dfrac{2}{x} = -3 \Longrightarrow x = -\dfrac{2}{3}; \left(-\dfrac{2}{3}, 0 \right)$

Symmetry: none

Horizontal asymptote: $y = 3$

Vertical asymptote: $x = 0$

65. $y = \dfrac{x^3}{\sqrt{x^2 - 4}}$

Domain: $(-\infty, -2), (2, \infty)$

Intercepts: none

Symmetry: origin

Horizontal asymptote: none

Vertical asymptotes: $x = \pm 2$ (discontinuities)

67. $f(x) = 5 - \dfrac{1}{x^2} = \dfrac{5x^2 - 1}{x^2}$

Domain: $(-\infty, 0), (0, \infty)$

$f'(x) = \dfrac{2}{x^3} \Longrightarrow$ No relative extrema

$f''(x) = -\dfrac{6}{x^4} \Longrightarrow$ No points of inflection

Vertical asymptote: $x = 0$

Horizontal asymptote: $y = 5$

69. $f(x) = \dfrac{x}{x^2 - 4}$

$f'(x) = \dfrac{(x^2 - 4) - x(2x)}{(x^2 - 4)^2}$

$\qquad = \dfrac{-(x^2 + 4)}{(x^2 - 4)^2} \neq 0$ for any x in the domain of f.

$f''(x) = \dfrac{(x^2 - 4)^2(-2x) + (x^2 + 4)(2)(x^2 - 4)(2x)}{(x^2 - 4)^2}$

$\qquad = \dfrac{2x(x^2 + 12)}{(x^2 - 4)^3} = 0$ when $x = 0$.

Since $f''(x) > 0$ on $(-2, 0)$ and $f''(x) < 0$ on $(0, 2)$, then $(0, 0)$ is a point of inflection.

Vertical asymptotes: $x = \pm 2$

Horizontal asymptote: $y = 0$

71. $f(x) = \dfrac{x - 2}{x^2 - 4x + 3} = \dfrac{x - 2}{(x - 1)(x - 3)}$

$f'(x) = \dfrac{(x^2 - 4x + 3) - (x - 2)(2x - 4)}{(x^2 - 4x + 3)^2} = \dfrac{-x^2 + 4x - 5}{(x^2 - 4x + 3)^2} \neq 0$

$f''(x) = \dfrac{(x^2 - 4x + 3)^2(-2x + 4) - (-x^2 + 4x - 5)(2)(x^2 - 4x + 3)(2x - 4)}{(x^2 - 4x + 3)^4}$

$\qquad = \dfrac{2(x^3 - 6x^2 + 15x - 14)}{(x^2 - 4x + 3)^3} = \dfrac{2(x - 2)(x^2 - 4x + 7)}{(x^2 - 4x + 3)^3} = 0$ when $x = 2$.

Since $f''(x) > 0$ on $(1, 2)$ and $f''(x) < 0$ on $(2, 3)$, then $(2, 0)$ is a point of inflection.

Vertical asymptotes: $x = 1, x = 3$

Horizontal asymptote: $y = 0$

73. $f(x) = \dfrac{3x}{\sqrt{4x^2 + 1}}$

$f'(x) = \dfrac{3}{(4x^2 + 1)^{3/2}} \Longrightarrow$ No relative extrema

$f''(x) = \dfrac{-36x}{(4x^2 + 1)^{5/2}} = 0$ when $x = 0$.

Point of inflection: $(0, 0)$

Horizontal asymptotes: $y = \pm\dfrac{3}{2}$

No vertical asymptotes

75. $g(x) = \sin\left(\dfrac{x}{x - 2}\right),\ 3 < x < \infty$

$g'(x) = \dfrac{-2\cos\left(\dfrac{x}{x - 2}\right)}{(x - 2)^2}$

Horizontal asymptote: $y = \sin(1)$

Relative maximum: $\dfrac{x}{x - 2} = \dfrac{\pi}{2} \Longrightarrow x = \dfrac{2\pi}{\pi - 2} \approx 5.5039$

No vertical asymptotes

77. $f(x) = \dfrac{x^3 - 3x^2 + 2}{x(x - 3)},\ g(x) = x + \dfrac{2}{x(x - 3)}$

(a)

(c)

The graph appears as the slant asymptote $y = x$.

(b) $f(x) = \dfrac{x^3 - 3x^2 + 2}{x(x - 3)}$

$= \dfrac{x^2(x - 3)}{x(x - 3)} + \dfrac{2}{x(x - 3)}$

$= x + \dfrac{2}{x(x - 3)} = g(x)$

79. $C = 0.5x + 500$

$\overline{C} = \dfrac{C}{x}$

$\overline{C} = 0.5 + \dfrac{500}{x}$

$\lim_{x \to \infty}\left(0.5 + \dfrac{500}{x}\right) = 0.5$

81. line: $mx - y + 4 = 0$

(a) $d = \dfrac{|Ax_1 + By_1 + C|}{\sqrt{A^2 + B^2}} = \dfrac{|m(3) - 1(1) + 4|}{\sqrt{m^2 + 1}}$

$= \dfrac{|3m + 3|}{\sqrt{m^2 + 1}}$

(b)

(c) $\lim\limits_{m \to \infty} d(m) = 3 = \lim\limits_{m \to -\infty} d(m)$

The line approaches the vertical line $x = 0$. Hence, the distance approaches 3.

85. Answers will vary. See page 195.

83. (a) $T_1(t) = -0.003t^2 + 0.677t + 26.564$

(b)

(c)

$T_2 = \dfrac{1451 + 86t}{58 + t}$

(d) $T_1(0) \approx 26.6$

$T_2(0) \approx 25.0$

(e) $\lim\limits_{t \to \infty} T_2 = \dfrac{86}{1} = 86$

(f) The limiting temperature is 86. T_1 has no horizontal asymptote.

87. False. Let $f(x) = \dfrac{2x}{\sqrt{x^2 + 2}}$. (See Exercise 2.)

Section 3.6   A Summary of Curve Sketching

1. f has constant negative slope. Matches (D)

3. The slope is periodic, and zero at $x = 0$. Matches (A)

5. (a) $f'(x) = 0$ for $x = -2$ and $x = 2$

f' is negative for $-2 < x < 2$ (decreasing function).

f' is positive for $x > 2$ and $x < -2$ (increasing function).

(b) $f''(x) = 0$ at $x = 0$ (Inflection point).

f'' is positive for $x > 0$ (Concave upwards).

f'' is negative for $x < 0$ (Concave downward).

(c) f' is increasing on $(0, \infty)$.   ($f'' > 0$)

(d) $f'(x)$ is minimum at $x = 0$. The rate of change of f at $x = 0$ is less than the rate of change of f for all other values of x.

7. $y = \dfrac{x^2}{x^2 + 3}$

$y' = \dfrac{6x}{(x^2 + 3)^2} = 0$ when $x = 0$.

$y'' = \dfrac{18(1 - x^2)}{(x^2 + 3)^3} = 0$ when $x = \pm 1$.

Horizontal asymptote: $y = 1$

	y	y'	y''	Conclusion
$-\infty < x < -1$		$-$	$-$	Decreasing, concave down
$x = -1$	$\frac{1}{4}$	$-$	0	Point of inflection
$-1 < x < 0$		$-$	$+$	Decreasing, concave up
$x = 0$	0	0	$+$	Relative minimum
$0 < x < 1$		$+$	$+$	Increasing, concave up
$x = 1$	$\frac{1}{4}$	$+$	0	Point of inflection
$1 < x < \infty$		$+$	$-$	Increasing, concave down

9. $y = \dfrac{1}{x - 2} - 3$

$y' = -\dfrac{1}{(x - 2)^2} < 0$ when $x \neq 2$.

$y'' = \dfrac{2}{(x - 2)^3}$

No relative extrema, no points of inflection

Intercepts: $\left(\dfrac{7}{3}, 0\right), \left(0, -\dfrac{7}{2}\right)$

Vertical asymptote: $x = 2$

Horizontal asymptote: $y = -3$

11. $y = \dfrac{2x}{x^2 - 1}$

$y' = \dfrac{-2(x^2 + 1)}{(x^2 - 1)^2} < 0$ if $x \neq \pm 1$.

$y'' = \dfrac{4x(x^2 + 3)}{(x^2 - 1)^3} = 0$ if $x = 0$.

Inflection point: $(0, 0)$

Intercept: $(0, 0)$

Vertical asymptote: $x = \pm 1$

Horizontal asymptote: $y = 0$

Symmetry with respect to the origin

13. $g(x) = x + \dfrac{4}{x^2 + 1}$

$g'(x) = 1 - \dfrac{8x}{(x^2 + 1)^2} = \dfrac{x^4 + 2x^2 - 8x + 1}{(x^2 + 1)^2} = 0$ when $x \approx 0.1292,\ 1.6085$

$g''(x) = \dfrac{8(3x^2 - 1)}{(x^2 + 1)^3} = 0$ when $x = \pm\dfrac{\sqrt{3}}{3}$

$g''(0.1292) < 0$, therefore, $(0.1292, 4.064)$ is relative maximum.

$g''(1.6085) > 0$, therefore, $(1.6085, 2.724)$ is a relative minimum.

Points of inflection: $\left(-\dfrac{\sqrt{3}}{3}, 2.423\right), \left(\dfrac{\sqrt{3}}{3}, 3.577\right)$

Intercepts: $(0, 4),\ (-1.3788, 0)$

Slant asymptote: $y = x$

15. $f(x) = \dfrac{x^2 + 1}{x} = x + \dfrac{1}{x}$

$f'(x) = 1 - \dfrac{1}{x^2} = 0$ when $x = \pm 1$.

$f''(x) = \dfrac{2}{x^3} \neq 0$

Relative maximum: $(-1, -2)$

Relative minimum: $(1, 2)$

Vertical asymptote: $x = 0$

Slant asymptote: $y = x$

17. $y = \dfrac{x^2 - 6x + 12}{x - 4} = x - 2 + \dfrac{4}{x - 4}$

$y' = 1 - \dfrac{4}{(x - 4)^2}$

$= \dfrac{(x - 2)(x - 6)}{(x - 4)^2} = 0$ when $x = 2, 6$.

$y'' = \dfrac{8}{(x - 4)^3}$

$y'' < 0$ when $x = 2$.

Therefore, $(2, -2)$ is a relative maximum.

$y'' > 0$ when $x = 6$.

Therefore, $(6, 6)$ is a relative minimum.

Vertical asymptote: $x = 4$

Slant asymptote: $y = x - 2$

19. $y = x\sqrt{4 - x}$,

Domain: $(-\infty, 4]$

$y' = \dfrac{8 - 3x}{2\sqrt{4 - x}} = 0$ when $x = \dfrac{8}{3}$ and undefined when $x = 4$.

$y'' = \dfrac{3x - 16}{4(4 - x)^{3/2}} = 0$ when $x = \dfrac{16}{3}$ and undefined when $x = 4$.

Note: $x = \dfrac{16}{3}$ is not in the domain.

	y	y'	y''	Conclusion
$-\infty < x < \dfrac{8}{3}$		+	−	Increasing, concave down
$x = \dfrac{8}{3}$	$\dfrac{16}{3\sqrt{3}}$	0	−	Relative maximum
$\dfrac{8}{3} < x < 4$		−	−	Decreasing, concave down
$x = 4$	0	Undefined	Undefined	Endpoint

21. $h(x) = x\sqrt{9 - x^2}$ Domain: $-3 \le x \le 3$

$h'(x) = \dfrac{9 - 2x^2}{\sqrt{9 - x^2}} = 0$ when $x = \pm\dfrac{3}{\sqrt{2}} = \pm\dfrac{3\sqrt{2}}{2}$

$h''(x) = \dfrac{x(2x^2 - 27)}{(9 - x^2)^{3/2}} = 0$ when $x = 0$

Relative maximum: $\left(\dfrac{3\sqrt{2}}{2}, \dfrac{9}{2}\right)$

Relative minimum: $\left(-\dfrac{3\sqrt{2}}{2}, -\dfrac{9}{2}\right)$

Intercepts: $(0, 0)$, $(\pm 3, 0)$

Symmetric with respect to the origin

Point of inflection: $(0, 0)$

23. $y = 3x^{2/3} - 2x$

$y' = 2x^{-1/3} - 2 = \dfrac{2(1 - x^{1/3})}{x^{1/3}}$

   $= 0$ when $x = 1$ and undefined when $x = 0$.

$y'' = \dfrac{-2}{3x^{4/3}} < 0$ when $x \ne 0$.

	y	y'	y''	Conclusion
$-\infty < x < 0$		−	−	Decreasing, concave down
$x = 0$	0	Undefined	Undefined	Relative minimum
$0 < x < 1$		+	−	Increasing, concave down
$x = 1$	1	0	−	Relative maximum
$1 < x < \infty$		−	−	Decreasing, concave down

25. $y = x^3 - 3x^2 + 3$

$y' = 3x^2 - 6x = 3x(x - 2) = 0$ when $x = 0$, $x = 2$

$y'' = 6x - 6 = 6(x - 1) = 0$ when $x = 1$

	y	y'	y''	Conclusion
$-\infty < x < 0$		$+$	$-$	Increasing, concave down
$x = 0$	3	0	$-$	Relative maximum
$0 < x < 1$		$-$	$-$	Decreasing, concave down
$x = 1$	1	$-$	0	Point of inflection
$1 < x < 2$		$-$	$+$	Decreasing, concave up
$x = 2$	-1	0	$+$	Relative minimum
$2 < x < \infty$		$+$	$+$	Increasing, concave up

27. $y = 2 - x - x^3$

$y' = -1 - 3x^2$

No critical numbers

$y'' = -6x = 0$ when $x = 0$.

	y	y'	y''	Conclusion
$-\infty < x < 0$		$-$	$+$	Decreasing, concave up
$x = 0$	2	$-$	0	Point of inflection
$0 < x < \infty$		$-$	$-$	Decreasing, concave down

29. $f(x) = 3x^3 - 9x + 1$

$f'(x) = 9x^2 - 9 = 9(x^2 - 1) = 0$ when $x = \pm 1$

$f''(x) = 18x = 0$ when $x = 0$

	$f(x)$	$f'(x)$	$f''(x)$	Conclusion
$-\infty < x < -1$		$+$	$-$	Increasing, concave down
$x = -1$	7	0	$-$	Relative maximum
$-1 < x < 0$		$-$	$-$	Decreasing, concave down
$x = 0$	1	$-$	0	Point of inflection
$0 < x < 1$		$-$	$+$	Decreasing, concave up
$x = 1$	-5	0	$+$	Relative minimum
$1 < x < \infty$		$+$	$+$	Increasing, concave up

31. $y = 3x^4 + 4x^3$

$y' = 12x^3 + 12x^2 = 12x^2(x + 1) = 0$ when $x = 0, x = -1$.

$y'' = 36x^2 + 24x = 12x(3x + 2) = 0$ when $x = 0, x = -\frac{2}{3}$.

	y	y'	y''	Conclusion
$-\infty < x < -1$		$-$	$+$	Decreasing, concave up
$x = -1$	-1	0	$+$	Relative minimum
$-1 < x < -\frac{2}{3}$		$+$	$+$	Increasing, concave up
$x = -\frac{2}{3}$	$-\frac{16}{27}$	$+$	0	Point of inflection
$-\frac{2}{3} < x < 0$		$+$	$-$	Increasing, concave down
$x = 0$	0	0	0	Point of inflection
$0 < x < \infty$		$+$	$+$	Increasing, concave up

33. $f(x) = x^4 - 4x^3 + 16x$

$f'(x) = 4x^3 - 12x^2 + 16 = 4(x + 1)(x - 2)^2 = 0$ when $x = -1, x = 2$.

$f''(x) = 12x^2 - 24x = 12x(x - 2) = 0$ when $x = 0, x = 2$.

	$f(x)$	$f'(x)$	$f''(x)$	Conclusion
$-\infty < x < -1$		$-$	$+$	Decreasing, concave up
$x = -1$	-11	0	$+$	Relative minimum
$-1 < x < 0$		$+$	$+$	Increasing, concave up
$x = 0$	0	$+$	0	Point of inflection
$0 < x < 2$		$+$	$-$	Increasing, concave down
$x = 2$	16	0	0	Point of inflection
$2 < x < \infty$		$+$	$+$	Increasing, concave up

35. $y = x^5 - 5x$

$y' = 5x^4 - 5 = 5(x^4 - 1) = 0$ when $x = \pm 1$.

$y'' = 20x^3 = 0$ when $x = 0$.

	y	y'	y''	Conclusion
$-\infty < x < -1$		$+$	$-$	Increasing, concave down
$x = -1$	4	0	$-$	Relative maximum
$-1 < x < 0$		$-$	$-$	Decreasing, concave down
$x = 0$	0	$-$	0	Point of inflection
$0 < x < 1$		$-$	$+$	Decreasing, concave up
$x = 1$	-4	0	$+$	Relative minimum
$1 < x < \infty$		$+$	$+$	Increasing, concave up

37. $y = |2x - 3|$

$y' = \dfrac{2(2x - 3)}{|2x - 3|}$ undefined at $x = \dfrac{3}{2}$.

$y'' = 0$

(0, 3)

$\left(\dfrac{3}{2}, 0\right)$

	y	y'	Conclusion
$-\infty < x < \frac{3}{2}$		$-$	Decreasing
$x = \frac{3}{2}$	0	Undefined	Relative minimum
$\frac{3}{2} < x < \infty$		$+$	Increasing

39. $y = \sin x - \dfrac{1}{18}\sin 3x,\ 0 \le x \le 2\pi$

$y' = \cos x - \dfrac{1}{6}\cos 3x = 0$ when $x = \dfrac{\pi}{2}, \dfrac{3\pi}{2}$.

$y'' = -\sin x + \dfrac{1}{2}\sin 3x = 0$ when $x = 0, \dfrac{\pi}{6}, \dfrac{5\pi}{6}, \pi, \dfrac{7\pi}{6}, \dfrac{11\pi}{6}$.

Relative maximum: $\left(\dfrac{\pi}{2}, \dfrac{19}{18}\right)$

Relative minimum: $\left(\dfrac{3\pi}{2}, -\dfrac{19}{18}\right)$

Inflection points: $\left(\dfrac{\pi}{6}, \dfrac{4}{9}\right), \left(\dfrac{5\pi}{6}, \dfrac{4}{9}\right), (\pi, 0), \left(\dfrac{7\pi}{6}, -\dfrac{4}{9}\right), \left(\dfrac{11\pi}{6}, -\dfrac{4}{9}\right)$

41. $y = 2x - \tan x,\ -\dfrac{\pi}{2} < x < \dfrac{\pi}{2}$

$y' = 2 - \sec^2 x = 0$ when $x = \pm\dfrac{\pi}{4}$.

$y'' = -2\sec^2 x \tan x = 0$ when $x = 0$.

Relative maximum: $\left(\dfrac{\pi}{4}, \dfrac{\pi}{2} - 1\right)$

Relative minimum: $\left(-\dfrac{\pi}{4}, 1 - \dfrac{\pi}{2}\right)$

Inflection point: $(0, 0)$

Vertical asymptotes: $x = \pm\dfrac{\pi}{2}$

43. $y = 2(\csc x + \sec x),\ 0 < x < \dfrac{\pi}{2}$

$y' = 2(\sec x \tan x - \csc x \cot x) = 0 \implies x = \pi/4$

Relative minimum: $\left(\dfrac{\pi}{4}, 4\sqrt{2}\right)$

Vertical asymptotes: $x = 0,\ x = \dfrac{\pi}{2}$

45. $g(x) = x \tan x$, $-\dfrac{3\pi}{2} < x < \dfrac{3\pi}{2}$

$g'(x) = \dfrac{x + \sin x \cos x}{\cos^2 x} = 0$ when $x = 0$

$g''(x) = \dfrac{2(\cos x + x \sin x)}{\cos^3 x}$

Vertical asymptotes: $x = -\dfrac{3\pi}{2}, -\dfrac{\pi}{2}, \dfrac{\pi}{2}, \dfrac{3\pi}{2}$

Intercepts: $(-\pi, 0), (0, 0), (\pi, 0)$

Symmetric with respect to y-axis.

Increasing on $\left(0, \dfrac{\pi}{2}\right)$ and $\left(\dfrac{\pi}{2}, \dfrac{3\pi}{2}\right)$

Points of inflection: $(\pm 2.80, 0)$

47. $f(x) = \dfrac{20x}{x^2 + 1} - \dfrac{1}{x} = \dfrac{19x^2 - 1}{x(x^2 + 1)}$

$x = 0$ vertical asymptote

$y = 0$ horizontal asymptote

Minimum: $(-1.10, -9.05)$

Maximum: $(1.10, 9.05)$

Points of inflection: $(-1.84, -7.86), (1.84, 7.86)$

49. $y = \dfrac{x}{\sqrt{x^2 + 7}}$

$(0, 0)$ point of inflection

$y = \pm 1$ horizontal asymptotes

51. f is cubic.

f' is quadratic.

f'' is linear.

53.

(any vertical translate of f will do)

55.

(any vertical translate of *f* will do)

57. Since the slope is negative, the function is decreasing on (2, 8), and hence $f(3) > f(5)$.

59. $f(x) = \dfrac{4(x-1)^2}{x^2 - 4x + 5}$

Vertical asymptote: none

Horizontal asymptote: $y = 4$

The graph crosses the horizontal asymptote $y = 4$. If a function has a vertical asymptote at $x = c$, the graph would not cross it since $f(c)$ is undefined.

61. $h(x) = \dfrac{6 - 2x}{3 - x}$

$= \dfrac{2(3 - x)}{3 - x} = \begin{cases} 2, & \text{if } x \neq 3 \\ \text{Undefined}, & \text{if } x = 3 \end{cases}$

The rational function is not reduced to lowest terms.

hole at $(3, 2)$

63. $f(x) = -\dfrac{x^2 - 3x - 1}{x - 2} = -x + 1 + \dfrac{3}{x - 2}$

The graph appears to approach the slant asymptote $y = -x + 1$.

65. $f(x) = \dfrac{\cos^2 \pi x}{\sqrt{x^2 + 1}}, \ (0, 4)$

(a)

On (0, 4) there seem to be 7 critical numbers:

0.5, 1.0, 1.5, 2.0, 2.5, 3.0, 3.5

(b) $f'(x) = \dfrac{-\cos \pi x (x \cos \pi x + 2\pi(x^2 + 1)\sin \pi x)}{(x^2 + 1)^{3/2}} = 0$

Critical numbers $\approx \dfrac{1}{2}, 0.97, \dfrac{3}{2}, 1.98, \dfrac{5}{2}, 2.98, \dfrac{7}{2}$.

The critical numbers where maxima occur appear to be integers in part (a), but approximating them using f' shows that they are not integers.

67. Vertical asymptote: $x = 5$

Horizontal asymptote: $y = 0$

$$y = \frac{1}{x - 5}$$

69. Vertical asymptote: $x = 5$

Slant asymptote: $y = 3x + 2$

$$y = 3x + 2 + \frac{1}{x - 5} = \frac{3x^2 - 13x - 9}{x - 5}$$

71. $f(x) = \dfrac{ax}{(x - b)^2}$

(a) The graph has a vertical asymptote at $x = b$. If $a > 0$, the graph approaches ∞ as $x \to b$. If $a < 0$, the graph approaches $-\infty$ as $x \to b$. The graph approaches its vertical asymptote faster as $|a| \to 0$.

(b) As b varies, the position of the vertical asymptote changes: $x = b$. Also, the coordinates of the minimum ($a > 0$) or maximum ($a < 0$) are changed.

73. $f(x) = \dfrac{3x^n}{x^4 + 1}$

(a) For n even, f is symmetric about the y-axis. For n odd, f is symmetric about the origin.

(b) The x-axis will be the horizontal asymptote if the degree of the numerator is less than 4. That is, $n = 0, 1, 2, 3$.

(c) $n = 4$ gives $y = 3$ as the horizontal asymptote.

(d) There is a slant asymptote $y = 3x$ if $n = 5$:

$$\frac{3x^5}{x^4 + 1} = 3x - \frac{3x}{x^4 + 1}.$$

(e)

n	0	1	2	3	4	5
M	1	2	3	2	1	0
N	2	3	4	5	2	3

75. (a)

(b) When $t = 10$, $N(10) \approx 2434$ bacteria.

(c) N is a maximum when $t \approx 7.2$ (seventh day).

(d) $N''(t) = 0$ for $t \approx 3.2$

(e) $\displaystyle\lim_{t \to \infty} N(t) = \frac{13,250}{7} \approx 1892.86$

Section 3.7   Optimization Problems

1. (a)

First Number, x	Second Number	Product, P
10	$110 - 10$	$10(110 - 10) = 1000$
20	$110 - 20$	$20(110 - 20) = 1800$
30	$110 - 30$	$30(110 - 30) = 2400$
40	$110 - 40$	$40(110 - 40) = 2800$
50	$110 - 50$	$50(110 - 50) = 3000$
60	$110 - 60$	$60(110 - 60) = 3000$

—CONTINUED—

1. —CONTINUED—

(b)

First Number, x	Second Number	Product, P
10	$110 - 10$	$10(110 - 10) = 1000$
20	$110 - 20$	$20(110 - 20) = 1800$
30	$110 - 30$	$30(110 - 30) = 2400$
40	$110 - 40$	$40(110 - 40) = 2800$
50	$110 - 50$	$50(110 - 50) = 3000$
60	$110 - 60$	$60(110 - 60) = 3000$
70	$110 - 70$	$70(110 - 70) = 2800$
80	$110 - 80$	$80(110 - 80) = 2400$
90	$110 - 90$	$90(110 - 90) = 1800$
100	$110 - 100$	$100(110 - 100) = 1000$

The maximum is attained near $x = 50$ and 60.

(c) $P = x(110 - x) = 110x - x^2$

(d)

The solution appears to be $x = 55$.

(e) $\dfrac{dP}{dx} = 110 - 2x = 0$ when $x = 55$.

$\dfrac{d^2P}{dx^2} = -2 < 0$

P is a maximum when $x = 110 - x = 55$.
The two numbers are 55 and 55.

3. Let x and y be two positive numbers such that $xy = 192$.

$$S = x + y = x + \frac{192}{x}$$

$$\frac{dS}{dx} = 1 - \frac{192}{x^2} = 0 \text{ when } x = \sqrt{192}.$$

$$\frac{d^2S}{dx^2} = \frac{384}{x^3} > 0 \text{ when } x = \sqrt{192}.$$

S is a minimum when $x = y = \sqrt{192}$.

5. Let x be a positive number.

$$S = x + \frac{1}{x}$$

$$\frac{dS}{dx} = 1 - \frac{1}{x^2} = 0 \text{ when } x = 1.$$

$$\frac{d^2S}{dx^2} = \frac{2}{x^3} > 0 \text{ when } x = 1.$$

The sum is a minimum when $x = 1$ and $1/x = 1$.

7. Let x be the length and y the width of the rectangle.

$$2x + 2y = 100$$
$$y = 50 - x$$
$$A = xy = x(50 - x)$$
$$\frac{dA}{dx} = 50 - 2x = 0 \text{ when } x = 25.$$
$$\frac{d^2A}{dx^2} = -2 < 0 \text{ when } x = 25.$$

A is maximum when $x = y = 25$ meters.

9. Let x be the length and y the width of the rectangle.

$$xy = 64$$
$$y = \frac{64}{x}$$
$$P = 2x + 2y = 2x + 2\left(\frac{64}{x}\right) = 2x + \frac{128}{x}$$
$$\frac{dP}{dx} = 2 - \frac{128}{x^2} = 0 \text{ when } x = 8.$$
$$\frac{d^2P}{dx^2} = \frac{256}{x^3} > 0 \text{ when } x = 8.$$

P is minimum when $x = y = 8$ feet.

11. $d = \sqrt{(x - 4)^2 + (\sqrt{x} - 0)^2}$

$\quad = \sqrt{x^2 - 7x + 16}$

Since d is smallest when the expression inside the radical is smallest, you need only find the critical numbers of

$\quad f(x) = x^2 - 7x + 16.$

$\quad f'(x) = 2x - 7 = 0$

$\quad\quad x = \frac{7}{2}$

By the First Derivative Test, the point nearest to $(4, 0)$ is $\left(7/2, \sqrt{7/2}\right)$.

13. $d = \sqrt{(x - 2)^2 + [x^2 - (1/2)]^2}$

$\quad = \sqrt{x^4 - 4x + (17/4)}$

Since d is smallest when the expression inside the radical is smallest, you need only find the critical numbers of

$\quad f(x) = x^4 - 4x + \frac{17}{4}.$

$\quad f'(x) = 4x^3 - 4 = 0$

$\quad\quad x = 1$

By the First Derivative Test, the point nearest to $\left(2, \frac{1}{2}\right)$ is $(1, 1)$.

15. $\dfrac{dQ}{dx} = kx(Q_0 - x) = kQ_0 x - kx^2$

$\quad \dfrac{d^2Q}{dx^2} = kQ_0 - 2kx$

$\quad\quad = k(Q_0 - 2x) = 0$ when $x = \dfrac{Q_0}{2}$.

$\quad \dfrac{d^3Q}{dx^3} = -2k < 0$ when $x = \dfrac{Q_0}{2}$.

dQ/dx is maximum when $x = Q_0/2$.

17. $xy = 180,000$ (see figure)

$\quad S = x + 2y = \left(x + \dfrac{360,000}{x}\right)$ where S is the length

of fence needed.

$\quad \dfrac{dS}{dx} = 1 - \dfrac{360,000}{x^2} = 0$ when $x = 600$.

$\quad \dfrac{d^2S}{dx^2} = \dfrac{720,000}{x^3} > 0$ when $x = 600$.

S is a minimum when $x = 600$ meters and $y = 300$ meters.

19. (a) $A = 4(\text{area of side}) + 2(\text{area of Top})$

(a) $A = 4(3)(11) + 2(3)(3) = 150$ square inches

(b) $A = 4(5)(5) + 2(5)(5) = 150$ square inches

(c) $A = 4(3.25)(6) + 2(6)(6) = 150$ square inches

(c) $S = 4xy + 2x^2 = 150 \Rightarrow y = \dfrac{150 - 2x^2}{4x}$

$\quad V = x^2 y = x^2 \left(\dfrac{150 - 2x^2}{4x}\right) = \dfrac{75}{2}x - \dfrac{1}{2}x^3$

$\quad V' = \dfrac{75}{2} - \dfrac{3}{2}x^2 = 0 \Rightarrow x = \pm 5$

(b) $V = (\text{length})(\text{width})(\text{height})$

(a) $V = (3)(3)(11) = 99$ cubic inches

(b) $V = (5)(5)(5) = 125$ cubic inches

(c) $V = (6)(6)(3.25) = 117$ cubic inches

By the First Derivative Test, $x = 5$ yields the maximum volume. Dimensions: $5 \times 5 \times 5$. (A cube!)

21. (a) $V = x(s - 2x)^2, 0 < x < \dfrac{s}{2}$

$\dfrac{dV}{dx} = 2x(s - 2x)(-2) + (s - 2x)^2$

$= (s - 2x)(s - 6x) = 0$ when $x = \dfrac{s}{2}, \dfrac{s}{6}$ ($s/2$ is not in the domain).

$\dfrac{d^2V}{dx^2} = 24x - 8s$

$\dfrac{d^2V}{dx^2} < 0$ when $x = \dfrac{s}{6}$.

$V = \dfrac{2s^3}{27}$ is maximum when $x = \dfrac{s}{6}$.

(b) If the length is doubled, $V = \dfrac{2}{27}(2s)^3 = 8\left(\dfrac{2}{27}s^3\right)$. Volume is increased by a factor of 8.

23. $16 = 2y + x + \pi\left(\dfrac{x}{2}\right)$

$32 = 4y + 2x + \pi x$

$y = \dfrac{32 - 2x - \pi x}{4}$

$A = xy + \dfrac{\pi}{2}\left(\dfrac{x}{2}\right)^2 = \left(\dfrac{32 - 2x - \pi x}{4}\right)x + \dfrac{\pi x^2}{8}$

$= 8x - \dfrac{1}{2}x^2 - \dfrac{\pi}{4}x^2 + \dfrac{\pi}{8}x^2$

$\dfrac{dA}{dx} = 8 - x - \dfrac{\pi}{2}x + \dfrac{\pi}{4}x = 8 - x\left(1 + \dfrac{\pi}{4}\right)$

$= 0$ when $x = \dfrac{8}{1 + (\pi/4)} = \dfrac{32}{4 + \pi}$.

$\dfrac{d^2A}{dx^2} = -\left(1 + \dfrac{\pi}{4}\right) < 0$ when $x = \dfrac{32}{4 + \pi}$

$y = \dfrac{32 - 2[32/(4 + \pi)] - \pi[32/(4 + \pi)]}{4} = \dfrac{16}{4 + \pi}$

The area is maximum when $y = \dfrac{16}{4 + \pi}$ feet and $x = \dfrac{32}{4 + \pi}$ feet.

25. (a) $\dfrac{y - 2}{0 - 1} = \dfrac{0 - 2}{x - 1}$

$y = 2 + \dfrac{2}{x - 1}$

$L = \sqrt{x^2 + y^2} = \sqrt{x^2 + \left(2 + \dfrac{2}{x - 1}\right)^2}$

$= \sqrt{x^2 + 4 + \dfrac{8}{x - 1} + \dfrac{4}{(x - 1)^2}},\ \ x > 1$

(b)

L is minimum when $x \approx 2.587$ and $L \approx 4.162$.

—CONTINUED—

25. —CONTINUED—

(c) Area $= A(x) = \frac{1}{2}xy = \frac{1}{2}x\left(2 + \frac{2}{x-1}\right) = x + \frac{x}{x-1}$

$A'(x) = 1 + \frac{(x-1) - x}{(x-1)^2} = 1 - \frac{1}{(x-1)^2} = 0$

$(x-1)^2 = 1$

$x - 1 = \pm 1$

$x = 0, 2$ (select $x = 2$)

Then $y = 4$ and $A = 4$.

Vertices: $(0, 0), (2, 0), (0, 4)$

27. $A = 2xy = 2x\sqrt{25 - x^2}$ (see figure)

$\frac{dA}{dx} = 2x\left(\frac{1}{2}\right)\left(\frac{-2x}{\sqrt{25 - x^2}}\right) + 2\sqrt{25 - x^2}$

$= 2\left(\frac{25 - 2x^2}{\sqrt{25 - x^2}}\right) = 0$ when $x = y = \frac{5\sqrt{2}}{2} \approx 3.54$.

By the First Derivative Test, the inscribed rectangle of maximum area has vertices

$\left(\pm\frac{5\sqrt{2}}{2}, 0\right), \left(\pm\frac{5\sqrt{2}}{2}, \frac{5\sqrt{2}}{2}\right)$.

Width: $\frac{5\sqrt{2}}{2}$; Length: $5\sqrt{2}$

29. $xy = 30 \Longrightarrow y = \frac{30}{x}$

$A = (x + 2)\left(\frac{30}{x} + 2\right)$ (see figure)

$\frac{dA}{dx} = (x + 2)\left(\frac{-30}{x^2}\right) + \left(\frac{30}{x} + 2\right) = \frac{2(x^2 - 30)}{x^2} = 0$ when $x = \sqrt{30}$.

$y = \frac{30}{\sqrt{30}} = \sqrt{30}$

By the First Derivative Test, the dimensions $(x + 2)$ by $(y + 2)$ are $\left(2 + \sqrt{30}\right)$ by $\left(2 + \sqrt{30}\right)$ (approximately 7.477 by 7.477). These dimensions yield a minimum area.

31. $V = \pi r^2 h = 22$ cubic inches or $h = \frac{22}{\pi r^2}$

(a)

Radius, r	Height	Surface Area	
0.2	$\dfrac{22}{\pi(0.2)^2}$	$2\pi(0.2)\left[0.2 + \dfrac{22}{\pi(0.2)^2}\right]$	≈ 220.3
0.4	$\dfrac{22}{\pi(0.4)^2}$	$2\pi(0.4)\left[0.4 + \dfrac{22}{\pi(0.4)^2}\right]$	≈ 111.0
0.6	$\dfrac{22}{\pi(0.6)^2}$	$2\pi(0.6)\left[0.6 + \dfrac{22}{\pi(0.6)^2}\right]$	≈ 75.6
0.8	$\dfrac{22}{\pi(0.8)^2}$	$2\pi(0.8)\left[0.8 + \dfrac{22}{\pi(0.8)^2}\right]$	≈ 59.0

—CONTINUED—

31. —CONTINUED—

(b)

Radius, r	Height	Surface Area
0.2	$\dfrac{22}{\pi(0.2)^2}$	$2\pi(0.2)\left[0.2 + \dfrac{22}{\pi(0.2)^2}\right] \approx 220.3$
0.4	$\dfrac{22}{\pi(0.4)^2}$	$2\pi(0.4)\left[0.4 + \dfrac{22}{\pi(0.4)^2}\right] \approx 111.0$
0.6	$\dfrac{22}{\pi(0.6)^2}$	$2\pi(0.6)\left[0.6 + \dfrac{22}{\pi(0.6)^2}\right] \approx 75.6$
0.8	$\dfrac{22}{\pi(0.8)^2}$	$2\pi(0.8)\left[0.8 + \dfrac{22}{\pi(0.8)^2}\right] \approx 59.0$
1.0	$\dfrac{22}{\pi(1.0)^2}$	$2\pi(1.0)\left[1.0 + \dfrac{22}{\pi(1.0)^2}\right] \approx 50.3$
1.2	$\dfrac{22}{\pi(1.2)^2}$	$2\pi(1.2)\left[1.2 + \dfrac{22}{\pi(1.2)^2}\right] \approx 45.7$
1.4	$\dfrac{22}{\pi(1.4)^2}$	$2\pi(1.4)\left[1.4 + \dfrac{22}{\pi(1.4)^2}\right] \approx 43.7$
1.6	$\dfrac{22}{\pi(1.6)^2}$	$2\pi(1.6)\left[1.6 + \dfrac{22}{\pi(1.6)^2}\right] \approx 43.6$
1.8	$\dfrac{22}{\pi(1.8)^2}$	$2\pi(1.8)\left[1.8 + \dfrac{22}{\pi(1.8)^2}\right] \approx 44.8$
2.0	$\dfrac{22}{\pi(2.0)^2}$	$2\pi(2.0)\left[2.0 + \dfrac{22}{\pi(2.0)^2}\right] \approx 47.1$

The minimum seems to be about 43.6 for $r = 1.6$.

(c) $S = 2\pi r^2 + 2\pi rh$

$$= 2\pi r(r + h) = 2\pi r\left[r + \frac{22}{\pi r^2}\right] = 2\pi r^2 + \frac{44}{r}$$

(d)

The minimum seems to be 43.46 for $r \approx 1.52$.

(e) $\dfrac{dS}{dr} = 4\pi r - \dfrac{44}{r^2} = 0$ when $r = \sqrt[3]{11/\pi} \approx 1.52$ in.

$$h = \frac{22}{\pi r^2} \approx 3.04 \text{ in.}$$

Note: Notice that

$$h = \frac{22}{\pi r^2} = \frac{22}{\pi(11/\pi)^{2/3}} = 2\left(\frac{11^{1/3}}{\pi^{1/3}}\right) = 2r.$$

33. Let x be the sides of the square ends and y the length of the package.

$$P = 4x + y = 108 \implies y = 108 - 4x$$
$$V = x^2 y = x^2(108 - 4x) = 108x^2 - 4x^3$$

$$\frac{dV}{dx} = 216x - 12x^2$$

$$= 12x(18 - x) = 0 \text{ when } x = 18.$$

$$\frac{d^2V}{dx^2} = 216 - 24x = -216 < 0 \text{ when } x = 18.$$

The volume is maximum when $x = 18$ inches and $y = 108 - 4(18) = 36$ inches.

35. $V = \dfrac{1}{3}\pi x^2 h = \dfrac{1}{3}\pi x^2\left(r + \sqrt{r^2 - x^2}\right)$ (see figure)

$$\frac{dV}{dx} = \frac{1}{3}\pi\left[\frac{-x^3}{\sqrt{r^2 - x^2}} + 2x\left(r + \sqrt{r^2 - x^2}\right)\right] = \frac{\pi x}{3\sqrt{r^2 - x^2}}\left(2r^2 + 2r\sqrt{r^2 - x^2} - 3x^2\right) = 0$$

$$2r^2 + 2r\sqrt{r^2 - x^2} - 3x^2 = 0$$

$$2r\sqrt{r^2 - x^2} = 3x^2 - 2r^2$$

$$4r^2(r^2 - x^2) = 9x^4 - 12x^2 r^2 + 4r^4$$

$$0 = 9x^4 - 8x^2 r^2 = x^2(9x^2 - 8r^2)$$

$$x = 0, \frac{2\sqrt{2}r}{3}$$

By the First Derivative Test, the volume is a maximum when

$$x = \frac{2\sqrt{2}r}{3} \text{ and } h = r + \sqrt{r^2 - x^2} = \frac{4r}{3}.$$

Thus, the maximum volume is

$$V = \frac{1}{3}\pi\left(\frac{8r^2}{9}\right)\left(\frac{4r}{3}\right) = \frac{32\pi r^3}{81} \text{ cubic units.}$$

37. No, there is no minimum area. If the sides are x and y, then $2x + 2y = 20 \Rightarrow y = 10 - x$.
The area is $A(x) = x(10 - x) = 10x - x^2$. This can be made arbitrarily small by selecting $x \approx 0$.

39. $V = 12 = \dfrac{4}{3}\pi r^3 + \pi r^2 h$

$$h = \frac{12 - (4/3)\pi r^3}{\pi r^2} = \frac{12}{\pi r^2} - \frac{4}{3}r$$

$$S = 4\pi r^2 + 2\pi rh = 4\pi r^2 + 2\pi r\left(\frac{12}{\pi r^2} - \frac{4}{3}r\right)$$

$$= 4\pi r^2 + \frac{24}{r} - \frac{8}{3}\pi r^2 = \frac{4}{3}\pi r^2 + \frac{24}{r}$$

$$\frac{dS}{dr} = \frac{8}{3}\pi r - \frac{24}{r^2} = 0 \text{ when } r = \sqrt[3]{9/\pi} \approx 1.42 \text{ cm.}$$

$$\frac{d^2S}{dr^2} = \frac{8}{3}\pi + \frac{48}{r^3} > 0 \text{ when } r = \sqrt[3]{9/\pi} \text{ cm.}$$

The surface area is minimum when $r = \sqrt[3]{9/\pi}$ cm and $h = 0$. The resulting solid is a sphere of radius $r \approx 1.42$ cm.

41. Let x be the length of a side of the square and y the length of a side of the triangle.

$$4x + 3y = 10$$

$$A = x^2 + \frac{1}{2}y\left(\frac{\sqrt{3}}{2}y\right)$$

$$= \frac{(10 - 3y)^2}{16} + \frac{\sqrt{3}}{4}y^2$$

$$\frac{dA}{dy} = \frac{1}{8}(10 - 3y)(-3) + \frac{\sqrt{3}}{2}y = 0$$

$$-30 + 9y + 4\sqrt{3}y = 0$$

$$y = \frac{30}{9 + 4\sqrt{3}}$$

$$\frac{d^2A}{dy^2} = \frac{9 + 4\sqrt{3}}{8} > 0$$

A is minimum when

$$y = \frac{30}{9 + 4\sqrt{3}} \text{ and } x = \frac{10\sqrt{3}}{9 + 4\sqrt{3}}.$$

43. Let S be the strength and k the constant of proportionality. Given $h^2 + w^2 = 24^2, h^2 = 24^2 - w^2$,

$$S = kwh^2$$

$$S = kw(576 - w^2) = k(576w - w^3)$$

$$\frac{dS}{dw} = k(576 - 3w^2) = 0 \text{ when } w = 8\sqrt{3}, h = 8\sqrt{6}.$$

$$\frac{d^2S}{dw^2} = -6kw < 0 \text{ when } w = 8\sqrt{3}.$$

These values yield a maximum.

45. $$R = \frac{v_0^2}{g} \sin 2\theta$$

$$\frac{dR}{d\theta} = \frac{2v_0^2}{g} \cos 2\theta = 0 \text{ when } \theta = \frac{\pi}{4}, \frac{3\pi}{4}.$$

$$\frac{d^2R}{d\theta^2} = -\frac{4v_0^2}{g} \sin 2\theta < 0 \text{ when } \theta = \frac{\pi}{4}.$$

By the Second Derivative Test, R is maximum when $\theta = \pi/4$.

47. $$\sin \alpha = \frac{h}{s} \Longrightarrow s = \frac{h}{\sin \alpha}, \ 0 < \alpha < \frac{\pi}{2}$$

$$\tan \alpha = \frac{h}{2} \Longrightarrow h = 2 \tan \alpha \Longrightarrow s = \frac{2 \tan \alpha}{\sin \alpha} = 2 \sec \alpha$$

$$I = \frac{k \sin \alpha}{s^2} = \frac{k \sin \alpha}{4 \sec^2 \alpha} = \frac{k}{4} \sin \alpha \cos^2 \alpha$$

$$\frac{dI}{d\alpha} = \frac{k}{4}\big[\sin \alpha(-2 \sin \alpha \cos \alpha) + \cos^2 \alpha(\cos \alpha)\big]$$

$$= \frac{k}{4} \cos \alpha[\cos^2 \alpha - 2 \sin^2 \alpha]$$

$$= \frac{k}{4} \cos \alpha[1 - 3 \sin^2 \alpha]$$

$$= 0 \text{ when } \alpha = \frac{\pi}{2}, \frac{3\pi}{2}, \text{ or when } \sin \alpha = \pm\frac{1}{\sqrt{3}}.$$

4 ft

Since α is acute, we have

$$\sin \alpha = \frac{1}{\sqrt{3}} \Longrightarrow h = 2 \tan \alpha = 2\left(\frac{1}{\sqrt{2}}\right) = \sqrt{2} \text{ feet.}$$

Since $(d^2I)/(d\alpha^2) = (k/4) \sin \alpha(9 \sin^2 \alpha - 7) < 0$ when $\sin \alpha = 1/\sqrt{3}$, this yields a maximum.

49. $$S = \sqrt{x^2 + 4}, L = \sqrt{1 + (3 - x)^2}$$

$$\text{Time} = T = \frac{\sqrt{x^2 + 4}}{2} + \frac{\sqrt{x^2 - 6x + 10}}{4}$$

$$\frac{dT}{dx} = \frac{x}{2\sqrt{x^2 + 4}} + \frac{x - 3}{4\sqrt{x^2 - 6x + 10}} = 0$$

$$\frac{x^2}{x^2 + 4} = \frac{9 - 6x + x^2}{4(x^2 - 6x + 10)}$$

$$x^4 - 6x^3 + 9x^2 + 8x - 12 = 0$$

You need to find the roots of this equation in the interval $[0, 3]$. By using a computer or graphics calculator, you can determine that this equation has only one root in this interval ($x = 1$). Testing at this value and at the endpoints, you see that $x = 1$ yields the minimum time. Thus, the man should row to a point 1 mile from the nearest point on the coast.

51. $T = \dfrac{\sqrt{x^2 + 4}}{v_1} + \dfrac{\sqrt{x^2 - 6x + 10}}{v_2}$

$\dfrac{dT}{dx} = \dfrac{x}{v_1\sqrt{x^2 + 4}} + \dfrac{x - 3}{v_2\sqrt{x^2 - 6x + 10}} = 0$

Since

$$\dfrac{x}{\sqrt{x^2 + 4}} = \sin\theta_1 \text{ and } \dfrac{x - 3}{\sqrt{x^2 - 6x + 10}} = -\sin\theta_2$$

we have

$$\dfrac{\sin\theta_1}{v_1} - \dfrac{\sin\theta_2}{v_2} = 0 \Rightarrow \dfrac{\sin\theta_1}{v_1} = \dfrac{\sin\theta_2}{v_2}.$$

Since

$$\dfrac{d^2T}{dx^2} = \dfrac{4}{v_1(x^2 + 4)^{3/2}} + \dfrac{1}{v_2(x^2 - 6x + 10)^{3/2}} > 0$$

this condition yields a minimum time.

53. $f(x) = 2 - 2\sin x$

(a) Distance from origin to y-intercept is 2.
Distance from origin to x-intercept is $\pi/2 \approx 1.57$.

(b) $d = \sqrt{x^2 + y^2} = \sqrt{x^2 + (2 - 2\sin x)^2}$

Minimum distance $= 0.9795$ at $x = 0.7967$.

(c) Let $f(x) = d^2(x) = x^2 + (2 - 2\sin x)^2$.

$f'(x) = 2x + 2(2 - 2\sin x)(-2\cos x)$

Setting $f'(x) = 0$, you obtain $x \approx 0.7967$, which corresponds to $d = 0.9795$.

55. $F\cos\theta = k(W - F\sin\theta)$

$F = \dfrac{kW}{\cos\theta + k\sin\theta}$

$\dfrac{dF}{d\theta} = \dfrac{-kW(k\cos\theta - \sin\theta)}{(\cos\theta + k\sin\theta)^2} = 0$

$k\cos\theta = \sin\theta \Rightarrow k = \tan\theta \Rightarrow \theta = \arctan k$

Since

$$\cos\theta + k\sin\theta = \dfrac{1}{\sqrt{k^2 + 1}} + \dfrac{k^2}{\sqrt{k^2 + 1}} = \sqrt{k^2 + 1},$$

the minimum force is

$$F = \dfrac{kW}{\cos\theta + k\sin\theta} = \dfrac{kW}{\sqrt{k^2 + 1}}.$$

57. (a)

Base 1	Base 2	Altitude	Area
8	$8 + 16 \cos 10°$	$8 \sin 10°$	≈ 22.1
8	$8 + 16 \cos 20°$	$8 \sin 20°$	≈ 42.5
8	$8 + 16 \cos 30°$	$8 \sin 30°$	≈ 59.7
8	$8 + 16 \cos 40°$	$8 \sin 40°$	≈ 72.7
8	$8 + 16 \cos 50°$	$8 \sin 50°$	≈ 80.5
8	$8 + 16 \cos 60°$	$8 \sin 60°$	≈ 83.1

(b)

Base 1	Base 2	Altitude	Area
8	$8 + 16 \cos 10°$	$8 \sin 10°$	≈ 22.1
8	$8 + 16 \cos 20°$	$8 \sin 20°$	≈ 42.5
8	$8 + 16 \cos 30°$	$8 \sin 30°$	≈ 59.7
8	$8 + 16 \cos 40°$	$8 \sin 40°$	≈ 72.7
8	$8 + 16 \cos 50°$	$8 \sin 50°$	≈ 80.5
8	$8 + 16 \cos 60°$	$8 \sin 60°$	≈ 83.1
8	$8 + 16 \cos 70°$	$8 \sin 70°$	≈ 80.7
8	$8 + 16 \cos 80°$	$8 \sin 80°$	≈ 74.0
8	$8 + 16 \cos 90°$	$8 \sin 90°$	$= 64.0$

The maximum cross-sectional area is approximately 83.1 square feet.

(c) $A = (a + b)\dfrac{h}{2}$

$= [8 + (8 + 16 \cos \theta)]\dfrac{8 \sin \theta}{2}$

$= 64(1 + \cos \theta)\sin \theta,\ 0° < \theta < 90°$

(d) $\dfrac{dA}{d\theta} = 64(1 + \cos \theta)\cos \theta + (-64 \sin \theta)\sin \theta$

$= 64(\cos \theta + \cos^2 \theta - \sin^2 \theta)$

$= 64(2 \cos^2 \theta + \cos \theta - 1)$

$= 64(2 \cos \theta - 1)(\cos \theta + 1)$

$= 0$ when $\theta = 60°, 180°, 300°$.

The maximum occurs when $\theta = 60°$.

(e)

59. $C = 100\left(\dfrac{200}{x^2} + \dfrac{x}{x + 30}\right),\ 1 \le x$

$C' = 100\left(-\dfrac{400}{x^3} + \dfrac{30}{(x + 30)^2}\right)$

Approximation: $x \approx 40.45$ units, or 4045 units

61. $S_1 = (4m - 1)^2 + (5m - 6)^2 + (10m - 3)^2$

$\dfrac{dS_1}{dm} = 2(4m - 1)(4) + 2(5m - 6)(5) + 2(10m - 3)(10) = 282m - 128 = 0$ when $m = \dfrac{64}{141}$.

Line: $y = \dfrac{64}{141}x$

$S = \left|4\left(\dfrac{64}{141}\right) - 1\right| + \left|5\left(\dfrac{64}{141}\right) - 6\right| + \left|10\left(\dfrac{64}{141}\right) - 3\right|$

$= \left|\dfrac{256}{141} - 1\right| + \left|\dfrac{320}{141} - 6\right| + \left|\dfrac{640}{141} - 3\right| = \dfrac{858}{141} \approx 6.1$ mi.

63. $S_3 = \dfrac{|4m - 1|}{\sqrt{m^2 + 1}} + \dfrac{|5m - 6|}{\sqrt{m^2 + 1}} + \dfrac{|10m - 3|}{\sqrt{m^2 + 1}}$

Using a graphing utility, you can see that the minimum occurs when $x \approx 0.3$.

Line: $y \approx 0.3x$

$S_3 = \dfrac{|4(0.3) - 1| + |5(0.3) - 6| + |10(0.3) - 3|}{\sqrt{(0.3)^2 + 1}} \approx 4.5$ mi.

Section 3.8   Newton's Method

1. $f(x) = x^2 - 3$

$f'(x) = 2x$

$x_1 = 1.7$

n	x_n	$f(x_n)$	$f'(x_n)$	$\dfrac{f(x_n)}{f'(x_n)}$	$x_n - \dfrac{f(x_n)}{f'(x_n)}$
1	1.7000	-0.1100	3.4000	-0.0324	1.7324
2	1.7324	0.0012	3.4648	0.0003	1.7321

3. $f(x) = \sin x$

$f'(x) = \cos x$

$x_1 = 3$

n	x_n	$f(x_n)$	$f'(x_n)$	$\dfrac{f(x_n)}{f'(x_n)}$	$x_n - \dfrac{f(x_n)}{f'(x_n)}$
1	3.0000	0.1411	-0.9900	-0.1425	3.1425
2	3.1425	-0.0009	-1.0000	0.0009	3.1416

5. $f(x) = x^3 + x - 1$

$f'(x) = 3x^2 + 1$

Approximation of the zero of f is 0.682.

n	x_n	$f(x_n)$	$f'(x_n)$	$\dfrac{f(x_n)}{f'(x_n)}$	$x_n - \dfrac{f(x_n)}{f'(x_n)}$
1	0.5000	-0.3750	1.7500	-0.2143	0.7143
2	0.7143	0.0788	2.5307	0.0311	0.6832
3	0.6832	0.0021	2.4003	0.0009	0.6823

7. $f(x) = 3\sqrt{x - 1} - x$

$f'(x) = \dfrac{3}{2\sqrt{x - 1}} - 1$

Approximation of the zero of f is 1.146.

Similarly, the other zero is approximately 7.854.

n	x_n	$f(x_n)$	$f'(x_n)$	$\dfrac{f(x_n)}{f'(x_n)}$	$x_n - \dfrac{f(x_n)}{f'(x_n)}$
1	1.2000	0.1416	2.3541	0.0602	1.1398
2	1.1398	-0.0181	3.0118	-0.0060	1.1458
3	1.1458	-0.0003	2.9284	-0.0001	1.1459

9. $f(x) = x^3 + 3$

$f'(x) = 3x^2$

Approximation of the zero of f is -1.442.

n	x_n	$f(x_n)$	$f'(x_n)$	$\dfrac{f(x_n)}{f'(x_n)}$	$x_n - \dfrac{f(x_n)}{f'(x_n)}$
1	-1.5000	-0.3750	6.7500	-0.0556	-1.4444
2	-1.4444	-0.0134	6.2589	-0.0021	-1.4423
3	-1.4423	-0.0003	6.2407	-0.0001	-1.4422

11. $f(x) = x^3 - 3.9x^2 + 4.79x - 1.881$

$f'(x) = 3x^2 - 7.8x + 4.79$

n	x_n	$f(x_n)$	$f'(x_n)$	$\dfrac{f(x_n)}{f'(x_n)}$	$x_n - \dfrac{f(x_n)}{f'(x_n)}$
1	0.5000	-0.3360	1.6400	-0.2049	0.7049
2	0.7049	-0.0921	0.7824	-0.1177	0.8226
3	0.8226	-0.0231	0.4037	-0.0573	0.8799
4	0.8799	-0.0045	0.2495	-0.0181	0.8980
5	0.8980	-0.0004	0.2048	-0.0020	0.9000
6	0.9000	0.0000	0.2000	0.0000	0.9000

Approximation of the zero of f is 0.900.

—CONTINUED—

11. —CONTINUED—

n	x_n	$f(x_n)$	$f'(x_n)$	$\dfrac{f(x_n)}{f'(x_n)}$	$x_n - \dfrac{f(x_n)}{f'(x_n)}$
1	1.1	0.0000	-0.1600	-0.0000	1.1000

Approximation of the zero of f is 1.100.

n	x_n	$f(x_n)$	$f'(x_n)$	$\dfrac{f(x_n)}{f'(x_n)}$	$x_n - \dfrac{f(x_n)}{f'(x_n)}$
1	1.9	0.0000	0.8000	0.0000	1.9000

Approximation of the zero of f is 1.900.

13. $f(x) = x + \sin(x + 1)$

$f'(x) = 1 + \cos(x + 1)$

Approximation of the zero of f is -0.489.

n	x_n	$f(x_n)$	$f'(x_n)$	$\dfrac{f(x_n)}{f'(x_n)}$	$x_n - \dfrac{f(x_n)}{f'(x_n)}$
1	-0.5000	-0.0206	1.8776	-0.0110	-0.4890
2	-0.4890	0.0000	1.8723	0.0000	-0.4890

15. $h(x) = f(x) - g(x) = 2x + 1 - \sqrt{x + 4}$

$h'(x) = 2 - \dfrac{1}{2\sqrt{x + 4}}$

Point of intersection of the graphs of f and g occurs when $x \approx 0.569$.

n	x_n	$h(x_n)$	$h'(x_n)$	$\dfrac{h(x_n)}{h'(x_n)}$	$x_n - \dfrac{h(x_n)}{h'(x_n)}$
1	0.6000	0.0552	1.7669	0.0313	0.5687
2	0.5687	-0.0001	1.7661	0.0000	0.5687

17. $h(x) = f(x) - g(x) = x - \tan x$

$h'(x) = 1 - \sec^2 x$

Point of intersection of the graphs of f and g occurs when $x \approx 4.493$.

n	x_n	$h(x_n)$	$h'(x_n)$	$\dfrac{h(x_n)}{h'(x_n)}$	$x_n - \dfrac{h(x_n)}{h'(x_n)}$
1	4.5000	-0.1373	-21.5048	0.0064	4.4936
2	4.4936	-0.0039	-20.2271	0.0002	4.4934

19. $f(x) = x^2 - a = 0$

$f'(x) = 2x$

$x_{i+1} = x_i - \dfrac{x_i^2 - a}{2x_i}$

$= \dfrac{2x_i^2 - x_i^2 + a}{2x_i} = \dfrac{x_i^2 + a}{2x_i} = \dfrac{x_i}{2} + \dfrac{a}{2x_i}$

21. $x_{i+1} = \dfrac{x_i^2 + 7}{2x_i}$

i	1	2	3	4	5
x_i	2.0000	2.7500	2.6477	2.6458	2.6458

$\sqrt{7} \approx 2.646$

23. $x_{i+1} = \dfrac{3x_i^4 + 6}{4x_i^3}$

i	1	2	3	4
x_i	1.5000	1.5694	1.5651	1.5651

$\sqrt[4]{6} \approx 1.565$

25. $f(x) = 1 + \cos x$

$f'(x) = -\sin x$

Approximation of the zero: 3.141

n	x_n	$f(x_n)$	$f'(x_n)$	$\dfrac{f(x_n)}{f'(x_n)}$	$x_n - \dfrac{f(x_n)}{f'(x_n)}$
1	3.0000	0.0100	-0.1411	-0.0709	3.0709
2	3.0709	0.0025	-0.0706	-0.0354	3.1063
3	3.1063	0.0006	-0.0353	-0.0176	3.1239
4	3.1239	0.0002	-0.0177	-0.0088	3.1327
5	3.1327	0.0000	-0.0089	-0.0044	3.1371
6	3.1371	0.0000	-0.0045	-0.0022	3.1393
7	3.1393	0.0000	-0.0023	-0.0011	3.1404
8	3.1404	0.0000	-0.0012	-0.0006	3.1410

27. $y = 2x^3 - 6x^2 + 6x - 1 = f(x)$

$y' = 6x^2 - 12x + 6 = f'(x)$

$x_1 = 1$

$f'(x) = 0$; therefore, the method fails.

n	x_n	$f(x_n)$	$f'(x_n)$
1	1	1	0

29. $y = -x^3 + 6x^2 - 10x + 6 = f(x)$

$y' = -3x^2 + 12x - 10 = f'(x)$

$x_1 = 2$

$x_2 = 1$

$x_3 = 2$

$x_4 = 1$ and so on.

Fails to converge

31. Answers will vary. See page 222.

Newton's Method uses tangent lines to approximate c such that $f(c) = 0$.

First, estimate an initial x_1 close to c (see graph).

Then determine x_2 by $x_2 = x_1 - \dfrac{f(x_1)}{f'(x_1)}$.

Calculate a third estimate by $x_3 = x_2 - \dfrac{f(x_2)}{f'(x_2)}$.

Continue this process until $|x_n - x_{n+1}|$ is within the desired accuracy.

Let x_{n+1} be the final approximation of c.

33. Let $g(x) = f(x) - x = \cos x - x$

$g'(x) = -\sin x - 1$.

The fixed point is approximately 0.74.

n	x_n	$g(x_n)$	$g'(x_n)$	$\dfrac{g(x_n)}{g'(x_n)}$	$x_n - \dfrac{g(x_n)}{g'(x_n)}$
1	1.0000	-0.4597	-1.8415	0.2496	0.7504
2	0.7504	-0.0190	-1.6819	0.0113	0.7391
3	0.7391	0.0000	-1.6736	0.0000	0.7391

35. $f(x) = x^3 - 3x^2 + 3$, $f'(x) = 3x^2 - 6x$

(a)

(b) $x_1 = 1$

$$x_2 = x_1 - \frac{f(x_1)}{f'(x_1)} \approx 1.333$$

Continuing, the zero is 1.347.

(c) $x_1 = \frac{1}{4}$

$$x_2 = x_1 - \frac{f(x_1)}{f'(x_1)} \approx 2.405$$

Continuing, the zero is 2.532.

(e) If the initial guess x_1 is not "close to" the desired zero of the function, the x-intercept of the tangent line may approximate another zero of the function.

(d)

The x-intercepts correspond to the values resulting from the first iteration of Newton's Method.

37. $f(x) = \frac{1}{x} - a = 0$

$$f'(x) = -\frac{1}{x^2}$$

$$x_{n+1} = x_n - \frac{(1/x_n) - a}{-1/x_n^2} = x_n + x_n^2\left(\frac{1}{x_n} - a\right) = x_n + x_n - x_n^2 a = 2x_n - x_n^2 a = x_n(2 - ax_n)$$

39. $f(x) = x \cos x$, $(0, \pi)$

$f'(x) = -x \sin x + \cos x = 0$

Letting $F(x) = f'(x)$, we can use Newton's Method as follows.

$[F'(x) = -2 \sin x + x \cos x]$

n	x_n	$F(x_n)$	$F'(x_n)$	$\dfrac{F(x_n)}{F'(x_n)}$	$x_n - \dfrac{F(x_n)}{F'(x_n)}$
1	0.9000	-0.0834	-2.1261	0.0392	0.8608
2	0.8608	-0.0010	-2.0778	0.0005	0.8603

Approximation to the critical number: 0.860

41. $y = f(x) = 4 - x^2, (1, 0)$

$d = \sqrt{(x - 1)^2 + (y - 0)^2} = \sqrt{(x - 1)^2 + (4 - x^2)^2} = \sqrt{x^4 - 7x^2 - 2x + 17}$

d is minimized when $D = x^4 - 7x^2 - 2x + 17$ is a minimum.

$g(x) = D' = 4x^3 - 14x - 2$

$g'(x) = 12x^2 - 14$

n	x_n	$g(x_n)$	$g'(x_n)$	$\dfrac{g(x_n)}{g'(x_n)}$	$x_n - \dfrac{g(x_n)}{g'(x_n)}$
1	2.0000	2.0000	34.0000	0.0588	1.9412
2	1.9412	0.0830	31.2191	0.0027	1.9385
3	1.9385	−0.0012	31.0934	0.0000	1.9385

$x \approx 1.939$

Point closest to $(1, 0)$ is $\approx (1.939, 0.240)$.

43.

$$\text{Minimize: } T = \frac{\text{Distance rowed}}{\text{Rate rowed}} + \frac{\text{Distance walked}}{\text{Rate walked}}$$

$$T = \frac{\sqrt{x^2 + 4}}{3} + \frac{\sqrt{x^2 - 6x + 10}}{4}$$

$$T' = \frac{x}{3\sqrt{x^2 + 4}} + \frac{x - 3}{4\sqrt{x^2 - 6x + 10}} = 0$$

$$4x\sqrt{x^2 - 6x + 10} = -3(x - 3)\sqrt{x^2 + 4}$$

$$16x^2(x^2 - 6x + 10) = 9(x - 3)^2(x^2 + 4)$$

$7x^4 - 42x^3 + 43x^2 + 216x - 324 = 0$

Let $f(x) = 7x^4 - 42x^3 + 43x^2 + 216x - 324$ and $f'(x) = 28x^3 - 126x^2 + 86x + 216$. Since $f(1) = -100$ and $f(2) = 56$, the solution is in the interval $(1, 2)$.

n	x_n	$f(x_n)$	$f'(x_n)$	$\dfrac{f(x_n)}{f'(x_n)}$	$x_n - \dfrac{f(x_n)}{f'(x_n)}$
1	1.7000	19.5887	135.6240	0.1444	1.5556
2	1.5556	−1.0480	150.2780	−0.0070	1.5626
3	1.5626	0.0014	49.5591	0.0000	1.5626

Approximation: $x \approx 1.563$ miles

45.

$$2,500,000 = -76x^3 + 4830x^2 - 320,000$$

$76x^3 - 4830x^2 + 2,820,000 = 0$

Let $f(x) = 76x^3 - 4830x^2 + 2,820,000$

$f'(x) = 228x^2 - 9660x$.

From the graph, choose $x_1 = 40$.

n	x_n	$f(x_n)$	$f'(x_n)$	$\dfrac{f(x_n)}{f'(x_n)}$	$x_n - \dfrac{f(x_n)}{f'(x_n)}$
1	40.0000	−44000.0000	−21600.0000	2.0370	37.9630
2	37.9630	17157.6209	−38131.4039	−0.4500	38.4130
3	38.4130	780.0914	−34642.2263	−0.0225	38.4355
4	38.4355	2.6308	−34465.3435	−0.0001	38.4356

The zero occurs when $x \approx 38.4356$ which corresponds to \$384,356.

47. False. Let $f(x) = (x^2 - 1)/(x - 1)$. $x = 1$ is a discontinuity. It is not a zero of $f(x)$. This statement would be true if $f(x) = p(x)/q(x)$ is given in **reduced** form.

49. True

51. $f(x) = \frac{1}{4}x^3 - 3x^2 + \frac{3}{4}x - 2$

$f'(x) = \frac{3}{4}x^2 - 6x + \frac{3}{4}$

Let $x_1 = 12$.

n	x_n	$f(x_n)$	$f'(x_n)$	$\dfrac{f(x_n)}{f'(x_n)}$	$x_n - \dfrac{f(x_n)}{f'(x_n)}$
1	12.0000	7.0000	36.7500	0.1905	11.8095
2	11.8095	0.2151	34.4912	0.0062	11.8033
3	11.8033	0.0015	34.4186	0.0000	11.8033

Approximation: $x \approx 11.803$

Section 3.9   Differentials

1. $f(x) = x^2$

$f'(x) = 2x$

Tangent line at $(2, 4)$: $y - f(2) = f'(2)(x - 2)$

$\qquad\qquad\qquad\qquad y - 4 = 4(x - 2)$

$\qquad\qquad\qquad\qquad\quad y = 4x - 4$

x	1.9	1.99	2	2.01	2.1
$f(x) = x^2$	3.6100	3.9601	4	4.0401	4.4100
$T(x) = 4x - 4$	3.6000	3.9600	4	4.0400	4.4000

3. $f(x) = x^5$

$f'(x) = 5x^4$

Tangent line at $(2, 32)$: $y - f(2) = f'(2)(x - 2)$

$\qquad\qquad\qquad\qquad\quad y - 32 = 80(x - 2)$

$\qquad\qquad\qquad\qquad\qquad\quad y = 80x - 128$

x	1.9	1.99	2	2.01	2.1
$f(x) = x^5$	24.7610	31.2080	32	32.8080	40.8410
$T(x) = 80x - 128$	24.0000	31.2000	32	32.8000	40.0000

5. $f(x) = \sin x$

$f'(x) = \cos x$

Tangent line at $(2, \sin 2)$:

$\qquad y - f(2) = f'(2)(x - 2)$

$\quad y - \sin 2 = (\cos 2)(x - 2)$

$\qquad\qquad y = (\cos 2)(x - 2) + \sin 2$

x	1.9	1.99	2	2.01	2.1
$f(x) = \sin x$	0.9463	0.9134	0.9093	0.9051	0.8632
$T(x) = (\cos 2)(x - 2) + \sin 2$	0.9509	0.9135	0.9093	0.9051	0.8677

7. $y = f(x) = \frac{1}{2}x^3, f'(x) = \frac{3}{2}x^2, x = 2, \Delta x = dx = 0.1$

$\qquad \Delta y = f(x + \Delta x) - f(x)$

$\qquad\qquad = f(2.1) - f(2)$

$\qquad\qquad = 0.6305$

$\qquad\qquad\qquad\qquad\qquad dy = f'(x)dx$

$\qquad\qquad\qquad\qquad\qquad\quad = f'(2)(0.1)$

$\qquad\qquad\qquad\qquad\qquad\quad = 6(0.1) = 0.6$

9. $y = f(x) = x^4 + 1, f'(x) = 4x^3, x = -1, \Delta x = dx = 0.01$

$$\Delta y = f(x + \Delta x) - f(x)$$
$$= f(-0.99) - f(-1)$$
$$= [(-0.99)^4 + 1] - [(-1)^4 + 1] \approx -0.0394$$

$$dy = f'(x)\,dx$$
$$= f'(-1)(0.01)$$
$$= (-4)(0.01) = -0.04$$

11. $y = 3x^2 - 4$

$dy = 6x\,dx$

13. $y = \dfrac{x + 1}{2x - 1}$

$$dy = \dfrac{-3}{(2x - 1)^2}\,dx$$

15. $y = x\sqrt{1 - x^2}$

$$dy = \left(x\dfrac{-x}{\sqrt{1 - x^2}} + \sqrt{1 - x^2}\right)dx = \dfrac{1 - 2x^2}{\sqrt{1 - x^2}}\,dx$$

17. $y = 2x - \cot^2 x$

$$dy = (2 + 2\cot x \csc^2 x)dx$$
$$= (2 + 2\cot x + 2\cot^3 x)dx$$

19. $y = \dfrac{1}{3}\cos\left(\dfrac{6\pi x - 1}{2}\right)$

$$dy = -\pi \sin\left(\dfrac{6\pi x - 1}{2}\right)dx$$

21. (a) $f(1.9) = f(2 - 0.1) \approx f(2) + f'(2)(-0.1)$
$$\approx 1 + (1)(-0.1) = 0.9$$
  (b) $f(2.04) = f(2 + 0.04) \approx f(2) + f'(2)(0.04)$
$$\approx 1 + (1)(0.04) = 1.04$$

23. (a) $f(1.9) = f(2 - 0.1) \approx f(2) + f'(2)(-0.1)$
$$\approx 1 + \left(-\tfrac{1}{2}\right)(-0.1) = 1.05$$
  (b) $f(2.04) = f(2 + 0.04) \approx f(2) + f'(2)(0.04)$
$$\approx 1 + \left(-\tfrac{1}{2}\right)(0.04) = 0.98$$

25. (a) $g(2.93) = g(3 - 0.07) \approx g(3) + g'(3)(-0.07)$
$$\approx 8 + \left(-\tfrac{1}{2}\right)(-0.07) = 8.035$$
  (b) $g(3.1) = g(3 + 0.1) \approx g(3) + g'(3)(0.1)$
$$\approx 8 + \left(-\tfrac{1}{2}\right)(0.1) = 7.95$$

27. (a) $g(2.93) = g(3 - 0.07) \approx g(3) + g'(3)(-0.07)$
$$\approx 8 + 0(-0.07) = 8$$
  (b) $g(3.1) = g(3 + 0.1) \approx g(3) + g'(3)(0.1)$
$$\approx 8 + 0(0.1) = 8$$

29. $A = x^2$

$x = 12$

$\Delta x = dx = \pm\dfrac{1}{64}$

$dA = 2x\,dx$

$\Delta A \approx dA = 2(12)\left(\pm\tfrac{1}{64}\right)$

$= \pm\dfrac{3}{8}$ square inches

31. $A = \pi r^2$

$r = 14$

$\Delta r = dr = \pm\dfrac{1}{4}$

$\Delta A \approx dA = 2\pi r\,dr = \pi(28)\left(\pm\tfrac{1}{4}\right)$

$= \pm 7\pi$ square inches

33. (a) $x = 15$ centimeter

$\Delta x = dx = \pm 0.05$ centimeters

$A = x^2$

$dA = 2x\,dx = 2(15)(\pm 0.05)$

$= \pm 1.5$ square centimeters

Percentage error:

$$\frac{dA}{A} = \frac{1.5}{(15)^2} = 0.00666\ldots = \frac{2}{3}\%$$

(b) $\dfrac{dA}{A} = \dfrac{2x\,dx}{x^2} = \dfrac{2\,dx}{x} \le 0.025$

$\dfrac{dx}{x} \le \dfrac{0.025}{2} = 0.0125 = 1.25\%$

35. $r = 6$ inches

$\Delta r = dr = \pm 0.02$ inches

(a) $V = \dfrac{4}{3}\pi r^3$

$dV = 4\pi r^2\,dr = 4\pi(6)^2(\pm 0.02) = \pm 2.88\pi$ cubic inches

(b) $S = 4\pi r^2$

$dS = 8\pi r\,dr = 8\pi(6)(\pm 0.02) = \pm 0.96\pi$ square inches

(c) Relative error: $\dfrac{dV}{V} = \dfrac{4\pi r^2\,dr}{(4/3)\pi r^3} = \dfrac{3dr}{r}$

$$= \frac{3}{6}(0.02) = 0.01 = 1\%$$

Relative error: $\dfrac{dS}{S} = \dfrac{8\pi r\,dr}{4\pi r^2} = \dfrac{2dr}{r}$

$$= \frac{2(0.02)}{6} = 0.00666\ldots = \frac{2}{3}\%$$

37. $V = \pi r^2 h = 40\pi r^2$, $r = 5$ cm, $h = 40$ cm, $dr = 0.2$ cm

$\Delta V \approx dV = 80\pi r\,dr = 80\pi(5)(0.2) = 80\pi$ cm^3

39. (a) $T = 2\pi\sqrt{L/g}$

$dT = \dfrac{\pi}{g\sqrt{L/g}}\,dL$

Relative error:

$$\frac{dT}{T} = \frac{(\pi\,dL)/\left(g\sqrt{L/g}\right)}{2\pi\sqrt{L/g}}$$

$$= \frac{dL}{2L}$$

$$= \frac{1}{2}\,(\text{relative error in } L)$$

$$= \frac{1}{2}(0.005) = 0.0025$$

Percentage error: $\dfrac{dT}{T}(100) = 0.25\% = \dfrac{1}{4}\%$

(b) $(0.0025)(3600)(24) = 216$ seconds

$= 3.6$ minutes

41. $\theta = 26°45' = 26.75°$

$d\theta = \pm 15' = \pm 0.25°$

(a) $h = 9.5\csc\theta$

$dh = -9.5\csc\theta\cot\theta\,d\theta$

$\dfrac{dh}{h} = -\cot\theta\,d\theta$

$\left|\dfrac{dh}{h}\right| = (\cot 26.75°)(0.25°)$

Converting to radians, $(\cot 0.4669)(0.0044)$

$\approx 0.0087 = 0.87\%$ (in radians).

(b) $\left|\dfrac{dh}{h}\right| = \cot\theta\,d\theta \le 0.02$

$\dfrac{d\theta}{\theta} \le \dfrac{0.02}{\theta(\cot\theta)} = \dfrac{0.02\tan\theta}{\theta}$

$\dfrac{d\theta}{\theta} \le \dfrac{0.02\tan 26.75°}{26.75°} \approx \dfrac{0.02\tan 0.4669}{0.4669}$

$\approx 0.0216 = 2.16\%$ (in radians)

43. $r = \dfrac{v_0{}^2}{32}(\sin 2\theta)$

$v_0 = 2200$ ft/sec

θ changes from $10°$ to $11°$

$dr = \dfrac{(2200)^2}{16}(\cos 2\theta)\,d\theta$

$\theta = 10\left(\dfrac{\pi}{180}\right)$

$d\theta = (11 - 10)\dfrac{\pi}{180}$

$\Delta r \approx dr$

$\quad = \dfrac{(2200)^2}{16}\cos\left(\dfrac{20\pi}{180}\right)\left(\dfrac{\pi}{180}\right) \approx 4961$ feet

$\quad \approx 4961$ feet

45. Let $f(x) = \sqrt{x}$, $x = 100$, $dx = -0.6$.

$f(x + \Delta x) \approx f(x) + f'(x)\,dx$

$\qquad\qquad = \sqrt{x} + \dfrac{1}{2\sqrt{x}}\,dx$

$f(x + \Delta x) = \sqrt{99.4}$

$\qquad\qquad \approx \sqrt{100} + \dfrac{1}{2\sqrt{100}}(-0.6) = 9.97$

Using a calculator: $\sqrt{99.4} \approx 9.96995$

47. Let $f(x) = \sqrt[4]{x}$, $x = 625$, $dx = -1$.

$f(x + \Delta x) \approx f(x) + f'(x)\,dx = \sqrt[4]{x} + \dfrac{1}{4\sqrt[4]{x^3}}\,dx$

$f(x + \Delta x) = \sqrt[4]{624} \approx \sqrt[4]{625} + \dfrac{1}{4\left(\sqrt[4]{625}\right)^3}(-1)$

$\qquad\qquad = 5 - \dfrac{1}{500} = 4.998$

Using a calculator, $\sqrt[4]{624} \approx 4.9980$.

49. Let $f(x) = \sqrt{x}$, $x = 4$, $dx = 0.02$, $f'(x) = 1/\left(2\sqrt{x}\right)$.

Then

$f(4.02) \approx f(4) + f'(4)\,dx$

$\sqrt{4.02} \approx \sqrt{4} + \dfrac{1}{2\sqrt{4}}(0.02) = 2 + \dfrac{1}{4}(0.02)$.

51. In general, when $\Delta x \to 0$, dy approaches Δy.

53. True

55. True

Review Exercises for Chapter 3

1. A number c in the domain of f is a critical number if $f'(c) = 0$ or f' is undefined at c.

3. $g(x) = 2x + 5\cos x$, $[0, 2\pi]$

$g'(x) = 2 - 5\sin x$

$\qquad = 0$ when $\sin x = \frac{2}{5}$.

Critical numbers: $x \approx 0.41$, $x \approx 2.73$

Left endpoint: $(0, 5)$

Critical number: $(0.41, 5.41)$

Critical number: $(2.73, 0.88)$ Minimum

Right endpoint: $(2\pi, 17.57)$ Maximum

5. Yes. $f(-3) = f(2) = 0$. f is continuous on $[-3, 2]$, differentiable on $(-3, 2)$.

$$f'(x) = (x + 3)(3x - 1) = 0 \text{ for } x = \tfrac{1}{3}.$$

$c = \tfrac{1}{3}$ satisfies $f'(c) - 0$.

7. $f(x) = 3 - |x - 4|$

(a)

$$f(1) = f(7) = 0$$

(b) f is not differentiable at $x = 4$.

9.
$$f(x) = x^{2/3}, 1 \le x \le 8$$

$$f'(x) = \frac{2}{3}x^{-1/3}$$

$$\frac{f(b) - f(a)}{b - a} = \frac{4 - 1}{8 - 1} = \frac{3}{7}$$

$$f'(c) = \frac{2}{3}c^{-1/3} = \frac{3}{7}$$

$$c = \left(\frac{14}{9}\right)^3 = \frac{2744}{729} \approx 3.764$$

11.
$$f(x) = x - \cos x, -\frac{\pi}{2} \le x \le \frac{\pi}{2}$$

$$f'(x) = 1 + \sin x$$

$$\frac{f(b) - f(a)}{b - a} = \frac{(\pi/2) - (-\pi/2)}{(\pi/2) - (-\pi/2)} = 1$$

$$f'(c) = 1 + \sin c = 1$$

$$c = 0$$

13.
$$f(x) = Ax^2 + Bx + C$$

$$f'(x) = 2Ax + B$$

$$\frac{f(x_2) - f(x_1)}{x_2 - x_1} = \frac{A(x_2^2 - x_1^2) + B(x_2 - x_1)}{x_2 - x_1}$$

$$= A(x_1 + x_2) + B$$

$$f'(c) = 2Ac + B = A(x_1 + x_2) + B$$

$$2Ac = A(x_1 + x_2)$$

$$c = \frac{x_1 + x_2}{2} = \text{Midpoint of } [x_1, x_2]$$

15. $f(x) = (x - 1)^2(x - 3)$

$$f'(x) = (x - 1)^2(1) + (x - 3)(2)(x - 1)$$

$$= (x - 1)(3x - 7)$$

Critical numbers: $x = 1$ and $x = \tfrac{7}{3}$

Interval:	$-\infty < x < 1$	$1 < x < \frac{7}{3}$	$\frac{7}{3} < x < \infty$
Sign of $f'(x)$:	$f'(x) > 0$	$f'(x) < 0$	$f'(x) > 0$
Conclusion:	Increasing	Decreasing	Increasing

17. $h(x) = \sqrt{x}(x - 3) = x^{3/2} - 3x^{1/2}$

Domain: $(0, \infty)$

$$h'(x) = \frac{3}{2}x^{1/2} - \frac{3}{2}x^{-1/2}$$

$$= \frac{3}{2}x^{-1/2}(x - 1) = \frac{3(x - 1)}{2\sqrt{x}}$$

Critical number: $x = 1$

Interval:	$0 < x < 1$	$1 < x < \infty$
Sign of $h'(x)$:	$h'(x) < 0$	$h'(x) > 0$
Conclusion:	Decreasing	Increasing

19. $h(t) = \frac{1}{4}t^4 - 8t$

$h'(t) = t^3 - 8 = 0$ when $t = 2$.

Relative minimum: $(2, -12)$

Test Interval:	$-\infty < t < 2$	$2 < t < \infty$
Sign of $h'(t)$:	$h'(t) < 0$	$h'(t) > 0$
Conclusion:	Decreasing	Increasing

21. $y = \frac{1}{3}\cos(12t) - \frac{1}{4}\sin(12t)$

$v = y' = -4\sin(12t) - 3\cos(12t)$

(a) When $t = \frac{\pi}{8}$, $y = \frac{1}{4}$ inch and $v = y' = 4$ inches/second.

(b) $y' = -4\sin(12t) - 3\cos(12t) = 0$ when $\dfrac{\sin(12t)}{\cos(12t)} = -\dfrac{3}{4} \Rightarrow \tan(12t) = -\dfrac{3}{4}$.

Therefore, $\sin(12t) = -\dfrac{3}{5}$ and $\cos(12t) = \dfrac{4}{5}$. The maximum displacement is

$$y = \left(\frac{1}{3}\right)\left(\frac{4}{5}\right) - \frac{1}{4}\left(-\frac{3}{5}\right) = \frac{5}{12} \text{ inch.}$$

(c) Period: $\dfrac{2\pi}{12} = \dfrac{\pi}{6}$

Frequency: $\dfrac{1}{\pi/6} = \dfrac{6}{\pi}$

23. $f(x) = x + \cos x$, $0 \le x \le 2\pi$

$f'(x) = 1 - \sin x$

$f''(x) = -\cos x = 0$ when $x = \dfrac{\pi}{2}, \dfrac{3\pi}{2}$.

Points of inflection: $\left(\dfrac{\pi}{2}, \dfrac{\pi}{2}\right), \left(\dfrac{3\pi}{2}, \dfrac{3\pi}{2}\right)$

Test Interval:	$0 < x < \dfrac{\pi}{2}$	$\dfrac{\pi}{2} < x < \dfrac{3\pi}{2}$	$\dfrac{3\pi}{2} < x < 2\pi$
Sign of $f''(x)$:	$f''(x) < 0$	$f''(x) > 0$	$f''(x) < 0$
Conclusion:	Concave downward	Concave upward	Concave downward

25. $g(x) = 2x^2(1 - x^2)$

$g'(x) = -4x(2x^2 - 1)$ Critical numbers: $x = 0, \pm\dfrac{1}{\sqrt{2}}$

$g''(x) = 4 - 24x^2$

$g''(0) = 4 > 0$ Relative minimum at $(0, 0)$

$g''\left(\pm\dfrac{1}{\sqrt{2}}\right) = -8 < 0$ Relative maximums at $\left(\pm\dfrac{1}{\sqrt{2}}, \dfrac{1}{2}\right)$

27.

29. The first derivative is positive and the second derivative is negative. The graph is increasing and is concave down.

31. (a) $D = 0.0034t^4 - 0.2352t^3 + 4.9423t^2 - 20.8641t + 94.4025$

(b)

(c) Maximum at $(21.9, 319.5)$ (≈ 1992)

Minimum at $(2.6, 69.6)$ (≈ 1972)

(d) Outlays increasing at greatest rate at the point of inflection $(9.8, 173.7)$ (≈ 1979)

33. $\lim\limits_{x \to \infty} \dfrac{2x^2}{3x^2 + 5} = \lim\limits_{x \to \infty} \dfrac{2}{3 + 5/x^2} = \dfrac{2}{3}$

35. $\lim\limits_{x \to \infty} \dfrac{5 \cos x}{x} = 0$, since $|5 \cos x| \le 5$.

37. $h(x) = \dfrac{2x + 3}{x - 4}$

Discontinuity: $x = 4$

$$\lim\limits_{x \to \infty} \frac{2x + 3}{x - 4} = \lim\limits_{x \to \infty} \frac{2 + (3/x)}{1 - (4/x)} = 2$$

Vertical asymptote: $x = 4$

Horizontal asymptote: $y = 2$

39. $f(x) = \dfrac{3}{x} - 2$

Discontinuity: $x = 0$

$$\lim\limits_{x \to \infty} \left(\frac{3}{x} - 2 \right) = -2$$

Vertical asymptote: $x = 0$

Horizontal asymptote: $y = -2$

41. $f(x) = x^3 + \dfrac{243}{x}$

Relative minimum: $(3, 108)$

Relative maximum: $(-3, -108)$

Vertical asymptote: $x = 0$

43. $f(x) = \dfrac{x - 1}{1 + 3x^2}$

Relative minimum: $(-0.155, -1.077)$

Relative maximum: $(2.155, 0.077)$

Horizontal asymptote: $y = 0$

45. $f(x) = 4x - x^2 = x(4 - x)$

Domain: $(-\infty, \infty)$; Range: $(-\infty, 4]$

$f'(x) = 4 - 2x = 0$ when $x = 2$.

$f''(x) = -2$

Therefore, $(2, 4)$ is a relative maximum.

Intercepts: $(0, 0), (4, 0)$

47. $f(x) = x\sqrt{16 - x^2}$, Domain: $[-4, 4]$, Range: $[-8, 8]$

Domain: $[-4, 4]$; Range: $[-8, 8]$

$f'(x) = \dfrac{16 - 2x^2}{\sqrt{16 - x^2}} = 0$ when $x = \pm 2\sqrt{2}$ and undefined when $x = \pm 4$.

$f''(x) = \dfrac{2x(x^2 - 24)}{(16 - x^2)^{3/2}}$

$\quad f''(-2\sqrt{2}) > 0$

Therefore, $\left(-2\sqrt{2}, -8\right)$ is a relative minimum.

$\quad f''(2\sqrt{2}) < 0$

Therefore, $\left(2\sqrt{2}, 8\right)$ is a relative maximum.

Point of inflection: $(0, 0)$

Intercepts: $(-4, 0), (0, 0), (4, 0)$

Symmetry with respect to origin

49. $f(x) = (x - 1)^3(x - 3)^2$

Domain: $(-\infty, \infty)$; Range: $(-\infty, \infty)$

$f'(x) = (x - 1)^2(x - 3)(5x - 11) = 0$ when $x = 1, \dfrac{11}{5}, 3.$

$f''(x) = 4(x - 1)(5x^2 - 22x + 23) = 0$ when $x = 1, \dfrac{11 \pm \sqrt{6}}{5}.$

$\quad f''(3) > 0$

Therefore, $(3, 0)$ is a relative minimum.

$\quad f''\left(\dfrac{11}{5}\right) < 0$

Therefore, $\left(\dfrac{11}{5}, \dfrac{3456}{3125}\right)$ is a relative maximum.

Points of inflection: $(1, 0), \left(\dfrac{11 - \sqrt{6}}{5}, 0.60\right), \left(\dfrac{11 + \sqrt{6}}{5}, 0.46\right)$

Intercepts: $(0, -9), (1, 0), (3, 0)$

51. $f(x) = x^{1/3}(x + 3)^{2/3}$

Domain: $(-\infty, \infty)$; Range: $(-\infty, \infty)$

$f'(x) = \dfrac{x + 1}{(x + 3)^{1/3}x^{2/3}} = 0$ when $x = -1$ and undefined when $x = -3, 0.$

$f''(x) = \dfrac{-2}{x^{5/3}(x + 3)^{4/3}}$ is undefined when $x = 0, -3.$

By the First Derivative Test $(-3, 0)$ is a relative maximum and $\left(-1, -\sqrt[3]{4}\right)$ is a relative minimum. $(0, 0)$ is a point of inflection.

Intercepts: $(-3, 0), (0, 0)$

53. $f(x) = \dfrac{x+1}{x-1}$

Domain: $(-\infty, 1), (1, \infty)$; Range: $(-\infty, 1), (1, \infty)$

$f'(x) = \dfrac{-2}{(x-1)^2} < 0$ if $x \neq 1$.

$f''(x) = \dfrac{4}{(x-1)^3}$

Horizontal asymptote: $y = 1$

Vertical asymptote: $x = 1$

Intercepts: $(-1, 0), (0, -1)$

55. $f(x) = \dfrac{4}{1+x^2}$

Domain: $(-\infty, \infty)$; Range: $(0, 4]$

$f'(x) = \dfrac{-8x}{(1+x^2)^2} = 0$ when $x = 0$.

$f''(x) = \dfrac{-8(1-3x^2)}{(1+x^2)^3} = 0$ when $x = \pm\dfrac{\sqrt{3}}{3}$.

$f''(0) < 0$

Therefore, $(0, 4)$ is a relative maximum.

Points of inflection: $\left(\pm\sqrt{3}/3, 3\right)$

Intercept: $(0, 4)$

Symmetric to the y-axis

Horizontal asymptote: $y = 0$

57. $f(x) = x^3 + x + \dfrac{4}{x}$

Domain: $(-\infty, 0), (0, \infty)$; Range: $(-\infty, -6], [6, \infty)$

$f'(x) = 3x^2 + 1 - \dfrac{4}{x^2} = \dfrac{3x^4 + x^2 - 4}{x^2} = \dfrac{(3x^2+4)(x^2-1)}{x^2} = 0$ when $x = \pm 1$.

$f''(x) = 6x + \dfrac{8}{x^3} = \dfrac{6x^4 + 8}{x^3} \neq 0$

$f''(-1) < 0$

Therefore, $(-1, -6)$ is a relative maximum.

$f''(1) > 0$

Therefore, $(1, 6)$ is a relative minimum.

Vertical asymptote: $x = 0$

Symmetric with respect to origin

59. $f(x) = |x^2 - 9|$

Domain: $(-\infty, \infty)$; Range: $[0, \infty)$

$f'(x) = \dfrac{2x(x^2 - 9)}{|x^2 - 9|} = 0$ when $x = 0$ and is undefined when $x = \pm 3$.

$f''(x) = \dfrac{2(x^2 - 9)}{|x^2 - 9|}$ is undefined at $x = \pm 3$.

$\qquad f''(0) < 0$

Therefore, $(0, 9)$ is a relative maximum.

Relative minima: $(\pm 3, 0)$

Intercepts: $(\pm 3, 0)$, $(0, 9)$

Symmetric to the y-axis

61. $f(x) = x + \cos x$

Domain: $[0, 2\pi]$; Range: $[1, 1 + 2\pi]$

$f'(x) = 1 - \sin x \geq 0$, f is increasing.

$f''(x) = -\cos x = 0$ when $x = \dfrac{\pi}{2}, \dfrac{3\pi}{2}$.

Points of inflection: $\left(\dfrac{\pi}{2}, \dfrac{\pi}{2}\right), \left(\dfrac{3\pi}{2}, \dfrac{3\pi}{2}\right)$

Intercept: $(0, 1)$

63. $x^2 + 4y^2 - 2x - 16y + 13 = 0$

(a) $(x^2 - 2x + 1) + 4(y^2 - 4y + 4) = -13 + 1 + 16$

$\qquad (x - 1)^2 + 4(y - 2)^2 = 4$

$\qquad \dfrac{(x - 1)^2}{4} + \dfrac{(y - 2)^2}{1} = 1$

The graph is an ellipse:

Maximum: $(1, 3)$

Minimum: $(1, 1)$

(b) $x^2 + 4y^2 - 2x - 16y + 13 = 0$

$\qquad 2x + 8y\dfrac{dy}{dx} - 2 - 16\dfrac{dy}{dx} = 0$

$\qquad\qquad \dfrac{dy}{dx}(8y - 16) = 2 - 2x$

$\qquad\qquad\quad \dfrac{dy}{dx} = \dfrac{2 - 2x}{8y - 16} = \dfrac{1 - x}{4y - 8}$

The critical numbers are $x = 1$ and $y = 2$. These correspond to the points $(1, 1)$, $(1, 3)$, $(2, -1)$, and $(2, 3)$.
Hence, the maximum is $(1, 3)$ and the minimum is $(1, 1)$.

65. Let $t = 0$ at noon.

$$L = d^2 = (100 - 12t)^2 + (-10t)^2 = 10,000 - 2400t + 244t^2$$

$$\frac{dL}{dt} = -2400 + 488t = 0 \text{ when } t = \frac{300}{61} \approx 4.92 \text{ hr.}$$

Ship A at $(40.98, 0)$; Ship B at $(0, -49.18)$

$$d^2 = 10,000 - 2400t + 244t^2$$

$$\approx 4098.36 \text{ when } t \approx 4.92 \approx 4\!:\!55 \text{ P.M.}.$$

$$d \approx 64 \text{ km}$$

67. We have points $(0, y)$, $(x, 0)$, and $(1, 8)$. Thus,

$$m = \frac{y - 8}{0 - 1} = \frac{0 - 8}{x - 1} \text{ or } y = \frac{8x}{x - 1}.$$

Let $f(x) = L^2 = x^2 + \left(\frac{8x}{x - 1}\right)^2.$

$$f'(x) = 2x + 128\left(\frac{x}{x - 1}\right)\left[\frac{(x - 1) - x}{(x - 1)^2}\right] = 0$$

$$x - \frac{64x}{(x - 1)^3} = 0$$

$$x[(x - 1)^3 - 64] = 0 \text{ when } x = 0, 5 \text{ (minimum).}$$

Vertices of triangle: $(0, 0)$, $(5, 0)$, $(0, 10)$

69. $A = $ (Average of bases)(Height)

$$= \left(\frac{x + s}{2}\right)\frac{\sqrt{3s^2 + 2sx - x^2}}{2} \text{ (see figure)}$$

$$\frac{dA}{dx} = \frac{1}{4}\left[\frac{(s - x)(s + x)}{\sqrt{3s^2 + 2sx - x^2}} + \sqrt{3s^2 + 2sx - x^2}\right]$$

$$= \frac{2(2s - x)(s + x)}{4\sqrt{3s^2 + 2sx - x^2}} = 0 \text{ when } x = 2s.$$

A is a maximum when $x = 2s$.

71. You can form a right triangle with vertices $(0, 0)$, $(x, 0)$ and $(0, y)$.
Assume that the hypotenuse of length L passes through $(4, 6)$.

$$m = \frac{y - 6}{0 - 4} = \frac{6 - 0}{4 - x} \text{ or } y = \frac{6x}{x - 4}$$

Let $f(x) = L^2 = x^2 + y^2 = x^2 + \left(\frac{6x}{x - 4}\right)^2.$

$$f'(x) = 2x + 72\left(\frac{x}{x - 4}\right)\left[\frac{-4}{(x - 4)^2}\right] = 0$$

$$x[(x - 4)^3 - 144] = 0 \text{ when } x = 0 \text{ or } x = 4 + \sqrt[3]{144}.$$

$$L \approx 14.05 \text{ feet}$$

73.

$$\csc \theta = \frac{L_1}{6} \text{ or } L_1 = 6 \csc \theta \qquad \text{(see figure)}$$

$$L = L_1 + L_2 = \sec \theta \frac{L_2}{9} \text{ or } L_2 = 9 \sec \theta = 6 \csc \theta + 9 \sec \theta$$

$$\frac{dL}{d\theta} = -6 \csc \theta \cot \theta + 9 \sec \theta \tan \theta = 0$$

$$\tan^3 \theta = \frac{2}{3} \Longrightarrow \tan \theta = \frac{\sqrt[3]{2}}{\sqrt[3]{3}}$$

$$\sec \theta = \sqrt{1 + \tan^2 \theta} = \sqrt{1 + \left(\frac{2}{3}\right)^{2/3}} = \frac{\sqrt{3^{2/3} + 2^{2/3}}}{3^{1/3}}$$

$$\csc \theta = \frac{\sec \theta}{\tan \theta} = \frac{\sqrt{3^{2/3} + 2^{2/3}}}{2^{1/3}}$$

$$L = 6 \frac{(3^{2/3} + 2^{2/3})^{1/2}}{2^{1/3}} + 9 \frac{(3^{2/3} + 2^{2/3})^{1/2}}{3^{1/3}} = 3(3^{2/3} + 2^{2/3})^{3/2} \text{ ft} \approx 21.07 \text{ ft (Compare to Exercise 72 using } a = 9 \text{ and } b = 6.)$$

75. Total cost = (Cost per hour)(Number of hours)

$$T = \left(\frac{v^2}{600} + 5\right)\left(\frac{110}{v}\right) = \frac{11v}{60} + \frac{550}{v}$$

$$\frac{dT}{dv} = \frac{11}{60} - \frac{550}{v^2} = \frac{11v^2 - 33{,}000}{60v^2}$$

$$= 0 \text{ when } v = \sqrt{3000} = 10\sqrt{30} \approx 54.8 \text{ mph.}$$

$$\frac{d^2T}{dv^2} = \frac{1100}{v^3} > 0 \text{ when } v = 10\sqrt{30} \text{ so this value yields a minimum.}$$

77. $f(x) = x^3 - 3x - 1$

From the graph you can see that $f(x)$ has three real zeros.

$$f'(x) = 3x^2 - 3$$

n	x_n	$f(x_n)$	$f'(x_n)$	$\dfrac{f(x_n)}{f'(x_n)}$	$x_n - \dfrac{f(x_n)}{f'(x_n)}$
1	-1.5000	0.1250	3.7500	0.0333	-1.5333
2	-1.5333	-0.0049	4.0530	-0.0012	-1.5321

n	x_n	$f(x_n)$	$f'(x_n)$	$\dfrac{f(x_n)}{f'(x_n)}$	$x_n - \dfrac{f(x_n)}{f'(x_n)}$
1	-0.5000	0.3750	-2.2500	-0.1667	-0.3333
2	-0.3333	-0.0371	-2.6667	0.0139	-0.3472
3	-0.3472	-0.0003	-2.6384	0.0001	-0.3473

n	x_n	$f(x_n)$	$f'(x_n)$	$\dfrac{f(x_n)}{f'(x_n)}$	$x_n - \dfrac{f(x_n)}{f'(x_n)}$
1	1.9000	0.1590	7.8300	0.0203	1.8797
2	1.8797	0.0024	7.5998	0.0003	1.8794

The three real zeros of $f(x)$ are $x \approx -1.532$, $x \approx -0.347$, and $x \approx 1.879$.

79. Find the zeros of $f(x) = x^4 - x - 3$.

$f'(x) = 4x^3 - 1$

From the graph you can see that $f(x)$ has two real zeros.

f changes sign in $[-2, -1]$.

n	x_n	$f(x_n)$	$f'(x_n)$	$\dfrac{f(x_n)}{f'(x_n)}$	$x_n - \dfrac{f(x_n)}{f'(x_n)}$
1	-1.2000	0.2736	-7.9120	-0.0346	-1.1654
2	-1.1654	0.0100	-7.3312	-0.0014	-1.1640

On the interval $[-2, -1]$: $x \approx -1.164$.

f changes sign in $[1, 2]$.

n	x_n	$f(x_n)$	$f'(x_n)$	$\dfrac{f(x_n)}{f'(x_n)}$	$x_n - \dfrac{f(x_n)}{f'(x_n)}$
1	1.5000	0.5625	12.5000	0.0450	1.4550
2	1.4550	0.0268	11.3211	0.0024	1.4526
3	1.4526	-0.0003	11.2602	0.0000	1.4526

On the interval $[1, 2]$; $x \approx 1.453$.

81. $y = x(1 - \cos x) = x - x \cos x$

$\dfrac{dy}{dx} = 1 + x \sin x - \cos x$

$dy = (1 + x \sin x - \cos x)\, dx$

83.
$$S = 4\pi r^2.\ dr = \Delta r = \pm 0.025$$
$$dS = 8\pi r\, dr = 8\pi(9)(\pm 0.025)$$
$$= \pm 1.8\pi \text{ square cm}$$

$$\frac{dS}{S}(100) = \frac{8\pi r\, dr}{4\pi r^2}(100) = \frac{2\, dr}{r}(100)$$
$$= \frac{2(\pm 0.025)}{9}(100) \approx \pm 0.56\%$$

$$V = \frac{4}{3}\pi r^3$$

$$dV = 4\pi r^2\, dr = 4\pi(9)^2(\pm 0.025)$$
$$= \pm 8.1\pi \text{ cubic cm}$$

$$\frac{dV}{V}(100) = \frac{4\pi r^2\, dr}{(4/3)\pi r^3}(100) = \frac{3\, dr}{r}(100)$$
$$= \frac{3(\pm 0.025)}{9}(100) \approx \pm 0.83\%$$

Problem Solving for Chapter 3

1. Assume $y_1 < d < y_2$. Let $g(x) = f(x) - d(x - a)$. g is continuous on $[a, b]$ and therefore has a minimum $(c, g(c))$ on $[a, b]$. The point c cannot be an endpoint of $[a, b]$ because

$g'(a) = f'(a) - d = y_1 - d < 0$

$g'(b) = f'(b) - d = y_2 - d > 0$

Hence, $a < c < b$ and $g'(c) = 0 \implies f'(c) = d$.

3. (a) For $a = -3, -2, -1, 0$, p has a relative maximum at $(0, 0)$.

For $a = 1, 2, 3$, p has a relative maximum at $(0, 0)$ and 2 relative minima.

(b) $p'(x) = 4ax^3 - 12x = 4x(ax^2 - 3) = 0 \implies x = 0, \pm\sqrt{\dfrac{3}{a}}$

$p''(x) = 12ax^2 - 12 = 12(ax^2 - 1)$

For $x = 0$, $p''(0) = -12 < 0 \implies p$ has a relative maximum at $(0, 0)$.

(c) If $a > 0$, $x = \pm\sqrt{\dfrac{3}{a}}$ are the remaining critical numbers.

$p''\left(\pm\sqrt{\dfrac{3}{a}}\right) = 12a\left(\dfrac{3}{a}\right) - 12 = 24 > 0 \implies p$ has relative minima for $a > 0$.

(d) $(0, 0)$ lies on $y = -3x^2$.

Let $x = \pm\sqrt{\dfrac{3}{a}}$. Then

$$p(x) = a\left(\dfrac{3}{a}\right)^2 - 6\left(\dfrac{3}{a}\right) = \dfrac{9}{a} - \dfrac{18}{a} = -\dfrac{9}{a}.$$

Thus, $y = -\dfrac{9}{a} = -3\left(\pm\sqrt{\dfrac{3}{a}}\right)^2 = -3x^2$ is satisfied by all the relative extrema of p.

5. $p(x) = x^4 + ax^2 + 1$

(a) $p'(x) = 4x^3 + 2ax = 2x(2x^2 + a)$

$p''(x) = 12x^2 + 2a$

For $a \geq 0$, there is one relative minimum at $(0, 1)$.

(b) For $a < 0$, there is a relative maximum at $(0, 1)$.

(c) For $a < 0$, there are two relative minima at $x = \pm\sqrt{-\dfrac{a}{2}}$.

(d) There are either 1 or 3 critical points. The above analysis shows that there cannot be exactly two relative extrema.

7. $f(x) = \dfrac{c}{x} + x^2$

$f'(x) = -\dfrac{c}{x^2} + 2x = 0 \implies \dfrac{c}{x^2} = 2x \implies x^3 = \dfrac{c}{2} \implies x = \sqrt[3]{\dfrac{c}{x}}$

$f''(x) = \dfrac{2c}{x^3} + 2$

If $c = 0$, $f(x) = x^2$ has a relative minimum, but no relative maximum.

If $c > 0$, $x = \sqrt[3]{\dfrac{c}{2}}$ is a relative minimum, because $f''\left(\sqrt[3]{\dfrac{c}{2}}\right) > 0$.

If $c < 0$, $x = \sqrt[3]{\dfrac{c}{2}}$ is a relative minimum too.

Answer: all c.

9. Set $\dfrac{f(b) - f(a) - f'(a)(b - a)}{(b - a)^2} = k.$

Define $F(x) = f(x) - f(a) - f'(a)(x - a) - k(x - a)^2.$

$F(a) = 0,\ F(b) = f(b) - f(a) - f'(a)(b - a) - k(b - a)^2 = 0$

F is continuous on $[a, b]$ and differentiable on (a, b).

There exists $c_1,\ a < c_1 < b$, satisfying $F'(c_1) = 0.$

$F'(x) = f'(x) - f'(a) - 2k(x - a)$ satisfies the hypothesis of Rolle's Theorem on $[a, c_1]$:
$$F'(a) = 0,\ F'(c_1) = 0.$$

There exists $c_2,\ a < c_2 < c_1$ satisfying $F''(c_2) = 0.$

Finally, $F''(x) = f''(x) - 2k$ and $F''(c_2) = 0$ implies that
$$k = \frac{f''(c_2)}{2}.$$

Thus, $k = \dfrac{f(b) - f(a) - f'(a)(b - a)}{(b - a)^2} = \dfrac{f''(c_2)}{2} \Rightarrow f(b) = f(a) + f'(a)(b - a) + \dfrac{1}{2}f''(c_2)(b - a)^2.$

11. $E(\phi) = \dfrac{\tan\phi(1 - 0.1\tan\phi)}{0.1 + \tan\phi} = \dfrac{10\tan\phi - \tan^2\phi}{1 + 10\tan\phi}$

$E'(\phi) = \dfrac{(1 + 10\tan\phi)(10\sec^2\phi - 2\tan\phi\sec^2\phi) - (10\tan\phi - \tan^2\phi)10\sec^2\phi}{(1 + 10\tan\phi)} = 0$

$\Rightarrow (1 + 10\tan\phi)(10\sec^2\phi - 2\tan\phi\sec^2\phi) = (10\tan\phi - \tan^2\phi)10\sec^2\phi$

$\Rightarrow 10\sec^2\phi - 2\tan\phi\sec^2\phi + 100\tan\phi\sec^2\phi - 20\tan^2\phi\sec^2\phi$
$$= 100\tan\phi\sec^2\phi - 10\tan^2\phi\sec^2\phi$$

$\Rightarrow 10 - 2\tan\phi = 10\tan^2\phi$

$\Rightarrow 10\tan^2\phi + 2\tan\phi - 10 = 0$

$\tan\phi = \dfrac{-2 \pm \sqrt{4 + 400}}{20} \approx 0.90499,\ -1.10499$

Using the positive value, $\phi \approx 0.7356$, or $42.14°$.

13. $v = -2400\pi\sin\theta$

$v' = -2400\pi\cos\theta = 0$

$\theta = \dfrac{\pi}{2} + 2n\pi, \dfrac{3\pi}{2} + 2n\pi,\ n$ an integer

15. The line has equation $\frac{x}{3} + \frac{y}{4} = 1$ or $y = -\frac{4}{3}x + 4$.

Rectangle:

$$\text{Area} = A = xy = x\left(-\frac{4}{3}x + 4\right) = -\frac{4}{3}x^2 + 4x.$$

$$A'(x) = -\frac{8}{3}x + 4 = 0 \implies \frac{8}{3}x = 4 \implies x = \frac{3}{2}$$

Dimensions: $\frac{3}{2} \times 2$ Calculus was helpful.

Circle: The distance from the center (r, r) to the line $\frac{x}{3} + \frac{y}{4} - 1 = 0$ must be r:

$$r = \frac{\left|\dfrac{r}{3} + \dfrac{r}{4} - 1\right|}{\sqrt{\dfrac{1}{9} + \dfrac{1}{16}}} = \frac{12}{5}\left|\frac{7r - 12}{12}\right| = \frac{|7r - 12|}{5}$$

$$5r = |7r - 12| \implies r = 1 \text{ or } r = 6.$$

Clearly, $r = 1$.

Semicircle: The center lies on the line $\frac{x}{3} + \frac{y}{4} = 1$ and satisfies $x = y = r$.

Thus $\frac{r}{3} + \frac{r}{4} = 1 \implies \frac{7}{12}r = 1 \implies r = \frac{12}{7}$. No calculus necessary.

17. $y = (1 + x^2)^{-1}$

$$y' = \frac{-2x}{(1 + x^2)^2}$$

$$y'' = \frac{2(3x^2 - 1)}{(x^2 + 1)^3} = 0 \implies x = \pm\frac{1}{\sqrt{3}} = \pm\frac{\sqrt{3}}{3}$$

$$y'': \quad \overset{+++}{\underset{-\frac{\sqrt{3}}{3}}{\rule{0pt}{0pt}}} \overset{----}{\rule{0pt}{0pt}} \overset{----}{\underset{0}{\rule{0pt}{0pt}}} \overset{+++}{\underset{\frac{\sqrt{3}}{3}}{\rule{0pt}{0pt}}}$$

The tangent line has greatest slope at $\left(-\frac{\sqrt{3}}{3}, \frac{3}{4}\right)$ and least slope at $\left(\frac{\sqrt{3}}{3}, \frac{3}{4}\right)$.

19. (a)

x	0.1	0.2	0.3	0.4	0.5	1.0
$\sin x$	0.09983	0.19867	0.29552	0.38942	0.47943	0.84147

$\sin x < x$

(b) Let $f(x) = \sin x$. For $x \geq 1$, $\sin x < 1$ is obviously true, so assume $0 < x < 1$. Then $f'(x) = \cos x$ and on $[0, x]$ you have by the Mean Value Theorem,

$$f'(c) = \frac{f(x) - f(0)}{x - 0}, \quad 0 < c < x$$

$$\cos(c) = \frac{\sin x}{x}$$

Hence, $\left|\dfrac{\sin x}{x}\right| = |\cos(c)| < 1$

$\implies |\sin x|, < |x|$

$\implies \sin x < x$

C H A P T E R 4
Integration

Section 4.1 Antiderivatives and Indefinite Integration **177**

Section 4.2 Area **182**

Section 4.3 Riemann Sums and Definite Integrals **188**

Section 4.4 The Fundamental Theorem of Calculus **192**

Section 4.5 Integration by Substitution **197**

Section 4.6 Numerical Integration **204**

Review Exercises **209**

Problem Solving **214**

CHAPTER 4
Integration

Section 4.1   Antiderivatives and Indefinite Integration

Solutions to Odd-Numbered Exercises

1. $\dfrac{d}{dx}\left(\dfrac{3}{x^3} + C\right) = \dfrac{d}{dx}(3x^{-3} + C) = -9x^{-4} = \dfrac{-9}{x^4}$

3. $\dfrac{d}{dx}\left(\dfrac{1}{3}x^3 - 4x + C\right) = x^2 - 4 = (x - 2)(x + 2)$

5. $\dfrac{dy}{dt} = 3t^2$

$y = t^3 + C$

Check: $\dfrac{d}{dt}[t^3 + C] = 3t^2$

7. $\dfrac{dy}{dx} = x^{3/2}$

$y = \dfrac{2}{5}x^{5/2} + C$

Check: $\dfrac{d}{dx}\left[\dfrac{2}{5}x^{5/2} + C\right] = x^{3/2}$

	Given	_Rewrite_	_Integrate_	_Simplify_
9.	$\displaystyle\int \sqrt[3]{x}\, dx$	$\displaystyle\int x^{1/3}\, dx$	$\dfrac{x^{4/3}}{4/3} + C$	$\dfrac{3}{4}x^{4/3} + C$
11.	$\displaystyle\int \dfrac{1}{x\sqrt{x}}\, dx$	$\displaystyle\int x^{-3/2}\, dx$	$\dfrac{x^{-1/2}}{-1/2} + C$	$-\dfrac{2}{\sqrt{x}} + C$
13.	$\displaystyle\int \dfrac{1}{2x^3}\, dx$	$\dfrac{1}{2}\displaystyle\int x^{-3}\, dx$	$\dfrac{1}{2}\left(\dfrac{x^{-2}}{-2}\right) + C$	$-\dfrac{1}{4x^2} + C$

15. $\displaystyle\int (x + 3)\,dx = \dfrac{x^2}{2} + 3x + C$

Check: $\dfrac{d}{dx}\left[\dfrac{x^2}{2} + 3x + C\right] = x + 3$

17. $\displaystyle\int (2x - 3x^2)\,dx = x^2 - x^3 + C$

Check: $\dfrac{d}{dx}[x^2 - x^3 + C] = 2x - 3x^2$

19. $\displaystyle\int (x^3 + 2)\, dx = \dfrac{1}{4}x^4 + 2x + C$

Check: $\dfrac{d}{dx}\left(\dfrac{1}{4}x^4 + 2x + C\right) = x^3 + 2$

21. $\displaystyle\int (x^{3/2} + 2x + 1)\, dx = \dfrac{2}{5}x^{5/2} + x^2 + x + C$

Check: $\dfrac{d}{dx}\left(\dfrac{2}{5}x^{5/2} + x^2 + x + C\right) = x^{3/2} + 2x + 1$

23. $\displaystyle\int \sqrt[3]{x^2}\, dx = \displaystyle\int x^{2/3}\, dx = \dfrac{x^{5/3}}{5/3} + C = \dfrac{3}{5}x^{5/3} + C$

Check: $\dfrac{d}{dx}\left(\dfrac{3}{5}x^{5/3} + C\right) = x^{2/3} = \sqrt[3]{x^2}$

25. $\displaystyle\int \dfrac{1}{x^3}\, dx = \displaystyle\int x^{-3}\, dx = \dfrac{x^{-2}}{-2} + C = -\dfrac{1}{2x^2} + C$

Check: $\dfrac{d}{dx}\left(-\dfrac{1}{2x^2} + C\right) = \dfrac{1}{x^3}$

27. $\int \dfrac{x^2 + x + 1}{\sqrt{x}}\, dx = \int (x^{3/2} + x^{1/2} + x^{-1/2})\, dx = \dfrac{2}{5}x^{5/2} + \dfrac{2}{3}x^{3/2} + 2x^{1/2} + C = \dfrac{2}{15}x^{1/2}(3x^2 + 5x + 15) + C$

Check: $\dfrac{d}{dx}\left(\dfrac{2}{5}x^{5/2} + \dfrac{2}{3}x^{3/2} + 2x^{1/2} + C\right) = x^{3/2} + x^{1/2} + x^{-1/2} = \dfrac{x^2 + x + 1}{\sqrt{x}}$

29. $\int (x + 1)(3x - 2)\, dx = \int (3x^2 + x - 2)\, dx$

$\qquad\qquad = x^3 + \dfrac{1}{2}x^2 - 2x + C$

Check: $\dfrac{d}{dx}\left(x^3 + \dfrac{1}{2}x^2 - 2x + C\right) = 3x^2 + x - 2$

$\qquad\qquad\qquad = (x + 1)(3x - 2)$

31. $\int y^2\sqrt{y}\, dy = \int y^{5/2}\, dy = \dfrac{2}{7}y^{7/2} + C$

Check: $\dfrac{d}{dy}\left(\dfrac{2}{7}y^{7/2} + C\right) = y^{5/2} = y^2\sqrt{y}$

33. $\int dx = \int 1\, dx = x + C$

Check: $\dfrac{d}{dx}(x + C) = 1$

35. $\int (2\sin x + 3\cos x)\, dx = -2\cos x + 3\sin x + C$

Check: $\dfrac{d}{dx}(-2\cos x + 3\sin x + C) = 2\sin x + 3\cos x$

37. $\int (1 - \csc t\cot t)\, dt = t + \csc t + C$

Check: $\dfrac{d}{dt}(t + \csc t + C) = 1 - \csc t\cot t$

39. $\int (\sec^2 \theta - \sin \theta)\, d\theta = \tan \theta + \cos \theta + C$

Check: $\dfrac{d}{d\theta}(\tan \theta + \cos \theta + C) = \sec^2 \theta - \sin \theta$

41. $\int (\tan^2 y + 1)\, dy = \int \sec^2 y\, dy = \tan y + C$

Check: $\dfrac{d}{dy}(\tan y + C) = \sec^2 y = \tan^2 y + 1$

43. $f(x) = \cos x$

45. $f'(x) = 2$

$\quad f(x) = 2x + C$

Answers will vary.

47. $f'(x) = 1 - x^2$

$\quad f(x) = x - \dfrac{x^3}{3} + C$

Answers will vary.

49. $\dfrac{dy}{dx} = 2x - 1,\ (1, 1)$

$\quad y = \int (2x - 1)\, dx = x^2 - x + C$

$\quad 1 = (1)^2 - (1) + C \implies C = 1$

$\quad y = x^2 - x + 1$

51. $\dfrac{dy}{dx} = \cos x,\ (0, 4)$

$y = \displaystyle\int \cos x\, dx = \sin x + C$

$4 = \sin 0 + C \Rightarrow C = 4$

$y = \sin x + 4$

53. (a) Answers will vary.

(b) $\dfrac{dy}{dx} = \dfrac{1}{2}x - 1,\ (4, 2)$

$y = \dfrac{x^2}{4} - x + C$

$2 = \dfrac{4^2}{4} - 4 + C$

$2 = C$

$y = \dfrac{x^2}{4} - x + 2$

55. $f'(x) = 4x, f(0) = 6$

$f(x) = \displaystyle\int 4x\, dx = 2x^2 + C$

$f(0) = 6 = 2(0)^2 + C \Rightarrow C = 6$

$f(x) = 2x^2 + 6$

57. $h'(t) = 8t^3 + 5, h(1) = -4$

$h(t) = \displaystyle\int (8t^3 + 5)dt = 2t^4 + 5t + C$

$h(1) = -4 = 2 + 5 + C \Rightarrow C = -11$

$h(t) = 2t^4 + 5t - 11$

59. $f''(x) = 2$

$f'(2) = 5$

$f(2) = 10$

$f'(x) = \displaystyle\int 2\, dx = 2x + C_1$

$f'(2) = 4 + C_1 = 5 \Rightarrow C_1 = 1$

$f'(x) = 2x + 1$

$f(x) = \displaystyle\int (2x + 1)\, dx = x^2 + x + C_2$

$f(2) = 6 + C_2 = 10 \Rightarrow C_2 = 4$

$f(x) = x^2 + x + 4$

61. $f''(x) = x^{-3/2}$

$f'(4) = 2$

$f(0) = 0$

$f'(x) = \displaystyle\int x^{-3/2}\, dx = -2x^{-1/2} + C_1 = -\dfrac{2}{\sqrt{x}} + C_1$

$f'(4) = -\dfrac{2}{2} + C_1 = 2 \Rightarrow C_1 = 3$

$f'(x) = -\dfrac{2}{\sqrt{x}} + 3$

$f(x) = \displaystyle\int (-2x^{-1/2} + 3)\, dx = -4x^{1/2} + 3x + C_2$

$f(0) = 0 + 0 + C_2 = 0 \Rightarrow C_2 = 0$

$f(x) = -4x^{1/2} + 3x = -4\sqrt{x} + 3x$

63. (a) $h(t) = \displaystyle\int (1.5t + 5)\, dt = 0.75t^2 + 5t + C$

$h(0) = 0 + 0 + C = 12 \Rightarrow C = 12$

$h(t) = 0.75t^2 + 5t + 12$

(b) $h(6) = 0.75(6)^2 + 5(6) + 12 = 69$ cm

65. $f(0) = -4$. Graph of f' is given.

(a) $f'(4) \approx -1.0$

(b) No. The slopes of the tangent lines are greater than 2 on $[0, 2]$. Therefore, f must increase more than 4 units on $[0, 4]$.

(c) No, $f(5) < f(4)$ because f is decreasing on $[4, 5]$.

(d) f is an maximum at $x = 3.5$ because $f'(3.5) \approx 0$ and the first derivative test.

(e) f is concave upward when f' is increasing on $(-\infty, 1)$ and $(5, \infty)$. f is concave downward on $(1, 5)$. Points of inflection at $x = 1, 5$.

(f) f'' is a minimum at $x = 3$.

(g)

67. $a(t) = -32$ ft/sec^2

$$v(t) = \int -32\, dt = -32t + C_1$$

$$v(0) = 60 = C_1$$

$$s(t) = \int (-32t + 60)dt = -16t^2 + 60t + C_2$$

$$s(0) = 6 = C_2$$

$$s(t) = -16t^2 + 60t + 6 \quad \text{Position function}$$

The ball reaches its maximim height when

$$v(t) = -32t + 60 = 0$$

$$32t = 60$$

$$t = \frac{15}{8} \text{ seconds}$$

$$s\left(\frac{15}{8}\right) = -16\left(\frac{15}{8}\right)^2 + 60\left(\frac{15}{8}\right) + 6 = 62.25 \text{ feet}$$

69. From Exercise 68, we have:

$$s(t) = -16t^2 + v_0 t$$

$$s'(t) = -32t + v_0 = 0 \text{ when } t = \frac{v_0}{32} = \text{ time to reach}$$

maximum height.

$$s\left(\frac{v_0}{32}\right) = -16\left(\frac{v_0}{32}\right)^2 + v_0\left(\frac{v_0}{32}\right) = 550$$

$$-\frac{v_0^2}{64} + \frac{v_0^2}{32} = 550$$

$$v_0^2 = 35,200$$

$$v_0 \approx 187.617 \text{ ft/sec}$$

71. $a(t) = -9.8$

$$v(t) = \int -9.8\, dt = -9.8t + C_1$$

$$v(0) = v_0 = C_1 \implies v(t) = -9.8t + v_0$$

$$f(t) = \int (-9.8t + v_0)\, dt = -4.9t^2 + v_0 t + C_2$$

$$f(0) = s_0 = C_2 \implies f(t) = -4.9t^2 + v_0 t + s_0$$

73. From Exercise 71, $f(t) = -4.9t^2 + 10t + 2$.

$$v(t) = -9.8t + 10 = 0 \quad (\text{Maximum height when } v = 0.)$$

$$9.8t = 10$$

$$t = \frac{10}{9.8}$$

$$f\left(\frac{10}{9.8}\right) \approx 7.1 \text{ m}$$

75. $a = -1.6$

$$v(t) = \int -1.6\, dt = -1.6t + v_0 = -1.6t, \text{ since the stone was dropped, } v_0 = 0.$$

$$s(t) = \int (-1.6t)\, dt = -0.8t^2 + s_0$$

$$s(20) = 0 \implies -0.8(20)^2 + s_0 = 0$$

$$s_0 = 320$$

Thus, the height of the cliff is 320 meters.

$$v(t) = -1.6t$$

$$v(20) = -32 \text{ m/sec}$$

77. $x(t) = t^3 - 6t^2 + 9t - 2 \qquad 0 \le t \le 5$

(a) $v(t) = x'(t) = 3t^2 - 12t + 9$

$\qquad = 3(t^2 - 4t + 3) = 3(t - 1)(t - 3)$

$a(t) = v'(t) = 6t - 12 = 6(t - 2)$

(b) $v(t) > 0$ when $0 < t < 1$ or $3 < t < 5$.

(c) $a(t) = 6(t - 2) = 0$ when $t = 2$.

$v(2) = 3(1)(-1) = -3$

79. $v(t) = \dfrac{1}{\sqrt{t}} = t^{-1/2} \qquad t > 0$

$x(t) = \displaystyle\int v(t)\, dt = 2t^{1/2} + C$

$x(1) = 4 = 2(1) + C \implies C = 2$

$x(t) = 2t^{1/2} + 2$ position function

$a(t) = v'(t) = -\dfrac{1}{2}t^{-3/2} = \dfrac{-1}{2t^{3/2}}$ acceleration

81. (a) $v(0) = 25 \text{ km/hr} = 25 \cdot \dfrac{1000}{3600} = \dfrac{250}{36} \text{ m/sec}$

$v(13) = 80 \text{ km/hr} = 80 \cdot \dfrac{1000}{3600} = \dfrac{800}{36} \text{ m/sec}$

$a(t) = a$ (constant acceleration)

$v(t) = at + C$

$v(0) = \dfrac{250}{36} \implies v(t) = at + \dfrac{250}{36}$

$v(13) = \dfrac{800}{36} = 13a + \dfrac{250}{36}$

$\dfrac{550}{36} = 13a$

$a = \dfrac{550}{468} = \dfrac{275}{234} \approx 1.175 \text{ m/sec}^2$

(b) $s(t) = a\dfrac{t^2}{2} + \dfrac{250}{36}t \quad (s(0) = 0)$

$s(13) = \dfrac{275}{234}\dfrac{(13)^2}{2} + \dfrac{250}{36}(13) \approx 189.58 \text{ m}$

83. Truck: $v(t) = 30$

$s(t) = 30t$ (Let $s(0) = 0$.)

Automobile: $a(t) = 6$

$v(t) = 6t$ (Let $v(0) = 0$.)

$s(t) = 3t^2$ (Let $s(0) = 0$.)

At the point where the automobile overtakes the truck:

$30t = 3t^2$

$0 = 3t^2 - 30t$

$0 = 3t(t - 10)$ when $t = 10$ sec.

(a) $s(10) = 3(10)^2 = 300 \text{ ft}$

(b) $v(10) = 6(10) = 60 \text{ ft/sec} \approx 41 \text{ mph}$

85. $\dfrac{(1 \text{ mi/hr})(5280 \text{ ft/mi})}{(3600 \text{ sec/hr})} = \dfrac{22}{15} \text{ ft/sec}$

(a)

t	0	5	10	15	20	25	30
V_1(ft/sec)	0	3.67	10.27	23.47	42.53	66	95.33
V_2(ft/sec)	0	30.8	55.73	74.8	88	93.87	95.33

(b) $V_1(t) = 0.1068t^2 - 0.0416t + 0.3679$

$V_2(t) = -0.1208t^2 + 6.7991t - 0.0707$

(c) $S_1(t) = \displaystyle\int V_1(t)\, dt = \dfrac{0.1068}{3}t^3 - \dfrac{0.0416}{2}t^2 + 0.3679t$

$S_2(t) = \displaystyle\int V_2(t)\, dt = -\dfrac{0.1208t^3}{3} + \dfrac{6.7991t^2}{2} - 0.0707t$

[In both cases, the constant of integration is 0 because $S_1(0) = S_2(0) = 0$]

$S_1(30) \approx 953.5$ feet

$S_2(30) \approx 1970.3$ feet

The second car was going faster than the first until the end.

87. $a(t) = k$

$v(t) = kt$

$s(t) = \dfrac{k}{2}t^2$ since $v(0) = s(0) = 0$.

At the time of lift-off, $kt = 160$ and $(k/2)t^2 = 0.7$. Since $(k/2)t^2 = 0.7$,

$$t = \sqrt{\dfrac{1.4}{k}}$$

$$v\left(\sqrt{\dfrac{1.4}{k}}\right) = k\sqrt{\dfrac{1.4}{k}} = 160$$

$$1.4k = 160^2 \implies k = \dfrac{160^2}{1.4}$$

$$\approx 18{,}285.714 \text{ mi/hr}^2$$

$$\approx 7.45 \text{ ft/sec}^2.$$

89. True

91. True

93. False. For example, $\displaystyle\int x \cdot x\,dx \neq \int x\,dx \cdot \int x\,dx$ because $\dfrac{x^3}{3} + C \neq \left(\dfrac{x^2}{2} + C_1\right)\left(\dfrac{x^2}{2} + C_2\right)$

95. $f'(x) = \begin{cases} 1, & 0 \leq x < 2 \\ 3x, & 2 \leq x \leq 5 \end{cases}$

$f(x) = \begin{cases} x + C_1, & 0 \leq x < 2 \\ \dfrac{3x^2}{2} + C_2, & 2 \leq x \leq 5 \end{cases}$

$f(1) = 3 \implies 1 + C_1 = 3 \implies C_1 = 2$

f is continuous: Values must agree at $x = 2$:

$4 = 6 + C_2 \implies C_2 = -2$

$f(x) = \begin{cases} x + 2, & 0 \leq x < 2 \\ \dfrac{3x^2}{2} - 2, & 2 \leq x \leq 5 \end{cases}$

The left and right hand derivatives at $x = 2$ do not agree. Hence f is not differentiable at $x = 2$.

Section 4.2   Area

1. $\displaystyle\sum_{i=1}^{5}(2i + 1) = 2\sum_{i=1}^{5}i + \sum_{i=1}^{5}1 = 2(1 + 2 + 3 + 4 + 5) + 5 = 35$

3. $\displaystyle\sum_{k=0}^{4}\dfrac{1}{k^2 + 1} = 1 + \dfrac{1}{2} + \dfrac{1}{5} + \dfrac{1}{10} + \dfrac{1}{17} = \dfrac{158}{85}$

5. $\displaystyle\sum_{k=1}^{4}c = c + c + c + c = 4c$

7. $\displaystyle\sum_{i=1}^{9}\dfrac{1}{3i}$

9. $\displaystyle\sum_{j=1}^{8}\left[5\left(\dfrac{j}{8}\right) + 3\right]$

11. $\displaystyle\dfrac{2}{n}\sum_{i=1}^{n}\left[\left(\dfrac{2i}{n}\right)^3 - \left(\dfrac{2i}{n}\right)\right]$

13. $\displaystyle\dfrac{3}{n}\sum_{i=1}^{n}\left[2\left(1 + \dfrac{3i}{n}\right)^2\right]$

15. $\displaystyle\sum_{i=1}^{20}2i = 2\sum_{i=1}^{20}i = 2\left[\dfrac{20(21)}{2}\right] = 420$

17. $\displaystyle\sum_{i=1}^{20}(i - 1)^2 = \sum_{i=1}^{19}i^2$

$$= \left[\dfrac{19(20)(39)}{6}\right] = 2470$$

19. $\displaystyle\sum_{i=1}^{15} i(i-1)^2 = \sum_{i=1}^{15} i^3 - 2\sum_{i=1}^{15} i^2 + \sum_{i=1}^{15} i$

$$= \frac{15^2(16)^2}{4} - 2\frac{15(16)(31)}{6} + \frac{15(16)}{2}$$

$$= 14{,}400 - 2{,}480 + 120$$

$$= 12{,}040$$

21. sum seq($x \boxed{\wedge} 2 + 3, x, 1, 20, 1$) = 2930   (*TI-82*)

$$\sum_{i=1}^{20}(i^2 + 3) = \frac{20(20+1)(2(20)+1)}{6} + 3(20)$$

$$= \frac{(20)(21)(41)}{6} + 60 = 2930$$

23. $S = \left[3 + 4 + \frac{9}{2} + 5\right](1) = \frac{33}{2} = 16.5$

$s = \left[1 + 3 + 4 + \frac{9}{2}\right](1) = \frac{25}{2} = 12.5$

25. $S = [3 + 3 + 5](1) = 11$

$s = [2 + 2 + 3](1) = 7$

27. $S(4) = \sqrt{\frac{1}{4}}\left(\frac{1}{4}\right) + \sqrt{\frac{1}{2}}\left(\frac{1}{4}\right) + \sqrt{\frac{3}{4}}\left(\frac{1}{4}\right) + \sqrt{1}\left(\frac{1}{4}\right) = \frac{1 + \sqrt{2} + \sqrt{3} + 2}{8} \approx 0.768$

$s(4) = 0\left(\frac{1}{4}\right) + \sqrt{\frac{1}{4}}\left(\frac{1}{4}\right) + \sqrt{\frac{1}{2}}\left(\frac{1}{4}\right) + \sqrt{\frac{3}{4}}\left(\frac{1}{4}\right) = \frac{1 + \sqrt{2} + \sqrt{3}}{8} \approx 0.518$

29. $S(5) = 1\left(\frac{1}{5}\right) + \frac{1}{6/5}\left(\frac{1}{5}\right) + \frac{1}{7/5}\left(\frac{1}{5}\right) + \frac{1}{8/5}\left(\frac{1}{5}\right) + \frac{1}{9/5}\left(\frac{1}{5}\right) = \frac{1}{5} + \frac{1}{6} + \frac{1}{7} + \frac{1}{8} + \frac{1}{9} \approx 0.746$

$s(5) = \frac{1}{6/5}\left(\frac{1}{5}\right) + \frac{1}{7/5}\left(\frac{1}{5}\right) + \frac{1}{8/5}\left(\frac{1}{5}\right) + \frac{1}{9/5}\left(\frac{1}{5}\right) + \frac{1}{2}\left(\frac{1}{5}\right) = \frac{1}{6} + \frac{1}{7} + \frac{1}{8} + \frac{1}{9} + \frac{1}{10} \approx 0.646$

31. $\displaystyle\lim_{n\to\infty}\left[\left(\frac{81}{n^4}\right)\frac{n^2(n+1)^2}{4}\right] = \frac{81}{4}\lim_{n\to\infty}\left[\frac{n^4 + 2n^3 + n^2}{n^4}\right] = \frac{81}{4}(1) = \frac{81}{4}$

33. $\displaystyle\lim_{n\to\infty}\left[\left(\frac{18}{n^2}\right)\frac{n(n+1)}{2}\right] = \frac{18}{2}\lim_{n\to\infty}\left[\frac{n^2 + n}{n^2}\right] = \frac{18}{2}(1) = 9$

35. $\displaystyle\sum_{i=1}^{n}\frac{2i+1}{n^2} = \frac{1}{n^2}\sum_{i=1}^{n}(2i+1) = \frac{1}{n^2}\left[2\frac{n(n+1)}{2} + n\right] = \frac{n+2}{n} = S(n)$

$S(10) = \frac{12}{10} = 1.2$

$S(100) = 1.02$

$S(1000) = 1.002$

$S(10{,}000) = 1.0002$

37. $\displaystyle\sum_{k=1}^{n}\frac{6k(k-1)}{n^3} = \frac{6}{n^3}\sum_{k=1}^{n}(k^2 - k) = \frac{6}{n^3}\left[\frac{n(n+1)(2n+1)}{6} - \frac{n(n+1)}{2}\right]$

$$= \frac{6}{n^2}\left[\frac{2n^2 + 3n + 1 - 3n - 3}{6}\right] = \frac{1}{n^2}[2n^2 - 2] = S(n)$$

$S(10) = 1.98$

$S(100) = 1.9998$

$S(1000) = 1.999998$

$S(10{,}000) = 1.99999998$

39. $\displaystyle\lim_{n\to\infty}\sum_{i=1}^{n}\left(\frac{16i}{n^2}\right) = \lim_{n\to\infty}\frac{16}{n^2}\sum_{i=1}^{n}i = \lim_{n\to\infty}\frac{16}{n^2}\left(\frac{n(n+1)}{2}\right) = \lim_{n\to\infty}\left[8\left(\frac{n^2+n}{n^2}\right)\right] = 8\lim_{n\to\infty}\left(1 + \frac{1}{n}\right) = 8$

41. $\displaystyle\lim_{n\to\infty}\sum_{i=1}^{n}\frac{1}{n^3}(i-1)^2 = \lim_{n\to\infty}\frac{1}{n^3}\sum_{i=1}^{n-1}i^2 = \lim_{n\to\infty}\frac{1}{n^3}\left[\frac{(n-1)(n)(2n-1)}{6}\right]$

$\displaystyle = \lim_{n\to\infty}\frac{1}{6}\left[\frac{2n^3-3n^2+n}{n^3}\right] = \lim_{n\to\infty}\left[\frac{1}{6}\left(\frac{2-(3/n)+(1/n^2)}{1}\right)\right] = \frac{1}{3}$

43. $\displaystyle\lim_{n\to\infty}\sum_{i=1}^{n}\left(1+\frac{i}{n}\right)\left(\frac{2}{n}\right) = 2\lim_{n\to\infty}\frac{1}{n}\left[\sum_{i=1}^{n}1+\frac{1}{n}\sum_{i=1}^{n}i\right] = 2\lim_{n\to\infty}\frac{1}{n}\left[n+\frac{1}{n}\left(\frac{n(n+1)}{2}\right)\right] = 2\lim_{n\to\infty}\left[1+\frac{n^2+n}{2n^2}\right] = 2\left(1+\frac{1}{2}\right) = 3$

45. (a)

(b) $\Delta x = \dfrac{2-0}{n} = \dfrac{2}{n}$

Endpoints:

$$0 < 1\left(\frac{2}{n}\right) < 2\left(\frac{2}{n}\right) < \cdots < (n-1)\left(\frac{2}{n}\right) < n\left(\frac{2}{n}\right) = 2$$

(c) Since $y = x$ is increasing, $f(m_i) = f(x_{i-1})$ on $[x_{i-1}, x_i]$.

$\displaystyle s(n) = \sum_{i=1}^{n}f(x_{i-1})\,\Delta x$

$\displaystyle = \sum_{i=1}^{n}f\left(\frac{2i-2}{n}\right)\left(\frac{2}{n}\right) = \sum_{i=1}^{n}\left[(i-1)\left(\frac{2}{n}\right)\right]\left(\frac{2}{n}\right)$

(d) $f(M_i) = f(x_i)$ on $[x_{i-1}, x_i]$

$\displaystyle S(n) = \sum_{i=1}^{n}f(x_i)\,\Delta x = \sum_{i=1}^{n}f\left(\frac{2i}{n}\right)\frac{2}{n} = \sum_{i=1}^{n}\left[i\left(\frac{2}{n}\right)\right]\left(\frac{2}{n}\right)$

(e)

x	5	10	50	100
$s(n)$	1.6	1.8	1.96	1.98
$S(n)$	2.4	2.2	2.04	2.02

(f) $\displaystyle\lim_{n\to\infty}\sum_{i=1}^{n}\left[(i-1)\left(\frac{2}{n}\right)\right]\left(\frac{2}{n}\right) = \lim_{n\to\infty}\frac{4}{n^2}\sum_{i=1}^{n}(i-1)$

$\displaystyle = \lim_{n\to\infty}\frac{4}{n^2}\left[\frac{n(n+1)}{2}-n\right]$

$\displaystyle = \lim_{n\to\infty}\left[\frac{2(n+1)}{n}-\frac{4}{n}\right] = 2$

$\displaystyle\lim_{n\to\infty}\sum_{i=1}^{n}\left[i\left(\frac{2}{n}\right)\right]\left(\frac{2}{n}\right) = \lim_{n\to\infty}\frac{4}{n^2}\sum_{i=1}^{n}i$

$\displaystyle = \lim_{n\to\infty}\left(\frac{4}{n^2}\right)\frac{n(n+1)}{2}$

$\displaystyle = \lim_{n\to\infty}\frac{2(n+1)}{n} = 2$

47. $y = -2x + 3$ on $[0, 1]$. $\left(\textbf{Note: }\Delta x = \dfrac{1-0}{n} = \dfrac{1}{n}\right)$

$\displaystyle s(n) = \sum_{i=1}^{n}f\left(\frac{i}{n}\right)\left(\frac{1}{n}\right) = \sum_{i=1}^{n}\left[-2\left(\frac{i}{n}\right)+3\right]\left(\frac{1}{n}\right)$

$\displaystyle = 3 - \frac{2}{n^2}\sum_{i=1}^{n}i = 3 - \frac{2(n+1)n}{2n^2} = 2 - \frac{1}{n}$

Area $= \displaystyle\lim_{n\to\infty}s(n) = 2$

49. $y = x^2 + 2$ on $[0, 1]$. $\left(\textbf{Note: }\Delta x = \dfrac{1}{n}\right)$

$\displaystyle S(n) = \sum_{i=1}^{n}f\left(\frac{i}{n}\right)\left(\frac{1}{n}\right) = \sum_{i=1}^{n}\left[\left(\frac{i}{n}\right)^2+2\right]\left(\frac{1}{n}\right)$

$\displaystyle = \left[\frac{1}{n^3}\sum_{i=1}^{n}i^2\right] + 2 = \frac{n(n+1)(2n+1)}{6n^3} + 2 = \frac{1}{6}\left(2+\frac{3}{n}+\frac{1}{n^2}\right) + 2$

Area $= \displaystyle\lim_{n\to\infty}S(n) = \frac{7}{3}$

51. $y = 16 - x^2$ on $[1, 3]$. $\left(\text{Note: } \Delta x = \dfrac{2}{n}\right)$

$$s(n) = \sum_{i=1}^{n} f\left(1 + \frac{2i}{n}\right)\left(\frac{2}{n}\right) = \sum_{i=1}^{n} \left[16 - \left(1 + \frac{2i}{n}\right)^2\right]\left(\frac{2}{n}\right)$$

$$= \frac{2}{n} \sum_{i=1}^{n} \left[15 - \frac{4i^2}{n^2} - \frac{4i}{n}\right]$$

$$= \frac{2}{n} \left[15n - \frac{4}{n^2}\frac{n(n+1)(2n+1)}{6} - \frac{4}{n}\frac{n(n+1)}{2}\right]$$

$$= 30 - \frac{8}{6n^2}(n+1)(2n+1) - \frac{4}{n}(n+1)$$

$$\text{Area} = \lim_{n \to \infty} s(n) = 30 - \frac{8}{3} - 4 = \frac{70}{3} = 23\frac{1}{3}$$

53. $y = 64 - x^3$ on $[1, 4]$. $\left(\text{Note: } \Delta x = \dfrac{4-1}{n} = \dfrac{3}{n}\right)$

$$s(n) = \sum_{i=1}^{n} f\left(1 + \frac{3i}{n}\right)\left(\frac{3}{n}\right) = \sum_{i=1}^{n} \left[64 - \left(1 + \frac{3i}{n}\right)^3\right]\left(\frac{3}{n}\right)$$

$$= \frac{3}{n} \sum_{i=1}^{n} \left[63 - \frac{27i^3}{n^3} - \frac{27i^2}{n^2} - \frac{9i}{n}\right]$$

$$= \frac{3}{n} \left[63n - \frac{27}{n^3}\frac{n^2(n+1)^2}{4} - \frac{27}{n^2}\frac{n(n+1)(2n+1)}{6} - \frac{9}{n}\frac{n(n+1)}{2}\right]$$

$$= 189 - \frac{81}{4n^2}(n+1)^2 - \frac{81}{6n^2}(n+1)(2n+1) - \frac{27}{2}\frac{n+1}{n}$$

$$\text{Area} = \lim_{n \to \infty} s(n) = 189 - \frac{81}{4} - 27 - \frac{27}{2} = \frac{513}{4} = 128.25$$

55. $y = x^2 - x^3$ on $[-1, 1]$. $\left(\text{Note: } \Delta x = \dfrac{1-(-1)}{n} = \dfrac{2}{n}\right)$

Again, $T(n)$ is neither an upper nor a lower sum.

$$T(n) = \sum_{i=1}^{n} f\left(-1 + \frac{2i}{n}\right)\left(\frac{2}{n}\right) = \sum_{i=1}^{n} \left[\left(-1 + \frac{2i}{n}\right)^2 - \left(-1 + \frac{2i}{n}\right)^3\right]\left(\frac{2}{n}\right)$$

$$= \sum_{i=1}^{n} \left[\left(1 - \frac{4i}{n} + \frac{4i^2}{n^2}\right) - \left(-1 + \frac{6i}{n} - \frac{12i^2}{n^2} + \frac{8i^3}{n^3}\right)\right]\left(\frac{2}{n}\right)$$

$$= \sum_{i=1}^{n} \left[2 - \frac{10i}{n} + \frac{16i^2}{n^2} - \frac{8i^3}{n^3}\right]\left(\frac{2}{n}\right) = \frac{4}{n}\sum_{i=1}^{n} 1 - \frac{20}{n^2}\sum_{i=1}^{n} i + \frac{32}{n^3}\sum_{i=1}^{n} i^2 - \frac{16}{n^4}\sum_{i=1}^{n} i^3$$

$$= \frac{4}{n}(n) - \frac{20}{n^2} \cdot \frac{n(n+1)}{2} + \frac{32}{n^3} \cdot \frac{n(n+1)(2n+1)}{6} - \frac{16}{n^4} \cdot \frac{n^2(n+1)^2}{4}$$

$$= 4 - 10\left(1 + \frac{1}{n}\right) + \frac{16}{3}\left(2 + \frac{3}{n} + \frac{1}{n^2}\right) - 4\left(1 + \frac{2}{n} + \frac{1}{n^2}\right)$$

$$\text{Area} = \lim_{n \to \infty} T(n) = 4 - 10 + \frac{32}{3} - 4 = \frac{2}{3}$$

57. $f(y) = 3y, 0 \le y \le 2$ $\left(\textbf{Note: } \Delta y = \dfrac{2-0}{n} = \dfrac{2}{n}\right)$

$$S(n) = \sum_{i=1}^{n} f(m_i)\,\Delta y = \sum_{i=1}^{n} f\left(\frac{2i}{n}\right)\left(\frac{2}{n}\right) = \sum_{i=1}^{n} 3\left(\frac{2i}{n}\right)\left(\frac{2}{n}\right)$$

$$= \frac{12}{n^2} \sum_{i=1}^{n} i = \left(\frac{12}{n^2}\right) \cdot \frac{n(n+1)}{2} = \frac{6(n+1)}{n} = 6 + \frac{6}{n}$$

$$\text{Area} = \lim_{n \to \infty} S(n) = \lim_{n \to \infty}\left(6 + \frac{6}{n}\right) = 6$$

59. $f(y) = y^2, 0 \le y \le 3$ $\left(\textbf{Note: } \Delta y = \dfrac{3-0}{n} = \dfrac{3}{n}\right)$

$$S(n) = \sum_{i=1}^{n} f\left(\frac{3i}{n}\right)\left(\frac{3}{n}\right) = \sum_{i=1}^{n}\left(\frac{3i}{n}\right)^2\left(\frac{3}{n}\right) = \frac{27}{n^3} \sum_{i=1}^{n} i^2$$

$$= \frac{27}{n^3} \cdot \frac{n(n+1)(2n+1)}{6} = \frac{9}{n^2}\left(\frac{2n^2 + 3n + 1}{2}\right) = 9 + \frac{27}{2n} + \frac{9}{2n^2}$$

$$\text{Area} = \lim_{n \to \infty} S(n) = \lim_{n \to \infty}\left(9 + \frac{27}{2n} + \frac{9}{2n^2}\right) = 9$$

61. $g(y) = 4y^2 - y^3, 1 \le y \le 3.$ $\left(\textbf{Note: } \Delta y = \dfrac{3-1}{n} = \dfrac{2}{n}\right)$

$$S(n) = \sum_{i=1}^{n} g\left(1 + \frac{2i}{n}\right)\left(\frac{2}{n}\right)$$

$$= \sum_{i=1}^{n}\left[4\left(1 + \frac{2i}{n}\right)^2 - \left(1 + \frac{2i}{n}\right)^3\right]\frac{2}{n}$$

$$= \frac{2}{n} \sum_{i=1}^{n} 4\left[1 + \frac{4i}{n} + \frac{4i^2}{n^2}\right] - \left[1 + \frac{6i}{n} + \frac{12i^2}{n^2} + \frac{8i^3}{n^3}\right]$$

$$= \frac{2}{n} \sum_{i=1}^{n}\left[3 + \frac{10i}{n} + \frac{4i^2}{n^2} - \frac{8i^3}{n^3}\right] = \frac{2}{n}\left[3n + \frac{10}{n}\frac{n(n+1)}{2} + \frac{4}{n^2}\frac{n(n+1)(2n+1)}{6} - \frac{8}{n^2}\frac{n^2(n+1)^2}{4}\right]$$

$$\text{Area} = \lim_{n \to \infty} S(n) = 6 + 10 + \frac{8}{3} - 4 = \frac{44}{3}$$

63. $f(x) = x^2 + 3, 0 \le x \le 2, n = 4$

Let $c_i = \dfrac{x_i + x_{i-1}}{2}$.

$$\Delta x = \frac{1}{2}, c_1 = \frac{1}{4}, c_2 = \frac{3}{4}, c_3 = \frac{5}{4}, c_4 = \frac{7}{4}$$

$$\text{Area} \approx \sum_{i=1}^{n} f(c_i)\,\Delta x = \sum_{i=1}^{4}[c_i^2 + 3]\left(\frac{1}{2}\right)$$

$$= \frac{1}{2}\left[\left(\frac{1}{16} + 3\right) + \left(\frac{9}{16} + 3\right) + \left(\frac{25}{16} + 3\right) + \left(\frac{49}{16} + 3\right)\right]$$

$$= \frac{69}{8}$$

65. $f(x) = \tan x, 0 \le x \le \dfrac{\pi}{4}, n = 4$

Let $c_i = \dfrac{x_i + x_{i-1}}{2}$.

$$\Delta x = \frac{\pi}{16}, c_1 = \frac{\pi}{32}, c_2 = \frac{3\pi}{32}, c_3 = \frac{5\pi}{32}, c_4 = \frac{7\pi}{32}$$

$$\text{Area} \approx \sum_{i=1}^{n} f(c_i)\,\Delta x = \sum_{i=1}^{4}(\tan c_i)\left(\frac{\pi}{16}\right)$$

$$= \frac{\pi}{16}\left(\tan\frac{\pi}{32} + \tan\frac{3\pi}{32} + \tan\frac{5\pi}{32} + \tan\frac{7\pi}{32}\right) \approx 0.345$$

67. $f(x) = \sqrt{x}$ on $[0, 4]$.

n	4	8	12	16	20
Approximate area	5.3838	5.3523	5.3439	5.3403	5.3384

(Exact value is 16/3)

69. $f(x) = \tan\left(\dfrac{\pi x}{8}\right)$ on $[1, 3]$.

n	4	8	12	16	20
Approximate area	2.2223	2.2387	2.2418	2.2430	2.2435

71. We can use the line $y = x$ bounded by $x = a$ and $x = b$. The sum of the areas of these inscribed rectangles is the lower sum.

The sum of the areas of these circumscribed rectangles is the upper sum.

We can see that the rectangles do not contain all of the area in the first graph and the rectangles in the second graph cover more than the area of the region.

The exact value of the area lies between these two sums.

73. (a)

Lower sum:
$$s(4) = 0 + 4 + 5\tfrac{1}{3} + 6 = 15\tfrac{1}{3} = \tfrac{46}{3} \approx 15.333$$

(b)

Upper sum:
$$S(4) = 4 + 5\tfrac{1}{3} + 6 + 6\tfrac{2}{5} = 21\tfrac{11}{15} = \tfrac{326}{15} \approx 21.733$$

(c)

(d) In each case, $\Delta x = 4/n$. The lower sum uses left endpoints, $(i - 1)(4/n)$. The upper sum uses right endpoints, $(i)(4/n)$. The Midpoint Rule uses midpoints, $\left(i - \tfrac{1}{2}\right)(4/n)$.

Midpoint Rule:
$$M(4) = 2\tfrac{2}{3} + 4\tfrac{4}{5} + 5\tfrac{5}{7} + 6\tfrac{2}{9} = \tfrac{6112}{315} \approx 19.403$$

(e)

n	4	8	20	100	200
$s(n)$	15.333	17.368	18.459	18.995	19.06
$S(n)$	21.733	20.568	19.739	19.251	19.188
$M(n)$	19.403	19.201	19.137	19.125	19.125

(f) $s(n)$ increases because the lower sum approaches the exact value as n increases. $S(n)$ decreases because the upper sum approaches the exact value as n increases. Because of the shape of the graph, the lower sum is always smaller than the exact value, whereas the upper sum is always larger.

75.

77. True. (Theorem 4.2 (2))

b. $A \approx 6$ square units

79. $f(x) = \sin x, \left[0, \dfrac{\pi}{2}\right]$

Let A_1 = area bounded by $f(x) = \sin x$, the x-axis, $x = 0$ and $x = \pi/2$. Let A_2 = area of the rectangle bounded by $y = 1$, $y = 0$, $x = 0$, and $x = \pi/2$. Thus, $A_2 = (\pi/2)(1) \approx 1.570796$.
In this program, the computer is generating N_2 pairs of random points in the rectangle whose area is represented by A_2. It is keeping track of how many of these points, N_1, lie in the region whose area is represented by A_1. Since the points are randomly generated, we assume that

$$\frac{A_1}{A_2} \approx \frac{N_1}{N_2} \implies A_1 \approx \frac{N_1}{N_2} A_2.$$

The larger N_2 is the better the approximation to A_1.

81. Suppose there are n rows and $n + 1$ columns in the figure. The stars on the left total $1 + 2 + \cdots + n$, as do the stars on the right. There are $n(n + 1)$ stars in total, hence

$$2[1 + 2 + \cdots + n] = n(n + 1)$$
$$1 + 2 + \cdots + n = \tfrac{1}{2}(n)(n + 1).$$

83. (a) $y = (-4.09 \times 10^{-5})x^3 + 0.016x^2 - 2.67x + 452.9$

(c) Using the integration capability of a graphing utility, you obtain

$$A \approx 76,897.5 \text{ ft}^2.$$

(b)

Section 4.3   Riemann Sums and Definite Integrals

1. $f(x) = \sqrt{x}, y = 0, x = 0, x = 3, c_i = \dfrac{3i^2}{n^2}$

$$\Delta x_i = \frac{3i^2}{n^2} - \frac{3(i - 1)^2}{n^2} = \frac{3}{n^2}(2i - 1)$$

$$\lim_{n \to \infty} \sum_{i=1}^{n} f(c_i)\Delta x_i = \lim_{n \to \infty} \sum_{i=1}^{n} \sqrt{\frac{3i^2}{n^2}} \frac{3}{n^2}(2i - 1)$$

$$= \lim_{n \to \infty} \frac{3\sqrt{3}}{n^3} \sum_{i=1}^{n} (2i^2 - i)$$

$$= \lim_{n \to \infty} \frac{3\sqrt{3}}{n^3}\left[2\frac{n(n + 1)(2n + 1)}{6} - \frac{n(n + 1)}{2}\right]$$

$$= \lim_{n \to \infty} 3\sqrt{3}\left[\frac{(n + 1)(2n + 1)}{3n^2} - \frac{n + 1}{2n^2}\right]$$

$$= 3\sqrt{3}\left[\frac{2}{3} - 0\right] = 2\sqrt{3} \approx 3.464$$

3. $y = 6$ on $[4, 10]$. $\left(\textbf{Note:} \; \Delta x = \dfrac{10 - 4}{n} = \dfrac{6}{n}, \|\Delta\| \to 0 \text{ as } n \to \infty\right)$

$$\sum_{i=1}^{n} f(c_i)\, \Delta x_i = \sum_{i=1}^{n} f\left(4 + \frac{6i}{n}\right)\left(\frac{6}{n}\right) = \sum_{i=1}^{n} 6\left(\frac{6}{n}\right) = \sum_{i=1}^{n} \frac{36}{n} = 36$$

$$\int_{4}^{10} 6\, dx = \lim_{n \to \infty} 36 = 36$$

5. $y = x^3$ on $[-1, 1]$. $\left(\textbf{Note:} \; \Delta x = \dfrac{1 - (-1)}{n} = \dfrac{2}{n}, \|\Delta\| \to 0 \text{ as } n \to \infty\right)$

$$\sum_{i=1}^{n} f(c_i)\, \Delta x_i = \sum_{i=1}^{n} f\left(-1 + \frac{2i}{n}\right)\left(\frac{2}{n}\right) = \sum_{i=1}^{n} \left(-1 + \frac{2i}{n}\right)^3\left(\frac{2}{n}\right) = \sum_{i=1}^{n} \left[-1 + \frac{6i}{n} - \frac{12i^2}{n^2} + \frac{8i^3}{n^3}\right]\left(\frac{2}{n}\right)$$

$$= -2 + \frac{12}{n^2}\sum_{i=1}^{n} i - \frac{24}{n^3}\sum_{i=1}^{n} i^2 + \frac{16}{n^4}\sum_{i=1}^{n} i^3$$

$$= -2 + 6\left(1 + \frac{1}{n}\right) - 4\left(2 + \frac{3}{n} + \frac{1}{n^2}\right) + 4\left(1 + \frac{2}{n} + \frac{1}{n^2}\right) = \frac{2}{n}$$

$$\int_{-1}^{1} x^3\, dx = \lim_{n \to \infty} \frac{2}{n} = 0$$

7. $y = x^2 + 1$ on $[1, 2]$. $\left(\textbf{Note:} \; \Delta x = \dfrac{2 - 1}{n} = \dfrac{1}{n}, \|\Delta\| \to 0 \text{ as } n \to \infty\right)$

$$\sum_{i=1}^{n} f(c_i)\, \Delta x_i = \sum_{i=1}^{n} f\left(1 + \frac{i}{n}\right)\left(\frac{1}{n}\right) = \sum_{i=1}^{n} \left[\left(1 + \frac{i}{n}\right)^2 + 1\right]\left(\frac{1}{n}\right) = \sum_{i=1}^{n} \left[1 + \frac{2i}{n} + \frac{i^2}{n^2} + 1\right]\left(\frac{1}{n}\right)$$

$$= 2 + \frac{2}{n^2}\sum_{i=1}^{n} i + \frac{1}{n^3}\sum_{i=1}^{n} i^2 = 2 + \left(1 + \frac{1}{n}\right) + \frac{1}{6}\left(2 + \frac{3}{n} + \frac{1}{n^2}\right) = \frac{10}{3} + \frac{3}{2n} + \frac{1}{6n^2}$$

$$\int_{1}^{2} (x^2 + 1)\, dx = \lim_{n \to \infty} \left(\frac{10}{3} + \frac{3}{2n} + \frac{1}{6n^2}\right) = \frac{10}{3}$$

9. $\displaystyle\lim_{\|\Delta\| \to 0} \sum_{i=1}^{n} (3c_i + 10)\, \Delta x_i = \int_{-1}^{5} (3x + 10)\, dx$

on the interval $[-1, 5]$.

11. $\displaystyle\lim_{\|\Delta\| \to 0} \sum_{i=1}^{n} \sqrt{c_i^2 + 4}\, \Delta x_i = \int_{0}^{3} \sqrt{x^2 + 4}\, dx$

on the interval $[0, 3]$.

13. $\displaystyle\int_{0}^{5} 3\, dx$

15. $\displaystyle\int_{-4}^{4} (4 - |x|)\, dx$

17. $\displaystyle\int_{-2}^{2} (4 - x^2)\, dx$

19. $\displaystyle\int_{0}^{\pi} \sin x\, dx$

21. $\displaystyle\int_{0}^{2} y^3\, dy$

23. Rectangle

$A = bh = 3(4)$

$A = \displaystyle\int_{0}^{3} 4\, dx = 12$

25. Triangle

$$A = \frac{1}{2}bh = \frac{1}{2}(4)(4)$$

$$A = \int_0^4 x \, dx = 8$$

27. Trapezoid

$$A = \frac{b_1 + b_2}{2}h = \left(\frac{5 + 9}{2}\right)2$$

$$A = \int_0^2 (2x + 5) \, dx = 14$$

29. Triangle

$$A = \frac{1}{2}bh = \frac{1}{2}(2)(1)$$

$$A = \int_{-1}^1 (1 - |x|) \, dx = 1$$

31. Semicircle

$$A = \frac{1}{2}\pi r^2 = \frac{1}{2}\pi(3)^2$$

$$A = \int_{-3}^3 \sqrt{9 - x^2} \, dx = \frac{9\pi}{2}$$

In Exercises 33–39, $\int_2^4 x^3 \, dx = 60, \int_2^4 x \, dx = 6, \int_2^4 dx = 2$

33. $\int_4^2 x \, dx = -\int_2^4 x \, dx = -6$

35. $\int_2^4 4x \, dx = 4\int_2^4 x \, dx = 4(6) = 24$

37. $\int_2^4 (x - 8) \, dx = \int_2^4 x \, dx - 8\int_2^4 dx = 6 - 8(2) = -10$

39. $\int_2^4 \left(\frac{1}{2}x^3 - 3x + 2\right) dx = \frac{1}{2}\int_2^4 x^3 \, dx - 3\int_2^4 x \, dx + 2\int_2^4 dx$

$$= \frac{1}{2}(60) - 3(6) + 2(2) = 16$$

41. (a) $\int_0^7 f(x) \, dx = \int_0^5 f(x) \, dx + \int_5^7 f(x) \, dx = 10 + 3 = 13$

(b) $\int_5^0 f(x) \, dx = -\int_0^5 f(x) \, dx = -10$

(c) $\int_5^5 f(x) \, dx = 0$

(d) $\int_0^5 3f(x) \, dx = 3\int_0^5 f(x) \, dx = 3(10) = 30$

43. (a) $\int_2^6 [f(x) + g(x)] \, dx = \int_2^6 f(x) \, dx + \int_2^6 g(x) \, dx$

$$= 10 + (-2) = 8$$

(b) $\int_2^6 [g(x) - f(x)] \, dx = \int_2^6 g(x) \, dx - \int_2^6 f(x) \, dx$

$$= -2 - 10 = -12$$

(c) $\int_2^6 2g(x) \, dx = 2\int_2^6 g(x) \, dx = 2(-2) = -4$

(d) $\int_2^6 3f(x) \, dx = 3\int_2^6 f(x) \, dx = 3(10) = 30$

45. (a) Quarter circle below x-axis: $-\frac{1}{4}\pi r^2 = -\frac{1}{4}\pi(2)^2 = -\pi$

(b) Triangle: $\frac{1}{2}bh = \frac{1}{2}(4)(2) = 4$

(c) Triangle + Semicircle below x-axis: $-\frac{1}{2}(2)(1) - \frac{1}{2}\pi(2)^2 = -(1 + 2\pi)$

(d) Sum of parts (b) and (c): $4 - (1 + 2\pi) = 3 - 2\pi$

(e) Sum of absolute values of (b) and (c): $4 + (1 + 2\pi) = 5 + 2\pi$

(f) Answer to (d) plus $2(10) = 20$: $(3 - 2\pi) + 20 = 23 - 2\pi$

47. The left endpoint approximation will be greater than the actual area: $>$

49. Because the curve is concave upward, the midpoint approximation will be less than the actual area: $<$

51. $f(x) = \dfrac{1}{x-4}$

is not integrable on the interval $[3, 5]$ because f has a discontinuity at $x = 4$.

53.

a. $A \approx 5$ square units

55.

d. $\displaystyle\int_0^1 2 \sin \pi x \, dx \approx \frac{1}{2}(1)(2) \approx 1$

57. $\displaystyle\int_0^3 x\sqrt{3-x}\, dx$

n	4	8	12	16	20
$L(n)$	3.6830	3.9956	4.0707	4.1016	4.1177
$M(n)$	4.3082	4.2076	4.1838	4.1740	4.1690
$R(n)$	3.6830	3.9956	4.0707	4.1016	4.1177

59. $\displaystyle\int_0^{\pi/2} \sin^2 x \, dx$

n	4	8	12	16	20
$L(n)$	0.5890	0.6872	0.7199	0.7363	0.7461
$M(n)$	0.7854	0.7854	0.7854	0.7854	0.7854
$R(n)$	0.9817	0.8836	0.8508	0.8345	0.8247

61. True

63. True

65. False

$$\int_0^2 (-x)\, dx = -2$$

67. $f(x) = x^2 + 3x$, $[0, 8]$

$x_0 = 0, x_1 = 1, x_2 = 3, x_3 = 7, x_4 = 8$

$\Delta x_1 = 1, \Delta x_2 = 2, \Delta x_3 = 4, \Delta x_4 = 1$

$c_1 = 1, c_2 = 2, c_3 = 5, c_4 = 8$

$$\sum_{i=1}^4 f(c_i)\, \Delta x = f(1)\, \Delta x_1 + f(2)\, \Delta x_2 + f(5)\, \Delta x_3 + f(8)\, \Delta x_4$$

$$= (4)(1) + (10)(2) + (40)(4) + (88)(1) = 272$$

69. $f(x) = \begin{cases} 1, & x \text{ is rational} \\ 0, & x \text{ is irrational} \end{cases}$

is not integrable on the interval $[0, 1]$. As $\|\Delta\| \to 0$, $f(c_i) = 1$ or $f(c_i) = 0$ in each subinterval since there are an infinite number of both rational and irrational numbers in any interval, no matter how small.

71. Let $f(x) = x^2$, $0 \le x \le 1$, and $\Delta x_i = 1/n$. The appropriate Riemann Sum is

$$\sum_{i=1}^{n} f(c_i)\Delta x_i = \sum_{i=1}^{n}\left(\frac{i}{n}\right)^2\frac{1}{n} = \frac{1}{n^3}\sum_{i=1}^{n}i^2.$$

$$\lim_{n\to\infty}\frac{1}{n^3}[1^2 + 2^2 + 3^2 + \cdots + n^2] = \lim_{n\to\infty}\frac{1}{n^3}\cdot\frac{n(2n + 1)(n + 1)}{6}$$

$$= \lim_{n\to\infty}\frac{2n^2 + 3n + 1}{6n^2} = \lim_{n\to\infty}\left(\frac{1}{3} + \frac{1}{2n} + \frac{1}{6n^2}\right) = \frac{1}{3}$$

Section 4.4 The Fundamental Theorem of Calculus

1. $f(x) = \dfrac{4}{x^2 + 1}$

$\displaystyle\int_0^{\pi}\frac{4}{x^2 + 1}\,dx$ is positive.

3. $f(x) = x\sqrt{x^2 + 1}$

$\displaystyle\int_{-2}^{2}x\sqrt{x^2 + 1}\,dx = 0$

5. $\displaystyle\int_0^1 2x\,dx = \left[x^2\right]_0^1 = 1 - 0 = 1$

7. $\displaystyle\int_{-1}^0 (x - 2)\,dx = \left[\frac{x^2}{2} - 2x\right]_{-1}^0 = 0 - \left(\frac{1}{2} + 2\right) = -\frac{5}{2}$

9. $\displaystyle\int_{-1}^1 (t^2 - 2)\,dt = \left[\frac{t^3}{3} - 2t\right]_{-1}^1 = \left(\frac{1}{3} - 2\right) - \left(-\frac{1}{3} + 2\right) = -\frac{10}{3}$

11. $\displaystyle\int_0^1 (2t - 1)^2\,dt = \int_0^1 (4t^2 - 4t + 1)\,dt = \left[\frac{4}{3}t^3 - 2t^2 + t\right]_0^1 = \frac{4}{3} - 2 + 1 = \frac{1}{3}$

13. $\displaystyle\int_1^2\left(\frac{3}{x^2} - 1\right)dx = \left[-\frac{3}{x} - x\right]_1^2 = \left(-\frac{3}{2} - 2\right) - (-3 - 1) = \frac{1}{2}$

15. $\displaystyle\int_1^4\frac{u - 2}{\sqrt{u}}\,du = \int_1^4 (u^{1/2} - 2u^{-1/2})\,du = \left[\frac{2}{3}u^{3/2} - 4u^{1/2}\right]_1^4 = \left[\frac{2}{3}(\sqrt{4})^3 - 4\sqrt{4}\right] - \left[\frac{2}{3} - 4\right] = \frac{2}{3}$

17. $\displaystyle\int_{-1}^1 (\sqrt[3]{t} - 2)\,dt = \left[\frac{3}{4}t^{4/3} - 2t\right]_{-1}^1 = \left(\frac{3}{4} - 2\right) - \left(\frac{3}{4} + 2\right) = -4$

19. $\displaystyle\int_0^1\frac{x - \sqrt{x}}{3}\,dx = \frac{1}{3}\int_0^1 (x - x^{1/2})\,dx = \frac{1}{3}\left[\frac{x^2}{2} - \frac{2}{3}x^{3/2}\right]_0^1 = \frac{1}{3}\left(\frac{1}{2} - \frac{2}{3}\right) = -\frac{1}{18}$

21. $\displaystyle\int_{-1}^0 (t^{1/3} - t^{2/3})\,dt = \left[\frac{3}{4}t^{4/3} - \frac{3}{5}t^{5/3}\right]_{-1}^0 = 0 - \left(\frac{3}{4} + \frac{3}{5}\right) = -\frac{27}{20}$

23. $\displaystyle\int_0^3 |2x - 3|\,dx = \int_0^{3/2} (3 - 2x)\,dx + \int_{3/2}^3 (2x - 3)\,dx \left(\text{split up the integral at the zero } x = \frac{3}{2}\right)$

$$= \left[3x - x^2\right]_0^{3/2} + \left[x^2 - 3x\right]_{3/2}^3 = \left(\frac{9}{2} - \frac{9}{4}\right) - 0 + (9 - 9) - \left(\frac{9}{4} - \frac{9}{2}\right) = 2\left(\frac{9}{2} - \frac{9}{4}\right) = \frac{9}{2}$$

25. $\displaystyle\int_0^3 |x^2 - 4|\, dx = \int_0^2 (4 - x^2)\, dx + \int_2^3 (x^2 - 4)\, dx$

$\qquad = \left[4x - \dfrac{x^3}{3} \right]_0^2 + \left[\dfrac{x^3}{3} - 4x \right]_2^3$

$\qquad = \left(8 - \dfrac{8}{3} \right) + (9 - 12) - \left(\dfrac{8}{3} - 8 \right)$

$\qquad = \dfrac{23}{3}$

27. $\displaystyle\int_0^\pi (1 + \sin x)\, dx = \left[x - \cos x \right]_0^\pi = (\pi + 1) - (0 - 1) = 2 + \pi$

29. $\displaystyle\int_{-\pi/6}^{\pi/6} \sec^2 x\, dx = \left[\tan x \right]_{-\pi/6}^{\pi/6} = \dfrac{\sqrt{3}}{3} - \left(-\dfrac{\sqrt{3}}{3} \right) = \dfrac{2\sqrt{3}}{3}$

31. $\displaystyle\int_{-\pi/3}^{\pi/3} 4 \sec \theta \tan \theta\, d\theta = \left[4 \sec \theta \right]_{-\pi/3}^{\pi/3} = 4(2) - 4(2) = 0$

33. $\displaystyle\int_0^3 10{,}000(t - 6)\, dt = 10{,}000 \left[\dfrac{t^2}{2} - 6t \right]_0^3 = -\$135{,}000$

35. $A = \displaystyle\int_0^1 (x - x^2)\, dx = \left[\dfrac{x^2}{2} - \dfrac{x^3}{3} \right]_0^1 = \dfrac{1}{6}$

37. $A = \displaystyle\int_0^3 (3 - x)\sqrt{x}\, dx = \int_0^3 (3x^{1/2} - x^{3/2})\, dx = \left[2x^{3/2} - \dfrac{2}{5}x^{5/2} \right]_0^3 = \left[\dfrac{x\sqrt{x}}{5}(10 - 2x) \right]_0^3 = \dfrac{12\sqrt{3}}{5}$

39. $A = \displaystyle\int_0^{\pi/2} \cos x\, dx = \left[\sin x \right]_0^{\pi/2} = 1$

41. Since $y \geq 0$ on $[0, 2]$,

$\qquad A = \displaystyle\int_0^2 (3x^2 + 1)\, dx = \left[x^3 + x \right]_0^2 = 8 + 2 = 10.$

43. Since $y \geq 0$ on $[0, 2]$,

$\qquad A = \displaystyle\int_0^2 (x^3 + x)\, dx = \left[\dfrac{x^4}{4} + \dfrac{x^2}{2} \right]_0^2 = 4 + 2 = 6.$

45. $\displaystyle\int_0^2 (x - 2\sqrt{x})\, dx = \left[\dfrac{x^2}{2} - \dfrac{4x^{3/2}}{3} \right]_0^2 = 2 - \dfrac{8\sqrt{2}}{3}$

$\qquad f(c)(2 - 0) = \dfrac{6 - 8\sqrt{2}}{3}$

$\qquad c - 2\sqrt{c} = \dfrac{3 - 4\sqrt{2}}{3}$

$\qquad c - 2\sqrt{c} + 1 = \dfrac{3 - 4\sqrt{2}}{3} + 1$

$\qquad \left(\sqrt{c} - 1 \right)^2 = \dfrac{6 - 4\sqrt{2}}{3}$

$\qquad \sqrt{c} - 1 = \pm \sqrt{\dfrac{6 - 4\sqrt{2}}{3}}$

$\qquad c = \left[1 \pm \sqrt{\dfrac{6 - 4\sqrt{2}}{3}} \right]^2$

$\qquad c \approx 0.4380 \text{ or } c \approx 1.7908$

47. $\int_{-\pi/4}^{\pi/4} 2\sec^2 x \, dx = \left[2\tan x\right]_{-\pi/4}^{\pi/4} = 2(1) - 2(-1) = 4$

$$f(c)\left[\frac{\pi}{4} - \left(-\frac{\pi}{4}\right)\right] = 4$$

$$2\sec^2 c = \frac{8}{\pi}$$

$$\sec^2 c = \frac{4}{\pi}$$

$$\sec c = \pm\frac{2}{\sqrt{\pi}}$$

$$c = \pm\mathrm{arcsec}\left(\frac{2}{\sqrt{\pi}}\right)$$

$$= \pm\arccos\frac{\sqrt{\pi}}{2} \approx \pm0.4817$$

49. $\dfrac{1}{2 - (-2)}\displaystyle\int_{-2}^{2}(4 - x^2)\,dx = \dfrac{1}{4}\left[4x - \dfrac{1}{3}x^3\right]_{-2}^{2} = \dfrac{1}{4}\left[\left(8 - \dfrac{8}{3}\right) - \left(-8 + \dfrac{8}{3}\right)\right] = \dfrac{8}{3}$

Average value $= \dfrac{8}{3}$

$4 - x^2 = \dfrac{8}{3}$ when $x^2 = 4 - \dfrac{8}{3}$ or $x = \pm\dfrac{2\sqrt{3}}{3} \approx \pm1.155$.

51. $\dfrac{1}{\pi - 0}\displaystyle\int_{0}^{\pi}\sin x \, dx = \left[-\dfrac{1}{\pi}\cos x\right]_{0}^{\pi} = \dfrac{2}{\pi}$

Average value $= \dfrac{2}{\pi}$

$$\sin x = \frac{2}{\pi}$$

$$x \approx 0.690, 2.451$$

53. If f is continuous on $[a, b]$ and $F'(x) = f(x)$ on $[a, b]$,

then $\displaystyle\int_{a}^{b} f(x)\,dx = F(b) - F(a)$.

55. $\displaystyle\int_{0}^{2} f(x)\,dx = -(\text{area of region } A) = -1.5$

57. $\displaystyle\int_{0}^{6} |f(x)|\,dx = -\int_{0}^{2} f(x)\,dx + \int_{2}^{6} f(x)\,dx = 1.5 + 5.0 = 6.5$

59. $\displaystyle\int_{0}^{6}[2 + f(x)]\,dx = \int_{0}^{6} 2\,dx + \int_{0}^{6} f(x)\,dx$

$$= 12 + 3.5 = 15.5$$

61. (a) $F(x) = k\sec^2 x$

$F(0) = k = 500$

$F(x) = 500\sec^2 x$

(b) $\dfrac{1}{\pi/3 - 0}\displaystyle\int_{0}^{\pi/3} 500\sec^2 x \, dx = \dfrac{1500}{\pi}\left[\tan x\right]_{0}^{\pi/3}$

$$= \frac{1500}{\pi}\left(\sqrt{3} - 0\right)$$

$$\approx 826.99 \text{ newtons}$$

$$\approx 827 \text{ newtons}$$

63. $\dfrac{1}{5 - 0}\displaystyle\int_{0}^{5}(0.1729t + 0.1522t^2 - 0.0374t^3)\,dt \approx \dfrac{1}{5}\left[0.08645t^2 + 0.05073t^3 - 0.00935t^4\right]_{0}^{5} \approx 0.5318 \text{ liter}$

65. (a)

The area above the *x*-axis equals the area below the *x*-axis. Thus, the average value is zero.

(b)

The average value of *S* appears to be *g*.

67. (a) $v = -8.61 \times 10^{-4}t^3 + 0.0782t^2 - 0.208t + 0.0952$

(b)

(c) $\int_0^{60} v(t)\,dt = \left[\dfrac{-8.61 \times 10^{-4}t^4}{4} + \dfrac{0.0782t^3}{3} - \dfrac{0.208t^2}{2} + 0.0952t\right]_0^{60} \approx 2476 \text{ meters}$

69. $F(x) = \int_0^x (t - 5)\,dt = \left[\dfrac{t^2}{2} - 5t\right]_0^x = \dfrac{x^2}{2} - 5x$

$F(2) = \dfrac{4}{2} - 5(2) = -8$

$F(5) = \dfrac{25}{2} - 5(5) = -\dfrac{25}{2}$

$F(8) = \dfrac{64}{2} - 5(8) = -8$

71. $F(x) = \int_1^x \dfrac{10}{v^2}\,dv = \int_1^x 10v^{-2}\,dv = \dfrac{-10}{v}\Big]_1^x$

$= -\dfrac{10}{x} + 10 = 10\left(1 - \dfrac{1}{x}\right)$

$F(2) = 10\left(\dfrac{1}{2}\right) = 5$

$F(5) = 10\left(\dfrac{4}{5}\right) = 8$

$F(8) = 10\left(\dfrac{7}{8}\right) = \dfrac{35}{4}$

73. $F(x) = \int_1^x \cos\theta\,d\theta = \sin\theta\Big]_1^x = \sin x - \sin 1$

$F(2) = \sin 2 - \sin 1 = 0.0678$

$F(5) = \sin 5 - \sin 1 \approx -1.8004$

$F(8) = \sin 8 - \sin 1 \approx 0.1479$

75. (a) $\int_0^x (t + 2)\,dt = \left[\dfrac{t^2}{2} + 2t\right]_0^x = \dfrac{1}{2}x^2 + 2x$

(b) $\dfrac{d}{dx}\left[\dfrac{1}{2}x^2 + 2x\right] = x + 2$

77. (a) $\int_8^x \sqrt[3]{t}\,dt = \left[\dfrac{3}{4}t^{4/3}\right]_8^x = \dfrac{3}{4}(x^{4/3} - 16) = \dfrac{3}{4}x^{4/3} - 12$

(b) $\dfrac{d}{dx}\left[\dfrac{3}{4}x^{4/3} - 12\right] = x^{1/3} = \sqrt[3]{x}$

79. (a) $\int_{x/4}^x \sec^2 t\,dt = \left[\tan t\right]_{x/4}^x = \tan x - 1$

(b) $\dfrac{d}{dx}[\tan x - 1] = \sec^2 x$

81. $F(x) = \int_{-2}^x (t^2 - 2t)\,dt$

$F'(x) = x^2 - 2x$

83. $F(x) = \int_{-1}^x \sqrt{t^4 + 1}\,dt$

$F'(x) = \sqrt{x^4 + 1}$

85. $F(x) = \int_0^x t\cos t\,dt$

$F'(x) = x\cos x$

87. $F(x) = \displaystyle\int_x^{x+2} (4t + 1)\, dt$

$\qquad = \left[2t^2 + t \right]_x^{x+2}$

$\qquad = [2(x + 2)^2 + (x + 2)] - [2x^2 + x]$

$\qquad = 8x + 10$

$\quad F'(x) = 8$

Alternate solution:

$F(x) = \displaystyle\int_x^{x+2} (4t + 1)\, dt$

$\qquad = \displaystyle\int_x^0 (4t + 1)\, dt + \displaystyle\int_0^{x+2} (4t + 1)\, dt$

$\qquad = -\displaystyle\int_0^x (4t + 1)\, dt + \displaystyle\int_0^{x+2} (4t + 1)\, dt$

$F'(x) = -(4x + 1) + 4(x + 2) + 1 = 8$

89. $F(x) = \displaystyle\int_0^{\sin x} \sqrt{t}\, dt = \left[\dfrac{2}{3} t^{3/2} \right]_0^{\sin x} = \dfrac{2}{3}(\sin x)^{3/2}$

$\quad F'(x) = (\sin x)^{1/2} \cos x = \cos x \sqrt{\sin x}$

Alternate solution

$F(x) = \displaystyle\int_0^{\sin x} \sqrt{t}\, dt$

$F'(x) = \sqrt{\sin x} \,\dfrac{d}{dx}(\sin x) = \sqrt{\sin x}(\cos x)$

91. $F(x) = \displaystyle\int_0^{x^3} \sin t^2\, dt$

$\quad F'(x) = \sin(x^3)^2 \cdot 3x^2 = 3x^2 \sin x^6$

93. $g(x) = \displaystyle\int_0^x f(t)\, dt$

$\quad g(0) = 0,\ g(1) \approx \dfrac{1}{2},\ g(2) \approx 1,\ g(3) \approx \dfrac{1}{2},\ g(4) = 0$

g has a relative maximum at $x = 2$.

95. (a) $C(x) = 5000\left(25 + 3\displaystyle\int_0^x t^{1/4}\, dt \right)$

$\qquad\quad = 5000\left(25 + 3\left[\dfrac{4}{5} t^{5/4} \right]_0^x \right)$

$\qquad\quad = 5000\left(25 + \dfrac{12}{5} x^{5/4} \right) = 1000(125 + 12x^{5/4})$

(b) $C(1) = 1000(125 + 12(1)) = \$137{,}000$

$\quad\ C(5) = 1000(125 + 12(5)^{5/4}) \approx \$214{,}721$

$\quad\ C(10) = 1000(125 + 12(10)^{5/4}) \approx \$338{,}394$

97. True

99. False; $\displaystyle\int_{-1}^1 x^{-2}\, dx = \int_{-1}^0 x^{-2}\, dx + \int_0^1 x^{-2}\, dx$

Each of these integrals is infinite. $f(x) = x^{-2}$ has a nonremovable discontinuity at $x = 0$.

101. $f(x) = \displaystyle\int_0^{1/x} \dfrac{1}{t^2 + 1}\, dt + \int_0^x \dfrac{1}{t^2 + 1}\, dt$

By the Second Fundamental Theorem of Calculus, we have

$\quad f'(x) = \dfrac{1}{(1/x)^2 + 1}\left(-\dfrac{1}{x^2} \right) + \dfrac{1}{x^2 + 1}$

$\qquad\quad = -\dfrac{1}{1 + x^2} + \dfrac{1}{x^2 + 1} = 0.$

Since $f'(x) = 0$, $f(x)$ must be constant.

103. $x(t) = t^3 - 6t^2 + 9t - 2$

$$x'(t) = 3t^2 - 12t + 9$$

$$= 3(t^2 - 4t + 3)$$

$$= 3(t - 3)(t - 1)$$

Total distance $= \displaystyle\int_0^5 |x'(t)|\,dt$

$$= \int_0^5 3|(t - 3)(t - 1)|\,dt$$

$$= 3\int_0^1 (t^2 - 4t + 3)\,dt - 3\int_1^3 (t^2 - 4t + 3)\,dt + 3\int_3^5 (t^2 - 4t + 3)\,dt$$

$$= 4 + 4 + 20$$

$$= 28 \text{ units}$$

105. Total distance $= \displaystyle\int_1^4 |x'(t)|\,dt$

$$= \int_1^4 |v(t)|\,dt$$

$$= \int_1^4 \frac{1}{\sqrt{t}}\,dt$$

$$= 2t^{1/2}\Big]_1^4$$

$$= 2(2 - 1) = 2 \text{ units}$$

Section 4.5   Integration by Substitution

$\displaystyle\int f(g(x))g'(x)\,dx$	$u = g(x)$	$du = g'(x)\,dx$
1. $\displaystyle\int (5x^2 + 1)^2(10x)\,dx$	$5x^2 + 1$	$10x\,dx$
3. $\displaystyle\int \frac{x}{\sqrt{x^2 + 1}}\,dx$	$x^2 + 1$	$2x\,dx$
5. $\displaystyle\int \tan^2 x \sec^2 x\,dx$	$\tan x$	$\sec^2 x\,dx$

7. $\displaystyle\int (1 + 2x)^4\, 2\,dx = \frac{(1 + 2x)^5}{5} + C$

Check: $\dfrac{d}{dx}\left[\dfrac{(1 + 2x)^5}{5} + C\right] = 2(1 + 2x)^4$

9. $\displaystyle\int (9 - x^2)^{1/2}(-2x)\,dx = \frac{(9 - x^2)^{3/2}}{3/2} + C = \frac{2}{3}(9 - x^2)^{3/2} + C$

Check: $\dfrac{d}{dx}\left[\dfrac{2}{3}(9 - x^2)^{3/2} + C\right] = \dfrac{2}{3} \cdot \dfrac{3}{2}(9 - x^2)^{1/2}(-2x) = \sqrt{9 - x^2}\,(-2x)$

11. $\int x^3(x^4 + 3)^2 \, dx = \frac{1}{4}\int (x^4 + 3)^2(4x^3) \, dx = \frac{1}{4}\frac{(x^4 + 3)^3}{3} + C = \frac{(x^4 + 3)^3}{12} + C$

Check: $\frac{d}{dx}\left[\frac{(x^4 + 3)^3}{12} + C\right] = \frac{3(x^4 + 3)^2}{12}(4x^3) = (x^4 + 3)^2(x^3)$

13. $\int x^2(x^3 - 1)^4 \, dx = \frac{1}{3}\int (x^3 - 1)^4(3x^2) \, dx = \frac{1}{3}\left[\frac{(x^3 - 1)^5}{5}\right] + C = \frac{(x^3 - 1)^5}{15} + C$

Check: $\frac{d}{dx}\left[\frac{(x^3 - 1)^5}{15} + C\right] = \frac{5(x^3 - 1)^4(3x^2)}{15} = x^2(x^3 - 1)^4$

15. $\int t\sqrt{t^2 + 2} \, dt = \frac{1}{2}\int (t^2 + 2)^{1/2}(2t) \, dt = \frac{1}{2}\frac{(t^2 + 2)^{3/2}}{3/2} + C = \frac{(t^2 + 2)^{3/2}}{3} + C$

Check: $\frac{d}{dt}\left[\frac{(t^2 + 2)^{3/2}}{3} + C\right] = \frac{3/2(t^2 + 2)^{1/2}(2t)}{3} = (t^2 + 2)^{1/2}t$

17. $\int 5x(1 - x^2)^{1/3} \, dx = -\frac{5}{2}\int (1 - x^2)^{1/3}(-2x) \, dx = -\frac{5}{2} \cdot \frac{(1 - x^2)^{4/3}}{4/3} + C = -\frac{15}{8}(1 - x^2)^{4/3} + C$

Check: $\frac{d}{dx}\left[-\frac{15}{8}(1 - x^2)^{4/3} + C\right] = -\frac{15}{8} \cdot \frac{4}{3}(1 - x^2)^{1/3}(-2x) = 5x(1 - x^2)^{1/3} = 5x\sqrt[3]{1 - x^2}$

19. $\int \frac{x}{(1 - x^2)^3} \, dx = -\frac{1}{2}\int (1 - x^2)^{-3}(-2x) \, dx = -\frac{1}{2}\frac{(1 - x^2)^{-2}}{-2} + C = \frac{1}{4(1 - x^2)^2} + C$

Check: $\frac{d}{dx}\left[\frac{1}{4(1 - x^2)^2} + C\right] = \frac{1}{4}(-2)(1 - x^2)^{-3}(-2x) = \frac{x}{(1 - x^2)^3}$

21. $\int \frac{x^2}{(1 + x^3)^2} \, dx = \frac{1}{3}\int (1 + x^3)^{-2}(3x^2) \, dx = \frac{1}{3}\left[\frac{(1 + x^3)^{-1}}{-1}\right] + C = -\frac{1}{3(1 + x^3)} + C$

Check: $\frac{d}{dx}\left[-\frac{1}{3(1 + x^3)} + C\right] = -\frac{1}{3}(-1)(1 + x^3)^{-2}(3x^2) = \frac{x^2}{(1 + x^3)^2}$

23. $\int \frac{x}{\sqrt{1 - x^2}} \, dx = -\frac{1}{2}\int (1 - x^2)^{-1/2}(-2x) \, dx = -\frac{1}{2}\frac{(1 - x^2)^{1/2}}{1/2} + C = -\sqrt{1 - x^2} + C$

Check: $\frac{d}{dx}[-(1 - x^2)^{1/2} + C] = -\frac{1}{2}(1 - x^2)^{-1/2}(-2x) = \frac{x}{\sqrt{1 - x^2}}$

25. $\int \left(1 + \frac{1}{t}\right)^3\left(\frac{1}{t^2}\right) dt = -\int \left(1 + \frac{1}{t}\right)^3\left(-\frac{1}{t^2}\right) dt = -\frac{[1 + (1/t)]^4}{4} + C$

Check: $\frac{d}{dt}\left[-\frac{[1 + (1/t)]^4}{4} + C\right] = -\frac{1}{4}(4)\left(1 + \frac{1}{t}\right)^3\left(-\frac{1}{t^2}\right) = \frac{1}{t^2}\left(1 + \frac{1}{t}\right)^3$

27. $\int \frac{1}{\sqrt{2x}} \, dx = \frac{1}{2}\int (2x)^{-1/2} \, 2 \, dx = \frac{1}{2}\left[\frac{(2x)^{1/2}}{1/2}\right] + C = \sqrt{2x} + C$

Check: $\frac{d}{dx}[\sqrt{2x} + C] = \frac{1}{2}(2x)^{-1/2}(2) = \frac{1}{\sqrt{2x}}$

29. $\displaystyle\int \frac{x^2 + 3x + 7}{\sqrt{x}}\,dx = \int (x^{3/2} + 3x^{1/2} + 7x^{-1/2})\,dx = \frac{2}{5}x^{5/2} + 2x^{3/2} + 14x^{1/2} + C = \frac{2}{5}\sqrt{x}(x^2 + 5x + 35) + C$

Check: $\displaystyle\frac{d}{dx}\left[\frac{2}{5}x^{5/2} + 2x^{3/2} + 14x^{1/2} + C\right] = \frac{x^2 + 3x + 7}{\sqrt{x}}$

31. $\displaystyle\int t^2\left(t - \frac{2}{t}\right)dt = \int (t^3 - 2t)\,dt = \frac{1}{4}t^4 - t^2 + C$

Check: $\displaystyle\frac{d}{dt}\left[\frac{1}{4}t^4 - t^2 + C\right] = t^3 - 2t = t^2\left(t - \frac{2}{t}\right)$

33. $\displaystyle\int (9 - y)\sqrt{y}\,dy = \int (9y^{1/2} - y^{3/2})\,dy = 9\left(\frac{2}{3}y^{3/2}\right) - \frac{2}{5}y^{5/2} + C = \frac{2}{5}y^{3/2}(15 - y) + C$

Check: $\displaystyle\frac{d}{dy}\left[\frac{2}{5}y^{3/2}(15 - y) + C\right] = \frac{d}{dy}\left[6y^{3/2} - \frac{2}{5}y^{5/2} + C\right] = 9y^{1/2} - y^{3/2} = (9 - y)\sqrt{y}$

35. $\displaystyle y = \int\left[4x + \frac{4x}{\sqrt{16 - x^2}}\right]dx$

$\displaystyle = 4\int x\,dx - 2\int (16 - x^2)^{-1/2}(-2x)\,dx$

$\displaystyle = 4\left(\frac{x^2}{2}\right) - 2\left[\frac{(16 - x^2)^{1/2}}{1/2}\right] + C$

$\displaystyle = 2x^2 - 4\sqrt{16 - x^2} + C$

37. $\displaystyle y = \int \frac{x + 1}{(x^2 + 2x - 3)^2}\,dx$

$\displaystyle = \frac{1}{2}\int (x^2 + 2x - 3)^{-2}(2x + 2)\,dx$

$\displaystyle = \frac{1}{2}\left[\frac{(x^2 + 2x - 3)^{-1}}{-1}\right] + C$

$\displaystyle = -\frac{1}{2(x^2 + 2x - 3)} + C$

39. (a)

(b) $\displaystyle\frac{dy}{dx} = x\sqrt{4 - x^2},\ (2, 2)$

$\displaystyle y = \int x\sqrt{4 - x^2}\,dx = -\frac{1}{2}\int (4 - x^2)^{1/2}(-2x\,dx)$

$\displaystyle = -\frac{1}{2}\cdot\frac{2}{3}(4 - x^2)^{3/2} + C = -\frac{1}{3}(4 - x^2)^{3/2} + C$

$\displaystyle (2, 2):\ 2 = -\frac{1}{3}(4 - 2^2)^{3/2} + C \implies C = 2$

$\displaystyle y = -\frac{1}{3}(4 - x^2)^{3/2} + 2$

41. $\displaystyle\int \pi \sin \pi x\,dx = -\cos \pi x + C$

43. $\displaystyle\int \sin 2x\,dx = \frac{1}{2}\int (\sin 2x)(2x)\,dx = -\frac{1}{2}\cos 2x + C$

45. $\displaystyle\int \frac{1}{\theta^2}\cos\frac{1}{\theta}\,d\theta = -\int \cos\frac{1}{\theta}\left(-\frac{1}{\theta^2}\right)d\theta = -\sin\frac{1}{\theta} + C$

47. $\displaystyle\int \sin 2x \cos 2x \, dx = \frac{1}{2}\int (\sin 2x)(2 \cos 2x) \, dx = \frac{1}{2}\frac{(\sin 2x)^2}{2} + C = \frac{1}{4}\sin^2 2x + C \quad \text{OR}$

$\displaystyle\int \sin 2x \cos 2x \, dx = -\frac{1}{2}\int (\cos 2x)(-2 \sin 2x) \, dx = -\frac{1}{2}\frac{(\cos 2x)^2}{2} + C_1 = -\frac{1}{4}\cos^2 2x + C_1 \quad \text{OR}$

$\displaystyle\int \sin 2x \cos 2x \, dx = \frac{1}{2}\int 2 \sin 2x \cos 2x \, dx = \frac{1}{2}\int \sin 4x \, dx = -\frac{1}{8}\cos 4x + C_2$

49. $\displaystyle\int \tan^4 x \sec^2 x \, dx = \frac{\tan^5 x}{5} + C = \frac{1}{5}\tan^5 x + C$

51. $\displaystyle\int \frac{\csc^2 x}{\cot^3 x} \, dx = -\int (\cot x)^{-3}(-\csc^2 x) \, dx$

$\displaystyle\qquad = -\frac{(\cot x)^{-2}}{-2} + C = \frac{1}{2 \cot^2 x} + C = \frac{1}{2}\tan^2 x + C = \frac{1}{2}(\sec^2 x - 1) + C = \frac{1}{2}\sec^2 x + C_1$

53. $\displaystyle\int \cot^2 x \, dx = \int (\csc^2 x - 1) \, dx = -\cot x - x + C$

55. $\displaystyle f(x) = \int \cos \frac{x}{2} \, dx = 2 \sin \frac{x}{2} + C$

$\qquad\qquad\qquad\qquad$ Since $f(0) = 3 = 2 \sin 0 + C, C = 3.$ Thus,

$$f(x) = 2 \sin \frac{x}{2} + 3.$$

57. $u = x + 2, x = u - 2, dx = du$

$\displaystyle\int x\sqrt{x + 2} \, dx = \int (u - 2)\sqrt{u} \, du$

$\displaystyle\qquad\qquad\qquad = \int (u^{3/2} - 2u^{1/2}) \, du$

$\displaystyle\qquad\qquad\qquad = \frac{2}{5}u^{5/2} - \frac{4}{3}u^{3/2} + C$

$\displaystyle\qquad\qquad\qquad = \frac{2u^{3/2}}{15}(3u - 10) + C$

$\displaystyle\qquad\qquad\qquad = \frac{2}{15}(x + 2)^{3/2}[3(x + 2) - 10] + C$

$\displaystyle\qquad\qquad\qquad = \frac{2}{15}(x + 2)^{3/2}(3x - 4) + C$

59. $u = 1 - x, x = 1 - u, dx = -du$

$\displaystyle\int x^2\sqrt{1 - x} \, dx = -\int (1 - u)^2\sqrt{u} \, du$

$\displaystyle\qquad\qquad\qquad = -\int (u^{1/2} - 2u^{3/2} + u^{5/2}) \, du$

$\displaystyle\qquad\qquad\qquad = -\left(\frac{2}{3}u^{3/2} - \frac{4}{5}u^{5/2} + \frac{2}{7}u^{7/2}\right) + C$

$\displaystyle\qquad\qquad\qquad = -\frac{2u^{3/2}}{105}(35 - 42u + 15u^2) + C$

$\displaystyle\qquad\qquad\qquad = -\frac{2}{105}(1 - x)^{3/2}[35 - 42(1 - x) + 15(1 - x)^2] + C$

$\displaystyle\qquad\qquad\qquad = -\frac{2}{105}(1 - x)^{3/2}(15x^2 + 12x + 8) + C$

61. $u = 2x - 1, x = \frac{1}{2}(u + 1), dx = \frac{1}{2} du$

$$\int \frac{x^2 - 1}{\sqrt{2x - 1}} dx = \int \frac{[(1/2)(u + 1)]^2 - 1}{\sqrt{u}} \frac{1}{2} du$$

$$= \frac{1}{8} \int u^{-1/2}[(u^2 + 2u + 1) - 4] du$$

$$= \frac{1}{8} \int (u^{3/2} + 2u^{1/2} - 3u^{-1/2}) du$$

$$= \frac{1}{8}\left(\frac{2}{5}u^{5/2} + \frac{4}{3}u^{3/2} - 6u^{1/2}\right) + C$$

$$= \frac{u^{1/2}}{60}(3u^2 + 10u - 45) + C$$

$$= \frac{\sqrt{2x - 1}}{60}[3(2x - 1)^2 + 10(2x - 1) - 45] + C$$

$$= \frac{1}{60}\sqrt{2x - 1}(12x^2 + 8x - 52) + C$$

$$= \frac{1}{15}\sqrt{2x - 1}(3x^2 + 2x - 13) + C$$

63. $u = x + 1, x = u - 1, dx = du$

$$\int \frac{-x}{(x + 1) - \sqrt{x + 1}} dx = \int \frac{-(u - 1)}{u - \sqrt{u}} du$$

$$= -\int \frac{(\sqrt{u} + 1)(\sqrt{u} - 1)}{\sqrt{u}(\sqrt{u} - 1)} du$$

$$= -\int (1 + u^{-1/2}) du$$

$$= -(u + 2u^{1/2}) + C$$

$$= -u - 2\sqrt{u} + C$$

$$= -(x + 1) - 2\sqrt{x + 1} + C$$

$$= -x - 2\sqrt{x + 1} - 1 + C$$

$$= -(x + 2\sqrt{x + 1}) + C_1$$

where $C_1 = -1 + C.$

65. Let $u = x^2 + 1, du = 2x\, dx.$

$$\int_{-1}^{1} x(x^2 + 1)^3 dx = \frac{1}{2}\int_{-1}^{1} (x^2 + 1)^3(2x)\, dx = \left[\frac{1}{8}(x^2 + 1)^4\right]_{-1}^{1} = 0$$

67. Let $u = x^3 + 1, du = 3x^2\, dx$

$$\int_{1}^{2} 2x^2\sqrt{x^3 + 1}\, dx = 2 \cdot \frac{1}{3}\int_{1}^{2} (x^3 + 1)^{1/2}(3x^2)\, dx$$

$$= \left[\frac{2}{3}\frac{(x^3 + 1)^{3/2}}{3/2}\right]_{1}^{2}$$

$$= \frac{4}{9}\left[(x^3 + 1)^{3/2}\right]_{1}^{2}$$

$$= \frac{4}{9}\left[27 - 2\sqrt{2}\right] = 12 - \frac{8}{9}\sqrt{2}$$

69. Let $u = 2x + 1$, $du = 2\,dx$.

$$\int_0^4 \frac{1}{\sqrt{2x+1}}\,dx = \frac{1}{2}\int_0^4 (2x+1)^{-1/2}(2)\,dx = \left[\sqrt{2x+1}\,\right]_0^4 = \sqrt{9} - \sqrt{1} = 2$$

71. Let $u = 1 + \sqrt{x}$, $du = \frac{1}{2\sqrt{x}}\,dx$.

$$\int_1^9 \frac{1}{\sqrt{x}(1+\sqrt{x})^2}\,dx = 2\int_1^9 (1+\sqrt{x})^{-2}\left(\frac{1}{2\sqrt{x}}\right)dx = \left[-\frac{2}{1+\sqrt{x}}\right]_1^9 = -\frac{1}{2} + 1 = \frac{1}{2}$$

73. $u = 2 - x$, $x = 2 - u$, $dx = -du$

When $x = 1$, $u = 1$. When $x = 2$, $u = 0$.

$$\int_1^2 (x-1)\sqrt{2-x}\,dx = \int_1^0 -[(2-u)-1]\sqrt{u}\,du = \int_1^0 (u^{3/2} - u^{1/2})\,du = \left[\frac{2}{5}u^{5/2} - \frac{2}{3}u^{3/2}\right]_1^0 = -\left[\frac{2}{5} - \frac{2}{3}\right] = \frac{4}{15}$$

75. $\displaystyle\int_0^{\pi/2} \cos\left(\frac{2}{3}x\right)dx = \left[\frac{3}{2}\sin\left(\frac{2}{3}x\right)\right]_0^{\pi/2} = \frac{3}{2}\left(\frac{\sqrt{3}}{2}\right) = \frac{3\sqrt{3}}{4}$

77. $u = x + 1$, $x = u - 1$, $dx = du$

When $x = 0$, $u = 1$. When $x = 7$, $u = 8$.

$$\text{Area} = \int_0^7 x\sqrt[3]{x+1}\,dx = \int_1^8 (u-1)\sqrt[3]{u}\,du$$

$$= \int_1^8 (u^{4/3} - u^{1/3})\,du = \left[\frac{3}{7}u^{7/3} - \frac{3}{4}u^{4/3}\right]_1^8 = \left(\frac{384}{7} - 12\right) - \left(\frac{3}{7} - \frac{3}{4}\right) = \frac{1209}{28}$$

79. $\displaystyle A = \int_0^{\pi} (2\sin x + \sin 2x)\,dx = -\left[2\cos x + \frac{1}{2}\cos 2x\right]_0^{\pi} = 4$

81. $\displaystyle \text{Area} = \int_{\pi/2}^{2\pi/3} \sec^2\left(\frac{x}{2}\right)dx = 2\int_{\pi/2}^{2\pi/3}\sec^2\left(\frac{x}{2}\right)\left(\frac{1}{2}\right)dx = \left[2\tan\left(\frac{x}{2}\right)\right]_{\pi/2}^{2\pi/3} = 2(\sqrt{3}-1)$

83. $\displaystyle\int_0^4 \frac{x}{\sqrt{2x+1}}\,dx \approx 3.333 = \frac{10}{3}$ 　**85.** $\displaystyle\int_3^7 x\sqrt{x-3}\,dx \approx 28.8 = \frac{144}{5}$ 　**87.** $\displaystyle\int_0^3 \left(\theta + \cos\frac{\theta}{6}\right)d\theta \approx 7.377$

89. $\displaystyle\int (2x-1)^2\,dx = \frac{1}{2}\int (2x-1)^2\,2\,dx = \frac{1}{6}(2x-1)^3 + C_1 = \frac{4}{3}x^3 - 2x^2 + x - \frac{1}{6} + C_1$

$$\int (2x-1)^2\,dx = \int (4x^2 - 4x + 1)\,dx = \frac{4}{3}x^3 - 2x^2 + x + C_2$$

They differ by a constant: $C_2 = C_1 - \frac{1}{6}$.

91. $f(x) = x^2(x^2 + 1)$ is even.

$$\int_{-2}^{2} x^2(x^2 + 1)\, dx = 2\int_{0}^{2}(x^4 + x^2)\, dx = 2\left[\frac{x^5}{5} + \frac{x^3}{3}\right]_{0}^{2}$$

$$= 2\left[\frac{32}{5} + \frac{8}{3}\right] = \frac{272}{15}$$

93. $f(x) = x(x^2 + 1)^3$ is odd.

$$\int_{-2}^{2} x(x^2 + 1)^3\, dx = 0$$

95. $\int_{0}^{2} x^2\, dx = \left[\frac{x^3}{3}\right]_{0}^{2} = \frac{8}{3}$; the function x^2 is an even function.

(a) $\int_{-2}^{0} x^2\, dx = \int_{0}^{2} x^2\, dx = \frac{8}{3}$

(b) $\int_{-2}^{2} x^2\, dx = 2\int_{0}^{2} x^2\, dx = \frac{16}{3}$

(c) $\int_{0}^{2} (-x^2)\, dx = -\int_{0}^{2} x^2\, dx = -\frac{8}{3}$

(d) $\int_{-2}^{0} 3x^2\, dx = 3\int_{0}^{2} x^2\, dx = 8$

97. $\int_{-4}^{4}(x^3 + 6x^2 - 2x - 3)\, dx = \int_{-4}^{4}(x^3 - 2x)\, dx + \int_{-4}^{4}(6x^2 - 3)\, dx = 0 + 2\int_{0}^{4}(6x^2 - 3)\, dx = 2\left[2x^3 - 3x\right]_{0}^{4} = 232$

99. Answers will vary. See "Guidelines for Making a Change of Variables" on page 292.

101. $f(x) = x(x^2 + 1)^2$ is odd. Hence, $\int_{-2}^{2} x(x^2 + 1)^2\, dx = 0$.

103. $\dfrac{dV}{dt} = \dfrac{k}{(t + 1)^2}$

$V(t) = \displaystyle\int \frac{k}{(t + 1)^2}\, dt = -\frac{k}{t + 1} + C$

$V(0) = -k + C = 500{,}000$

$V(1) = -\dfrac{1}{2}k + C = 400{,}000$

Solving this system yields $k = -200{,}000$ and $C = 300{,}000$. Thus,

$$V(t) = \frac{200{,}000}{t + 1} + 300{,}000.$$

When $t = 4$, $V(4) = \$340{,}000$.

105. $\dfrac{1}{b - a}\displaystyle\int_{a}^{b}\left[74.50 + 43.75 \sin \frac{\pi t}{6}\right] dt = \frac{1}{b - a}\left[74.50t - \frac{262.5}{\pi}\cos\frac{\pi t}{6}\right]_{a}^{b}$

(a) $\dfrac{1}{3}\left[74.50t - \dfrac{262.5}{\pi}\cos\dfrac{\pi t}{6}\right]_{0}^{3} = \dfrac{1}{3}\left(223.5 + \dfrac{262.5}{\pi}\right) \approx 102.352$ thousand units

(b) $\dfrac{1}{3}\left[74.50t - \dfrac{262.5}{\pi}\cos\dfrac{\pi t}{6}\right]_{3}^{6} = \dfrac{1}{3}\left(447 + \dfrac{262.5}{\pi} - 223.5\right) \approx 102.352$ thousand units

(c) $\dfrac{1}{12}\left[74.50t - \dfrac{262.5}{\pi}\cos\dfrac{\pi t}{6}\right]_{0}^{12} = \dfrac{1}{12}\left(894 - \dfrac{262.5}{\pi} + \dfrac{262.5}{\pi}\right) = 74.5$ thousand units

107. $\dfrac{1}{b-a}\displaystyle\int_a^b [2\sin(60\pi t) + \cos(120\pi t)]\, dt = \dfrac{1}{b-a}\left[-\dfrac{1}{30\pi}\cos(60\pi t) + \dfrac{1}{120\,\pi}\sin(120\pi t) \right]_a^b$

(a) $\dfrac{1}{(1/60)-0}\left[-\dfrac{1}{30\pi}\cos(60\pi t) + \dfrac{1}{120\pi}\sin(120\pi t) \right]_0^{1/60} = 60\left[\left(\dfrac{1}{30\pi}+0\right) - \left(-\dfrac{1}{30\pi}\right) \right] = \dfrac{4}{\pi} \approx 1.273 \text{ amps}$

(b) $\dfrac{1}{(1/240)-0}\left[-\dfrac{1}{30\pi}\cos(60\pi t) + \dfrac{1}{120\pi}\sin(120\pi t) \right]_0^{1/240} = 240\left[\left(-\dfrac{1}{30\sqrt{2}\,\pi}+\dfrac{1}{120\pi}\right) - \left(-\dfrac{1}{30\pi}\right) \right]$

$= \dfrac{2}{\pi}\left(5 - 2\sqrt{2}\right) \approx 1.382 \text{ amps}$

(c) $\dfrac{1}{(1/30)-0}\left[-\dfrac{1}{30\pi}\cos(60\pi t) + \dfrac{1}{120\pi}\sin(120\pi t) \right]_0^{1/30} = 30\left[\left(\dfrac{1}{30\pi}\right) - \left(-\dfrac{1}{30\pi}\right) \right] = 0 \text{ amps}$

109. False

$$\int (2x+1)^2\, dx = \frac{1}{2}\int (2x+1)^2\, 2\, dx = \frac{1}{6}(2x+1)^3 + C$$

111. True

$$\int_{-10}^{10} (ax^3 + bx^2 + cx + d)\, dx = \underbrace{\int_{-10}^{10}(ax^3+cx)\,dx}_{\text{Odd}} + \underbrace{\int_{-10}^{10}(bx^2+d)\,dx}_{\text{Even}} = 0 + 2\int_0^{10}(bx^2+d)\,dx$$

113. True

$$4\int \sin x \cos x\, dx = 2\int \sin 2x\, dx = -\cos 2x + C$$

115. Let $u = x + h$, then $du = dx$. When $x = a$, $u = a + h$. When $x = b$, $u = b + h$. Thus,

$$\int_a^b f(x+h)\, dx = \int_{a+h}^{b+h} f(u)\, du = \int_{a+h}^{b+h} f(x)\, dx.$$

Section 4.6   Numerical Integration

1. Exact: $\displaystyle\int_0^2 x^2\, dx = \left[\frac{1}{3}x^3\right]_0^2 = \frac{8}{3} \approx 2.6667$

Trapezoidal: $\displaystyle\int_0^2 x^2\, dx \approx \frac{1}{4}\left[0 + 2\left(\frac{1}{2}\right)^2 + 2(1)^2 + 2\left(\frac{3}{2}\right)^2 + (2)^2 \right] = \frac{11}{4} = 2.7500$

Simpson's: $\displaystyle\int_0^2 x^2\, dx \approx \frac{1}{6}\left[0 + 4\left(\frac{1}{2}\right)^2 + 2(1)^2 + 4\left(\frac{3}{2}\right)^2 + (2)^2 \right] = \frac{8}{3} \approx 2.6667$

3. Exact: $\displaystyle\int_0^2 x^3\, dx = \left[\frac{x^4}{4}\right]_0^2 = 4.000$

Trapezoidal: $\displaystyle\int_0^2 x^3\, dx \approx \frac{1}{4}\left[0 + 2\left(\frac{1}{2}\right)^3 + 2(1)^3 + 2\left(\frac{3}{2}\right)^3 + (2)^3 \right] = \frac{17}{4} = 4.2500$

Simpson's: $\displaystyle\int_0^2 x^3\, dx \approx \frac{1}{6}\left[0 + 4\left(\frac{1}{2}\right)^3 + 2(1)^3 + 4\left(\frac{3}{2}\right)^3 + (2)^3 \right] = \frac{24}{6} = 4.0000$

5. Exact:
$$\int_0^2 x^3\,dx = \left[\frac{1}{4}x^4\right]_0^2 = 4.0000$$

Trapezoidal:
$$\int_0^2 x^3\,dx \approx \frac{1}{8}\left[0 + 2\left(\frac{1}{4}\right)^3 + 2\left(\frac{2}{4}\right)^3 + 2\left(\frac{3}{4}\right)^3 + 2(1)^3 + 2\left(\frac{5}{4}\right)^3 + 2\left(\frac{6}{4}\right)^3 + 2\left(\frac{7}{4}\right)^3 + 8\right] = 4.0625$$

Simpson's:
$$\int_0^2 x^3\,dx \approx \frac{1}{12}\left[0 + 4\left(\frac{1}{4}\right)^3 + 2\left(\frac{2}{4}\right)^3 + 4\left(\frac{3}{4}\right)^3 + 2(1)^3 + 4\left(\frac{5}{4}\right)^3 + 2\left(\frac{6}{4}\right)^3 + 4\left(\frac{7}{4}\right)^3 + 8\right] = 4.0000$$

7. Exact:
$$\int_4^9 \sqrt{x}\,dx = \left[\frac{2}{3}x^{3/2}\right]_4^9 = 18 - \frac{16}{3} = \frac{38}{3} \approx 12.6667$$

Trapezoidal:
$$\int_4^9 \sqrt{x}\,dx \approx \frac{5}{16}\left[2 + 2\sqrt{\frac{37}{8}} + 2\sqrt{\frac{21}{4}} + 2\sqrt{\frac{47}{8}} + 2\sqrt{\frac{26}{4}} + 2\sqrt{\frac{57}{8}} + 2\sqrt{\frac{31}{4}} + 2\sqrt{\frac{67}{8}} + 3\right]$$
$$\approx 12.6640$$

Simpson's:
$$\int_4^9 \sqrt{x}\,dx \approx \frac{5}{24}\left[2 + 4\sqrt{\frac{37}{8}} + \sqrt{21} + 4\sqrt{\frac{47}{8}} + \sqrt{26} + 4\sqrt{\frac{57}{8}} + \sqrt{31} + 4\sqrt{\frac{67}{8}} + 3\right] \approx 12.6667$$

9. Exact:
$$\int_1^2 \frac{1}{(x+1)^2}\,dx = \left[-\frac{1}{x+1}\right]_1^2 = -\frac{1}{3} + \frac{1}{2} = \frac{1}{6} \approx 0.1667$$

Trapezoidal:
$$\int_1^2 \frac{1}{(x+1)^2}\,dx \approx \frac{1}{8}\left[\frac{1}{4} + 2\left(\frac{1}{((5/4)+1)^2}\right) + 2\left(\frac{1}{((3/2)+1)^2}\right) + 2\left(\frac{1}{((7/4)+1)^2}\right) + \frac{1}{9}\right]$$
$$= \frac{1}{8}\left(\frac{1}{4} + \frac{32}{81} + \frac{8}{25} + \frac{32}{121} + \frac{1}{9}\right) \approx 0.1676$$

Simpson's:
$$\int_1^2 \frac{1}{(x+1)^2}\,dx \approx \frac{1}{12}\left[\frac{1}{4} + 4\left(\frac{1}{((5/4)+1)^2}\right) + 2\left(\frac{1}{((3/2)+1)^2}\right) + 4\left(\frac{1}{((7/4)+1)^2}\right) + \frac{1}{9}\right]$$
$$= \frac{1}{12}\left(\frac{1}{4} + \frac{64}{81} + \frac{8}{25} + \frac{64}{121} + \frac{1}{9}\right) \approx 0.1667$$

11. Trapezoidal:
$$\int_0^2 \sqrt{1+x^3}\,dx \approx \frac{1}{4}\left[1 + 2\sqrt{1+(1/8)} + 2\sqrt{2} + 2\sqrt{1+(27/8)} + 3\right] \approx 3.283$$

Simpson's:
$$\int_0^2 \sqrt{1+x^3}\,dx \approx \frac{1}{6}\left[1 + 4\sqrt{1+(1/8)} + 2\sqrt{2} + 4\sqrt{1+(27/8)} + 3\right] \approx 3.240$$

Graphing utility: 3.241

13.
$$\int_0^1 \sqrt{x}\sqrt{1-x}\,dx = \int_0^1 \sqrt{x(1-x)}\,dx$$

Trapezoidal:
$$\int_0^1 \sqrt{x(1-x)}\,dx \approx \frac{1}{8}\left[0 + 2\sqrt{\frac{1}{4}\left(1-\frac{1}{4}\right)} + 2\sqrt{\frac{1}{2}\left(1-\frac{1}{2}\right)} + 2\sqrt{\frac{3}{4}\left(1-\frac{3}{4}\right)}\right] \approx 0.342$$

Simpson's:
$$\int_0^1 \sqrt{x(1-x)}\,dx \approx \frac{1}{12}\left[0 + 4\sqrt{\frac{1}{4}\left(1-\frac{1}{4}\right)} + 2\sqrt{\frac{1}{2}\left(1-\frac{1}{2}\right)} + 4\sqrt{\frac{3}{4}\left(1-\frac{3}{4}\right)}\right] \approx 0.372$$

Graphing utility: 0.393

15. Trapezoidal: $\displaystyle\int_0^{\sqrt{\pi/2}} \cos(x^2)\,dx \approx \frac{\sqrt{\pi/2}}{8}\left[\cos 0 + 2\cos\left(\frac{\sqrt{\pi/2}}{4}\right)^2 + 2\cos\left(\frac{\sqrt{\pi/2}}{2}\right)^2 + 2\cos\left(\frac{\sqrt{\pi/2}}{4}\right)^2 + \cos\left(\sqrt{\frac{\pi}{2}}\right)^2\right]$

≈ 0.957

Simpson's: $\displaystyle\int_0^{\sqrt{\pi/2}} \cos(x^2)\,dx \approx \frac{\sqrt{\pi/2}}{12}\left[\cos 0 + 4\cos\left(\frac{\sqrt{\pi/2}}{4}\right)^2 + 2\cos\left(\frac{\sqrt{\pi/2}}{2}\right)^2 + 4\cos\left(\frac{\sqrt{\pi/2}}{4}\right)^2 + \cos\left(\sqrt{\frac{\pi}{2}}\right)^2\right]$

≈ 0.978

Graphing utility: 0.977

17. Trapezoidal: $\displaystyle\int_1^{1.1} \sin x^2\,dx \approx \frac{1}{80}[\sin(1) + 2\sin(1.025)^2 + 2\sin(1.05)^2 + 2\sin(1.075)^2 + \sin(1.1)^2] \approx 0.089$

Simpson's: $\displaystyle\int_1^{1.1} \sin x^2\,dx \approx \frac{1}{120}[\sin(1) + 4\sin(1.025)^2 + 2\sin(1.05)^2 + 4\sin(1.075)^2 + \sin(1.1)^2] \approx 0.089$

Graphing utility: 0.089

19. Trapezoidal: $\displaystyle\int_0^{\pi/4} x\tan x\,dx \approx \frac{\pi}{32}\left[0 + 2\left(\frac{\pi}{16}\right)\tan\left(\frac{\pi}{16}\right) + 2\left(\frac{2\pi}{16}\right)\tan\left(\frac{2\pi}{16}\right) + 2\left(\frac{3\pi}{16}\right)\tan\left(\frac{3\pi}{16}\right) + \frac{\pi}{4}\right] \approx 0.194$

Simpson's: $\displaystyle\int_0^{\pi/4} x\tan x\,dx \approx \frac{\pi}{48}\left[0 + 4\left(\frac{\pi}{16}\right)\tan\left(\frac{\pi}{16}\right) + 2\left(\frac{2\pi}{16}\right)\tan\left(\frac{2\pi}{16}\right) + 4\left(\frac{3\pi}{16}\right)\tan\left(\frac{3\pi}{16}\right) + \frac{\pi}{4}\right] \approx 0.186$

Graphing utility: 0.186

21. (a)

The Trapezoidal Rule overestimates the area if the graph of the integrand is concave up.

23. $f(x) = x^3$

$f'(x) = 3x^2$

$f''(x) = 6x$

$f'''(x) = 6$

$f^{(4)}(x) = 0$

(a) Trapezoidal: Error $\leq \dfrac{(2-0)^3}{12(4^2)}(12) = 0.5$ since

$f''(x)$ is maximum in $[0,2]$ when $x = 2$.

(b) Simpson's: Error $\leq \dfrac{(2-0)^5}{180(4^4)}(0) = 0$ since

$f^{(4)}(x) = 0$.

25. $f''(x) = \dfrac{2}{x^3}$ in $[1,3]$.

(a) $|f''(x)|$ is maximum when $x = 1$ and $|f''(1)| = 2$.

Trapezoidal: Error $\leq \dfrac{2^3}{12n^2}(2) < 0.00001$, $n^2 > 133{,}333.33$, $n > 365.15$; let $n = 366$.

$f^{(4)}(x) = \dfrac{24}{x^5}$ in $[1,3]$

(b) $|f^{(4)}(x)|$ is maximum when $x = 1$ and $|f^{(4)}(1)| = 24$.

Simpson's: Error $\leq \dfrac{2^5}{180n^4}(24) < 0.00001$, $n^4 > 426{,}666.67$, $n > 25.56$; let $n = 26$.

27. $f(x) = \sqrt{1 + x}$

(a) $f''(x) = -\dfrac{1}{4(1 + x)^{3/2}}$ in $[0, 2]$.

$|f''(x)|$ is maximum when $x = 0$ and $|f''(0)| = \dfrac{1}{4}$.

Trapezoidal: Error $\leq \dfrac{8}{12n^2}\left(\dfrac{1}{4}\right) < 0.00001$, $n^2 > 16{,}666.67$, $n > 129.10$; let $n = 130$.

(b) $f^{(4)}(x) = \dfrac{-15}{16(1 + x)^{7/2}}$ in $[0, 2]$

$|f^{(4)}(x)|$ is maximum when $x = 0$ and $|f^{(4)}(0)| = \dfrac{15}{16}$.

Simpson's: Error $\leq \dfrac{32}{180n^4}\left(\dfrac{15}{16}\right) < 0.00001$, $n^4 > 16{,}666.67$, $n > 11.36$; let $n = 12$.

29. $f(x) = \tan(x^2)$

(a) $f''(x) = 2\sec^2(x^2)[1 + 4x^2\tan(x^2)]$ in $[0, 1]$.

$|f''(x)|$ is maximum when $x = 1$ and $|f''(1)| \approx 49.5305$.

Trapezoidal: Error $\leq \dfrac{(1 - 0)^3}{12n^2}(49.5305) < 0.00001$, $n^2 > 412{,}754.17$, $n > 642.46$; let $n = 643$.

(b) $f^{(4)}(x) = 8\sec^2(x^2)[12x^2 + (3 + 32x^4)\tan(x^2) + 36x^2\tan^2(x^2) + 48x^4\tan^3(x^2)]$ in $[0, 1]$

$|f^{(4)}(x)|$ is maximum when $x = 1$ and $|f^{(4)}(1)| \approx 9184.4734$.

Simpson's: Error $\leq \dfrac{(1 - 0)^5}{180n^4}(9184.4734) < 0.00001$, $n^4 > 5{,}102{,}485.22$, $n > 47.53$; let $n = 48$.

31. Let $f(x) = Ax^3 + Bx^2 + Cx + D$. Then $f^{(4)}(x) = 0$.

Simpson's: Error $\leq \dfrac{(b - a)^5}{180n^4}(0) = 0$

Therefore, Simpson's Rule is exact when approximating the integral of a cubic polynomial.

Example: $\displaystyle\int_0^1 x^3\, dx = \dfrac{1}{6}\left[0 + 4\left(\dfrac{1}{2}\right)^3 + 1\right] = \dfrac{1}{4}$

This is the exact value of the integral.

33. $f(x) = \sqrt{2 + 3x^2}$ on $[0, 4]$.

n	$L(n)$	$M(n)$	$R(n)$	$T(n)$	$S(n)$
4	12.7771	15.3965	18.4340	15.6055	15.4845
8	14.0868	15.4480	16.9152	15.5010	15.4662
10	14.3569	15.4544	16.6197	15.4883	15.4658
12	14.5386	15.4578	16.4242	15.4814	15.4657
16	14.7674	15.4613	16.1816	15.4745	15.4657
20	14.9056	15.4628	16.0370	15.4713	15.4657

35. $f(x) = \sin\sqrt{x}$ on $[0, 4]$.

n	$L(n)$	$M(n)$	$R(n)$	$T(n)$	$S(n)$
4	2.8163	3.5456	3.7256	3.2709	3.3996
8	3.1809	3.5053	3.6356	3.4083	3.4541
10	3.2478	3.4990	3.6115	3.4296	3.4624
12	3.2909	3.4952	3.5940	3.4425	3.4674
16	3.3431	3.4910	3.5704	3.4568	3.4730
20	3.3734	3.4888	3.5552	3.4643	3.4759

37. $A = \displaystyle\int_0^{\pi/2} \sqrt{x}\cos x \, dx$

Simpson's Rule: $n = 14$

$$\int_0^{\pi/2} \sqrt{x}\cos x \, dx \approx \frac{\pi}{84}\left[\sqrt{0}\cos 0 + 4\sqrt{\frac{\pi}{28}}\cos\frac{\pi}{28} + 2\sqrt{\frac{\pi}{14}}\cos\frac{\pi}{14} + 4\sqrt{\frac{3\pi}{28}}\cos\frac{3\pi}{28} + \cdots + \sqrt{\frac{\pi}{2}}\cos\frac{\pi}{2}\right]$$

$$\approx 0.701$$

39. $W = \displaystyle\int_0^5 100x\sqrt{125 - x^3}\, dx$

Simpson's Rule: $n = 12$

$$\int_0^5 100x\sqrt{125 - x^3}\, dx \approx \frac{5}{3(12)}\left[0 + 400\left(\frac{5}{12}\right)\sqrt{125 - \left(\frac{5}{12}\right)^3} + 200\left(\frac{10}{12}\right)\sqrt{125 - \left(\frac{10}{12}\right)^3}\right.$$

$$\left. + 400\left(\frac{15}{12}\right)\sqrt{125 - \left(\frac{15}{12}\right)^3} + \cdots + 0\right] \approx 10{,}233.58 \text{ ft} \cdot \text{lb}$$

41. $\displaystyle\int_0^{1/2} \frac{6}{\sqrt{1 - x^2}}\, dx$ Simpson's Rule, $n = 6$

$$\pi \approx \frac{\left(\frac{1}{2} - 0\right)}{3(6)}\left[6 + 4(6.0209) + 2(6.0851) + 4(6.1968) + 2(6.3640) + 4(6.6002) + 6.9282\right]$$

$$\approx \frac{1}{36}[113.098] \approx 3.1416$$

43. Area $\approx \dfrac{1000}{2(10)}[125 + 2(125) + 2(120) + 2(112) + 2(90) + 2(90) + 2(95) + 2(88) + 2(75) + 2(35)] = 89{,}250 \text{ sq m}$

45. $\displaystyle\int_0^t \sin\sqrt{x}\, dx = 2, \; n = 10$

By trial and error, we obtain $t \approx 2.477$.

Review Exercises for Chapter 4

1.

3. $\int (2x^2 + x - 1)\, dx = \frac{2}{3}x^3 + \frac{1}{2}x^2 - x + C$

5. $\int \frac{x^3 + 1}{x^2}\, dx = \int \left(x + \frac{1}{x^2} \right) dx = \frac{1}{2}x^2 - \frac{1}{x} + C$

7. $\int (4x - 3\sin x)\, dx = 2x^2 + 3\cos x + C$

9. $f'(x) = -2x, \ (-1, 1)$

$f(x) = \int -2x\, dx = -x^2 + C$

When $x = -1$:

$\qquad y = -1 + C = 1$

$\qquad C = 2$

$\qquad y = 2 - x^2$

11. $\quad a(t) = a$

$\qquad v(t) = \int a\, dt = at + C_1$

$\qquad v(0) = 0 + C_1 = 0$ when $C_1 = 0$.

$\qquad v(t) = at$

$\qquad s(t) = \int at\, dt = \frac{a}{2}t^2 + C_2$

$\qquad s(0) = 0 + C_2 = 0$ when $C_2 = 0$.

$\qquad s(t) = \frac{a}{2}t^2$

$\quad s(30) = \frac{a}{2}(30)^2 = 3600$ or

$\qquad a = \frac{2(3600)}{(30)^2} = 8 \text{ ft/sec}^2.$

$\quad v(30) = 8(30) = 240 \text{ ft/sec}$

13. $a(t) = -32$

$\quad v(t) = -32t + 96$

$\quad s(t) = -16t^2 + 96t$

(a) $v(t) = -32t + 96 = 0$ when $t = 3$ sec.

(b) $s(3) = -144 + 288 = 144$ ft

(c) $v(t) = -32t + 96 = \frac{96}{2}$ when $t = \frac{3}{2}$ sec.

(d) $s\left(\frac{3}{2} \right) = -16\left(\frac{9}{4} \right) + 96\left(\frac{3}{2} \right) = 108$ ft

15. (a) $\sum\limits_{i=1}^{10} (2i - 1)$

(b) $\sum\limits_{i=1}^{n} i^3$

(c) $\sum\limits_{i=1}^{10} (4i + 2)$

17. $y = \dfrac{10}{x^2 + 1}, \Delta x = \dfrac{1}{2}, n = 4$

$$S(n) = S(4) = \frac{1}{2}\left[\frac{10}{1} + \frac{10}{(1/2)^2 + 1} + \frac{10}{(1)^2 + 1} + \frac{10}{(3/2)^2 + 1}\right]$$

$$\approx 13.0385$$

$$s(n) = s(4) = \frac{1}{2}\left[\frac{10}{(1/2)^2 + 1} + \frac{10}{1 + 1} + \frac{10}{(3/2)^2 + 1} + \frac{10}{2^2 + 1}\right]$$

$$\approx 9.0385$$

$9.0385 < \text{Area of Region} < 13.0385$

19. $y = 6 - x, \Delta x = \dfrac{4}{n},$ right endpoints

$$\text{Area} = \lim_{n\to\infty} \sum_{i=1}^{n} f(c_i)\,\Delta x$$

$$= \lim_{n\to\infty} \sum_{i=1}^{n} \left(6 - \frac{4i}{n}\right)\frac{4}{n}$$

$$= \lim_{n\to\infty} \frac{4}{n}\left[6n - \frac{4}{n}\frac{n(n+1)}{2}\right]$$

$$= \lim_{n\to\infty} \left[24 - 8\frac{n+1}{n}\right] = 24 - 8 = 16$$

21. $y = 5 - x^2, \Delta x = \dfrac{3}{n}$

$$\text{Area} = \lim_{n\to\infty} \sum_{i=1}^{n} f(c_i)\,\Delta x$$

$$= \lim_{n\to\infty} \sum_{i=1}^{n} \left[5 - \left(-2 + \frac{3i}{n}\right)^2\right]\left(\frac{3}{n}\right)$$

$$= \lim_{n\to\infty} \frac{3}{n}\sum_{i=1}^{n}\left[1 + \frac{12i}{n} - \frac{9i^2}{n^2}\right]$$

$$= \lim_{n\to\infty} \frac{3}{n}\left[n + \frac{12}{n}\frac{n(n+1)}{2} - \frac{9}{n^2}\frac{n(n+1)(2n+1)}{6}\right]$$

$$= \lim_{n\to\infty} \left[3 + 18\frac{n+1}{n} - \frac{9}{2}\frac{(n+1)(2n+1)}{n^2}\right]$$

$$= 3 + 18 - 9 = 12$$

23. $x = 5y - y^2, 2 \le y \le 5, \Delta y = \dfrac{3}{n}$

$$\text{Area} = \lim_{n\to\infty} \sum_{i=1}^{n}\left[5\left(2 + \frac{3i}{n}\right) - \left(2 + \frac{3i}{n}\right)^2\right]\left(\frac{3}{n}\right)$$

$$= \lim_{n\to\infty} \frac{3}{n}\sum_{i=1}^{n}\left[10 + \frac{15i}{n} - 4 - 12\frac{i}{n} - \frac{9i^2}{n^2}\right]$$

$$= \lim_{n\to\infty} \frac{3}{n}\sum_{i=1}^{n}\left[6 + \frac{3i}{n} - \frac{9i^2}{n^2}\right]$$

$$= \lim_{n\to\infty} \frac{3}{n}\left[6n + \frac{3}{n}\frac{n(n+1)}{2} - \frac{9}{n^2}\frac{n(n+1)(2n+1)}{6}\right]$$

$$= \left[18 + \frac{9}{2} - 9\right] = \frac{27}{2}$$

25. $\displaystyle\lim_{\|\Delta\|\to\infty}\sum_{i=1}^{n}(2c_i - 3)\,\Delta xi = \int_4^6 (2x - 3)\,dx$

27.

$$\int_0^5 (5 - |x - 5|)\,dx = \int_0^5 (5 - (5 - x))\,dx = \int_0^5 x\,dx = \frac{25}{2}$$

(triangle)

29. (a) $\displaystyle\int_2^6 [f(x) + g(x)]\,dx = \int_2^6 f(x)\,dx + \int_2^6 g(x)\,dx = 10 + 3 = 13$

(b) $\displaystyle\int_2^6 [f(x) - g(x)]\,dx = \int_2^6 f(x)\,dx - \int_2^6 g(x)\,dx = 10 - 3 = 7$

(c) $\displaystyle\int_2^6 [2f(x) - 3g(x)]\,dx = 2\int_2^6 f(x)\,dx - 3\int_2^6 g(x)\,dx = 2(10) - 3(3) = 11$

(d) $\displaystyle\int_2^6 5f(x)\,dx = 5\int_2^6 f(x)\,dx = 5(10) = 50$

31. $\displaystyle\int_1^8 \left(\sqrt[3]{x} + 1\right)dx = \left[\frac{3}{4}x^{4/3} + x\right]_1^8 = \left[\frac{3}{4}(16) + 8\right] - \left[\frac{3}{4} + 1\right] = \frac{73}{4}$ (c)

33. $\displaystyle\int_0^4 (2 + x)\,dx = \left[2x + \frac{x^2}{2}\right]_0^4 = 8 + \frac{16}{2} = 16$ **35.** $\displaystyle\int_{-1}^1 (4t^3 - 2t)\,dt = \left[t^4 - t^2\right]_{-1}^1 = 0$

37. $\displaystyle\int_4^9 x\sqrt{x}\,dx = \int_4^9 x^{3/2}\,dx = \left[\frac{2}{5}x^{5/2}\right]_4^9 = \frac{2}{5}\left[(\sqrt{9})^5 - (\sqrt{4})^5\right] = \frac{2}{5}(243 - 32) = \frac{422}{5}$

39. $\displaystyle\int_0^{3\pi/4} \sin\theta\,d\theta = \left[-\cos\theta\right]_0^{3\pi/4} = -\left(-\frac{\sqrt{2}}{2}\right) + 1 = 1 + \frac{\sqrt{2}}{2} = \frac{\sqrt{2} + 2}{2}$

41. $\displaystyle\int_1^3 (2x - 1)\,dx = \left[x^2 - x\right]_1^3 = 6$

43. $\displaystyle\int_3^4 (x^2 - 9)\,dx = \left[\frac{x^3}{3} - 9x\right]_3^4$

$$= \left(\frac{64}{3} - 36\right) - (9 - 27)$$

$$= \frac{64}{3} - \frac{54}{3} = \frac{10}{3}$$

45. $\int_0^1 (x - x^3)\,dx = \left[\dfrac{x^2}{2} - \dfrac{x^4}{4}\right]_0^1 = \dfrac{1}{2} - \dfrac{1}{4} = \dfrac{1}{4}$

47. Area $= \int_1^9 \dfrac{4}{\sqrt{x}}\,dx = \left[\dfrac{4x^{1/2}}{(1/2)}\right]_1^9 = 8(3 - 1) = 16$

49. $\dfrac{1}{9 - 4}\int_4^9 \dfrac{1}{\sqrt{x}}\,dx = \left[\dfrac{1}{5}\,2\sqrt{x}\right]_4^9 = \dfrac{2}{5}(3 - 2) = \dfrac{2}{5}$ Average value

$\dfrac{2}{5} = \dfrac{1}{\sqrt{x}}$

$\sqrt{x} = \dfrac{5}{2}$

$x = \dfrac{25}{4}$

51. $F'(x) = x^2\sqrt{1 + x^3}$

53. $F'(x) = x^2 + 3x + 2$

55. $\int (x^2 + 1)^3\,dx = \int (x^6 + 3x^4 + 3x^2 + 1)\,dx = \dfrac{x^7}{7} + \dfrac{3}{5}x^5 + x^3 + x + C$

57. $u = x^3 + 3,\, du = 3x^2\,dx$

$\int \dfrac{x^2}{\sqrt{x^3 + 3}}\,dx = \int (x^3 + 3)^{-1/2}\,x^2\,dx = \dfrac{1}{3}\int (x^3 + 3)^{-1/2}\,3x^2\,dx = \dfrac{2}{3}(x^3 + 3)^{1/2} + C$

59. $u = 1 - 3x^2,\, du = -6x\,dx$

$\int x(1 - 3x^2)^4\,dx = -\dfrac{1}{6}\int (1 - 3x^2)^4(-6x\,dx) = -\dfrac{1}{30}(1 - 3x^2)^5 + C = \dfrac{1}{30}(3x^2 - 1)^5 + C$

61. $\int \sin^3 x \cos x\,dx = \dfrac{1}{4}\sin^4 x + C$

63. $\int \dfrac{\sin\theta}{\sqrt{1 - \cos\theta}}\,d\theta = \int (1 - \cos\theta)^{-1/2}\sin\theta\,d\theta = 2(1 - \cos\theta)^{1/2} + C = 2\sqrt{1 - \cos\theta} + C$

65. $\int \tan^n x \sec^2 x\,dx = \dfrac{\tan^{n+1} x}{n + 1} + C,\, n \neq -1$

67. $\int (1 + \sec\pi x)^2 \sec\pi x \tan\pi x\,dx = \dfrac{1}{\pi}\int (1 + \sec\pi x)^2(\pi\sec\pi x \tan\pi x)\,dx = \dfrac{1}{3\pi}(1 + \sec\pi x)^3 + C$

69. $\int_{-1}^2 x(x^2 - 4)\,dx = \dfrac{1}{2}\int_{-1}^2 (x^2 - 4)(2x)\,dx = \dfrac{1}{2}\dfrac{(x^2 - 4)^2}{2}\Big]_{-1}^2 = \dfrac{1}{4}[0 - 9] = -\dfrac{9}{4}$

71. $\int_0^3 \frac{1}{\sqrt{1+x}}\,dx = \int_0^3 (1+x)^{-1/2}\,dx = \left[2(1+x)^{1/2}\right]_0^3 = 4 - 2 = 2$

73. $u = 1 - y, y = 1 - u, dy = -du$

When $y = 0$, $u = 1$. When $y = 1$, $u = 0$.

$$2\pi\int_0^1 (y+1)\sqrt{1-y}\,dy = 2\pi\int_1^0 -[(1-u)+1]\sqrt{u}\,du$$

$$= 2\pi\int_1^0 (u^{3/2} - 2u^{1/2})\,du = 2\pi\left[\frac{2}{5}u^{5/2} - \frac{4}{3}u^{3/2}\right]_1^0 = \frac{28\pi}{15}$$

75. $\int_0^\pi \cos\left(\frac{x}{2}\right)dx = 2\int_0^\pi \cos\left(\frac{x}{2}\right)\frac{1}{2}\,dx = \left[2\sin\left(\frac{x}{2}\right)\right]_0^\pi = 2$

77. $u = 1 - x, x = 1 - u, dx = -du$

When $x = a$, $u = 1 - a$. When $x = b$, $u = 1 - b$.

$$P_{a,b} = \int_a^b \frac{15}{4}x\sqrt{1-x}\,dx = \frac{15}{4}\int_{1-a}^{1-b} -(1-u)\sqrt{u}\,du$$

$$= \frac{15}{4}\int_{1-a}^{1-b} (u^{3/2} - u^{1/2})\,du = \frac{15}{4}\left[\frac{2}{5}u^{5/2} - \frac{2}{3}u^{3/2}\right]_{1-a}^{1-b} = \frac{15}{4}\left[\frac{2u^{3/2}}{15}(3u-5)\right]_{1-a}^{1-b} = \left[-\frac{(1-x)^{3/2}}{2}(3x+2)\right]_a^b$$

(a) $P_{0.50,0.75} = \left[-\frac{(1-x)^{3/2}}{2}(3x+2)\right]_{0.50}^{0.75} = 0.353 = 35.3\%$

(b) $P_{0,b} = \left[-\frac{(1-x)^{3/2}}{2}(3x+2)\right]_0^b = -\frac{(1-b)^{3/2}}{2}(3b+2) + 1 = 0.5$

$(1-b)^{3/2}(3b+2) = 1$

$b \approx 0.586 = 58.6\%$

79. $p = 1.20 + 0.04t$

$$C = \frac{15,000}{M}\int_t^{t+1} p\,ds = \frac{15,000}{M}\int_t^{t+1} (1.20 + 0.04s)\,ds$$

(a) 2000 corresponds to $t = 10$.

$$C = \frac{15,000}{M}\int_{10}^{11} [1.20 + 0.04t]\,dt$$

$$= \frac{15,000}{M}\left[1.20t + 0.02t^2\right]_{10}^{11} = \frac{24,300}{M}$$

(b) 2005 corresponds to $t = 15$.

$$C = \frac{15,000}{M}\left[1.20t + 0.02t^2\right]_{15}^{16} = \frac{27,300}{M}$$

81. Trapezoidal Rule ($n = 4$): $\int_1^2 \frac{1}{1+x^3}\,dx \approx \frac{1}{8}\left[\frac{1}{1+1^3} + \frac{2}{1+(1.25)^3} + \frac{2}{1+(1.5)^3} + \frac{2}{1+(1.75)^3} + \frac{1}{1+2^3}\right] \approx 0.257$

Simpson's Rule ($n = 4$): $\int_1^2 \frac{1}{1+x^3}\,dx \approx \frac{1}{12}\left[\frac{1}{1+1^3} + \frac{4}{1+(1.25)^3} + \frac{2}{1+(1.5)^3} + \frac{4}{1+(1.75)^3} + \frac{1}{1+2^3}\right] \approx 0.254$

Graphing utility: 0.254

83. Trapezoidal Rule ($n = 4$): $\int_0^{\pi/2} \sqrt{x}\cos x\,dx \approx 0.637$

Simpson's Rule ($n = 4$): 0.685

Graphing Utility: 0.704

85. (a) $R < I < T < L$

(b) $S(4) = \frac{4-0}{3(4)}[f(0) + 4f(1) + 2f(2) + 4f(3) + f(4)]$

$\approx \frac{1}{3}\left[4 + 4(2) + 2(1) + 4\left(\frac{1}{2}\right) + \frac{1}{4}\right] \approx 5.417$

Problem Solving for Chapter 4

1. (a) $L(1) = \int_1^1 \frac{1}{t}\,dt = 0$

(b) $L'(x) = \frac{1}{x}$ by the Second Fundamental Theorem of Calculus.

$L'(1) = 1$

(c) $L(x) = 1 = \int_1^x \frac{1}{t}\,dt$ for $x \approx 2.718$

$\int_1^{2.718} \frac{1}{t}\,dt = 0.999896$ (**Note:** The exact value of x is e, the base of the natural logarithm function.)

(d) We first show that $\int_1^{x_1} \frac{1}{t}\,dt = \int_{1/x_1}^1 \frac{1}{t}\,dt$.

To see this, let $u = \frac{t}{x_1}$ and $du = \frac{1}{x_1}\,dt$.

Then $\int_1^{x_1} \frac{1}{t}\,dt = \int_{1/x_1}^1 \frac{1}{ux_1}(x_1\,du) = \int_{1/x_1}^1 \frac{1}{u}\,du = \int_{1/x_1}^1 \frac{1}{t}\,dt$.

Now, $L(x_1 x_2) = \int_1^{x_1 x_2} \frac{1}{t}\,dt = \int_{1/x_1}^{x_2} \frac{1}{u}\,du \left(\text{using } u = \frac{t}{x_1}\right)$

$= \int_{1/x_1}^1 \frac{1}{u}\,du + \int_1^{x_2} \frac{1}{u}\,du$

$= \int_1^{x_1} \frac{1}{u}\,du + \int_1^{x_2} \frac{1}{u}\,du$

$= L(x_1) + L(x_2)$.

3. $S(x) = \int_0^x \sin\left(\frac{\pi t^2}{2}\right)dt$

(a)

(b)

The zeros of $y = \sin \dfrac{\pi x^2}{2}$ correspond to the relative extrema of $S(x)$.

(c) $S'(x) = \sin \dfrac{\pi x^2}{2} = 0 \;\Rightarrow\; \dfrac{\pi x^2}{2} = n\pi \;\Rightarrow\; x^2 = 2n \;\Rightarrow\; x = \sqrt{2n}$, n integer.

Relative maximum at $x = \sqrt{2} \approx 1.4142$ and $x = \sqrt{6} \approx 2.4495$

Relative minimum at $x = 2$ and $x = \sqrt{8} \approx 2.8284$

(d) $S''(x) = \cos\left(\dfrac{\pi x^2}{2}\right)(\pi x) = 0 \;\Rightarrow\; \dfrac{\pi x^2}{2} = \dfrac{\pi}{2} + n\pi \;\Rightarrow\; x^2 = 1 + 2n \;\Rightarrow\; x = \sqrt{1+2n}$, n integer

Points of inflection at $x = 1$, $\sqrt{3}$, $\sqrt{5}$, and $\sqrt{7}$.

5. (a)

(b)

x	0	1	2	3	4	5	6	7	8
$F(x)$	0	$-\dfrac{1}{2}$	-2	$-\dfrac{7}{2}$	-4	$-\dfrac{7}{2}$	-2	$\dfrac{1}{4}$	3

(c) $f(x) = \begin{cases} -x, & 0 \le x < 2 \\ x - 4, & 2 \le x < 6 \\ \dfrac{1}{2}x - 1, & 6 \le x \le 8 \end{cases}$

$F(x) = \displaystyle\int_0^x f(t)\,dt = \begin{cases} (-x^2/2), & 0 \le x < 2 \\ (x^2/2) - 4x + 4, & 2 \le x < 6 \\ (1/4)x^2 - x - 5, & 6 \le x \le 8 \end{cases}$

$F'(x) = f(x)$. F is decreasing on $(0, 4)$ and increasing on $(4, 8)$. Therefore, the minimum is -4 at $x = 4$, and the maximum is 3 at $x = 8$.

(d) $F''(x) = f'(x) = \begin{cases} -1, & 0 < x < 2 \\ 1, & 2 < x < 6 \\ \dfrac{1}{2}, & 6 < x < 8 \end{cases}$

$x = 2$ is a point of inflection, whereas $x = 6$ is not.

7. (a) $\displaystyle\int_{-1}^1 \cos x\,dx \approx \cos\left(-\frac{1}{\sqrt{3}}\right) + \cos\left(\frac{1}{\sqrt{3}}\right) = 2\cos\left(\frac{1}{\sqrt{3}}\right) \approx 1.6758$

$\displaystyle\int_{-1}^1 \cos x\,dx = \sin x\Big]_{-1}^1 = 2\sin(1) \approx 1.6829$

Error. $|1.6829 - 1.6758| = 0.0071$

(b) $\displaystyle\int_{-1}^1 \frac{1}{1 + x^2}\,dx \approx \frac{1}{1 + (1/3)} + \frac{1}{1 + (1/3)} = \frac{3}{2}$

(Note: exact answer is $\pi/2 \approx 1.5708$)

(c) Let $p(x) = ax^3 + bx^2 + cx + d$.

$\displaystyle\int_{-1}^1 p(x)\,dx = \left[\frac{ax^4}{4} + \frac{bx^3}{3} + \frac{cx^2}{2} + dx\right]_{-1}^1 = \frac{2b}{3} + 2d$

$p\left(-\frac{1}{\sqrt{3}}\right) + p\left(\frac{1}{\sqrt{3}}\right) = \left(\frac{b}{3} + d\right) + \left(\frac{b}{3} + d\right) = \frac{2b}{3} + 2d$

9. Consider $F(x) = [f(x)]^2 \implies F'(x) = 2f(x)f'(x)$. Thus,

$\displaystyle\int_a^b f(x)f'(x)\,dx = \int_a^b \frac{1}{2}F'(x)\,dx$

$= \left[\frac{1}{2}F(x)\right]_a^b$

$= \frac{1}{2}[F(b) - F(a)]$

$= \frac{1}{2}[f(b)^2 - f(a)^2]$

11. Consider $\displaystyle\int_0^1 x^5\,dx = \frac{x^6}{6}\Big]_0^1 = \frac{1}{6}$.

The corresponding Riemann Sum using right endpoints is

$S(n) = \dfrac{1}{n}\left[\left(\dfrac{1}{n}\right)^5 + \left(\dfrac{2}{n}\right)^5 + \cdots + \left(\dfrac{n}{n}\right)^5\right]$

$= \dfrac{1}{n^6}[1^5 + 2^5 + \cdots + n^5]$

Thus, $\displaystyle\lim_{n\to\infty} S(n) = \lim_{n\to\infty} \frac{1^5 + 2^5 + \cdots + n^5}{n^6} = \frac{1}{6}$.

13. By Theorem 4.8, $0 < f(x) \leq M \implies \int_a^b f(x)\, dx \leq \int_a^b M\, dx = M(b - a)$.

Similarly, $m \leq f(x) \implies m(b - a) = \int_a^b m\, dx \leq \int_a^b f(x)\, dx$.

Thus, $m(b - a) \leq \int_a^b f(x)\, dx \leq M(b - a)$. On the interval $[0, 1]$, $1 \leq \sqrt{1 + x^4} \leq \sqrt{2}$ and $b - a = 1$.

Thus, $1 \leq \int_0^1 \sqrt{1 + x^4}\, dx \leq \sqrt{2}$. $\left(\textbf{Note: } \int_0^1 \sqrt{1 + x^4}\, dx \approx 1.0894 \right)$

15. Since $-|f(x)| \leq f(x) \leq |f(x)|$,

$$-\int_a^b |f(x)|\, dx \leq \int_a^b f(x)\, dx \leq \int_a^b |f(x)|\, dx \implies \left| \int_a^b f(x)\, dx \right| \leq \int_a^b |f(x)|\, dx.$$

17. $\dfrac{1}{365} \displaystyle\int_0^{365} 100{,}000 \left[1 + \sin \dfrac{2\pi(t - 60)}{365} \right] dt = \dfrac{100{,}000}{365} \left[t - \dfrac{365}{2\pi} \cos \dfrac{2\pi(t - 60)}{365} \right]_0^{365} = 100{,}000 \text{ lbs.}$

CHAPTER 5
Logarithmic, Exponential, and Other Transcendental Functions

Section 5.1   The Natural Logarithmic Function: Differentiation 218

Section 5.2   The Natural Logarithmic Function: Integration 223

Section 5.3   Inverse Functions 227

Section 5.4   Exponential Functions: Differentiation and Integration . . 233

Section 5.5   Bases Other than *e* and Applications 240

Section 5.6   Differential Equations: Growth and Decay 246

Section 5.7   Differential Equations: Separation of Variables 251

Section 5.8   Inverse Trigonometric Functions: Differentiation 259

Section 5.9   Inverse Trigonometric Functions: Integration 263

Section 5.10 Hyperbolic Functions 267

Review Exercises 272

Problem Solving 278

CHAPTER 5
Logarithmic, Exponential, and Other Transcendental Functions

Section 5.1   The Natural Logarithmic Function: Differentiation
Solutions to Odd-Numbered Exercises

1. Simpson's Rule: $n = 10$

x	0.5	1.5	2	2.5	3	3.5	4
$\int_1^x \frac{1}{t}\,dt$	-0.6932	0.4055	0.6932	0.9163	1.0987	1.2529	1.3865

Note: $\displaystyle\int_1^{0.5} \frac{1}{t}\,dt = -\int_{0.5}^1 \frac{1}{t}\,dt$

3. (a) $\ln 45 \approx 3.8067$

   (b) $\displaystyle\int_1^{45} \frac{1}{t}\,dt \approx 3.8067$

5. (a) $\ln 0.8 \approx -0.2231$

   (b) $\displaystyle\int_1^{0.8} \frac{1}{t}\,dt \approx -0.2231$

7. $f(x) = \ln x + 2$

Vertical shift 2 units upward

Matches (b)

9. $f(x) = \ln(x - 1)$

Horizontal shift 1 unit to the right

Matches (a)

11. $f(x) = 3 \ln x$

Domain: $x > 0$

13. $f(x) = \ln 2x$

Domain: $x > 0$

15. $f(x) = \ln(x - 1)$

Domain: $x > 1$

17. (a) $\ln 6 = \ln 2 + \ln 3 \approx 1.7917$

   (b) $\ln \frac{2}{3} = \ln 2 - \ln 3 \approx -0.4055$

   (c) $\ln 81 = \ln 3^4 = 4 \ln 3 \approx 4.3944$

   (d) $\ln \sqrt{3} = \ln 3^{1/2} = \frac{1}{2} \ln 3 \approx 0.5493$

19. $\ln \frac{2}{3} = \ln 2 - \ln 3$

21. $\ln \frac{xy}{z} = \ln x + \ln y - \ln z$

23. $\ln \sqrt[3]{a^2 + 1} = \ln(a^2 + 1)^{1/3} = \frac{1}{3} \ln(a^2 + 1)$

25. $\ln\left(\dfrac{x^2-1}{x^3}\right)^3 = 3[\ln(x^2-1) - \ln x^3]$

$\qquad\qquad\qquad = 3[\ln(x+1) + \ln(x-1) - 3\ln x]$

27. $\ln z(z-1)^2 = \ln z + \ln(z-1)^2$

$\qquad\qquad\quad = \ln z + 2\ln(z-1)$

29. $\ln(x-2) - \ln(x+2) = \ln\dfrac{x-2}{x+2}$

31. $\dfrac{1}{3}[2\ln(x+3) + \ln x - \ln(x^2-1)] = \dfrac{1}{3}\ln\dfrac{x(x+3)^2}{x^2-1} = \ln\sqrt[3]{\dfrac{x(x+3)^2}{x^2-1}}$

33 . $2\ln 3 - \dfrac{1}{2}\ln(x^2+1) = \ln 9 - \ln\sqrt{x^2+1} = \ln\dfrac{9}{\sqrt{x^2+1}}$

35.

37. $\displaystyle\lim_{x\to 3^+}\ln(x-3) = -\infty$

39. $\displaystyle\lim_{x\to 2^-}\ln[x^2(3-x)] = \ln 4 \approx 1.3863$

41. $y = \ln x^3 = 3\ln x$

$\qquad y' = \dfrac{3}{x}$

$\qquad$ At $(1, 0)$, $y' = 3$.

43. $y = \ln x^2 = 2\ln x$

$\qquad y' = \dfrac{2}{x}$

$\qquad$ At $(1, 0)$, $y' = 2$.

45. $g(x) = \ln x^2 = 2\ln x$

$\qquad g'(x) = \dfrac{2}{x}$

47. $y = (\ln x)^4$

$\qquad \dfrac{dy}{dx} = 4(\ln x)^3\left(\dfrac{1}{x}\right) = \dfrac{4(\ln x)^3}{x}$

49. $y = \ln\left[x\sqrt{x^2-1}\right] = \ln x + \dfrac{1}{2}\ln(x^2-1)$

$\qquad \dfrac{dy}{dx} = \dfrac{1}{x} + \dfrac{1}{2}\left(\dfrac{2x}{x^2-1}\right) = \dfrac{2x^2-1}{x(x^2-1)}$

51. $f(x) = \ln\dfrac{x}{x^2+1} = \ln x - \ln(x^2+1)$

$\qquad f'(x) = \dfrac{1}{x} - \dfrac{2x}{x^2+1} = \dfrac{1-x^2}{x(x^2+1)}$

53. $g(t) = \dfrac{\ln t}{t^2}$

$\qquad g'(t) = \dfrac{t^2(1/t) - 2t\ln t}{t^4} = \dfrac{1 - 2\ln t}{t^3}$

55. $y = \ln(\ln x^2)$

$\qquad \dfrac{dy}{dx} = \dfrac{1}{\ln x^2}\dfrac{d}{dx}(\ln x^2) = \dfrac{(2x/x^2)}{\ln x^2} = \dfrac{2}{x\ln x^2} = \dfrac{1}{x\ln x}$

57. $y = \ln\sqrt{\dfrac{x+1}{x-1}} = \dfrac{1}{2}[\ln(x+1) - \ln(x-1)]$

$\qquad \dfrac{dy}{dx} = \dfrac{1}{2}\left[\dfrac{1}{x+1} - \dfrac{1}{x-1}\right] = \dfrac{1}{1-x^2}$

59. $f(x) = \ln\dfrac{\sqrt{4+x^2}}{x} = \dfrac{1}{2}\ln(4+x^2) - \ln x$

$\qquad f'(x) = \dfrac{x}{4+x^2} - \dfrac{1}{x} = \dfrac{-4}{x(x^2+4)}$

61. $y = \dfrac{-\sqrt{x^2+1}}{x} + \ln\left(x + \sqrt{x^2+1}\right)$

$\dfrac{dy}{dx} = \dfrac{-x\left(x/\sqrt{x^2+1}\right) + \sqrt{x^2+1}}{x^2} + \left(\dfrac{1}{x + \sqrt{x^2+1}}\right)\left(1 + \dfrac{x}{\sqrt{x^2+1}}\right)$

$= \dfrac{1}{x^2\sqrt{x^2+1}} + \left(\dfrac{1}{x + \sqrt{x^2+1}}\right)\left(\dfrac{\sqrt{x^2+1} + x}{\sqrt{x^2+1}}\right) = \dfrac{1}{x^2\sqrt{x^2+1}} + \dfrac{1}{\sqrt{x^2+1}} = \dfrac{1 + x^2}{x^2\sqrt{x^2+1}} = \dfrac{\sqrt{x^2+1}}{x^2}$

63. $y = \ln|\sin x|$

$\dfrac{dy}{dx} = \dfrac{\cos x}{\sin x} = \cot x$

65. $y = \ln\left|\dfrac{\cos x}{\cos x - 1}\right|$

$= \ln|\cos x| - \ln|\cos x - 1|$

$\dfrac{dy}{dx} = \dfrac{-\sin x}{\cos x} - \dfrac{-\sin x}{\cos x - 1} = -\tan x + \dfrac{\sin x}{\cos x - 1}$

67. $y = \ln\left|\dfrac{-1 + \sin x}{2 + \sin x}\right|$

$= \ln|-1 + \sin x| - \ln|2 + \sin x|$

$\dfrac{dy}{dx} = \dfrac{\cos x}{-1 + \sin x} - \dfrac{\cos x}{2 + \sin x}$

$= \dfrac{3\cos x}{(\sin x - 1)(\sin x + 2)}$

69. $f(x) = \sin 2x \ln x^2 = 2 \sin 2x \ln x$

$f'(x) = (2 \sin 2x)\left(\dfrac{1}{x}\right) + 4 \cos 2x \ln x$

$= \dfrac{2}{x}(\sin 2x + 2x \cos 2x \ln x)$

$= \dfrac{2}{x}(\sin 2x + x \cos 2x \ln x^2)$

71. (a) $y = 3x^2 - \ln x,\quad (1, 3)$

$\dfrac{dy}{dx} = 6x - \dfrac{1}{x}$

When $x = 1$, $\dfrac{dy}{dx} = 5$.

Tangent line: $y - 3 = 5(x - 1)$

$y = 5x - 2$

$0 = 5x - y - 2$

(b)

73. $x^2 - 3 \ln y + y^2 = 10$

$2x - \dfrac{3}{y}\dfrac{dy}{dx} + 2y\dfrac{dy}{dx} = 0$

$2x = \dfrac{dy}{dx}\left(\dfrac{3}{y} - 2y\right)$

$\dfrac{dy}{dx} = \dfrac{2x}{(3/y) - 2y} = \dfrac{2xy}{3 - 2y^2}$

75. $y = 2(\ln x) + 3$

$y' = \dfrac{2}{x}$

$y'' = -\dfrac{2}{x^2}$

$xy'' + y' = x\left(-\dfrac{2}{x^2}\right) + \dfrac{2}{x} = 0$

77. $y = \dfrac{x^2}{2} - \ln x$

Domain: $x > 0$

$y' = x - \dfrac{1}{x} = \dfrac{(x+1)(x-1)}{x} = 0$ when $x = 1$.

$y'' = 1 + \dfrac{1}{x^2} > 0$

Relative minimum: $\left(1, \tfrac{1}{2}\right)$

79. $y = x \ln x$

Domain: $x > 0$

$y' = x\left(\dfrac{1}{x}\right) + \ln x = 1 + \ln x = 0$ when $x = e^{-1}$.

$y'' = \dfrac{1}{x} > 0$

Relative minimum: $(e^{-1}, -e^{-1})$

81. $y = \dfrac{x}{\ln x}$

Domain: $0 < x < 1, x > 1$

$y' = \dfrac{(\ln x)(1) - (x)(1/x)}{(\ln x)^2} = \dfrac{\ln x - 1}{(\ln x)^2} = 0$ when $x = e$.

$y'' = \dfrac{(\ln x)^2(1/x) - (\ln x - 1)(2/x)\ln x}{(\ln x)^4} = \dfrac{2 - \ln x}{x(\ln x)^3} = 0$ when $x = e^2$.

Relative minimum: (e, e)

Point of inflection: $(e^2, e^2/2)$

83. $f(x) = \ln x, \quad f(1) = 0$

$f'(x) = \dfrac{1}{x}, \quad f'(1) = 1$

$f''(x) = -\dfrac{1}{x^2}, \quad f''(1) = -1$

$P_1(x) = f(1) + f'(1)(x - 1) = x - 1, \quad P_1(1) = 0$

$P_2(x) = f(1) + f'(1)(x - 1) + \dfrac{1}{2}f''(1)(x - 1)^2$

$\qquad = (x - 1) - \dfrac{1}{2}(x - 1)^2, \quad P_2(1) = 0$

$P_1'(x) = 1, \quad P_1'(1) = 1$

$P_2'(x) = 1 - (x - 1) = 2 - x, \quad P_2'(1) = 1$

$P_2''(x) = -1, \quad P_2''(1) = -1$

The values of f, P_1, P_2, and their first derivatives agree at $x = 1$. The values of the second derivatives of f and P_2 agree at $x = 1$.

85. Find x such that $\ln x = -x$.

$f(x) = (\ln x) + x = 0$

$f'(x) = \dfrac{1}{x} + 1$

$x_{n+1} = x_n - \dfrac{f(x_n)}{f'(x_n)} = x_n\left[\dfrac{1 - \ln x_n}{1 + x_n}\right]$

n	1	2	3
x_n	0.5	0.5644	0.5671
$f(x_n)$	-0.1931	-0.0076	-0.0001

Approximate root: $x = 0.567$

87. $y = x\sqrt{x^2 - 1}$

$\ln y = \ln x + \dfrac{1}{2}\ln(x^2 - 1)$

$\dfrac{1}{y}\left(\dfrac{dy}{dx}\right) = \dfrac{1}{x} + \dfrac{x}{x^2 - 1}$

$\dfrac{dy}{dx} = y\left[\dfrac{2x^2 - 1}{x(x^2 - 1)}\right] = \dfrac{2x^2 - 1}{\sqrt{x^2 - 1}}$

89.
$$y = \frac{x^2\sqrt{3x-2}}{(x-1)^2}$$

$$\ln y = 2 \ln x + \frac{1}{2}\ln(3x-2) - 2\ln(x-1)$$

$$\frac{1}{y}\left(\frac{dy}{dx}\right) = \frac{2}{x} + \frac{3}{2(3x-2)} - \frac{2}{x-1}$$

$$\frac{dy}{dx} = y\left[\frac{3x^2 - 15x + 8}{2x(3x-2)(x-1)}\right]$$

$$= \frac{3x^3 - 15x^2 + 8x}{2(x-1)^3\sqrt{3x-2}}$$

91.
$$y = \frac{x(x-1)^{3/2}}{\sqrt{x+1}}$$

$$\ln y = \ln x + \frac{3}{2}\ln(x-1) - \frac{1}{2}\ln(x+1)$$

$$\frac{1}{y}\left(\frac{dy}{dx}\right) = \frac{1}{x} + \frac{3}{2}\left(\frac{1}{x-1}\right) - \frac{1}{2}\left(\frac{1}{x+1}\right)$$

$$\frac{dy}{dx} = \frac{y}{2}\left[\frac{2}{x} + \frac{3}{x-1} - \frac{1}{x+1}\right]$$

$$= \frac{y}{2}\left[\frac{4x^2 + 4x - 2}{x(x^2-1)}\right] = \frac{(2x^2 + 2x - 1)\sqrt{x-1}}{(x+1)^{3/2}}$$

93. Answers will vary. See Theorem 5.1 and 5.2.

95. $\ln e^x = x$ because $f(x) = \ln x$ and $g(x) = e^x$ are inverse functions.

97. (a) $f(1) \neq f(3)$

(b) $f'(x) = 1 - \dfrac{2}{x} = 0$ for $x = 2$.

99.
$$\beta = 10 \log_{10}\left(\frac{I}{10^{-16}}\right) = \frac{10}{\ln 10}\ln\left(\frac{I}{10^{-16}}\right) = \frac{10}{\ln 10}[\ln I + 16 \ln 10] = 160 + 10 \log_{10} I$$

$$\beta(10^{-10}) = \frac{10}{\ln 10}[\ln 10^{-10} + 16 \ln 10] = \frac{10}{\ln 10}[-10 \ln 10 + 16 \ln 10] = \frac{10}{\ln 10}[6 \ln 10] = 60 \text{ decibels}$$

101. (a) You get an error message because $\ln h$ does not exist for $h = 0$.

(b) Reversing the data, you obtain

$$h = 0.8627 - 6.4474 \ln p.$$

(c)

(d) If $p = 0.75$, $h \approx 2.72$ km.

(e) If $h = 13$ km, $p \approx 0.15$ atmosphere.

(f) $h = 0.8627 - 6.4474 \ln p$

$$1 = -6.4474\frac{1}{p}\frac{dp}{dh} \quad \text{(implicit differentiation)}$$

$$\frac{dp}{dh} = \frac{p}{-6.4474}$$

For $h = 5$, $p = 0.55$ and $dp/dh = -0.0853$ atmos/km.

For $h = 20$, $p = 0.06$ and $dp/dh = -0.00931$ atmos/km.

As the altitude increases, the rate of change of pressure decreases.

103. (a) $f(x) = \ln x$, $g(x) = \sqrt{x}$

$$f'(x) = \frac{1}{x}, g'(x) = \frac{1}{2\sqrt{x}}$$

For $x > 4$, $g'(x) > f'(x)$. g is increasing at a faster rate than f for "large" values of x.

(b) $f(x) = \ln x$, $g(x) = \sqrt[4]{x}$

$$f'(x) = \frac{1}{x}, g'(x) = \frac{1}{4\sqrt[4]{x^3}}$$

For $x > 256$, $g'(x) > f'(x)$. g is increasing at a faster rate than f for "large" values of x. $f(x) = \ln x$ increases very slowly for "large" values of x.

105. False

$$\ln x + \ln 25 = \ln(25x) \neq \ln(x + 25)$$

Section 5.2   The Natural Logarithmic Function: Integration

1. $\displaystyle\int \frac{5}{x}\,dx = 5\int \frac{1}{x}\,dx = 5\ln|x| + C$

3. $u = x + 1,\ du = dx$

$$\int \frac{1}{x + 1}\,dx = \ln|x + 1| + C$$

5. $u = 3 - 2x,\ du = -2\,dx$

$$\int \frac{1}{3 - 2x}\,dx = -\frac{1}{2}\int \frac{1}{3 - 2x}(-2)\,dx$$

$$= -\frac{1}{2}\ln|3 - 2x| + C$$

7. $u = x^2 + 1,\ du = 2x\,dx$

$$\int \frac{x}{x^2 + 1}\,dx = \frac{1}{2}\int \frac{1}{x^2 + 1}(2x)\,dx$$

$$= \frac{1}{2}\ln(x^2 + 1) + C$$

$$= \ln\sqrt{x^2 + 1} + C$$

9. $\displaystyle\int \frac{x^2 - 4}{x}\,dx = \int \left(x - \frac{4}{x}\right)dx$

$$= \frac{x^2}{2} - 4\ln|x| + C$$

11. $u = x^3 + 3x^2 + 9x,\ du = 3(x^2 + 2x + 3)\,dx$

$$\int \frac{x^2 + 2x + 3}{x^3 + 3x^2 + 9x}\,dx = \frac{1}{3}\int \frac{3(x^2 + 2x + 3)}{x^3 + 3x^2 + 9x}\,dx$$

$$= \frac{1}{3}\ln|x^3 + 3x^2 + 9x| + C$$

13. $\displaystyle\int \frac{x^2 - 3x + 2}{x + 1}\,dx = \int \left(x - 4 + \frac{6}{x + 1}\right)dx$

$$= \frac{x^2}{2} - 4x + 6\ln|x + 1| + C$$

15. $\displaystyle\int \frac{x^3 - 3x^2 + 5}{x - 3}\,dx = \int \left(x^2 + \frac{5}{x - 3}\right)dx$

$$= \frac{x^3}{3} + 5\ln|x - 3| + C$$

17. $\displaystyle\int \frac{x^4 + x - 4}{x^2 + 2}\,dx = \int \left(x^2 - 2 + \frac{x}{x^2 + 2}\right)dx$

$$= \frac{x^3}{3} - 2x + \frac{1}{2}\ln(x^2 + 2) + C$$

19. $u = \ln x,\ du = \dfrac{1}{x}\,dx$

$$\int \frac{(\ln x)^2}{x}\,dx = \frac{1}{3}(\ln x)^3 + C$$

21. $u = x + 1,\ du = dx$

$$\int \frac{1}{\sqrt{x + 1}}\,dx = \int (x + 1)^{-1/2}\,dx$$

$$= 2(x + 1)^{1/2} + C$$

$$= 2\sqrt{x + 1} + C$$

23. $\displaystyle\int \frac{2x}{(x - 1)^2}\,dx = \int \frac{2x - 2 + 2}{(x - 1)^2}\,dx$

$$= \int \frac{2(x - 1)}{(x - 1)^2}\,dx + 2\int \frac{1}{(x - 1)^2}\,dx$$

$$= 2\int \frac{1}{x - 1}\,dx + 2\int \frac{1}{(x - 1)^2}\,dx$$

$$= 2\ln|x - 1| - \frac{2}{(x - 1)} + C$$

25. $u = 1 + \sqrt{2x}, du = \dfrac{1}{\sqrt{2x}} dx \implies (u - 1)\, du = dx$

$$\int \frac{1}{1 + \sqrt{2x}} dx = \int \frac{(u - 1)}{u} du = \int \left(u - \frac{1}{u} \right) du$$

$$= u - \ln|u| + C_1$$

$$= \left(1 + \sqrt{2x} \right) - \ln\left| 1 + \sqrt{2x} \right| + C_1$$

$$= \sqrt{2x} - \ln\left(1 + \sqrt{2x} \right) + C$$

where $C = C_1 + 1$.

27. $u = \sqrt{x} - 3, du = \dfrac{1}{2\sqrt{x}} dx \implies 2(u + 3)\, du = dx$

$$\int \frac{\sqrt{x}}{\sqrt{x} - 3} dx = 2 \int \frac{(u + 3)^2}{u} du = 2 \int \frac{u^2 + 6u + 9}{u} du = 2 \int \left(u + 6 + \frac{9}{u} \right) du$$

$$= 2 \left[\frac{u^2}{2} + 6u + 9 \ln|u| \right] + C_1 = u^2 + 12u + 18 \ln|u| + C_1$$

$$= \left(\sqrt{x} - 3 \right)^2 + 12\left(\sqrt{x} - 3 \right) + 18 \ln\left| \sqrt{x} - 3 \right| + C_1$$

$$= x + 6\sqrt{x} + 18 \ln\left| \sqrt{x} - 3 \right| + C \text{ where } C = C_1 - 27.$$

29. $\displaystyle\int \frac{\cos \theta}{\sin \theta} d\theta = \ln|\sin \theta| + C$

$\quad (u = \sin \theta, du = \cos \theta\, d\theta)$

31. $\displaystyle\int \csc 2x\, dx = \frac{1}{2} \int (\csc 2x)(2)\, dx$

$$= -\frac{1}{2} \ln|\csc 2x + \cot 2x| + C$$

33. $\displaystyle\int \frac{\cos t}{1 + \sin t} dt = \ln|1 + \sin t| + C$

35. $\displaystyle\int \frac{\sec x \tan x}{\sec x - 1} dx = \ln|\sec x - 1| + C$

37. $y = \displaystyle\int \frac{3}{2 - x} dx$

$$= -3 \int \frac{1}{x - 2} dx$$

$$= -3 \ln|x - 2| + C$$

$(1, 0)$: $0 = -3 \ln|1 - 2| + C \implies C = 0$

$y = -3 \ln|x - 2|$

39. $s = \displaystyle\int \tan(2\theta)\, d\theta$

$$= \frac{1}{2} \int \tan(2\theta)(2\, d\theta)$$

$$= -\frac{1}{2} \ln|\cos 2\theta| + C$$

$(0, 2)$: $2 = -\dfrac{1}{2} \ln|\cos (0)| + C \implies C = 2$

$$s = -\frac{1}{2} \ln|\cos 2\theta| + 2$$

41. $\dfrac{dy}{dx} = \dfrac{1}{x + 2}, (0, 1)$

(a)

(b) $y = \displaystyle\int \frac{1}{x + 2} dx = \ln|x + 2| + C$

$\quad\quad y(0) = 1 \implies 1 = \ln 2 + C \implies C = 1 - \ln 2$

$\quad\quad$ Hence, $y = \ln|x + 2| + 1 - \ln 2 = \ln\left| \dfrac{x + 2}{2} \right| + 1.$

43. $\displaystyle\int_0^4 \frac{5}{3x+1}\,dx = \left[\frac{5}{3}\ln|3x+1|\right]_0^4$

$\qquad = \dfrac{5}{3}\ln 13 \approx 4.275$

45. $u = 1 + \ln x,\; du = \dfrac{1}{x}\,dx$

$\displaystyle\int_1^e \frac{(1+\ln x)^2}{x}\,dx = \left[\frac{1}{3}(1+\ln x)^3\right]_1^e = \frac{7}{3}$

47. $\displaystyle\int_0^2 \frac{x^2-2}{x+1}\,dx = \int_0^2\left(x - 1 - \frac{1}{x+1}\right)dx$

$\qquad = \left[\dfrac{1}{2}x^2 - x - \ln|x+1|\right]_0^2 = -\ln 3$

49. $\displaystyle\int_1^2 \frac{1-\cos\theta}{\theta - \sin\theta}\,d\theta = \Big[\ln|\theta - \sin\theta|\Big]_1^2$

$\qquad = \ln\left|\dfrac{2-\sin 2}{1-\sin 1}\right| \approx 1.929$

51. $-\ln|\cos x| + C = \ln\left|\dfrac{1}{\cos x}\right| + C = \ln|\sec x| + C$

53. $\ln|\sec x + \tan x| + C = \ln\left|\dfrac{(\sec x + \tan x)(\sec x - \tan x)}{(\sec x - \tan x)}\right| + C = \ln\left|\dfrac{\sec^2 x - \tan^2 x}{\sec x - \tan x}\right| + C$

$\qquad = \ln\left|\dfrac{1}{\sec x - \tan x}\right| + C = -\ln|\sec x - \tan x| + C$

55. $\displaystyle\int \frac{1}{1+\sqrt{x}}\,dx = 2\left(1 + \sqrt{x}\right) - 2\ln\left(1+\sqrt{x}\right) + C_1$

$\qquad = 2\left[\sqrt{x} - \ln\left(1+\sqrt{x}\right)\right] + C \text{ where } C = C_1 + 2.$

57. $\displaystyle\int \cos(1-x)\,dx = -\sin(1-x) + C$

59. $\displaystyle\int_{\pi/4}^{\pi/2} (\csc x - \sin x)\,dx = \left[-\ln|\csc x + \cot x| + \cos x\right]_{\pi/4}^{\pi/2} = \ln\left(\sqrt{2}+1\right) - \frac{\sqrt{2}}{2} \approx 0.174$

Note: In Exercises 61 and 63, you can use the Second Fundamental Theorem of Calculus or integrate the function.

61. $F(x) = \displaystyle\int_1^x \frac{1}{t}\,dt$

$F'(x) = \dfrac{1}{x}$

63. $F(x) = \displaystyle\int_x^{3x} \frac{1}{t}\,dt = \int_1^{3x} \frac{1}{t}\,dt - \int_1^x \frac{1}{t}\,dt$

$F'(x) = \dfrac{3}{3x} - \dfrac{1}{x} = 0$

65.

$A \approx 1.25$
Matches (d)

67. $A = \displaystyle\int_1^4 \frac{x^2+4}{x}\,dx = \int_1^4\left(x + \frac{4}{x}\right)dx$

$\qquad = \left[\dfrac{x^2}{2} + 4\ln x\right]_1^4 = (8 + 4\ln 4) - \dfrac{1}{2}$

$\qquad = \dfrac{15}{2} + 8\ln 2 \approx 13.045 \text{ square units}$

69. $\int_0^2 2 \sec \frac{\pi x}{6} \, dx = \frac{12}{\pi} \int_0^2 \sec\left(\frac{\pi x}{6}\right)\frac{\pi}{6} \, dx$

$= \left[\frac{12}{\pi} \ln\left|\sec \frac{\pi x}{6} + \tan \frac{\pi x}{6}\right|\right]_0^2$

$= \frac{12}{\pi} \ln\left|\sec \frac{\pi}{3} + \tan \frac{\pi}{3}\right| - \frac{12}{\pi} \ln|1 + 0|$

$= \frac{12}{\pi} \ln\left(2 + \sqrt{3}\right) \approx 5.03041$

71. Power Rule

73. Substitution: $(u = x^2 + 4)$ and Log Rule

75. Divide the polynomials:

$\frac{x^2}{x + 1} = x - 1 + \frac{1}{x + 1}$

77. Average value $= \frac{1}{4 - 2}\int_2^4 \frac{8}{x^2} \, dx = 4\int_2^4 x^{-2} \, dx$

$= \left[-4\frac{1}{x}\right]_2^4$

$= -4\left(\frac{1}{4} - \frac{1}{2}\right) = 1$

79. Average value $= \frac{1}{e - 1}\int_1^e \frac{\ln x}{x} \, dx = \frac{1}{e - 1}\left[\frac{(\ln x)^2}{2}\right]_1^e$

$= \frac{1}{e - 1}\left(\frac{1}{2}\right)$

$= \frac{1}{2e - 2} \approx 0.291$

81. $P(t) = \int \frac{3000}{1 + 0.25t} \, dt = (3000)(4)\int \frac{0.25}{1 + 0.25t} \, dt = 12{,}000 \ln|1 + 0.25t| + C$

$P(0) = 12{,}000 \ln|1 + 0.25(0)| + C = 1000$

$C = 1000$

$P(t) = 12{,}000 \ln|1 + 0.25t| + 1000 = 1000[12 \ln|1 + 0.25t| + 1]$

$P(3) = 1000[12(\ln 1.75) + 1] \approx 7715$

83. $\frac{1}{50 - 40}\int_{40}^{50} \frac{90{,}000}{400 + 3x} \, dx = \left[3000 \ln|400 + 3x|\right]_{40}^{50} \approx \$168.27$

85. (a) $2x^2 - y^2 = 8$

$y^2 = 2x^2 - 8$

$y_1 = \sqrt{2x^2 - 8}$

$y_2 = -\sqrt{2x^2 - 8}$

(b) $y^2 = e^{-\int(1/x)dx} = e^{-\ln x + C} = e^{\ln(1/x)}(e^C) = \frac{1}{x}k$

Let $k = 4$ and graph $y^2 = \frac{4}{x}$ $\left(\begin{array}{l} y_1 = 2/\sqrt{x} \\ y_2 = -2/\sqrt{x} \end{array}\right)$

(c) In part (a), $\quad 2x^2 - y^2 = 8$

$4x - 2yy' = 0$

$y' = \frac{2x}{y}.$

In part (b), $\quad y^2 = \frac{4}{x} = 4x^{-1}$

$2yy' = \frac{-4}{x^2}$

$y' = \frac{-2}{yx^2} = \frac{-2y}{y^2 x^2} = \frac{-2y}{4x} = \frac{-y}{2x}.$

Using a graphing utility the graphs intersect at $(2.214, 1.344)$. The slopes are 3.295 and $-0.304 = (-1)/3.295$, respectively.

87. False

$$\frac{1}{2}(\ln x) = \ln(x^{1/2}) \neq (\ln x)^{1/2}$$

89. True

$$\int \frac{1}{x}\,dx = \ln|x| + C_1$$

$$= \ln|x| + \ln|C| = \ln|Cx|, \ \ C \neq 0$$

Section 5.3   Inverse Functions

1. (a) $f(x) = 5x + 1$

$$g(x) = \frac{x - 1}{5}$$

$$f(g(x)) = f\left(\frac{x - 1}{5}\right) = 5\left(\frac{x - 1}{5}\right) + 1 = x$$

$$g((f(x)) = g(5x + 1) = \frac{(5x + 1) - 1}{5} = x$$

(b)

3. (a) $f(x) = x^3$

$$g(x) = \sqrt[3]{x}$$

$$f(g(x)) = f(\sqrt[3]{x}) = (\sqrt[3]{x})^3 = x$$

$$g(f(x)) = g(x^3) = \sqrt[3]{x^3} = x$$

(b)

5. (a) $f(x) = \sqrt{x - 4}$

$$g(x) = x^2 + 4, \ x \geq 0$$

$$f(g(x)) = f(x^2 + 4)$$

$$= \sqrt{(x^2 + 4) - 4} = \sqrt{x^2} = x$$

$$g(f(x)) = g(\sqrt{x - 4})$$

$$= (\sqrt{x - 4})^2 + 4 = x - 4 + 4 = x$$

(b)

7. (a) $f(x) = \frac{1}{x}$

$$g(x) = \frac{1}{x}$$

$$f(g(x)) = \frac{1}{1/x} = x$$

$$g(f(x)) = \frac{1}{1/x} = x$$

(b)

9. Matches (c)

11. Matches (a)

13. $f(x) = \frac{3}{4}x + 6$

One-to-one; has an inverse

15. $f(\theta) = \sin \theta$

Not one-to-one; does not have an inverse

17. $h(s) = \frac{1}{s - 2} - 3$

One-to-one; has an inverse

19. $f(x) = \ln x$

One-to-one; has an inverse

21. $g(x) = (x + 5)^3$

One-to-one; has an inverse

23. $f(x) = (x + a)^3 + b$

$f'(x) = 3(x + a)^2 \geq 0$ for all x.

f is increasing on $(-\infty, \infty)$. Therefore, f is strictly monotonic and has an inverse.

25. $f(x) = \frac{x^4}{4} - 2x^2$

$f'(x) = x^3 - 4x = 0$ when $x = 0, 2, -2$.

f is not strictly monotonic on $(-\infty, \infty)$. Therefore, f does not have an inverse.

27. $f(x) = 2 - x - x^3$

$f'(x) = -1 - 3x^2 < 0$ for all x.

f is decreasing on $(-\infty, \infty)$. Therefore, f is strictly monotonic and has an inverse.

29. $f(x) = 2x - 3 = y$

$x = \frac{y + 3}{2}$

$y = \frac{x + 3}{2}$

$f^{-1}(x) = \frac{x + 3}{2}$

31. $f(x) = x^5 = y$

$x = \sqrt[5]{y}$

$y = \sqrt[5]{x}$

$f^{-1}(x) = \sqrt[5]{x} = x^{1/5}$

33. $f(x) = \sqrt{x} = y$

$x = y^2$

$y = x^2$

$f^{-1}(x) = x^2, \ x \geq 0$

35. $f(x) = \sqrt{4 - x^2} = y, \ 0 \le x \le 2$

$\quad\quad x = \sqrt{4 - y^2}$

$\quad\quad y = \sqrt{4 - x^2}$

$\quad f^{-1}(x) = \sqrt{4 - x^2}, \ 0 \le x \le 2$

37. $f(x) = \sqrt[3]{x - 1} = y$

$\quad\quad x = y^3 + 1$

$\quad\quad y = x^3 + 1$

$\quad f^{-1}(x) = x^3 + 1$

The graphs of f and f^{-1} are reflections of each other across the line $y = x$.

39. $f(x) = x^{2/3} = y, \ x \ge 0$

$\quad\quad x = y^{3/2}$

$\quad\quad y = x^{3/2}$

$\quad f^{-1}(x) = x^{3/2}, \ x \ge 0$

The graphs of f and f^{-1} are reflections of each other across the line $y = x$.

41. $f(x) = \dfrac{x}{\sqrt{x^2 + 7}} = y$

$\quad\quad\quad\quad\quad x = \dfrac{\sqrt{7}y}{\sqrt{1 - y^2}}$

$\quad\quad y = \dfrac{\sqrt{7}x}{\sqrt{1 - x^2}}$

$\quad f^{-1}(x) = \dfrac{\sqrt{7}x}{\sqrt{1 - x^2}}, \ -1 < x < 1$

The graphs of f and f^{-1} are reflections of each other across the line $y = x$.

43.

x	1	2	3	4
$f^{-1}(x)$	0	1	2	4

45. (a) Let x be the number of pounds of the commodity costing 1.25 per pound. Since there are 50 pounds total, the amount of the second commodity is $50 - x$. The total cost is

$$y = 1.25x + 1.60(50 - x)$$

$$= -0.35x + 80 \quad\quad 0 \le x \le 50.$$

(b) We find the inverse of the original function:

$$y = -0.35x + 80$$

$$0.35x = 80 - y$$

$$x = \frac{100}{35}(80 - y)$$

Inverse: $y = \frac{100}{35}(80 - x) = \frac{20}{7}(80 - x)$.

x represents cost and y represents pounds.

(c) Domain of inverse is $62.5 \le x \le 80$.

(d) If $x = 73$ in the inverse function, $y = \frac{100}{35}(80 - 73) = \frac{100}{5} = 20$ pounds.

47. $f(x) = (x - 4)^2$ on $[4, \infty)$

$f'(x) = 2(x - 4) > 0$ on $(4, \infty)$

f is increasing on $[4, \infty)$. Therefore, f is strictly monotonic and has an inverse.

49. $f(x) = \dfrac{4}{x^2}$ on $(0, \infty)$

$f'(x) = -\dfrac{8}{x^3} < 0$ on $(0, \infty)$

f is decreasing on $(0, \infty)$. Therefore, f is strictly monotonic and has an inverse.

51. $f(x) = \cos x$ on $[0, \pi]$

$f'(x) = -\sin x < 0$ on $(0, \pi)$

f is decreasing on $[0, \pi]$. Therefore, f is strictly monotonic and has an inverse.

53. $f(x) = \dfrac{x}{x^2 - 4} = y$ on $(-2, 2)$

$x^2 y - 4y = x$

$x^2 y - x - 4y = 0$

$a = y, \, b = -1, \, c = -4y$

$x = \dfrac{1 \pm \sqrt{1 - 4(y)(-4y)}}{2y} = \dfrac{1 \pm \sqrt{1 + 16y^2}}{2y}$

$y = f^{-1}(x) = \begin{cases} \left(1 - \sqrt{1 + 16x^2}\right)/2x, & \text{if } x \neq 0 \\ 0, & \text{if } x = 0 \end{cases}$

Domain of f^{-1}: all x

Range of f^{-1}: $-2 < y < 2$

The graphs of f and f^{-1} are reflections of each other across the line $y = x$.

55. (a), (b)

(c) Yes, f is one-to-one and has an inverse. The inverse relation is an inverse function.

57. (a), (b)

(c) g is not one-to-one and does not have an inverse. The inverse relation is not an inverse function.

59. $f(x) = \sqrt{x - 2}$, Domain: $x \geq 2$

$f'(x) = \dfrac{1}{2\sqrt{x - 2}} > 0$ for $x > 2$.

f is one-to-one; has an inverse

$\sqrt{x - 2} = y$

$x - 2 = y^2$

$x = y^2 + 2$

$y = x^2 + 2$

$f^{-1}(x) = x^2 + 2, \, x \geq 0$

61. $f(x) = |x - 2|, \, x \leq 2$

$\quad = -(x - 2)$

$\quad = 2 - x$

f is one-to-one; has an inverse

$2 - x = y$

$2 - y = x$

$f^{-1}(x) = 2 - x, \, x \geq 0$

63. $f(x) = (x - 3)^2$ is one-to-one for $x \geq 3$.

$(x - 3)^2 = y$

$x - 3 = \sqrt{y}$

$x = \sqrt{y} + 3$

$y = \sqrt{x} + 3$

$f^{-1}(x) = \sqrt{x} + 3, \, x \geq 0$

(Answer is not unique)

65. $f(x) = |x + 3|$ is one-to-one for $x \geq -3$.

$x + 3 = y$

$x = y - 3$

$y = x - 3$

$f^{-1}(x) = x - 3, \, x \geq 0$

(Answer is not unique)

67. Yes, the volume is an increasing function, and hence one-to-one. The inverse function gives the time t corresponding to the volume V.

69. No, $C(t)$ is not one-to-one because long distance costs are step functions. A call lasting 2.1 minutes costs the same as one lasting 2.2 minutes.

71.
$$f(x) = x^3 + 2x - 1, \quad f(1) = 2 = a$$
$$f'(x) = 3x^2 + 2$$
$$(f^{-1})'(2) = \frac{1}{f'(f^{-1}(2))} = \frac{1}{f'(1)} = \frac{1}{3(1)^2 + 2} = \frac{1}{5}$$

73.
$$f(x) = \sin x, \quad f\left(\frac{\pi}{6}\right) = \frac{1}{2} = a$$
$$f'(x) = \cos x$$
$$(f^{-1})'\left(\frac{1}{2}\right) = \frac{1}{f'(f^{-1}(1/2))} = \frac{1}{f'(\pi/6)} = \frac{1}{\cos(\pi/6)}$$
$$= \frac{1}{\sqrt{3}/2} = \frac{2\sqrt{3}}{3}$$

75.
$$f(x) = x^3 - \frac{4}{x}, \quad f(2) = 6 = a$$
$$f'(x) = 3x^2 + \frac{4}{x^2}$$
$$(f^{-1})'(6) = \frac{1}{f'(f^{-1}(6))} = \frac{1}{f'(2)} = \frac{1}{3(2)^2 + (4/2^2)} = \frac{1}{13}$$

77. (a) Domain f = Domain $f^{-1} = (-\infty, \infty)$

(b) Range f = Range $f^{-1} = (-\infty, \infty)$

(c)

(d)
$$f(x) = x^3, \quad \left(\frac{1}{2}, \frac{1}{8}\right)$$
$$f'(x) = 3x^2$$
$$f'\left(\frac{1}{2}\right) = \frac{3}{4}$$
$$f^{-1}(x) = \sqrt[3]{x}, \quad \left(\frac{1}{8}, \frac{1}{2}\right)$$
$$(f^{-1})'(x) = \frac{1}{3\sqrt[3]{x^2}}$$
$$(f^{-1})'\left(\frac{1}{8}\right) = \frac{4}{3}$$

79. (a) Domain $f = [4, \infty)$, Domain $f^{-1} = [0, \infty)$

(b) Range $f = [0, \infty)$, Range $f^{-1} = [4, \infty)$

(c)

(d)
$$f(x) = \sqrt{x - 4}, \quad (5, 1)$$
$$f'(x) = \frac{1}{2\sqrt{x - 4}}$$
$$f'(5) = \frac{1}{2}$$
$$f^{-1}(x) = x^2 + 4, \quad (1, 5)$$
$$(f^{-1})'(x) = 2x$$
$$(f^{-1})'(1) = 2$$

81. $x = y^3 - 7y^2 + 2$

$$1 = 3y^2 \frac{dy}{dx} - 14y \frac{dy}{dx}$$

$\dfrac{dy}{dx} = \dfrac{1}{3y^2 - 14y}$. At $(-4, 1)$, $\dfrac{dy}{dx} = \dfrac{1}{3 - 14} = \dfrac{-1}{11}$.

Alternate solution: let $f(x) = x^3 - 7x^2 + 2$.

Then $f'(x) = 3x^2 - 14x$ and $f'(1) = -11$.

Hence, $\dfrac{dy}{dx} = \dfrac{1}{-11} = \dfrac{-1}{11}$.

In Exercises 83 and 85, use the following.

$$f(x) = \tfrac{1}{8}x - 3 \text{ and } g(x) = x^3$$

$$f^{-1}(x) = 8(x + 3) \text{ and } g^{-1}(x) = \sqrt[3]{x}$$

83. $(f^{-1} \circ g^{-1})(1) = f^{-1}(g^{-1}(1)) = f^{-1}(1) = 32$

85. $(f^{-1} \circ f^{-1})(6) = f^{-1}(f^{-1}(6)) = f^{-1}(72) = 600$

In Exercises 87 and 89, use the following.

$$f(x) = x + 4 \text{ and } g(x) = 2x - 5$$

$$f^{-1}(x) = x - 4 \text{ and } g^{-1}(x) = \frac{x + 5}{2}$$

87. $(g^{-1} \circ f^{-1})(x) = g^{-1}(f^{-1}(x))$

$$= g^{-1}(x - 4)$$

$$= \frac{(x - 4) + 5}{2}$$

$$= \frac{x + 1}{2}$$

89. $(f \circ g)(x) = f(g(x))$

$$= f(2x - 5)$$

$$= (2x - 5) + 4$$

$$= 2x - 1$$

Hence, $(f \circ g)^{-1}(x) = \dfrac{x + 1}{2}$

(Note: $(f \circ g)^{-1} = g^{-1} \circ f^{-1}$)

91.   Answers will vary. See page 335 and Example 3.

93. $y = x^2$ on $(-\infty, \infty)$ does not have an inverse.

95. f is not one-to-one because many different x-values yield the same y-value.

Example: $f(0) = f(\pi) = 0$

Not continuous at $\dfrac{(2n - 1)\pi}{2}$, where n is an integer

97.   Let $(f \circ g)(x) = y$ then $x = (f \circ g)^{-1}(y)$. Also,

$$(f \circ g)(x) = y$$

$$f(g(x)) = y$$

$$g(x) = f^{-1}(y)$$

$$x = g^{-1}(f^{-1}(y))$$

$$= (g^{-1} \circ f^{-1})(y)$$

Since f and g are one-to-one functions,
$(f \circ g)^{-1} = g^{-1} \circ f^{-1}$.

99.   Suppose $g(x)$ and $h(x)$ are both inverses of $f(x)$. Then the graph of $f(x)$ contains the point (a, b) if and only if the graphs of $g(x)$ and $h(x)$ contain the point (b, a). Since the graphs of $g(x)$ and $h(x)$ are the same, $g(x) = h(x)$. Therefore, the inverse of $f(x)$ is unique.

101. False

Let $f(x) = x^2$.

103. True

105. Not true

Let $f(x) = \begin{cases} x, & 0 \le x \le 1 \\ 1 - x, & 1 < x \le 2 \end{cases}$.

f is one-to-one, but not strictly monotonic.

107. $f(x) = \displaystyle\int_2^x \dfrac{dt}{\sqrt{1 + t^4}}$, $f(2) = 0$

$f'(x) = \dfrac{1}{\sqrt{1 + x^4}}$

$(f^{-1})'(0) = \dfrac{1}{f'(2)} = \dfrac{1}{1/\sqrt{17}} = \sqrt{17}$

Section 5.4   Exponential Functions: Differentiation and Integration

1. $e^0 = 1$

$\ln 1 = 0$

3. $\ln 2 = 0.6931\ldots$

$e^{0.6931\ldots} = 2$

5. $e^{\ln x} = 4$

$x = 4$

7. $e^x = 12$

$x = \ln 12 \approx 2.485$

9. $9 - 2e^x = 7$

$2e^x = 2$

$e^x = 1$

$x = 0$

11. $50e^{-x} = 30$

$e^{-x} = \dfrac{3}{5}$

$-x = \ln\left(\dfrac{3}{5}\right)$

$x = \ln\left(\dfrac{5}{3}\right) \approx 0.511$

13. $\ln x = 2$

$x = e^2 \approx 7.389$

15. $\ln(x - 3) = 2$

$x - 3 = e^2$

$x = 3 + e^2 \approx 10.389$

17. $\ln\sqrt{x + 2} = 1$

$\sqrt{x + 2} = e^1 = e$

$x + 2 = e^2$

$x = e^2 - 2 \approx 5.389$

19. $y = e^{-x}$

21. $y = e^{-x^2}$

Symmetric with respect to the y-axis

Horizontal asymptote: $y = 0$

23. (a)

Horizontal shift 2 units to the right

(b)

A reflection in the *x*-axis and a vertical shrink

(c)

Vertical shift 3 units upward and a reflection in the *y*-axis

25. $y = Ce^{ax}$

Horizontal asymptote: $y = 0$

Matches (c)

27. $y = C(1 - e^{-ax})$

Vertical shift C units

Reflection in both the *x*- and *y*-axes

Matches (a)

29. $f(x) = e^{2x}$

$g(x) = \ln \sqrt{x} = \dfrac{1}{2} \ln x$

31. $f(x) = e^x - 1$

$g(x) = \ln(x + 1)$

33.

As $x \to \infty$, the graph of f approaches the graph of g.

$$\lim_{x \to \infty} \left(1 + \frac{0.5}{x}\right)^x = e^{0.5}$$

35. $\left(1 + \dfrac{1}{1,000,000}\right)^{1,000,000} \approx 2.718280469$

$e \approx 2.718281828$

$e > \left(1 + \dfrac{1}{1,000,000}\right)^{1,000,000}$

37. (a) $y = e^{3x}$

$y' = 3e^{3x}$

At $(0, 1)$, $y' = 3$.

(b) $y = e^{-3x}$

$y' = -3e^{-3x}$

At $(0, 1)$, $y' = -3$.

39. $f(x) = e^{2x}$

$f'(x) = 2e^{2x}$

41. $f(x) = e^{-2x + x^2}$

$\dfrac{dy}{dx} = 2(x - 1)e^{-2x + x^2}$

43. $y = e^{\sqrt{x}}$

$\dfrac{dy}{dx} = \dfrac{e^{\sqrt{x}}}{2\sqrt{x}}$

45. $g(t) = (e^{-t} + e^t)^3$

$g'(t) = 3(e^{-t} + e^t)^2(e^t - e^{-t})$

47. $y = \ln e^{x^2} = x^2$

$\dfrac{dy}{dx} = 2x$

49. $y = \ln(1 + e^{2x})$

$\dfrac{dy}{dx} = \dfrac{2e^{2x}}{1 + e^{2x}}$

51. $y = \dfrac{2}{e^x + e^{-x}} = 2(e^x + e^{-x})^{-1}$

$\dfrac{dy}{dx} = -2(e^x + e^{-x})^{-2}(e^x - e^{-x})$

$\phantom{\dfrac{dy}{dx}} = \dfrac{-2(e^x - e^{-x})}{(e^x + e^{-x})^2}$

53. $y = x^2 e^x - 2xe^x + 2e^x = e^x(x^2 - 2x + 2)$

$\dfrac{dy}{dx} = e^x(2x - 2) + e^x(x^2 - 2x + 2) = x^2 e^x$

55. $f(x) = e^{-x} \ln x$

$f'(x) = e^{-x}\left(\dfrac{1}{x}\right) - e^{-x} \ln x = e^{-x}\left(\dfrac{1}{x} - \ln x\right)$

57. $y = e^x(\sin x + \cos x)$

$\dfrac{dy}{dx} = e^x(\cos x - \sin x) + (\sin x + \cos x)(e^x)$

$\phantom{\dfrac{dy}{dx}} = e^x(2 \cos x) = 2e^x \cos x$

59. $\qquad xe^y - 10x + 3y = 0$

$xe^y \dfrac{dy}{dx} + e^y - 10 + 3\dfrac{dy}{dx} = 0$

$\dfrac{dy}{dx}(xe^y + 3) = 10 - e^y$

$\dfrac{dy}{dx} = \dfrac{10 - e^y}{xe^y + 3}$

61. $f(x) = (3 + 2x)e^{-3x}$

$f'(x) = (3 + 2x)(-3e^{-3x}) + 2e^{-3x}$

$ = (-7 - 6x)e^{-3x}$

$f''(x) = (-7 - 6x)(-3e^{-3x}) - 6e^{-3x}$

$ = 3(6x + 5)e^{-3x}$

63. $\qquad y = e^x\left(\cos\sqrt{2}x + \sin\sqrt{2}x\right)$

$y' = e^x\left(-\sqrt{2}\sin\sqrt{2}x + \sqrt{2}\cos\sqrt{2}x\right) + e^x\left(\cos\sqrt{2}x + \sin\sqrt{2}x\right)$

$ = e^x\left[\left(1 + \sqrt{2}\right)\cos\sqrt{2}x + \left(1 - \sqrt{2}\right)\sin\sqrt{2}x\right]$

$y'' = e^x\left[-\left(\sqrt{2} + 2\right)\sin\sqrt{2}x + \left(\sqrt{2} - 2\right)\cos\sqrt{2}x\right] + e^x\left[\left(1 + \sqrt{2}\right)\cos\sqrt{2}x + \left(1 - \sqrt{2}\right)\sin\sqrt{2}x\right]$

$ = e^x\left[\left(-1 - 2\sqrt{2}\right)\sin\sqrt{2}x + \left(-1 + 2\sqrt{2}\right)\cos\sqrt{2}x\right]$

$-2y' + 3y = -2e^x\left[\left(1 + \sqrt{2}\right)\cos\sqrt{2}x + \left(1 - \sqrt{2}\right)\sin\sqrt{2}x\right] + 3e^x\left[\cos\sqrt{2}x + \sin\sqrt{2}x\right]$

$ = e^x\left[\left(1 - 2\sqrt{2}\right)\cos\sqrt{2}x + \left(1 + 2\sqrt{2}\right)\sin\sqrt{2}x\right] = -y''$

Therefore, $-2y' + 3y = -y'' \implies y'' - 2y' + 3y = 0$.

65. $f(x) = \dfrac{e^x + e^{-x}}{2}$

$f'(x) = \dfrac{e^x - e^{-x}}{2} = 0$ when $x = 0$.

$f''(x) = \dfrac{e^x + e^{-x}}{2} > 0$

Relative minimum: $(0, 1)$

67. $g(x) = \dfrac{1}{\sqrt{2\pi}}e^{-(x-2)^2/2}$

$g'(x) = \dfrac{-1}{\sqrt{2\pi}}(x - 2)e^{-(x-2)^2/2}$

$g''(x) = \dfrac{1}{\sqrt{2\pi}}(x - 1)(x - 3)e^{-(x-2)^2/2}$

Relative maximum: $\left(2, \dfrac{1}{\sqrt{2\pi}}\right) \approx (2, 0.399)$

Points of inflection: $\left(1, \dfrac{1}{\sqrt{2\pi}}e^{-1/2}\right), \left(3, \dfrac{1}{\sqrt{2\pi}}e^{-1/2}\right) \approx (1, 0.242), (3, 0.242)$

69. $f(x) = x^2 e^{-x}$

$\quad f'(x) = -x^2 e^{-x} + 2xe^{-x} = xe^{-x}(2 - x) = 0$ when $x = 0, 2$.

$\quad f''(x) = -e^{-x}(2x - x^2) + e^{-x}(2 - 2x)$

$\qquad = e^{-x}(x^2 - 4x + 2) = 0$ when $x = 2 \pm \sqrt{2}$.

Relative minimum: $(0, 0)$

Relative maximum: $(2, 4e^{-2})$

$\quad x = 2 \pm \sqrt{2}$

$\quad y = (2 \pm \sqrt{2})^2 e^{-(2 \pm \sqrt{2})}$

Points of inflection: $(3.414, 0.384), (0.586, 0.191)$

71. $g(t) = 1 + (2 + t)e^{-t}$

$\quad g'(t) = -(1 + t)e^{-t}$

$\quad g''(t) = te^{-t}$

Relative maximum: $(-1, 1 + e) \approx (-1, 3.718)$

Point of inflection: $(0, 3)$

73. $A = (\text{base})(\text{height}) = 2xe^{-x^2}$

$\quad \dfrac{dA}{dx} = -4x^2 e^{-x^2} + 2e^{-x^2}$

$\qquad = 2e^{-x^2}(1 - 2x^2) = 0$ when $x = \dfrac{\sqrt{2}}{2}$.

$\quad A = \sqrt{2}e^{-1/2}$

75. $y = \dfrac{L}{1 + ae^{-x/b}}, a > 0, b > 0, L > 0$

$y' = \dfrac{-L\left(-\dfrac{a}{b}e^{-x/b}\right)}{(1 + ae^{-x/b})^2} = \dfrac{\dfrac{aL}{b}e^{-x/b}}{(1 + ae^{-x/b})^2}$

$y'' = \dfrac{(1 + ae^{-x/b})^2 \left(\dfrac{-aL}{b^2}e^{-x/b}\right) - \left(\dfrac{aL}{b}e^{-x/b}\right)2(1 + ae^{-x/b})\left(\dfrac{-a}{b}e^{-x/b}\right)}{(1 + ae^{-x/b})^4}$

$\quad = \dfrac{(1 + ae^{-x/b})\left(\dfrac{-aL}{b^2}e^{-x/b}\right) + 2\left(\dfrac{aL}{b}e^{-x/b}\right)\left(\dfrac{a}{b}e^{-x/b}\right)}{(1 + ae^{-x/b})^3}$

$\quad = \dfrac{Lae^{-x/b}[ae^{-x/b} - 1]}{(1 + ae^{-x/b})^3 b^2}$

$y'' = 0$ if $ae^{-x/b} = 1 \Longrightarrow \dfrac{-x}{b} = \ln\left(\dfrac{1}{a}\right) \Longrightarrow x = b \ln a$

$y(b \ln a) = \dfrac{L}{1 + ae^{-(b \ln a)/b}} = \dfrac{L}{1 + a(1/a)} = \dfrac{L}{2}$

Therefore, the y-coordinate of the inflection point is $L/2$.

77. $e^{-x} = x \implies f(x) = x - e^{-x}$

$f'(x) = 1 + e^{-x}$

$x_{n+1} = x_n - \dfrac{f(x_n)}{f'(x_n)} = x_n - \dfrac{x_n - e^{-x_n}}{1 + e^{-x_n}}$

$x_1 = 1$

$x_2 = x_1 - \dfrac{f(x_1)}{f'(x_1)} \approx 0.5379$

$x_3 = x_2 - \dfrac{f(x_2)}{f'(x_2)} \approx 0.5670$

$x_4 = x_3 - \dfrac{f(x_3)}{f'(x_3)} \approx 0.5671$

We approximate the root of f to be $x = 0.567$.

79. (a)

(b) When x increases without bound, $1/x$ approaches zero, and $e^{1/x}$ approaches 1. Therefore, $f(x)$ approaches $2/(1 + 1) = 1$. Thus, $f(x)$ has a horizontal asymptote at $y = 1$. As x approaches zero from the right, $1/x$ approaches ∞, $e^{1/x}$ approaches ∞ and $f(x)$ approaches zero. As x approaches zero from the left, $1/x$ approaches $-\infty$, $e^{1/x}$ approaches zero, and $f(x)$ approaches 2. The limit does not exist since the left limit does not equal the right limit. Therefore, $x = 0$ is a nonremovable discontinuity.

81.

h	0	5	10	15	20
P	10,332	5,583	2,376	1,240	517
$\ln P$	9.243	8.627	7.773	7.123	6.248

(a)

$y = -0.1499h + 9.3018$ is the regression line for data $(h, \ln P)$.

(c)

(b) $\ln P = ah + b$

$P = e^{ah+b} = e^b e^{ah}$

$P = Ce^{ah}, C = e^b$

For our data, $a = -0.1499$ and $C = e^{9.3018} = 10,957.7$

$P = 10,957.7e^{-0.1499h}$

(d) $\dfrac{dP}{dh} = (10,957.71)(-0.1499)e^{-0.1499h}$

$= -1642.56e^{-0.1499h}$

For $h = 5, \dfrac{dP}{dh} = -776.3$. For $h = 18, \dfrac{dP}{dh} \approx -110.6$.

83. $f(x) = e^{x/2}, f(0) = 1$

$f'(x) = \dfrac{1}{2}e^{x/2}, f'(0) = \dfrac{1}{2}$

$f''(x) = \dfrac{1}{4}e^{x/2}, f''(0) = \dfrac{1}{4}$

$P_1(x) = 1 + \dfrac{1}{2}(x - 0) = \dfrac{x}{2} + 1, P_1(0) = 1$

$P_1'(x) = \dfrac{1}{2}, P_1'(0) = \dfrac{1}{2}$

$P_2(x) = 1 + \dfrac{1}{2}(x - 0) + \dfrac{1}{8}(x - 0)^2 = \dfrac{x^2}{8} + \dfrac{x}{2} + 1, P_2(0) = 1$

$P_2'(x) = \dfrac{1}{4}x + \dfrac{1}{2}, P_2'(0) = \dfrac{1}{2}$

$P_2''(x) = \dfrac{1}{4}, P_2''(0) = \dfrac{1}{4}$

The values of f, P_1, P_2 and their first derivatives agree at $x = 0$. The values of the second derivatives of f and P_2 agree at $x = 0$.

85. (a) $y = e^x$

$y_1 = 1 + x$

(b) $y = e^x$

$y_2 = 1 + x + \left(\dfrac{x^2}{2}\right)$

(c) $y = e^x$

$y_3 = 1 + x + \dfrac{x^2}{2} + \dfrac{x^3}{6}$

87. Let $u = 5x$, $du = 5\,dx$.

$$\int e^{5x}\, 5\,dx = e^{5x} + C$$

89. Let $u = -2x$, $du = -2\,dx$.

$$\int_0^1 e^{-2x}\,dx = -\frac{1}{2}\int_0^1 e^{-2x}(-2)\,dx = \left[-\frac{1}{2}e^{-2x}\right]_0^1$$

$$= \frac{1}{2}(1 - e^{-2}) = \frac{e^2 - 1}{2e^2}$$

91. $\displaystyle\int xe^{-x^2}dx = -\frac{1}{2}\int e^{-x^2}(-2x)dx = -\frac{1}{2}e^{-x^2} + C$

93. $\displaystyle\int \frac{e^{\sqrt{x}}}{\sqrt{x}}dx = 2\int e^{\sqrt{x}}\left(\frac{1}{2\sqrt{x}}\right)dx = 2e^{\sqrt{x}} + C$

95. Let $u = 1 + e^{-x}$, $du = -e^{-x}\,dx$.

$$\int \frac{e^{-x}}{1 + e^{-x}}\,dx = -\int \frac{-e^{-x}}{1 + e^{-x}}\,dx = -\ln(1 + e^{-x}) + C = \ln\left(\frac{e^x}{e^x + 1}\right) + C = x - \ln(e^x + 1) + C$$

97. Let $u = \dfrac{3}{x}$, $du = -\dfrac{3}{x^2}\,dx$.

$$\int_1^3 \frac{e^{3/x}}{x^2}\,dx = -\frac{1}{3}\int_1^3 e^{3/x}\left(-\frac{3}{x^2}\right)dx$$

$$= \left[-\frac{1}{3}e^{3/x}\right]_1^3 = \frac{e}{3}(e^2 - 1)$$

99. Let $u = 1 - e^x$, $du = -e^x\,dx$.

$$\int e^x\sqrt{1 - e^x}\,dx = -\int (1 - e^x)^{1/2}(-e^x)\,dx$$

$$= -\frac{2}{3}(1 - e^x)^{3/2} + C$$

101. Let $u = e^x - e^{-x}$, $du = (e^x + e^{-x})\,dx$.

$$\int \frac{e^x + e^{-x}}{e^x - e^{-x}}\,dx = \ln|e^x - e^{-x}| + C$$

103. $\displaystyle\int \frac{5 - e^x}{e^{2x}}\,dx = \int 5e^{-2x}\,dx - \int e^{-x}\,dx$

$$= -\frac{5}{2}e^{-2x} + e^{-x} + C$$

105. $\displaystyle\int e^{\sin \pi x}\cos \pi x\, dx = \frac{1}{\pi}\int e^{\sin \pi x}(\pi \cos \pi x)\, dx$

$$= \frac{1}{\pi}e^{\sin \pi x} + C$$

107. $\displaystyle\int e^{-x}\tan(e^{-x})\,dx = -\int [\tan(e^{-x})](-e^{-x})\,dx$

$$= \ln|\cos(e^{-x})| + C$$

109. Let $u = ax^2$, $du = 2ax\,dx$. (Assume $a \neq 0$)

$$y = \int xe^{ax^2}\,dx$$

$$= \frac{1}{2a}\int e^{ax^2}(2ax)\,dx = \frac{1}{2a}e^{ax^2} + C$$

111. $f'(x) = \int \frac{1}{2}(e^x + e^{-x})\,dx = \frac{1}{2}(e^x - e^{-x}) + C_1$

$f'(0) = C_1 = 0$

$f(x) = \int \frac{1}{2}(e^x - e^{-x})\,dx = \frac{1}{2}(e^x + e^{-x}) + C_2$

$f(0) = 1 + C_2 = 1 \implies C_2 = 0$

$f(x) = \frac{1}{2}(e^x + e^{-x})$

113. (a)

(b) $\dfrac{dy}{dx} = 2e^{-x/2}, \quad (0, 1)$

$$y = \int 2e^{-x/2}\,dx = -4\int e^{-x/2}\left(-\frac{1}{2}\,dx\right)$$

$$= -4e^{-x/2} + C$$

$(0, 1): \ 1 = -4e^0 + C = -4 + C \implies C = 5$

$y = -4e^{-x/2} + 5$

115. $\displaystyle\int_0^5 e^x\,dx = \left[e^x\right]_0^5 = e^5 - 1 \approx 147.413$

117. $\displaystyle\int_0^{\sqrt{6}} xe^{-x^2/4}\,dx = \left[-2e^{-x^2/4}\right]_0^{\sqrt{6}}$

$$= -2e^{-3/2} + 2 \approx 1.554$$

119. (a) $f(u - v) = e^{u-v} = (e^u)(e^{-v}) = \dfrac{e^u}{e^v} = \dfrac{f(u)}{f(v)}$

(b) $f(kx) = e^{kx} = (e^x)^k = [f(x)]^k$.

121. $0.0665\displaystyle\int_{48}^{60} e^{-0.0139(t-48)^2}\,dt$

Graphing Utility: $0.4772 = 47.72\%$

123. $\displaystyle\int_0^x e^t\,dt \geq \int_0^x 1\,dt$

$\left[e^t\right]_0^x \geq \left[t\right]_0^x$

$e^x - 1 \geq x \implies e^x \geq 1 + x \text{ for } x \geq 0$

125. $f(x) = e^x$. Domain is $(-\infty, \infty)$ and range is $(0, \infty)$.
f is continuous, increasing, one-to-one, and concave upwards on its entire domain.

$$\lim_{x \to -\infty} e^x = 0 \text{ and } \lim_{x \to \infty} e^x = \infty.$$

127. Yes. $f(x) = Ce^x$, C a constant.

129. $e^{-x} > 0 \implies \displaystyle\int_0^2 e^{-x}\,dx > 0.$

131. $f(x) = \dfrac{\ln x}{x}$

(a) $f'(x) = \dfrac{1 - \ln x}{x^2} = 0$ when $x = e$.

On $(0, e), f'(x) > 0 \implies f$ is increasing.

On $(e, \infty), f'(x) < 0 \implies f$ is decreasing.

(b) For $e \leq A < B$, we have:

$$\dfrac{\ln A}{A} > \dfrac{\ln B}{B}$$

$$B \ln A > A \ln B$$

$$\ln A^B > \ln B^A$$

$$A^B > B^A.$$

(c) Since $e < \pi$, from part (b) we have $e^\pi > \pi^e$.

Section 5.5   Bases Other than *e* and Applications

1. $y = \left(\dfrac{1}{2}\right)^{t/3}$

At $t_0 = 6, y = \left(\dfrac{1}{2}\right)^{6/3} = \dfrac{1}{4}$

3. $y = \left(\dfrac{1}{2}\right)^{t/7}$

At $t_0 = 10, y = \left(\dfrac{1}{2}\right)^{10/7} \approx 0.3715$

5. $\log_2 \dfrac{1}{8} = \log_2 2^{-3} = -3$

7. $\log_7 1 = 0$

9. (a) $\quad 2^3 = 8$

$\quad\quad \log_2 8 = 3$

(b) $\quad 3^{-1} = \dfrac{1}{3}$

$\quad\quad \log_3 \dfrac{1}{3} = -1$

11. (a) $\log_{10} 0.01 = -2$

$\quad\quad 10^{-2} = 0.01$

(b) $\log_{0.5} 8 = -3$

$\quad\quad 0.5^{-3} = 8$

$\quad\quad \left(\dfrac{1}{2}\right)^{-3} = 8$

13. $y = 3^x$

x	-2	-1	0	1	2
y	$\dfrac{1}{9}$	$\dfrac{1}{3}$	1	3	9

15. $y = \left(\dfrac{1}{3}\right)^x = 3^{-x}$

x	-2	-1	0	1	2
y	9	3	1	$\dfrac{1}{3}$	$\dfrac{1}{9}$

17. $h(x) = 5^{x-2}$

x	-1	0	1	2	3
y	$\dfrac{1}{125}$	$\dfrac{1}{25}$	$\dfrac{1}{5}$	1	5

19. (a) $\log_{10} 1000 = x$

$10^x = 1000$

$x = 3$

(b) $\log_{10} 0.1 = x$

$10^x = 0.1$

$x = -1$

21. (a) $\log_3 x = -1$

$3^{-1} = x$

$x = \frac{1}{3}$

(b) $\log_2 x = -4$

$2^{-4} = x$

$x = \frac{1}{16}$

23. (a) $x^2 - x = \log_5 25$

$x^2 - x = \log_5 5^2 = 2$

$x^2 - x - 2 = 0$

$(x + 1)(x - 2) = 0$

$x = -1 \text{ OR } x = 2$

(b) $3x + 5 = \log_2 64$

$3x + 5 = \log_2 2^6 = 6$

$3x = 1$

$x = \frac{1}{3}$

25. $3^{2x} = 75$

$2x \ln 3 = \ln 75$

$x = \left(\frac{1}{2}\right)\dfrac{\ln 75}{\ln 3} \approx 1.965$

27. $2^{3-z} = 625$

$(3 - z)\ln 2 = \ln 625$

$3 - z = \dfrac{\ln 625}{\ln 2}$

$z = 3 - \dfrac{\ln 625}{\ln 2} \approx -6.288$

29. $\left(1 + \dfrac{0.09}{12}\right)^{12t} = 3$

$12t \ln\left(1 + \dfrac{0.09}{12}\right) = \ln 3$

$t = \left(\dfrac{1}{12}\right)\dfrac{\ln 3}{\ln\left(1 + \dfrac{0.09}{12}\right)} \approx 12.253$

31. $\log_2(x - 1) = 5$

$x - 1 = 2^5 = 32$

$x = 33$

33. $\log_3 x^2 = 4.5$

$x^2 = 3^{4.5}$

$x = \pm\sqrt{3^{4.5}} \approx \pm 11.845$

35. $g(x) = 6(2^{1-x}) - 25$

Zero: $x \approx -1.059$

37. $h(s) = 32 \log_{10}(s - 2) + 15$

Zero: $s \approx 2.340$

39. $f(x) = 4^x$

$g(x) = \log_4 x$

x	-2	-1	0	$\frac{1}{2}$	1
$f(x)$	$\frac{1}{16}$	$\frac{1}{4}$	1	2	4

x	$\frac{1}{16}$	$\frac{1}{4}$	1	2	4
$g(x)$	-2	-1	0	$\frac{1}{2}$	1

41. $f(x) = 4^x$

$f'(x) = (\ln 4)\, 4^x$

43. $y = 5^{x-2}$

$\dfrac{dy}{dx} = (\ln 5)\, 5^{x-2}$

45. $g(t) = t^2\, 2^t$

$g'(t) = t^2 (\ln 2)\, 2^t + (2t)\, 2^t$

$\quad = t\, 2^t\, (t \ln 2 + 2)$

$\quad = 2^t\, t(2 + t \ln 2)$

47. $h(\theta) = 2^{-\theta} \cos \pi\theta$

$h'(\theta) = 2^{-\theta}(-\pi \sin \pi\theta) - (\ln 2) 2^{-\theta} \cos \pi\theta$

$\quad = -2^{-\theta}[(\ln 2) \cos \pi\theta + \pi \sin \pi\theta]$

49. $y = \log_3 x$

$\dfrac{dy}{dx} = \dfrac{1}{x \ln 3}$

51. $f(x) = \log_2 \dfrac{x^2}{x-1}$

$\quad = 2 \log_2 x - \log_2 (x-1)$

$f'(x) = \dfrac{2}{x \ln 2} - \dfrac{1}{(x-1) \ln 2}$

$\quad = \dfrac{x-2}{(\ln 2)x(x-1)}$

53. $y = \log_5 \sqrt{x^2 - 1} = \dfrac{1}{2} \log_5 (x^2 - 1)$

$\dfrac{dy}{dx} = \dfrac{1}{2} \cdot \dfrac{2x}{(x^2-1)\ln 5} = \dfrac{x}{(x^2-1)\ln 5}$

55. $g(t) = \dfrac{10 \log_4 t}{t} = \dfrac{10}{\ln 4}\left(\dfrac{\ln t}{t}\right)$

$g'(t) = \dfrac{10}{\ln 4}\left[\dfrac{t(1/t) - \ln t}{t^2}\right]$

$\quad = \dfrac{10}{t^2 \ln 4}[1 - \ln t] = \dfrac{5}{t^2 \ln 2}(1 - \ln t)$

57. $\qquad y = x^{2/x}$

$\ln y = \dfrac{2}{x} \ln x$

$\dfrac{1}{y}\left(\dfrac{dy}{dx}\right) = \dfrac{2}{x}\left(\dfrac{1}{x}\right) + \ln x\left(-\dfrac{2}{x^2}\right) = \dfrac{2}{x^2}(1 - \ln x)$

$\dfrac{dy}{dx} = \dfrac{2y}{x^2}(1 - \ln x) = 2x^{(2/x)-2}(1 - \ln x)$

59. $\qquad y = (x-2)^{x+1}$

$\ln y = (x+1) \ln(x-2)$

$\dfrac{1}{y}\left(\dfrac{dy}{dx}\right) = (x+1)\left(\dfrac{1}{x-2}\right) + \ln(x-2)$

$\dfrac{dy}{dx} = y\left[\dfrac{x+1}{x-2} + \ln(x-2)\right]$

$\quad = (x-2)^{x+1}\left[\dfrac{x+1}{x-2} + \ln(x-2)\right]$

61. $\displaystyle\int 3^x\, dx = \dfrac{3^x}{\ln 3} + C$

63. $\int_{-1}^{2} 2^x \, dx = \left[\dfrac{2^x}{\ln 2}\right]_{-1}^{2}$

$= \dfrac{1}{\ln 2}\left[4 - \dfrac{1}{2}\right]$

$= \dfrac{7}{2 \ln 2} = \dfrac{7}{\ln 4}$

65. $\int x\left(5^{-x^2}\right) dx = -\dfrac{1}{2}\int 5^{-x^2}(-2x) \, dx$

$= -\left(\dfrac{1}{2}\right)\dfrac{5^{-x^2}}{\ln 5} + C$

$= \dfrac{-1}{2 \ln 5}\left(5^{-x^2}\right) + C$

67. $\int \dfrac{3^{2x}}{1 + 3^{2x}} \, dx,\ u = 1 + 3^{2x},\ du = 2(\ln 3)3^{2x} \, dx$

$\dfrac{1}{2 \ln 3}\int \dfrac{(2 \ln 3)3^{2x}}{1 + 3^{2x}} \, dx = \dfrac{1}{2 \ln 3}\ln(1 + 3^{2x}) + C$

69. $\dfrac{dy}{dx} = 0.4^{x/3},\ \left(0, \dfrac{1}{2}\right)$

$y = \int 0.4^{x/3} \, dx = 3\int 0.4^{x/3}\left(\dfrac{1}{3} \, dx\right)$

$= \dfrac{3}{\ln 0.4} 0.4^{x/3} + C = 3(\ln 2.5)(0.4)^{x/3} + C$

$\dfrac{1}{2} = 3(\ln 2.5) + C \implies C = \dfrac{1}{2} - 3 \ln 2.5$

$y = 3\,(\ln 2.5)(0.4)^{x/3} + \dfrac{1}{2} - 3 \ln 2.5$

$= \dfrac{3(1 - 0.4^{x/3})}{\ln 2.5} + \dfrac{1}{2}$

(a) (b)

71. Answers will vary. Example: Growth and decay problems

73.

x	1	2	8
y	0	1	3

(a) y is an exponential function of x: False

(b) y is a logarithmic function of x: True; $y = \log_2 x$

(c) x is an exponential function of y: True, $2^y = x$

(d) y is a linear function of x: False

75. $f(x) = \log_2 x \implies f'(x) = \dfrac{1}{x \ln 2}$

$g(x) = x^x \implies g'(x) = x^x(1 + \ln x)$

Note: Let $y = g(x)$. Then: $\ln y = \ln x^x = x \ln x$

$\dfrac{1}{y} y' = x \cdot \dfrac{1}{x} + \ln x$

$y' = y(1 + \ln x)$

$y' = x^x(1 + \ln x) = g'(x)$

$h(x) = x^2 \implies h'(x) = 2x$

$k(x) = 2^x \implies k'(x) = (\ln 2)2^x$

From greatest to smallest rate of growth:

$g(x),\ k(x),\ h(x),\ f(x)$

77. $C(t) = P(1.05)^t$

(a) $C(10) = 24.95(1.05)^{10}$

$\approx \$40.64$

(b) $\dfrac{dC}{dt} = P(\ln 1.05)(1.05)^t$

When $t = 1,\ \dfrac{dC}{dt} \approx 0.051P.$

When $t = 8,\ \dfrac{dC}{dt} \approx 0.072P.$

(c) $\dfrac{dC}{dt} = (\ln 1.05)[P(1.05)^t]$

$= (\ln 1.05)C(t)$

The constant of proportionality is $\ln 1.05$.

79. $P = \$1000$, $r = 3\frac{1}{2}\% = 0.035$, $t = 10$

$A = 1000\left(1 + \dfrac{0.035}{n}\right)^{10n}$

$A = 1000e^{(0.035)(10)} = 1419.07$

n	1	2	4	12	365	Continuous
A	1410.60	1414.78	1416.91	1418.34	1419.04	1419.07

81. $P = \$1000$, $r = 5\% = 0.05$, $t = 30$

$A = 1000\left(1 + \dfrac{0.05}{n}\right)^{30n}$

$A = 1000e^{(0.05)30} = 4481.69$

n	1	2	4	12	365	Continuous
A	4321.94	4399.79	4440.21	4467.74	4481.23	4481.69

83. $100,000 = Pe^{0.05t} \Rightarrow P = 100,000e^{-0.05t}$

t	1	10	20	30	40	50
P	95,122.94	60,653.07	36,787.94	22,313.02	13,533.53	8208.50

85. $100,000 = P\left(1 + \dfrac{0.05}{12}\right)^{12t} \Rightarrow P = 100,000\left(1 + \dfrac{0.05}{12}\right)^{-12t}$

t	1	10	20	30	40	50
P	95,132.82	60,716.10	36,864.45	22,382.66	13,589.88	8251.24

87. (a) $A = 20,000\left(1 + \dfrac{0.06}{365}\right)^{(365)(8)} \approx \$32,320.21$

(b) $A = \$30,000$

(c) $A = 8000\left(1 + \dfrac{0.06}{365}\right)^{(365)(8)} + 20,000\left(1 + \dfrac{0.06}{365}\right)^{(365)(4)}$

$\approx \$12,928.09 + 25,424.48 = \$38,352.57$

(d) $A = 9000\left[\left(1 + \dfrac{0.06}{365}\right)^{(365)(8)} + \left(1 + \dfrac{0.06}{365}\right)^{(365)(4)} + 1\right]$

$\approx \$34,985.11$

Take option (c).

89. (a) $\lim\limits_{t \to \infty} 6.7e^{(-48.1)/t} = 6.7e^0 = 6.7$ million ft^3

(b) $V' = \dfrac{322.27}{t^2} e^{-(48.1)/t}$

$V'(20) \approx 0.073$ million ft^3/yr

$V'(60) \approx 0.040$ million ft^3/yr

91. $y = \dfrac{300}{3 + 17e^{-0.0625x}}$

(a)

(b) If $x = 2$ (2000 egg masses), $y \approx 16.67 \approx 16.7\%$.

(c) If $y = 66.67\%$, then $x \approx 38.8$ or 38,800 egg masses.

(d) $y = 300(3 + 17e^{-0.0625x})^{-1}$

$y' = \dfrac{318.75e^{-0.0625x}}{(3 + 17e^{-0.0625x})^2}$

$y'' = \dfrac{19.921875e^{-0.0625x}(17e^{-0.0625x} - 3)}{(3 + 17e^{-0.0625x})^3}$

$17e^{-0.0625x} - 3 = 0 \Rightarrow x \approx 27.8$ or 27,800 egg masses.

93. (a) $B = 4.7539(6.7744)^d = 4.7539e^{1.9132d}$

(b)

(c) $B'(d) = 9.0952e^{1.9132d}$

$B'(0.8) \approx 42.03$ tons/inch

$B'(1.5) \approx 160.38$ tons/inch

95. (a) $\displaystyle\int_0^4 f(t)\,dt \approx 5.67$

$\displaystyle\int_0^4 g(t)\,dt \approx 5.67$

$\displaystyle\int_0^4 h(t)\,dt \approx 5.67$

(b)

(c) The functions appear to be equal: $f(t) = g(t) = h(t)$

Analytically,

$$f(t) = 4\left(\frac{3}{8}\right)^{2t/3} = 4\left[\left(\frac{3}{8}\right)^{2/3}\right]^t = 4\left(\frac{9^{1/3}}{4}\right)^t = g(t)$$

$$h(t) = 4e^{-0.653886t} = 4[e^{-0.653886}]^t \approx 4(0.52002)^t$$

$$g(t) = 4\left(\frac{9^{1/3}}{4}\right)^t \approx 4(0.52002)^t$$

No. The definite integrals over a given interval may be equal when the functions are not equal.

97. $P = \displaystyle\int_0^{10} 2000e^{-0.06t}\,dt$

$= \left[\dfrac{2000}{-0.06}e^{-0.06t}\right]_0^{10}$

$\approx \$15,039.61$

99.

t	0	1	2	3	4
y	1200	720	432	259.20	155.52

$y = C(k^t)$

When $t = 0$, $y = 1200 \implies C = 1200$.

$y = 1200(k^t)$

$\dfrac{720}{1200} = 0.6,\ \dfrac{432}{720} = 0.6,\ \dfrac{259.20}{432} = 0.6,\ \dfrac{155.52}{259.20} = 0.6$

Let $k = 0.6$.

$y = 1200(0.6)^t$

101. False. e is an irrational number.

103. True.

$f(g(x)) = 2 + e^{\ln(x-2)}$

$= 2 + x - 2 = x$

$g(f(x)) = \ln(2 + e^x - 2)$

$= \ln e^x = x$

105. True.

$\dfrac{d}{dx}[e^x] = e^x$ and $\dfrac{d}{dx}[e^{-x}] = -e^{-x}$

$e^x = e^{-x}$ when $x = 0$.

$(e^0)(-e^{-0}) = -1$

107.

$$\frac{dy}{dt} = \frac{8}{25} y \left(\frac{5}{4} - y \right), \, y(0) = 1$$

$$\frac{dy}{y[(5/4) - y]} = \frac{8}{25} dt \implies \frac{4}{5} \int \left(\frac{1}{y} + \frac{1}{(5/4) - y} \right) dy = \int \frac{8}{25} dt \implies$$

$$\ln y - \ln\left(\frac{5}{4} - y \right) = \frac{2}{5} t + C$$

$$\ln\left(\frac{y}{(5/4) - y} \right) = \frac{2}{5} t + C$$

$$\frac{y}{(5/4) - y} = e^{(2/5)t + C} = C_1 e^{(2/5)t}$$

$$y(0) = 1 \implies C_1 = 4 \implies 4e^{(2/5)t} = \frac{y}{(5/4) - y}$$

$$\implies 4e^{(2/5)t}\left(\frac{5}{4} - y \right) = y \implies 5e^{(2/5)t} = 4e^{(2/5)t} y + y = (4e^{(2/5)t} + 1)y$$

$$\implies y = \frac{5e^{(2/5)t}}{4e^{(2/5)t} + 1} = \frac{5}{4 + e^{-0.4t}} = \frac{1.25}{1 + 0.25 e^{-0.4t}}$$

Section 5.6   Differential Equations: Growth and Decay

1. $\dfrac{dy}{dx} = x + 2$

$$y = \int (x + 2)dx = \frac{x^2}{2} + 2x + C$$

3. $\dfrac{dy}{dx} = y + 2$

$$\frac{dy}{y + 2} = dx$$

$$\int \frac{1}{y + 2} dy = \int dx$$

$$\ln|y + 2| = x + C_1$$

$$y + 2 = e^{x + C_1} = Ce^x$$

$$y = Ce^x - 2$$

5. $y' = \dfrac{5x}{y}$

$$yy' = 5x$$

$$\int yy' \, dx = \int 5x \, dx$$

$$\int y \, dy = \int 5x \, dx$$

$$\frac{1}{2} y^2 = \frac{5}{2} x^2 + C_1$$

$$y^2 - 5x^2 = C$$

7. $y' = \sqrt{x}\, y$

$$\frac{y'}{y} = \sqrt{x}$$

$$\int \frac{y'}{y} \, dx = \int \sqrt{x} \, dx$$

$$\int \frac{dy}{y} = \int \sqrt{x} \, dx$$

$$\ln|y| = \frac{2}{3} x^{3/2} + C_1$$

$$y = e^{(2/3)x^{3/2} + C_1}$$

$$= e^{C_1} e^{(2/3)x^{3/2}}$$

$$= Ce^{(2/3)x^{3/2}}$$

9. $(1 + x^2)y' - 2xy = 0$

$$y' = \frac{2xy}{1 + x^2}$$

$$\frac{y'}{y} = \frac{2x}{1 + x^2}$$

$$\int \frac{y'}{y}\, dx = \int \frac{2x}{1 + x^2}\, dx$$

$$\int \frac{dy}{y} = \int \frac{2x}{1 + x^2}\, dx$$

$$\ln|y| = \ln(1 + x^2) + C_1$$

$$\ln|y| = \ln(1 + x^2) + \ln C$$

$$\ln|y| = \ln\left[C(1 + x^2)\right]$$

$$y = C(1 + x^2)$$

11. $\dfrac{dQ}{dt} = \dfrac{k}{t^2}$

$$\int \frac{dQ}{dt}\, dt = \int \frac{k}{t^2}\, dt$$

$$\int dQ = -\frac{k}{t} + C$$

$$Q = -\frac{k}{t} + C$$

13. $\dfrac{dN}{ds} = k(250 - s)$

$$\int \frac{dN}{ds}\, ds = \int k(250 - s)\, ds$$

$$\int dN = -\frac{k}{2}(250 - s)^2 + C$$

$$N = -\frac{k}{2}(250 - s)^2 + C$$

15. (a)

(b) $\dfrac{dy}{dx} = x(6 - y),\quad (0,0)$

$$\frac{dy}{y - 6} = -x\, dx$$

$$\ln|y - 6| = \frac{-x^2}{2} + C$$

$$y - 6 = e^{-x^2/2 + C} = C_1 e^{-x^2/2}$$

$$y = 6 + C_1 e^{-x^2/2}$$

$(0,0):\ 0 = 6 + C_1 \implies C_1 = -6 \implies y = 6 - 6e^{-x^2/2}$

17. $\dfrac{dy}{dt} = \dfrac{1}{2}t,\quad (0, 10)$

$$\int dy = \int \frac{1}{2}t\, dt$$

$$y = \frac{1}{4}t^2 + C$$

$$10 = \frac{1}{4}(0)^2 + C \implies C = 10$$

$$y = \frac{1}{4}t^2 + 10$$

19. $\dfrac{dy}{dt} = -\dfrac{1}{2}y,\quad (0, 10)$

$$\int \frac{dy}{y} = \int -\frac{1}{2}\, dt$$

$$\ln|y| = -\frac{1}{2}t + C_1$$

$$y = e^{-(t/2) + C_1} = e^{C_1} e^{-t/2} = Ce^{-t/2}$$

$$10 = Ce^0 \implies C = 10$$

$$y = 10e^{-t/2}$$

21. $\dfrac{dy}{dx} = ky$

$y = Ce^{kx}$ (Theorem 5.16)

$(0, 4):\ 4 = Ce^0 = C$

$(3, 10):\ 10 = 4e^{3k} \implies k = \dfrac{1}{3}\ln\left(\dfrac{5}{2}\right)$

When $x = 6$, $y = 4e^{1/3\,\ln(5/2)(6)} = 4e^{\ln(5/2)^2}$

$$= 4\left(\frac{5}{2}\right)^2 = 25$$

23. $\dfrac{dV}{dt} = kV$

$V = Ce^{kt}$ (Theorem 5.16)

$(0, 20{,}000):\ C = 20{,}000$

$(4, 12{,}500):\ 12{,}500 = 20{,}000e^{4k} \implies k = \dfrac{1}{4}\ln\left(\dfrac{5}{8}\right)$

When $t = 6$, $V = 20{,}000e^{1/4\,\ln(5/8)(6)} = 20{,}000e^{\ln(5/8)^{3/2}}$

$$= 20{,}000\left(\frac{5}{8}\right)^{3/2} \approx 9882.118$$

25. $y = Ce^{kt}$, $\left(0, \dfrac{1}{2}\right)$, $(5, 5)$

$C = \dfrac{1}{2}$

$y = \dfrac{1}{2} e^{kt}$

$5 = \dfrac{1}{2} e^{5k}$

$k = \dfrac{\ln 10}{5} \approx 0.4605$

$y = \dfrac{1}{2} e^{0.4605t}$

27. $y = Ce^{kt}$, $(1, 1)$, $(5, 5)$

$1 = Ce^k$

$5 = Ce^{5k}$

$5Ce^k = Ce^{5k}$

$5e^k = e^{5k}$

$5 = e^{4k}$

$k = \dfrac{\ln 5}{4} \approx 0.4024$

$y = Ce^{0.4024t}$

$1 = Ce^{0.4024}$

$C \approx 0.6687$

$y = 0.6687e^{0.4024t}$

29. A differential equation in x and y is an equation that involves x, y and derivatives of y.

31. $\dfrac{dy}{dx} = \dfrac{1}{2}xy$

$\dfrac{dy}{dx} > 0$ when $xy > 0$. Quadrants I and III.

33. Since the initial quantity is 10 grams, $y = 10e^{[\ln(1/2)/1620]t}$. When $t = 1000$, $y = 10e^{[\ln(1/2)/1620](1000)} \approx 6.52$ grams. When $t = 10,000$, $y = 10e^{[\ln(1/2)/1620](10,000)} \approx 0.14$ gram.

35. Since $y = Ce^{[\ln(1/2)/1620]t}$, we have $0.5 = Ce^{[\ln(1/2)/1620](10,000)} \implies C \approx 36.07$.

Initial quantity: 36.07 grams.

When $t = 1000$, we have $y = Ce^{[\ln(1/2)/1620](1000)} \approx 23.51$ grams.

37. Since the initial quantity is 5 grams, we have $y = 5.0e^{[\ln(1/2)/5730]t}$.

When $t = 1000$, $y \approx 4.43$ g.

When $t = 10,000$, $y \approx 1.49$ g.

39. Since $y = Ce^{[\ln(1/2)/24,360]t}$, we have $2.1 = Ce^{[\ln(1/2)/24,360](1000)} \implies C \approx 2.16$. Thus, the initial quantity is 2.16 grams. When $t = 10,000$, $y = 2.16e^{[\ln(1/2)/24,360](10,000)} \approx 1.63$ grams.

41. Since $\dfrac{dy}{dx} = ky$, $y = Ce^{kt}$ or $y = y_0e^{kt}$.

$\dfrac{1}{2}y_0 = y_0e^{1620k}$

$k = \dfrac{-\ln 2}{1620}$

$y = y_0e^{-(\ln 2)t/1620}$.

When $t = 100$, $y = y_0e^{-(\ln 2)/16.2} \approx y_0(0.9581)$.
Therefore, 95.81% of the present amount still exists.

43. Since $A = 1000e^{0.06t}$, the time to double is given by $2000 = 1000e^{0.06t}$ and we have

$2 = e^{0.06t}$

$\ln 2 = 0.06t$

$t = \dfrac{\ln 2}{0.06} \approx 11.55$ years.

Amount after 10 years: $A = 1000e^{(0.06)(10)} \approx \$1822.12$

45. Since $A = 750e^{rt}$ and $A = 1500$ when $t = 7.75$, we have the following.

$$1500 = 750e^{7.75r}$$

$$r = \frac{\ln 2}{7.75} \approx 0.0894 = 8.94\%$$

Amount after 10 years: $A = 750e^{0.0894(10)} \approx \$1833.67$

47. Since $A = 500e^{rt}$ and $A = 1292.85$ when $t = 10$, we have the following.

$$1292.85 = 500e^{10r}$$

$$r = \frac{\ln(1292.85/500)}{10} \approx 0.0950 = 9.50\%$$

The time to double is given by

$$1000 = 500e^{0.0950t}$$

$$t = \frac{\ln 2}{0.095} \approx 7.30 \text{ years.}$$

49. $500{,}000 = P\left(1 + \dfrac{0.075}{12}\right)^{(12)(20)}$

$$P = 500{,}000\left(1 + \frac{0.075}{12}\right)^{-240}$$

$$\approx \$112{,}087.09$$

51. $500{,}000 = P\left(1 + \dfrac{0.08}{12}\right)^{(12)(35)}$

$$P = 500{,}000\left(1 + \frac{0.08}{12}\right)^{-420}$$

$$= \$30{,}688.87$$

53. (a) $2000 = 1000(1 + 0.07)^t$

$$2 = 1.07^t$$

$$\ln 2 = t \ln 1.07$$

$$t = \frac{\ln 2}{\ln 1.07} \approx 10.24 \text{ years}$$

(b) $2000 = 1000\left(1 + \dfrac{0.07}{12}\right)^{12t}$

$$2 = \left(1 + \frac{0.007}{12}\right)^{12t}$$

$$\ln 2 = 12t \ln\left(1 + \frac{0.07}{12}\right)$$

$$t = \frac{\ln 2}{12 \ln(1 + (0.07/12))} \approx 9.93 \text{ years}$$

(c) $2000 = 1000\left(1 + \dfrac{0.07}{365}\right)^{365t}$

$$2 = \left(1 + \frac{0.07}{365}\right)^{365t}$$

$$\ln 2 = 365t \ln\left(1 + \frac{0.07}{365}\right)$$

$$t = \frac{\ln 2}{365 \ln(1 + (0.07/365))} \approx 9.90 \text{ years}$$

(d) $2000 = 1000e^{(0.07)t}$

$$2 = e^{0.07t}$$

$$\ln 2 = 0.07t$$

$$t = \frac{\ln 2}{0.07} \approx 9.90 \text{ years}$$

55. (a) $2000 = 1000(1 + 0.085)^t$

$$2 = 1.085^t$$

$$\ln 2 = t \ln 1.085$$

$$t = \frac{\ln 2}{\ln 1.085} \approx 8.50 \text{ years}$$

(b) $2000 = 1000\left(1 + \dfrac{0.085}{12}\right)^{12t}$

$$2 = \left(1 + \frac{0.085}{12}\right)^{12t}$$

$$\ln 2 = 12t \ln\left(1 + \frac{0.085}{12}\right)$$

$$t = \frac{1}{12}\frac{\ln 2}{\ln\left(1 + \dfrac{0.085}{12}\right)} \approx 8.18 \text{ years}$$

(c) $2000 = 1000\left(1 + \dfrac{0.085}{365}\right)^{365t}$

$$2 = \left(1 + \frac{0.085}{365}\right)^{365t}$$

$$\ln 2 = 365t \ln\left(1 + \frac{0.085}{365}\right)$$

$$t = \frac{1}{365}\frac{\ln 2}{\ln\left(1 + \dfrac{0.085}{365}\right)} \approx 8.16 \text{ years}$$

(d) $2000 = 1000e^{0.085t}$

$$2 = e^{0.085t}$$

$$\ln 2 = 0.085t$$

$$t = \frac{\ln 2}{0.085} \approx 8.15 \text{ years}$$

57. $P = Ce^{kt} = Ce^{-0.009t}$

$P(-1) = 8.2 = Ce^{-0.009(-1)} \Rightarrow C = 8.1265$

$P = 8.1265e^{-0.009t}$

$P(10) \approx 7.43$ or 7,430,000 people in 2010

59. $P = Ce^{kt} = Ce^{0.036t}$

$P(-1) = 4.6 = Ce^{0.036(-1)} \Rightarrow C = 4.7686$

$P = 4.7686e^{0.036t}$

$P(10) \approx 6.83$ or 6,830,000 people in 2010

61. If $k < 0$, the population decreases.

If $k > 0$, the population increases.

63. $P = Ce^{kx}$, (0, 760), (1000, 672.71)

$C = 760$

$672.71 = 760e^{1000x}$

$x = \dfrac{\ln(672.71/760)}{1000} \approx -0.000122$

$P \approx 760e^{-0.000122x}$

When $x = 3000$, $P \approx 527.06$ mm Hg.

65. (a) $19 = 30(1 - e^{20k})$

$30e^{20k} = 11$

$k = \dfrac{\ln(11/30)}{20} \approx -0.0502$

$N \approx 30(1 - e^{-0.0502t})$

(b) $25 = 30(1 - e^{-0.0502t})$

$e^{-0.0502t} = \dfrac{1}{6}$

$t = \dfrac{-\ln 6}{-0.0502} \approx 36$ days

67. $S = Ce^{k/t}$

(a) $S = 5$ when $t = 1$

$5 = Ce^k$

$\lim_{t \to \infty} Ce^{k/t} = C = 30$

$5 = 30e^k$

$k = \ln\frac{1}{6} \approx -1.7918$

$S \approx 30e^{-1.7918/t}$

(b) When $t = 5$, $S \approx 20.9646$ which is 20,965 units.

(c)

69. $A(t) = V(t)e^{-0.10t} = 100,000e^{0.8\sqrt{t}}e^{-0.10t} = 100,000e^{0.8\sqrt{t} - 0.10t}$

$\dfrac{dA}{dt} = 100,000\left(\dfrac{0.4}{\sqrt{t}} - 0.10\right)e^{0.8\sqrt{t} - 0.10t} = 0$ when 16.

The timber should be harvested in the year 2014, (1998 + 16). **Note:** You could also use a graphing utility to graph $A(t)$ and find the maximum of $A(t)$. Use the viewing rectangle $0 \le x \le 30$ and $0 \le y \le 600,000$.

71. $\beta(I) = 10 \log_{10} \dfrac{I}{I_0}, I_0 = 10^{-16}$

(a) $\beta(10^{-14}) = 10 \log_{10} \dfrac{10^{-14}}{10^{-16}} = 20$ decibels

(b) $\beta(10^{-9}) = 10 \log_{10} \dfrac{10^{-9}}{10^{-16}} = 70$ decibels

(c) $\beta(10^{-6.5}) = 10 \log_{10} \dfrac{10^{-6.5}}{10^{-16}} = 95$ decibels

(d) $\beta(10^{-4}) = 10 \log_{10} \dfrac{10^{-4}}{10^{-16}} = 120$ decibels

73. $R = \dfrac{\ln I - \ln I_0}{\ln 10}, I = e^{R \ln 10} = 10^R$

(a) $8.3 = \dfrac{\ln I - \ln I_0}{\ln 10}$

$I = 10^{8.3} \approx 199,526,231.5$

(b) $2R = \dfrac{\ln I - \ln I_0}{\ln 10}$

$I = e^{2R \ln 10} = e^{2R \ln 10} = (e^{R \ln 10})^2 = (10^R)^2$

Increases by a factor of $e^{2R \ln 10}$ or 10^R.

(c) $\dfrac{dR}{dI} = \dfrac{\ln I - \ln I_0}{\ln 10}$

75. False. If $y = Ce^{kt}$, $y' = Cke^{kt} \neq$ constant.   **77.** True

Section 5.7   Differential Equations: Separation of Variables

1. Differential equation: $y' = 4y$

Solution: $y = Ce^{4x}$

Check: $y' = 4Ce^{4x} = 4y$

3. Differential equation: $y'' + y = 0$

Solution: $y = C_1 \cos x + C_2 \sin x$

Check: $y' = -C_1 \sin x + C_2 \cos x$

$y'' = -C_1 \cos x - C_2 \sin x$

$y'' + y = -C_1 \cos x - C_2 \sin x + C_1 \cos x + C_2 \sin x = 0$

5. Differential Equation:

$y'' + y = \tan x$

$y = -\cos x \ln|\sec x + \tan x|$

$y' = (-\cos x)\dfrac{1}{\sec x + \tan x}(\sec x \cdot \tan x + \sec^2 x) + \sin x \ln|\sec x + \tan x|$

$= \dfrac{(-\cos x)}{\sec x + \tan x}(\sec x)(\tan x + \sec x) + \sin x \ln|\sec x + \tan x|$

$= -1 + \sin x \ln|\sec x + \tan x|$

$y'' = (\sin x)\dfrac{1}{\sec x + \tan x}(\sec x \cdot \tan x + \sec^2 x) + \cos x \ln|\sec x + \tan x|$

$= (\sin x)(\sec x) + \cos x \ln|\sec x + \tan x|$

Substituting,

$y'' + y = (\sin x)(\sec x) + \cos x \ln|\sec x + \tan x| - \cos x \ln|\sec x + \tan x|$

$= \tan x.$

In Exercises 7–11, the differential equation is $y^{(4)} - 16y = 0$.

7. $y = 3 \cos x$

$y^{(4)} = 3 \cos x$

$y^{(4)} - 16y = -45 \cos x \neq 0,$

No.

9. $y = e^{-2x}$

$y^{(4)} = 16e^{-2x}$

$y^{(4)} - 16y = 16e^{-2x} - 16e^{-2x} = 0,$

Yes.

11. $y = C_1 e^{2x} + C_2 e^{-2x} + C_3 \sin 2x + C_4 \cos 2x$

$y^{(4)} = 16C_1 e^{2x} + 16C_2 e^{-2x} + 16C_3 \sin 2x + 16C_4 \cos 2x$

$y^{(4)} - 16y = 0,$

Yes.

In 13–17, the differential equation is $xy' - 2y = x^3e^x$.

13. $y = x^2$, $y' = 2x$

$xy' - 2y = x(2x) - 2(x^2) = 0 \neq x^3e^x$

No.

15. $y = x^2(2 + e^x)$, $y' = x^2(e^x) + 2x(2 + e^x)$

$xy' - 2y = x[x^2e^x + 2xe^x + 4x] - 2[x^2e^x + 2x^2] = x^3e^x$,

Yes.

17. $y = \ln x$, $y' = \dfrac{1}{x}$

$xy' - 2y = x\left(\dfrac{1}{x}\right) - 2 \ln x \neq x^3e^x$,   No.

19. $y = Ce^{kx}$

$\dfrac{dy}{dx} = Cke^{kx}$

Since $dy/dx = 0.07y$, we have $Cke^{kx} = 0.07Ce^{kx}$.
Thus, $k = 0.07$.

C cannot be determined.

21. $y^2 = Cx^3$ passes through $(4, 4)$

$16 = C(64) \implies C = \frac{1}{4}$

Particular solution: $y^2 = \frac{1}{4} x^3$ or $4y^2 = x^3$

23. Differential equation: $4yy' - x = 0$

General solution: $4y^2 - x^2 = C$

Particular solutions: $C = 0$, Two intersecting lines
        $C = \pm 1$, $C = \pm 4$, Hyperbolas

25. Differential equation: $y' + 2y = 0$

General Solution: $y = Ce^{-2x}$

$y' + 2y = C(-2)e^{-2x} + 2(Ce^{-2x}) = 0$

Initial condition: $y(0) = 3$, $3 = Ce^0 = C$

Particular solution: $y = 3e^{-2x}$

27. Differential equation: $y'' + 9y = 0$

General solution: $y = C_1 \sin 3x + C_2 \cos 3x$

$y' = 3C_1 \cos 3x - 3C_2 \sin 3x$,

$y'' = -9C_1 \sin 3x - 9C_2 \cos 3x$

$y'' + 9y = (-9C_1 \sin 3x - 9C_2 \cos 3x) +$
$\qquad\qquad 9(C_1 \sin 3x + C_2 \cos 3x) = 0$

Initial conditions: $y\left(\dfrac{\pi}{6}\right) = 2$, $y'\left(\dfrac{\pi}{6}\right) = 1$

$2 = C_1 \sin\left(\dfrac{\pi}{2}\right) + C_2 \cos\left(\dfrac{\pi}{2}\right) \implies C_1 = 2$

$y' = 3C_1 \cos 3x - 3C_2 \sin 3x$

$1 = 3C_1 \cos\left(\dfrac{\pi}{2}\right) - 3C_2 \sin\left(\dfrac{\pi}{2}\right)$

$\qquad = -3C_2 \implies C_2 = -\dfrac{1}{3}$

Particular solution: $y = 2 \sin 3x - \dfrac{1}{3} \cos 3x$

29. Differential equation: $x^2y'' - 3xy' + 3y = 0$

General solution: $y = C_1x + C_2x^3$

$y' = C_1 + 3C_2x^2, \ y'' = 6C_2x$

$x^2y'' - 3xy' + 3y = x^2(6C_2x) - 3x(C_1 + 3C_2x^2) +$

$\qquad\qquad 3(C_1x + C_2x^3) = 0$

Initial conditions: $y(2) = 0, \ y'(2) = 4$

$0 = 2C_1 + 8C_2$

$y' = C_1 + 3C_2x^2$

$4 = C_1 + 12C_2$

$\left.\begin{array}{l} C_1 + 4C_2 = 0 \\ C_1 + 12C_2 = 4 \end{array}\right\}\ C_2 = \dfrac{1}{2},\ C_1 = -2$

Particular solution: $y = -2x + \dfrac{1}{2}x^3$

31. $\dfrac{dy}{dx} = 3x^2$

$y = \displaystyle\int 3x^2\,dx = x^3 + C$

33. $\dfrac{dy}{dx} = \dfrac{x}{1 + x^2}$

$y = \displaystyle\int \dfrac{x}{1 + x^2}\,dx = \dfrac{1}{2}\ln(1 + x^2) + C$

$(u = 1 + x^2, du = 2x\,dx)$

35. $\dfrac{dy}{dx} = \dfrac{x - 2}{x} = 1 - \dfrac{2}{x}$

$y = \displaystyle\int \left[1 - \dfrac{2}{x}\right]dx$

$\quad = x - 2\ln|x| + C = x - \ln x^2 + C$

37. $\dfrac{dy}{dx} = \sin 2x$

$y = \displaystyle\int \sin 2x\,dx = -\dfrac{1}{2}\cos 2x + C$

$(u = 2x, du = 2\,dx)$

39. $\dfrac{dy}{dx} = x\sqrt{x - 3}$ Let $u = \sqrt{x - 3}$, then $x = u^2 + 3$ and $dx = 2u\,du$.

$y = \displaystyle\int x\sqrt{x - 3}\,dx = \int (u^2 + 3)(u)(2u)du$

$\quad = 2\displaystyle\int (u^4 + 3u^2)\,du = 2\left(\dfrac{u^5}{5} + u^3\right) + C = \dfrac{2}{5}(x - 3)^{5/2} + 2(x - 3)^{3/2} + C$

41. $\dfrac{dy}{dx} = xe^{x^2}$

$y = \displaystyle\int xe^{x^2}\,dx = \dfrac{1}{2}e^{x^2} + C$

$(u = x^2,\ du = 2x\,dx)$

43. $\quad\dfrac{dy}{dx} = \dfrac{x}{y}$

$\displaystyle\int y\,dy = \int x\,dx$

$\dfrac{y^2}{2} = \dfrac{x^2}{2} + C_1$

$y^2 - x^2 = C$

45. $\quad\dfrac{dr}{ds} = 0.05r$

$\displaystyle\int \dfrac{dr}{r} = \int 0.05\,ds$

$\ln|r| = 0.05s + C_1$

$r = e^{0.05s + C_1} = Ce^{0.05s}$

47. $(2 + x)y' = 3y$

$\displaystyle\int \dfrac{dy}{y} = \int \dfrac{3}{2 + x}\,dx$

$\ln|y| = 3\ln|2 + x| + \ln C = \ln|C(2 + x)^3|$

$y = C(2 + x)^3$

49. $yy' = \sin x$

$$\int y \, dy = \int \sin x \, dx$$

$$\frac{y^2}{2} = -\cos x + C_1$$

$$y^2 = -2 \cos x + C$$

51. $\sqrt{1 - 4x^2} \dfrac{dy}{dx} = x$

$$dy = \frac{x}{\sqrt{1 - 4x^2}} \, dx$$

$$\int dy = \int \frac{x}{\sqrt{1 - 4x^2}} \, dx$$

$$= -\frac{1}{8} \int (1 - 4x^2)^{-1/2}(-8x \, dx)$$

$$y = -\frac{1}{4}(1 - 4x^2)^{1/2} + C$$

53. $y \ln x - xy' = 0$

$$\int \frac{dy}{y} = \int \frac{\ln x}{x} \, dx \quad \left(u = \ln x, \, du = \frac{dx}{x} \right)$$

$$\ln|y| = \frac{1}{2}(\ln x)^2 + C_1$$

$$y = e^{(1/2)(\ln x)^2 + C_1} = Ce^{(\ln x)^2/2}$$

55. $yy' - e^x = 0$

$$\int y \, dy = \int e^x \, dx$$

$$\frac{y^2}{2} = e^x + C_1$$

$$y^2 = 2e^x + C$$

Initial condition: $y(0) = 4$, $16 = 2 + C$, $C = 14$

Particular solution: $y^2 = 2e^x + 14$

57. $y(x + 1) + y' = 0$

$$\int \frac{dy}{y} = -\int (x + 1) \, dx$$

$$\ln|y| = -\frac{(x + 1)^2}{2} + C_1$$

$$y = Ce^{-(x+1)^2/2}$$

Initial condition: $y(-2) = 1$, $1 = Ce^{-1/2}$, $C = e^{1/2}$

Particular solution: $y = e^{[1 - (x+1)^2]/2} = e^{-(x^2 + 2x)/2}$

59. $y(1 + x^2) \dfrac{dy}{dx} = x(1 + y^2)$

$$\frac{y}{1 + y^2} \, dy = \frac{x}{1 + x^2} \, dx$$

$$\frac{1}{2} \ln(1 + y^2) = \frac{1}{2} \ln(1 + x^2) + C_1$$

$$\ln(1 + y^2) = \ln(1 + x^2) + \ln C = \ln[C(1 + x^2)]$$

$$1 + y^2 = C(1 + x^2)$$

$$y(0) = \sqrt{3}: \ 1 + 3 = C \Rightarrow C = 4$$

$$1 + y^2 = 4(1 + x^2)$$

$$y^2 = 3 + 4x^2$$

61. $\dfrac{du}{dv} = uv \sin v^2$

$$\int \frac{du}{u} = \int v \sin v^2 \, dv$$

$$\ln|u| = -\frac{1}{2} \cos v^2 + C_1$$

$$u = Ce^{-(\cos v^2)/2}$$

Initial condition: $u(0) = 1$, $C = \dfrac{1}{e^{-1/2}} = e^{1/2}$

Particular solution: $u = e^{(1 - \cos v^2)/2}$

63. $dP - kP \, dt = 0$

$$\int \frac{dP}{P} = k \int dt$$

$$\ln|P| = kt + C_1$$

$$P = Ce^{kt}$$

Initial condition: $P(0) = P_0$, $P_0 = Ce^0 = C$

Particular solution: $P = P_0 e^{kt}$

65.
$$\frac{dy}{dx} = \frac{-9x}{16y}$$

$$\int 16y \, dy = -\int 9x \, dx$$

$$8y^2 = \frac{-9}{2}x^2 + C$$

Initial condition: $y(1) = 1$, $8 = -\frac{9}{2} + C$, $C = \frac{25}{2}$

Particular solution: $8y^2 = \frac{-9}{2}x^2 + \frac{25}{2}$,

$$16y^2 + 9x^2 = 25$$

67.
$$m = \frac{dy}{dx} = \frac{0-y}{(x+2)-x} = -\frac{y}{2}$$

$$\int \frac{dy}{y} = \int -\frac{1}{2} \, dx$$

$$\ln|y| = -\frac{1}{2}x + C_1$$

$$y = Ce^{-x/2}$$

69. $f(x, y) = x^3 - 4xy^2 + y^3$

$f(tx, ty) = t^3 x^3 - 4t\,xt^2 y^2 + t^3 y^3$

$\qquad = t^3(x^3 - 4xy^2 + y^3)$

Homogeneous of degree 3

71. $f(x, y) = \dfrac{x^2 y^2}{\sqrt{x^2 + y^2}}$

$$f(tx, ty) = \frac{t^4 x^2 y^2}{\sqrt{t^2 x^2 + t^2 y^2}} = t^3 \frac{x^2 y^2}{\sqrt{x^2 + y^2}}$$

Homogeneous of degree 3

73. $f(x, y) = 2 \ln xy$

$f(tx, ty) = 2 \ln[tx \, ty]$

$\qquad = 2 \ln[t^2 xy] = 2(\ln t^2 + \ln xy)$

Not homogeneous

75. $f(x, y) = 2 \ln \dfrac{x}{y}$

$$f(tx, ty) = 2 \ln \frac{tx}{ty} = 2 \ln \frac{x}{y}$$

Homogeneous degree 0

77.
$$y' = \frac{x + y}{2x}, y = vx$$

$$v + x\frac{dv}{dx} = \frac{x + vx}{2x}$$

$$x\frac{dv}{dx} = \frac{1+v}{2} - v = \frac{1-v}{2}$$

$$2\int \frac{dv}{1-v} = \int \frac{dx}{x}$$

$$-\ln(1-v)^2 = \ln|x| + \ln C = \ln|Cx|$$

$$\frac{1}{(1-v)^2} = |Cx|$$

$$\frac{1}{[1 - (y/x)]^2} = |Cx|$$

$$\frac{x^2}{(x-y)^2} = |Cx|$$

$$|x| = C(x-y)^2$$

79.
$$y' = \frac{x - y}{x + y}, y = vx$$

$$v + x\frac{dv}{dx} = \frac{x - xv}{x + xv}$$

$$v \, dx + x \, dv = \frac{1-v}{1+v} dx$$

$$x \, dv = \left(\frac{1-v}{1+v} - v\right) dx = \frac{1 - 2v - v^2}{1+v} dx$$

$$\int \frac{v+1}{v^2 + 2v - 1} dv = -\int \frac{dx}{x}$$

$$\frac{1}{2}\ln|v^2 + 2v - 1| = -\ln|x| + \ln C_1 = \ln\left|\frac{C_1}{x}\right|$$

$$|v^2 + 2v - 1| = \frac{C}{x^2}$$

$$\left|\frac{y^2}{x^2} + 2\frac{y}{x} - 1\right| = \frac{C}{x^2}$$

$$|y^2 + 2xy - x^2| = C$$

81. $\quad y' = \dfrac{xy}{x^2 - y^2},\ y = vx$

$$v + x\frac{dv}{dx} = \frac{x^2 v}{x^2 - x^2 v^2}$$

$$v\,dx + x\,dv = \frac{v}{1 - v^2}\,dx$$

$$x\,dv = \left(\frac{v}{1 - v^2} - v\right)dx = \left(\frac{v^3}{1 - v^2}\right)dx$$

$$\int \frac{1 - v^2}{v^3}\,dv = \int \frac{dx}{x}$$

$$-\frac{1}{2v^2} - \ln|v| = \ln|x| + \ln C_1 = \ln|C_1 x|$$

$$\frac{-1}{2v^2} = \ln|C_1 x v|$$

$$\frac{-x^2}{2y^2} = \ln|C_1 y|$$

$$y = Ce^{-x^2/2y^2}$$

83. $\quad x\,dy - (2xe^{-y/x} + y)\,dx = 0,\ y = vx$

$$x(v\,dx + x\,dv) - (2xe^{-v} + vx)\,dx = 0$$

$$\int e^v\,dv = \int \frac{2}{x}\,dx$$

$$e^v = \ln C_1 x^2$$

$$e^{y/x} = \ln C_1 + \ln x^2$$

$$e^{y/x} = C + \ln x^2$$

Initial condition: $y(1) = 0,\ 1 = C$

Particular solution: $e^{y/x} = 1 + \ln x^2$

85. $\quad \left(x \sec \dfrac{y}{x} + y\right)dx - x\,dy = 0,\ y = vx$

$$(x \sec v + xv)dx - x(v\,dx + x\,dv) = 0$$

$$(\sec v + v)\,dx = v\,dx + x\,dv$$

$$\int \cos v\,dv = \int \frac{dx}{x}$$

$$\sin v = \ln x + \ln C_1$$

$$x = Ce^{\sin v}$$

$$= Ce^{\sin(y/x)}$$

Initial condition: $y(1) = 0,\ 1 = Ce^0 = C$

Particular solution: $x = e^{\sin(y/x)}$

87. $\quad \dfrac{dy}{dx} = x$

$$y = \int x\,dx = \frac{1}{2}x^2 + C$$

89. $\dfrac{dy}{dx} = 4 - y$

$$\int \frac{dy}{4 - y} = \int dx$$

$$\ln|4 - y| = -x + C_1$$

$$4 - y = e^{-x + C_1}$$

$$y = 4 + Ce^{-x}$$

91. $\dfrac{dy}{dx} = 0.5y,\ y(0) = 6$

93. $\dfrac{dy}{dx} = 0.02y(10 - y),\ y(0) = 2$

95. $\dfrac{dy}{dt} = ky,\ y = Ce^{kt}$

Initial conditions: $y(0) = y_0$

$$y(1620) = \frac{y_0}{2}$$

$$C = y_0$$

$$\frac{y_0}{2} = y_0 e^{1620k}$$

$$k = \frac{\ln(1/2)}{1620}$$

Particular solution: $y = y_0 e^{-t(\ln 2)/1620}$

When $t = 25$, $y \approx 0.989 y_0$, $y = 98.9\%$ of y_0.

97. $\dfrac{dy}{dx} = k(y - 4)$

The direction field satisfies $(dy/dx) = 0$ along $y = 4$; but not along $y = 0$. Matches (a).

99. $\dfrac{dy}{dx} = ky(y - 4)$

The direction field satisfies $(dy/dx) = 0$ along $y = 0$ and $y = 4$. Matches (c).

101.

$$\frac{dw}{dt} = k(1200 - w)$$

$$\int \frac{dw}{1200 - w} = \int k\,dt$$

$$\ln|1200 - w| = -kt + C_1$$

$$1200 - w = e^{-kt + C_1} = Ce^{-kt}$$

$$w = 1200 - Ce^{-kt}$$

$w(0) = 60 = 1200 - C \Rightarrow C = 1200 - 60 = 1140$

$w = 1200 - 1140e^{-kt}$

(a)

(b) $k = 0.8$: $t = 1.31$ years

$\quad k = 0.9$: $t = 1.16$ years

$\quad k = 1.0$: $t = 1.05$ years

(c) Maximum weight: 1200 pounds

$\quad \lim\limits_{t \to \infty} w = 1200$

103. (a)

$$\frac{dv}{dt} = k(W - v)$$

$$\int \frac{dv}{W - v} = \int k\,dt$$

$$-\ln|W - v| = kt + C_1$$

$$v = W - Ce^{-kt}$$

Initial conditions:

$\quad W = 20$, $v = 0$ when $t = 0$, and

$\quad v = 5$ when $t = 1$.

$\quad C = 20$, $k = -\ln(3/4)$

Particular solution:

$$v = 20(1 - e^{\ln(3/4)t}) = 20\left(1 - \left(\frac{3}{4}\right)^t\right)$$

or

$$v \approx 20(1 - e^{-0.2877t})$$

(b) $s = \displaystyle\int 20(1 - e^{-0.2877t})\,dt$

$\quad \approx 20[t + 3.4761e^{-0.2877t}] + C$

Since $s(0) = 0$, $C \approx -69.5$ and we have
$s \approx 20t + 69.5(e^{-0.2877t} - 1)$.

105. Given family (circles): $x^2 + y^2 = C$

$$2x + 2yy' = 0$$

$$y' = -\frac{x}{y}$$

Orthogonal trajectory (lines): $y' = \frac{y}{x}$

$$\int \frac{dy}{y} = \int \frac{dx}{x}$$

$$\ln|y| = \ln|x| + \ln K$$

$$y = Kx$$

107. Given family (parabolas): $x^2 = Cy$

$$2x = Cy'$$

$$y' = \frac{2x}{C} = \frac{2x}{x^2/y} = \frac{2y}{x}$$

Orthogonal trajectory (ellipses): $y' = -\frac{x}{2y}$

$$2\int y\, dy = -\int x\, dx$$

$$y^2 = -\frac{x^2}{2} + K_1$$

$$x^2 + 2y^2 = K$$

109. Given family: $y^2 = Cx^3$

$$2yy' = 3Cx^2$$

$$y' = \frac{3Cx^2}{2y} = \frac{3x^2}{2y}\left(\frac{y^2}{x^3}\right) = \frac{3y}{2x}$$

Orthogonal trajectory (ellipses): $y' = -\frac{2x}{3y}$

$$3\int y\, dy = -2\int x\, dx$$

$$\frac{3y^2}{2} = -x^2 + K_1$$

$$3y^2 + 2x^2 = K$$

111. A general solution of order n has n arbitrary constants while in a particular solution initial conditions are given in order to solve for all these constants.

113. $M(x, y)dx + N(x, y)dy = 0$, where M and N are homogeneous functions of the same degree.

See Example 7a.

115. False. Consider Example 2. $y = x^3$ is a solution to $xy' - 3y = 0$, but $y = x^3 + 1$ is not a solution.

117. False

$$f(tx, ty) = t^2x^2 + t^2xy + 2$$

$$\neq t^2 f(x, y)$$

Section 5.8   Inverse Trigonometric Functions: Differentiation

1. $y = \arcsin x$

(a)

x	-1	-0.8	-0.6	-0.4	-0.2	0	0.2	0.4	0.6	0.8	1
y	-1.571	-0.927	-0.644	-0.412	-0.201	0	0.201	0.412	0.644	0.927	1.571

(b)

(c)

(d) Symmetric about origin:
$$\arcsin(-x) = -\arcsin x$$

Intercept: $(0, 0)$

3. False.

$$\arccos \frac{1}{2} = \frac{\pi}{3}$$

since the range is $[0, \pi]$.

5. $\arcsin \dfrac{1}{2} = \dfrac{\pi}{6}$

7. $\arccos \dfrac{1}{2} = \dfrac{\pi}{3}$

9. $\arctan \dfrac{\sqrt{3}}{3} = \dfrac{\pi}{6}$

11. $\operatorname{arccsc}\left(-\sqrt{2}\right) = -\dfrac{\pi}{4}$

13. $\arccos(-0.8) \approx 2.50$

15. $\operatorname{arcsec}(1.269) = \arccos\left(\dfrac{1}{1.269}\right)$

≈ 0.66

17. (a) $\sin\left(\arctan \frac{3}{4}\right) = \frac{3}{5}$

(b) $\sec\left(\arcsin \frac{4}{5}\right) = \frac{5}{3}$

19. (a) $\cot\left[\arcsin\left(-\dfrac{1}{2}\right)\right] = \cot\left(-\dfrac{\pi}{6}\right) = -\sqrt{3}$

(b) $\csc\left[\arctan\left(-\dfrac{5}{12}\right)\right] = -\dfrac{13}{5}$

21. $y = \cos(\arcsin 2x)$

$\theta = \arcsin 2x$

$y = \cos \theta = \sqrt{1 - 4x^2}$

23. $y = \sin(\operatorname{arcsec} x)$

$\theta = \operatorname{arcsec} x,\ 0 \le \theta \le \pi,\ \theta \ne \dfrac{\pi}{2}$

$y = \sin \theta = \dfrac{\sqrt{x^2 - 1}}{|x|}$

The absolute value bars on x are necessary because of
the restriction $0 \le \theta \le \pi$, $\theta \ne \pi/2$, and $\sin \theta$ for this
domain must always be nonnegative.

25. $y = \tan\left(\operatorname{arcsec}\dfrac{x}{3}\right)$

$\theta = \operatorname{arcsec}\dfrac{x}{3}$

$y = \tan\theta = \dfrac{\sqrt{x^2 - 9}}{3}$

27. $y = \csc\left(\arctan\dfrac{x}{\sqrt{2}}\right)$

$\theta = \arctan\dfrac{x}{\sqrt{2}}$

$y = \csc\theta = \dfrac{\sqrt{x^2 + 2}}{x}$

29. $\sin(\arctan 2x) = \dfrac{2x}{\sqrt{1 + 4x^2}}$

Asymptotes: $y = \pm 1$

$\arctan 2x = \theta$

$\tan\theta = 2x$

$\sin\theta = \dfrac{2x}{\sqrt{1 + 4x^2}}$

31. $\arcsin(3x - \pi) = \frac{1}{2}$

$3x - \pi = \sin\left(\frac{1}{2}\right)$

$x = \frac{1}{3}\left[\sin\left(\frac{1}{2}\right) + \pi\right] \approx 1.207$

33. $\arcsin\sqrt{2x} = \arccos\sqrt{x}$

$\sqrt{2x} = \sin\left(\arccos\sqrt{x}\right)$

$\sqrt{2x} = \sqrt{1 - x}, \, 0 \le x \le 1$

$2x = 1 - x$

$3x = 1$

$x = \dfrac{1}{3}$

35. (a) $\operatorname{arccsc} x = \arcsin\dfrac{1}{x}, \, x \ge 1$

Let $y = \operatorname{arccsc} x$. Then for

$$-\frac{\pi}{2} \le y < 0 \text{ and } 0 < y \le \frac{\pi}{2},$$

$\csc y = x \implies \sin y = 1/x$. Thus, $y = \arcsin(1/x)$.
Therefore, $\operatorname{arccsc} x = \arcsin(1/x)$.

(b) $\arctan x + \arctan\dfrac{1}{x} = \dfrac{\pi}{2}, \, x > 0$

Let $y = \arctan x + \arctan(1/x)$. Then,

$$\tan y = \frac{\tan(\arctan x) + \tan[\arctan(1/x)]}{1 - \tan(\arctan x)\tan[\arctan(1/x)]}$$

$$= \frac{x + (1/x)}{1 - x(1/x)}$$

$$= \frac{x + (1/x)}{0} \text{ (which is undefined).}$$

Thus, $y = \pi/2$. Therefore, $\arctan x + \arctan(1/x) = \pi/2$.

37. $f(x) = \arcsin(x - 1)$

$x - 1 = \sin y$

$x = 1 + \sin y$

Domain: $[0, 2]$

Range: $\left[-\dfrac{\pi}{2}, \dfrac{\pi}{2}\right]$

$f(x)$ is the graph of $\arcsin x$ shifted 1 unit to the right.

39. $f(x) = \operatorname{arcsec} 2x$

$2x = \sec y$

$x = \dfrac{1}{2}\sec y$

Domain: $\left(-\infty, -\dfrac{1}{2}\right], \left[\dfrac{1}{2}, \infty\right)$

Range: $\left[0, \dfrac{\pi}{2}\right), \left(\dfrac{\pi}{2}, \pi\right]$

41. $f(x) = 2 \arcsin(x - 1)$

$$f'(x) = \frac{2}{\sqrt{1 - (x-1)^2}} = \frac{2}{\sqrt{2x - x^2}}$$

43. $g(x) = 3 \arccos \dfrac{x}{2}$

$$g'(x) = \frac{-3(1/2)}{\sqrt{1 - (x^2/4)}} = \frac{-3}{\sqrt{4 - x^2}}$$

45. $f(x) = \arctan \dfrac{x}{a}$

$$f'(x) = \frac{1/a}{1 + (x^2/a^2)} = \frac{a}{a^2 + x^2}$$

47. $g(x) = \dfrac{\arcsin 3x}{x}$

$$g'(x) = \frac{x\left(3/\sqrt{1 - 9x^2}\right) - \arcsin 3x}{x^2}$$

$$= \frac{3x - \sqrt{1 - 9x^2}\,\arcsin 3x}{x^2\sqrt{1 - 9x^2}}$$

49. $h(t) = \sin(\arccos t) = \sqrt{1 - t^2}$

$$h'(t) = \frac{1}{2}(1 - t^2)^{-1/2}(-2t) = \frac{-t}{\sqrt{1 - t^2}}$$

51. $y = x \arccos x - \sqrt{1 - x^2}$

$$y' = \arccos x - \frac{x}{\sqrt{1 - x^2}} - \frac{1}{2}(1 - x^2)^{-1/2}(-2x)$$

$$= \arccos x$$

53. $y = \dfrac{1}{2}\left(\dfrac{1}{2}\ln\dfrac{x + 1}{x - 1} + \arctan x\right)$

$$= \frac{1}{4}[\ln(x + 1) - \ln(x - 1)] + \frac{1}{2}\arctan x$$

$$\frac{dy}{dx} = \frac{1}{4}\left(\frac{1}{x + 1} - \frac{1}{x - 1}\right) + \frac{1/2}{1 + x^2} = \frac{1}{1 - x^4}$$

55. $y = x \arcsin x + \sqrt{1 - x^2}$

$$\frac{dy}{dx} = x\left(\frac{1}{\sqrt{1 - x^2}}\right) + \arcsin x - \frac{x}{\sqrt{1 - x^2}} = \arcsin x$$

57. $y = 8 \arcsin \dfrac{x}{4} - \dfrac{x\sqrt{16 - x^2}}{2}$

$$y' = 2\frac{1}{\sqrt{1 - (x/4)^2}} - \frac{\sqrt{16 - x^2}}{2} - \frac{x}{4}(16 - x^2)^{-1/2}(-2x)$$

$$= \frac{8}{\sqrt{16 - x^2}} - \frac{\sqrt{16 - x^2}}{2} + \frac{x^2}{2\sqrt{16 - x^2}}$$

$$= \frac{16 - (16 - x^2) + x^2}{2\sqrt{16 - x^2}} = \frac{x^2}{\sqrt{16 - x^2}}$$

59. $y = \arctan x + \dfrac{x}{1 + x^2}$

$$y' = \frac{1}{1 + x^2} + \frac{(1 + x^2) - x(2x)}{(1 + x^2)^2}$$

$$= \frac{(1 + x^2) + (1 - x^2)}{(1 + x^2)^2}$$

$$= \frac{2}{(1 + x^2)^2}$$

61. $f(x) = \arcsin x, \ a = \dfrac{1}{2}$

$$f'(x) = \frac{1}{\sqrt{1 - x^2}}$$

$$f''(x) = \frac{x}{(1 - x^2)^{3/2}}$$

$$P_1(x) = f\left(\frac{1}{2}\right) + f'\left(\frac{1}{2}\right)\left(x - \frac{1}{2}\right) = \frac{\pi}{6} + \frac{2\sqrt{3}}{3}\left(x - \frac{1}{2}\right)$$

$$P_2(x) = f\left(\frac{1}{2}\right) + f'\left(\frac{1}{2}\right)\left(x - \frac{1}{2}\right) + \frac{1}{2}f''\left(\frac{1}{2}\right)\left(x - \frac{1}{2}\right)^2 = \frac{\pi}{6} + \frac{2\sqrt{3}}{3}\left(x - \frac{1}{2}\right) + \frac{2\sqrt{3}}{9}\left(x - \frac{1}{2}\right)^2$$

63. $f(x) = \text{arcsec } x - x$

$$f'(x) = \frac{1}{|x|\sqrt{x^2-1}} - 1$$

$$= 0 \text{ when } |x|\sqrt{x^2-1} = 1.$$

$$x^2(x^2-1) = 1$$

$$x^4 - x^2 - 1 = 0 \text{ when } x^2 = \frac{1+\sqrt{5}}{2} \text{ or}$$

$$x = \pm\sqrt{\frac{1+\sqrt{5}}{2}} = \pm 1.272$$

Relative maximum: $(1.272, -0.606)$

Relative minimum: $(-1.272, 3.747)$

65. $f(x) = \arctan x - \arctan(x-4)$

$$f'(x) = \frac{1}{1+x^2} - \frac{1}{1+(x-4)^2} = 0$$

$$1 + x^2 = 1 + (x-4)^2$$

$$0 = -8x + 16$$

$$x = 2$$

By the First Derivative Test, $(2, 2.214)$ is a relative maximum.

67. The trigonometric functions are not one-to-one on $(-\infty, \infty)$, so their domains must be restricted to intervals on which they are one-to-one.

69. $y = \text{arccot } x, \; 0 < y < \pi$

$$x = \cot y$$

$$\tan y = \frac{1}{x}$$

So, graph the function

$$y = \arctan\left(\frac{1}{x}\right) \text{ for } x > 0 \text{ and } y = \arctan\left(\frac{1}{x}\right) + \pi \text{ for } x < 0.$$

71. (a) $\cot\theta = \dfrac{x}{5}$

$$\theta = \text{arccot}\left(\frac{x}{5}\right)$$

(b) $\dfrac{d\theta}{dt} = \dfrac{-\dfrac{1}{5}}{1 + \left(\dfrac{x}{5}\right)^2}\dfrac{dx}{dt} = \dfrac{-5}{x^2 + 25}\dfrac{dx}{dt}$

If $\dfrac{dx}{dt} = -400$ and $x = 10$, $\dfrac{d\theta}{dt} = 16$ rad/hr.

If $\dfrac{dx}{dt} = -400$ and $x = 3$, $\dfrac{d\theta}{dt} \approx 58.824$ rad/hr.

73. (a) $h(t) = -16t^2 + 256$

$$-16t^2 + 256 = 0 \text{ when } t = 4 \text{ sec.}$$

(b) $\tan\theta = \dfrac{h}{500} = \dfrac{-16t^2 + 256}{500}$

$$\theta = \arctan\left[\frac{16}{500}(-t^2 + 16)\right]$$

$$\frac{d\theta}{dt} = \frac{-8t/125}{1 + [(4/125)(-t^2+16)]^2} = \frac{-1000t}{15,625 + 16(16-t^2)^2}$$

When $t = 1$, $d\theta/dt \approx -0.0520$ rad/sec.

When $t = 2$, $d\theta/dt \approx -0.1116$ rad/sec.

75. $\tan(\arctan x + \arctan y) = \dfrac{\tan(\arctan x) + \tan(\arctan y)}{1 - \tan(\arctan x)\tan(\arctan y)} = \dfrac{x + y}{1 - xy}, xy \neq 1$

Therefore,

$$\arctan x + \arctan y = \arctan\left(\frac{x + y}{1 - xy}\right), xy \neq 1.$$

Let $x = \frac{1}{2}$ and $y = \frac{1}{3}$.

$$\arctan\left(\frac{1}{2}\right) + \arctan\left(\frac{1}{3}\right) = \arctan\frac{(1/2) + (1/3)}{1 - [(1/2) \cdot (1/3)]} = \arctan\frac{5/6}{1 - (1/6)} = \arctan\frac{5/6}{5/6} = \arctan 1 = \frac{\pi}{4}$$

77. $f(x) = kx + \sin x$

$f'(x) = k + \cos x \geq 0$ for $k \geq 1$

$f'(x) = k + \cos x \leq 0$ for $k \leq -1$

Therefore, $f(x) = kx + \sin x$ is strictly monotonic and has an inverse for $k \leq -1$ or $k \geq 1$.

79. True

$$\frac{d}{dx}[\arctan x] = \frac{1}{1 + x^2} > 0 \text{ for all } x.$$

81. True

$$\frac{d}{dx}[\arctan(\tan x)] = \frac{\sec^2 x}{1 + \tan^2 x} = \frac{\sec^2 x}{\sec^2 x} = 1$$

Section 5.9   Inverse Trigonometric Functions: Integration

1. $\displaystyle\int \frac{5}{\sqrt{9 - x^2}}\, dx = 5\arcsin\left(\frac{x}{3}\right) + C$

3. Let $u = 3x$, $du = 3\, dx$.

$$\int_0^{1/6} \frac{1}{\sqrt{1 - 9x^2}}\, dx = \frac{1}{3}\int_0^{1/6} \frac{1}{\sqrt{1 - (3x)^2}}(3)\, dx = \left[\frac{1}{3}\arcsin(3x)\right]_0^{1/6} = \frac{\pi}{18}$$

5. $\displaystyle\int \frac{7}{16 + x^2}\, dx = \frac{7}{4}\arctan\left(\frac{x}{4}\right) + C$

7. Let $u = 2x$, $du = 2\, dx$.

$$\int_0^{\sqrt{3}/2} \frac{1}{1 + 4x^2}\, dx = \frac{1}{2}\int_0^{\sqrt{3}/2} \frac{2}{1 + (2x)^2}\, dx = \left[\frac{1}{2}\arctan(2x)\right]_0^{\sqrt{3}/2} = \frac{\pi}{6}$$

9. $\displaystyle\int \frac{1}{x\sqrt{4x^2 - 1}}\, dx = \int \frac{2}{2x\sqrt{(2x)^2 - 1}}\, dx = \text{arcsec}|2x| + C$

11. $\displaystyle\int \frac{x^3}{x^2 + 1}\, dx = \int \left[x - \frac{x}{x^2 + 1}\right] dx = \int x\, dx - \frac{1}{2}\int \frac{2x}{x^2 + 1}\, dx = \frac{1}{2}x^2 - \frac{1}{2}\ln(x^2 + 1) + C$ (Use long division.)

13. $\displaystyle\int \frac{1}{\sqrt{1 - (x + 1)^2}}\, dx = \arcsin(x + 1) + C$

15. Let $u = t^2$, $du = 2t\, dt$.

$$\int \frac{t}{\sqrt{1 - t^4}}\, dt = \frac{1}{2}\int \frac{1}{\sqrt{1 - (t^2)^2}}(2t)\, dt = \frac{1}{2}\arcsin(t^2) + C$$

17. Let $u = \arcsin x$, $du = \dfrac{1}{\sqrt{1 - x^2}}\, dx$.

$$\int_0^{1/\sqrt{2}} \frac{\arcsin x}{\sqrt{1 - x^2}}\, dx = \left[\frac{1}{2}\arcsin^2 x\right]_0^{1/\sqrt{2}} = \frac{\pi^2}{32} \approx 0.308$$

19. Let $u = 1 - x^2$, $du = -2x\, dx$.

$$\int_{-1/2}^0 \frac{x}{\sqrt{1 - x^2}}\, dx = -\frac{1}{2}\int_{-1/2}^0 (1 - x^2)^{-1/2}(-2x)\, dx$$

$$= \left[-\sqrt{1 - x^2}\right]_{-1/2}^0 = \frac{\sqrt{3} - 2}{2}$$

$$\approx -0.134$$

21. Let $u = e^{2x}$, $du = 2e^{2x}\, dx$.

$$\int \frac{e^{2x}}{4 + e^{4x}}\, dx = \frac{1}{2}\int \frac{2e^{2x}}{4 + (e^{2x})^2}\, dx = \frac{1}{4}\arctan\frac{e^{2x}}{2} + C$$

23. Let $u = \cos x$, $du = -\sin x\, dx$.

$$\int_{\pi/2}^{\pi} \frac{\sin x}{1 + \cos^2 x}\, dx = -\int_{\pi/2}^{\pi} \frac{-\sin x}{1 + \cos^2 x}\, dx$$

$$= \left[-\arctan(\cos x)\right]_{\pi/2}^{\pi} = \frac{\pi}{4}$$

25. $\displaystyle\int \frac{1}{\sqrt{x}\,\sqrt{1 - x}}\, dx$, $u = \sqrt{x}$, $x = u^2$, $dx = 2u\, du$

$$\int \frac{1}{u\sqrt{1 - u^2}}(2u\, du) = 2\int \frac{du}{\sqrt{1 - u^2}} = 2\arcsin u + C$$

$$= 2\arcsin\sqrt{x} + C$$

27. $\displaystyle\int \frac{x - 3}{x^2 + 1}\, dx = \frac{1}{2}\int \frac{2x}{x^2 + 1}\, dx - 3\int \frac{1}{x^2 + 1}\, dx$

$$= \frac{1}{2}\ln(x^2 + 1) - 3\arctan x + C$$

29. $\displaystyle\int \frac{x + 5}{\sqrt{9 - (x - 3)^2}}\, dx = \int \frac{(x - 3)}{\sqrt{9 - (x - 3)^2}}\, dx + \int \frac{8}{\sqrt{9 - (x - 3)^2}}\, dx$

$$= -\sqrt{9 - (x - 3)^2} - 8\arcsin\left(\frac{x - 3}{3}\right) + C$$

$$= -\sqrt{6x - x^2} + 8\arcsin\left(\frac{x}{3} - 1\right) + C$$

31. $\displaystyle\int_0^2 \frac{dx}{x^2 - 2x + 2} = \int_0^2 \frac{1}{1 + (x - 1)^2}\, dx = \left[\arctan(x - 1)\right]_0^2 = \frac{\pi}{2}$

33. $\displaystyle\int \frac{2x}{x^2 + 6x + 13}\, dx = \int \frac{2x + 6}{x^2 + 6x + 13}\, dx - 6\int \frac{1}{x^2 + 6x + 13}\, dx = \int \frac{2x + 6}{x^2 + 6x + 13}\, dx - 6\int \frac{1}{4 + (x + 3)^2}\, dx$

$$= \ln|x^2 + 6x + 13| - 3\arctan\left(\frac{x + 3}{2}\right) + C$$

35. $\displaystyle\int \frac{1}{\sqrt{-x^2 - 4x}}\, dx = \int \frac{1}{\sqrt{4 - (x + 2)^2}}\, dx = \arcsin\left(\frac{x + 2}{2}\right) + C$

37. Let $u = -x^2 - 4x$, $du = (-2x - 4)\, dx$.

$$\int \frac{x + 2}{\sqrt{-x^2 - 4x}}\, dx = -\frac{1}{2}\int (-x^2 - 4x)^{-1/2}(-2x - 4)\, dx = -\sqrt{-x^2 - 4x} + C$$

39. $\displaystyle\int_2^3 \frac{2x - 3}{\sqrt{4x - x^2}}\, dx = \int_2^3 \frac{2x - 4}{\sqrt{4x - x^2}}\, dx + \int_2^3 \frac{1}{\sqrt{4x - x^2}}\, dx = -\int_2^3 (4x - x^2)^{-1/2}(4 - 2x)\, dx + \int_2^3 \frac{1}{\sqrt{4 - (x - 2)^2}}\, dx$

$$= \left[-2\sqrt{4x - x^2} + \arcsin\left(\frac{x - 2}{2}\right)\right]_2^3 = 4 - 2\sqrt{3} + \frac{\pi}{6} \approx 1.059$$

41. Let $u = x^2 + 1$, $du = 2x\,dx$.

$$\int \frac{x}{x^4 + 2x^2 + 2}\,dx = \frac{1}{2}\int \frac{2x}{(x^2 + 1)^2 + 1}\,dx = \frac{1}{2}\arctan(x^2 + 1) + C$$

43. Let $u = \sqrt{e^t - 3}$. Then $u^2 + 3 = e^t$, $2u\,du = e^t\,dt$, and $\dfrac{2u\,du}{u^2 + 3} = dt$.

$$\int \sqrt{e^t - 3}\,dt = \int \frac{2u^2}{u^2 + 3}\,du = \int 2\,du - \int 6\,\frac{1}{u^2 + 3}\,du$$

$$= 2u - 2\sqrt{3}\arctan\frac{u}{\sqrt{3}} + C = 2\sqrt{e^t - 3} - 2\sqrt{3}\arctan\sqrt{\frac{e^t - 3}{3}} + C$$

45. A perfect square trinomial is an expression in x with three terms that factor as a perfect square.

Example: $x^2 + 6x + 9 = (x + 3)^2$

47. (a) $\displaystyle\int \frac{1}{\sqrt{1 - x^2}}\,dx = \arcsin x + C$, $u = x$ (b) $\displaystyle\int \frac{x}{\sqrt{1 - x^2}}\,dx = -\sqrt{1 - x^2} + C$, $u = 1 - x^2$

  (c) $\displaystyle\int \frac{1}{x\sqrt{1 - x^2}}\,dx$ cannot be evaluated using the basic integration rules.

49. (a) $\displaystyle\int \sqrt{x - 1}\,dx = \frac{2}{3}(x - 1)^{3/2} + C$, $u = x - 1$

  (b) Let $u = \sqrt{x - 1}$. Then $x = u^2 + 1$ and $dx = 2u\,du$.

$$\int x\sqrt{x - 1}\,dx = \int (u^2 + 1)(u)(2u)\,du = 2\int (u^4 + u^2)\,du = 2\left(\frac{u^5}{5} + \frac{u^3}{3}\right) + C$$

$$= \frac{2}{15}u^3(3u^2 + 5) + C = \frac{2}{15}(x - 1)^{3/2}[3(x - 1) + 5] + C = \frac{2}{15}(x - 1)^{3/2}(3x + 2) + C$$

  (c) Let $u = \sqrt{x - 1}$. Then $x = u^2 + 1$ and $dx = 2u\,du$.

$$\int \frac{x}{\sqrt{x - 1}}\,dx = \int \frac{u^2 + 1}{u}(2u)\,du = 2\int (u^2 + 1)\,du = 2\left(\frac{u^3}{3} + u\right) + C = \frac{2}{3}u(u^2 + 3) + C = \frac{2}{3}\sqrt{x - 1}(x + 2) + C$$

Note: In (b) and (c), substitution was necessary *before* the basic integration rules could be used.

51. (a)

(0, 0)

  (b) $\dfrac{dy}{dx} = \dfrac{3}{1 + x^2}$, $(0, 0)$

$$y = 3\int \frac{dx}{1 + x^2} = 3\arctan x + C$$

$(0, 0)$: $0 = 3\arctan(0) + C \implies C = 0$

$$y = 3\arctan x$$

53. $\dfrac{dy}{dx} = \dfrac{10}{x\sqrt{x^2 - 1}}$, $y(3) = 0$

55. $A = \displaystyle\int_1^3 \frac{1}{x^2 - 2x + 1 + 4}\,dx = \int_1^3 \frac{1}{(x - 1)^2 + 2^2}\,dx$

$$= \left[\frac{1}{2}\arctan\left(\frac{x - 1}{2}\right)\right]_1^3 = \frac{1}{2}\arctan(1) = \frac{\pi}{8} \approx 0.3927$$

57. Area ≈ (1)(1) = 1

Matches (c)

59. (a) $\int_0^1 \frac{4}{1 + x^2}\, dx = \left[4 \arctan x \right]_0^1 = 4 \arctan 1 - 4 \arctan 0 = 4\left(\frac{\pi}{4}\right) - 4(0) = \pi$

(b) Let $n = 6$.

$4\int_0^1 \frac{1}{1 + x^2}\, dx \approx 4\left(\frac{1}{18}\right)\left[1 + \frac{4}{1 + (1/36)} + \frac{2}{1 + (1/9)} + \frac{4}{1 + (1/4)} + \frac{2}{1 + (4/9)} + \frac{4}{1 + (25/36)} + \frac{1}{2} \right] \approx 3.1415918$

(c) 3.1415927

61. (a) $\dfrac{d}{dx}\left[\arcsin\left(\dfrac{u}{a}\right) + C \right] = \dfrac{1}{\sqrt{1 - (u^2/a^2)}}\left(\dfrac{u'}{a}\right) = \dfrac{u'}{\sqrt{a^2 - u^2}}$

Thus, $\displaystyle\int \frac{du}{\sqrt{a^2 - u^2}} = \arcsin\left(\frac{u}{a}\right) + C.$

(b) $\dfrac{d}{dx}\left[\dfrac{1}{a}\arctan \dfrac{u}{a} + C \right] = \dfrac{1}{a}\left[\dfrac{u'/a}{1 + (u/a)^2} \right] = \dfrac{1}{a^2}\left[\dfrac{u'}{(a^2 + u^2)/a^2} \right] = \dfrac{u'}{a^2 + u^2}$

Thus, $\displaystyle\int \frac{du}{a^2 + u^2} = \int \frac{u'}{a^2 + u^2}\, dx = \frac{1}{a}\arctan \frac{u}{a} + C.$

(c) Assume $u > 0$.

$\dfrac{d}{dx}\left[\dfrac{1}{a}\operatorname{arcsec} \dfrac{u}{a} + C \right] = \dfrac{1}{a}\left[\dfrac{u'/a}{(u/a)\sqrt{(u/a)^2 - 1}} \right] = \dfrac{1}{a}\left[\dfrac{u'}{u\sqrt{(u^2 - a^2)/a^2}} \right] = \dfrac{u'}{u\sqrt{u^2 - a^2}}.$

The case $u < 0$ is handled in a similar manner.

Thus, $\displaystyle\int \frac{du}{u\sqrt{u^2 - a^2}} = \int \frac{u'}{u\sqrt{u^2 - a^2}}\, dx = \frac{1}{a}\operatorname{arcsec} \frac{|u|}{a} + C.$

63. (a) $v(t) = -32t + 500$

(b) $s(t) = \displaystyle\int v(t)\, dt = \int (-32t + 500)\, dt$

$\qquad = -16t^2 + 500t + C$

$s(0) = -16(0) + 500(0) + C = 0 \implies C = 0$

$s(t) = -16t^2 + 500t$

When the object reaches its maximum height, $v(t) = 0$.

$v(t) = -32t + 500 = 0$

$-32t = -500$

$t = 15.625$

$s(15.625) = -16(15.625)^2 + 500(15.625)$

$= 3906.25$ ft (Maximum height)

—CONTINUED—

63. —CONTINUED—

(c) $$\int \frac{1}{32 + kv^2}\, dv = -\int dt$$

$$\frac{1}{\sqrt{32k}} \arctan\left(\sqrt{\frac{k}{32}}v\right) = -t + C_1$$

$$\arctan\left(\sqrt{\frac{k}{32}}v\right) = -\sqrt{32k}\,t + C$$

$$\sqrt{\frac{k}{32}}v = \tan\left(C - \sqrt{32k}\,t\right)$$

$$v = \sqrt{\frac{32}{k}}\,\tan\left(C - \sqrt{32k}\,t\right)$$

When $t = 0$, $v = 500$, $C = \arctan\left(500\sqrt{k/32}\right)$, and we have

$$v(t) = \sqrt{\frac{32}{k}}\,\tan\left[\arctan\left(500\sqrt{\frac{k}{32}}\right) - \sqrt{32k}\,t\right].$$

(d) When $k = 0.001$, $v(t) = \sqrt{32{,}000}\,\tan\left[\arctan\left(500\sqrt{0.00003125}\right) - \sqrt{0.032}\,t\right]$.

$v(t) = 0$ when $t_0 \approx 6.86$ sec.

(e) $$h = \int_0^{6.86} \sqrt{32{,}000}\,\tan\left[\arctan\left(500\sqrt{0.00003125}\right) - \sqrt{0.032}\,t\right] dt$$

Simpson's Rule: $n = 10$; $h \approx 1088$ feet

(f) Air resistance lowers the maximum height.

Section 5.10   Hyperbolic Functions

1. (a) $\sinh 3 = \dfrac{e^3 - e^{-3}}{2} \approx 10.018$

(b) $\tanh(-2) = \dfrac{\sinh(-2)}{\cosh(-2)} = \dfrac{e^{-2} - e^2}{e^{-2} + e^2} \approx -0.964$

3. (a) $\operatorname{csch}(\ln 2) = \dfrac{2}{e^{\ln 2} - e^{-\ln 2}} = \dfrac{2}{2 - (1/2)} = \dfrac{4}{3}$

(b) $\coth(\ln 5) = \dfrac{\cosh(\ln 5)}{\sinh(\ln 5)} = \dfrac{e^{\ln 5} + e^{-\ln 5}}{e^{\ln 5} - e^{-\ln 5}}$

$$= \dfrac{5 + (1/5)}{5 - (1/5)} = \dfrac{13}{12}$$

5. (a) $\cosh^{-1}(2) = \ln\left(2 + \sqrt{3}\right) \approx 1.317$

(b) $\operatorname{sech}^{-1}\left(\dfrac{2}{3}\right) = \ln\left(\dfrac{1 + \sqrt{1 - (4/9)}}{2/3}\right) \approx 0.962$

7. $\tanh^2 x + \operatorname{sech}^2 x = \left(\dfrac{e^x - e^{-x}}{e^x + e^{-x}}\right)^2 + \left(\dfrac{2}{e^x + e^{-x}}\right)^2 = \dfrac{e^{2x} - 2 + e^{-2x} + 4}{(e^x + e^{-x})^2} = \dfrac{e^{2x} + 2 + e^{-2x}}{e^{2x} + 2 + e^{-2x}} = 1$

9. $\sinh x \cosh y + \cosh x \sinh y = \left(\dfrac{e^x - e^{-x}}{2}\right)\left(\dfrac{e^y + e^{-y}}{2}\right) + \left(\dfrac{e^x + e^{-x}}{2}\right)\left(\dfrac{e^y - e^{-y}}{2}\right)$

$$= \frac{1}{4}[e^{x+y} - e^{-x+y} + e^{x-y} - e^{-(x+y)} + e^{x+y} + e^{-x+y} - e^{x-y} - e^{-(x+y)}]$$

$$= \frac{1}{4}[2(e^{x+y} - e^{-(x+y)})] = \frac{e^{(x+y)} - e^{-(x+y)}}{2} = \sinh(x + y)$$

11. $3 \sinh x + 4 \sinh^3 x = \sinh x(3 + 4 \sinh^2 x) = \left(\dfrac{e^x - e^{-x}}{2}\right)\left[3 + 4\left(\dfrac{e^x - e^{-x}}{2}\right)^2\right]$

$$= \left(\frac{e^x - e^{-x}}{2}\right)[3 + e^{2x} - 2 + e^{-2x}] = \frac{1}{2}(e^x - e^{-x})(e^{2x} + e^{-2x} + 1)$$

$$= \frac{1}{2}[e^{3x} + e^{-x} + e^x - e^x - e^{-3x} - e^{-x}] = \frac{e^{3x} - e^{-3x}}{2} = \sinh(3x)$$

13. $\sinh x = \dfrac{3}{2}$

$$\cosh^2 x - \left(\frac{3}{2}\right)^2 = 1 \implies \cosh^2 x = \frac{13}{4} \implies \cosh x = \frac{\sqrt{13}}{2}$$

$$\tanh x = \frac{3/2}{\sqrt{13}/2} = \frac{3\sqrt{13}}{13}$$

$$\operatorname{csch} x = \frac{1}{3/2} = \frac{2}{3}$$

$$\operatorname{sech} x = \frac{1}{\sqrt{13}/2} = \frac{2\sqrt{13}}{13}$$

$$\coth x = \frac{1}{3/\sqrt{13}} = \frac{\sqrt{13}}{3}$$

15. $y = \sinh(1 - x^2)$

$y' = -2x \cosh(1 - x^2)$

17. $f(x) = \ln(\sinh x)$

$$f'(x) = \frac{1}{\sinh x}(\cosh x) = \coth x$$

19. $y = \ln\left(\tanh \dfrac{x}{2}\right)$

$$y' = \frac{1/2}{\tanh(x/2)} \operatorname{sech}^2\left(\frac{x}{2}\right) = \frac{1}{2 \sinh(x/2) \cosh(x/2)}$$

$$= \frac{1}{\sinh x} = \operatorname{csch} x$$

21. $h(x) = \dfrac{1}{4} \sinh(2x) - \dfrac{x}{2}$

$$h'(x) = \frac{1}{2} \cosh(2x) - \frac{1}{2} = \frac{\cosh(2x) - 1}{2} = \sinh^2 x$$

23. $f(t) = \arctan(\sinh t)$

$$f'(t) = \frac{1}{1 + \sinh^2 t}(\cosh t)$$

$$= \frac{\cosh t}{\cosh^2 t} = \operatorname{sech} t$$

25. Let $y = g(x)$.

$$y = x^{\cosh x}$$

$$\ln y = \cosh x \ln x$$

$$\frac{1}{y}\left(\frac{dy}{dx}\right) = \frac{\cosh x}{x} + \sinh x \ln x$$

$$\frac{dy}{dx} = \frac{y}{x}[\cosh x + x(\sinh x) \ln x]$$

$$= \frac{x^{\cosh x}}{x}[\cosh x + x(\sinh x) \ln x]$$

27. $y = (\cosh x - \sinh x)^2$

$y' = 2(\cosh x - \sinh x)(\sinh x - \cosh x)$

$\quad = -2(\cosh x - \sinh x)^2 = -2e^{-2x}$

29. $f(x) = \sin x \sinh x - \cos x \cosh x$, $-4 \leq x \leq 4$

$f'(x) = \sin x \cosh x + \cos x \sinh x - \cos x \sinh x + \sin x \cosh x$

$\quad = 2 \sin x \cosh x = 0$ when $x = 0, \pm \pi$.

Relative maxima: $(\pm \pi, \cosh \pi)$

Relative minimum: $(0, -1)$

31. $g(x) = x \operatorname{sech} x = \dfrac{x}{\cosh x}$

Relative maximum: $(1.20, 0.66)$

Relative minimum: $(-1.20, -0.66)$

33. $y = a \sinh x$

$y' = a \cosh x$

$y'' = a \sinh x$

$y''' = a \cosh x$

Therefore, $y''' - y' = 0$.

35. $f(x) = \tanh x$ $\qquad f(1) = \tanh(1) \approx 0.7616$

$f'(x) = \operatorname{sech}^2 x$ $\qquad f'(1) = \dfrac{1}{\cosh^2(1)} \approx 0.4200$

$f''(x) = -2 \operatorname{sech}^2 x \cdot \tanh x \qquad f''(1) \approx -0.6397$

$P_1(x) = f(1) + f'(1)(x - 1) = 0.7616 + 0.42(x - 1)$

$P_2(x) = 0.7616 + 0.42(x - 1) - \dfrac{0.6397}{2}(x - 1)^2$

37. (a) $y = 10 + 15 \cosh \dfrac{x}{15}$, $-15 \leq x \leq 15$

(b) At $x = \pm 15$, $y = 10 + 15 \cosh(1) \approx 33.146$.

At $x = 0$, $y = 10 + 15 \cosh(0) = 25$.

(c) $y' = \sinh \dfrac{x}{15}$. At $x = 15$, $y' = \sinh(1) \approx 1.175$

39. Let $u = 1 - 2x$, $du = -2\,dx$.

$\displaystyle \int \sinh(1 - 2x)\,dx = -\frac{1}{2}\int \sinh(1 - 2x)(-2)\,dx$

$\qquad\qquad\qquad\qquad = -\dfrac{1}{2}\cosh(1 - 2x) + C$

41. Let $u = \cosh(x - 1)$, $du = \sinh(x - 1)\,dx$.

$\displaystyle \int \cosh^2(x - 1)\sinh(x - 1)\,dx = \frac{1}{3}\cosh^3(x - 1) + C$

43. Let $u = \sinh x$, $du = \cosh x\, dx$.

$$\int \frac{\cosh x}{\sinh x}\, dx = \ln|\sinh x| + C$$

45. Let $u = \dfrac{x^2}{2}$, $du = x\, dx$.

$$\int x \operatorname{csch}^2 \frac{x^2}{2}\, dx = \int \left(\operatorname{csch}^2 \frac{x^2}{2}\right) x\, dx = -\coth \frac{x^2}{2} + C$$

47. Let $u = \dfrac{1}{x}$, $du = -\dfrac{1}{x^2}\, dx$.

$$\int \frac{\operatorname{csch}(1/x)\,\coth(1/x)}{x^2}\, dx = -\int \operatorname{csch}\frac{1}{x} \coth\frac{1}{x}\left(-\frac{1}{x^2}\right) dx = \operatorname{csch}\frac{1}{x} + C$$

49. $\displaystyle\int_0^4 \frac{1}{25 - x^2}\, dx = \frac{1}{10}\int \frac{1}{5 - x}\, dx + \frac{1}{10}\int \frac{1}{5 + x}\, dx = \left[\frac{1}{10}\ln\left|\frac{5 + x}{5 - x}\right|\right]_0^4 = \frac{1}{10}\ln 9 = \frac{1}{5}\ln 3$

51. Let $u = 2x$, $du = 2\, dx$.

$$\int_0^{\sqrt{2}/4} \frac{2}{\sqrt{1 - 4x^2}}\, dx = \int_0^{\sqrt{2}/4} \frac{1}{\sqrt{1 - (2x)^2}}(2)\, dx = \Big[\arcsin(2x)\Big]_0^{\sqrt{2}/4} = \frac{\pi}{4}$$

53. Let $u = x^2$, $du = 2x\, dx$.

$$\int \frac{x}{x^4 + 1}\, dx = \frac{1}{2}\int \frac{2x}{(x^2)^2 + 1}\, dx = \frac{1}{2}\arctan(x^2) + C$$

55. $y = \cosh^{-1}(3x)$

$$y' = \frac{3}{\sqrt{9x^2 - 1}}$$

57. $y = \sinh^{-1}(\tan x)$

$$y' = \frac{1}{\sqrt{\tan^2 x + 1}}(\sec^2 x) = |\sec x|$$

59. $y = \coth^{-1}(\sin 2x)$

$$y' = \frac{1}{1 - \sin^2 2x}(2\cos 2x) = 2\sec 2x$$

61. $y = 2x \sinh^{-1}(2x) - \sqrt{1 + 4x^2}$

$$y' = 2x\left(\frac{2}{\sqrt{1 + 4x^2}}\right) + 2\sinh^{-1}(2x) - \frac{4x}{\sqrt{1 + 4x^2}} = 2\sinh^{-1}(2x)$$

63. See page 395.

65. $y = a \operatorname{sech}^{-1} \dfrac{x}{a} - \sqrt{a^2 - x^2}$, $a > 0$

$$\frac{dy}{dx} = \frac{-1}{(x/a)\sqrt{1 - (x^2/a^2)}} + \frac{x}{\sqrt{a^2 - x^2}} = \frac{-a^2}{x\sqrt{a^2 - x^2}} + \frac{x}{\sqrt{a^2 - x^2}} = \frac{x^2 - a^2}{x\sqrt{a^2 - x^2}} = \frac{-\sqrt{a^2 - x^2}}{x}$$

67. $\displaystyle\int \frac{1}{\sqrt{1 + e^{2x}}}\, dx = \int \frac{e^x}{e^x\sqrt{1 + (e^x)^2}}\, dx = -\operatorname{csch}^{-1}(e^x) + C = -\ln\left(\frac{1 + \sqrt{1 + e^{2x}}}{e^x}\right) + C$

69. Let $u = \sqrt{x}$, $du = \dfrac{1}{2\sqrt{x}}\, dx$.

$$\int \frac{1}{\sqrt{x}\sqrt{1 + x}}\, dx = 2\int \frac{1}{\sqrt{1 + \left(\sqrt{x}\right)^2}}\left(\frac{1}{2\sqrt{x}}\right) dx = 2\sinh^{-1}\sqrt{x} + C = 2\ln\left(\sqrt{x} + \sqrt{1 + x}\right) + C$$

71. $\displaystyle\int \frac{-1}{4x - x^2}\, dx = \int \frac{1}{(x - 2)^2 - 4}\, dx = \frac{1}{4}\ln\left|\frac{(x - 2) - 2}{(x - 2) + 2}\right| = \frac{1}{4}\ln\left|\frac{x - 4}{x}\right| + C$

73. $\displaystyle\int \frac{1}{1 - 4x - 2x^2}\,dx = \int \frac{1}{3 - 2(x + 1)^2}\,dx = \frac{-1}{\sqrt{2}}\int \frac{\sqrt{2}}{\left[\sqrt{2}(x + 1)\right]^2 - \left(\sqrt{3}\right)^2}\,dx$

$\displaystyle = \frac{-1}{2\sqrt{6}}\ln\left|\frac{\sqrt{2}(x + 1) - \sqrt{3}}{\sqrt{2}(x + 1) + \sqrt{3}}\right| + C = \frac{1}{2\sqrt{6}}\ln\left|\frac{\sqrt{2}(x + 1) + \sqrt{3}}{\sqrt{2}(x + 1) - \sqrt{3}}\right| + C$

75. Let $u = 4x - 1,\, du = 4\,dx.$

$\displaystyle y = \int \frac{1}{\sqrt{80 + 8x - 16x^2}}\,dx = \frac{1}{4}\int \frac{4}{\sqrt{81 - (4x - 1)^2}}\,dx = \frac{1}{4}\arcsin\left(\frac{4x - 1}{9}\right) + C$

77. $\displaystyle y = \int \frac{x^3 - 21x}{5 + 4x - x^2}\,dx = \int\left(-x - 4 + \frac{20}{5 + 4x - x^2}\right)dx = \int(-x - 4)\,dx + 20\int \frac{1}{3^2 - (x - 2)^2}\,dx$

$\displaystyle = -\frac{x^2}{2} - 4x + \frac{20}{6}\ln\left|\frac{3 + (x - 2)}{3 - (x - 2)}\right| + C = -\frac{x^2}{2} - 4x + \frac{10}{3}\ln\left|\frac{1 + x}{5 - x}\right| + C = \frac{-x^2}{2} - 4x - \frac{10}{3}\ln\left|\frac{5 - x}{x + 1}\right| + C$

79. $\displaystyle A = 2\int_0^4 \operatorname{sech}\frac{x}{2}\,dx$

$\displaystyle = 2\int_0^4 \frac{2}{e^{x/2} + e^{-x/2}}\,dx$

$\displaystyle = 4\int_0^4 \frac{e^{x/2}}{(e^{x/2})^2 + 1}\,dx$

$\displaystyle = \left[8\arctan(e^{x/2})\right]_0^4$

$\displaystyle = 8\arctan(e^2) - 2\pi \approx 5.207$

81. $\displaystyle A = \int_0^2 \frac{5x}{\sqrt{x^4 + 1}}\,dx$

$\displaystyle = \frac{5}{2}\int_0^2 \frac{2x}{\sqrt{(x^2)^2 + 1}}\,dx$

$\displaystyle = \left[\frac{5}{2}\ln\left(x^2 + \sqrt{x^4 + 1}\right)\right]_0^2$

$\displaystyle = \frac{5}{2}\ln\left(4 + \sqrt{17}\right) \approx 5.237$

83. $\displaystyle\int \frac{3k}{16}\,dt = \int \frac{1}{x^2 - 12x + 32}\,dx$

$\displaystyle \frac{3kt}{16} = \int \frac{1}{(x - 6)^2 - 4}\,dx = \frac{1}{2(2)}\ln\left|\frac{(x - 6) - 2}{(x - 6) + 2}\right| + C = \frac{1}{4}\ln\left|\frac{x - 8}{x - 4}\right| + C$

When $x = 0$: $t = 0$

$$C = -\frac{1}{4}\ln(2)$$

When $x = 1$: $t = 10$

$$\frac{30k}{16} = \frac{1}{4}\ln\left|\frac{-7}{-3}\right| - \frac{1}{4}\ln(2) = \frac{1}{4}\ln\left(\frac{7}{6}\right)$$

$$k = \frac{2}{15}\ln\left(\frac{7}{6}\right)$$

When $t = 20$: $\displaystyle\left(\frac{3}{16}\right)\left(\frac{2}{15}\right)\ln\left(\frac{7}{6}\right)(20) = \frac{1}{4}\ln\frac{x - 8}{2x - 8}$

$$\ln\left(\frac{7}{6}\right)^2 = \ln\frac{x - 8}{2x - 8}$$

$$\frac{49}{36} = \frac{x - 8}{2x - 8}$$

$$62x = 104$$

$$x = \frac{104}{62} = \frac{52}{31} \approx 1.677\text{ kg}$$

85. As k increases, the time required for the object to reach the ground increases.

87. $y = \cosh x = \dfrac{e^x + e^{-x}}{2}$

$y' = \dfrac{e^x - e^{-x}}{2} = \sinh x$

89. $\quad y = \cosh^{-1} x$

$\cosh y = x$

$(\sinh y)(y') = 1$

$y' = \dfrac{1}{\sinh y} = \dfrac{1}{\sqrt{\cosh^2 y - 1}} = \dfrac{1}{\sqrt{x^2 - 1}}$

91. $y = \operatorname{sech} x = \dfrac{2}{e^x + e^{-x}}$

$y' = -2(e^x + e^{-x})^{-2}(e^x - e^{-x}) = \left(\dfrac{-2}{e^x + e^{-x}}\right)\left(\dfrac{e^x - e^{-x}}{e^x + e^{-x}}\right) = -\operatorname{sech} x \tanh x$

Review Exercises for Chapter 5

1. $f(x) = \ln x + 3$

Vertical shift 3 units upward
Vertical asymptote: $x = 0$

3. $\ln \sqrt[5]{\dfrac{4x^2 - 1}{4x^2 + 1}} = \dfrac{1}{5}\ln\dfrac{(2x-1)(2x+1)}{4x^2 + 1} = \dfrac{1}{5}[\ln(2x - 1) + \ln(2x + 1) - \ln(4x^2 + 1)]$

5. $\ln 3 + \dfrac{1}{3}\ln(4 - x^2) - \ln x = \ln 3 + \ln\sqrt[3]{4 - x^2} - \ln x = \ln\left(\dfrac{3\sqrt[3]{4 - x^2}}{x}\right)$

7. $\ln\sqrt{x + 1} = 2$

$\sqrt{x + 1} = e^2$

$x + 1 = e^4$

$x = e^4 - 1 \approx 53.598$

9. $g(x) = \ln\sqrt{x} = \dfrac{1}{2}\ln x$

$g'(x) = \dfrac{1}{2x}$

11. $f(x) = x\sqrt{\ln x}$

$f'(x) = \left(\dfrac{x}{2}\right)(\ln x)^{-1/2}\left(\dfrac{1}{x}\right) + \sqrt{\ln x}$

$\quad = \dfrac{1}{2\sqrt{\ln x}} + \sqrt{\ln x} = \dfrac{1 + 2\ln x}{2\sqrt{\ln x}}$

13. $y = \dfrac{1}{b^2}\left[\ln(a + bx) + \dfrac{a}{a + bx}\right]$

$\dfrac{dy}{dx} = \dfrac{1}{b^2}\left[\dfrac{b}{a + bx} - \dfrac{ab}{(a + bx)^2}\right] = \dfrac{x}{(a + bx)^2}$

15. $y = -\dfrac{1}{a}\ln\left(\dfrac{a + bx}{x}\right) = -\dfrac{1}{a}[\ln(a + bx) - \ln x]$

$\dfrac{dy}{dx} = -\dfrac{1}{a}\left(\dfrac{b}{a + bx} - \dfrac{1}{x}\right) = \dfrac{1}{x(a + bx)}$

17. $u = 7x - 2,\ du = 7\,dx$

$\displaystyle\int\dfrac{1}{7x - 2}\,dx = \dfrac{1}{7}\int\dfrac{1}{7x - 2}(7)\,dx = \dfrac{1}{7}\ln|7x - 2| + C$

19. $\int \dfrac{\sin x}{1 + \cos x}\, dx = -\int \dfrac{-\sin x}{1 + \cos x}\, dx$

$$= -\ln|1 + \cos x| + C$$

21. $\displaystyle\int_{1}^{4} \dfrac{x + 1}{x}\, dx = \int_{1}^{4}\left(1 + \dfrac{1}{x}\right) dx = \left[x + \ln|x|\right]_{1}^{4} = 3 + \ln 4$

23. $\displaystyle\int_{0}^{\pi/3} \sec \theta \, d\theta = \left[\ln|\sec \theta + \tan \theta|\right]_{0}^{\pi/3} = \ln\left(2 + \sqrt{3}\right)$

25. (a)
$$f(x) = \tfrac{1}{2}x - 3$$
$$y = \tfrac{1}{2}x - 3$$
$$2(y + 3) = x$$
$$2(x + 3) = y$$
$$f^{-1}(x) = 2x + 6$$

(b)

(c) $f^{-1}(f(x)) = f^{-1}\left(\tfrac{1}{2}x - 3\right) = 2\left(\tfrac{1}{2}x - 3\right) + 6 = x$

$f(f^{-1}(x)) = f(2x + 6) = \tfrac{1}{2}(2x + 6) - 3 = x$

27. (a)
$$f(x) = \sqrt{x + 1}$$
$$y = \sqrt{x + 1}$$
$$y^2 - 1 = x$$
$$x^2 - 1 = y$$
$$f^{-1}(x) = x^2 - 1, \; x \geq 0$$

(b)

(c) $f^{-1}(f(x)) = f^{-1}\left(\sqrt{x + 1}\right) = \sqrt{(x^2 - 1)^2} - 1 = x$

$f(f^{-1}(x)) = f(x^2 - 1) = \sqrt{(x^2 - 1) + 1}$

$\qquad\qquad\quad = \sqrt{x^2} = x \text{ for } x \geq 0.$

29. (a)
$$f(x) = \sqrt[3]{x + 1}$$
$$y = \sqrt[3]{x + 1}$$
$$y^3 - 1 = x$$
$$x^3 - 1 = y$$
$$f^{-1}(x) = x^3 - 1$$

(b)

(c) $f^{-1}(f(x)) = f^{-1}\left(\sqrt[3]{x + 1}\right) = \left(\sqrt[3]{x + 1}\right)^3 - 1 = x$

$f(f^{-1}(x)) = f(x^3 - 1) = \sqrt[3]{(x^3 - 1) + 1} = x$

31.
$$f(x) = x^3 + 2$$
$$f^{-1}(x) = (x - 2)^{1/3}$$
$$(f^{-1})\,'(x) = \tfrac{1}{3}(x - 2)^{-2/3}$$
$$(f^{-1})\,'(-1) = \tfrac{1}{3}(-1 - 2)^{-2/3} = \dfrac{1}{3(-3)^{2/3}}$$
$$= \dfrac{1}{3^{5/3}} \approx 0.160$$

33.
$$f(x) = \tan x$$
$$f\left(\dfrac{\pi}{6}\right) = \dfrac{\sqrt{3}}{3}$$
$$f'(x) = \sec^2 x$$
$$f'\left(\dfrac{\pi}{6}\right) = \dfrac{4}{3}$$
$$(f^{-1})\,'\left(\dfrac{\sqrt{3}}{3}\right) = \dfrac{1}{f'(\pi/6)} = \dfrac{3}{4}$$

35. (a) $f(x) = \ln \sqrt{x}$

$y = \ln \sqrt{x}$

$e^y = \sqrt{x}$

$e^{2y} = x$

$e^{2x} = y$

$f^{-1}(x) = e^{2x}$

(b)

(c) $f^{-1}(f(x)) = f^{-1}(\ln \sqrt{x}) = e^{2\ln \sqrt{x}} = e^{\ln x} = x$

$f(f^{-1}(x)) = f(e^{2x}) = \ln \sqrt{e^{2x}} = \ln e^x = x$

37. $y = e^{-x/2}$

39. $f(x) = \ln(e^{-x^2}) = -x^2$

$f'(x) = -2x$

41. $g(t) = t^2 e^t$

$g'(t) = t^2 e^t + 2t e^t = t e^t(t + 2)$

43. $y = \sqrt{e^{2x} + e^{-2x}}$

$y' = \frac{1}{2}(e^{2x} + e^{-2x})^{-1/2}(2e^{2x} - 2e^{-2x}) = \frac{e^{2x} - e^{-2x}}{\sqrt{e^{2x} + e^{-2x}}}$

45. $g(x) = \dfrac{x^2}{e^x}$

$g'(x) = \dfrac{e^x(2x) - x^2 e^x}{e^{2x}} = \dfrac{x(2 - x)}{e^x}$

47. $y(\ln x) + y^2 = 0$

$y\left(\dfrac{1}{x}\right) + (\ln x)\left(\dfrac{dy}{dx}\right) + 2y\left(\dfrac{dy}{dx}\right) = 0$

$(2y + \ln x)\dfrac{dy}{dx} = \dfrac{-y}{x}$

$\dfrac{dy}{dx} = \dfrac{-y}{x(2y + \ln x)}$

49. Let $u = -3x^2$, $du = -6x\, dx$.

$\displaystyle\int xe^{-3x^2}\, dx = -\frac{1}{6}\int e^{-3x^2}(-6x)\, dx = -\frac{1}{6}e^{-3x^2} + C$

51. $\displaystyle\int \dfrac{e^{4x} - e^{2x} + 1}{e^x}\, dx = \int (e^{3x} - e^x + e^{-x})\, dx$

$= \dfrac{1}{3}e^{3x} - e^x - e^{-x} + C$

$= \dfrac{e^{4x} - 3e^{2x} - 3}{3e^x} + C$

53. $\displaystyle\int xe^{1-x^2}\, dx = -\frac{1}{2}\int e^{1-x^2}(-2x)\, dx$

$= -\dfrac{1}{2}e^{1-x^2} + C$

55. Let $u = e^x - 1$, $du = e^x\, dx$.

$\displaystyle\int \dfrac{e^x}{e^x - 1}\, dx = \ln|e^x - 1| + C$

57. $y = e^x(a \cos 3x + b \sin 3x)$

$y' = e^x(-3a \sin 3x + 3b \cos 3x) + e^x(a \cos 3x + b \sin 3x)$

$= e^x[(-3a + b) \sin 3x + (a + 3b) \cos 3x]$

$y'' = e^x[3(-3a + b) \cos 3x - 3(a + 3b) \sin 3x] + e^x[(-3a + b) \sin 3x + (a + 3b) \cos 3x]$

$= e^x[(-6a - 8b) \sin 3x + (-8a + 6b) \cos 3x]$

$y'' - 2y' + 10y = e^x\{[(-6a - 8b) - 2(-3a + b) + 10b] \sin 3x + [(-8a + 6b) - 2(a + 3b) + 10a] \cos 3x\} = 0$

59. Area $= \int_0^4 xe^{-x^2}\,dx = \left[-\frac{1}{2}e^{-x^2} \right]_0^4 = -\frac{1}{2}(e^{-16} - 1) \approx 0.500$

61. $y = 3^{x/2}$

63. $y = \log_2(x - 1)$

65. $f(x) = 3^{x-1}$

$f'(x) = 3^{x-1} \ln 3$

67. $y = x^{2x+1}$

$\ln y = (2x + 1) \ln x$

$\dfrac{y'}{y} = \dfrac{2x + 1}{x} + 2 \ln x$

$y' = y\left(\dfrac{2x + 1}{x} + 2 \ln x \right) = x^{2x+1}\left(\dfrac{2x + 1}{x} + 2 \ln x \right)$

69. $g(x) = \log_3 \sqrt{1 - x} = \dfrac{1}{2} \log_3(1 - x)$

$g'(x) = \left(\dfrac{1}{2} \right) \dfrac{-1}{(1 - x) \ln 3} = \dfrac{1}{2(x - 1) \ln 3}$

71. $\displaystyle\int (x + 1)5^{(x+1)^2}\,dx = \left(\dfrac{1}{2} \right) \dfrac{1}{\ln 5} 5^{(x+1)^2} + C$

73. (a) $\ y = x^a$

$\quad\quad y' = ax^{a-1}$

    (b) $\ y = a^x$

$\quad\quad y' = (\ln a)a^x$

    (c) $\quad y = x^x$

$\quad\quad \ln y = x \ln x$

$\quad\quad \dfrac{1}{y} y' = x \cdot \dfrac{1}{x} + (1) \ln x$

$\quad\quad\quad y' = y(1 + \ln x)$

$\quad\quad\quad y' = x^x(1 + \ln x)$

    (d) $\ y = a^a$

$\quad\quad y' = 0$

75. $10{,}000 = Pe^{(0.07)(15)}$

$\quad\quad P = \dfrac{10{,}000}{e^{1.05}} \approx \$3499.38$

77. $\dfrac{dP}{dh} = kP, P(0) = 30$

$\quad\quad P(h) = 30e^{kh}$

$\quad P(18{,}000) = 30e^{18{,}000k} = 15$

$\quad\quad\quad k = \dfrac{\ln(1/2)}{18{,}000} = \dfrac{-\ln 2}{18{,}000}$

$\quad\quad P(h) = 30e^{-(h \ln 2)/18{,}000}$

$\quad P(35{,}000) = 30e^{-(35{,}000 \ln 2)/18{,}000} \approx 7.79 \text{ inches}$

79. $\quad P = Ce^{0.015t}$

$\quad\quad 2C = Ce^{0.015t}$

$\quad\quad\ \ 2 = e^{0.015t}$

$\quad \ln 2 = 0.015t$

$\quad\quad\ \ t = \dfrac{\ln 2}{0.015} \approx 46.21 \text{ years}$

81. $\dfrac{dy}{dx} = \dfrac{x^2 + 3}{x}$

$\quad \displaystyle\int dy = \int \left(x + \dfrac{3}{x} \right) dx$

$\quad\quad y = \dfrac{x^2}{2} + 3 \ln|x| + C$

83. $y' - 2xy = 0$

$$\frac{dy}{dx} = 2xy$$

$$\int \frac{1}{y}\, dy = \int 2x\, dx$$

$$\ln|y| = x^2 + C_1$$

$$e^{x^2 + C_1} = y$$

$$y = Ce^{x^2}$$

85. $\dfrac{dy}{dx} = \dfrac{x^2 + y^2}{2xy}$ (homogeneous differential equation)

$$(x^2 + y^2)\, dx - 2xy\, dy = 0$$

Let $y = vx$, $dy = x\, dv + v\, dx$.

$$(x^2 + v^2 x^2)\, dx - 2x(vx)(x\, dv + v\, dx) = 0$$

$$(x^2 + v^2 x^2 - 2x^2 v^2)\, dx - 2x^3 v\, dv = 0$$

$$(x^2 - x^2 v^2)\, dx = 2x^3 v\, dv$$

$$(1 - v^2)\, dx = 2xv\, dv$$

$$\int \frac{dx}{x} = \int \frac{2v}{1 - v^2}\, dv$$

$$\ln|x| = -\ln|1 - v^2| + C_1 = -\ln|1 - v^2| + \ln C$$

$$x = \frac{C}{1 - v^2} = \frac{C}{1 - (y/x)^2} = \frac{Cx^2}{x^2 - y^2}$$

$$1 = \frac{Cx}{x^2 - y^2} \quad \text{or} \quad C_1 = \frac{x}{x^2 - y^2}$$

87. $y = C_1 x + C_2 x^3$

$y' = C_1 + 3C_2 x^2$

$y'' = 6C_2 x$

$x^2 y'' - 3xy' + 3y = x^2(6C_2 x) - 3x(C_1 + 3C_2 x^2) + (C_1 x = C_2 x^3)$

$$\qquad = 6C_2 x^3 - 3C_1 x - 9C_2 x^3 + 3C_1 x + 3C_2 x^3 = 0$$

$x = 2, y = 0$: $0 = 2C_1 + 8C_2 \implies C_1 = -4C_2$

$x = 2, y' = 4$: $4 = C_1 + 12C_2$

$$4 = (-4C_2) + 12C_2 = 8C_2 \implies C_2 = \frac{1}{2}, C_1 = -2$$

$y = -2x + \dfrac{1}{2}x^3$

89. $f(x) = 2\arctan(x + 3)$

91. (a)　　　Let $\theta = \arcsin \dfrac{1}{2}$

$$\sin \theta = \frac{1}{2}$$

$$\sin\left(\arcsin \frac{1}{2}\right) = \sin \theta = \frac{1}{2}.$$

(b)　　　Let $\theta = \arcsin \dfrac{1}{2}$

$$\sin \theta = \frac{1}{2}$$

$$\cos\left(\arcsin \frac{1}{2}\right) = \cos \theta = \frac{\sqrt{3}}{2}.$$

93. $y = \tan(\arcsin x) = \dfrac{x}{\sqrt{1 - x^2}}$

$$y' = \frac{(1 - x^2)^{1/2} + x^2(1 - x^2)^{-1/2}}{1 - x^2} = (1 - x^2)^{-3/2}$$

95. $y = x\,\text{arcsec}\,x$

$$y' = \frac{x}{|x|\sqrt{x^2 - 1}} + \text{arcsec}\,x$$

97. $y = x(\arcsin x)^2 - 2x + 2\sqrt{1 - x^2}\,\arcsin x$

$$y' = \frac{2x \arcsin x}{\sqrt{1 - x^2}} + (\arcsin x)^2 - 2 + \frac{2\sqrt{1 - x^2}}{\sqrt{1 - x^2}} - \frac{2x}{\sqrt{1 - x^2}}\,\arcsin x = (\arcsin x)^2$$

99. Let $u = e^{2x}, du = 2e^{2x}\,dx$.

$$\int \frac{1}{e^{2x} + e^{-2x}}\,dx = \int \frac{e^{2x}}{1 + e^{4x}}\,dx = \frac{1}{2}\int \frac{1}{1 + (e^{2x})^2}(2e^{2x})\,dx = \frac{1}{2}\arctan(e^{2x}) + C$$

101. Let $u = x^2, du = 2x\,dx$.

$$\int \frac{x}{\sqrt{1 - x^4}}\,dx = \frac{1}{2}\int \frac{1}{\sqrt{1 - (x^2)^2}}(2x)\,dx = \frac{1}{2}\arcsin x^2 + C$$

103. Let $u = 16 + x^2, du = 2x\,dx$.

$$\int \frac{x}{16 + x^2}\,dx = \frac{1}{2}\int \frac{1}{16 + x^2}(2x)\,dx = \frac{1}{2}\ln(16 + x^2) + C$$

105. Let $u = \arctan\left(\dfrac{x}{2}\right), du = \dfrac{2}{4 + x^2}\,dx$.

$$\int \frac{\arctan(x/2)}{4 + x^2}\,dx = \frac{1}{2}\int \left(\arctan \frac{x}{2}\right)\left(\frac{2}{4 + x^2}\right)\,dx = \frac{1}{4}\left(\arctan \frac{x}{2}\right)^2 + C$$

107. $\displaystyle \int \frac{dy}{\sqrt{A^2 - y^2}} = \int \sqrt{\frac{k}{m}}\,dt$

$$\arcsin\left(\frac{y}{A}\right) = \sqrt{\frac{k}{m}}\,t + C$$

Since $y = 0$ when $t = 0$, you have $C = 0$. Thus,

$$\sin\left(\sqrt{\frac{k}{m}}\,t\right) = \frac{y}{A}$$

$$y = A\sin\left(\sqrt{\frac{k}{m}}\,t\right)$$

109. $y = 2x - \cosh\sqrt{x}$

$$y' = 2 - \frac{1}{2\sqrt{x}}\left(\sinh \sqrt{x}\right) = 2 - \frac{\sinh\sqrt{x}}{2\sqrt{x}}$$

111. Let $u = x^2, du = 2x\,dx$.

$$\int \frac{x}{\sqrt{x^4 - 1}}\,dx = \frac{1}{2}\int \frac{1}{\sqrt{(x^2)^2 - 1}}(2x)\,dx = \frac{1}{2}\ln\left(x^2 + \sqrt{x^4 - 1}\right) + C$$

Problem Solving for Chapter 5

1. $\tan \theta_1 = \dfrac{3}{x}$

$\tan \theta_2 = \dfrac{6}{10 - x}$

Minimize $\theta_1 + \theta_2$:

$$f(x) = \theta_1 + \theta_2 = \arctan\!\left(\frac{3}{x}\right) + \arctan\!\left(\frac{6}{10 - x}\right)$$

$$f'(x) = \frac{1}{1 + \dfrac{9}{x^2}}\left(\frac{-3}{x^2}\right) + \frac{1}{1 + \dfrac{36}{(10 - x)^2}}\left(\frac{6}{(10 - x)^2}\right) = 0$$

$$\frac{3}{x^2 + 9} = \frac{6}{(10 - x)^2 + 36}$$

$$(10 - x)^2 + 36 = 2(x^2 + 9)$$

$$100 - 20x + x^2 + 36 = 2x^2 + 18$$

$$x^2 + 20x - 118 = 0$$

$$x = \frac{-20 \pm \sqrt{20^2 - 4(-118)}}{2} = -10 \pm \sqrt{218}$$

$a = -10 + \sqrt{218} \approx 4.7648 \quad f(a) \approx 1.4153$

$\theta = \pi - (\theta_1 + \theta_2) \approx 1.7263 \quad \text{or} \quad 98.9°$

Endpoints: $a = 0: \ \theta \approx 1.0304$

$\qquad\qquad a = 10: \ \theta \approx 1.2793$

Maximum is 1.7263 at $a = -10 + \sqrt{218} \approx 4.7648$.

3. $f(x) = \sin(\ln x)$

(a) Domain: $x > 0 \quad \text{or} \quad (0, \infty)$

(b) $f(x) = 1 = \sin(\ln x) \implies \ln x = \dfrac{\pi}{2} + 2k\pi$.

Two values are $x = e^{\pi/2}, \ e^{(\pi/2) + 2\pi}$.

(c) $f(x) = -1 = \sin(\ln x) \implies \ln x = \dfrac{3\pi}{2} + 2k\pi$.

Two values are $x = e^{-\pi/2}, \ e^{3\pi/2}$.

(d) Since the range of the sine function is $[-1, 1]$, parts (b) and (c) show that the range of f is $[-1, 1]$.

(e) $f'(x) = \dfrac{1}{x}\cos(\ln x)$

$$f'(x) = 0 \implies \cos(\ln x) = 0 \implies \ln x = \frac{\pi}{2} + k\pi \implies$$

$$x = e^{\pi/2} \text{ on } [1, 10]$$

$\left.\begin{array}{l} f(e^{\pi/2}) = 1 \\[4pt] f(1) = 0 \\[4pt] f(10) \approx 0.7440 \end{array}\right\}$ Maximum is 1 at $x = e^{\pi/2} \approx 4.8105$

(f)

$\lim\limits_{x \to 0^+} f(x)$ seems to be $-\dfrac{1}{2}$. (This in incorrect.)

(g) For the points $x = e^{\pi/2}, \, e^{-3\pi/2}, \, e^{-7\pi/2}, \ldots$

we have $f(x) = 1$.

For the points $x = e^{-\pi/2}, \, e^{-5\pi/2}, \, e^{-9\pi/2}, \ldots$

we have $f(x) = -1$.

That is, as $x \to 0^+$, there is an infinite number of points where $f(x) = 1$, and an infinite number where $f(x) = -1$. Thus $\lim\limits_{x \to 0^+} \sin(\ln x)$ does not exist.

You can verifiy this by graphing $f(x)$ on small intervals close to the origin.

5. (a) $\dfrac{\text{Area sector}}{\text{Area circle}} = \dfrac{t}{2\pi} \Rightarrow \text{Area sector} = \dfrac{t}{2\pi}(\pi) = \dfrac{t}{2}$

(b) Area $AOP = \dfrac{1}{2}(\text{base})(\text{height}) - \displaystyle\int_1^{\cosh t} \sqrt{x^2 - 1}\, dx$

$A(t) = \dfrac{1}{2}\cosh t \cdot \sinh t - \displaystyle\int_1^{\cosh t} \sqrt{x^2 - 1}\, dx$

$A'(t) = \dfrac{1}{2}[\cosh^2 t + \sinh^2 t] - \sqrt{\cosh^2 t - 1}\,\sinh t$

$ = \dfrac{1}{2}[\cosh^2 t + \sinh^2 t] - \sinh^2 t$

$ = \dfrac{1}{2}[\cosh^2 t - \sinh^2 t] = \dfrac{1}{2}$

$A(t) = \dfrac{1}{2}t + C.$ But, $A(0) = C = 0 \Rightarrow C = 0$

Thus, $A(t) = \dfrac{1}{2}t$ or $t = 2A(t).$

7. $y = \ln x$

$y' = \dfrac{1}{x}$

$y - b = \dfrac{1}{a}(x - a)$

$y = \dfrac{1}{a}x + b - 1$ Tangent line

If $x = 0,\ c = b - 1.$ Thus, $b - c = b - (b - 1) = 1.$

9. Let $u = 1 + \sqrt{x},\ \sqrt{x} = u - 1,\ x = u^2 - 2u + 1,$

$dx = (2u - 2)du.$

Area $= \displaystyle\int_1^4 \dfrac{1}{\sqrt{x} + x}\, dx = \int_2^3 \dfrac{2u - 2}{(u - 1) + (u^2 - 2u + 1)}\, du$

$\phantom{\text{Area}} = \displaystyle\int_2^3 \dfrac{2(u - 1)}{u^2 - u}\, du$

$\phantom{\text{Area}} = \displaystyle\int_2^3 \dfrac{2}{u}\, du$

$\phantom{\text{Area}} = \Big[\, 2 \ln|u| \,\Big]_2^3$

$\phantom{\text{Area}} = 2 \ln 3 - 2 \ln 2 = 2 \ln\!\left(\dfrac{3}{2}\right)$

$\phantom{\text{Area}} \approx 0.8109$

11. (a) $\dfrac{dy}{dt} = y^{1.01}$

$\displaystyle\int y^{-1.01}\, dy = \int dt$

$\dfrac{y^{-0.01}}{-0.01} = t + C_1$

$\dfrac{1}{y^{0.01}} = -0.01t + C$

$y^{0.01} = \dfrac{1}{C - 0.01t}$

$y = \dfrac{1}{(C - 0.01t)^{100}}$

$y(0) = 1: \ 1 = \dfrac{1}{C^{100}} \Rightarrow C = 1$

Hence, $y = \dfrac{1}{(1 - 0.01t)^{100}}.$

For $T = 100,\ \displaystyle\lim_{t \to T^-} y = \infty.$

(b) $\displaystyle\int y^{-(1+\varepsilon)}\, dy = \int k\, dt$

$\dfrac{y^{-\varepsilon}}{-\varepsilon} = kt + C_1$

$y^{-\varepsilon} = -\varepsilon kt + C$

$y = \dfrac{1}{(C - \varepsilon kt)^{1/\varepsilon}}$

$y(0) = y_0 = \dfrac{1}{C^{1/\varepsilon}} \Rightarrow C^{1/\varepsilon} = \dfrac{1}{y_0} \Rightarrow C = \left(\dfrac{1}{y_0}\right)^{\varepsilon}$

Hence, $y = \dfrac{1}{\left(\dfrac{1}{y_0^{\varepsilon}} - \varepsilon kt\right)^{1/\varepsilon}}.$

For $t \to \dfrac{1}{y_0^{\varepsilon}\varepsilon k},\ y \to \infty.$

13. Since $\dfrac{dy}{dt} = k(y - 20)$,

$$\int \frac{1}{y - 20}\, dy = \int k\, dt$$

$$\ln|y - 20| = kt + C$$

$$y = Ce^{kt} + 20.$$

When $t = 0$, $y = 72$. Therefore, $C = 52$.

When $t = 1$, $y = 48$. Therefore, $48 = 52e^k + 20$, $e^k = (28/52) = (7/13)$, and $k = \ln(7/13)$. Thus, $y = 52e^{[\ln(7/13)]t} + 20$.

When $t = 5$, $y = 52e^{5\ln(7/13)} + 20 \approx 22.35°$.

15. (a) $\dfrac{dS}{dt} = k_1 S(L - S)$

$S = \dfrac{L}{1 + Ce^{-kt}}$ is a solution because

$$\frac{dS}{dt} = -L(1 + Ce^{-kt})^{-2}(-Cke^{-kt})$$

$$= \frac{LCke^{-kt}}{(1 + Ce^{-kt})^2}$$

$$= \left(\frac{k}{L}\right)\frac{L}{1 + Ce^{-kt}} \cdot \frac{CLe^{-kt}}{1 + Ce^{-kt}}$$

$$= \left(\frac{k}{L}\right)\frac{L}{1 + Ce^{-kt}} \cdot \left(L - \frac{L}{1 + Ce^{-kt}}\right)$$

$$= k_1 S(L - S), \text{ where } k_1 = \frac{k}{L}.$$

$L = 100$. Also, $S = 10$ when $t = 0 \Rightarrow C = 9$. And, $S = 20$ when $t = 1 \Rightarrow k = -\ln(4/9)$.

Particular Solution. $S = \dfrac{100}{1 + 9e^{\ln(4/9)t}}$

$$= \frac{100}{1 + 9e^{-0.8109t}}$$

(c)

(b) $\dfrac{dS}{dt} = k_1 S(100 - S)$

$$\frac{d^2S}{dt^2} = k_1\left[S\left(-\frac{dS}{dt}\right) + (100 - S)\frac{dS}{dt}\right]$$

$$= k_1(100 - 2S)\frac{dS}{dt}$$

$$= 0 \text{ when } S = 50 \text{ or } \frac{dS}{dt} = 0.$$

Choosing $S = 50$, we have:

$$50 = \frac{100}{1 + 9e^{\ln(4/9)t}}$$

$$2 = 1 + 9e^{\ln(4/9)t}$$

$$\frac{\ln(1/9)}{\ln(4/9)} = t$$

$$t \approx 2.7 \text{ months (This is the inflection point)}$$

(d)

(e) Sales will decrease toward the line $S = L$.

C H A P T E R   6
Applications of Integration

Section 6.1   Area of a Region Between Two Curves 282

Section 6.2   Volume: The Disk Method 289

Section 6.3   Volume: The Shell Method 297

Section 6.4   Arc Length and Surfaces of Revolution 302

Section 6.5   Work 307

Section 6.6   Moments, Centers of Mass, and Centroids 310

Section 6.7   Fluid Pressure and Fluid Force 317

Review Exercises 320

Problem Solving 326

CHAPTER 6
Applications of Integration

Section 6.1   Area of a Region Between Two Curves

Solutions to Odd-Numbered Exercises

1. $A = \int_0^6 [0 - (x^2 - 6x)] \, dx = -\int_0^6 (x^2 - 6x) \, dx$

3. $A = \int_0^3 [(-x^2 + 2x + 3) - (x^2 - 4x + 3)] \, dx = \int_0^3 (-2x^2 + 6x) \, dx$

5. $A = 2\int_{-1}^0 3(x^3 - x) \, dx = 6\int_{-1}^0 (x^3 - x) \, dx \quad$ or $\quad -6\int_0^1 (x^3 - x) \, dx$

7. $\int_0^4 \left[(x + 1) - \dfrac{x}{2} \right] dx$

9. $\int_0^6 \left[4(2^{-x/3}) - \dfrac{x}{6} \right] dx$

11. $\int_{-\pi/3}^{\pi/3} [2 - \sec x] \, dx$

13. $f(x) = x + 1$

$g(x) = (x - 1)^2$

$A \approx 4$

Matches (d)

15. $A = \int_0^2 \left[\left(\dfrac{1}{2}x^3 + 2 \right) - (x + 1) \right] dx$

$= \int_0^2 \left(\dfrac{1}{2}x^3 - x + 1 \right) dx$

$= \left[\dfrac{x^4}{8} - \dfrac{x^2}{2} + x \right]_0^2$

$= \left(\dfrac{16}{8} - \dfrac{4}{2} + 2 \right) - 0 = 2$

17. The points of intersection are given by:

$$x^2 - 4x = 0$$

$$x(x - 4) = 0 \quad \text{when} \quad x = 0, 4$$

$$A = \int_0^4 [g(x) - f(x)] \, dx$$

$$= -\int_0^4 (x^2 - 4x) \, dx$$

$$= -\left[\frac{x^3}{3} - 2x^2 \right]_0^4$$

$$= \frac{32}{3}$$

19. The points of intersection are given by:

$$x^2 + 2x + 1 = 3x + 3$$

$$(x - 2)(x + 1) = 0 \quad \text{when} \quad x = -1, 2$$

$$A = \int_{-1}^2 [g(x) - f(x)] \, dx$$

$$= \int_{-1}^2 [(3x + 3) - (x^2 + 2x + 1)] \, dx$$

$$= \int_{-1}^2 (2 + x - x^2) \, dx$$

$$= \left[2x + \frac{x^2}{2} - \frac{x^3}{3} \right]_{-1}^2 = \frac{9}{2}$$

21. The points of intersection are given by:

$$x = 2 - x \quad \text{and} \quad x = 0 \quad \text{and} \quad 2 - x = 0$$

$$x = 1 \qquad\qquad x = 0 \qquad\qquad x = 2$$

$$A = \int_0^1 [(2 - y) - (y)] \, dy = \left[2y - y^2 \right]_0^1 = 1$$

Note that if we integrate with respect to x, we need two integrals. Also, note that the region is a triangle.

23. The points of intersection are given by:

$$\sqrt{3x} + 1 = x + 1$$

$$\sqrt{3x} = x \quad \text{when} \quad x = 0, 3$$

$$A = \int_0^3 [f(x) - g(x)] \, dx$$

$$= \int_0^3 [(\sqrt{3x} + 1) - (x + 1)] \, dx$$

$$= \int_0^3 [(3x)^{1/2} - x] \, dx$$

$$= \left[\frac{2}{9} (3x)^{3/2} - \frac{x^2}{2} \right]_0^3 = \frac{3}{2}$$

25. The points of intersection are given by:

$$y^2 = y + 2$$

$$(y - 2)(y + 1) = 0 \quad \text{when} \quad y = -1, 2$$

$$A = \int_{-1}^2 [g(y) - f(y)] \, dy$$

$$= \int_{-1}^2 [(y + 2) - y^2] \, dy$$

$$= \left[2y + \frac{y^2}{2} - \frac{y^3}{3} \right]_{-1}^2 = \frac{9}{2}$$

27. $A = \displaystyle\int_{-1}^{2} [f(y) - g(y)]\,dy$

$= \displaystyle\int_{-1}^{2} [(y^2 + 1) - 0]\,dy$

$= \left[\dfrac{y^3}{3} + y \right]_{-1}^{2} = 6$

29. $y = \dfrac{10}{x} \implies x = \dfrac{10}{y}$

$A = \displaystyle\int_{2}^{10} \dfrac{10}{y}\,dy$

$= \Big[10 \ln y \Big]_{2}^{10}$

$= 10(\ln 10 - \ln 2)$

$= 10 \ln 5 \approx 16.0944$

31. The points of intersection are given by:

$x^3 - 3x^2 + 3x = x^2$

$x(x - 1)(x - 3) = 0 \quad \text{when} \quad x = 0, 1, 3$

$A = \displaystyle\int_{0}^{1} [f(x) - g(x)]\,dx + \int_{1}^{3} [g(x) - f(x)]\,dx$

$= \displaystyle\int_{0}^{1} [(x^3 - 3x^2 + 3x) - x^2]\,dx + \int_{1}^{3} [x^2 - (x^3 - 3x^2 + 3x)]\,dx$

$= \displaystyle\int_{0}^{1} (x^3 - 4x^2 + 3x)\,dx + \int_{1}^{3} (-x^3 + 4x^2 - 3x)\,dx$

$= \left[\dfrac{x^4}{4} - \dfrac{4}{3}x^3 + \dfrac{3}{2}x^2 \right]_{0}^{1} + \left[\dfrac{-x^4}{4} + \dfrac{4}{3}x^3 - \dfrac{3}{2}x^2 \right]_{1}^{3} = \dfrac{5}{12} + \dfrac{8}{3} = \dfrac{37}{12}$

Numerical Approximation: $0.417 + 2.667 \approx 3.083$

33. The points of intersection are given by:

$x^2 - 4x + 3 = 3 + 4x - x^2$

$2x(x - 4) = 0 \quad \text{when} \quad x = 0, 4$

$A = \displaystyle\int_{0}^{4} [(3 + 4x - x^2) - (x^2 - 4x + 3)]\,dx$

$= \displaystyle\int_{0}^{4} (-2x^2 + 8x)\,dx$

$= \left[-\dfrac{2x^3}{3} + 4x^2 \right]_{0}^{4} = \dfrac{64}{3}$

Numerical Approximation: 21.333

35. $f(x) = x^4 - 4x^2$, $g(x) = x^2 - 4$

The points of intersection are given by:

$$x^4 - 4x^2 = x^2 - 4$$

$$x^4 - 5x^2 + 4 = 0$$

$$(x^2 - 4)(x^2 - 1) = 0 \quad \text{when} \quad x = \pm 2, \pm 1$$

By symmetry,

$$A = 2\int_0^1 [(x^4 - 4x^2) - (x^2 - 4)]\,dx + 2\int_1^2 [(x^2 - 4) - (x^4 - 4x^2)]\,dx$$

$$= 2\int_0^1 (x^4 - 5x^2 + 4)\,dx + 2\int_1^2 (-x^4 + 5x^2 - 4)\,dx$$

$$= 2\left[\frac{x^5}{5} - \frac{5x^3}{3} + 4x\right]_0^1 + 2\left[-\frac{x^5}{5} + \frac{5x^3}{3} - 4x\right]_1^2$$

$$= 2\left[\frac{1}{5} - \frac{5}{3} + 4\right] + 2\left[\left(-\frac{32}{5} + \frac{40}{3} - 8\right) - \left(-\frac{1}{5} + \frac{5}{3} - 4\right)\right] = 8.$$

Numerical Approximation: $5.067 + 2.933 = 8.0$

37. The points of intersection are given by:

$$\frac{1}{1 + x^2} = \frac{x^2}{2}$$

$$x^4 + x^2 - 2 = 0$$

$$(x^2 + 2)(x^2 - 1) = 0$$

$$x = \pm 1$$

$$A = 2\int_0^1 [f(x) - g(x)]\,dx$$

$$= 2\int_0^1 \left[\frac{1}{1 + x^2} - \frac{x^2}{2}\right]\,dx$$

$$= 2\left[\arctan x - \frac{x^3}{6}\right]_0^1$$

$$= 2\left(\frac{\pi}{4} - \frac{1}{6}\right) = \frac{\pi}{2} - \frac{1}{3} \approx 1.237$$

Numerical Approximation: 1.237

39. $\sqrt{1 + x^3} \leq \frac{1}{2}x + 2$ on $[0, 2]$

Numerical approximation: 1.759

$$A = \int_0^2 \left[\frac{1}{2}x + 2 - \sqrt{1 + x^3}\right]\,dx \approx 1.759$$

41. $A = 2\int_0^{\pi/3} [f(x) - g(x)]\,dx$

$$= 2\int_0^{\pi/3} (2\sin x - \tan x)\,dx$$

$$= 2\left[-2\cos x + \ln|\cos x|\right]_0^{\pi/3}$$

$$= 2(1 - \ln 2) \approx 0.614$$

43. $A = \int_0^{2\pi} [(2 - \cos x) - \cos x] \, dx$

$= 2\int_0^{2\pi} (1 - \cos x) \, dx$

$= 2\Big[x - \sin x\Big]_0^{2\pi} = 4\pi \approx 12.566$

45. $A = \int_0^1 [xe^{-x^2} - 0] \, dx$

$= \Big[-\frac{1}{2}e^{-x^2}\Big]_0^1 = \frac{1}{2}\Big(1 - \frac{1}{e}\Big) \approx 0.316$

47. $A = \int_0^{\pi} [(2 \sin x + \sin 2x) - 0] \, dx$

$= \Big[-2 \cos x - \frac{1}{2} \cos 2x\Big]_0^{\pi} = 4.0$

49. $A = \int_1^3 \Big[\frac{1}{x^2}e^{1/x} - 0\Big] \, dx$

$= \Big[-e^{1/x}\Big]_1^3 = e - e^{1/3} \approx 1.323$

51. (a) $y = \sqrt{\dfrac{x^3}{4 - x}}$, $y = 0$, $x = 3$

(b) $A = \int_0^3 \sqrt{\dfrac{x^3}{4 - x}} \, dx$,

No, it cannot be evaluated by hand.

(c) 4.7721

53. $F(x) = \int_0^x \Big(\frac{1}{2}t + 1\Big) \, dt = \Big[\frac{t^2}{4} + t\Big]_0^x = \frac{x^2}{4} + x$

(a) $F(0) = 0$

(b) $F(2) = \dfrac{2^2}{4} + 2 = 3$

(c) $F(6) = \dfrac{6^2}{4} + 6 = 15$

55. $F(\alpha) = \displaystyle\int_{-1}^{\alpha} \cos\frac{\pi\theta}{2}\,d\theta = \left[\frac{2}{\pi}\sin\frac{\pi\theta}{2}\right]_{-1}^{\alpha} = \frac{2}{\pi}\sin\frac{\pi\alpha}{2} + \frac{2}{\pi}$

(a) $F(-1) = 0$

(b) $F(0) = \dfrac{2}{\pi} \approx 0.6366$

(c) $F\left(\dfrac{1}{2}\right) = \dfrac{2 + \sqrt{2}}{\pi} \approx 1.0868$

57. $A = \displaystyle\int_0^c \left[\left(\frac{b-a}{c}y + a\right) - \frac{b}{c}y\right]dy$

$= \displaystyle\int_0^c \left(-\frac{a}{c}y + a\right)dy$

$= \left[-\frac{a}{2c}y^2 + ay\right]_0^c$

$= -\dfrac{ac}{2} + ac = \dfrac{ac}{2} \quad \left(= \dfrac{1}{2}(\text{base})(\text{height})\right)$

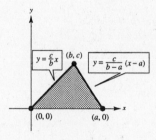

59. $f(x) = x^3$

$f'(x) = 3x^2$

At $(1, 1)$, $f'(1) = 3$.

Tangent line:

$\quad y - 1 = 3(x - 1)$ or $y = 3x - 2$

The tangent line intersects $f(x) = x^3$ at $x = -2$.

$\quad A = \displaystyle\int_{-2}^{1}\left[x^3 - (3x - 2)\right]dx = \left[\frac{x^4}{4} - \frac{3x^2}{2} + 2x\right]_{-2}^{1} = \frac{27}{4}$

61. The variable is y.

63. $x^4 - 2x^2 + 1 \le 1 - x^2$ on $[-1, 1]$

$A = \displaystyle\int_{-1}^{1}\left[(1 - x^2) - (x^4 - 2x^2 + 1)\right]dx$

$= \displaystyle\int_{-1}^{1}(x^2 - x^4)\,dx$

$= \left[\frac{x^3}{3} - \frac{x^5}{5}\right]_{-1}^{1} = \frac{4}{15}$

You can use a single integral because $x^4 - 2x^2 + 1 \le 1 - x^2$ on $[-1, 1]$.

65. Offer 2 is better because the accumulated salary (area under the curve) is larger.

67. $A = \int_{-3}^{3} (9 - x^2) \, dx = 36$

$$\int_{-\sqrt{9-b}}^{\sqrt{9-b}} [(9 - x^2) - b] \, dx = 18$$

$$\int_{0}^{\sqrt{9-b}} [(9 - b) - x^2] \, dx = 9$$

$$\left[(9 - b)x - \frac{x^3}{3} \right]_{0}^{\sqrt{9-b}} = 9$$

$$\frac{2}{3} (9 - b)^{3/2} = 9$$

$$(9 - b)^{3/2} = \frac{27}{2}$$

$$9 - b = \frac{9}{\sqrt[3]{4}}$$

$$b = 9 - \frac{9}{\sqrt[3]{4}} \approx 3.330$$

69. $\lim\limits_{\|\Delta\| \to 0} \sum\limits_{i=1}^{n} (x_i - x_i^2) \, \Delta x$

where $x_i = \dfrac{i}{n}$ and $\Delta x = \dfrac{1}{n}$ is the same as

$$\int_{0}^{1} (x - x^2) \, dx = \left[\frac{x^2}{2} - \frac{x^3}{3} \right]_{0}^{1} = \frac{1}{6}.$$

71. $\int_{0}^{5} [(7.21 + 0.58t) - (7.21 + 0.45t)] \, dt = \int_{0}^{5} 0.13t \, dt = \left[\frac{0.13t^2}{2} \right]_{0}^{5} = \$1.625 \text{ billion}$

73. (a) $y_1 = (275.0675)(1.0537)^t = (275.0675)e^{0.0523t}$ (b) $y_2 = (239.9407)(1.0417)^t = (239.9407)e^{0.0408t}$

(c) $\int_{10}^{15} (y_1 - y_2) \, dt \approx 649.5$ billion dollars

(d) No, model $y_1 > y_2$ forever because $1.0537 > 1.0417$.

No, these models are not accurate. According to news reports, $E > R$ eventually.

75. The total area is 8 times the area of the shaded region to the right. A point (x, y) is on the upper boundary of the region if

$$\sqrt{x^2 + y^2} = 2 - y$$
$$x^2 + y^2 = 4 - 4y + y^2$$
$$x^2 = 4 - 4y$$
$$4y = 4 - x^2$$
$$y = 1 - \frac{x^2}{4}.$$

We now determine where this curve intersects the line $y = x$.

$$x = 1 - \frac{x^2}{4}$$
$$x^2 + 4x - 4 = 0$$
$$x = \frac{-4 \pm \sqrt{16 + 16}}{2} = -2 \pm 2\sqrt{2} \implies x = -2 + 2\sqrt{2}$$

Total area $= 8 \displaystyle\int_0^{-2+2\sqrt{2}} \left(1 - \frac{x^2}{4} - x\right) dx$

$$= 8 \left[x - \frac{x^3}{12} - \frac{x^2}{2}\right]_0^{-2+2\sqrt{2}} = \frac{16}{3}(4\sqrt{2} - 5) \approx 8(0.4379) = 3.503$$

77. (a) $A = 2 \left[\displaystyle\int_0^5 \left(1 - \frac{1}{3}\sqrt{5 - x}\right) dx + \int_5^{5.5} (1 - 0)\, dx \right]$

$$= 2\left(\left[x + \frac{2}{9}(5 - x)^{3/2}\right]_0^5 + \left[x\right]_5^{5.5}\right) = 2\left(5 - \frac{10\sqrt{5}}{9} + 5.5 - 5\right) \approx 6.031 \text{ m}^2$$

(b) $V = 2A \approx 2(6.031) \approx 12.062 \text{ m}^3$ (c) $5000\,V \approx 5000(12.062) = 60{,}310$ pounds

79. True

81. False. Let $f(x) = x$ and $g(x) = 2x - x^2$. f and g intersect at $(1, 1)$, the midpoint of $[0, 2]$. But

$$\int_a^b [f(x) - g(x)]\, dx = \int_0^2 [x - (2x - x^2)]\, dx = \frac{2}{3} \ne 0.$$

Section 6.2   Volume: The Disk Method

1. $V = \pi \displaystyle\int_0^1 (-x + 1)^2\, dx = \pi \int_0^1 (x^2 - 2x + 1)\, dx = \pi \left[\frac{x^3}{3} - x^2 + x\right]_0^1 = \frac{\pi}{3}$

3. $V = \pi \displaystyle\int_1^4 \left(\sqrt{x}\right)^2 dx = \pi \int_1^4 x\, dx = \pi \left[\frac{x^2}{2}\right]_1^4 = \frac{15\pi}{2}$

5. $V = \pi \displaystyle\int_0^1 [(x^2)^2 - (x^3)^2]\, dx = \pi \int_0^1 (x^4 - x^6)\, dx = \pi \left[\frac{x^5}{5} - \frac{x^7}{7}\right]_0^1 = \frac{2\pi}{35}$

7. $y = x^2 \implies x = \sqrt{y}$ **9.** $y = x^{2/3} \implies x = y^{3/2}$

$V = \pi \displaystyle\int_0^4 \left(\sqrt{y}\right)^2 dy = \pi \int_0^4 y\, dy$ $V = \pi \displaystyle\int_0^1 (y^{3/2})^2\, dy = \pi \int_0^1 y^3\, dy = \pi \left[\frac{y^4}{4}\right]_0^1 = \frac{\pi}{4}$

$= \pi \left[\frac{y^2}{2}\right]_0^4 = 8\pi$

11. $y = \sqrt{x}$, $y = 0$, $x = 4$

(a) $R(x) = \sqrt{x}$, $r(x) = 0$

$$V = \pi \int_0^4 \left(\sqrt{x}\right)^2 dx$$

$$= \pi \int_0^4 x \, dx = \left[\frac{\pi}{2}x^2\right]_0^4 = 8\pi$$

(b) $R(y) = 4$, $r(y) = y^2$

$$V = \pi \int_0^2 (16 - y^4) \, dy$$

$$= \pi \left[16y - \frac{1}{5}y^5\right]_0^2 = \frac{128\pi}{5}$$

(c) $R(y) = 4 - y^2$, $r(y) = 0$

$$V = \pi \int_0^2 (4 - y^2)^2 \, dy$$

$$= \pi \int_0^2 (16 - 8y^2 + y^4) \, dy$$

$$= \pi \left[16y - \frac{8}{3}y^3 + \frac{1}{5}y^5\right]_0^2 = \frac{256\pi}{15}$$

(d) $R(y) = 6 - y^2$, $r(y) = 2$

$$V = \pi \int_0^2 \left[(6 - y^2)^2 - 4\right] dy$$

$$= \pi \int_0^2 (32 - 12y^2 + y^4) \, dy$$

$$= \pi \left[32y - 4y^3 + \frac{1}{5}y^5\right]_0^2 = \frac{192\pi}{5}$$

13. $y = x^2$, $y = 4x - x^2$ intersect at $(0, 0)$ and $(2, 4)$.

(a) $R(x) = 4x - x^2 \quad r(x) = x^2$

$$V = \pi \int_0^2 \left[(4x - x^2)^2 - x^4\right] dx$$

$$= \pi \int_0^2 (16x^2 - 8x^3) \, dx$$

$$= \pi \left[\frac{16}{3}x^3 - 2x^4\right]_0^2 = \frac{32\pi}{3}$$

(b) $R(x) = 6 - x^2$, $r(x) = 6 - (4x - x^2)$

$$V = \pi \int_0^2 \left[(6 - x^2)^2 - (6 - 4x + x^2)^2\right] dx$$

$$= 8\pi \int_0^2 (x^3 - 5x^2 + 6x) \, dx$$

$$= 8\pi \left[\frac{x^4}{4} - \frac{5}{3}x^3 + 3x^2\right]_0^2 = \frac{64\pi}{3}$$

15. $R(x) = 4 - x$, $r(x) = 1$

$$V = \pi \int_0^3 \left[(4 - x)^2 - (1)^2 \right] dx$$

$$= \pi \int_0^3 (x^2 - 8x + 15)\, dx$$

$$= \pi \left[\frac{x^3}{3} - 4x^2 + 15x \right]_0^3 = 18\pi$$

17. $R(x) = 4$, $r(x) = 4 - \dfrac{1}{1 + x}$

$$V = \pi \int_0^3 \left[4^2 - \left(4 - \frac{1}{1 + x} \right)^2 \right] dx$$

$$= \pi \int_0^3 \left[\frac{8}{1 + x} - \frac{1}{(1 + x)^2} \right] dx$$

$$= \pi \left[8 \ln(1 + x) + \frac{1}{1 + x} \right]_0^3$$

$$= \pi \left[8 \ln 4 + \frac{1}{4} - 1 \right]$$

$$= \left(8 \ln 4 - \frac{3}{4} \right) \pi \approx 32.485$$

19. $R(y) = 6 - y$, $r(y) = 0$

$$V = \pi \int_0^4 (6 - y)^2\, dy$$

$$= \pi \int_0^4 (y^2 - 12y + 36)\, dy$$

$$= \pi \left[\frac{y^3}{3} - 6y^2 + 36y \right]_0^4$$

$$= \frac{208\pi}{3}$$

21. $R(y) = 6 - y^2$, $r(y) = 2$

$$V = \pi \int_{-2}^2 \left[(6 - y^2)^2 - (2)^2 \right] dy$$

$$= 2\pi \int_0^2 (y^4 - 12y^2 + 32)\, dy$$

$$= 2\pi \left[\frac{y^5}{5} - 4y^3 + 32y \right]_0^2$$

$$= \frac{384\pi}{5}$$

23. $R(x) = \dfrac{1}{\sqrt{x + 1}}$, $r(x) = 0$

$$V = \pi \int_0^3 \left(\frac{1}{\sqrt{x + 1}} \right)^2 dx$$

$$= \pi \int_0^3 \frac{1}{x + 1}\, dx$$

$$= \left[\pi \ln|x + 1| \right]_0^3 = \pi \ln 4$$

25. $R(x) = \dfrac{1}{x}$, $r(x) = 0$

$$V = \pi \int_1^4 \left(\frac{1}{x}\right)^2 dx$$

$$= \pi \left[-\frac{1}{x} \right]_1^4$$

$$= \frac{3\pi}{4}$$

27. $R(x) = e^{-x}$, $r(x) = 0$

$$V = \pi \int_0^1 (e^{-x})^2 dx$$

$$= \pi \int_0^1 e^{-2x} dx$$

$$= \left[-\frac{\pi}{2} e^{-2x} \right]_0^1$$

$$= \frac{\pi}{2}(1 - e^{-2}) \approx 1.358$$

29.
$$x^2 + 1 = -x^2 + 2x + 5$$
$$x^2 - x - 2 = 0$$
$$(x - 2)(x + 1) = 0$$

$(-1, 2)$, $(2, 5)$ are points of intersection.

$$V = \pi \int_0^2 [(5 + 2x - x^2)^2 - (x^2 + 1)^2)] \, dx + \pi \int_2^3 [(x^2 + 1)^2 - (5 + 2x - x^2)^2] \, dx$$

$$= \pi \int_0^2 (-4x^3 - 8x^2 + 20x + 24) \, dx + \pi \int_2^3 (4x^3 + 8x^2 - 20x - 24) \, dx$$

$$= \pi \left[-x^4 - \frac{8}{3}x^3 + 10x^2 + 24x \right]_0^2 + \pi \left[x^4 + \frac{8}{3}x^3 - 10x^2 - 24x \right]_2^3$$

$$= \pi \frac{152}{3} + \pi \frac{125}{3} = \frac{277\pi}{3}$$

31. $y = 6 - 3x \implies x = \dfrac{1}{3}(6 - y)$

$$V = \pi \int_0^6 \left[\frac{1}{3}(6 - y) \right]^2 dy$$

$$= \frac{\pi}{9} \int_0^6 [36 - 12y + y^2] \, dy$$

$$= \frac{\pi}{9} \left[36y - 6y^2 + \frac{y^3}{3} \right]_0^6$$

$$= \frac{\pi}{9} \left[216 - 216 + \frac{216}{3} \right]$$

$$= 8\pi \left(= \frac{1}{3}\pi r^2 h \text{ volume of cone} \right)$$

33. $V = \pi \displaystyle\int_0^\pi [\sin x]^2 \, dx \approx 4.9348$

35. $V = \pi \int_0^2 [e^{-x^2}]^2 \, dx \approx 1.9686$

37. $V = \pi \int_{-1}^2 [e^{x/2} + e^{-x/2}]^2 \, dx \approx 49.0218$

39. $A \approx 3$

Matches (a)

41. Disk Method:

$$V = \pi \int_a^b [R(x)]^2 \, dx \quad \text{or} \quad V = \pi \int_c^d [R(y)]^2 \, dy$$

Washer Method:

$$V = \pi \int_a^b ([R(x)]^2 - [r(x)]^2) \, dx \quad \text{or}$$

$$V = \pi \int_c^d ([R(y)]^2 - [r(y)]^2) \, dy$$

43.

The volumes are the same because the solid has been translated horizontally.

45. $R(x) = \dfrac{1}{2}x, \ r(x) = 0$

$$V = \pi \int_0^6 \frac{1}{4}x^2 \, dx$$

$$= \left[\frac{\pi}{12}x^3 \right]_0^6 = 18\pi$$

Note: $V = \dfrac{1}{3}\pi r^2 h$

$$= \frac{1}{3}\pi(3^2)6$$

$$= 18\pi$$

47. $R(x) = \sqrt{r^2 - x^2}, \ r(x) = 0$

$$V = \pi \int_{-r}^r (r^2 - x^2) \, dx$$

$$= 2\pi \int_0^r (r^2 - x^2) \, dx$$

$$= 2\pi \left[r^2 x - \frac{1}{3}x^3 \right]_0^r$$

$$= 2\pi \left(r^3 - \frac{1}{3}r^3 \right) = \frac{4}{3}\pi r^3$$

49. $x = r - \dfrac{r}{H}y = r\left(1 - \dfrac{y}{H}\right)$, $R(y) = r\left(1 - \dfrac{y}{H}\right)$, $r(y) = 0$

$$V = \pi \int_0^h \left[r\left(1 - \dfrac{y}{H}\right) \right]^2 dy = \pi r^2 \int_0^h \left(1 - \dfrac{2}{H}y + \dfrac{1}{H^2}y^2\right) dy$$

$$= \pi r^2 \left[y - \dfrac{1}{H}y^2 + \dfrac{1}{3H^2}y^3 \right]_0^h$$

$$= \pi r^2 \left(h - \dfrac{h^2}{H} + \dfrac{h^3}{3H^2} \right)$$

$$= \pi r^2 h \left(1 - \dfrac{h}{H} + \dfrac{h^2}{3H^2} \right)$$

51. $V = \pi \int_0^2 \left(\dfrac{1}{8}x^2 \sqrt{2 - x}\right)^2 dx = \dfrac{\pi}{64} \int_0^2 x^4(2 - x)\, dx = \dfrac{\pi}{64} \left[\dfrac{2x^5}{5} - \dfrac{x^6}{6} \right]_0^2 = \dfrac{\pi}{30}$

53. (a) $R(x) = \dfrac{3}{5}\sqrt{25 - x^2}$, $r(x) = 0$ (b) $R(y) = \dfrac{5}{3}\sqrt{9 - y^2}$, $r(y) = 0$, $x \geq 0$

$$V = \dfrac{9\pi}{25} \int_{-5}^5 (25 - x^2)\, dx \qquad\qquad V = \dfrac{25\pi}{9} \int_0^3 (9 - y^2)\, dy$$

$$= \dfrac{18\pi}{25} \int_0^5 (25 - x^2)\, dx \qquad\qquad = \dfrac{25\pi}{9} \left[9y - \dfrac{y^3}{3} \right]_0^3 = 50\pi$$

$$= \dfrac{18\pi}{25} \left[25x - \dfrac{x^3}{3} \right]_0^5 = 60\pi$$

55. Total volume: $V = \dfrac{4\pi(50)^3}{3} = \dfrac{500{,}000}{3}$ ft^3

Volume of water in the tank:

$$\pi \int_{-50}^{y_0} \left(\sqrt{2500 - y^2}\right)^2 dy = \pi \int_{-50}^{y_0} (2500 - y^2)\, dy$$

$$= \pi \left[2500y - \dfrac{y^3}{3} \right]_{-50}^{y_0}$$

$$= \pi \left(2500y_0 - \dfrac{y_0^3}{3} + \dfrac{250{,}000}{3} \right)$$

When the tank is one-fourth of its capacity:

$$\dfrac{1}{4}\left(\dfrac{500{,}000\pi}{3}\right) = \pi \left(2500y_0 - \dfrac{y_0^3}{3} + \dfrac{250{,}00}{3} \right)$$

$$125{,}000 = 7500y_0 - y_0^3 + 250{,}000$$

$$y_0^3 - 7500y_0 - 125{,}000 = 0$$

$$y_0 \approx -17.36$$

Depth: $-17.36 - (-50) = 32.64$ feet

When the tank is three-fourths of its capacity the depth is $100 - 32.64 = 67.36$ feet.

57. (a) $\pi \displaystyle\int_0^h r^2 \, dx$ (ii)

is the volume of a right circular cylinder with radius r and height h.

(b) $\pi \displaystyle\int_{-b}^{b} \left(a \sqrt{1 - \dfrac{x^2}{b^2}} \right)^2 dx$ (iv)

is the volume of an ellipsoid with axes $2a$ and $2b$.

$y = a \sqrt{1 - \dfrac{x^2}{b^2}}$

(c) $\pi \displaystyle\int_{-r}^{r} \left(\sqrt{r^2 - x^2} \right)^2 dx$ (iii)

is the volume of a sphere with radius r.

$y = \sqrt{r^2 - x^2}$

(d) $\pi \displaystyle\int_0^h \left(\dfrac{rx}{h} \right)^2 dx$ (i)

is the volume of a right circular cone with the radius of the base as r and height h.

$y = \dfrac{r}{h} x$

(e) $\pi \displaystyle\int_{-r}^{r} \left[\left(R + \sqrt{r^2 - x^2} \right)^2 - \left(R - \sqrt{r^2 - x^2} \right)^2 \right] dx$ (v)

is the volume of a torus with the radius of its circular cross section as r and the distance from the axis of the torus to the center of its cross section as R.

$R + \sqrt{r^2 - x^2}$

R

$R - \sqrt{r^2 - x^2}$

59.

Base of Cross Section $= (x + 1) - (x^2 - 1) = 2 + x - x^2$

(a) $A(x) = b^2 = (2 + x - x^2)^2$

$\qquad = 4 + 4x - 3x^2 - 2x^3 + x^4$

$V = \displaystyle\int_{-1}^{2} (4 + 4x - 3x^2 - 2x^3 + x^4) \, dx$

$\quad = \left[4x + 2x^2 - x^3 - \dfrac{1}{2}x^4 + \dfrac{1}{5}x^5 \right]_{-1}^{2} = \dfrac{81}{10}$

$2 + x - x^2$

$2 + x - x^2$

(b) $A(x) = bh = (2 + x - x^2)1$

$V = \displaystyle\int_{-1}^{2} (2 + x - x^2) \, dx = \left[2x + \dfrac{x^2}{2} - \dfrac{x^3}{3} \right]_{-1}^{2} = \dfrac{9}{2}$

$2 + x - x^2$

1

61.

Base of Cross Section $= 1 - \sqrt[3]{y}$

(b) $A(y) = \dfrac{1}{2}\pi r^2 = \dfrac{1}{2}\pi\left(\dfrac{1 - \sqrt[3]{y}}{2}\right)^2 = \dfrac{1}{8}\pi(1 - \sqrt[3]{y})^2$

$V = \dfrac{1}{8}\pi\displaystyle\int_0^1 (1 - \sqrt[3]{y})^2\, dy = \dfrac{\pi}{8}\left(\dfrac{1}{10}\right) = \dfrac{\pi}{80}$

(c) $A(y) = \dfrac{1}{2}bh = \dfrac{1}{2}(1 - \sqrt[3]{y})\left(\dfrac{\sqrt{3}}{2}\right)(1 - \sqrt[3]{y})$

$\qquad = \dfrac{\sqrt{3}}{4}(1 - \sqrt[3]{y})^2$

$V = \dfrac{\sqrt{3}}{4}\displaystyle\int_0^1 (1 - \sqrt[3]{y})^2\, dy = \dfrac{\sqrt{3}}{4}\left(\dfrac{1}{10}\right) = \dfrac{\sqrt{3}}{40}$

(d) $A(y) = \dfrac{1}{2}\pi ab = \dfrac{\pi}{2}(2)(1 - \sqrt[3]{y})\dfrac{1 - \sqrt[3]{y}}{2}$

$\qquad = \dfrac{\pi}{2}(1 - \sqrt[3]{y})^2$

$V = \dfrac{\pi}{2}\displaystyle\int_0^1 (1 - \sqrt[3]{y})^2\, dy = \dfrac{\pi}{2}\left(\dfrac{1}{10}\right) = \dfrac{\pi}{20}$

(a) $A(y) = b^2 = \left(1 - \sqrt[3]{y}\right)^2$

$V = \displaystyle\int_0^1 \left(1 - \sqrt[3]{y}\right)^2\, dy$

$\quad = \displaystyle\int_0^1 (1 - 2y^{1/3} + y^{2/3})\, dy$

$\quad = \left[y - \dfrac{3}{2}y^{4/3} + \dfrac{3}{5}y^{5/3}\right]_0^1 = \dfrac{1}{10}$

63. Let $A_1(x)$ and $A_2(x)$ equal the areas of the cross sections of the two solids for $a \le x \le b$. Since $A_1(x) = A_2(x)$, we have

$$V_1 = \int_a^b A_1(x)\, dx = \int_a^b A_2(x)\, dx = V_2$$

Thus, the volumes are the same.

65. $\dfrac{4}{3}\pi(25 - r^2)^{3/2} = \dfrac{1}{2}\left(\dfrac{4}{3}\right)\pi(125)$

$\qquad (25 - r^2)^{3/2} = \dfrac{125}{2}$

$\qquad\qquad 25 - r^2 = \left(\dfrac{125}{2}\right)^{2/3}$

$\qquad 25 - \dfrac{25}{(2^{2/3})} = r^2$

$\qquad 25(1 - 2^{-2/3}) = r^2$

$\qquad\qquad r = 5\sqrt{1 - 2^{-2/3}} \approx 3.0415$

67. (a) Since the cross sections are isosceles right triangles:

$$A(x) = \dfrac{1}{2}bh = \dfrac{1}{2}\left(\sqrt{r^2 - y^2}\right)\left(\sqrt{r^2 - y^2}\right) = \dfrac{1}{2}(r^2 - y^2)$$

$$V = \dfrac{1}{2}\int_{-r}^r (r^2 - y^2)\, dy = \int_0^r (r^2 - y^2)\, dy = \left[r^2 y - \dfrac{y^3}{3}\right]_0^r = \dfrac{2}{3}r^3$$

(b) $A(x) = \dfrac{1}{2}bh = \dfrac{1}{2}\sqrt{r^2 - y^2}\left(\sqrt{r^2 - y^2}\tan\theta\right) = \dfrac{\tan\theta}{2}(r^2 - y^2)$

$$V = \dfrac{\tan\theta}{2}\int_{-r}^r (r^2 - y^2)\, dy = \tan\theta\int_0^r (r^2 - y^2)\, dy = \tan\theta\left[r^2 y - \dfrac{y^3}{3}\right]_0^r = \dfrac{2}{3}r^3\tan\theta$$

As $\theta \to 90°$, $V \to \infty$.

Section 6.3   Volume: The Shell Method

1. $p(x) = x$

$h(x) = x$

$V = 2\pi \displaystyle\int_0^2 x(x)\,dx = \left[\dfrac{2\pi x^3}{3}\right]_0^2 = \dfrac{16\pi}{3}$

3. $p(x) = x$

$h(x) = \sqrt{x}$

$V = 2\pi \displaystyle\int_0^4 x\sqrt{x}\,dx$

$= 2\pi \displaystyle\int_0^4 x^{3/2}\,dx$

$= \left[\dfrac{4\pi}{5}x^{5/2}\right]_0^4 = \dfrac{128\pi}{5}$

5. $p(x) = x$

$h(x) = x^2$

$V = 2\pi \displaystyle\int_0^2 x^3\,dx$

$= \left[\dfrac{\pi}{2}x^4\right]_0^2 = 8\pi$

7. $p(x) = x$

$h(x) = (4x - x^2) - x^2 = 4x - 2x^2$

$V = 2\pi \displaystyle\int_0^2 x(4x - 2x^2)\,dx$

$= 4\pi \displaystyle\int_0^2 (2x^2 - x^3)\,dx$

$= 4\pi \left[\dfrac{2}{3}x^3 - \dfrac{1}{4}x^4\right]_0^2 = \dfrac{16\pi}{3}$

9. $p(x) = x$

$h(x) = 4 - (4x - x^2) = x^2 - 4x + 4$

$V = 2\pi \displaystyle\int_0^2 (x^3 - 4x^2 + 4x)\,dx$

$= 2\pi \left[\dfrac{x^4}{4} - \dfrac{4}{3}x^3 + 2x^2\right]_0^2 = \dfrac{8\pi}{3}$

11. $p(x) = x$

$h(x) = \dfrac{1}{\sqrt{2\pi}}e^{-x^2/2}$

$V = 2\pi \displaystyle\int_0^1 x\left(\dfrac{1}{\sqrt{2\pi}}e^{-x^2/2}\right)dx$

$= \sqrt{2\pi} \displaystyle\int_0^1 e^{-x^2/2}x\,dx$

$= \left[-\sqrt{2\pi}\,e^{-x^2/2}\right]_0^1 = \sqrt{2\pi}\left(1 - \dfrac{1}{\sqrt{e}}\right) \approx 0.986$

13. $p(y) = y$

$h(y) = 2 - y$

$V = 2\pi \displaystyle\int_0^2 y(2 - y)\,dy$

$= 2\pi \displaystyle\int_0^2 (2y - y^2)\,dy$

$= 2\pi \left[y^2 - \dfrac{y^3}{3}\right]_0^2 = \dfrac{8\pi}{3}$

15. $p(y) = y$ and $h(y) = 1$ if $0 \le y < \dfrac{1}{2}$.

$p(y) = y$ and $h(y) = \dfrac{1}{y} - 1$ if $\dfrac{1}{2} \le y \le 1$.

$V = 2\pi \displaystyle\int_0^{1/2} y \, dy + 2\pi \int_{1/2}^1 (1 - y) \, dy$

$= 2\pi \left[\dfrac{y^2}{2}\right]_0^{1/2} + 2\pi \left[y - \dfrac{y^2}{2}\right]_{1/2}^1 = \dfrac{\pi}{4} + \dfrac{\pi}{4} = \dfrac{\pi}{2}$

17. $p(x) = 4 - x$

$h(x) = 4x - x^2 - x^2 = 4x - 2x^2$

$V = 2\pi \displaystyle\int_0^2 (4 - x)(4x - 2x^2) \, dx$

$= 2\pi(2) \displaystyle\int_0^2 (x^3 - 6x^2 + 8x) \, dx$

$= 4\pi \left[\dfrac{x^4}{4} - 2x^3 + 4x^2\right]_0^2 = 16\pi$

19. $p(x) = 5 - x$

$h(x) = 4x - x^2$

$V = 2\pi \displaystyle\int_0^4 (5 - x)(4x - x^2) \, dx$

$= 2\pi \displaystyle\int_0^4 (x^3 - 9x^2 + 20x) \, dx$

$= 2\pi \left[\dfrac{x^4}{4} - 3x^3 + 10x^2\right]_0^4 = 64\pi$

21. (a) **Disk**

$R(x) = x^3$

$r(x) = 0$

$V = \pi \displaystyle\int_0^2 x^6 \, dx = \pi \left[\dfrac{x^7}{7}\right]_0^2 = \dfrac{128\pi}{7}$

(b) **Shell**

$p(x) = x$

$h(x) = x^3$

$V = 2\pi \displaystyle\int_0^2 x^4 \, dx = 2\pi \left[\dfrac{x^5}{5}\right]_0^2 = \dfrac{64\pi}{5}$

(c) **Shell**

$p(x) = 4 - x$

$h(x) = x^3$

$V = 2\pi \displaystyle\int_0^2 (4 - x)x^3 \, dx$

$= 2\pi \displaystyle\int_0^2 (4x^3 - x^4) \, dx$

$= 2\pi \left[x^4 - \dfrac{1}{5}x^5\right]_0^2 = \dfrac{96\pi}{5}$

23. (a) **Shell**

$$p(y) = y$$

$$h(y) = (a^{1/2} - y^{1/2})^2$$

$$V = 2\pi \int_0^a y(a - 2a^{1/2}y^{1/2} + y)\,dy$$

$$= 2\pi \int_0^a (ay - 2a^{1/2}y^{3/2} + y^2)\,dy$$

$$= 2\pi \left[\frac{a}{2}y^2 - \frac{4a^{1/2}}{5}y^{5/2} + \frac{y^3}{3} \right]_0^a$$

$$= 2\pi \left[\frac{a^3}{2} - \frac{4a^3}{5} + \frac{a^3}{3} \right] = \frac{\pi a^3}{15}$$

(b) Same as part (a) by symmetry

25. $V = 2\pi \displaystyle\int_c^d p(y)h(y)\,dy$ or $V = 2\pi \displaystyle\int_a^b p(x)h(x)\,dx$

27. $\pi \displaystyle\int_1^5 (x - 1)\,dx = \pi \int_1^5 \left(\sqrt{x - 1} \right)^2 dx$

This integral represents the volume of the solid generated by revolving the region bounded by $y = \sqrt{x - 1}$, $y = 0$, and $x = 5$ about the x-axis by using the Disk Method.

$$2\pi \int_0^2 y[5 - (y^2 + 1)]\,dy$$

represents this same volume by using the Shell Method.

Disk Method

(c) **Shell**

$$p(x) = a - x$$

$$h(x) = (a^{1/2} - x^{1/2})^2$$

$$V = 2\pi \int_0^a (a - x)(a^{1/2} - x^{1/2})^2\,dx$$

$$= 2\pi \int_0^a (a^2 - 2a^{3/2}x^{1/2} + 2a^{1/2}x^{3/2} - x^2)\,dx$$

$$= 2\pi \left[a^2x - \frac{4}{3}a^{3/2}x^{3/2} + \frac{4}{5}a^{1/2}x^{5/2} - \frac{1}{3}x^3 \right]_0^a = \frac{4\pi a^3}{15}$$

29. (a)

(b) $x^{4/3} + y^{4/3} = 1$, $x = 0$, $y = 0$

$$y = (1 - x^{4/3})^{3/4}$$

$$V = 2\pi \int_0^1 x(1 - x^{4/3})^{3/4}\,dx \approx 1.5056$$

31. (a)

$$y = \sqrt[3]{(x-2)^2(x-6)^2}$$

(b) $V = 2\pi \int_2^6 x \sqrt[3]{(x-2)^2(x-6)^2}\, dx \approx 187.249$

33. $y = 2e^{-x}, \ y = 0, \ x = 0, \ x = 2$

Volume ≈ 7.5

Matches (d)

35. $p(x) = x$

$$h(x) = 2 - \frac{1}{2}x^2$$

$$V = 2\pi \int_0^2 x\left(2 - \frac{1}{2}x^2\right) dx = 2\pi \int_0^2 \left(2x - \frac{1}{2}x^3\right) dx = 2\pi \left[x^2 - \frac{1}{8}x^4\right]_0^2 = 4\pi \ (\text{total volume})$$

Now find x_0 such that

$$\pi = 2\pi \int_0^{x_0} \left(2x - \frac{1}{2}x^3\right) dx$$

$$1 = 2\left[x^2 - \frac{1}{8}x^4\right]_0^{x_0}$$

$$1 = 2x_0^2 - \frac{1}{4}x_0^4$$

$$x_0^4 - 8x_0^2 + 4 = 0$$

$$x_0^2 = 4 \pm 2\sqrt{3} \quad (\text{Quadratic Formula})$$

Take $x_0 = \sqrt{4 - 2\sqrt{3}}$ since the other root is too large.

Diameter: $2\sqrt{4 - 2\sqrt{3}} \approx 1.464$

37. $V = 4\pi \int_{-1}^1 (2 - x)\sqrt{1 - x^2}\, dx$

$$= 8\pi \int_{-1}^1 \sqrt{1 - x^2}\, dx - 4\pi \int_{-1}^1 x\sqrt{1 - x^2}\, dx$$

$$= 8\pi \left(\frac{\pi}{2}\right) + 2\pi \int_{-1}^1 x(1 - x^2)^{1/2}(-2)\, dx$$

$$= 4\pi^2 + \left[2\pi\left(\frac{2}{3}\right)(1 - x^2)^{3/2}\right]_{-1}^1 = 4\pi^2$$

39. Disk Method

$$R(y) = \sqrt{r^2 - y^2}$$

$$r(y) = 0$$

$$V = \pi \int_{r-h}^r (r^2 - y^2)\, dy$$

$$= \pi \left[r^2 y - \frac{y^3}{3}\right]_{r-h}^r = \frac{1}{3}\pi h^2(3r - h)$$

41. (a) $2\pi \int_0^r hx\left(1 - \dfrac{x}{r}\right) dx$ (ii)

is the volume of a right circular cone with the radius of the base as r and height h.

(b) $2\pi \int_{-r}^r (R - x)\left(2\sqrt{r^2 - x^2}\right) dx$ (v)

is the volume of a torus with the radius of its circular cross section as r and the distance from the axis of the torus to the center of its cross section as R.

(c) $2\pi \int_0^r 2x\sqrt{r^2 - x^2}\, dx$ (iii) is the volume of a sphere with radius r.

(d) $2\pi \int_0^r hx\, dx$ (i) is the volume of a right circular cylinder with a radius of r and a height of h.

(e) $2\pi \int_0^b 2ax\sqrt{1 - (x^2/b^2)}\, dx$ (iv)

is the volume of an ellipsoid with axes $2a$ and $2b$.

43. (a) $V = 2\pi \displaystyle\int_0^{200} xf(x)\, dx$

$\approx \dfrac{2\pi(200)}{3(8)}[0 + 4(25)(19) + 2(50)(19) + 4(75)(17) + 2(100)15 + 4(125)(14) + 2(150)(10) + 4(175)(6) + 0]$

$\approx 1,366,593$ cubic feet

(b) $d = -0.000561x^2 + 0.0189x + 19.39$

(c) $V \approx 2\pi \displaystyle\int_0^{200} xd(x)\, dx \approx 2\pi(213,800) = 1,343,345$ cubic feet

(d) Number gallons $\approx V(7.48) = 10,048,221$ gallons

Section 6.4   Arc Length and Surfaces of Revolution

1. (0, 0), (5, 12)

(a) $d = \sqrt{(5-0)^2 + (12-0)^2} = 13$

(b) $y = \dfrac{12}{5}x$

$y' = \dfrac{12}{5}$

$s = \displaystyle\int_0^5 \sqrt{1 + \left(\dfrac{12}{5}\right)^2}\, dx = \left[\dfrac{13}{5}x\right]_0^5 = 13$

3. $y = \dfrac{2}{3}x^{3/2} + 1$

$y' = x^{1/2}, [0, 1]$

$s = \displaystyle\int_0^1 \sqrt{1 + x}\, dx$

$= \left[\dfrac{2}{3}(1 + x)^{3/2}\right]_0^1$

$= \dfrac{2}{3}\left(\sqrt{8} - 1\right) \approx 1.219$

5. $y = \dfrac{3}{2}x^{2/3}$

$y' = \dfrac{1}{x^{1/3}}, [1, 8]$

$s = \displaystyle\int_1^8 \sqrt{1 + \left(\dfrac{1}{x^{1/3}}\right)^2}\, dx$

$= \displaystyle\int_1^8 \sqrt{\dfrac{x^{2/3} + 1}{x^{2/3}}}\, dx$

$= \dfrac{3}{2}\displaystyle\int_1^8 \sqrt{x^{2/3} + 1}\left(\dfrac{2}{3x^{1/3}}\right)\, dx$

$= \dfrac{3}{2}\left[\dfrac{2}{3}(x^{2/3} + 1)^{3/2}\right]_1^8$

$= 5\sqrt{5} - 2\sqrt{2} \approx 8.352$

7. $y = \dfrac{x^4}{8} + \dfrac{1}{4x^2}$

$y' = \dfrac{1}{2}x^3 - \dfrac{1}{2x^3}, [1, 2]$

$1 + (y')^2 = \left(\dfrac{1}{2}x^3 + \dfrac{1}{2x^3}\right)^2, [1, 2]$

$s = \displaystyle\int_a^b \sqrt{1 + (y')^2}\, dx$

$= \displaystyle\int_1^2 \left(\dfrac{1}{2}x^3 + \dfrac{1}{2x^3}\right)\, dx$

$= \left[\dfrac{1}{8}x^4 - \dfrac{1}{4x^2}\right]_1^2 = \dfrac{33}{16} \approx 2.063$

9. $y = \ln(\sin x), \left[\dfrac{\pi}{4}, \dfrac{3\pi}{4}\right]$

$y' = \dfrac{1}{\sin x}\cos x = \cot x$

$1 + (y')^2 = 1 + \cot^2 x = \csc^2 x$

$s = \displaystyle\int_{\pi/4}^{3\pi/4} \csc x\, dx$

$= \left[\ln|\csc x - \cot x|\right]_{\pi/4}^{3\pi/4}$

$= \ln\left(\sqrt{2} + 1\right) - \ln\left(\sqrt{2} - 1\right) \approx 1.763$

11. (a) $y = 4 - x^2, 0 \le x \le 2$

(b) $y' = -2x$

$1 + (y')^2 = 1 + 4x^2$

$L = \displaystyle\int_0^2 \sqrt{1 + 4x^2}\, dx$

(c) $L \approx 4.647$

13. (a) $y = \dfrac{1}{x}, 1 \le x \le 3$

(b) $y' = -\dfrac{1}{x^2}$

$1 + (y')^2 = 1 + \dfrac{1}{x^4}$

$L = \displaystyle\int_1^3 \sqrt{1 + \dfrac{1}{x^4}}\, dx$

(c) $L \approx 2.147$

15. (a) $y = \sin x, 0 \le x \le \pi$

(b) $y' = \cos x$

$1 + (y')^2 = 1 + \cos^2 x$

$L = \displaystyle\int_0^\pi \sqrt{1 + \cos^2 x}\, dx$

(c) $L \approx 3.820$

17. (a) $x = e^{-y}, 0 \le y \le 2$

$y = -\ln x$

$1 \ge x \ge e^{-2} \approx 0.135$

(b) $y' = -\dfrac{1}{x}$

$1 + (y')^2 = 1 + \dfrac{1}{x^2}$

$L = \displaystyle\int_{e^{-2}}^1 \sqrt{1 + \dfrac{1}{x^2}}\, dx$

(c) $L \approx 2.221$

Alternatively, you can do all the computations with respect to y.

(a) $x = e^{-y} \quad 0 \le y \le 2$

(b) $\dfrac{dx}{dy} = -e^{-y}$

$1 + \left(\dfrac{dx}{dy}\right)^2 = 1 + e^{-2y}$

$L = \displaystyle\int_0^2 \sqrt{1 + e^{-2y}}\, dy$

(c) $L \approx 2.221$

19. (a) $y = 2 \arctan x, 0 \le x \le 1$

(b) $y' = \dfrac{2}{1 + x^2}$

$L = \displaystyle\int_0^1 \sqrt{1 + \dfrac{4}{(1 + x^2)^2}}\, dx$

(c) $L \approx 1.871$

21. $\displaystyle\int_0^2 \sqrt{1 + \left[\frac{d}{dx}\left(\frac{5}{x^2 + 1}\right)\right]^2}\, dx$

$s \approx 5$

Matches (b)

23. $y = x^3$, $[0, 4]$

(a) $d = \sqrt{(4 - 0)^2 + (64 - 0)^2} \approx 64.125$

(b) $d = \sqrt{(1 - 0)^2 + (1 - 0)^2} + \sqrt{(2 - 1)^2 + (8 - 1)^2} + \sqrt{(3 - 2)^2 + (27 - 8)^2} + \sqrt{(4 - 3)^2 + (64 - 27)^2}$

≈ 64.525

(c) $s = \displaystyle\int_0^4 \sqrt{1 + (3x^2)^2}\, dx = \int_0^4 \sqrt{1 + 9x^4}\, dx \approx 64.666$

(d) 64.672

25. (a)

(c) $y_1' = 1$, $L_1 = \displaystyle\int_0^4 \sqrt{2}\, dx \approx 5.657$

$y_2' = \dfrac{3}{4}x^{1/2}$, $L_2 = \displaystyle\int_0^4 \sqrt{1 + \frac{9x}{16}}\, dx \approx 5.759$

$y_3' = \dfrac{1}{2}x$, $L_3 = \displaystyle\int_0^4 \sqrt{1 + \frac{x^2}{4}}\, dx \approx 5.916$

$y_4' = \dfrac{5}{16}x^{3/2}$, $L_4 = \displaystyle\int_0^4 \sqrt{1 + \frac{25}{256}x^3}\, dx \approx 6.063$

(b) y_1, y_2, y_3, y_4

27. $y = \dfrac{1}{3}\left[x^{3/2} - 3x^{1/2} + 2\right]$

When $x = 0$, $y = \frac{2}{3}$. Thus, the fleeting object has traveled $\frac{2}{3}$ units when it is caught.

$$y' = \frac{1}{3}\left[\frac{3}{2}x^{1/2} - \frac{3}{2}x^{-1/2}\right] = \left(\frac{1}{2}\right)\frac{x - 1}{x^{1/2}}$$

$$1 + (y')^2 = 1 + \frac{(x - 1)^2}{4x} = \frac{(x + 1)^2}{4x}$$

$$s = \int_0^1 \frac{x + 1}{2x^{1/2}}\, dx = \frac{1}{2}\int_0^1 (x^{1/2} + x^{-1/2})\, dx = \frac{1}{2}\left[\frac{2}{3}x^{3/2} + 2x^{1/2}\right]_0^1 = \frac{4}{3} = 2\left(\frac{2}{3}\right)$$

The pursuer has traveled twice the distance that the fleeing object has traveled when it is caught.

29. $\qquad y = 20 \cosh\dfrac{x}{20}$, $-20 \le x \le 20$

$y' = \sinh\dfrac{x}{20}$

$1 + (y')^2 = 1 + \sinh^2\dfrac{x}{20} = \cosh^2\dfrac{x}{20}$

$$L = \int_{-20}^{20} \cosh\frac{x}{20}\, dx = 2\int_0^{20} \cosh\frac{x}{20}\, dx = 2(20)\sinh\frac{x}{20}\bigg]_0^{20}$$

$$= 40\sinh(1) \approx 47.008 \text{ m.}$$

31.

$$y = \sqrt{9 - x^2}$$

$$y' = \frac{-x}{\sqrt{9 - x^2}}$$

$$1 + (y')^2 = \frac{9}{9 - x^2}$$

$$s = \int_0^2 \sqrt{\frac{9}{9 - x^2}}\, dx$$

$$= \int_0^2 \frac{3}{\sqrt{9 - x^2}}\, dx$$

$$= \left[3 \arcsin \frac{x}{3} \right]_0^2$$

$$= 3\left(\arcsin \frac{2}{3} - \arcsin 0 \right)$$

$$= 3 \arcsin \frac{2}{3} \approx 2.1892$$

33. $y = \dfrac{x^3}{3}$

$$y' = x^2,\ [0, 3]$$

$$S = 2\pi \int_0^3 \frac{x^3}{3} \sqrt{1 + x^4}\, dx$$

$$= \frac{\pi}{6} \int_0^3 (1 + x^4)^{1/2}(4x^3)\, dx$$

$$= \left[\frac{\pi}{9}(1 + x^4)^{3/2} \right]_0^3$$

$$= \frac{\pi}{9}\left(82\sqrt{82} - 1 \right) \approx 258.85$$

35.

$$y = \frac{x^3}{6} + \frac{1}{2x}$$

$$y' = \frac{x^2}{2} - \frac{1}{2x^2}$$

$$1 + (y')^2 = \left(\frac{x^2}{2} + \frac{1}{2x^2} \right)^2,\ [1, 2]$$

$$S = 2\pi \int_1^2 \left(\frac{x^3}{6} + \frac{1}{2x} \right)\left(\frac{x^2}{2} + \frac{1}{2x^2} \right)\, dx$$

$$= 2\pi \int_1^2 \left(\frac{x^5}{12} + \frac{x}{3} + \frac{1}{4x^3} \right)\, dx$$

$$= 2\pi \left[\frac{x^6}{72} + \frac{x^2}{6} - \frac{1}{8x^2} \right]_1^2 = \frac{47\pi}{16}$$

37. $y = \sqrt[3]{x} + 2$

$$y' = \frac{1}{3x^{2/3}},\ [1, 8]$$

$$S = 2\pi \int_1^8 x \sqrt{1 + \frac{1}{9x^{4/3}}}\, dx$$

$$= \frac{2\pi}{3} \int_1^8 x^{1/3} \sqrt{9x^{4/3} + 1}\, dx$$

$$= \frac{\pi}{18} \int_1^8 (9x^{4/3} + 1)^{1/2}(12x^{1/3})\, dx$$

$$= \left[\frac{\pi}{27}(9x^{4/3} + 1)^{3/2} \right]_1^8$$

$$= \frac{\pi}{27}\left(145\sqrt{145} - 10\sqrt{10} \right) \approx 199.48$$

39. $y = \sin x$

$$y' = \cos x,\ [0, \pi]$$

$$S = 2\pi \int_0^\pi \sin x \sqrt{1 + \cos^2 x}\, dx$$

$$\approx 14.4236$$

41. A rectifiable curve is one that has a finite arc length.

43. The precalculus formula is the surface area formula for the lateral surface of the frustum of a right circular cone. The representative element is

$$2\pi f(d_i)\sqrt{\Delta x_i^2 + \Delta y_i^2} = 2\pi f(d_i) \sqrt{1 + \left(\frac{\Delta y_i}{\Delta x_i} \right)^2}\, \Delta x_i.$$

45.
$$y = \frac{hx}{r}$$

$$y' = \frac{h}{r}$$

$$1 + (y')^2 = \frac{r^2 + h^2}{r^2}$$

$$S = 2\pi \int_0^r x \sqrt{\frac{r^2 + h^2}{r^2}} \, dx$$

$$= \left[\frac{2\pi \sqrt{r^2 + h^2}}{r} \left(\frac{x^2}{2} \right) \right]_0^r = \pi r \sqrt{r^2 + h^2}$$

47.
$$y = \sqrt{9 - x^2}$$

$$y' = \frac{-x}{\sqrt{9 - x^2}}$$

$$\sqrt{1 + (y')^2} = \frac{3}{\sqrt{9 - x^2}}$$

$$S = 2\pi \int_0^2 \frac{3x}{\sqrt{9 - x^2}} \, dx$$

$$= -3\pi \int_0^2 \frac{-2x}{\sqrt{9 - x^2}} \, dx$$

$$= \left[-6\pi \sqrt{9 - x^2} \right]_0^2$$

$$= 6\pi \left(3 - \sqrt{5} \right) \approx 14.40$$

See figure in Exercise 48.

49.
$$y = \frac{1}{3} x^{1/2} - x^{3/2}$$

$$y' = \frac{1}{6} x^{-1/2} - \frac{3}{2} x^{1/2} = \frac{1}{6} (x^{-1/2} - 9x^{1/2})$$

$$1 + (y')^2 = 1 + \frac{1}{36} (x^{-1} - 18 + 81x) = \frac{1}{36} (x^{-1/2} + 9x^{1/2})^2$$

$$S = 2\pi \int_0^{1/3} \left(\frac{1}{3} x^{1/2} - x^{3/2} \right) \sqrt{\frac{1}{36} (x^{-1/2} + 9x^{1/2})^2} \, dx = \frac{2\pi}{6} \int_0^{1/3} \left(\frac{1}{3} x^{1/2} - x^{3/2} \right) (x^{-1/2} + 9x^{1/2}) \, dx$$

$$= \frac{\pi}{3} \int_0^{1/3} \left(\frac{1}{3} + 2x - 9x^2 \right) dx = \frac{\pi}{3} \left[\frac{1}{3} x + x^2 - 3x^3 \right]_0^{1/3} = \frac{\pi}{27} \text{ ft}^2 \approx 0.1164 \text{ ft}^2 \approx 16.8 \text{ in}^2$$

Amount of glass needed: $V = \frac{\pi}{27} \left(\frac{0.015}{12} \right) \approx 0.00015 \text{ ft}^3 \approx 0.25 \text{ in}^3$

51. (a) $y = f(x) = 0.0000001953x^4 - 0.0001804x^3 + 0.0496x^2 - 4.8323x + 536.9270$

(b) Area $= \displaystyle\int_0^{400} f(x) \, dx \approx 131{,}734.5$ square feet

≈ 3.0 acres (1 acre = 43,560 square feet)

(Answers will vary.)

(c) $L = \displaystyle\int_0^{400} \sqrt{1 + f'(x)^2} \, dx \approx 794.9$ feet

(Answers will vary.)

53. (a) $V = \pi \displaystyle\int_1^b \frac{1}{x^2} \, dx = \left[-\frac{\pi}{x} \right]_1^b = \pi \left(1 - \frac{1}{b} \right)$

(b) $S = 2\pi \displaystyle\int_1^b \frac{1}{x} \sqrt{1 + \left(-\frac{1}{x^2} \right)^2} \, dx$

$$= 2\pi \int_1^b \frac{1}{x} \sqrt{1 + \frac{1}{x^4}} \, dx$$

$$= 2\pi \int_1^b \frac{\sqrt{x^4 + 1}}{x^3} \, dx$$

—CONTINUED—

53. —CONTINUED—

(c) $\displaystyle\lim_{b\to\infty} V = \lim_{b\to\infty}\ \pi\left(1 - \frac{1}{b}\right) = \pi$

(d) Since

$$\frac{\sqrt{x^4 + 1}}{x^3} > \frac{\sqrt{x^4}}{x^3} = \frac{1}{x} > 0 \text{ on } [1, b]$$

we have

$$\int_1^b \frac{\sqrt{x^4 + 1}}{x^3}\, dx > \int_1^b \frac{1}{x}\, dx = \Big[\ln x\Big]_1^b = \ln b$$

and $\displaystyle\lim_{b\to\infty} \ln b \to \infty$. Thus,

$$\lim_{b\to\infty} 2\pi \int_1^b \frac{\sqrt{x^4 + 1}}{x^3}\, dx = \infty.$$

55. (a) Area of circle with radius L: $A = \pi L^2$

Area of sector with central angle θ (in radians)

$$S = \frac{\theta}{2\pi} A = \frac{\theta}{2\pi}(\pi L^2) = \frac{1}{2}L^2\theta$$

(b) Let s be the arc length of the sector, which is the circumference of the base of the cone. Here, $s = L\theta = 2\pi r$, and you have

$$S = \frac{1}{2}L^2\theta = \frac{1}{2}L^2\left(\frac{s}{L}\right) = \frac{1}{2}Ls = \frac{1}{2}L(2\pi r) = \pi r L$$

(c) The lateral surface area of the frustum is the difference of the large cone and the small one.

$$S = \pi r_2(L + L_1) - \pi r_1 L_1$$

$$= \pi r_2 L + \pi L_1(r_2 - r_1)$$

By similar triangles, $\dfrac{L + L_1}{r_2} = \dfrac{L_1}{r_1} \implies Lr_1 = L_1(r_2 - r_1)$

Hence,

$$S = \pi r_2 L + \pi L_1(r_2 - r_1) = \pi r_2 L + \pi L r_1$$

$$= \pi L(r_1 + r_2).$$

Section 6.5   Work

1. $W = Fd = (100)(10) = 1000 \text{ ft} \cdot \text{lb}$

3. $W = Fd = (112)(4) = 448 \text{ joules (newton-meters)}$

5. Work equals force times distance, $W = FD$.

7. Since the work equals the area under the force function, you have $(c) < (d) < (a) < (b)$.

9. $F(x) = kx$

$5 = k(4)$

$k = \dfrac{5}{4}$

$W = \displaystyle\int_0^7 \frac{5}{4}x\, dx = \left[\frac{5}{8}x^2\right]_0^7$

$= \dfrac{245}{8} \text{ in} \cdot \text{lb}$

$= 30.625 \text{ in} \cdot \text{lb} \approx 2.55 \text{ ft} \cdot \text{lb}$

11. $F(x) = kx$

$250 = k(30) \implies k = \dfrac{25}{3}$

$W = \displaystyle\int_{20}^{50} F(x)\, dx = \int_{20}^{50} \frac{25}{3}x\, dx = \frac{25x^2}{6}\bigg]_{20}^{50}$

$= 8750 \text{ n} \cdot \text{cm} = 87.5 \text{ joules or Nm}$

13. $F(x) = kx$

$20 = k(9)$

$k = \dfrac{20}{9}$

$W = \displaystyle\int_0^{12} \dfrac{20}{9}x\, dx = \left[\dfrac{10}{9}x^2\right]_0^{12} = 160 \text{ in} \cdot \text{lb} = \dfrac{40}{3}\,\text{ft}\cdot\text{lb}$

15. $W = 18 = \displaystyle\int_0^{1/3} kx\, dx = \dfrac{kx^2}{2}\Big]_0^{1/3} = \dfrac{k}{18} \implies k = 324$

$W = \displaystyle\int_{1/3}^{7/12} 324x\, dx = 162x^2\Big]_{1/3}^{7/12} = 37.125 \text{ ft}\cdot\text{lbs}$

$\left[\textbf{Note:}\ 4 \text{ inches} = \tfrac{1}{3}\text{ foot}\right]$

17. Assume that Earth has a radius of 4000 miles.

$F(x) = \dfrac{k}{x^2}$

$5 = \dfrac{k}{(4000)^2}$

$k = 80{,}000{,}000$

$F(x) = \dfrac{80{,}000{,}000}{x^2}$

(a) $W = \displaystyle\int_{4000}^{4100} \dfrac{80{,}000{,}000}{x^2}\, dx = \left[\dfrac{-80{,}000{,}000}{x}\right]_{4000}^{4100} \approx 487.8 \text{ mi}\cdot\text{tons}$

$\approx 5.15 \times 10^9 \text{ ft}\cdot\text{lb}$

(b) $W = \displaystyle\int_{4000}^{4300} \dfrac{80{,}000{,}000}{x^2}\, dx \approx 1395.3 \text{ mi}\cdot\text{ton}$

$\approx 1.47 \times 10^{10} \text{ ft}\cdot\text{lb}$

19. Assume that the earth has a radius of 4000 miles.

$F(x) = \dfrac{k}{x^2}$

$10 = \dfrac{k}{(4000)^2}$

$k = 160{,}000{,}000$

$F(x) = \dfrac{160{,}000{,}000}{x^2}$

(a) $W = \displaystyle\int_{4000}^{15{,}000} \dfrac{160{,}000{,}000}{x^2}\, dx = \left[-\dfrac{160{,}000{,}000}{x}\right]_{4000}^{15{,}000} \approx -10{,}666.667 + 40{,}000$

$= 29{,}333.333 \text{ mi}\cdot\text{ton}$

$\approx 2.93 \times 10^4 \text{ mi}\cdot\text{ton}$

$\approx 3.10 \times 10^{11} \text{ ft}\cdot\text{lb}$

(b) $W = \displaystyle\int_{4000}^{26{,}000} \dfrac{160{,}000{,}000}{x^2}\, dx = \left[-\dfrac{160{,}000{,}000}{x}\right]_{4000}^{26{,}000} \approx -6{,}153.846 + 40{,}000$

$= 33{,}846.154 \text{ mi}\cdot\text{ton}$

$\approx 3.38 \times 10^4 \text{ mi}\cdot\text{ton}$

$\approx 3.57 \times 10^{11} \text{ ft}\cdot\text{lb}$

21. Weight of each layer: $62.4(20)\,\Delta y$

Distance: $4 - y$

(a) $W = \displaystyle\int_2^4 62.4(20)(4 - y)\, dy = \left[4992y - 624y^2\right]_2^4 = 2496 \text{ ft}\cdot\text{lb}$

(b) $W = \displaystyle\int_0^4 62.4(20)(4 - y)\, dy = \left[4992y - 624y^2\right]_0^4 = 9984 \text{ ft}\cdot\text{lb}$

23. Volume of disk: $\pi(2)^2\,\Delta y = 4\pi\,\Delta y$

Weight of disk of water: $9800(4\pi)\,\Delta y$

Distance the disk of water is moved: $5 - y$

$W = \displaystyle\int_0^4 (5 - y)(9800)4\pi\, dy = 39{,}200\pi\displaystyle\int_0^4 (5 - y)\, dy$

$= 39{,}200\pi\left[5y - \dfrac{y^2}{2}\right]_0^4$

$= 39{,}200\pi(12) = 470{,}400\pi \text{ newton–meters}$

25. Volume of disk: $\pi\left(\frac{2}{3}y\right)^2 \Delta y$

Weight of disk: $62.4\pi\left(\frac{2}{3}y\right)^2 \Delta y$

Distance: $6 - y$

$$W = \frac{4(62.4)\pi}{9}\int_0^6 (6 - y)y^2\,dy = \frac{4}{9}(62.4)\pi\left[2y^3 - \frac{1}{4}y^4\right]_0^6 = 2995.2\pi \text{ ft} \cdot \text{lb}$$

27. Volume of disk: $\pi\left(\sqrt{36 - y^2}\right)^2 \Delta y$

Weight of disk: $62.4\pi(36 - y^2)\Delta y$

Distance: y

$$W = 62.4\pi\int_0^6 y(36 - y^2)\,dy$$

$$= 62.4\pi\int_0^6 (36y - y^3)\,dy = 62.4\pi\left[18y^2 - \frac{1}{4}y^4\right]_0^6$$

$$= 20{,}217.6\pi \text{ ft} \cdot \text{lb}$$

29. Volume of layer: $V = lwh = 4(2)\sqrt{(9/4) - y^2}\,\Delta y$

Weight of layer: $W = 42(8)\sqrt{(9/4) - y^2}\,\Delta y$

Distance: $\frac{13}{2} - y$

$$W = \int_{-1.5}^{1.5} 42(8)\sqrt{(9/4) - y^2}\left(\frac{13}{2} - y\right)dy$$

$$= 336\left[\frac{13}{2}\int_{-1.5}^{1.5}\sqrt{(9/4) - y^2}\,dy - \int_{-1.5}^{1.5}\sqrt{(9/4) - y^2}\,y\,dy\right]$$

The second integral is zero since the integrand is odd and the limits of integration are symmetric to the origin. The first integral represents the area of a semicircle of radius $\frac{3}{2}$. Thus, the work is

$$W = 336\left(\frac{13}{2}\right)\pi\left(\frac{3}{2}\right)^2\left(\frac{1}{2}\right) = 2457\pi \text{ ft} \cdot \text{lb}$$

31. Weight of section of chain: $3\,\Delta y$

Distance: $15 - y$

$$W = 3\int_0^{15} (15 - y)\,dy$$

$$= \left[-\frac{3}{2}(15 - y)^2\right]_0^{15}$$

$$= 337.5 \text{ ft} \cdot \text{lb}$$

33. The lower 5 feet of chain are raised 10 feet with a constant force.

$$W_1 = 3(5)(10) = 150 \text{ ft} \cdot \text{lb}$$

The top 10 feet of chain are raised with a variable force.

Weight per section: $3\,\Delta y$

Distance: $10 - y$

$$W_2 = 3\int_0^{10} (10 - y)\,dy \;=\; \left[-\frac{3}{2}(10 - y)^2\right]_0^{10}$$

$$= 150 \text{ ft} \cdot \text{lb}$$

$$W = W_1 + W_2 = 300 \text{ ft} \cdot \text{lb}$$

35. Weight of section of chain: $3\,\Delta y$

Distance: $15 - 2y$

$$W = 3\int_0^{7.5}(15 - 2y)\,dy = \left[-\frac{3}{4}(15 - 2y)^2\right]_0^{7.5}$$

$$= \frac{3}{4}(15)^2 = 168.75 \text{ ft} \cdot \text{lb}$$

37. Work to pull up the ball: $W_1 = 500(15) = 7500$ ft $\cdot$ lb

Work to wind up the top 15 feet of cable: force is variable

Weight per section: $1\,\Delta y$

Distance: $15 - x$

$$W_2 = \int_0^{15}(15 - x)\,dx = \left[-\frac{1}{2}(15 - x)^2\right]_0^{15}$$

$$= 112.5 \text{ ft} \cdot \text{lb}$$

Work to lift the lower 25 feet of cable with a constant force:

$$W_3 = (1)(25)(15) = 375 \text{ ft} \cdot \text{lb}$$

$$W = W_1 + W_2 + W_3 = 7500 + 112.5 + 375$$

$$= 7987.5 \text{ ft} \cdot \text{lb}$$

39. $$p = \frac{k}{V}$$

$$1000 = \frac{k}{2}$$

$$k = 2000$$

$$W = \int_2^3 \frac{2000}{V}\,dV = \Big[2000\ln|V|\Big]_2^3$$

$$= 2000\ln\!\left(\frac{3}{2}\right) \approx 810.93 \text{ ft} \cdot \text{lb}$$

41. $$F(x) = \frac{k}{(2 - x)^2}$$

$$W = \int_{-2}^1 \frac{k}{(2 - x)^2}\,dx = \left[\frac{k}{2 - x}\right]_{-2}^1 = k\!\left(1 - \frac{1}{4}\right)$$

$$= \frac{3k}{4}\text{(units of work)}$$

43. $$W = \int_0^5 1000[1.8 - \ln(x + 1)]\,dx \approx 3249.44 \text{ ft} \cdot \text{lb}$$

45. $$W = \int_0^5 100x\sqrt{125 - x^3}\,dx \approx 10{,}330.3 \text{ ft} \cdot \text{lb}$$

Section 6.6   Moments, Centers of Mass, and Centroids

1. $\bar{x} = \dfrac{6(-5) + 3(1) + 5(3)}{6 + 3 + 5} = -\dfrac{6}{7}$

3. $\bar{x} = \dfrac{1(7) + 1(8) + 1(12) + 1(15) + 1(18)}{1 + 1 + 1 + 1 + 1} = 12$

5. (a) $\bar{x} = \dfrac{(7 + 5) + (8 + 5) + (12 + 5) + (15 + 5) + (18 + 5)}{5} = 17 = 12 + 5$

   (b) $\bar{x} = \dfrac{12(-6 - 3) + 1(-4 - 3) + 6(-2 - 3) + 3(0 - 3) + 11(8 - 3)}{12 + 1 + 6 + 3 + 11} = \dfrac{-99}{33} = -3$

7. $50x = 75(L - x) = 75(10 - x)$

$50x = 750 - 75x$

$125x = 750$

$x = 6$ feet

9. $\bar{x} = \dfrac{5(2) + 1(-3) + 3(1)}{5 + 1 + 3} = \dfrac{10}{9}$

$\bar{y} = \dfrac{5(2) + 1(1) + 3(-4)}{5 + 1 + 3} = -\dfrac{1}{9}$

$(\bar{x}, \bar{y}) = \left(\dfrac{10}{9}, -\dfrac{1}{9}\right)$

11. $\bar{x} = \dfrac{3(-2) + 4(-1) + 2(7) + 1(0) + 6(-3)}{3 + 4 + 2 + 1 + 6} = -\dfrac{7}{8}$

$\bar{y} = \dfrac{3(-3) + 4(0) + 2(1) + 1(0) + 6(0)}{3 + 4 + 2 + 1 + 6} = -\dfrac{7}{16}$

$(\bar{x}, \bar{y}) = \left(-\dfrac{7}{8}, -\dfrac{7}{16}\right)$

13. $m = \rho \displaystyle\int_0^4 \sqrt{x}\, dx = \left[\dfrac{2\rho}{3} x^{3/2}\right]_0^4 = \dfrac{16\rho}{3}$

$M_x = \rho \displaystyle\int_0^4 \dfrac{\sqrt{x}}{2}\left(\sqrt{x}\right) dx = \left[\rho\dfrac{x^2}{4}\right]_0^4 = 4\rho$

$\bar{y} = \dfrac{M_x}{m} = 4\rho\left(\dfrac{3}{16\rho}\right) = \dfrac{3}{4}$

$M_y = \rho \displaystyle\int_0^4 x\sqrt{x}\, dx = \left[\rho\dfrac{2}{5}x^{5/2}\right]_0^4 = \dfrac{64\rho}{5}$

$\bar{x} = \dfrac{M_y}{m} = \dfrac{64\rho}{5}\left(\dfrac{3}{16\rho}\right) = \dfrac{12}{5}$

$(\bar{x}, \bar{y}) = \left(\dfrac{12}{5}, \dfrac{3}{4}\right)$

15. $m = \rho \displaystyle\int_0^1 (x^2 - x^3)\, dx = \rho\left[\dfrac{x^3}{3} - \dfrac{x^4}{4}\right]_0^1 = \dfrac{\rho}{12}$

$M_x = \rho \displaystyle\int_0^1 \dfrac{(x^2 + x^3)}{2}(x^2 - x^3)\, dx = \dfrac{\rho}{2}\displaystyle\int_0^1 (x^4 - x^6)\, dx = \dfrac{\rho}{2}\left[\dfrac{x^5}{5} - \dfrac{x^7}{7}\right]_0^1 = \dfrac{\rho}{35}$

$\bar{y} = \dfrac{M_x}{m} = \dfrac{\rho}{35}\left(\dfrac{12}{\rho}\right) = \dfrac{12}{35}$

$M_y = \rho \displaystyle\int_0^1 x(x^2 - x^3)\, dx = \rho\displaystyle\int_0^1 (x^3 - x^4)\, dx = \rho\left[\dfrac{x^4}{4} - \dfrac{x^5}{5}\right]_0^1 = \dfrac{\rho}{20}$

$\bar{x} = \dfrac{M_y}{m} = \dfrac{\rho}{20}\left(\dfrac{12}{\rho}\right) = \dfrac{3}{5}$

$(\bar{x}, \bar{y}) = \left(\dfrac{3}{5}, \dfrac{12}{35}\right)$

17. $m = \rho \displaystyle\int_0^3 [(-x^2 + 4x + 2) - (x + 2)]\, dx = -\rho\left[\dfrac{x^3}{3} + \dfrac{3x^2}{2}\right]_0^3 = \dfrac{9\rho}{2}$

$M_x = \rho \displaystyle\int_0^3 \left[\dfrac{(-x^2 + 4x + 2) + (x + 2)}{2}\right][(-x^2 + 4x + 2) - (x + 2)]\, dx$

$\quad = \dfrac{\rho}{2}\displaystyle\int_0^3 (-x^2 + 5x + 4)(-x^2 + 3x)\, dx = \dfrac{\rho}{2}\displaystyle\int_0^3 (x^4 - 8x^3 + 11x^2 + 12x)\, dx$

$\quad = \dfrac{\rho}{2}\left[\dfrac{x^5}{5} - 2x^4 + \dfrac{11x^3}{3} + 6x^2\right]_0^3 = \dfrac{99\rho}{5}$

$\bar{y} = \dfrac{M_x}{m} = \dfrac{99\rho}{5}\left(\dfrac{2}{9\rho}\right) = \dfrac{22}{5}$

$M_y = \rho \displaystyle\int_0^3 x[(-x^2 + 4x - 2) - (x + 2)]\, dx = \rho\displaystyle\int_0^3 (-x^3 + 3x^2)\, dx = \rho\left[-\dfrac{x^4}{4} + x^3\right]_0^3 = \dfrac{27\rho}{4}$

$\bar{x} = \dfrac{M_y}{m} = \dfrac{27\rho}{4}\left(\dfrac{2}{9\rho}\right) = \dfrac{3}{2}$

$(\bar{x}, \bar{y}) = \left(\dfrac{3}{2}, \dfrac{22}{5}\right)$

19. $m = \rho \int_0^8 x^{2/3}\, dx = \rho\left[\frac{3}{5}x^{5/3}\right]_0^8 = \frac{96\rho}{5}$

$M_x = \rho \int_0^8 \frac{x^{2/3}}{2}(x^{2/3})\, dx = \frac{\rho}{2}\left[\frac{3}{7}x^{7/3}\right]_0^8 = \frac{192\rho}{7}$

$\bar{y} = \frac{M_x}{m} = \frac{192\rho}{7}\left(\frac{5}{96\rho}\right) = \frac{10}{7}$

$M_y = \rho \int_0^8 x(x^{2/3})\, dx = \rho\left[\frac{3}{8}x^{8/3}\right]_0^8 = 96\rho$

$\bar{x} = \frac{M_y}{m} = 96\rho\left(\frac{5}{96\rho}\right) = 5$

$(\bar{x}, \bar{y}) = \left(5, \frac{10}{7}\right)$

21. $m = 2\rho \int_0^2 (4 - y^2)\, dy = 2\rho\left[4y - \frac{y^3}{3}\right]_0^2 = \frac{32\rho}{3}$

$M_y = 2\rho \int_0^2 \left(\frac{4 - y^2}{2}\right)(4 - y^2)\, dy = \rho\left[16y - \frac{8}{3}y^3 + \frac{y^5}{5}\right]_0^2 = \frac{256\rho}{15}$

$\bar{x} = \frac{M_y}{m} = \frac{256\rho}{15}\left(\frac{3}{32\rho}\right) = \frac{8}{5}$

By symmetry, M_x and $\bar{y} = 0$.

$(\bar{x}, \bar{y}) = \left(\frac{8}{5}, 0\right)$

23. $m = \rho \int_0^3 [(2y - y^2) - (-y)]\, dy = \rho\left[\frac{3y^2}{2} - \frac{y^3}{3}\right]_0^3 = \frac{9\rho}{2}$

$M_y = \rho \int_0^3 \frac{[(2y - y^2) + (-y)]}{2}[(2y - y^2) - (-y)]\, dy = \frac{\rho}{2}\int_0^3 (y - y^2)(3y - y^2)\, dy$

$= \frac{\rho}{2}\int_0^3 (y^4 - 4y^3 + 3y^2)\, dy = \frac{\rho}{2}\left[\frac{y^5}{5} - y^4 + y^3\right]_0^3 = -\frac{27\rho}{10}$

$\bar{x} = \frac{M_y}{m} = -\frac{27\rho}{10}\left(\frac{2}{9\rho}\right) = -\frac{3}{5}$

$M_x = \rho \int_0^3 y[(2y - y^2) - (-y)]\, dy = \rho\int_0^3 (3y^2 - y^3)\, dy = \rho\left[y^3 - \frac{y^4}{4}\right]_0^3 = \frac{27\rho}{4}$

$\bar{y} = \frac{M_x}{m} = \frac{27\rho}{4}\left(\frac{2}{9\rho}\right) = \frac{3}{2}$

$(\bar{x}, \bar{y}) = \left(-\frac{3}{5}, \frac{3}{2}\right)$

25. $A = \int_0^1 (x - x^2)\, dx = \left[\frac{1}{2}x^2 - \frac{x^3}{3}\right]_0^1 = \frac{1}{6}$

$M_x = \frac{1}{2}\int_0^1 (x^2 - x^4)\, dx = \frac{1}{2}\left[\frac{x^3}{3} - \frac{x^5}{5}\right]_0^1 = \frac{1}{2}\left(\frac{1}{3} - \frac{1}{5}\right) = \frac{1}{15}$

$M_y = \int_0^1 (x^2 - x^3)\, dx = \left[\frac{x^3}{3} - \frac{x^4}{4}\right]_0^1 = \left(\frac{1}{3} - \frac{1}{4}\right) = \frac{1}{12}$

27. $A = \int_0^3 (2x + 4)\,dx = \left[x^2 + 4x\right]_0^3 = 9 + 12 = 21$

$M_x = \frac{1}{2}\int_0^3 (2x + 4)^2\,dx = \int_0^3 (2x^2 + 8x + 8)\,dx = \left[\frac{2x^3}{3} + 4x^2 + 8x\right]_0^3 = 18 + 36 + 24 = 78$

$M_y = \int_0^3 (2x^2 + 4x)\,dx = \left[\frac{2x^3}{3} + 2x^2\right]_0^3 = 18 + 18 = 36$

29. $m = \rho \int_0^5 10x\sqrt{125 - x^3}\,dx \approx 1033.0\rho$

$M_x = \rho \int_0^5 \left(\frac{10x\sqrt{125 - x^3}}{2}\right)\left(10x\sqrt{125 - x^3}\right)dx = 50\rho \int_0^5 x^2(125 - x^3)\,dx = \frac{3{,}124{,}375\rho}{24} \approx 130{,}208\rho$

$M_y = \rho \int_0^5 10x^2\sqrt{125 - x^3}\,dx = -\frac{10\rho}{3}\int_0^5 \sqrt{125 - x^3}(-3x^2)\,dx = \frac{12{,}500\sqrt{5}\rho}{9} \approx 3105.6\rho$

$\bar{x} = \frac{M_y}{m} \approx 3.0$

$\bar{y} = \frac{M_x}{m} \approx 126.0$

Therefore, the centroid is (3.0, 126.0).

31. $m = \rho \int_{-20}^{20} 5\sqrt[3]{400 - x^2}\,dx \approx 1239.76\rho$

$M_x = \rho \int_{-20}^{20} \frac{5\sqrt[3]{400 - x^2}}{2}\left(5\sqrt[3]{400 - x^2}\right)dx$

$= \frac{25\rho}{2}\int_{-20}^{20} (400 - x^2)^{2/3}\,dx \approx 20064.27$

$\bar{y} = \frac{M_x}{m} \approx 16.18$

$\bar{x} = 0$ by symmetry. Therefore, the centroid is (0, 16.2).

33. $A = \frac{1}{2}(2a)c = ac$

$\frac{1}{A} = \frac{1}{ac}$

$\bar{x} = \left(\frac{1}{ac}\right)\frac{1}{2}\int_0^c \left[\left(\frac{b - a}{c}y + a\right)^2 - \left(\frac{b + a}{c}y - a\right)^2\right]dy$

$= \frac{1}{2ac}\int_0^c \left[\frac{4ab}{c}y - \frac{4ab}{c^2}y^2\right]dy$

$= \frac{1}{2ac}\left[\frac{2ab}{c}y^2 - \frac{4ab}{3c^2}y^3\right]_0^c = \frac{1}{2ac}\left(\frac{2}{3}abc\right) = \frac{b}{3}$

$\bar{y} = \frac{1}{ac}\int_0^c y\left[\left(\frac{b - a}{c}y + a\right) - \left(\frac{b + a}{c}y - a\right)\right]dy$

$= \frac{1}{ac}\int_0^c y\left(-\frac{2a}{c}y + 2a\right)dy = \frac{2}{c}\int_0^c \left(y - \frac{y^2}{c}\right)dy$

$= \frac{2}{c}\left[\frac{y^2}{2} - \frac{y^3}{3c}\right]_0^c = \frac{c}{3}$

$(\bar{x}, \bar{y}) = \left(\frac{b}{3}, \frac{c}{3}\right)$

In Exercise 73 of Section P.2, you found that $(b/3, c/3)$ is the point of intersection of the medians.

35. $A = \dfrac{c}{2}(a + b)$

$\dfrac{1}{A} = \dfrac{2}{c(a + b)}$

$\bar{x} = \dfrac{2}{c(a + b)} \displaystyle\int_0^c x\left(\dfrac{b - a}{c}x + a\right) dx = \dfrac{2}{c(a + b)} \int_0^c \left(\dfrac{b - a}{c}x^2 + ax\right) dx = \dfrac{2}{c(a + b)}\left[\dfrac{b - a}{c}\dfrac{x^3}{3} + \dfrac{ax^2}{2}\right]_0^c$

$\quad = \dfrac{2}{c(a + b)}\left[\dfrac{(b - a)c^2}{3} + \dfrac{ac^2}{2}\right] = \dfrac{2}{c(a + b)}\left[\dfrac{2bc^2 - 2ac^2 + 3ac^2}{6}\right] = \dfrac{c(2b + a)}{3(a + b)} = \dfrac{(a + 2b)c}{3(a + b)}$

$\bar{y} = \dfrac{2}{c(a + b)}\dfrac{1}{2}\displaystyle\int_0^c \left(\dfrac{b - a}{c}x + a\right)^2 dx = \dfrac{1}{c(a + b)}\int_0^c \left[\left(\dfrac{b - a}{c}\right)^2 x^2 + \dfrac{2a(b - a)}{c}x + a^2\right] dx$

$\quad = \dfrac{1}{c(a + b)}\left[\left(\dfrac{b - a}{c}\right)^2 \dfrac{x^3}{3} + \dfrac{2a(b - a)}{c}\dfrac{x^2}{2} + a^2 x\right]_0^c = \dfrac{1}{c(a + b)}\left[\dfrac{(b - a)^2 c}{3} + ac(b - a) + a^2 c\right]$

$\quad = \dfrac{1}{3c(a + b)}[(b^2 - 2ab + a^2)c + 3ac(b - a) + 3a^2 c]$

$\quad = \dfrac{1}{3(a + b)}[b^2 - 2ab + a^2 + 3ab - 3a^2 + 3a^2] = \dfrac{a^2 + ab + b^2}{3(a + b)}$

Thus, $(\bar{x}, \bar{y}) = \left(\dfrac{(a + 2b)c}{3(a + b)}, \dfrac{a^2 + ab + b^2}{3(a + b)}\right)$.

The one line passes through $(0, a/2)$ and $(c, b/2)$. It's equation is $y = \dfrac{b - a}{2c}x + \dfrac{a}{2}$.

The other line passes through $(0, -b)$ and $(c, a + b)$. It's equation is $y = \dfrac{a + 2b}{c}x - b$.

$(\bar{x}, \bar{y})$ is the point of intersection of these two lines.

37. $\bar{x} = 0$ by symmetry

$\quad A = \dfrac{1}{2}\pi ab$

$\quad \dfrac{1}{A} = \dfrac{2}{\pi ab}$

$\quad \bar{y} = \dfrac{2}{\pi ab}\dfrac{1}{2}\displaystyle\int_{-a}^a \left(\dfrac{b}{a}\sqrt{a^2 - x^2}\right)^2 dx$

$\quad = \dfrac{1}{\pi ab}\left(\dfrac{b^2}{a^2}\right)\left[a^2 x - \dfrac{x^3}{3}\right]_{-a}^a = \dfrac{b}{\pi a^3}\left[\dfrac{4a^3}{3}\right] = \dfrac{4b}{3\pi}$

$(\bar{x}, \bar{y}) = \left(0, \dfrac{4b}{3\pi}\right)$

39. (a)

(b) $\bar{x} = 0$ by symmetry

(c) $M_y = \displaystyle\int_{-\sqrt{b}}^{\sqrt{b}} x(b - x^2)\, dx = 0$ because $bx - x^3$ is odd

(d) $\bar{y} > \dfrac{b}{2}$ since there is more area above $y = \dfrac{b}{2}$ than below

(e) $M_x = \displaystyle\int_{-\sqrt{b}}^{\sqrt{b}} \dfrac{(b + x^2)(b - x^2)}{2}\, dx$

$\quad = \displaystyle\int_{-\sqrt{b}}^{\sqrt{b}} \dfrac{b^2 - x^4}{2}\, dx = \dfrac{1}{2}\left[b^2 x - \dfrac{x^5}{5}\right]_{-\sqrt{b}}^{\sqrt{b}}$

$\quad = b^2\sqrt{b} - \dfrac{b^2\sqrt{b}}{5} = \dfrac{4b^2\sqrt{b}}{5}$

$\quad A = \displaystyle\int_{-\sqrt{b}}^{\sqrt{b}} (b - x^2)\, dx = \left[bx - \dfrac{x^3}{3}\right]_{-\sqrt{b}}^{\sqrt{b}}$

$\quad = \left(b\sqrt{b} - \dfrac{b\sqrt{b}}{3}\right)2 = 4\dfrac{b\sqrt{b}}{3}$

$\quad \bar{y} = \dfrac{M_x}{A} = \dfrac{4b^2\sqrt{b}/5}{4b\sqrt{b}/3} = \dfrac{3}{5}b.$

41. (a) $\bar{x} = 0$ by symmetry

$$A = 2\int_0^{40} f(x)\, dx = \frac{2(40)}{3(4)}[30 + 4(29) + 2(26) + 4(20) + 0] = \frac{20}{3}(278) = \frac{5560}{3}$$

$$M_x = \int_{-40}^{40} \frac{f(x)^2}{2}\, dx = \frac{40}{3(4)}[30^2 + 4(29)^2 + 2(26)^2 + 4(20)^2 + 0] = \frac{10}{3}(7216) = \frac{72160}{3}$$

$$\bar{y} = \frac{M_x}{A} = \frac{72160/3}{5560/3} = \frac{72160}{5560} \approx 12.98$$

$(\bar{x}, \bar{y}) = (0, 12.98)$

(b) $y = (-1.02 \times 10^{-5})x^4 - 0.0019x^2 + 29.28$

(c) $\bar{y} = \frac{M_x}{A} \approx \frac{23697.68}{1843.54} \approx 12.85$

$(\bar{x}, \bar{y}) = (0, 12.85)$

43. Centroids of the given regions: $(1, 0)$ and $(3, 0)$

Area: $A = 4 + \pi$

$$\bar{x} = \frac{4(1) + \pi(3)}{4 + \pi} = \frac{4 + 3\pi}{4 + \pi}$$

$$\bar{y} = \frac{4(0) + \pi(0)}{4 + \pi} = 0$$

$(\bar{x}, \bar{y}) = \left(\frac{4 + 3\pi}{4 + \pi}, 0\right) \approx (1.88, 0)$

45. Centroids of the given regions: $\left(0, \frac{3}{2}\right)$, $(0, 5)$, and $\left(0, \frac{15}{2}\right)$

Area: $A = 15 + 12 + 7 = 34$

$$\bar{x} = \frac{15(0) + 12(0) + 7(0)}{34} = 0$$

$$\bar{y} = \frac{15(3/2) + 12(5) + 7(15/2)}{34} = \frac{135}{34}$$

$(\bar{x}, \bar{y}) = \left(0, \frac{135}{34}\right)$

47. Centroids of the given regions: $(1, 0)$ and $(3, 0)$

Mass: $4 + 2\pi$

$$\bar{x} = \frac{4(1) + 2\pi(3)}{4 + 2\pi} = \frac{2 + 3\pi}{2 + \pi}$$

$$\bar{y} = 0$$

$(\bar{x}, \bar{y}) = \left(\frac{2 + 3\pi}{2 + \pi}, 0\right) \approx (2.22, 0)$

49. $V = 2\pi r A = 2\pi(5)(16\pi) = 160\pi^2 \approx 1579.14$

51. $A = \dfrac{1}{2}(4)(4) = 8$

$$\bar{y} = \left(\dfrac{1}{8}\right)\dfrac{1}{2}\int_0^4 (4+x)(4-x)\,dx = \dfrac{1}{16}\left[16x - \dfrac{x^3}{3}\right]_0^4 = \dfrac{8}{3}$$

$$r = \bar{y} = \dfrac{8}{3}$$

$$V = 2\pi r A = 2\pi\left(\dfrac{8}{3}\right)(8) = \dfrac{128\pi}{3} \approx 134.04$$

53. $m = m_1 + \cdots + m_n$

$M_y = m_1 x_1 + \cdots + m_n x_n$

$M_x = m_1 y_1 + \cdots + m_n y_n$

$\bar{x} = \dfrac{M_y}{m},\ \bar{y} = \dfrac{M_x}{m}$

55. (a) Yes. $(\bar{x}, \bar{y}) = \left(\dfrac{5}{6}, \dfrac{5}{18} + 2\right) = \left(\dfrac{5}{6}, \dfrac{41}{18}\right)$

(b) Yes. $(\bar{x}, \bar{y}) = \left(\dfrac{5}{6} + 2, \dfrac{5}{18}\right) = \left(\dfrac{17}{6}, \dfrac{5}{18}\right)$

(c) Yes. $(\bar{x}, \bar{y}) = \left(\dfrac{5}{6}, -\dfrac{5}{18}\right)$

(d) No.

57. The surface area of the sphere is $S = 4\pi r^2$. The arc length of C is $s = \pi r$. The distance traveled by the centroid is

$$d = \dfrac{S}{s} = \dfrac{4\pi r^2}{\pi r} = 4r.$$

This distance is also the circumference of the circle of radius y.

$$d = 2\pi y$$

Thus, $2\pi y = 4r$ and we have $y = 2r/\pi$. Therefore, the centroid of the semicircle $y = \sqrt{r^2 - x^2}$ is $(0, 2r/\pi)$.

59. $A = \displaystyle\int_0^1 x^n\,dx = \left[\dfrac{x^{n+1}}{n+1}\right]_0^1 = \dfrac{1}{n+1}$

$m = \rho A = \dfrac{\rho}{n+1}$

$M_x = \dfrac{\rho}{2}\displaystyle\int_0^1 (x^n)^2\,dx = \left[\dfrac{\rho}{2}\cdot\dfrac{x^{2n+1}}{2n+1}\right]_0^1 = \dfrac{\rho}{2(2n+1)}$

$M_y = \rho\displaystyle\int_0^1 x(x^n)\,dx = \left[\rho\cdot\dfrac{x^{n+2}}{n+2}\right]_0^1 = \dfrac{\rho}{n+2}$

$\bar{x} = \dfrac{M_y}{m} = \dfrac{n+1}{n+2}$

$\bar{y} = \dfrac{M_x}{m} = \dfrac{n+1}{2(2n+1)} = \dfrac{n+1}{4n+2}$

Centroid: $\left(\dfrac{n+1}{n+2}, \dfrac{n+1}{4n+2}\right)$

As $n \to \infty$, $(\bar{x}, \bar{y}) \to \left(1, \dfrac{1}{4}\right)$.

The graph approaches the x-axis and the line $x = 1$ as $n \to \infty$.

Section 6.7   Fluid Pressure and Fluid Force

1. $F = PA = [62.4(5)](3) = 936 \text{ lb}$

3. $F = 62.4(h + 2)(6) - (62.4)(h)(6)$

$\qquad = 62.4(2)(6) = 748.8 \text{ lb}$

5. $h(y) = 3 - y$

$L(y) = 4$

$F = 62.4 \int_0^3 (3 - y)(4)\, dy$

$\quad = 249.6 \int_0^3 (3 - y)\, dy$

$\quad = 249.6 \left[3y - \dfrac{y^2}{2} \right]_0^3 = 1123.2 \text{ lb}$

7. $h(y) = 3 - y$

$L(y) = 2 \left(\dfrac{y}{3} + 1 \right)$

$F = 2(62.4) \int_0^3 (3 - y)\left(\dfrac{y}{3} + 1 \right) dy$

$\quad = 124.8 \int_0^3 \left(3 - \dfrac{y^2}{3} \right) dy$

$\quad = 124.8 \left[3y - \dfrac{y^3}{9} \right]_0^3 = 748.8 \text{ lb}$

9. $h(y) = 4 - y$

$L(y) = 2\sqrt{y}$

$F = 2(62.4) \int_0^4 (4 - y)\sqrt{y}\, dy$

$\quad = 124.8 \int_0^4 (4y^{1/2} - y^{3/2})\, dy$

$\quad = 124.8 \left[\dfrac{8y^{3/2}}{3} - \dfrac{2y^{5/2}}{5} \right]_0^4 = 1064.96 \text{ lb}$

11. $h(y) = 4 - y$

$L(y) = 2$

$F = 9800 \int_0^2 2(4 - y)\, dy$

$\quad = 9800 \left[8y - y^2 \right]_0^2 = 117,600 \text{ Newtons}$

13. $h(y) = 12 - y$

$L(y) = 6 - \dfrac{2y}{3}$

$F = 9800 \displaystyle\int_0^9 (12 - y)\left(6 - \dfrac{2y}{3}\right) dy$

$= 9800\left[72y - 7y^2 + \dfrac{2y^3}{9}\right]_0^9 = 2{,}381{,}400 \text{ Newtons}$

15. $h(y) = 2 - y$

$L(y) = 10$

$F = 140.7 \displaystyle\int_0^2 (2 - y)(10) \, dy$

$= 1407 \displaystyle\int_0^2 (2 - y) \, dy$

$= 1407\left[2y - \dfrac{y^2}{2}\right]_0^2 = 2814 \text{ lb}$

17. $h(y) = 4 - y$

$L(y) = 6$

$F = 140.7 \displaystyle\int_0^4 (4 - y)(6) \, dy$

$= 844.2 \displaystyle\int_0^4 (4 - y) \, dy$

$= 844.2\left[4y - \dfrac{y^2}{2}\right]_0^4 = 6753.6 \text{ lb}$

19. $h(y) = -y$

$L(y) = 2\left(\dfrac{1}{2}\right)\sqrt{9 - 4y^2}$

$F = 42 \displaystyle\int_{-3/2}^0 (-y)\sqrt{9 - 4y^2} \, dy$

$= \dfrac{42}{8} \displaystyle\int_{-3/2}^0 (9 - 4y^2)^{1/2}(-8y) \, dy$

$= \left[\left(\dfrac{21}{4}\right)\left(\dfrac{2}{3}\right)(9 - 4y^2)^{3/2}\right]_{-3/2}^0 = 94.5 \text{ lb}$

21. $h(y) = k - y$

$L(y) = 2\sqrt{r^2 - y^2}$

$F = w \displaystyle\int_{-r}^r (k - y)\sqrt{r^2 - y^2}\,(2) \, dy$

$= w\left[2k \displaystyle\int_{-r}^r \sqrt{r^2 - y^2} \, dy + \displaystyle\int_{-r}^r \sqrt{r^2 - y^2}\,(-2y) \, dy\right]$

The second integral is zero since its integrand is odd and the limits of integration are symmetric to the origin. The first integral is the area of a semicircle with radius r.

$F = w\left[(2k)\dfrac{\pi r^2}{2} + 0\right] = wk\pi r^2$

23. $h(y) = k - y$

$L(y) = b$

$$F = w \int_{-h/2}^{h/2} (k - y)b \, dy$$

$$= wb \left[ky - \frac{y^2}{2} \right]_{-h/2}^{h/2} = wb(hk) = wkhb$$

27. $h(y) = 4 - y$

$$F = 62.4 \int_0^4 (4 - y)L(y) \, dy$$

Using Simpson's Rule with $n = 8$ we have:

$$F \approx 62.4 \left(\frac{4 - 0}{3(8)} \right) [0 + 4(3.5)(3) + 2(3)(5) + 4(2.5)(8) + 2(2)(9) + 4(1.5)(10) + 2(1)(10.25) + 4(0.5)(10.5) + 0]$$

$$= 3010.8 \text{ lb}$$

29. $h(y) = 12 - y$

$L(y) = 2(4^{2/3} - y^{2/3})^{3/2}$

$$F = 62.4 \int_0^4 2(12 - y)(4^{2/3} - y^{2/3})^{3/2} \, dy$$

$$\approx 6448.73 \text{ lb}$$

25. From Exercise 23:

$$F = 64(15)(1)(1) = 960 \text{ lb}$$

31. (a) If the fluid force is one half of 1123.2 lb, and the height of the water is b, then

$$h(y) = b - y$$

$$L(y) = 4$$

$$F = 62.4 \int_0^b (b - y)(4) \, dy = \frac{1}{2}(1123.2)$$

$$\int_0^b (b - y) \, dy = 2.25$$

$$\left[by - \frac{y^2}{2} \right]_0^b = 2.25$$

$$b^2 - \frac{b^2}{2} = 2.25$$

$$b^2 = 4.5 \implies b \approx 2.12 \text{ ft.}$$

(b) The pressure increases with increasing depth.

33. $F = Fw = w \int_c^d h(y)L(y) \, dy$, see page 471.

Review Exercises for Chapter 6

1. $A = \int_1^5 \frac{1}{x^2}\, dx = \left[-\frac{1}{x} \right]_1^5 = \frac{4}{5}$

3. $A = \int_{-1}^1 \frac{1}{x^2 + 1}\, dx$

$= \left[\arctan x \right]_{-1}^1$

$= \frac{\pi}{4} - \left(-\frac{\pi}{4} \right) = \frac{\pi}{2}$

5. $A = 2 \int_0^1 (x - x^3)\, dx$

$= 2 \left[\frac{1}{2}x^2 - \frac{1}{4}x^4 \right]_0^1$

$= \frac{1}{2}$

7. $A = \int_0^2 (e^2 - e^x)\, dx$

$= \left[xe^2 - e^x \right]_0^2$

$= e^2 + 1$

9. $A = \int_{\pi/4}^{5\pi/4} (\sin x - \cos x)\, dx$

$= \left[-\cos x - \sin x \right]_{\pi/4}^{5\pi/4}$

$= \left(\frac{1}{\sqrt{2}} + \frac{1}{\sqrt{2}} \right) - \left(-\frac{1}{\sqrt{2}} - \frac{1}{\sqrt{2}} \right)$

$= \frac{4}{\sqrt{2}} = 2\sqrt{2}$

11. $A = \int_0^8 \left[(3 + 8x - x^2) - (x^2 - 8x + 3) \right] dx$

$= \int_0^8 (16x - 2x^2)\, dx$

$= \left[8x^2 - \frac{2}{3}x^3 \right]_0^8 = \frac{512}{3} \approx 170.667$

13. $y = \left(1 - \sqrt{x}\right)^2$

$$A = \int_0^1 \left(1 - \sqrt{x}\right)^2 dx$$

$$= \int_0^1 (1 - 2x^{1/2} + x)\, dx$$

$$= \left[x - \frac{4}{3}x^{3/2} + \frac{1}{2}x^2\right]_0^1 = \frac{1}{6} \approx 0.1667$$

15. $x = y^2 - 2y \Rightarrow x + 1 = (y - 1)^2 \Rightarrow y = 1 \pm \sqrt{x + 1}$

$$A = \int_{-1}^0 \left[\left(1 + \sqrt{x + 1}\right) - \left(1 - \sqrt{x + 1}\right)\right] dx = \int_{-1}^0 2\sqrt{x + 1}\, dx$$

$$A = \int_0^2 \left[0 - (y^2 - 2y)\right] dy = \int_0^2 (2y - y^2)\, dy = \left[y^2 - \frac{1}{3}y^3\right]_0^2 = \frac{4}{3}$$

17. $A = \int_0^2 \left[1 - \left(1 - \frac{x}{2}\right)\right] dx + \int_2^3 \left[1 - (x - 2)\right] dx$

$$= \int_0^2 \frac{x}{2}\, dx + \int_2^3 (3 - x)\, dx$$

$$y = 1 - \frac{x}{2} \Rightarrow x = 2 - 2y$$

$$y = x - 2 \Rightarrow x = y + 2,\ y = 1$$

$$A = \int_0^1 \left[(y + 2) - (2 - 2y)\right] dy$$

$$= \int_0^1 3y\, dy = \left[\frac{3}{2}y^2\right]_0^1 = \frac{3}{2}$$

19. Job 1 is better. The salary for Job 1 is greater than the salary for Job 2 for all the years except the first and 10th years.

21. (a) **Disk**

$$V = \pi \int_0^4 x^2\, dx = \left[\frac{\pi x^3}{3}\right]_0^4 = \frac{64\pi}{3}$$

(b) **Shell**

$$V = 2\pi \int_0^4 x^2\, dx = \left[\frac{2\pi}{3}x^3\right]_0^4 = \frac{128\pi}{3}$$

—CONTINUED—

21. —CONTINUED—

(c) **Shell**

$$V = 2\pi \int_0^4 (4 - x)x \, dx$$

$$= 2\pi \int_0^4 (4x - x^2) \, dx$$

$$= 2\pi \left[2x^2 - \frac{x^3}{3} \right]_0^4 = \frac{64\pi}{3}$$

(d) **Shell**

$$V = 2\pi \int_0^4 (6 - x)x \, dx$$

$$= 2\pi \int_0^4 (6x - x^2) \, dx$$

$$= 2\pi \left[3x^2 - \frac{1}{3}x^3 \right]_0^4 = \frac{160\pi}{3}$$

23. (a) **Shell**

$$V = 4\pi \int_0^4 x\left(\frac{3}{4}\right)\sqrt{16 - x^2} \, dx$$

$$= \left[3\pi\left(-\frac{1}{2}\right)\left(\frac{2}{3}\right)(16 - x^2)^{3/2} \right]_0^4 = 64\pi$$

(b) **Disk**

$$V = 2\pi \int_0^4 \left[\frac{3}{4}\sqrt{16 - x^2} \right]^2 dx$$

$$= \frac{9\pi}{8} \left[16x - \frac{x^3}{3} \right]_0^4 = 48\pi$$

25. Shell

$$V = 2\pi \int_0^1 \frac{x}{x^4 + 1} \, dx$$

$$= \pi \int_0^1 \frac{(2x)}{(x^2)^2 + 1} \, dx$$

$$= \left[\pi \arctan(x^2) \right]_0^1$$

$$= \pi \left[\frac{\pi}{4} - 0 \right] = \frac{\pi^2}{4}$$

27. Shell

$$u = \sqrt{x-2}$$

$$x = u^2 + 2$$

$$dx = 2u\,du$$

$$V = 2\pi \int_2^6 \frac{x}{1+\sqrt{x-2}}\,dx = 4\pi \int_0^2 \frac{(u^2+2)u}{1+u}\,du$$

$$= 4\pi \int_0^2 \frac{u^3+2u}{1+u}\,du = 4\pi \int_0^2 \left(u^2 - u + 3 - \frac{3}{1+u}\right)du$$

$$= 4\pi \left[\frac{1}{3}u^3 - \frac{1}{2}u^2 + 3u - 3\ln(1+u)\right]_0^2 = \frac{4\pi}{3}(20 - 9\ln 3) \approx 42.359$$

29. Since $y \le 0$, $A = -\displaystyle\int_{-1}^0 x\sqrt{x+1}\,dx.$

$$u = x + 1$$

$$x = u - 1$$

$$dx = du$$

$$A = -\int_0^1 (u-1)\sqrt{u}\,du = -\int_0^1 (u^{3/2} - u^{1/2})\,du$$

$$= -\left[\frac{2}{5}u^{5/2} - \frac{2}{3}u^{3/2}\right]_0^1 = \frac{4}{15}$$

31. From Exercise 23(a) we have: $V = 64\pi$ ft^3

$$\frac{1}{4}V = 16\pi$$

Disk: $\pi \displaystyle\int_{-3}^{y_0} \frac{16}{9}(9 - y^2)\,dy = 16\pi$

$$\frac{1}{9}\int_{-3}^{y_0} (9 - y^2)\,dy = 1$$

$$\left[9y - \frac{1}{3}y^3\right]_{-3}^{y_0} = 9$$

$$\left(9y_0 - \frac{1}{3}y_0^3\right) - (-27 + 9) = 9$$

$$y_0^3 - 27y_0 - 27 = 0$$

By Newton's Method, $y_0 \approx -1.042$ and the depth of the

gasoline is $3 - 1.042 = 1.958$ ft.

33. $f(x) = \dfrac{4}{5}x^{5/4}$

$$f'(x) = x^{1/4}$$

$$1 + [f'(x)]^2 = 1 + \sqrt{x}$$

$$u = 1 + \sqrt{x}$$

$$x = (u-1)^2$$

$$dx = 2(u-1)\,du$$

$$s = \int_0^4 \sqrt{1+\sqrt{x}}\,dx = 2\int_1^3 \sqrt{u}(u-1)\,du$$

$$= 2\int_1^3 (u^{3/2} - u^{1/2})\,du$$

$$= 2\left[\frac{2}{5}u^{5/2} - \frac{2}{3}u^{3/2}\right]_1^3 = \frac{4}{15}\left[u^{3/2}(3u-5)\right]_1^3$$

$$= \frac{8}{15}\left(1 + 6\sqrt{3}\right) \approx 6.076$$

35. $y = 300\cosh\left(\dfrac{x}{2000}\right) - 280,\ -2000 \le x \le 2000$

$$y' = \frac{3}{20}\sinh\left(\frac{x}{2000}\right)$$

$$s = \int_{-2000}^{2000} \sqrt{1 + \left[\frac{3}{20}\sinh\left(\frac{x}{2000}\right)\right]^2}\,dx$$

$$= \frac{1}{20}\int_{-2000}^{2000} \sqrt{400 + 9\sinh^2\left(\frac{x}{2000}\right)}\,dx$$

$$\approx 4018.2 \text{ ft (by Simpson's Rule or graphing utility)}$$

37. $y = \dfrac{3}{4}x$

$$y' = \dfrac{3}{4}$$

$$1 + (y')^2 = \dfrac{25}{16}$$

$$S = 2\pi \int_0^4 \left(\dfrac{3}{4}x\right) \sqrt{\dfrac{25}{16}}\, dx = \left[\left(\dfrac{15\pi}{8}\right)\dfrac{x^2}{2}\right]_0^4 = 15\pi$$

(4, 3)

39. $F = kx$

$$4 = k(1)$$

$$F = 4x$$

$$W = \int_0^5 4x\, dx = \left[2x^2\right]_0^5$$

$$= 50 \text{ in} \cdot \text{lb} \approx 4.167 \text{ ft} \cdot \text{lb}$$

41. Volume of disk: $\pi\left(\dfrac{1}{3}\right)^2 \Delta y$

Weight of disk: $62.4\pi\left(\dfrac{1}{3}\right)^2 \Delta y$

Distance: $175 - y$

$$W = \dfrac{62.4\pi}{9}\int_0^{150}(175 - y)\, dy = \dfrac{62.4\pi}{9}\left[175y - \dfrac{y^2}{2}\right]_0^{150}$$

$$= 104{,}000\pi \text{ ft} \cdot \text{lb} \approx 163.4 \text{ ft} \cdot \text{ton}$$

43. Weight of section of chain: $5\,\Delta x$

Distance moved: $10 - x$

$$W = 5\int_0^{10}(10 - x)\, dx = \left[-\dfrac{5}{2}(10 - x)^2\right]_0^{10}$$

$$= 250 \text{ ft} \cdot \text{lb}$$

45. $W = \displaystyle\int_a^b F(x)\, dx$

$$80 = \int_0^4 ax^2 dx = \dfrac{ax^3}{3}\bigg]_0^4 = \dfrac{64}{3}a$$

$$a = \dfrac{3(80)}{64} = \dfrac{15}{4} = 3.75$$

47. $A = \displaystyle\int_0^a \left(\sqrt{a} - \sqrt{x}\right)^2 dx = \int_0^a \left(a - 2\sqrt{a}\,x^{1/2} + x\right) dx = \left[ax - \dfrac{4}{3}\sqrt{a}\,x^{3/2} + \dfrac{1}{2}x^2\right]_0^a = \dfrac{a^2}{6}$

$$\dfrac{1}{A} = \dfrac{6}{a^2}$$

$$\bar{x} = \dfrac{6}{a^2}\int_0^a x\left(\sqrt{a} - \sqrt{x}\right)^2 dx = \dfrac{6}{a^2}\int_0^a \left(ax - 2\sqrt{a}\,x^{3/2} + x^2\right) dx = \dfrac{a}{5}$$

$$\bar{y} = \left(\dfrac{6}{a^2}\right)\dfrac{1}{2}\int_0^a \left(\sqrt{a} - \sqrt{x}\right)^4 dx$$

$$= \dfrac{3}{a^2}\int_0^a \left(a^2 - 4a^{3/2}x^{1/2} + 6ax - 4a^{1/2}x^{3/2} + x^2\right) dx$$

$$= \dfrac{3}{a^2}\left[a^2 x - \dfrac{8}{3}a^{3/2}x^{3/2} + 3ax^2 - \dfrac{8}{5}a^{1/2}x^{5/2} + \dfrac{1}{3}x^3\right]_0^a = \dfrac{a}{5}$$

$$(\bar{x}, \bar{y}) = \left(\dfrac{a}{5}, \dfrac{a}{5}\right)$$

$(\bar{x}, \bar{y})$

49. By symmetry, $x = 0$.

$$A = 2 \int_0^1 (a^2 - x^2) \, dx = 2\left[a^2x - \frac{x^3}{3} \right]_0^a = \frac{4a^3}{3}$$

$$\frac{1}{A} = \frac{3}{4a^3}$$

$$\bar{y} = \left(\frac{3}{4a^3} \right) \frac{1}{2} \int_{-a}^a (a^2 - x^2)^2 \, dx$$

$$= \frac{6}{8a^3} \int_0^a (a^4 - 2a^2x^2 + x^4) \, dx$$

$$= \frac{6}{8a^3} \left[a^4x - \frac{2a^2}{3}x^3 + \frac{1}{5}x^5 \right]_0^a$$

$$= \frac{6}{8a^3} \left(a^5 - \frac{2}{3}a^5 + \frac{1}{5}a^5 \right) = \frac{2a^2}{5}$$

$$(\bar{x}, \bar{y}) = \left(0, \frac{2a^2}{5} \right)$$

51. $\bar{y} = 0$ by symmetry

For the trapezoid:

$$m = [(4)(6) - (1)(6)]\rho = 18\rho$$

$$M_y = \rho \int_0^6 x\left[\left(\frac{1}{6}x + 1 \right) - \left(-\frac{1}{6}x - 1 \right) \right] dx$$

$$= \rho \int_0^6 \left(\frac{1}{3}x^2 + 2x \right) dx = \rho\left[\frac{x^3}{9} + x^2 \right]_0^6 = 60\rho$$

For the semicircle:

$$m = \left(\frac{1}{2} \right)(\pi)(2)^2\rho = 2\pi\rho$$

$$M_y = \rho \int_6^8 x\left[\sqrt{4 - (x-6)^2} - \left(-\sqrt{4 - (x-6)^2} \right) \right] dx = 2\rho \int_6^8 x\sqrt{4 - (x-6)^2} \, dx$$

Let $u = x - 6$, then $x = u + 6$ and $dx = du$. When $x = 6$, $u = 0$. When $x = 8$, $u = 2$.

$$M_y = 2\rho \int_0^2 (u + 6)\sqrt{4 - u^2} \, du = 2\rho \int_0^2 u\sqrt{4 - u^2} \, du + 12\rho \int_0^2 \sqrt{4 - u^2} \, du$$

$$= 2\rho\left[\left(-\frac{1}{2} \right)\left(\frac{2}{3} \right)(4 - u^2)^{3/2} \right]_0^2 + 12\rho\left[\frac{\pi(2)^2}{4} \right] = \frac{16\rho}{3} + 12\pi\rho = \frac{4\rho(4 + 9\pi)}{3}$$

Thus, we have:

$$\bar{x}(18\rho + 2\pi\rho) = 60\rho + \frac{4\rho(4 + 9\pi)}{3}$$

$$\bar{x} = \frac{180\rho + 4\rho(4 + 9\pi)}{3} \cdot \frac{1}{2\rho(9 + \pi)} = \frac{2(9\pi + 49)}{3(\pi + 9)}$$

The centroid of the blade is $\left(\dfrac{2(9\pi + 49)}{3(\pi + 9)}, 0 \right)$.

53. Let D = surface of liquid; ρ = weight per cubic volume.

$$F = \rho \int_c^d (D - y)[f(y) - g(y)]\, dy$$

$$= \rho\left[\int_c^d D[f(y) - g(y)]\, dy - \int_c^d y[f(y) - g(y)]\, dy\right]$$

$$= \rho\left[\int_c^d [f(y) - g(y)]\, dy\right]\left[D - \frac{\int_c^d y[f(y) - g(y)]\, dy}{\int_c^d [f(y) - g(y)]\, dy}\right]$$

$$= \rho(\text{Area})(D - \bar{y})$$

$$= \rho(\text{Area})(\text{depth of centroid})$$

Problem Solving for Chapter 6

1. $T = \dfrac{1}{2}c(c^2) = \dfrac{1}{2}c^3$

$$R = \int_0^c (cx - x^2)\, dx = \left[\frac{cx^2}{2} - \frac{x^3}{3}\right]_0^c = \frac{c^3}{2} - \frac{c^3}{3} = \frac{c^3}{6}$$

$$\lim_{c \to 0^+} \frac{T}{R} = \lim_{c \to 0^+} \frac{\frac{1}{2}c^3}{\frac{1}{6}c^3} = 3$$

3. (a) $\dfrac{1}{2}V = \displaystyle\int_0^1 \left[\pi\left(2 + \sqrt{1 - y^2}\right)^2 - \pi\left(2 - \sqrt{1 - y^2}\right)^2\right] dy$

$$= \pi\int_0^1 \left[\left(4 + 4\sqrt{1 - y^2} + (1 - y^2)\right) - \left(4 - 4\sqrt{1 - y^2} + (1 - y^2)\right)\right] dy$$

$$= 8\pi\int_0^1 \sqrt{1 - y^2}\, dy \quad \text{(Integral represents 1/4 (area of circle))}$$

$$= 8\pi\left(\frac{\pi}{4}\right) = 2\pi^2 \implies V = 4\pi^2$$

(b) $(x - R)^2 + y^2 = r^2 \implies x = R \pm \sqrt{r^2 - y^2}$

$$\frac{1}{2}V = \int_0^r \left[\pi\left(R + \sqrt{r^2 - y^2}\right)^2 - \pi\left(R - \sqrt{r^2 - y^2}\right)^2\right] dy$$

$$= \pi\int_0^r 4R\sqrt{r^2 - y^2}\, dy$$

$$= \pi(4R)\frac{1}{4}\pi r^2 = \pi^2 r^2 R$$

$$V = 2\pi^2 r^2 R$$

5. $V = 2(2\pi)\displaystyle\int_{\sqrt{r^2 - (h^2/4)}}^r x\sqrt{r^2 - x^2}\, dx$

$$= -2\pi\left[\frac{2}{3}(r^2 - x^2)^{3/2}\right]_{\sqrt{r^2 - (h^2/4)}}^r$$

$$= \frac{-4\pi}{3}\left[-\frac{h^3}{8}\right] = \frac{\pi h^3}{6} \text{ which does not depend on } r!$$

7. (a) Tangent at A: $y = x^3$, $y' = 3x^2$

$$y - 1 = 3(x - 1)$$
$$y = 3x - 2$$

To find point B: $\quad x^3 = 3x - 2$

$$x^3 - 3x + 2 = 0$$
$$(x - 1)^2(x + 2) = 0 \implies B = (-2, -8)$$

Tangent at B: $\quad y = x^3$, $y' = 3x^2$

$$y + 8 = 12(x + 2)$$
$$y = 12x + 16$$

To find point C: $\quad x^3 = 12x + 16$

$$x^3 - 12x - 16 = 0$$
$$(x + 2)^2(x - 4) = 0 \implies C = (4, 64)$$

Area of $R = \displaystyle\int_{-2}^{1} (x^3 - 3x + 2)\, dx = \dfrac{27}{4}$

Area of $S = \displaystyle\int_{-2}^{4} (12x + 16 - x^3)\, dx = 108$

Area of $S = 16(\text{area of } R)$ $\quad \left[\dfrac{\text{area } S}{\text{area } R} = 16 \right]$

(b) Tangent at $A(a, a^3)$: $y - a^3 = 3a^2(x - a)$

$$y = 3a^2x - 2a^3$$

To find point B: $\quad x^3 - 3a^2x + 2a^3 = 0$

$$(x - a)^2(x + 2a) = 0 \implies$$
$$B = (-2a, -8a^3)$$

Tangent at B: $\quad y + 8a^3 = 12a^2(x + 2a)$

$$y = 12a^2x + 16a^3$$

To find point C: $\quad x^3 - 12a^2x - 16a^3 = 0$

$$(x + 2a)^2(x - 4a) = 0 \implies$$
$$C = (4a, 64a^3)$$

Area of $R = \displaystyle\int_{-2a}^{a} [x^3 - 3a^2x + 2a^3]\, dx = \dfrac{27}{4}a^4$

Area of $S = \displaystyle\int_{-2a}^{4a} [12a^2x + 16a^3 - x^3]\, dx = 108a^4$

Area of $S = 16(\text{area of } R)$

9. $s(x) = \displaystyle\int_{\alpha}^{x} \sqrt{1 + f'(t)^2}\, dt$

(a) $s'(x) = \dfrac{ds}{dx} = \sqrt{1 + f'(x)^2}$

(b) $ds = \sqrt{1 + f'(x)^2}\, dx$

$$(ds)^2 = [1 + f'(x)^2](dx)^2 = \left[1 + \left(\dfrac{dy}{dx} \right)^2 \right](dx)^2 = (dx)^2 + (dy)^2$$

(c) $s(x) = \displaystyle\int_{1}^{x} \sqrt{1 + \left(\dfrac{3}{2}t^{1/2} \right)^2}\, dt = \int_{1}^{x} \sqrt{1 + \dfrac{9}{4}t}\, dt$

(d) $s(2) = \displaystyle\int_{1}^{2} \sqrt{1 + \dfrac{9}{4}t}\, dt = \left[\dfrac{8}{27}\left(1 + \dfrac{9}{4}t \right)^{3/2} \right]_{1}^{2} = \dfrac{22}{27}\sqrt{22} - \dfrac{13}{27}\sqrt{13} \approx 2.0858$

This is the length of the curve $y = x^{3/2}$ from $x = 1$ to $x = 2$.

11. (a) $\bar{y} = 0$ by symmetry

$$M_y = \int_{1}^{6} x\left(\dfrac{1}{x^3} - \left(-\dfrac{1}{x^3} \right) \right) dx = \int_{1}^{6} \dfrac{2}{x^2}\, dx = \left[-2\dfrac{1}{x} \right]_{1}^{6} = \dfrac{5}{3}$$

$$m = 2\int_{1}^{6} \dfrac{1}{x^3}\, dx = \left[-\dfrac{1}{x^2} \right]_{1}^{6} = \dfrac{35}{36}$$

$$\bar{x} = \dfrac{5/3}{35/36} = \dfrac{12}{7} \qquad (\bar{x}, \bar{y}) = \left(\dfrac{12}{7}, 0 \right)$$

(b) $m = 2\displaystyle\int_{1}^{b} \dfrac{1}{x^3}\, dx = \dfrac{b^2 - 1}{b^2}$

$M_y = 2\displaystyle\int_{1}^{6} \dfrac{1}{x^2}\, dx = \dfrac{2(b - 1)}{b}$

$\bar{x} = \dfrac{2(b - 1)/b}{(b^2 - 1)/b^2} = \dfrac{2b}{b + 1} \qquad (\bar{x}, \bar{y}) = \left(\dfrac{2b}{b + 1}, 0 \right)$

(c) $\displaystyle\lim_{b \to \infty} \bar{x} = \lim_{b \to \infty} \dfrac{2b}{b + 1} = 2 \qquad (\bar{x}, \bar{y}) = (2, 0)$

13. (a) $W = \text{area} = 2 + 4 + 6 = 12$

(b) $W = \text{area} = 3 + (1 + 1) + 2 + \dfrac{1}{2} = 7\dfrac{1}{2}$

15. Point of equilibrium: $50 - 0.5x = 0.125x$

$$x = 80, p = 10$$

$(P_0, x_0) = (10, 80)$

Consumer surplus $= \displaystyle\int_0^{80} [(50 - 0.5x) - 10]\, dx = 1600$

Producer surplus $= \displaystyle\int_0^{80} [10 - 0.125x]\, dx = 400$

17. We use Exercise 23 in Section 6.7:

(a) Wall at shallow end

From Exercise 23: $F = 62.4(2)(4)(20) = 9984\,\text{lb}$

(b) Wall at deep end

From Exercise 23: $F = 62.4(4)(8)(20) = 39{,}936\,\text{lb}$

(c) Side wall

From Exercise 23: $F_1 = 62.4(2)(4)(40) = 19{,}968\,\text{lb}$

$$F_2 = 62.4 \int_0^4 (8 - y)(10y)\, dy$$

$$= 624 \int_0^4 (8y - y^2)\, dy = 624\left[4y^2 - \frac{y^3}{3}\right]_0^4$$

$$= 26{,}624\,\text{lb}$$

Total force: $F_1 + F_2 = 46{,}592\,\text{lb}$

C H A P T E R   7
Integration Techniques, L'Hôpital's Rule, and Improper Integrals

Section 7.1   Basic Integration Rules **330**

Section 7.2   Integration by Parts **335**

Section 7.3   Trigonometric Integrals **345**

Section 7.4   Trigonometric Substitution **354**

Section 7.5   Partial Fractions **364**

Section 7.6   Integration by Tables and Other Integration Techniques . . **370**

Section 7.7   Indeterminate Forms and L'Hôpital's Rule **376**

Section 7.8   Improper Integrals **382**

Review Exercises **388**

Problem Solving **394**

CHAPTER 7
Integration Techniques, L'Hôpital's Rule, and Improper Integrals

Section 7.1   Basic Integration Rules

Solutions to Odd-Numbered Exercises

1. (a) $\dfrac{d}{dx}\left[2\sqrt{x^2+1}+C\right]=2\left(\dfrac{1}{2}\right)(x^2+1)^{-1/2}(2x)=\dfrac{2x}{\sqrt{x^2+1}}$

  (b) $\dfrac{d}{dx}\left[\sqrt{x^2+1}+C\right]=\dfrac{1}{2}(x^2+1)^{-1/2}(2x)=\dfrac{x}{\sqrt{x^2+1}}$

  (c) $\dfrac{d}{dx}\left[\dfrac{1}{2}\sqrt{x^2+1}+C\right]=\dfrac{1}{2}\left(\dfrac{1}{2}\right)(x^2+1)^{-1/2}(2x)=\dfrac{x}{2\sqrt{x^2+1}}$

  (d) $\dfrac{d}{dx}\left[\ln(x^2+1)+C\right]=\dfrac{2x}{x^2+1}$

  $\displaystyle\int\dfrac{x}{\sqrt{x^2+1}}\,dx$ matches (b).

3. (a) $\dfrac{d}{dx}\left[\ln\sqrt{x^2+1}+C\right]=\dfrac{1}{2}\left(\dfrac{2x}{x^2+1}\right)=\dfrac{x}{x^2+1}$

  (b) $\dfrac{d}{dx}\left[\dfrac{2x}{(x^2+1)^2}+C\right]=\dfrac{(x^2+1)^2(2)-(2x)(2)(x^2+1)(2x)}{(x^2+1)^4}=\dfrac{2(1-3x^2)}{(x^2+1)^3}$

  (c) $\dfrac{d}{dx}\left[\arctan x+C\right]=\dfrac{1}{1+x^2}$

  (d) $\dfrac{d}{dx}\left[\ln(x^2+1)+C\right]=\dfrac{2x}{x^2+1}$

  $\displaystyle\int\dfrac{1}{x^2+1}\,dx$ matches (c).

5. $\displaystyle\int(3x-2)^4\,dx$

$u=3x-2,\,du=3\,dx,\,n=4$

Use $\displaystyle\int u^n\,du.$

7. $\displaystyle\int\dfrac{1}{\sqrt{x}(1-2\sqrt{x})}\,dx$

$u=1-2\sqrt{x},\,du=-\dfrac{1}{\sqrt{x}}\,dx$

Use $\displaystyle\int\dfrac{du}{u}.$

9. $\displaystyle\int\dfrac{3}{\sqrt{1-t^2}}\,dt$

$u=t,\,du=dt,\,a=1$

Use $\displaystyle\int\dfrac{du}{\sqrt{a^2-u^2}}$

11. $\displaystyle\int t\sin t^2\,dt$

$u=t^2,\,du=2t\,dt$

Use $\displaystyle\int\sin u\,du.$

13. $\displaystyle\int(\cos x)e^{\sin x}\,dx$

$u=\sin x,\,du=\cos x\,dx$

Use $\displaystyle\int e^u\,du.$

15. Let $u = -2x + 5$, $du = -2\,dx$.

$$\int (-2x + 5)^{3/2}\,dx = -\frac{1}{2}\int (-2x + 5)^{3/2}(-2)\,dx$$

$$= -\frac{1}{5}(-2x + 5)^{5/2} + C$$

17. Let $u = z - 4$, $du = dz$

$$\int \frac{5}{(z-4)^5}\,dz = 5\int (z-4)^{-5}\,dz = 5\frac{(z-4)^{-4}}{-4} + C$$

$$= \frac{-5}{4(z-4)^4} + C$$

19. Let $u = t^3 - 1$, $du = 3t^2\,dt$.

$$\int t^2\sqrt[3]{t^3 - 1}\,dt = \frac{1}{3}\int (t^3 - 1)^{1/3}(3t^2)\,dt$$

$$= \frac{1}{3}\frac{(t^3 - 1)^{4/3}}{4/3} + C$$

$$= \frac{(t^3 - 1)^{4/3}}{4} + C$$

21. $\displaystyle\int \left[v + \frac{1}{(3v - 1)^3} \right] dv = \int v\,dv + \frac{1}{3}\int (3v - 1)^{-3}(3)\,dv$

$$= \frac{1}{2}v^2 - \frac{1}{6(3v - 1)^2} + C$$

23. Let $u = -t^3 + 9t + 1$, $du = (-3t^2 + 9)\,dt = -3(t^2 - 3)\,dt$.

$$\int \frac{t^2 - 3}{-t^3 + 9t + 1}\,dt = -\frac{1}{3}\int \frac{-3(t^2 - 3)}{-t^3 + 9t + 1}\,dt = -\frac{1}{3}\ln|-t^3 + 9t + 1| + C$$

25. $\displaystyle\int \frac{x^2}{x - 1}\,dx = \int (x + 1)\,dx + \int \frac{1}{x - 1}\,dx$

$$= \frac{1}{2}x^2 + x + \ln|x - 1| + C$$

27. Let $u = 1 + e^x$, $du = e^x\,dx$.

$$\int \frac{e^x}{1 + e^x}\,dx = \ln(1 + e^x) + C$$

29. $\displaystyle\int (1 + 2x^2)^2\,dx = \int (4x^4 + 4x^2 + 1)\,dx = \frac{4}{5}x^5 + \frac{4}{3}x^3 + x + C = \frac{x}{15}(12x^4 + 20x^2 + 15) + C$

31. Let $u = 2\pi x^2$, $du = 4\pi x\,dx$.

$$\int x(\cos 2\pi x^2)\,dx = \frac{1}{4\pi}\int (\cos 2\pi x^2)(4\pi x)\,dx$$

$$= \frac{1}{4\pi}\sin 2\pi x^2 + C$$

33. Let $u = \pi x$, $du = \pi\,dx$.

$$\int \csc(\pi x)\cot(\pi x)\,dx = \frac{1}{\pi}\int \csc(\pi x)\cot(\pi x)\pi\,dx = -\frac{1}{\pi}\csc(\pi x) + C$$

35. Let $u = 5x$, $du = 5\,dx$.

$$\int e^{5x}\,dx = \frac{1}{5}\int e^{5x}(5)\,dx = \frac{1}{5}e^{5x} + C$$

37. Let $u = 1 + e^x$, $du = e^x\,dx$.

$$\int \frac{2}{e^{-x} + 1}\,dx = 2\int \left(\frac{1}{e^{-x} + 1}\right)\left(\frac{e^x}{e^x}\right)\,dx$$

$$= 2\int \frac{e^x}{1 + e^x}\,dx = 2\ln(1 + e^x) + C$$

39. $\displaystyle\int \frac{\ln x^2}{x}\,dx = 2\int (\ln x)\frac{1}{x}\,dx = 2\,\frac{(\ln x)^2}{2} + C = (\ln x)^2 + C$

41. $\displaystyle\int \frac{1 + \sin x}{\cos x}\,dx = \int (\sec x + \tan x)\,dx = \ln|\sec x + \tan x| + \ln|\sec x| + C = \ln|\sec x(\sec x + \tan x)| + C$

43.
$$\frac{1}{\cos\theta - 1} = \frac{1}{\cos\theta - 1}\cdot\frac{\cos\theta + 1}{\cos\theta + 1} = \frac{\cos\theta + 1}{\cos^2\theta - 1} = \frac{\cos\theta + 1}{-\sin^2\theta}$$

$$= -\csc\theta\cdot\cot\theta - \csc^2\theta$$

$$\int \frac{1}{\cos\theta - 1}\,d\theta = \int(-\csc\theta\cot\theta - \csc^2\theta)\,d\theta$$

$$= \csc\theta + \cot\theta + C$$

$$= \frac{1}{\sin\theta} + \frac{\cos\theta}{\sin\theta} + C$$

$$= \frac{1 + \cos\theta}{\sin\theta} + C$$

45.
$$\int \frac{3z + 2}{z^2 + 9}\,dz = \frac{3}{2}\int \frac{2z}{z^2 + 9}\,dz + 2\int \frac{dz}{z^2 + 9}$$

$$= \frac{3}{2}\ln(z^2 + 9) + \frac{2}{3}\arctan\!\left(\frac{z}{3}\right) + C$$

47. Let $u = 2t - 1$, $du = 2\,dt$.
$$\int \frac{-1}{\sqrt{1 - (2t - 1)^2}}\,dt = -\frac{1}{2}\int \frac{2}{\sqrt{1 - (2t - 1)^2}}\,dt$$

$$= -\frac{1}{2}\arcsin(2t - 1) + C$$

49. Let $u = \cos\!\left(\dfrac{2}{t}\right)$, $du = \dfrac{2\sin(2/t)}{t^2}\,dt$.

$$\int \frac{\tan(2/t)}{t^2}\,dt = \frac{1}{2}\int \frac{1}{\cos(2/t)}\left[\frac{2\sin(2/t)}{t^2}\right]dt$$

$$= \frac{1}{2}\ln\left|\cos\!\left(\frac{2}{t}\right)\right| + C$$

51. $\displaystyle\int \frac{3}{\sqrt{6x - x^2}}\,dx = 3\int \frac{1}{\sqrt{9 - (x - 3)^2}}\,dx = 3\arcsin\!\left(\frac{x - 3}{3}\right) + C$

53. $\displaystyle\int \frac{4}{4x^2 + 4x + 65}\,dx = \int \frac{1}{[x + (1/2)]^2 + 16}\,dx = \frac{1}{4}\arctan\!\left[\frac{x + (1/2)}{4}\right] + C = \frac{1}{4}\arctan\!\left(\frac{2x + 1}{8}\right) + C$

55. $\dfrac{ds}{dt} = \dfrac{t}{\sqrt{1 - t^4}}$, $\left(0, -\dfrac{1}{2}\right)$

(a)

(b) $u = t^2$, $du = 2t\,dt$

$$\int \frac{t}{\sqrt{1 - t^4}}\,dt = \frac{1}{2}\int \frac{2t}{\sqrt{1 - (t^2)^2}}\,dt = \frac{1}{2}\arcsin t^2 + C$$

$$\left(0, -\frac{1}{2}\right)\!: \quad -\frac{1}{2} = \frac{1}{2}\arcsin 0 + C \implies C = -\frac{1}{2}$$

$$s = \frac{1}{2}\arcsin t^2 - \frac{1}{2}$$

57.

$y = 3e^{0.2x}$

59. $y = \displaystyle\int (1 + e^x)^2 \, dx = \int (e^{2x} + 2e^x + 1) \, dx$

$\qquad = \dfrac{1}{2}e^{2x} + 2e^x + x + C$

61. $\dfrac{dy}{dx} = \dfrac{\sec^2 x}{4 + \tan^2 x}$

Let $u = \tan x,\ du = \sec^2 x \, dx$.

$y = \displaystyle\int \dfrac{\sec^2 x}{4 + \tan^2 x} \, dx = \dfrac{1}{2} \arctan\!\left(\dfrac{\tan x}{2}\right) + C$

63. Let $u = 2x,\ du = 2 \, dx$.

$\displaystyle\int_0^{\pi/4} \cos 2x \, dx = \dfrac{1}{2}\int_0^{\pi/4} \cos 2x (2) \, dx$

$\qquad = \left[\dfrac{1}{2}\sin 2x\right]_0^{\pi/4} = \dfrac{1}{2}$

65. Let $u = -x^2,\ du = -2x \, dx$.

$\displaystyle\int_0^1 xe^{-x^2} dx = -\dfrac{1}{2}\int_0^1 e^{-x^2}(-2x)\, dx = \left[-\dfrac{1}{2}e^{-x^2}\right]_0^1$

$\qquad = \dfrac{1}{2}(1 - e^{-1}) \approx 0.316$

67. Let $u = x^2 + 9,\ du = 2x \, dx$.

$\displaystyle\int_0^4 \dfrac{2x}{\sqrt{x^2 + 9}}\, dx = \int_0^4 (x^2 + 9)^{-1/2}(2x)\, dx$

$\qquad = \left[2\sqrt{x^2 + 9}\right]_0^4 = 4$

69. Let $u = 3x,\ du = 3 \, dx$.

$\displaystyle\int_0^{2/\sqrt{3}} \dfrac{1}{4 + 9x^2}\, dx = \dfrac{1}{3}\int_0^{2/\sqrt{3}} \dfrac{3}{4 + (3x)^2}\, dx$

$\qquad = \left[\dfrac{1}{6}\arctan\!\left(\dfrac{3x}{2}\right)\right]_0^{2/\sqrt{3}}$

$\qquad = \dfrac{\pi}{18} \approx 0.175$

71. $\displaystyle\int \dfrac{1}{x^2 + 4x + 13}\, dx = \dfrac{1}{3}\arctan\!\left(\dfrac{x + 2}{3}\right) + C$

The antiderivatives are vertical translations of each other.

73. $\displaystyle\int \dfrac{1}{1 + \sin\theta}\, d\theta = \tan\theta - \sec\theta + C \left(\text{or } \dfrac{-2}{1 + \tan(\theta/2)}\right)$

The antiderivatives are vertical translations of each other.

75. Power Rule: $\displaystyle\int u^n \, du = \dfrac{u^{n+1}}{n + 1} + C,\ n \neq -1$.

$\qquad u = x^2 + 1,\ n = 3$

77. Log Rule: $\displaystyle\int \dfrac{du}{u} = \ln|u| + C,\ u = x^2 + 1$.

79. The are equivalent because

$\qquad e^{x + C_1} = e^x \cdot e^{C_1} = Ce^x,\ C = e^{C_1}$

81. $\sin x + \cos x = a \sin(x + b)$

$\sin x + \cos x = a \sin x \cos b + a \cos x \sin b$

$\sin x + \cos x = (a \cos b) \sin x + (a \sin b) \cos x$

Equate coefficients of like terms to obtain the following.

$\quad 1 = a \cos b \quad$ and $\quad 1 = a \sin b$

Thus, $a = 1/\cos b$. Now, substitute for a in $1 = a \sin b$.

$$1 = \left(\frac{1}{\cos b}\right) \sin b$$

$$1 = \tan b \implies b = \frac{\pi}{4}$$

Since $b = \frac{\pi}{4}$, $a = \dfrac{1}{\cos(\pi/4)} = \sqrt{2}$. Thus, $\sin x + \cos x = \sqrt{2} \sin\left(x + \dfrac{\pi}{4}\right)$.

$$\int \frac{dx}{\sin x + \cos x} = \int \frac{dx}{\sqrt{2}\,\sin(x + (\pi/4))} = \frac{1}{\sqrt{2}} \int \csc\left(x + \frac{\pi}{4}\right) dx = -\frac{1}{\sqrt{2}} \ln\left|\csc\left(x + \frac{\pi}{4}\right) + \cot\left(x + \frac{\pi}{4}\right)\right| + C$$

83. $\displaystyle\int_0^2 \frac{4x}{x^2 + 1}\, dx \approx 3$

Matches (a).

85. Let $u = 1 - x^2$, $du = -2x\, dx$.

$$A = 4\int_0^1 x\sqrt{1 - x^2}\, dx$$

$$= -2\int_0^1 (1 - x^2)^{1/2}(-2x)\, dx$$

$$= \left[-\frac{4}{3}(1 - x^2)^{3/2}\right]_0^1 = \frac{4}{3}$$

87. $\displaystyle\int_0^{1/a} (x - ax^2)\, dx = \left[\frac{1}{2}x^2 - \frac{a}{3}x^3\right]_0^{1/a}$

$$= \frac{1}{6a^2}$$

Let $\dfrac{1}{6a^2} = \dfrac{2}{3}$, $12a^2 = 3$, $a = \dfrac{1}{2}$.

89. **(a) Shell Method:**

Let $u = -x^2$, $du = -2x\, dx$.

$$V = 2\pi\int_0^1 xe^{-x^2}\, dx$$

$$= -\pi\int_0^1 e^{-x^2}(-2x)\, dx$$

$$= \left[-\pi e^{-x^2}\right]_0^1$$

$$= \pi(1 - e^{-1}) \approx 1.986$$

(b) Shell Method:

$$V = 2\pi\int_0^b xe^{-x^2}\, dx$$

$$= \left[-\pi e^{-x^2}\right]_0^b$$

$$= \pi(1 - e^{-b^2}) = \frac{4}{3}$$

$$e^{-b^2} = \frac{3\pi - 4}{3\pi}$$

$$b = \sqrt{\ln\left(\frac{3\pi}{3\pi - 4}\right)}$$

$$\approx 0.743$$

91. $A = \displaystyle\int_0^4 \frac{5}{\sqrt{25 - x^2}}\, dx = \left[5 \arcsin \frac{x}{5}\right]_0^4 = 5 \arcsin \frac{4}{5}$

$\bar{x} = \dfrac{1}{A} \displaystyle\int_0^4 x\left(\frac{5}{\sqrt{25 - x^2}}\right) dx$

$\phantom{\bar{x}} = \dfrac{1}{5 \arcsin(4/5)}\left(-\frac{5}{2}\right)\displaystyle\int_0^4 (25 - x^2)^{-1/2}(-2x)\, dx$

$\phantom{\bar{x}} = \dfrac{1}{5 \arcsin(4/5)}(-5)\left[(25 - x^2)^{1/2}\right]_0^4$

$\phantom{\bar{x}} = -\dfrac{1}{\arcsin(4/5)}[3 - 5]$

$\phantom{\bar{x}} = \dfrac{2}{\arcsin(4/5)} \approx 2.157$

93. $y = \tan(\pi x)$

$\quad y' = \pi \sec^2(\pi x)$

$1 + (y')^2 = 1 + \pi^2 \sec^4(\pi x)$

$\ s = \displaystyle\int_0^{1/4} \sqrt{1 + \pi^2 \sec^4(\pi x)}\, dx$

$\ \approx 1.0320$

Section 7.2   Integration by Parts

1. $\dfrac{d}{dx}[\sin x - x \cos x] = \cos x - (-x \sin x + \cos x) = x \sin x.$ Matches (b)

3. $\dfrac{d}{dx}[x^2 e^x - 2xe^x + 2e^x] = x^2 e^x + 2xe^x - 2xe^x - 2e^x + 2e^x = x^2 e^x.$ Matches (c)

5. $\displaystyle\int xe^{2x}\, dx$

$u = x,\, dv = e^{2x}\, dx$

7. $\displaystyle\int (\ln x)^2\, dx$

$u = (\ln x)^2,\, dv = dx$

9. $\displaystyle\int x \sec^2 x\, dx$

$u = x,\, dv = \sec^2 x\, dx$

11. $dv = e^{-2x}\, dx \implies v = \displaystyle\int e^{-2x}\, dx = -\frac{1}{2}e^{-2x}$

$\ u = x \qquad \implies du = dx$

$\displaystyle\int xe^{-2x}\, dx = -\frac{1}{2}xe^{-2x} - \displaystyle\int -\frac{1}{2}e^{-2x}\, dx$

$\phantom{\displaystyle\int xe^{-2x}\, dx} = -\frac{1}{2}xe^{-2x} - \frac{1}{4}e^{-2x} + C = \frac{-1}{4e^{2x}}(2x + 1) + C$

13. Use integration by parts three times.

(1) $dv = e^x \, dx \implies v = \int e^x \, dx = e^x$ (2) $dv = e^x \, dx \implies v = \int e^x \, dx = e^x$ (3) $dv = e^x \, dx \implies v = \int e^x \, dx = e^x$

$u = x^3 \implies du = 3x^2 \, dx$ $u = x^2 \implies du = 2x \, dx$ $u = x \implies du = dx$

$$\int x^3 e^x \, dx = x^3 e^x - 3 \int x^2 e^x \, dx = x^3 e^x - 3x^2 e^x + 6 \int x e^x \, dx$$

$$= x^3 e^x - 3x^2 e^x + 6x e^x - 6e^x + C = e^x(x^3 - 3x^2 + 6x - 6) + C$$

15. $\int x^2 e^{x^3} \, dx = \dfrac{1}{3} \int e^{x^3}(3x^2) \, dx = \dfrac{1}{3} e^{x^3} + C$

17. $dv = t \, dt \implies v = \int t \, dt = \dfrac{t^2}{2}$

$u = \ln(t + 1) \implies du = \dfrac{1}{t + 1} \, dt$

$$\int t \ln(t + 1) \, dt = \frac{t^2}{2} \ln(t + 1) - \frac{1}{2} \int \frac{t^2}{t + 1} \, dt$$

$$= \frac{t^2}{2} \ln(t + 1) - \frac{1}{2} \int \left(t - 1 + \frac{1}{t + 1} \right) dt$$

$$= \frac{t^2}{2} \ln(t + 1) - \frac{1}{2} \left[\frac{t^2}{2} - t + \ln(t + 1) \right] + C$$

$$= \frac{1}{4} \left[2(t^2 - 1) \ln|t + 1| - t^2 + 2t \right] + C$$

19. Let $u = \ln x, \, du = \dfrac{1}{x} \, dx$.

$$\int \frac{(\ln x)^2}{x} \, dx = \int (\ln x)^2 \left(\frac{1}{x} \right) dx = \frac{(\ln x)^3}{3} + C$$

21. $dv = \dfrac{1}{(2x + 1)^2} \, dx \implies v = \int (2x + 1)^{-2} \, dx$

$$= -\frac{1}{2(2x + 1)}$$

$u = x e^{2x} \implies du = (2x e^{2x} + e^{2x}) \, dx$

$$= e^{2x}(2x + 1) \, dx$$

$$\int \frac{x e^{2x}}{(2x + 1)^2} \, dx = -\frac{x e^{2x}}{2(2x + 1)} + \int \frac{e^{2x}}{2} \, dx$$

$$= \frac{-x e^{2x}}{2(2x + 1)} + \frac{e^{2x}}{4} + C$$

$$= \frac{e^{2x}}{4(2x + 1)} + C$$

23. Use integration by parts twice.

(1) $dv = e^x \, dx \implies v = \int e^x \, dx = e^x$ (2) $dv = e^x \, dx \implies v = \int e^x \, dx = e^x$

$u = x^2 \implies du = 2x \, dx$ $u = x \implies du = dx$

$$\int (x^2 - 1)e^x \, dx = \int x^2 e^x \, dx - \int e^x \, dx = x^2 e^x - 2 \int x e^x \, dx - e^x$$

$$= x^2 e^x - 2 \left[x e^x - \int e^x \, dx \right] - e^x = x^2 e^x - 2x e^x + e^x + C = (x - 1)^2 e^x + C$$

25. $dv = \sqrt{x-1}\,dx \;\Rightarrow\; v = \displaystyle\int (x-1)^{1/2}dx = \frac{2}{3}(x-1)^{3/2}$

$u = x \qquad\qquad \Rightarrow du = dx$

$\displaystyle\int x\sqrt{x-1}\,dx = \frac{2}{3}x(x-1)^{3/2} - \frac{2}{3}\int (x-1)^{3/2}\,dx$

$\qquad\qquad\qquad = \frac{2}{3}x(x-1)^{3/2} - \frac{4}{15}(x-1)^{5/2} + C$

$\qquad\qquad\qquad = \frac{2(x-1)^{3/2}}{15}(3x+2) + C$

27. $dv = \cos x\,dx \;\Rightarrow\; v = \displaystyle\int \cos x\,dx = \sin x$

$u = x \qquad \Rightarrow du = dx$

$\displaystyle\int x\cos x\,dx = x\sin x - \int \sin x\,dx = x\sin x + \cos x + C$

29. Use integration by parts three times.

(1) $u = x^3,\, du = 3x^2,\, dv = \sin x\,dx,\, v = -\cos x$

$\displaystyle\int x^3 \sin dx = -x^3\cos x + 3\int x^2 \cos x\,dx$

(2) $u = x^2,\, du = 2x\,dx,\, dv = \cos x\,dx,\, v = \sin x$

$\displaystyle\int x^3 \sin x\,dx = -x^3\cos x + 3\left[x^2\sin x - 2\int x\sin x\,dx \right]$

$\qquad\qquad\qquad = -x^3\cos x + 3x^2\sin x - 6\int x\sin x\,dx$

(3) $u = x,\, du = dx,\, dv = \sin x\,dx,\, v = -\cos x$

$\displaystyle\int x^3 \sin x\,dx = -x^3\cos x + 3x^2\sin x - 6\left[-x\cos x + \int \cos x\,dx \right]$

$\qquad\qquad\qquad = -x^3\cos x + 3x^2\sin x + 6x\cos x - 6\sin x + C$

31. $u = t,\, du = dt,\, dv = \csc t\cot t\,dt,\, v = -\csc t$

$\displaystyle\int t\csc t\cot t\,dt = -t\csc t + \int \csc t\,dt$

$\qquad\qquad\qquad = -t\csc t - \ln|\csc t + \cot t| + C$

33. $dv = dx \qquad \Rightarrow\; v = \displaystyle\int dx = x$

$u = \arctan x \Rightarrow du = \dfrac{1}{1+x^2}\,dx$

$\displaystyle\int \arctan x\,dx = x\arctan x - \int \frac{x}{1+x^2}\,dx$

$\qquad\qquad\qquad = x\arctan x - \frac{1}{2}\ln(1+x^2) + C$

35. Use integration by parts twice.

(1) $dv = e^{2x}dx \;\Rightarrow\; v = \displaystyle\int e^{2x}dx = \frac{1}{2}e^{2x}$

$u = \sin x \;\Rightarrow\; du = \cos x\,dx$

(2) $dv = e^{2x}dx \;\Rightarrow\; v = \displaystyle\int e^{2x}dx = \frac{1}{2}e^{2x}$

$u = \cos x \;\Rightarrow\; du = -\sin x\,dx$

$\displaystyle\int e^{2x}\sin x\,dx = \frac{1}{2}e^{2x}\sin x - \frac{1}{2}\int e^{2x}\cos x\,dx = \frac{1}{2}e^{2x}\sin x - \frac{1}{2}\left(\frac{1}{2}e^{2x}\cos x + \frac{1}{2}\int e^{2x}\sin x\,dx \right)$

$\displaystyle\frac{5}{4}\int e^{2x}\sin x\,dx = \frac{1}{2}e^{2x}\sin x - \frac{1}{4}e^{2x}\cos x$

$\displaystyle\int e^{2x}\sin x\,dx = \frac{1}{5}e^{2x}(2\sin x - \cos x) + C$

37. $y' = xe^{x^2}$

$$y = \int xe^{x^2}\, dx = \frac{1}{2}e^{x^2} + C$$

39. Use integration by parts twice.

(1) $dv = \dfrac{1}{\sqrt{2 + 3t}}\, dt \implies v = \int (2 + 3t)^{-1/2}\, dt = \dfrac{2}{3}\sqrt{2 + 3t}$

$\quad u = t^2 \qquad\qquad \implies du = 2t\, dt$

(2) $dv = \sqrt{2 + 3t}\, dt \implies v = \int (2 + 3t)^{1/2}\, dt = \dfrac{2}{9}(2 + 3t)^{3/2}$

$\quad u = t \qquad\qquad \implies du = dt$

$$
\begin{aligned}
y &= \int \frac{t^2}{\sqrt{2 + 3t}}\, dt = \frac{2t^2\sqrt{2 + 3t}}{3} - \frac{4}{3}\int t\sqrt{2 + 3t}\, dt \\
&= \frac{2t^2\sqrt{2 + 3t}}{3} - \frac{4}{3}\left[\frac{2t}{9}(2 + 3t)^{3/2} - \frac{2}{9}\int (2 + 3t)^{3/2}\, dt\right] \\
&= \frac{2t^2\sqrt{2 + 3t}}{3} - \frac{8t}{27}(2 + 3t)^{3/2} + \frac{16}{405}(2 + 3t)^{5/2} + C \\
&= \frac{2\sqrt{2 + 3t}}{405}(27t^2 - 24t + 32) + C
\end{aligned}
$$

41. $(\cos y)y' = 2x$

$$\int \cos y\, dy = \int 2x\, dx$$

$$\sin y = x^2 + C$$

43. (a)

(b) $\quad \dfrac{dy}{dx} = x\sqrt{y}\cos x,\ (0, 4)$

$$\int \frac{dy}{\sqrt{y}} = \int x\cos x\, dx$$

$$\int y^{-1/2}\, dy = \int x\cos x\, dx \qquad (u = x,\, du = dx,\, dv = \cos x\, dx,\, v = \sin x)$$

$$2y^{1/2} = x\sin x - \int \sin x\, dx$$

$$\qquad\quad = x\sin x + \cos x + C$$

$(0, 4)$: $2(4)^{1/2} = 0 + 1 + C \implies C = 3$

$$2\sqrt{y} = x\sin x + \cos x + 3$$

45. $\dfrac{dy}{dx} = \dfrac{x}{y}e^{x/8},\ y(0) = 2$

47. $u = x, du = dx, dv = e^{-x/2} dx, v = -2e^{-x/2}$

$$\int xe^{-x/2} dx = -2xe^{-x/2} + \int 2e^{-x/2} dx = -2xe^{-x/2} - 4e^{-x/2} + C$$

Thus, $\displaystyle\int_0^4 xe^{-x/2} dx = \left[-2xe^{-x/2} - 4e^{-x/2} \right]_0^4$

$$= -8e^{-2} - 4e^{-2} + 4$$

$$= -12e^{-2} + 4 \approx 2.376.$$

49. See Exercise 27.

$$\int_0^{\pi/2} x \cos x \, dx = \left[x \sin x + \cos x \right]_0^{\pi/2} = \frac{\pi}{2} - 1$$

51. $u = \arccos x, du = -\dfrac{1}{\sqrt{1-x^2}} dx, dv = dx, v = x$

$$\int \arccos x \, dx = x \arccos x + \int \frac{x}{\sqrt{1-x^2}} dx = x \arccos x - \sqrt{1-x^2} + C$$

Thus, $\displaystyle\int_0^{1/2} \arccos x = \left[x \arccos x - \sqrt{1-x^2} \right]_0^{1/2}$

$$= \frac{1}{2} \arccos\left(\frac{1}{2}\right) - \sqrt{\frac{3}{4}} + 1$$

$$= \frac{\pi}{6} - \frac{\sqrt{3}}{2} + 1 \approx 0.658.$$

53. Use integration by parts twice.

(1) $dv = e^x dx \implies v = \displaystyle\int e^x dx = e^x$ (2) $dv = e^x dx \implies v = \displaystyle\int e^x dx = e^x$

   $u = \sin x \implies du = \cos x \, dx$ $u = \cos x \implies du = -\sin x \, dx$

$$\int e^x \sin x \, dx = e^x \sin x - \int e^x \cos x \, dx = e^x \sin x - e^x \cos x - \int e^x \sin x \, dx$$

$$2 \int e^x \sin x \, dx = e^x (\sin x - \cos x)$$

$$\int e^x \sin x \, dx = \frac{e^x}{2} (\sin x - \cos x) + C$$

Thus, $\displaystyle\int_0^1 e^x \sin x \, dx = \left[\frac{e^x}{2}(\sin x - \cos x) \right]_0^1 = \frac{e}{2}(\sin 1 - \cos 1) + \frac{1}{2} = \frac{e(\sin 1 - \cos 1)}{}$

55. $dv = x^2 dx, v = \dfrac{x^3}{3}, u = \ln x, du = \dfrac{1}{x} dx$

$$\int x^2 \ln x \, dx = \frac{x^3}{3} \ln x - \int \frac{x^3}{3}\left(\frac{1}{x}\right) dx$$

$$= \frac{x^3}{3} \ln x - \frac{1}{3} \int x^2 dx$$

Hence, $\displaystyle\int_1^2 x^2 \ln x \, dx = \left[\frac{x^3}{3} \ln x - \frac{1}{9} x^3 \right.$

$$= \frac{8}{3} \ln 2 - \frac{8}{9} +$$

57. $dv = x\,dx,\ v = \dfrac{x^2}{2},\ u = \operatorname{arcsec} x,\ du = \dfrac{1}{x\sqrt{x^2-1}}\,dx$

$$\int x \operatorname{arcsec} x\,dx = \frac{x^2}{2}\operatorname{arcsec} x - \int \frac{x^2/2}{x\sqrt{x^2-1}}\,dx$$

$$= \frac{x^2}{2}\operatorname{arcsec} x - \frac{1}{4}\int \frac{2x}{\sqrt{x^2-1}}\,dx$$

$$= \frac{x^2}{2}\operatorname{arcsec} x - \frac{1}{2}\sqrt{x^2-1} + C$$

Hence,

$$\int_2^4 x \operatorname{arcsec} x\,dx = \left[\frac{x^2}{2}\operatorname{arcsec} x - \frac{1}{2}\sqrt{x^2-1}\right]_2^4$$

$$= \left(8\operatorname{arcsec} 4 - \frac{\sqrt{15}}{2}\right) - \left(\frac{2\pi}{3} - \frac{\sqrt{3}}{2}\right)$$

$$= 8\operatorname{arcsec} 4 - \frac{\sqrt{15}}{2} + \frac{\sqrt{3}}{2} - \frac{2\pi}{3}$$

$$\approx 7.380.$$

59. $\displaystyle\int x^2 e^{2x}\,dx = x^2\left(\frac{1}{2}e^{2x}\right) - (2x)\left(\frac{1}{4}e^{2x}\right) + 2\left(\frac{1}{8}e^{2x}\right) + C$

$$= \frac{1}{2}x^2 e^{2x} - \frac{1}{2}x e^{2x} + \frac{1}{4}e^{2x} + C$$

$$= \frac{1}{4}e^{2x}(2x^2 - 2x + 1) + C$$

Alternate signs	u and its derivatives	v' and its antiderivatives
$+$	x^2	e^{2x}
$-$	$2x$	$\frac{1}{2}e^{2x}$
$+$	2	$\frac{1}{4}e^{2x}$
$-$	0	$\frac{1}{8}e^{2x}$

61. $\displaystyle\int x^3 \sin x\,dx = x^3(-\cos x) - 3x^2(-\sin x) + 6x\cos x - 6\sin x + C$

$$= -x^3\cos x + 3x^2\sin x + 6x\cos x - 6\sin x + C$$

$$= (3x^2 - 6)\sin x - (x^3 - 6x)\cos x + C$$

Alternate signs	u and its derivatives	v' and its antiderivatives
$+$	x^3	$\sin x$
$-$	$3x^2$	$-\cos x$
$+$	$6x$	$-\sin x$
$-$	6	$\cos x$
$+$	0	$\sin x$

63. $\displaystyle\int x \sec^2 x\,dx = x\tan x + \ln|\cos x| + C$

Alternate signs	u and its derivatives	v' and its antiderivatives
$+$	x	$\sec^2 x$
$-$	1	$\tan x$
$+$	0	$-\ln|\cos x|$

71. based on the
(Su

67. No. Substitution.

69. Yes. $u = x^2,\ dv = e^{2x}\,dx$

75. $\displaystyle\int_0^{\pi/2} e^{-x}$

73. $\displaystyle\int t^3 e^{-4t}\,dt = -\frac{e^{-4t}}{128}(32t^3 + 24t^2 + 12t + 3) + C$

$\dfrac{1}{13}(2e^{-\pi} + 3) \approx 0.2374$

77. (a) $dv = \sqrt{2x-3}\,dx \implies v = \int (2x-3)^{1/2}\,dx = \frac{1}{3}(2x-3)^{3/2}$

$\qquad u = 2x \qquad\qquad \implies du = 2\,dx$

$\qquad \int 2x\sqrt{2x-3}\,dx = \frac{2}{3}x(2x-3)^{3/2} - \frac{2}{3}\int (2x-3)^{3/2}\,dx$

$\qquad\qquad = \frac{2}{3}x(2x-3)^{3/2} - \frac{2}{15}(2x-3)^{5/2} + C$

$\qquad\qquad = \frac{2}{15}(2x-3)^{3/2}(3x+3) + C = \frac{2}{5}(2x-3)^{3/2}(x+1) + C$

(b) $u = 2x-3 \implies x = \dfrac{u+3}{2}$ and $dx = \dfrac{1}{2}\,du$

$\qquad \int 2x\sqrt{2x-3}\,dx = \int 2\left(\frac{u+3}{2}\right)u^{1/2}\left(\frac{1}{2}\right)du = \frac{1}{2}\int (u^{3/2}+3u^{1/2})\,du = \frac{1}{2}\left[\frac{2}{5}u^{5/2}+2u^{3/2}\right] + C$

$\qquad\qquad = \frac{1}{5}u^{3/2}(u+5) + C = \frac{1}{5}(2x-3)^{3/2}[(2x-3)+5] + C = \frac{2}{5}(2x-3)^{3/2}(x+1) + C$

79. (a) $dv = \dfrac{x}{\sqrt{4+x^2}}\,dx \implies v = \int (4+x^2)^{-1/2}x\,dx = \sqrt{4+x^2}$

$\qquad u = x^2 \implies du = 2x\,dx$

$\qquad \int \dfrac{x^3}{\sqrt{4+x^2}}\,dx = x^2\sqrt{4+x^2} - 2\int x\sqrt{4+x^2}\,dx$

$\qquad\qquad = x^2\sqrt{4+x^2} - \frac{2}{3}(4+x^2)^{3/2} + C = \frac{1}{3}\sqrt{4+x^2}(x^2-8) + C$

(b) $u = 4+x^2 \implies x^2 = u-4$ and $2x\,dx = du \implies x\,dx = \dfrac{1}{2}\,du$

$\qquad \int \dfrac{x^3}{\sqrt{4+x^2}}\,dx = \int \dfrac{x^2}{\sqrt{4+x^2}}x\,dx = \int \dfrac{u-4}{\sqrt{u}}\frac{1}{2}\,du$

$\qquad\qquad = \frac{1}{2}\int (u^{1/2}-4u^{-1/2})\,du = \frac{1}{2}\left(\frac{2}{3}u^{3/2}-8u^{1/2}\right) + C$

$\qquad\qquad = \frac{1}{3}u^{1/2}(u-12) + C = \frac{1}{3}\sqrt{4+x^2}[(4+x^2)-12] + C = \frac{1}{3}\sqrt{4+x^2}(x^2-8) + C$

81. $n = 0$: $\int \ln x\,dx = x(\ln x - 1) + C$

$n = 1$: $\int x\ln x\,dx = \dfrac{x^2}{4}(2\ln x - 1) + C$

$n = 2$: $\int x^2\ln x\,dx = \dfrac{x^3}{9}(3\ln x - 1) + C$

$n = 3$: $\int x^3\ln x\,dx = \dfrac{x^4}{16}(4\ln x - 1) + C$

$n = 4$: $\int x^4\ln x\,dx = \dfrac{x^5}{25}(5\ln x - 1) + C$

In general, $\int x^n\ln x\,dx = \dfrac{x^{n+1}}{(n+1)^2}[(n+1)\ln x - 1] + C$. (See Exercise 85.)

83. $dv = \sin x\, dx \implies v = -\cos x$

$\quad u = x^n \qquad \implies du = nx^{n-1}\, dx$

$\displaystyle\int x^n \sin x\, dx = -x^n \cos x + n\int x^{n-1} \cos x\, dx$

85. $dv = x^n\, dx \implies v = \dfrac{x^{n+1}}{n+1}$

$\quad u = \ln x \implies du = \dfrac{1}{x}\, dx$

$\displaystyle\int x^n \ln x\, dx = \frac{x^{n+1}}{n+1}\ln x - \int \frac{x^n}{n+1}\, dx$

$\displaystyle\qquad\qquad = \frac{x^{n+1}}{n+1}\ln x - \frac{x^{n+1}}{(n+1)^2} + C$

$\displaystyle\qquad\qquad = \frac{x^{n+1}}{(n+1)^2}[(n+1)\ln x - 1] + C$

87. Use integration by parts twice.

(1) $dv = e^{ax}\, dx \implies v = \dfrac{1}{a}e^{ax}$

$\quad u = \sin bx \implies du = b\cos bx\, dx$

(2) $dv = e^{ax}\, dx \implies v = \dfrac{1}{a}e^{ax}$

$\quad u = \cos bx \implies du = -b\sin bx\, dx$

$\displaystyle\int e^{ax}\sin bx\, dx = \frac{e^{ax}\sin bx}{a} - \frac{b}{a}\int e^{ax}\cos bx\, dx$

$\displaystyle\qquad = \frac{e^{ax}\sin bx}{a} - \frac{b}{a}\left[\frac{e^{ax}\cos bx}{a} + \frac{b}{a}\int e^{ax}\sin bx\, dx\right] = \frac{e^{ax}\sin bx}{a} - \frac{b}{a^2}e^{ax}\cos bx - \frac{b^2}{a^2}\int e^{ax}\sin bx\, dx$

Therefore, $\left(1 + \dfrac{b^2}{a^2}\right)\displaystyle\int e^{ax}\sin bx\, dx = \frac{e^{ax}(a\sin bx - b\cos bx)}{a^2}$

$\displaystyle\qquad\qquad\int e^{ax}\sin bx\, dx = \frac{e^{ax}(a\sin bx - b\cos bx)}{a^2 + b^2} + C.$

89. $n = 3$ (Use formula in Exercise 85.)

$\displaystyle\int x^3 \ln x\, dx = \frac{x^4}{16}[4\ln x - 1] + C$

91. $a = 2, b = 3$ (Use formula in Exercise 88.)

$\displaystyle\int e^{2x}\cos 3x\, dx = \frac{e^{2x}(2\cos 3x + 3\sin 3x)}{13} + C$

93. $dv = e^{-x}\, dx \implies v = -e^{-x}$

$\quad u = x \qquad \implies du = dx$

$\displaystyle A = \int_0^4 xe^{-x}\, dx = \left[-xe^{-x}\right]_0^4 + \int_0^4 e^{-x}\, dx = \frac{-4}{e^4} - \left[e^{-x}\right]_0^4$

$\displaystyle\quad = 1 - \frac{5}{e^4} \approx 0.908$

95. $\displaystyle A = \int_0^1 e^{-x}\sin(\pi x)\, dx$

$\displaystyle\quad = \left[\frac{e^{-x}(-\sin \pi x - \pi\cos \pi x)}{1 + \pi^2}\right]_0^1$

$\displaystyle\quad = \frac{1}{1 + \pi^2}\left(\frac{\pi}{e} + \pi\right) = \frac{\pi}{1 + \pi^2}\left(\frac{1}{e} + 1\right)$

$\quad \approx 0.395$ (See Exercise 87.)

97. (a) $A = \displaystyle\int_1^e \ln x \, dx = \Big[-x + x \ln x \Big]_1^e = 1$ (See Exercise 4.)

(b) $R(x) = \ln x, \, r(x) = 0$

$$V = \pi \int_1^e (\ln x)^2 \, dx$$

$$= \pi \Big[x(\ln x)^2 - 2x \ln x + 2x \Big]_1^e \text{ (Use integration by parts twice, see Exercise 7.)}$$

$$= \pi(e - 2) \approx 2.257$$

(c) $p(x) = x, \, h(x) = \ln x$

$$V = 2\pi \int_1^e x \ln x \, dx = 2\pi \Big[\frac{x^2}{4}(-1 + 2 \ln x) \Big]_1^e$$

$$= \frac{(e^2 + 1)\pi}{2} \approx 13.177 \text{ (See Exercise 85.)}$$

(d) $\displaystyle \bar{x} = \frac{\int_1^e x \ln x \, dx}{1} = \frac{e^2 + 1}{4} \approx 2.097$

$\displaystyle \bar{y} = \frac{\frac{1}{2}\int_1^e (\ln x)^2 dx}{1} = \frac{e - 2}{2} \approx 0.359$

$\displaystyle (\bar{x}, \bar{y}) = \left(\frac{e^2 + 1}{4}, \frac{e - 2}{2} \right) \approx (2.097, 0.359)$

99. Average value $= \displaystyle \frac{1}{\pi} \int_0^\pi e^{-4t}(\cos 2t + 5 \sin 2t) \, dt$

$$= \frac{1}{\pi} \Big[e^{-4t}\left(\frac{-4 \cos 2t + 2 \sin 2t}{20} \right) + 5e^{-4t}\left(\frac{-4 \sin 2t - 2 \cos 2t}{20} \right) \Big]_0^\pi \text{ (From Exercises 87 and 88)}$$

$$= \frac{7}{10\pi}(1 - e^{-4\pi}) \approx 0.223$$

101. $c(t) = 100{,}000 + 4000t, \, r = 5\%, \, t_1 = 10$

$$P = \int_0^{10} (100{,}000 + 4000t)e^{-0.05t} \, dt = 4000 \int_0^{10} (25 + t)e^{-0.05t} \, dt$$

Let $u = 25 + t, \, dv = e^{-0.05t} dt, \, du = dt, \, v = -\dfrac{100}{5} e^{-0.05t}$

$$P = 4000 \left\{ \Big[(25 + t)\left(-\frac{100}{5} e^{-0.05t} \right) \Big]_0^{10} + \frac{100}{5} \int_0^{10} e^{-0.05t} \, dt \right\}$$

$$= 4000 \left\{ \Big[(25 + t)\left(-\frac{100}{5} e^{-0.05t} \right) \Big]_0^{10} - \Big[\frac{10{,}000}{25} e^{-0.05t} \Big]_0^{10} \right\} \approx \$931{,}265$$

103. $\displaystyle \int_{-\pi}^{\pi} x \sin nx \, dx = \Big[-\frac{x}{n} \cos nx + \frac{1}{n^2} \sin nx \Big]_{-\pi}^{\pi}$

$$= -\frac{\pi}{n} \cos \pi n - \frac{\pi}{n} \cos(-\pi n)$$

$$= -\frac{2\pi}{n} \cos \pi n$$

$$= \begin{cases} -(2\pi/n), & \text{if } n \text{ is even} \\ (2\pi/n), & \text{if } n \text{ is odd} \end{cases}$$

105. Let $u = x$, $dv = \sin\left(\dfrac{n\pi}{2}x\right)dx$, $du = dx$, $v = -\dfrac{2}{n\pi}\cos\left(\dfrac{n\pi}{2}x\right)$.

$$I_1 = \int_0^1 x\sin\left(\frac{n\pi}{2}x\right)dx = \left[\frac{-2x}{n\pi}\cos\left(\frac{n\pi}{2}x\right)\right]_0^1 + \frac{2}{n\pi}\int_0^1 \cos\left(\frac{n\pi}{2}x\right)dx$$

$$= -\frac{2}{n\pi}\cos\left(\frac{n\pi}{2}\right) + \left[\left(\frac{2}{n\pi}\right)^2 \sin\left(\frac{n\pi}{2}x\right)\right]_0^1$$

$$= -\frac{2}{n\pi}\cos\left(\frac{n\pi}{2}\right) + \left(\frac{2}{n\pi}\right)^2 \sin\left(\frac{n\pi}{2}\right)$$

Let $u = (-x + 2)$, $dv = \sin\left(\dfrac{n\pi}{2}x\right)dx$, $du = -dx$, $v = -\dfrac{2}{n\pi}\cos\left(\dfrac{n\pi}{2}x\right)$.

$$I_2 = \int_1^2 (-x+2)\sin\left(\frac{n\pi}{2}x\right)dx = \left[\frac{-2(-x+2)}{n\pi}\cos\left(\frac{n\pi}{2}x\right)\right]_1^2 - \frac{2}{n\pi}\int_1^2 \cos\left(\frac{n\pi}{2}x\right)dx$$

$$= \frac{2}{n\pi}\cos\left(\frac{n\pi}{2}\right) - \left[\left(\frac{2}{n\pi}\right)^2 \sin\left(\frac{n\pi}{2}x\right)\right]_1^2$$

$$= \frac{2}{n\pi}\cos\left(\frac{n\pi}{2}\right) + \left(\frac{2}{n\pi}\right)^2 \sin\left(\frac{n\pi}{2}\right)$$

$$h(I_1 + I_2) = b_n = h\left[\left(\frac{2}{n\pi}\right)^2 \sin\left(\frac{n\pi}{2}\right) + \left(\frac{2}{n\pi}\right)^2 \sin\left(\frac{n\pi}{2}\right)\right] = \frac{8h}{(n\pi)^2}\sin\left(\frac{n\pi}{2}\right)$$

107. Shell Method:

$$V = 2\pi\int_a^b xf(x)\,dx$$

$$dv = x\,dx \implies v = \frac{x^2}{2}$$

$$u = f(x) \implies du = f'(x)\,dx$$

$$V = 2\pi\left[\frac{x^2}{2}f(x) - \int\frac{x^2}{2}f'(x)\,dx\right]_a^b$$

$$= \pi\left[(b^2f(b) - a^2f(a)) - \int_a^b x^2 f'(x)\,dx\right]$$

Disk Method:

$$V = \pi\int_0^{f(a)} (b^2 - a^2)\,dy + \pi\int_{f(a)}^{f(b)} [b^2 - [f^{-1}(y)]^2]\,dy$$

$$= \pi(b^2 - a^2)f(a) + \pi b^2(f(b) - f(a)) - \pi\int_{f(a)}^{f(b)} [f^{-1}(y)]^2\,dy$$

$$= \pi\left[(b^2f(b) - a^2f(a)) - \int_{f(a)}^{f(b)} [f^{-1}(y)]^2\,dy\right]$$

Since $x = f^{-1}(y)$, we have $f(x) = y$ and $f'(x)dx = dy$. When $y = f(a)$, $x = a$. When $y = f(b)$, $x = b$. Thus,

$$\int_{f(a)}^{f(b)} [f^{-1}(y)]^2\,dy = \int_a^b x^2 f'(x)\,dx$$

and the volumes are the same.

109. $f'(x) = xe^{-x}$

(a) $f(x) = \displaystyle\int xe^{-x}\,dx = -xe^{-x} - e^{-x} + C$

(b)

(Parts: $u = x$, $dv = e^{-x}\,dx$)

$f(0) = 0 = -1 + C \implies C = 1$

$f(x) = -xe^{-x} - e^{-x} + 1$

(c) You obtain the points

n	x_n	y_n
0	0	0
1	0.05	0
2	0.10	2.378×10^{-3}
3	0.15	0.0069
4	0.20	0.0134
⋮	⋮	⋮
80	4.0	0.9064

(d) You obtain the points

n	x_n	y_n
0	0	0
1	0.1	0
2	0.2	0.0090484
3	0.3	0.025423
4	0.4	0.047648
⋮	⋮	⋮
40	4.0	0.9039

(e) $f(4) = 0.9084$

The approximations are tangent line approximations. The results in (c) are better because Δx is smaller.

Section 7.3   Trigonometric Integrals

1. $f(x) = \sin^4 x + \cos^4 x$

(a) $\sin^4 x + \cos^4 x = \left(\dfrac{1 - \cos 2x}{2}\right)^2 + \left(\dfrac{1 + \cos 2x}{2}\right)^2$

$\qquad = \dfrac{1}{4}[1 - 2\cos 2x + \cos^2 2x + 1 + 2\cos 2x + \cos^2 2x]$

$\qquad = \dfrac{1}{4}\left[2 + 2\dfrac{1 + \cos 4x}{2}\right]$

$\qquad = \dfrac{1}{4}[3 + \cos 4x]$

(b) $\sin^4 x + \cos^4 x = (\sin^2 x)^2 + \cos^4 x$

$\qquad = (1 - \cos^2 x)^2 + \cos^4 x$

$\qquad = 1 - 2\cos^2 x + 2\cos^4 x$

(c) $\sin^4 x + \cos^4 x = \sin^4 x + 2\sin^2 x \cos^2 x + \cos^4 x - 2\sin^2 x \cos^2 x$

$\qquad = (\sin^2 x + \cos^2 x)^2 - 2\sin^2 x \cos^2 x$

$\qquad = 1 - 2\sin^2 x \cos^2 x$

—CONTINUED—

1. —CONTINUED—

(d) $1 - 2\sin^2 x \cos^2 x = 1 - (2\sin x \cos x)(\sin x \cos x)$

$$= 1 - (\sin 2x)\left(\frac{1}{2}\sin 2x\right)$$

$$= 1 - \frac{1}{2}\sin^2(2x)$$

(e) Four ways. There is often more than one way to rewrite a trigonometric expression.

3. Let $u = \cos x$, $du = -\sin x\, dx$.

$$\int \cos^3 x \sin x\, dx = -\int \cos^3 x(-\sin x)\, dx$$

$$= -\frac{1}{4}\cos^4 x + C$$

5. Let $u = \sin 2x$, $du = 2\cos 2x\, dx$.

$$\int \sin^5 2x \cos 2x\, dx = \frac{1}{2}\int \sin^5 2x(2\cos 2x)\, dx$$

$$= \frac{1}{12}\sin^6 2x + C$$

7. Let $u = \cos x$, $du = -\sin x\, dx$.

$$\int \sin^5 x \cos^2 x\, dx = \int \sin x(1 - \cos^2 x)^2 \cos^2 x\, dx$$

$$= -\int (\cos^2 x - 2\cos^4 x + \cos^6 x)(-\sin x)\, dx = \frac{-1}{3}\cos^3 x + \frac{2}{5}\cos^5 x - \frac{1}{7}\cos^7 x + C$$

9. $\displaystyle\int \cos^3 \theta \sqrt{\sin \theta}\, d\theta = \int \cos\theta(1 - \sin^2\theta)(\sin\theta)^{1/2}\, d\theta$

$$= \int \left[(\sin\theta)^{1/2} - (\sin\theta)^{5/2}\right]\cos\theta\, d\theta$$

$$= \frac{2}{3}(\sin\theta)^{3/2} - \frac{2}{7}(\sin\theta)^{7/2} + C$$

11. $\displaystyle\int \cos^2 3x\, dx = \int \frac{1 + \cos 6x}{2}\, dx$

$$= \frac{1}{2}\left(x + \frac{1}{6}\sin 6x\right) + C$$

$$= \frac{1}{12}(6x + \sin 6x) + C$$

13. $\displaystyle\int \sin^2 \alpha \cdot \cos^2 \alpha\, d\alpha = \int \frac{1 - \cos 2\alpha}{2} \cdot \frac{1 + \cos 2\alpha}{2}\, d\alpha$

$$= \frac{1}{4}\int (1 - \cos^2 2\alpha)\, d\alpha$$

$$= \frac{1}{4}\int \left(1 - \frac{1 + \cos 4\alpha}{2}\right)\, d\alpha$$

$$= \frac{1}{8}\int (1 - \cos 4\alpha)\, d\alpha$$

$$= \frac{1}{8}\left[\alpha - \frac{1}{4}\sin 4\alpha\right] + C$$

$$= \frac{1}{32}[4\alpha - \sin 4\alpha] + C$$

15. Integration by parts.

$$dv = \sin^2 x \, dx = \frac{1 - \cos 2x}{2} \implies v = \frac{x}{2} - \frac{\sin 2x}{4} = \frac{1}{4}(2x - \sin 2x)$$

$$u = x \implies du = dx$$

$$\int x \sin^2 x \, dx = \frac{1}{4} x(2x - \sin 2x) - \frac{1}{4} \int (2x - \sin 2x) \, dx$$

$$= \frac{1}{4}x(2x - \sin 2x) - \frac{1}{4}\left(x^2 + \frac{1}{2}\cos 2x\right) + C = \frac{1}{8}(2x^2 - 2x \sin 2x - \cos 2x) + C$$

17. Let $u = \sin x$, $du = \cos x \, dx$.

$$\int_0^{\pi/2} \cos^3 x \, dx = \int_0^{\pi/2} (1 - \sin^2 x) \cos x \, dx$$

$$= \left[\sin x - \frac{1}{3} \sin^3 x \right]_0^{\pi/2} = \frac{2}{3}$$

19. Let $u = \sin x$, $du = \cos x \, dx$.

$$\int_0^{\pi/2} \cos^7 x \, dx = \int_0^{\pi/2} (1 - \sin^2 x)^3 \cos x \, dx = \int_0^{\pi/2} (1 - 3\sin^2 x + 3\sin^4 x - \sin^6 x) \cos x \, dx$$

$$= \left[\sin x - \sin^3 x + \frac{3}{5} \sin^5 x - \frac{1}{7} \sin^7 x \right]_0^{\pi/2} = \frac{16}{35}$$

21. $\displaystyle\int \sec(3x) \, dx = \frac{1}{3} \ln|\sec 3x + \tan 3x| + C$

23. $\displaystyle\int \sec^4 5x \, dx = \int (1 + \tan^2 5x) \sec^2 5x \, dx$

$$= \frac{1}{5}\left(\tan 5x + \frac{\tan^3 5x}{3} \right) + C$$

$$= \frac{\tan 5x}{15}(3 + \tan^2 5x) + C$$

25. $dv = \sec^2 \pi x \, dx \implies v = \dfrac{1}{\pi} \tan \pi x$

$$u = \sec \pi x \implies du = \pi \sec \pi x \tan \pi x \, dx$$

$$\int \sec^3 \pi x \, dx = \frac{1}{\pi} \sec \pi x \tan \pi x - \int \sec \pi x \tan^2 \pi x \, dx = \frac{1}{\pi} \sec \pi x \tan \pi x - \int \sec \pi x (\sec^2 \pi x - 1) \, dx$$

$$2 \int \sec^3 \pi x \, dx = \frac{1}{\pi}(\sec \pi x \tan \pi x + \ln|\sec \pi x + \tan \pi x|) + C_1$$

$$\int \sec^3 \pi x \, dx = \frac{1}{2\pi}(\sec \pi x \tan \pi x + \ln|\sec \pi x + \tan \pi x|) + C$$

27. $\displaystyle\int \tan^5 \frac{x}{4} \, dx = \int \left(\sec^2 \frac{x}{4} - 1 \right) \tan^3 \frac{x}{4} \, dx$

$$= \int \tan^3 \frac{x}{4} \sec^2 \frac{x}{4} \, dx - \int \tan^3 \frac{x}{4} \, dx$$

$$= \tan^4 \frac{x}{4} - \int \left(\sec^2 \frac{x}{4} - 1 \right) \tan \frac{x}{4} \, dx$$

$$= \tan^4 \frac{x}{4} - 2 \tan^2 \frac{x}{4} - 4 \ln\left| \cos \frac{x}{4} \right| + C$$

29. $u = \tan x$, $du = \sec^2 x \, dx$

$$\int \sec^2 x \tan x \, dx = \frac{1}{2} \tan^2 x + C$$

31. $\int \tan^2 x \sec^2 x \, dx = \dfrac{\tan^3 x}{3} + C$

33. $\int \sec^6 4x \tan 4x \, dx = \dfrac{1}{4} \int \sec^5 4x(4 \sec 4x \tan 4x) \, dx$

$$= \dfrac{\sec^6 4x}{24} + C$$

35. Let $u = \sec x$, $du = \sec x \tan x \, dx$.

$$\int \sec^3 x \tan x \, dx = \int \sec^2 x(\sec x \tan x) \, dx$$

$$= \dfrac{1}{3} \sec^3 x + C$$

37. $\int \dfrac{\tan^2 x}{\sec x} \, dx = \int \dfrac{(\sec^2 x - 1)}{\sec x} \, dx$

$$= \int (\sec x - \cos x) \, dx$$

$$= \ln|\sec x + \tan x| - \sin x + C$$

39. $r = \int \sin^4(\pi\theta) \, d\theta = \dfrac{1}{4} \int [1 - \cos(2\pi\theta)]^2 \, d\theta$

$$= \dfrac{1}{4} \int [1 - 2\cos(2\pi\theta) + \cos^2(2\pi\theta)] \, d\theta$$

$$= \dfrac{1}{4} \int \left[1 - 2\cos(2\pi\theta) + \dfrac{1 + \cos(4\pi\theta)}{2} \right] d\theta$$

$$= \dfrac{1}{4} \left[\theta - \dfrac{1}{\pi} \sin(2\pi\theta) + \dfrac{\theta}{2} + \dfrac{1}{8\pi} \sin(4\pi\theta) \right] + C$$

$$= \dfrac{1}{32\pi} [12\pi\theta - 8\sin(2\pi\theta) + \sin(4\pi\theta)] + C$$

41. $y = \int \tan^3 3x \sec 3x \, dx$

$$= \int (\sec^2 3x - 1)\sec 3x \tan 3x \, dx$$

$$= \dfrac{1}{3} \int \sec^2 3x(3 \sec 3x \tan 3x) \, dx - \dfrac{1}{3} \int 3 \sec 3x \tan 3x \, dx$$

$$= \dfrac{1}{9} \sec^3 3x - \dfrac{1}{3} \sec 3x + C$$

43. (a)

(b) $\dfrac{dy}{dx} = \sin^2 x$, $(0, 0)$

$$y = \int \sin^2 x \, dx = \int \dfrac{1 - \cos 2x}{2} \, dx$$

$$= \dfrac{1}{2} x - \dfrac{\sin 2x}{4} + C$$

$(0, 0)$: $0 = C$, $y = \dfrac{1}{2} x - \dfrac{\sin 2x}{4}$

45. $\dfrac{dy}{dx} = \dfrac{3 \sin x}{y}$, $y(0) = 2$

47. $\int \sin 3x \cos 2x \, dx = \dfrac{1}{2} \int (\sin 5x + \sin x) \, dx$

$$= \dfrac{-1}{2} \left(\dfrac{1}{5} \cos 5x + \cos x \right) + C$$

$$= \dfrac{-1}{10} (\cos 5x + 5 \cos x) + C$$

49. $\int \sin\theta \sin 3\theta \, d\theta = \dfrac{1}{2} \int (\cos 2\theta - \cos 4\theta) \, d\theta$

$$= \dfrac{1}{2} \left(\dfrac{1}{2} \sin 2\theta - \dfrac{1}{4} \sin 4\theta \right) + C$$

$$= \dfrac{1}{8} (2 \sin 2\theta - \sin 4\theta) + C$$

51. $\int \cot^3 2x \, dx = \int (\csc^2 2x - 1) \cot 2x \, dx$

$$= -\dfrac{1}{2} \int \cot 2x(-2\csc^2 2x) \, dx - \dfrac{1}{2} \int \dfrac{2 \cos 2x}{\sin 2x} \, dx$$

$$= -\dfrac{1}{4} \cot^2 2x - \dfrac{1}{2} \ln|\sin 2x| + C$$

$$= \dfrac{1}{4} (\ln|\csc^2 2x| - \cot^2 2x) + C$$

53. Let $u = \cot \theta$, $du = -\csc^2 \theta \, d\theta$.

$$\int \csc^4 \theta \, d\theta = \int \csc^2 \theta (1 + \cot^2 \theta) \, d\theta$$

$$= \int \csc^2 \theta \, d\theta + \int \csc^2 \theta \cot^2 \theta \, d\theta$$

$$= -\cot \theta - \frac{1}{3} \cot^3 \theta + C$$

55. $\displaystyle \int \frac{\cot^2 t}{\csc t} \, dt = \int \frac{\csc^2 t - 1}{\csc t} \, dt$

$$= \int (\csc t - \sin t) dt$$

$$= \ln|\csc t - \cot t| + \cos t + C$$

57. $\displaystyle \int \frac{1}{\sec x \tan x} \, dx = \int \frac{\cos^2 x}{\sin x} \, dx = \int \frac{1 - \sin^2 x}{\sin x} \, dx$

$$= \int (\csc x - \sin x) \, dx$$

$$= \ln|\csc x - \cot x| + \cos x + C$$

59. $\displaystyle \int (\tan^4 t - \sec^4 t) \, dt = \int (\tan^2 t + \sec^2 t)(\tan^2 t - \sec^2 t) \, dt \qquad (\tan^2 t - \sec^2 t = -1)$

$$= -\int (\tan^2 t + \sec^2 t) \, dt = -\int (2 \sec^2 t - 1) \, dt = -2 \tan t + t + C$$

61. $\displaystyle \int_{-\pi}^{\pi} \sin^2 x \, dx = 2 \int_0^{\pi} \frac{1 - \cos 2x}{2} \, dx$

$$= \left[x - \frac{1}{2} \sin 2x \right]_0^{\pi} = \pi$$

63. $\displaystyle \int_0^{\pi/4} \tan^3 x \, dx = \int_0^{\pi/4} (\sec^2 x - 1) \tan x \, dx$

$$= \int_0^{\pi/4} \sec^2 x \tan x \, dx - \int_0^{\pi/4} \frac{\sin x}{\cos x} \, dx$$

$$= \left[\frac{1}{2} \tan^2 x + \ln|\cos x| \right]_0^{\pi/4}$$

$$= \frac{1}{2}(1 - \ln 2)$$

65. Let $u = 1 + \sin t$, $du = \cos t \, dt$.

$$\int_0^{\pi/2} \frac{\cos t}{1 + \sin t} \, dt = \left[\ln|1 + \sin t| \right]_0^{\pi/2} = \ln 2$$

67. Let $u = \sin x$, $du = \cos x \, dx$.

$$\int_{-\pi/2}^{\pi/2} \cos^3 x \, dx = 2 \int_0^{\pi/2} (1 - \sin^2 x) \cos x \, dx$$

$$= 2 \left[\sin x - \frac{1}{3} \sin^3 x \right]_0^{\pi/2} = \frac{4}{3}$$

69. $\displaystyle \int \cos^4 \frac{x}{2} \, dx = \frac{1}{16} [6x + 8 \sin x + \sin 2x] + C$

71. $\displaystyle \int \sec^5 \pi x \, dx = \frac{1}{4\pi} \left\{ \sec^3 \pi x \tan \pi x + \frac{3}{2} [\sec \pi x \tan \pi x + \ln|\sec \pi x + \tan \pi x|] \right\} + C$

73. $\int \sec^5 \pi x \tan \pi x \, dx = \frac{1}{5\pi} \sec^5 \pi x + C$

75. $\int_0^{\pi/4} \sin 2\theta \sin 3\theta \, d\theta = \frac{1}{2}\left[\sin \theta - \frac{1}{5} \sin 5\theta \right]_0^{\pi/4} = \frac{3\sqrt{2}}{10}$

77. $\int_0^{\pi/2} \sin^4 x \, dx = \frac{1}{4}\left[\frac{3x}{2} - \sin 2x + \frac{1}{8} \sin 4x \right]_0^{\pi/2}$

$$= \frac{3\pi}{16}$$

79. (a) Save one sine factor and convert the remaining sine factors to cosine. Then expand and integrate.

(b) Save one cosine factor and convert the remaining cosine factors to sine. Then expand and integrate.

(c) Make repeated use of the power reducing formula to convert the integrand to odd powers of the cosine.

81. (a) Let $u = \tan 3x$, $du = 3 \sec^2 3x \, dx$.

$$\int \sec^4 3x \tan^3 3x \, dx = \int \sec^2 3x \tan^3 3x \sec^2 3x \, dx$$

$$= \frac{1}{3} \int (\tan^2 3x + 1) \tan^3 3x (3 \sec^2 3x) \, dx$$

$$= \frac{1}{3} \int (\tan^5 3x + \tan^3 3x)(3 \sec^2 3x) \, dx$$

$$= \frac{\tan^6 3x}{18} + \frac{\tan^4 3x}{12} + C_1$$

Or let $u = \sec 3x$, $du = 3 \sec 3x \tan 3x \, dx$.

$$\int \sec^4 3x \tan^3 3x \, dx = \int \sec^3 3x \tan^2 3x \sec 3x \tan 3x \, dx$$

$$= \frac{1}{3} \int \sec^3 3x (\sec^2 3x - 1)(3 \sec 3x \tan 3x) \, dx$$

$$= \frac{\sec^6 3x}{18} - \frac{\sec^4 3x}{12} + C$$

(b)

(c) $\dfrac{\sec^6 3x}{18} - \dfrac{\sec^4 3x}{12} + C = \dfrac{(1 + \tan^2 3x)^3}{18} - \dfrac{(1 + \tan^2 3x)^2}{12} + C$

$$= \frac{1}{18} \tan^6 3x + \frac{1}{6} \tan^4 3x + \frac{1}{6} \tan^2 3x + \frac{1}{18} - \frac{1}{12} \tan^4 3x - \frac{1}{6} \tan^2 3x - \frac{1}{12} + C$$

$$= \frac{\tan^6 3x}{18} + \frac{\tan^4 3x}{12} + \left(\frac{1}{18} - \frac{1}{12} \right) + C$$

$$= \frac{\tan^6 3x}{18} + \frac{\tan^4 3x}{12} + C_2$$

83. $A = \int_0^1 \sin^2(\pi x) \, dx$

$$= \int_0^1 \frac{1 - \cos(2\pi x)}{2} \, dx$$

$$= \left[\frac{x}{2} - \frac{1}{4\pi} \sin(2\pi x) \right]_0^1$$

$$= \frac{1}{2}$$

85. (a) $V = \pi \int_0^\pi \sin^2 x \, dx = \dfrac{\pi}{2} \int_0^\pi (1 - \cos 2x) \, dx = \dfrac{\pi}{2} \left[x - \dfrac{1}{2} \sin 2x \right]_0^\pi = \dfrac{\pi^2}{2}$

(b) $A = \int_0^\pi \sin x \, dx = \left[-\cos x \right]_0^\pi = 1 + 1 = 2$

Let $u = x, \, dv = \sin x \, dx, \, du = dx, \, v = -\cos x$.

$$\bar{x} = \dfrac{1}{A} \int_0^\pi x \sin x \, dx = \dfrac{1}{2} \left[\left[-x \cos x \right]_0^\pi + \int_0^\pi \cos x \, dx \right] = \dfrac{1}{2} \left[-x \cos x + \sin x \right]_0^\pi = \dfrac{\pi}{2}$$

$$\bar{y} = \dfrac{1}{2A} \int_0^\pi \sin^2 x \, dx$$

$$= \dfrac{1}{8} \int_0^\pi (1 - \cos 2x) \, dx$$

$$= \dfrac{1}{8} \left[x - \dfrac{1}{2} \sin 2x \right]_0^\pi = \dfrac{\pi}{8}$$

$$(\bar{x}, \bar{y}) = \left(\dfrac{\pi}{2}, \dfrac{\pi}{8} \right)$$

87. $dv = \sin x \, dx \implies v = -\cos x$

$u = \sin^{n-1} x \implies du = (n - 1)\sin^{n-2} x \cos x \, dx$

$$\int \sin^n x \, dx = -\sin^{n-1} x \cos x + (n - 1) \int \sin^{n-2} x \cos^2 x \, dx$$

$$= -\sin^{n-1} x \cos x + (n - 1) \int \sin^{n-2} x (1 - \sin^2 x) \, dx$$

$$= -\sin^{n-1} x \cos x + (n - 1) \int \sin^{n-2} x \, dx - (n - 1) \int \sin^n x \, dx$$

Therefore, $n \int \sin^n x \, dx = -\sin^{n-1} x \cos x + (n - 1) \int \sin^{n-2} x \, dx$

$$\int \sin^n x \, dx = \dfrac{-\sin^{n-1} x \cos x}{n} + \dfrac{n - 1}{n} \int \sin^{n-2} x \, dx.$$

89. Let $u = \sin^{n-1} x, \, du = (n - 1)\sin^{n-2} x \cos x \, dx, \, dv = \cos^m x \sin x \, dx, \, v = \dfrac{-\cos^{m+1} x}{m + 1}$.

$$\int \cos^m x \sin^n x \, dx = \dfrac{-\sin^{n-1} x \cos^{m+1} x}{m + 1} + \dfrac{n - 1}{m + 1} \int \sin^{n-2} x \cos^{m+2} x \, dx$$

$$= \dfrac{-\sin^{n-1} x \cos^{m+1} x}{m + 1} + \dfrac{n - 1}{m + 1} \int \sin^{n-2} x \cos^m x (1 - \sin^2 x) \, dx$$

$$= \dfrac{-\sin^{n-1} x \cos^{m+1} x}{m + 1} + \dfrac{n - 1}{m + 1} \int \sin^{n-2} x \cos^m x \, dx - \dfrac{n - 1}{m + 1} \int \sin^n x \cos^m x \, dx$$

$$\dfrac{m + n}{m + 1} \int \cos^m x \sin^n x \, dx = \dfrac{-\sin^{n-1} x \cos^{m+1} x}{m + 1} + \dfrac{n - 1}{m + 1} \int \sin^{n-2} x \cos^m x \, dx$$

$$\int \cos^m x \sin^n x \, dx = \dfrac{-\cos^{m+1} x \sin^{n-1}}{m + n} + \dfrac{n - 1}{m + n} \int \cos^m x \sin^{n-2} x \, dx$$

91. $\int \sin^5 x \, dx = -\dfrac{\sin^4 x \cos x}{5} + \dfrac{4}{5} \int \sin^3 x \, dx$

$\qquad\qquad = -\dfrac{\sin^4 x \cos x}{5} + \dfrac{4}{5}\left[-\dfrac{\sin^2 x \cos x}{3} + \dfrac{2}{3} \int \sin x \, dx \right]$

$\qquad\qquad = -\dfrac{1}{5} \sin^4 x \cos x - \dfrac{4}{15} \sin^2 x \cos x - \dfrac{8}{15} \cos x + C$

$\qquad\qquad = -\dfrac{\cos x}{15}[3 \sin^4 x + 4 \sin^2 x + 8] + C$

93. $\int \sec^4\left(\dfrac{2\pi x}{5}\right) dx = \dfrac{5}{2\pi} \int \sec^4\left(\dfrac{2\pi x}{5}\right)\dfrac{2\pi}{5} \, dx$

$\qquad\qquad = \dfrac{5}{2\pi}\left[\dfrac{1}{3} \sec^2\left(\dfrac{2\pi x}{5}\right)\tan\left(\dfrac{2\pi x}{5}\right) + \dfrac{2}{3} \int \sec^2\left(\dfrac{2\pi x}{5}\right)\dfrac{2\pi}{5} \, dx \right]$

$\qquad\qquad = \dfrac{5}{6\pi}\left[\sec^2\left(\dfrac{2\pi x}{5}\right)\tan\left(\dfrac{2\pi x}{5}\right) + 2\tan\left(\dfrac{2\pi x}{5}\right) \right] + C$

$\qquad\qquad = \dfrac{5}{6\pi}\tan\left(\dfrac{2\pi x}{5}\right)\left[\sec^2\left(\dfrac{2\pi x}{5}\right) + 2 \right] + C$

95. (a) $f(t) = a_0 + a_1 \cos\dfrac{\pi t}{6} + b_1 \sin\dfrac{\pi t}{6}$ where:

$\qquad a_0 = \dfrac{1}{12} \displaystyle\int_0^{12} f(t) \, dt$

$\qquad a_1 = \dfrac{1}{6} \displaystyle\int_0^{12} f(t) \cos\dfrac{\pi t}{6} \, dt$

$\qquad b_1 = \dfrac{1}{6} \displaystyle\int_0^{12} f(t) \sin\dfrac{\pi t}{6} \, dt$

$a_0 \approx \dfrac{12-0}{3(12)^2}[30.9 + 4(32.2) + 2(41.1) + 4(53.7) + 2(64.6) + 4(74.0) + 2(78.2) + 4(77.0) + 2(71.0) +$

$\qquad\qquad 4(60.1) + 2(47.1) + 4(35.7) + 30.9] \approx 55.46$

$a_1 \approx \dfrac{12-0}{6(3)(12)}\left[30.9 \cos 0 + 4\left(32.2 \cos\dfrac{\pi}{6}\right) + 2\left(41.1 \cos\dfrac{\pi}{3}\right) + 4\left(53.7 \cos\dfrac{\pi}{2}\right) + 2\left(64.6 \cos\dfrac{2\pi}{3}\right) + \right.$

$\qquad\qquad 4\left(74.0 \cos\dfrac{5\pi}{6}\right) + 2(78.2 \cos \pi) + 4\left(77.0 \cos\dfrac{7\pi}{6}\right) + 2\left(71.0 \cos\dfrac{4\pi}{3}\right) +$

$\qquad\qquad \left. 4\left(60.1 \cos\dfrac{3\pi}{2}\right) + 2\left(47.1 \cos\dfrac{5\pi}{3}\right) + 4\left(35.7 \cos\dfrac{11\pi}{6}\right) + 30.9 \cos 2\pi \right] \approx -23.88$

$b_1 \approx \dfrac{12-0}{6(3)(12)}\left[30.9 \sin 0 + 4\left(32.2 \sin\dfrac{\pi}{6}\right) + 2\left(41.1 \sin\dfrac{\pi}{3}\right) + 4\left(53.7 \sin\dfrac{\pi}{2}\right) + 2\left(64.6 \sin\dfrac{2\pi}{3}\right) + \right.$

$\qquad\qquad 4\left(74.0 \sin\dfrac{5\pi}{6}\right) + 2(78.2 \sin \pi) + 4\left(77.0 \sin\dfrac{7\pi}{6}\right) + 2\left(71.0 \sin\dfrac{4\pi}{3}\right) +$

$\qquad\qquad \left. 4\left(60.1 \sin\dfrac{3\pi}{2}\right) + 2\left(47.1 \sin\dfrac{5\pi}{3}\right) + 4\left(35.7 \sin\dfrac{11\pi}{6}\right) + 30.9 \sin 2\pi \right] \approx -3.34$

$\qquad H(t) \approx 55.46 - 23.88 \cos\dfrac{\pi t}{6} - 3.34 \sin\dfrac{\pi t}{6}$

—CONTINUED—

95. **—CONTINUED—**

(b) $a_0 \approx \dfrac{12 - 0}{3(12)^2}[18.0 + 4(17.7) + 2(25.8) + 4(36.1) + 2(45.4) + 4(55.2) + 2(59.9) + 4(59.4) + 2(53.1) +$

$$4(43.2) + 2(34.3) + 4(24.2) + 18.0] \approx 39.34$$

$a_1 \approx \dfrac{12 - 0}{6(3)(12)}\left[18.0 \cos 0 + 4\left(17.7 \cos \dfrac{\pi}{6}\right) + 2\left(25.8 \cos \dfrac{\pi}{3}\right) + 4\left(36.1 \cos \dfrac{\pi}{2}\right) + 2\left(45.4 \cos \dfrac{2\pi}{3}\right) +\right.$

$$4\left(55.2 \cos \dfrac{5\pi}{6}\right) + 2(59.9 \cos \pi) + 4\left(59.4 \cos \dfrac{7\pi}{6}\right) + 2\left(53.1 \cos \dfrac{4\pi}{3}\right) +$$

$$\left. 4\left(43.2 \cos \dfrac{3\pi}{2}\right) + 2\left(34.3 \cos \dfrac{5\pi}{3}\right) + 4\left(24.2 \cos \dfrac{11\pi}{6}\right) + 18 \cos 2\pi\right] \approx -20.78$$

$b_1 \approx \dfrac{12 - 0}{6(3)(12)}\left[18.0 \sin 0 + 4\left(17.7 \sin \dfrac{\pi}{6}\right) + 2\left(25.8 \sin \dfrac{\pi}{3}\right) + 4\left(36.1 \sin \dfrac{\pi}{2}\right) + 2\left(45.4 \sin \dfrac{2\pi}{3}\right) +\right.$

$$4\left(55.2 \sin \dfrac{5\pi}{6}\right) + 2(59.9 \sin \pi) + 4\left(59.4 \sin \dfrac{7\pi}{6}\right) + 2\left(53.1 \sin \dfrac{4\pi}{3}\right) +$$

$$\left. 4\left(43.2 \sin \dfrac{3\pi}{2}\right) + 2\left(34.3 \sin \dfrac{5\pi}{3}\right) + 4\left(24.2 \sin \dfrac{11\pi}{6}\right) + 18 \sin 2\pi\right] \approx -4.33$$

$$L(t) \approx 39.34 - 20.78 \cos \dfrac{\pi t}{6} - 4.33 \sin \dfrac{\pi t}{6}$$

(c) The difference between the maximum and minimum temperatures is greatest in the summer.

97. $\displaystyle\int_{-\pi}^{\pi} \cos(mx) \cos(nx)\, dx = \dfrac{1}{2}\left[\dfrac{\sin(m + n)x}{m + n} + \dfrac{\sin(m - n)x}{m - n}\right]_{-\pi}^{\pi} = 0, \; (m \neq n)$

$\displaystyle\int_{-\pi}^{\pi} \sin(mx) \sin(nx)\, dx = \dfrac{1}{2}\int_{-\pi}^{\pi} [\cos(m - n)x - \cos(m + n)x]\, dx$

$$= \dfrac{1}{2}\left[\dfrac{\sin(m - n)x}{m - n} - \dfrac{\sin(m + n)x}{m + n}\right]_{-\pi}^{\pi} = 0, \; (m \neq n)$$

$\displaystyle\int_{-\pi}^{\pi} \sin(mx) \cos(nx)\, dx = \dfrac{1}{2}\int_{-\pi}^{\pi} [\sin(m + n)x + \sin(m - n)x]\, dx$

$$= -\dfrac{1}{2}\left[\dfrac{\cos(m + n)x}{m + n} + \dfrac{\cos(m - n)x}{m - n}\right]_{-\pi}^{\pi}, \, (m \neq n)$$

$$= -\dfrac{1}{2}\left[\left(\dfrac{\cos(m + n)\pi}{m + n} + \dfrac{\cos(m - n)\pi}{m - n}\right) - \left(\dfrac{\cos(m + n)(-\pi)}{m + n} + \dfrac{\cos(m - n)(-\pi)}{m - n}\right)\right]$$

$$= 0, \text{ since } \cos(-\theta) = \cos\theta.$$

$\displaystyle\int_{-\pi}^{\pi} \sin(mx) \cos(mx)\, dx = \dfrac{1}{m} \dfrac{\sin^2(mx)}{2}\Big]_{-\pi}^{\pi} = 0$

Section 7.4   Trigonometric Substitution

1. $\dfrac{d}{dx}\left[4\ln\left|\dfrac{\sqrt{x^2+16}-4}{x}\right| + \sqrt{x^2+16} + C\right] = \dfrac{d}{dx}\left[4\ln\left|\sqrt{x^2+16}-4\right| - 4\ln|x| + \sqrt{x^2+16} + C\right]$

$$= 4\left[\dfrac{x/\sqrt{x^2+16}}{\sqrt{x^2+16}-4}\right] - \dfrac{4}{x} + \dfrac{x}{\sqrt{x^2+16}}$$

$$= \dfrac{4x}{\sqrt{x^2+16}\left(\sqrt{x^2+16}-4\right)} - \dfrac{4}{x} + \dfrac{x}{\sqrt{x^2+16}}$$

$$= \dfrac{4x^2 - 4\sqrt{x^2+16}\left(\sqrt{x^2+16}-4\right) + x^2\left(\sqrt{x^2+16}-4\right)}{x\sqrt{x^2+16}\left(\sqrt{x^2+16}-4\right)}$$

$$= \dfrac{4x^2 - 4(x^2+16) + 16\sqrt{x^2+16} + x^2\sqrt{x^2+16} - 4x^2}{x\sqrt{x^2+16}\left(\sqrt{x^2+16}-4\right)}$$

$$= \dfrac{\sqrt{x^2+16}(x^2+16) - 4(x^2+16)}{x\sqrt{x^2+16}\left(\sqrt{x^2+16}-4\right)}$$

$$= \dfrac{(x^2+16)\left(\sqrt{x^2+16}-4\right)}{x\sqrt{x^2+16}\left(\sqrt{x^2+16}-4\right)} = \dfrac{\sqrt{x^2+16}}{x}$$

Indefinite integral: $\displaystyle\int \dfrac{\sqrt{x^2+16}}{x}\,dx$ Matches (b)

3. $\dfrac{d}{dx}\left[8\arcsin\dfrac{x}{4} - \dfrac{x\sqrt{16-x^2}}{2} + C\right] = 8\dfrac{1/4}{\sqrt{1-(x/4)^2}} - \dfrac{x(1/2)(16-x^2)^{-1/2}(-2x) + \sqrt{16-x^2}}{2}$

$$= \dfrac{8}{\sqrt{16-x^2}} + \dfrac{x^2}{2\sqrt{16-x^2}} - \dfrac{\sqrt{16-x^2}}{2}$$

$$= \dfrac{16}{2\sqrt{16-x^2}} + \dfrac{x^2}{2\sqrt{16-x^2}} - \dfrac{(16-x^2)}{2\sqrt{16-x^2}} = \dfrac{x^2}{\sqrt{16-x^2}}$$

Matches (a)

5. Let $x = 5\sin\theta$, $dx = 5\cos\theta\,d\theta$, $\sqrt{25-x^2} = 5\cos\theta$.

$$\int \dfrac{1}{(25-x^2)^{3/2}}\,dx = \int \dfrac{5\cos\theta}{(5\cos\theta)^3}\,d\theta$$

$$= \dfrac{1}{25}\int \sec^2\theta\,d\theta$$

$$= \dfrac{1}{25}\tan\theta + C$$

$$= \dfrac{x}{25\sqrt{25-x^2}} + C$$

7. Same substitution as in Exercise 5

$$\int \dfrac{\sqrt{25-x^2}}{x}\,dx = \int \dfrac{25\cos^2\theta\,d\theta}{5\sin\theta} = 5\int \dfrac{1-\sin^2\theta}{\sin\theta}\,d\theta = 5\int (\csc\theta - \sin\theta)\,d\theta$$

$$= 5[\ln|\csc\theta - \cot\theta| + \cos\theta] + C = 5\ln\left|\dfrac{5-\sqrt{25-x^2}}{x}\right| + \sqrt{25-x^2} + C$$

9. Let $x = 2 \sec \theta$, $dx = 2 \sec \theta \tan \theta\, d\theta$, $\sqrt{x^2 - 4} = 2 \tan \theta$.

$$\int \frac{1}{\sqrt{x^2 - 4}}\, dx = \int \frac{2 \sec \theta \tan \theta\, d\theta}{2 \tan \theta} = \int \sec \theta\, d\theta = \ln|\sec \theta + \tan\theta| + C_1$$

$$= \ln \left| \frac{x}{2} + \frac{\sqrt{x^2 - 4}}{2} \right| + C_1$$

$$= \ln \left| x + \sqrt{x^2 - 4} \right| - \ln 2 + C_1 = \ln \left| x + \sqrt{x^2 - 4} \right| + C$$

11. Same substitution as in Exercise 9

$$\int x^3 \sqrt{x^2 - 4}\, dx = \int (8 \sec^3 \theta)(2 \tan \theta)(2 \sec \theta \tan \theta)\, d\theta = 32 \int \tan^2 \theta \sec^4 \theta\, d\theta$$

$$= 32 \int \tan^2 \theta (1 + \tan^2 \theta) \sec^2 \theta\, d\theta = 32 \left(\frac{\tan^3 \theta}{3} + \frac{\tan^5 \theta}{5} \right) + C$$

$$= \frac{32}{15} \tan^3 \theta [5 + 3 \tan^2 \theta] + C = \frac{32}{15} \frac{(x^2 - 4)^{3/2}}{8} \left[5 + 3 \frac{(x^2 - 4)}{4} \right] + C$$

$$= \frac{1}{15}(x^2 - 4)^{3/2}[20 + 3(x^2 - 4)] + C = \frac{1}{15}(x^2 - 4)^{3/2}(3x^2 + 8) + C$$

13. Let $x = \tan \theta$, $dx = \sec^2 \theta\, d\theta$, $\sqrt{1 + x^2} = \sec \theta$.

$$\int x \sqrt{1 + x^2}\, dx = \int \tan \theta (\sec \theta) \sec^2 \theta\, d\theta = \frac{\sec^3 \theta}{3} + C = \frac{1}{3}(1 + x^2)^{3/2} + C$$

Note: This integral could have been evaluated with the Power Rule.

15. Same substitution as in Exercise 13

$$\int \frac{1}{(1 + x^2)^2}\, dx = \int \frac{1}{\left(\sqrt{1 + x^2} \right)^4}\, dx$$

$$= \int \frac{\sec^2 \theta\, d\theta}{\sec^4 \theta}$$

$$= \int \cos^2 \theta\, d\theta = \frac{1}{2} \int (1 + \cos 2\theta)\, d\theta$$

$$= \frac{1}{2} \left[\theta + \frac{\sin 2\theta}{2} \right]$$

$$= \frac{1}{2} [\theta + \sin \theta \cos \theta] + C$$

$$= \frac{1}{2} \left[\arctan x + \left(\frac{x}{\sqrt{1 + x^2}} \right) \left(\frac{1}{\sqrt{1 + x^2}} \right) \right] + C$$

$$= \frac{1}{2} \left[\arctan x + \frac{x}{1 + x^2} \right] + C$$

17. Let $u = 3x$, $a = 2$, and $du = 3\, dx$.

$$\int \sqrt{4 + 9x^2}\, dx = \frac{1}{3} \int \sqrt{(2)^2 + (3x)^2}\, 3\, dx$$

$$= \frac{1}{3} \left(\frac{1}{2} \right) \left(3x \sqrt{4 + 9x^2} + 4 \ln \left| 3x + \sqrt{4 + 9x^2} \right| \right) + C$$

$$= \frac{1}{2} x \sqrt{4 + 9x^2} + \frac{2}{3} \ln \left| 3x + \sqrt{4 + 9x^2} \right| + C$$

19. $\int \dfrac{x}{\sqrt{x^2 + 9}}\, dx = \dfrac{1}{2} \int (x^2 + 9)^{-1/2}(2x)\, dx$

$\qquad\qquad\qquad = \sqrt{x^2 + 9} + C$

(Power Rule)

21. $\int \dfrac{1}{\sqrt{16 - x^2}}\, dx = \arcsin\left(\dfrac{x}{4}\right) + C$

23. Let $x = 2 \sin \theta$, $dx = 2 \cos \theta\, d\theta$, $\sqrt{4 - x^2} = 2 \cos \theta$.

$$\int \sqrt{16 - 4x^2}\, dx = 2 \int \sqrt{4 - x^2}\, dx$$

$$= 2 \int 2 \cos \theta (2 \cos \theta\, d\theta)$$

$$= 8 \int \cos^2 \theta\, d\theta$$

$$= 4 \int (1 + \cos 2\theta)\, d\theta$$

$$= 4 \left[\theta + \dfrac{1}{2} \sin 2\theta \right] + C$$

$$= 4\theta + 4 \sin \theta \cos \theta + C$$

$$= 4 \arcsin\left(\dfrac{x}{2}\right) + x\sqrt{4 - x^2} + C$$

25. Let $x = 3 \sec \theta$, $dx = 3 \sec \theta \tan \theta\, d\theta$,

$\sqrt{x^2 - 9} = 3 \tan \theta$.

$$\int \dfrac{1}{\sqrt{x^2 - 9}}\, dx = \int \dfrac{3 \sec \theta \tan \theta\, d\theta}{3 \tan \theta}$$

$$= \int \sec \theta\, d\theta$$

$$= \ln|\sec \theta + \tan \theta| + C_1$$

$$= \ln\left| \dfrac{x}{3} + \dfrac{\sqrt{x^2 - 9}}{3} \right| + C_1$$

$$= \ln\left| x + \sqrt{x^2 - 9} \right| + C$$

27. Let $x = \sin \theta$, $dx = \cos \theta\, d\theta$, $\sqrt{1 - x^2} = \cos \theta$.

$$\int \dfrac{\sqrt{1 - x^2}}{x^4}\, dx = \int \dfrac{\cos \theta (\cos \theta\, d\theta)}{\sin^4 \theta}$$

$$= \int \cot^2 \theta \csc^2 \theta\, d\theta$$

$$= -\dfrac{1}{3} \cot^3 \theta + C$$

$$= \dfrac{-(1 - x^2)^{3/2}}{3x^3} + C$$

29. Same substitutions as in Exercise 28

$$\int \dfrac{1}{x\sqrt{4x^2 + 9}}\, dx = \int \dfrac{(3/2) \sec^2 \theta\, d\theta}{(3/2) \tan \theta\, 3\sec \theta}$$

$$= \dfrac{1}{3} \int \csc \theta\, d\theta = -\dfrac{1}{3} \ln|\csc \theta + \cot \theta| + C = -\dfrac{1}{3} \ln\left| \dfrac{\sqrt{4x^2 + 9} + 3}{2x} \right| + C$$

31. Let $x = \sqrt{5}\tan\theta,\ dx = \sqrt{5}\sec^2\theta\,d\theta,\ x^2 + 5 = 5\sec^2\theta.$

$$\int \frac{-5x}{(x^2+5)^{3/2}}\,dx = \int \frac{-5\sqrt{5}\tan\theta}{(5\sec^2\theta)^{3/2}}\sqrt{5}\sec^2\theta\,d\theta$$

$$= -\sqrt{5}\int \frac{\tan\theta}{\sec\theta}\,d\theta$$

$$= -\sqrt{5}\int \sin\theta\,d\theta$$

$$= \sqrt{5}\cos\theta + C$$

$$= \sqrt{5}\,\frac{\sqrt{5}}{\sqrt{x^2+5}} + C$$

$$= \frac{5}{\sqrt{x^2+5}} + C$$

33. Let $u = 1 + e^{2x},\ du = 2e^{2x}\,dx.$

$$\int e^{2x}\sqrt{1+e^{2x}}\,dx = \frac{1}{2}\int (1+e^{2x})^{1/2}(2e^{2x})\,dx = \frac{1}{3}(1+e^{2x})^{3/2} + C$$

35. Let $e^x = \sin\theta,\ e^x\,dx = \cos\theta\,d\theta,\ \sqrt{1-e^{2x}} = \cos\theta.$

$$\int e^x\sqrt{1-e^{2x}}\,dx = \int \cos^2\theta\,d\theta$$

$$= \frac{1}{2}\int (1 + \cos 2\theta)\,d\theta$$

$$= \frac{1}{2}\left[\theta + \frac{\sin 2\theta}{2}\right]$$

$$= \frac{1}{2}(\theta + \sin\theta\cos\theta) + C = \frac{1}{2}\left(\arcsin e^x + e^x\sqrt{1-e^{2x}}\right) + C$$

37. Let $x = \sqrt{2}\tan\theta,\ dx = \sqrt{2}\sec^2\theta\,d\theta,\ x^2 + 2 = 2\sec^2\theta.$

$$\int \frac{1}{4 + 4x^2 + x^4}\,dx = \int \frac{1}{(x^2+2)^2}\,dx$$

$$= \int \frac{\sqrt{2}\sec^2\theta\,d\theta}{4\sec^4\theta}$$

$$= \frac{\sqrt{2}}{4}\int \cos^2\theta\,d\theta$$

$$= \frac{\sqrt{2}}{4}\left(\frac{1}{2}\right)\int (1 + \cos 2\theta)\,d\theta$$

$$= \frac{\sqrt{2}}{8}\left(\theta + \frac{1}{2}\sin 2\theta\right) + C$$

$$= \frac{\sqrt{2}}{8}(\theta + \sin\theta\cos\theta) + C$$

$$= \frac{1}{4}\left[\frac{x}{x^2+2} + \frac{1}{\sqrt{2}}\arctan\frac{x}{\sqrt{2}}\right] + C$$

39. Since $x > \dfrac{1}{2}$,

$$u = \text{arcsec } 2x, \;\Rightarrow\; du = \frac{1}{x\sqrt{4x^2 - 1}}\, dx, \; dv = dx \;\Rightarrow\; v = x$$

$$\int \text{arcsec } 2x\, dx = x\, \text{arcsec } 2x - \int \frac{1}{\sqrt{4x^2 - 1}}\, dx$$

$$2x = \sec \theta,\; dx = \frac{1}{2}\sec\theta \tan\theta\, d\theta,\; \sqrt{4x^2 - 1} = \tan\theta$$

$$\int \text{arcsec } 2x\, dx = x\, \text{arcsec } 2x - \int \frac{(1/2)\sec\theta\tan\theta\, d\theta}{\tan\theta} = x\, \text{arcsec } 2x - \frac{1}{2}\int \sec\theta\, d\theta$$

$$= x\, \text{arcsec } 2x - \frac{1}{2}\ln|\sec\theta + \tan\theta| + C = x\, \text{arcsec } 2x - \frac{1}{2}\ln\left|2x + \sqrt{4x^2 - 1}\right| + C.$$

41. $\displaystyle \int \frac{1}{\sqrt{4x - x^2}}\, dx = \int \frac{1}{\sqrt{4 - (x - 2)^2}}\, dx = \arcsin\!\left(\frac{x - 2}{2}\right) + C$

43. Let $x + 2 = 2\tan\theta,\; dx = 2\sec^2\theta\, d\theta,\; \sqrt{(x + 2)^2 + 4} = 2\sec\theta.$

$$\int \frac{x}{\sqrt{x^2 + 4x + 8}}\, dx = \int \frac{x}{\sqrt{(x + 2)^2 + 4}}\, dx = \int \frac{(2\tan\theta - 2)(2\sec^2\theta)\, d\theta}{2\sec\theta}$$

$$= 2\int (\tan\theta - 1)(\sec\theta)\, d\theta$$

$$= 2[\sec\theta - \ln|\sec\theta + \tan\theta|] + C_1$$

$$= 2\left[\frac{\sqrt{(x + 2)^2 + 4}}{2} - \ln\left|\frac{\sqrt{(x + 2)^2 + 4}}{2} + \frac{x + 2}{2}\right|\right] + C_1$$

$$= \sqrt{x^2 + 4x + 8} - 2\left[\ln\left|\sqrt{x^2 + 4x + 8} + (x + 2)\right| - \ln 2\right] + C_1$$

$$= \sqrt{x^2 + 4x + 8} - 2\ln\left|\sqrt{x^2 + 4x + 8} + (x + 2)\right| + C$$

45. Let $t = \sin\theta,\; dt = \cos\theta\, d\theta,\; 1 - t^2 = \cos^2\theta.$

(a) $\displaystyle \int \frac{t^2}{(1 - t^2)^{3/2}}\, dt = \int \frac{\sin^2\theta \cos\theta\, d\theta}{\cos^3\theta}$

$$= \int \tan^2\theta\, d\theta$$

$$= \int (\sec^2\theta - 1)\, d\theta$$

$$= \tan\theta - \theta + C$$

$$= \frac{t}{\sqrt{1 - t^2}} - \arcsin t + C$$

Thus, $\displaystyle \int_0^{\sqrt{3}/2} \frac{t^2}{(1 - t^2)^{3/2}}\, dt = \left[\frac{t}{\sqrt{1 - t^2}} - \arcsin t\right]_0^{\sqrt{3}/2} = \frac{\sqrt{3}/2}{\sqrt{1/4}} - \arcsin\frac{\sqrt{3}}{2} = \sqrt{3} - \frac{\pi}{3} \approx 0.685.$

(b) When $t = 0,\; \theta = 0.$ When $t = \sqrt{3}/2,\; \theta = \pi/3.$ Thus,

$$\int_0^{\sqrt{3}/2} \frac{t^2}{(1 - t^2)^{3/2}}\, dt = \left[\tan\theta - \theta\right]_0^{\pi/3} = \sqrt{3} - \frac{\pi}{3} \approx 0.685.$$

47. (a) Let $x = 3 \tan \theta$, $dx = 3 \sec^2 \theta \, d\theta$, $\sqrt{x^2 + 9} = 3 \sec \theta$.

$$\int \frac{x^3}{\sqrt{x^2 + 9}} \, dx = \int \frac{(27 \tan^3 \theta)(3 \sec^2 \theta \, d\theta)}{3 \sec \theta}$$

$$= 27 \int (\sec^2 \theta - 1) \sec \theta \tan \theta \, d\theta$$

$$= 27 \left[\frac{1}{3} \sec^3 \theta - \sec \theta \right] + C = 9[\sec^3 \theta - 3 \sec \theta] + C$$

$$= 9 \left[\left(\frac{\sqrt{x^2 + 9}}{3} \right)^3 - 3 \left(\frac{\sqrt{x^2 + 9}}{3} \right) \right] + C = \frac{1}{3}(x^2 + 9)^{3/2} - 9\sqrt{x^2 + 9} + C$$

Thus, $\displaystyle\int_0^3 \frac{x^3}{\sqrt{x^2 + 9}} \, dx = \left[\frac{1}{3}(x^2 + 9)^{3/2} - 9\sqrt{x^2 + 9} \right]_0^3$

$$= \left(\frac{1}{3}\left(54\sqrt{2} \right) - 27\sqrt{2} \right) - (9 - 27)$$

$$= 18 - 9\sqrt{2} = 9\left(2 - \sqrt{2} \right) \approx 5.272.$$

(b) When $x = 0$, $\theta = 0$. When $x = 3$, $\theta = \pi/4$. Thus,

$$\int_0^3 \frac{x^3}{\sqrt{x^2 + 9}} \, dx = 9 \left[\sec^3 \theta - 3 \sec \theta \right]_0^{\pi/4} = 9\left(2\sqrt{2} - 3\sqrt{2} \right) - 9(1 - 3) = 9\left(2 - \sqrt{2} \right) \approx 5.272.$$

49. (a) Let $x = 3 \sec \theta$, $dx = 3 \sec \theta \tan \theta \, d\theta$, $\sqrt{x^2 - 9} = 3 \tan \theta$.

$$\int \frac{x^2}{\sqrt{x^2 - 9}} \, dx = \int \frac{9 \sec^2 \theta}{3 \tan \theta} 3 \sec \theta \tan \theta \, d\theta$$

$$= 9 \int \sec^3 \theta \, d\theta$$

$$= 9 \left[\frac{1}{2} \sec \theta \tan \theta + \frac{1}{2} \int \sec \theta \, d\theta \right] \qquad \text{(7.3 Exercise 90)}$$

$$= \frac{9}{2}[\sec \theta \tan \theta + \ln|\sec \theta + \tan \theta|]$$

$$= \frac{9}{2}\left[\frac{x}{3} \cdot \frac{\sqrt{x^2 - 9}}{3} + \ln\left| \frac{x}{3} + \frac{\sqrt{x^2 - 9}}{3} \right| \right]$$

Hence,

$$\int_4^6 \frac{x^2}{\sqrt{x^2 - 9}} \, dx = \frac{9}{2}\left[\frac{x\sqrt{x^2 - 9}}{9} + \ln\left| \frac{x}{3} + \frac{\sqrt{x^2 - 9}}{3} \right| \right]_4^6$$

$$= \frac{9}{2}\left[\left(\frac{6\sqrt{27}}{9} + \ln\left| 2 + \frac{\sqrt{27}}{3} \right| \right) - \left(\frac{4\sqrt{7}}{9} + \ln\left| \frac{4}{3} + \frac{\sqrt{7}}{3} \right| \right) \right]$$

$$= 9\sqrt{3} - 2\sqrt{7} + \frac{9}{2}\left(\ln\left(\frac{6 + \sqrt{27}}{3} \right) - \ln\left(\frac{4 + \sqrt{7}}{3} \right) \right)$$

$$= 9\sqrt{3} - 2\sqrt{7} + \frac{9}{2} \ln\left(\frac{6 + 3\sqrt{3}}{4 + \sqrt{7}} \right)$$

$$= 9\sqrt{3} - 2\sqrt{7} + \frac{9}{2} \ln\left(\frac{\left(4 - \sqrt{7} \right)\left(2 + \sqrt{3} \right)}{3} \right) \approx 12.644.$$

—CONTINUED—

49. —CONTINUED—

(b) When $x = 4$, $\theta = \text{arcsec}\left(\dfrac{4}{3}\right)$.

When $x = 6$, $\theta = \text{arcsec}(2) = \dfrac{\pi}{3}$.

$$\int_4^6 \frac{x^2}{\sqrt{x^2-9}}\, dx = \frac{9}{2}\Big[\sec\theta\tan\theta + \ln|\sec\theta + \tan\theta|\Big]_{\text{arcsec}(4/3)}^{\pi/3}$$

$$= \frac{9}{2}\Big[2\cdot\sqrt{3} + \ln\big|2 + \sqrt{3}\big|\Big] - \frac{9}{2}\Big[\frac{4}{3}\frac{\sqrt{7}}{3} + \ln\Big|\frac{4}{3} + \frac{\sqrt{7}}{3}\Big|\Big]$$

$$= 9\sqrt{3} - 2\sqrt{7} + \frac{9}{2}\ln\left(\frac{6 + 3\sqrt{3}}{4 + \sqrt{7}}\right) \approx 12.644$$

51. $\displaystyle\int \frac{x^2}{\sqrt{x^2 + 10x + 9}}\, dx = \frac{1}{2}\sqrt{x^2 + 10x + 9}\,(x - 15) + 33\ln\big|(x + 5) + \sqrt{x^2 + 10x + 9}\big| + C$

53. $\displaystyle\int \frac{x^2}{\sqrt{x^2 - 1}}\, dx = \frac{1}{2}\big(x\sqrt{x^2 - 1} + \ln\big|x + \sqrt{x^2 - 1}\big|\big) + C$ **55.** (a) $u = a\sin\theta$ (b) $u = a\tan\theta$ (c) $u = a\sec\theta$

57. $\displaystyle A = 4\int_0^a \frac{b}{a}\sqrt{a^2 - x^2}\, dx$

$$= \frac{4b}{a}\int_0^a \sqrt{a^2 - x^2}\, dx$$

$$= \left[\frac{4b}{a}\left(\frac{1}{2}\right)\left(a^2\arcsin\frac{x}{a} + x\sqrt{a^2 - x^2}\right)\right]_0^a$$

$$= \frac{2b}{a}\left(a^2\left(\frac{\pi}{2}\right)\right)$$

$$= \pi ab$$

Note: See Theorem 7.2 for $\int\sqrt{a^2 - x^2}\, dx$.

59. $x^2 + y^2 = a^2$

$$x = \pm\sqrt{a^2 - y^2}$$

$$A = 2\int_h^a \sqrt{a^2 - y^2}\, dy = \left[a^2\arcsin\left(\frac{y}{a}\right) + y\sqrt{a^2 - y^2}\right]_h^a \qquad \text{(Theorem 7.2)}$$

$$= \left(a^2\frac{\pi}{2}\right) - \left(a^2\arcsin\left(\frac{h}{a}\right) + h\sqrt{a^2 - h^2}\right)$$

$$= \frac{a^2\pi}{2} - a^2\arcsin\left(\frac{h}{a}\right) - h\sqrt{a^2 - h^2}$$

61. Let $x - 3 = \sin\theta$, $dx = \cos\theta\, d\theta$, $\sqrt{1 - (x - 3)^2} = \cos\theta$.

Shell Method:

$$V = 4\pi\int_2^4 x\sqrt{1 - (x - 3)^2}\, dx$$

$$= 4\pi\int_{-\pi/2}^{\pi/2} (3 + \sin\theta)\cos^2\theta\, d\theta$$

$$= 4\pi\left[\frac{3}{2}\int_{-\pi/2}^{\pi/2}(1 + \cos 2\theta)\, d\theta + \int_{-\pi/2}^{\pi/2}\cos^2\theta\sin\theta\, d\theta\right]$$

$$= 4\pi\left[\frac{3}{2}\left(\theta + \frac{1}{2}\sin 2\theta\right) - \frac{1}{3}\cos^3\theta\right]_{-\pi/2}^{\pi/2} = 6\pi^2$$

63. $y = \ln x, y' = \dfrac{1}{x}, 1 + (y')^2 = 1 + \dfrac{1}{x^2} = \dfrac{x^2+1}{x^2}$

Let $x = \tan \theta, dx = \sec^2 \theta\, d\theta, \sqrt{x^2+1} = \sec \theta.$

$$\begin{aligned}
s &= \int_1^5 \sqrt{\frac{x^2+1}{x^2}}\, dx = \int_1^5 \frac{\sqrt{x^2+1}}{x}\, dx \\
&= \int_a^b \frac{\sec \theta}{\tan \theta} \sec^2 \theta\, d\theta = \int_a^b \frac{\sec \theta}{\tan \theta}(1 + \tan^2 \theta)\, d\theta \\
&= \int_a^b (\csc \theta + \sec \theta \tan \theta)\, d\theta \\
&= \left[-\ln|\csc \theta + \cot \theta| + \sec \theta \right]_a^b \\
&= \left[-\ln\left|\frac{\sqrt{x^2+1}}{x} + \frac{1}{x}\right| + \sqrt{x^2+1}\, \right]_1^5 \\
&= \left[-\ln\left(\frac{\sqrt{26}+1}{5}\right) + \sqrt{26} \right] - \left[-\ln(\sqrt{2}+1) + \sqrt{2} \right] \\
&= \ln\left[\frac{5(\sqrt{2}+1)}{\sqrt{26}+1}\right] + \sqrt{26} - \sqrt{2} \approx 4.367 \ \text{ or } \ln\left[\frac{\sqrt{26}-1}{5(\sqrt{2}-1)}\right] + \sqrt{26} - \sqrt{2}
\end{aligned}$$

65. Length of one arch of sine curve: $y = \sin x, y' = \cos x$

$$L_1 = \int_0^\pi \sqrt{1 + \cos^2 x}\, dx$$

Length of one arch of cosine curve: $y = \cos x, y' = -\sin x$

$$\begin{aligned}
L_2 &= \int_{-\pi/2}^{\pi/2} \sqrt{1 + \sin^2 x}\, dx \\
&= \int_{-\pi/2}^{\pi/2} \sqrt{1 + \cos^2\left(x - \frac{\pi}{2}\right)}\, dx \qquad u = x - \frac{\pi}{2}, du = dx \\
&= \int_{-\pi}^0 \sqrt{1 + \cos^2 u}\, du \\
&= \int_0^\pi \sqrt{1 + \cos^2 u}\, du = L_1
\end{aligned}$$

67. (a)

(b) $y = 0$ for $x = 200$ (range)

(c) $y = x - 0.005x^2, y' = 1 - 0.01x, 1 + (y')^2 = 1 + (1 - 0.01x)^2$

Let $u = 1 - 0.01x, du = -0.01\, dx, a = 1.$ (See Theorem 7.2.)

$$\begin{aligned}
s &= \int_0^{200} \sqrt{1 + (1 - 0.01x)^2}\, dx = -100 \int_0^{200} \sqrt{(1 - 0.01x)^2 + 1}\,(-0.01)\, dx \\
&= -50\left[(1 - 0.01x)\sqrt{(1 - 0.01x)^2 + 1} + \ln\left|(1 - 0.01x) + \sqrt{(1 - 0.01x)^2 + 1}\,\right| \right]_0^{200} \\
&= -50\left[\left(-\sqrt{2} + \ln\left|-1 + \sqrt{2}\,\right|\right) - \left(\sqrt{2} + \ln\left|1 + \sqrt{2}\,\right|\right) \right] \\
&= 100\sqrt{2} + 50 \ln\left(\frac{\sqrt{2}+1}{\sqrt{2}-1}\right) \approx 229.559
\end{aligned}$$

69. Let $x = 3 \tan \theta$, $dx = 3 \sec^2 \theta \, d\theta$, $\sqrt{x^2 + 9} = 3 \sec \theta$.

$$A = 2 \int_0^4 \frac{3}{\sqrt{x^2 + 9}} \, dx = 6 \int_0^4 \frac{dx}{\sqrt{x^2 + 9}} = 6 \int_a^b \frac{3 \sec^2 \theta \, d\theta}{3 \sec \theta}$$

$$= 6 \int_a^b \sec \theta \, d\theta = \left[6 \ln|\sec \theta + \tan \theta| \right]_a^b = \left[6 \ln \left| \frac{\sqrt{x^2 + 9} + x}{3} \right| \right]_0^4 = 6 \ln 3$$

$\bar{x} = 0$ (by symmetry)

$$\bar{y} = \frac{1}{2} \left(\frac{1}{A} \right) \int_{-4}^4 \left(\frac{3}{\sqrt{x^2 + 9}} \right)^2 dx$$

$$= \frac{9}{12 \ln 3} \int_{-4}^4 \frac{1}{x^2 + 9} \, dx$$

$$= \frac{3}{4 \ln 3} \left[\frac{1}{3} \arctan \frac{x}{3} \right]_{-4}^4$$

$$= \frac{2}{4 \ln 3} \arctan \frac{4}{3} \approx 0.422$$

$$(\bar{x}, \bar{y}) = \left(0, \frac{1}{2 \ln 3} \arctan \frac{4}{3} \right) \approx (0, 0.422)$$

71. $y = x^2$, $\qquad y' = 2x$, $\qquad 1 + (y')^2 = 1 + 4x^2$

$2x = \tan \theta$, $dx = \frac{1}{2} \sec^2 \theta \, d\theta$, $\sqrt{1 + 4x^2} = \sec \theta$

(For $\int \sec^5 \theta \, d\theta$ and $\int \sec^3 \theta \, d\theta$, see Exercise 90 in Section 7.3)

$$S = 2\pi \int_0^{\sqrt{2}} x^2 \sqrt{1 + 4x^2} \, dx = 2\pi \int_a^b \left(\frac{\tan \theta}{2} \right)^2 (\sec \theta) \left(\frac{1}{2} \sec^2 \theta \right) d\theta$$

$$= \frac{\pi}{4} \int_a^b \sec^3 \theta \tan^2 \theta \, d\theta = \frac{\pi}{4} \left[\int_a^b \sec^5 \theta \, d\theta - \int_a^b \sec^3 \theta \, d\theta \right]$$

$$= \frac{\pi}{4} \left\{ \frac{1}{4} \left[\sec^3 \theta \tan \theta + \frac{3}{2} (\sec \theta \tan \theta + \ln|\sec \theta + \tan \theta|) \right] - \frac{1}{2} (\sec \theta \tan \theta + \ln|\sec \theta + \tan \theta|) \right\}_a^b$$

$$= \frac{\pi}{4} \left[\frac{1}{4} [(1 + 4x^2)^{3/2}(2x)] - \frac{1}{8} [(1 + 4x^2)^{1/2}(2x) + \ln \left| \sqrt{1 + 4x^2} + 2x \right|] \right]_0^{\sqrt{2}}$$

$$= \frac{\pi}{4} \left[\frac{54 \sqrt{2}}{4} - \frac{6 \sqrt{2}}{8} - \frac{1}{8} \ln \left(3 + 2 \sqrt{2} \right) \right]$$

$$= \frac{\pi}{4} \left(\frac{51 \sqrt{2}}{4} - \frac{\ln \left(3 + 2 \sqrt{2} \right)}{8} \right) = \frac{\pi}{32} \left[102 \sqrt{2} - \ln \left(3 + 2 \sqrt{2} \right) \right] \approx 13.989$$

73. (a) Area of representative rectangle: $2 \sqrt{1 - y^2} \, \Delta y$

Pressure: $2(62.4)(3 - y) \sqrt{1 - y^2} \, \Delta y$

$$F = 124.8 \int_{-1}^1 (3 - y) \sqrt{1 - y^2} \, dy$$

$$= 124.8 \left[3 \int_{-1}^1 \sqrt{1 - y^2} \, dy - \int_{-1}^1 y \sqrt{1 - y^2} \, dy \right]$$

$$= 124.8 \left[\frac{3}{2} \left(\arcsin y + y \sqrt{1 - y^2} \right) + \frac{1}{2} \left(\frac{2}{3} \right) (1 - y^2)^{3/2} \right]_{-1}^1$$

$$= (62.4)3[\arcsin 1 - \arcsin(-1)] = 187.2 \pi \text{ lb}$$

(b) $F = 124.8 \int_{-1}^1 (d - y) \sqrt{1 - y^2} \, dy = 124.8 d \int_{-1}^1 \sqrt{1 - y^2} \, dy - 124.8 \int_{-1}^1 y \sqrt{1 - y^2} \, dy$

$$= 124.8 \left(\frac{d}{2} \right) \left[\arcsin y + y \sqrt{1 - y^2} \right]_{-1}^1 - 124.8(0) = 62.4 \pi d \text{ lb}$$

75. (a) $m = \dfrac{dy}{dx} = \dfrac{y - \left(y + \sqrt{144 - x^2}\right)}{x - 0}$

$= -\dfrac{\sqrt{144 - x^2}}{x}$

(b) $y = -\displaystyle\int \dfrac{\sqrt{144 - x^2}}{x}\, dx$

Let $x = 12 \sin\theta$, $dx = 12 \cos\theta\, d\theta$, $\sqrt{144 - x^2} = 12 \cos\theta$.

$y = -\displaystyle\int \dfrac{12 \cos\theta}{12 \sin\theta} 12 \cos\theta\, d\theta = -12 \int \dfrac{1 - \sin^2\theta}{\sin\theta}\, d\theta$

$= -12 \displaystyle\int (\csc\theta - \sin\theta)\, d\theta = -12 \ln|\csc\theta - \cot\theta| - 12 \cos\theta + C$

$= -12 \ln\left|\dfrac{12}{x} - \dfrac{\sqrt{144 - x^2}}{x}\right| - 12\left(\dfrac{\sqrt{144 - x^2}}{12}\right) + C$

$= -12 \ln\left|\dfrac{12 - \sqrt{144 - x^2}}{x}\right| - \sqrt{144 - x^2} + C$

When $x = 12$, $y = 0 \Rightarrow C = 0$. Thus, $y = -12 \ln\left(\dfrac{12 - \sqrt{144 - x^2}}{x}\right) - \sqrt{144 - x^2}$.

Note: $\dfrac{12 - \sqrt{144 - x^2}}{x} > 0$ for $0 < x \le 12$

(c) Vertical asymptote: $x = 0$

(d) $y + \sqrt{144 - x^2} = 12 \Rightarrow y = 12 - \sqrt{144 - x^2}$

Thus,

$$12 - \sqrt{144 - x^2} = -12 \ln\left(\dfrac{12 - \sqrt{144 - x^2}}{x}\right) - \sqrt{144 - x^2}$$

$$-1 = \ln\left(\dfrac{12 - \sqrt{144 - x^2}}{x}\right)$$

$$xe^{-1} = 12 - \sqrt{144 - x^2}$$

$$(xe^{-1} - 12)^2 = \left(-\sqrt{144 - x^2}\right)^2$$

$$x^2 e^{-2} - 24xe^{-1} + 144 = 144 - x^2$$

$$x^2(e^{-2} + 1) - 24xe^{-1} = 0$$

$$x[x(e^{-2} + 1) - 24e^{-1}] = 0$$

$$x = 0 \text{ or } x = \dfrac{24e^{-1}}{e^{-2} + 1} \approx 7.77665.$$

Therefore,

$$s = \int_{7.77665}^{12} \sqrt{1 + \left(-\dfrac{\sqrt{144 - x^2}}{x}\right)^2}\, dx = \int_{7.77665}^{12} \sqrt{\dfrac{x^2 + (144 - x^2)}{x^2}}\, dx$$

$$= \int_{7.77665}^{12} \dfrac{12}{x}\, dx = \left[12 \ln|x|\right]_{7.77665}^{12} = 12(\ln 12 - \ln 7.77665) \approx 5.2 \text{ meters.}$$

77. True

$$\int \frac{dx}{\sqrt{1 - x^2}} = \int \frac{\cos \theta \, d\theta}{\cos \theta} = \int d\theta$$

79. False

$$\int_0^{\sqrt{3}} \frac{dx}{(\sqrt{1 + x^2})^3} = \int_0^{\pi/3} \frac{\sec^2 \theta \, d\theta}{\sec^3 \theta} = \int_0^{\pi/3} \cos \theta \, d\theta$$

81. Let $u = a \sin \theta$, $du = a \cos \theta \, d\theta$, $\sqrt{a^2 - u^2} = a \cos \theta$.

$$\int \sqrt{a^2 - u^2} \, du = \int a^2 \cos^2 \theta \, d\theta = a^2 \int \frac{1 + \cos 2\theta}{2} \, d\theta$$

$$= \frac{a^2}{2} \left(\theta + \frac{1}{2} \sin 2\theta \right) + C = \frac{a^2}{2} (\theta + \sin \theta \cos \theta) + C$$

$$= \frac{a^2}{2} \left[\arcsin \frac{u}{a} + \left(\frac{u}{a} \right) \left(\frac{\sqrt{a^2 + u^2}}{a} \right) \right] + C = \frac{1}{2} \left[a^2 \arcsin \frac{u}{a} + u\sqrt{a^2 - u^2} \right] + C$$

Let $u = a \sec \theta$, $du = a \sec \theta \tan \theta \, d\theta$, $\sqrt{u^2 - a^2} = a \tan \theta$.

$$\int \sqrt{u^2 - a^2} \, du = \int a \tan \theta (a \sec \theta \tan \theta) \, d\theta = a^2 \int \tan^2 \theta \sec \theta \, d\theta$$

$$= a^2 \int (\sec^2 \theta - 1) \sec \theta \, d\theta = a^2 \int (\sec^3 \theta - \sec \theta) \, d\theta$$

$$= a^2 \left[\frac{1}{2} \sec \theta \tan \theta + \frac{1}{2} \int \sec \theta \, d\theta \right] - a^2 \int \sec \theta \, d\theta = a^2 \left[\frac{1}{2} \sec \theta \tan \theta - \frac{1}{2} \ln|\sec \theta + \tan \theta| \right]$$

$$= \frac{a^2}{2} \left[\frac{u}{a} \cdot \frac{\sqrt{u^2 - a^2}}{a} - \ln \left| \frac{u}{a} + \frac{\sqrt{u^2 - a^2}}{a} \right| \right] + C_1$$

$$= \frac{1}{2} \left[u\sqrt{u^2 - a^2} - a^2 \ln|u + \sqrt{u^2 - a^2}| \right] + C$$

Let $u = a \tan \theta$, $du = a \sec^2 \theta \, d\theta$, $\sqrt{u^2 + a^2} = a \sec \theta \, d\theta$.

$$\int \sqrt{u^2 + a^2} \, du = \int (a \sec \theta)(a \sec^2 \theta) \, d\theta$$

$$= a^2 \int \sec^3 \theta \, d\theta = a^2 \left[\frac{1}{2} \sec \theta \tan \theta + \frac{1}{2} \ln|\sec \theta + \tan \theta| \right] + C_1$$

$$= \frac{a^2}{2} \left[\frac{\sqrt{u^2 + a^2}}{a} \cdot \frac{u}{a} + \ln \left| \frac{\sqrt{u^2 + a^2}}{a} + \frac{u}{a} \right| \right] + C_1 = \frac{1}{2} \left[u\sqrt{u^2 + a^2} + a^2 \ln|u + \sqrt{u^2 + a^2}| \right] + C$$

Section 7.5 Partial Fractions

1. $\dfrac{5}{x^2 - 10x} = \dfrac{5}{x(x - 10)} = \dfrac{A}{x} + \dfrac{B}{x - 10}$

3. $\dfrac{2x - 3}{x^3 + 10x} = \dfrac{2x - 3}{x(x^2 + 10)} = \dfrac{A}{x} + \dfrac{Bx + C}{x^2 + 10}$

5. $\dfrac{16x}{x^3 - 10x^2} = \dfrac{16x}{x^2(x - 10)} = \dfrac{A}{x} + \dfrac{B}{x^2} + \dfrac{C}{x - 10}$

7. $\dfrac{1}{x^2 - 1} = \dfrac{1}{(x + 1)(x - 1)} = \dfrac{A}{x + 1} + \dfrac{B}{x - 1}$

$$1 = A(x - 1) + B(x + 1)$$

When $x = -1$, $1 = -2A$, $A = -\frac{1}{2}$.

When $x = 1$, $1 = 2B$, $B = \frac{1}{2}$.

$$\int \frac{1}{x^2 - 1} \, dx = -\frac{1}{2} \int \frac{1}{x + 1} \, dx + \frac{1}{2} \int \frac{1}{x - 1} \, dx$$

$$= -\frac{1}{2} \ln|x + 1| + \frac{1}{2} \ln|x - 1| + C$$

$$= \frac{1}{2} \ln \left| \frac{x - 1}{x + 1} \right| + C$$

9. $\dfrac{3}{x^2 + x - 2} = \dfrac{3}{(x-1)(x+2)} = \dfrac{A}{x-1} + \dfrac{B}{x+2}$

$$3 = A(x+2) + B(x-1)$$

When $x = 1, 3 = 3A, A = 1$.
When $x = -2, 3 = -3B, B = -1$.

$$\int \frac{3}{x^2 + x - 2}\,dx = \int \frac{1}{x-1}\,dx - \int \frac{1}{x+2}\,dx$$

$$= \ln|x-1| - \ln|x+2| + C$$

$$= \ln\left|\frac{x-1}{x+2}\right| + C$$

11. $\dfrac{5-x}{2x^2 + x - 1} = \dfrac{5-x}{(2x-1)(x+1)} = \dfrac{A}{2x-1} + \dfrac{B}{x+1}$

$$5 - x = A(x+1) + B(2x-1)$$

When $x = \frac{1}{2}, \frac{9}{2} = \frac{3}{2}A, A = 3$.

When $x = -1, 6 = -3B, B = -2$.

$$\int \frac{5-x}{2x^2 + x - 1}\,dx = 3\int \frac{1}{2x-1}\,dx - 2\int \frac{1}{x+1}\,dx$$

$$= \frac{3}{2}\ln|2x-1| - 2\ln|x+1| + C$$

13. $\dfrac{x^2 + 12x + 12}{x(x+2)(x-2)} = \dfrac{A}{x} + \dfrac{B}{x+2} + \dfrac{C}{x-2}$

$$x^2 + 12x + 12 = A(x+2)(x-2) + Bx(x-2) + Cx(x+2)$$

When $x = 0, 12 = -4A, A = -3$. When $x = -2, -8 = 8B, B = -1$. When $x = 2, 40 = 8C, C = 5$.

$$\int \frac{x^2 + 12x + 12}{x^3 - 4x}\,dx = 5\int \frac{1}{x-2}\,dx - \int \frac{1}{x+2}\,dx - 3\int \frac{1}{x}\,dx$$

$$= 5\ln|x-2| - \ln|x+2| - 3\ln|x| + C$$

15. $\dfrac{2x^3 - 4x^2 - 15x + 5}{x^2 - 2x - 8} = 2x + \dfrac{x+5}{(x-4)(x+2)} = 2x + \dfrac{A}{x-4} + \dfrac{B}{x+2}$

$$x + 5 = A(x+2) + B(x-4)$$

When $x = 4, 9 = 6A, A = \frac{3}{2}$. When $x = -2, 3 = -6B, B = -\frac{1}{2}$.

$$\int \frac{2x^3 - 4x^2 - 15x + 5}{x^2 - 2x - 8}\,dx = \int \left[2x + \frac{3/2}{x-4} - \frac{1/2}{x+2}\right]dx$$

$$= x^2 + \frac{3}{2}\ln|x-4| - \frac{1}{2}\ln|x+2| + C$$

17. $\dfrac{4x^2 + 2x - 1}{x^2(x+1)} = \dfrac{A}{x} + \dfrac{B}{x^2} + \dfrac{C}{x+1}$

$$4x^2 + 2x - 1 = Ax(x+1) + B(x+1) + Cx^2$$

When $x = 0, B = -1$. When $x = -1, C = 1$. When $x = 1, A = 3$.

$$\int \frac{4x^2 + 2x - 1}{x^3 + x^2}\,dx = \int \left[\frac{3}{x} - \frac{1}{x^2} + \frac{1}{x+1}\right]dx = 3\ln|x| + \frac{1}{x} + \ln|x+1| + C$$

$$= \frac{1}{x} + \ln|x^4 + x^3| + C$$

19. $\dfrac{x^2 + 3x - 4}{x^3 - 4x^2 + 4x} = \dfrac{x^2 + 3x - 4}{x(x-2)^2} = \dfrac{A}{x} + \dfrac{B}{(x-2)} + \dfrac{C}{(x-2)^2}$

$$x^2 + 3x - 4 = A(x-2)^2 + Bx(x-2) + Cx$$

When $x = 0, -4 = 4A \implies A = -1$. When $x = 2, 6 = 2C \implies C = 3$. When $x = 1, 0 = -1 - B + 3 \implies B = 2$.

$$\int \frac{x^2 + 3x - 4}{x^3 - 4x^2 + 4x}\,dx = \int \frac{-1}{x}\,dx + \int \frac{2}{(x-2)}\,dx + \int \frac{3}{(x-2)^2}\,dx$$

$$= -\ln|x| + 2\ln|x-2| - \frac{3}{(x-2)} + C$$

21. $\dfrac{x^2 - 1}{x(x^2 + 1)} = \dfrac{A}{x} + \dfrac{Bx + C}{x^2 + 1}$

$$x^2 - 1 = A(x^2 + 1) + (Bx + C)x$$

When $x = 0$, $A = -1$. When $x = 1$, $0 = -2 + B + C$. When $x = -1$, $0 = -2 + B - C$.
Solving these equations we have $A = -1$, $B = 2$, $C = 0$.

$$\int \frac{x^2 - 1}{x^3 + x}\,dx = -\int \frac{1}{x}\,dx + \int \frac{2x}{x^2 + 1}\,dx$$

$$= \ln|x^2 + 1| - \ln|x| + C$$

$$= \ln\left|\frac{x^2 + 1}{x}\right| + C$$

23. $\dfrac{x^2}{x^4 - 2x^2 - 8} = \dfrac{A}{x - 2} + \dfrac{B}{x + 2} + \dfrac{Cx + D}{x^2 + 2}$

$$x^2 = A(x + 2)(x^2 + 2) + B(x - 2)(x^2 + 2) + (Cx + D)(x + 2)(x - 2)$$

When $x = 2$, $4 = 24A$. When $x = -2$, $4 = -24B$. When $x = 0$, $0 = 4A - 4B - 4D$, and when $x = 1$,
$1 = 9A - 3B - 3C - 3D$. Solving these equations we have $A = \frac{1}{6}$, $B = -\frac{1}{6}$, $C = 0$, $D = \frac{1}{3}$.

$$\int \frac{x^2}{x^4 - 2x^2 - 8}\,dx = \frac{1}{6}\left[\int \frac{1}{x - 2}\,dx - \int \frac{1}{x + 2}\,dx + 2\int \frac{1}{x^2 + 2}\,dx\right]$$

$$= \frac{1}{6}\left[\ln\left|\frac{x - 2}{x + 2}\right| + \sqrt{2}\arctan\frac{x}{\sqrt{2}}\right] + C$$

25. $\dfrac{x}{(2x - 1)(2x + 1)(4x^2 + 1)} = \dfrac{A}{2x - 1} + \dfrac{B}{2x + 1} + \dfrac{Cx + D}{4x^2 + 1}$

$$x = A(2x + 1)(4x^2 + 1) + B(2x - 1)(4x^2 + 1) + (Cx + D)(2x - 1)(2x + 1)$$

When $x = \frac{1}{2}$, $\frac{1}{2} = 4A$. When $x = -\frac{1}{2}$, $-\frac{1}{2} = -4B$. When $x = 0$, $0 = A - B - D$, and when $x = 1$,
$1 = 15A + 5B + 3C + 3D$. Solving these equations we have $A = \frac{1}{8}$, $B = \frac{1}{8}$, $C = -\frac{1}{2}$, $D = 0$.

$$\int \frac{x}{16x^4 - 1}\,dx = \frac{1}{8}\left[\int \frac{1}{2x - 1}\,dx + \int \frac{1}{2x + 1}\,dx - 4\int \frac{x}{4x^2 + 1}\,dx\right]$$

$$= \frac{1}{16}\ln\left|\frac{4x^2 - 1}{4x^2 + 1}\right| + C$$

27. $\dfrac{x^2 + 5}{(x + 1)(x^2 - 2x + 3)} = \dfrac{A}{x + 1} + \dfrac{Bx + C}{x^2 - 2x + 3}$

$$x^2 + 5 = A(x^2 - 2x + 3) + (Bx + C)(x + 1)$$

$$= (A + B)x^2 + (-2A + B + C)x + (3A + C)$$

When $x = -1$, $A = 1$. By equating coefficients of like terms, we have $A + B = 1$, $-2A + B + C = 0$,
$3A + C = 5$. Solving these equations we have $A = 1$, $B = 0$, $C = 2$.

$$\int \frac{x^2 + 5}{x^3 - x^2 + x + 3}\,dx = \int \frac{1}{x + 1}\,dx + 2\int \frac{1}{(x - 1)^2 + 2}\,dx$$

$$= \ln|x + 1| + \sqrt{2}\arctan\left(\frac{x - 1}{\sqrt{2}}\right) + C$$

29. $\dfrac{3}{(2x+1)(x+2)} = \dfrac{A}{2x+1} + \dfrac{B}{x+2}$

$$3 = A(x+2) + B(2x+1)$$

When $x = -\frac{1}{2}$, $A = 2$. When $x = -2$, $B = -1$.

$$\int_0^1 \frac{3}{2x^2+5x+2}\,dx = \int_0^1 \frac{2}{2x+1}\,dx - \int_0^1 \frac{1}{x+2}\,dx$$

$$= \Big[\ln|2x+1| - \ln|x+2|\Big]_0^1$$

$$= \ln 2$$

31. $\dfrac{x+1}{x(x^2+1)} = \dfrac{A}{x} + \dfrac{Bx+C}{x^2+1}$

$$x+1 = A(x^2+1) + (Bx+C)x$$

When $x = 0$, $A = 1$. When $x = 1$, $2 = 2A + B + C$. When $x = -1$, $0 = 2A + B - C$. Solving these equations we have $A = 1$, $B = -1$, $C = 1$.

$$\int_1^2 \frac{x+1}{x(x^2+1)}\,dx = \int_1^2 \frac{1}{x}\,dx - \int_1^2 \frac{x}{x^2+1}\,dx + \int_1^2 \frac{1}{x^2+1}\,dx$$

$$= \left[\ln|x| - \frac{1}{2}\ln(x^2+1) + \arctan x\right]_1^2$$

$$= \frac{1}{2}\ln\frac{8}{5} - \frac{\pi}{4} + \arctan 2$$

$$\approx 0.557$$

33. $\displaystyle\int \frac{3x\,dx}{x^2-6x+9} = 3\ln|x-3| - \frac{9}{x-3} + C$

$(4, 0)$: $3\ln|4-3| - \dfrac{9}{4-3} + C = 0 \Rightarrow C = 9$

35. $\displaystyle\int \frac{x^2+x+2}{(x^2+2)^2}\,dx = \frac{\sqrt{2}}{2}\arctan\frac{x}{\sqrt{2}} - \frac{1}{2(x^2+2)} + C$

$(0, 1)$: $0 - \dfrac{1}{4} + C = 1 \Rightarrow C = \dfrac{5}{4}$

37. $\displaystyle\int \frac{2x^2-2x+3}{x^3-x^2-x-2}\,dx = \ln|x-2| + \frac{1}{2}\ln|x^2+x+1| - \sqrt{3}\arctan\!\left(\frac{2x+1}{\sqrt{3}}\right) + C$

$(3, 10)$: $0 + \dfrac{1}{2}\ln 13 - \sqrt{3}\arctan\dfrac{7}{\sqrt{3}} + C = 10 \Rightarrow C = 10 - \dfrac{1}{2}\ln 13 + \sqrt{3}\arctan\dfrac{7}{\sqrt{3}}$

39. $\displaystyle\int \frac{1}{x^2-4}\,dx = \frac{1}{4}\ln\left|\frac{x-2}{x+2}\right| + C$

$(6, 4)$: $\dfrac{1}{4}\ln\left|\dfrac{4}{8}\right| + C = 4 \Rightarrow C = 4 - \dfrac{1}{4}\ln\dfrac{1}{2} = 4 + \dfrac{1}{4}\ln 2$

41. Let $u = \cos x$ $du = -\sin x\,dx$.

$$\frac{1}{u(u-1)} = \frac{A}{u} + \frac{B}{u-1}$$

$$1 = A(u-1) + Bu$$

When $u = 0, A = -1$. When $u = 1, B = 1, u = \cos x$, $du = -\sin x\,dx$.

$$\int \frac{\sin x}{\cos x(\cos x - 1)}\,dx = -\int \frac{1}{u(u-1)}\,du$$

$$= \int \frac{1}{u}\,du - \int \frac{1}{u-1}\,du$$

$$= \ln|u| - \ln|u-1| + C$$

$$= \ln\left|\frac{u}{u-1}\right| + C$$

$$= \ln\left|\frac{\cos x}{\cos x - 1}\right| + C$$

43. $\displaystyle\int \frac{3\cos x}{\sin^2 x + \sin x - 2}\,dx = 3\int \frac{1}{u^2 + u - 2}\,du$

$$= \ln\left|\frac{u-1}{u+2}\right| + C$$

$$= \ln\left|\frac{-1 + \sin x}{2 + \sin x}\right| + C$$

(From Exercise 9 with $u = \sin x$, $du = \cos x\,dx$)

45. Let $u = e^x$, $du = e^x\,dx$.

$$\frac{1}{(u-1)(u+4)} = \frac{A}{u-1} + \frac{B}{u+4}$$

$$1 = A(u+4) + B(u-1)$$

When $u = 1, A = \frac{1}{5}$. When $u = -4, B = -\frac{1}{5}, u = e^x$, $du = e^x\,dx$.

$$\int \frac{e^x}{(e^x - 1)(e^x + 4)}\,dx = \int \frac{1}{(u-1)(u+4)}\,du$$

$$= \frac{1}{5}\left(\int \frac{1}{u-1}\,du - \int \frac{1}{u+4}\,du\right)$$

$$= \frac{1}{5}\ln\left|\frac{u-1}{u+4}\right| + C$$

$$= \frac{1}{5}\ln\left|\frac{e^x - 1}{e^x + 4}\right| + C$$

47. $\displaystyle\frac{1}{x(a + bx)} = \frac{A}{x} + \frac{B}{a + bx}$

$$1 = A(a + bx) + Bx$$

When $x = 0, 1 = aA \Longrightarrow A = 1/a$.
When $x = -a/b, 1 = -(a/b)B \Longrightarrow B = -b/a$.

$$\int \frac{1}{x(a + bx)}\,dx = \frac{1}{a}\int\left(\frac{1}{x} - \frac{b}{a + bx}\right)\,dx$$

$$= \frac{1}{a}\bigl(\ln|x| - \ln|a + bx|\bigr) + C$$

$$= \frac{1}{a}\ln\left|\frac{x}{a + bx}\right| + C$$

49. $\displaystyle\frac{x}{(a + bx)^2} = \frac{A}{a + bx} + \frac{B}{(a + bx)^2}$

$$x = A(a + bx) + B$$

When $x = -a/b, B = -a/b$.
When $x = 0, 0 = aA + B \Longrightarrow A = 1/b$.

$$\int \frac{x}{(a + bx)^2}\,dx = \int\left(\frac{1/b}{a + bx} + \frac{-a/b}{(a + bx)^2}\right)\,dx$$

$$= \frac{1}{b}\int \frac{1}{a + bx}\,dx - \frac{a}{b}\int \frac{1}{(a + bx)^2}\,dx$$

$$= \frac{1}{b^2}\ln|a + bx| + \frac{a}{b^2}\left(\frac{1}{a + bx}\right) + C$$

$$= \frac{1}{b^2}\left(\frac{a}{a + bx} + \ln|a + bx|\right) + C$$

51. $\displaystyle\frac{dy}{dx} = \frac{6}{4 - x^2}$, $y(0) = 3$

53. Dividing x^3 by $x - 5$.

55. (a) Substitution: $u = x^2 + 2x - 8$

(b) Partial fractions

(c) Trigonometric substitution (tan) or inverse tangent rule

57. Average Cost $= \dfrac{1}{80 - 75} \displaystyle\int_{75}^{80} \dfrac{124p}{(10 + p)(100 - p)}\, dp$

$\qquad = \dfrac{1}{5} \displaystyle\int_{75}^{80} \left(\dfrac{-124}{(10 + p)11} + \dfrac{1240}{(100 - p)11} \right) dp$

$\qquad = \dfrac{1}{5} \left[\dfrac{-124}{11} \ln(10 + p) - \dfrac{1240}{11} \ln(100 - p) \right]_{75}^{80}$

$\qquad \approx \dfrac{1}{5}(24.51) = 4.9$

Approximately \$490,000.

59. $A = \displaystyle\int_{1}^{3} \dfrac{10}{x(x^2 + 1)}\, dx \approx 3$

Matches (c)

61. $\qquad \dfrac{1}{(x + 1)(n - x)} = \dfrac{A}{x + 1} + \dfrac{B}{n - x}, A = B = \dfrac{1}{n + 1}$

$\dfrac{1}{n + 1} \displaystyle\int \left(\dfrac{1}{x + 1} + \dfrac{1}{n - x} \right) dx = kt + C$

$\qquad \dfrac{1}{n + 1} \ln\left| \dfrac{x + 1}{n - x} \right| = kt + C$

When $t = 0, x = 0, C = \dfrac{1}{n + 1} \ln\dfrac{1}{n}$.

$\qquad \dfrac{1}{n + 1} \ln\left| \dfrac{x + 1}{n - x} \right| = kt + \dfrac{1}{n + 1} \ln\dfrac{1}{n}$

$\dfrac{1}{n + 1} \left[\ln\left| \dfrac{x + 1}{n - x} \right| - \ln\dfrac{1}{n} \right] = kt$

$\qquad \ln\dfrac{nx + n}{n - x} = (n + 1)kt$

$\qquad \dfrac{nx + n}{n - x} = e^{(n + 1)kt}$

$\qquad x = \dfrac{n\left[e^{(n + 1)kt} - 1 \right]}{n + e^{(n + 1)kt}}$ **Note:** $\displaystyle\lim_{t \to \infty} x = n$

63. $\dfrac{x}{1+x^4} = \dfrac{Ax+B}{x^2+\sqrt{2}\,x+1} + \dfrac{Cx+D}{x^2-\sqrt{2}\,x+1}$

$$x = (Ax+B)(x^2-\sqrt{2}\,x+1) + (Cx+D)(x^2+\sqrt{2}\,x+1)$$

$$= (A+C)x^3 + (B+D-\sqrt{2}\,A+\sqrt{2}\,C)x^2 + (A+C-\sqrt{2}\,B+\sqrt{2}\,D)x + (B+D)$$

$0 = A + C \implies C = -A$

$\left.\begin{array}{l} 0 = B + D - \sqrt{2}\,A + \sqrt{2}\,C \\[2pt] 1 = A + C - \sqrt{2}\,B + \sqrt{2}\,D \end{array}\right\}$ $\begin{array}{l} -2\sqrt{2}\,A = 0 \implies A = 0 \text{ and } C = 0 \\[4pt] -2\sqrt{2}\,B = 1 \implies B = -\dfrac{\sqrt{2}}{4} \text{ and } D = \dfrac{\sqrt{2}}{4} \end{array}$

$0 = B + D \implies D = -B$

Thus,

$$\int_0^1 \frac{x}{1+x^4}\,dx = \int_0^1 \left[\frac{-\sqrt{2}/4}{x^2+\sqrt{2}\,x+1} + \frac{\sqrt{2}/4}{x^2-\sqrt{2}\,x+1} \right] dx$$

$$= \frac{\sqrt{2}}{4} \int_0^1 \left[\frac{-1}{\left[x+\left(\sqrt{2}/2\right)\right]^2 + (1/2)} + \frac{1}{\left[x-\left(\sqrt{2}/2\right)\right]^2 + (1/2)} \right] dx$$

$$= \frac{\sqrt{2}}{4} \cdot \frac{1}{1/\sqrt{2}} \left[-\arctan\!\left(\frac{x+\left(\sqrt{2}/2\right)}{1/\sqrt{2}} \right) + \arctan\!\left(\frac{x-\left(\sqrt{2}/2\right)}{1/\sqrt{2}} \right) \right]_0^1$$

$$= \frac{1}{2}\left[-\arctan\!\left(\sqrt{2}\,x+1\right) + \arctan\!\left(\sqrt{2}\,x-1\right) \right]_0^1$$

$$= \frac{1}{2}\left[\left(-\arctan(\sqrt{2}+1) + \arctan(\sqrt{2}-1)\right) - \left(-\arctan 1 + \arctan(-1)\right) \right]$$

$$= \frac{1}{2}\left[\arctan(\sqrt{2}-1) - \arctan(\sqrt{2}+1) + \frac{\pi}{4} + \frac{\pi}{4} \right].$$

Since $\arctan x - \arctan y = \arctan[(x-y)/(1+xy)]$, we have:

$$\int_0^1 \frac{x}{1+x^4}\,dx = \frac{1}{2}\left[\arctan\!\left(\frac{(\sqrt{2}-1)-(\sqrt{2}+1)}{1+(\sqrt{2}-1)(\sqrt{2}+1)} \right) + \frac{\pi}{2} \right] = \frac{1}{2}\left[\arctan\!\left(\frac{-2}{2}\right) + \frac{\pi}{2} \right] = \frac{1}{2}\left[-\frac{\pi}{4} + \frac{\pi}{2} \right] = \frac{\pi}{8}$$

Section 7.6   Integration by Tables and Other Integration Techniques

1. By Formula 6: $\displaystyle\int \frac{x^2}{1+x}\,dx = -\frac{x}{2}(2-x) + \ln|1+x| + C$

3. By Formula 26: $\displaystyle\int e^x\sqrt{1+e^{2x}}\,dx = \frac{1}{2}\left[e^x\sqrt{e^{2x}+1} + \ln\left|e^x + \sqrt{e^{2x}+1}\right| \right] + C$

$u = e^x,\ du = e^x\,dx$

5. By Formula 44: $\displaystyle\int \frac{1}{x^2\sqrt{1-x^2}}\,dx = -\frac{\sqrt{1-x^2}}{x} + C$

7. By Formulas 50 and 48: $\displaystyle\int \sin^4(2x)\,dx = \frac{1}{2}\int \sin^4(2x)(2)\,dx$

$$= \frac{1}{2}\left[\frac{-\sin^3(2x)\cos(2x)}{4} + \frac{3}{4}\int \sin^2(2x)(2)\,dx\right]$$

$$= \frac{1}{2}\left[\frac{-\sin^3(2x)\cos(2x)}{4} + \frac{3}{8}(2x - \sin 2x \cos 2x)\right] + C$$

$$= \frac{1}{16}(6x - 3\sin 2x \cos 2x - 2\sin^3 2x \cos 2x) + C$$

9. By Formula 57: $\displaystyle\int \frac{1}{\sqrt{x}(1 - \cos\sqrt{x})}\,dx = 2\int \frac{1}{1 - \cos\sqrt{x}}\left(\frac{1}{2\sqrt{x}}\right)dx$

$$= -2\left(\cot\sqrt{x} + \csc\sqrt{x}\right) + C$$

$$u = \sqrt{x},\, du = \frac{1}{2\sqrt{x}}\,dx$$

11. By Formula 84:

$$\int \frac{1}{1 + e^{2x}}\,dx = x - \frac{1}{2}\ln(1 + e^{2x}) + C$$

13. By Formula 89:

$$\int x^3 \ln x\,dx = \frac{x^4}{16}(4\ln|x| - 1) + C$$

15. (a) By Formulas 83 and 82: $\displaystyle\int x^2 e^x\,dx = x^2 e^x - 2\int xe^x\,dx$

$$= x^2 e^x - 2[(x - 1)e^x + C_1]$$

$$= x^2 e^x - 2xe^x + 2e^x + C$$

(b) Integration by parts: $u = x^2,\, du = 2x\,dx,\, dv = e^x\,dx,\, v = e^x$

$$\int x^2 e^x\,dx = x^2 e^x - \int 2xe^x\,dx$$

Parts again: $u = 2x,\, du = 2\,dx,\, dv = e^x\,dx,\, v = e^x$

$$\int x^2 e^x\,dx = x^2 e^x - \left[2xe^x - \int 2e^x\,dx\right] = x^2 e^x - 2xe^x + 2e^x + C$$

17. (a) By Formula: 12, $a = b = 1,\, u = x$, and

$$\int \frac{1}{x^2(x + 1)}\,dx = \frac{-1}{1}\left(\frac{1}{x} + \frac{1}{1}\ln\left|\frac{x}{1 + x}\right|\right) + C$$

$$= \frac{-1}{x} - \ln\left|\frac{x}{1 + x}\right| + C$$

$$= \frac{-1}{x} + \ln\left|\frac{x + 1}{x}\right| + C$$

(b) Partial fractions:

$$\frac{1}{x^2(x + 1)} = \frac{A}{x} + \frac{B}{x^2} + \frac{C}{x + 1}$$

$$1 = Ax(x + 1) + B(x + 1) + Cx^2$$

$x = 0$: $1 = B$

$x = -1$: $1 = C$

$x = 1$: $1 = 2A + 2 + 1 \Rightarrow A = -1$

$$\int \frac{1}{x^2(x + 1)}\,dx = \int\left[\frac{-1}{x} + \frac{1}{x^2} + \frac{1}{x + 1}\right]dx$$

$$= -\ln|x| - \frac{1}{x} + \ln|x + 1| + C$$

$$= -\frac{1}{x} - \ln\left|\frac{x}{x + 1}\right| + C$$

19. By Formula 81: $\int xe^{x^2} = \dfrac{1}{2}e^{x^2} + C$

21. By Formula 79: $\int x \arcsec(x^2 + 1)\, dx = \dfrac{1}{2}\int \arcsec(x^2 + 1)(2x)\, dx$

$$= \dfrac{1}{2}\Big[(x^2 + 1)\arcsec(x^2 + 1) - \ln\big((x^2 + 1) + \sqrt{x^4 + 2x^2}\big)\Big] + C$$

$u = x^2 + 1,\, du = 2x\, dx$

23. By Formula 89: $\int x^2 \ln x\, dx = \dfrac{x^3}{9}(-1 + 3\ln|x|) + C$

25. By Formula 35: $\int \dfrac{1}{x^2\sqrt{x^2 - 4}}\, dx = \dfrac{\sqrt{x^2 - 4}}{4x} + C$

27. By Formula 4: $\int \dfrac{2x}{(1 - 3x)^2}\, dx = 2\int \dfrac{x}{(1 - 3x)^2}\, dx = \dfrac{2}{9}\left(\ln|1 - 3x| + \dfrac{1}{1 - 3x}\right) + C$

29. By Formula 76:

$$\int e^x \arccos e^x\, dx = e^x \arccos e^x - \sqrt{1 - e^{2x}} + C$$

$u = e^x,\, du = e^x\, dx$

31. By Formula 73:

$$\int \dfrac{x}{1 - \sec x^2}\, dx = \dfrac{1}{2}\int \dfrac{2x}{1 - \sec x^2}\, dx$$

$$= \dfrac{1}{2}(x^2 + \cot x^2 + \csc x^2) + C$$

33. By Formula 23: $\int \dfrac{\cos x}{1 + \sin^2 x}\, dx = \arctan(\sin x) + C$

$u = \sin x,\, du = \cos x\, dx$

35. By Formula 14: $\int \dfrac{\cos \theta}{3 + 2\sin \theta + \sin^2 \theta}\, d\theta = \dfrac{\sqrt{2}}{2}\arctan\!\left(\dfrac{1 + \sin \theta}{\sqrt{2}}\right) + C$

$u = \sin \theta,\, du = \cos \theta\, d\theta$

37. By Formula 35: $\int \dfrac{1}{x^2\sqrt{2 + 9x^2}}\, dx = 3\int \dfrac{3}{(3x)^2\sqrt{(\sqrt{2})^2 + (3x)^2}}\, dx$

$$= -\dfrac{3\sqrt{2 + 9x^2}}{6x} + C$$

$$= -\dfrac{\sqrt{2 + 9x^2}}{2x} + C$$

39. By Formulas 54 and 55:

$$\int t^3 \cos t\, dt = t^3 \sin t - 3\int t^2 \sin t\, dt$$

$$= t^3 \sin t - 3\left[-t^2 \cos t + 2\int t \cos t\, dt\right]$$

$$= t^3 \sin t + 3t^2 \cos t - 6\left[t \sin t - \int \sin t\, dt\right]$$

$$= t^3 \sin t + 3t^2 \cos t - 6t \sin t - 6\cos t + C$$

41. By Formula 3: $\displaystyle\int \frac{\ln x}{x(3 + 2 \ln x)}\, dx = \frac{1}{4}\big(2 \ln|x| - 3 \ln|3 + 2 \ln||x|\big) + C$

$$u = \ln x,\ du = \frac{1}{x}\, dx$$

43. By Formulas 1, 25, and 33: $\displaystyle\int \frac{x}{(x^2 - 6x + 10)^2}\, dx = \frac{1}{2}\int \frac{2x - 6 + 6}{(x^2 - 6x + 10)^2}\, dx$

$$= \frac{1}{2}\int (x^2 - 6x + 10)^{-2}(2x - 6)\, dx + 3\int \frac{1}{[(x - 3)^2 + 1]^2}\, dx$$

$$= -\frac{1}{2(x^2 - 6x + 10)} + \frac{3}{2}\left[\frac{x - 3}{x^2 - 6x + 10} + \arctan(x - 3)\right] + C$$

$$= \frac{3x - 10}{2(x^2 - 6x + 10)} + \frac{3}{2}\arctan(x - 3) + C$$

45. By Formula 31: $\displaystyle\int \frac{x}{\sqrt{x^4 - 6x^2 + 5}}\, dx = \frac{1}{2}\int \frac{2x}{\sqrt{(x^2 - 3)^2 - 4}}\, dx$

$$= \frac{1}{2}\ln|x^2 - 3 + \sqrt{x^4 - 6x^2 + 5}| + C$$

$$u = x^2 - 3,\ du = 2x\, dx$$

47. $\displaystyle\int \frac{x^3}{\sqrt{4 - x^2}}\, dx = \int \frac{8 \sin^3 \theta(2 \cos \theta\, d\theta)}{2 \cos \theta}$

$$= 8\int (1 - \cos^2 \theta) \sin \theta\, d\theta$$

$$= 8\int [\sin \theta - \cos^2 \theta(\sin \theta)]\, d\theta$$

$$= -8 \cos \theta + \frac{8 \cos^3 \theta}{3} + C$$

$$= \frac{-\sqrt{4 - x^2}}{3}(x^2 + 8) + C$$

$$x = 2 \sin \theta,\ dx = 2 \cos \theta\, d\theta,\ \sqrt{4 - x^2} = 2 \cos \theta$$

49. By Formula 8: $\displaystyle\int \frac{e^{3x}}{(1 + e^x)^3}\, dx = \int \frac{(e^x)^2}{(1 + e^x)^3}(e^x)\, dx$

$$= \frac{2}{1 + e^x} - \frac{1}{2(1 + e^x)^2} + \ln|1 + e^x| + C$$

$$u = e^x,\ du = e^x\, dx$$

51. $\displaystyle\frac{u^2}{(a + bu)^2} = \frac{1}{b^2} - \frac{(2a/b)u + (a^2/b^2)}{(a + bu)^2} = \frac{1}{b^2} + \frac{A}{a + bu} + \frac{B}{(a + bu)^2}$

$$-\frac{2a}{b}u - \frac{a^2}{b^2} = A(a + bu) + B = (aA + B) + bAu$$

Equating the coefficients of like terms we have $aA + B = -a^2/b^2$ and $bA = -2a/b$. Solving these equations we have $A = -2a/b^2$ and $B = a^2/b^2$.

$$\int \frac{u^2}{(a + bu)^2}\, du = \frac{1}{b^2}\int du - \frac{2a}{b^2}\left(\frac{1}{b}\right)\int \frac{1}{a + bu}b\, du + \frac{a^2}{b^2}\left(\frac{1}{b}\right)\int \frac{1}{(a + bu)^2}b\, du = \frac{1}{b^2}u - \frac{2a}{b^3}\ln|a + bu| - \frac{a^2}{b^3}\left(\frac{1}{a + bu}\right) + C$$

$$= \frac{1}{b^3}\left(bu - \frac{a^2}{a + bu} - 2a \ln|a + bu|\right) + C$$

53. When we have $u^2 + a^2$:

$$u = a \tan \theta$$

$$du = a \sec^2 \theta \, d\theta$$

$$u^2 + a^2 = a^2 \sec^2 \theta$$

$$\int \frac{1}{(u^2 + a^2)^{3/2}} \, du = \int \frac{a \sec^2 \theta \, d\theta}{a^3 \sec^3 \theta}$$

$$= \frac{1}{a^2} \int \cos \theta \, d\theta$$

$$= \frac{1}{a^2} \sin \theta + C$$

$$= \frac{u}{a^2 \sqrt{u^2 + a^2}} + C$$

When we have $u^2 - a^2$:

$$u = a \sec \theta$$

$$du = a \sec \theta \tan \theta \, d\theta$$

$$u^2 - a^2 = a^2 \tan^2 \theta$$

$$\int \frac{1}{(u^2 - a^2)^{3/2}} \, du = \int \frac{a \sec \theta \tan \theta \, d\theta}{a^3 \tan^3 \theta}$$

$$= \frac{1}{a^2} \int \frac{\cos \theta}{\sin^2 \theta} \, d\theta$$

$$= -\frac{1}{a^2} \csc \theta + C$$

$$= \frac{-u}{a^2 \sqrt{u^2 - a^2}} + C$$

55. $\displaystyle \int (\arctan u) \, du = u \arctan u - \frac{1}{2} \int \frac{2u}{1 + u^2} \, du$

$$= u \arctan u - \frac{1}{2} \ln(1 + u^2) + C$$

$$= u \arctan u - \ln\sqrt{1 + u^2} + C$$

$$w = \arctan u, \, dv = du, \, dw = \frac{du}{1 + u^2}, \, v = u$$

57. $\displaystyle \int \frac{1}{x^{3/2}\sqrt{1 - x}} \, dx = \frac{-2\sqrt{1 - x}}{\sqrt{x}} + C$

$$\left(\frac{1}{2}, 5 \right): \frac{-2\sqrt{1/2}}{\sqrt{1/2}} + C = 5 \Rightarrow C = 7$$

$$y = \frac{-2\sqrt{1 - x}}{\sqrt{x}} + 7$$

59. $\displaystyle \int \frac{1}{(x^2 - 6x + 10)^2} \, dx = \frac{1}{2}\left[\tan^{-1}(x - 3) + \frac{x - 3}{x^2 - 6x + 10} \right] + C$

$$(3, 0): \frac{1}{2}\left[0 + \frac{0}{10} \right] + C = 0 \Rightarrow C = 0$$

$$y = \frac{1}{2}\left[\tan^{-1}(x - 3) + \frac{x - 3}{x^2 - 6x + 10} \right]$$

61. $\displaystyle \int \frac{1}{\sin \theta \tan \theta} \, d\theta = -\csc \theta + C$

$$\left(\frac{\pi}{4}, 2 \right): -\frac{2}{\sqrt{2}} + C = 2 \Rightarrow C = 2 + \sqrt{2}$$

$$y = -\csc \theta + 2 + \sqrt{2}$$

63. $\displaystyle\int \frac{1}{2 - 3\sin\theta}\,d\theta = \int \left[\frac{\dfrac{2\,du}{1 + u^2}}{2 - 3\left(\dfrac{2u}{1 + u^2}\right)}\right]$

$\displaystyle = \int \frac{2}{2(1 + u^2) - 6u}\,du$

$\displaystyle = \int \frac{1}{u^2 - 3u + 1}\,du$

$\displaystyle = \int \frac{1}{\left(u - \dfrac{3}{2}\right)^2 - \dfrac{5}{4}}\,du$

$\displaystyle = \frac{1}{\sqrt{5}} \ln\left|\frac{\left(u - \dfrac{3}{2}\right) - \dfrac{\sqrt{5}}{2}}{\left(u - \dfrac{3}{2}\right) + \dfrac{\sqrt{5}}{2}}\right| + C$

$\displaystyle = \frac{1}{\sqrt{5}} \ln\left|\frac{2u - 3 - \sqrt{5}}{2u - 3 + \sqrt{5}}\right| + C$

$\displaystyle = \frac{1}{\sqrt{5}} \ln\left|\frac{2\tan\left(\dfrac{\theta}{2}\right) - 3 - \sqrt{5}}{2\tan\left(\dfrac{\theta}{2}\right) - 3 + \sqrt{5}}\right| + C$

$u = \tan\dfrac{\theta}{2}$

65. $\displaystyle\int_0^{\pi/2} \frac{1}{1 + \sin\theta + \cos\theta}\,d\theta = \int_0^1 \left[\frac{\dfrac{2\,du}{1 + u^2}}{1 + \dfrac{2u}{1 + u^2} + \dfrac{1 - u^2}{1 + u^2}}\right]$

$\displaystyle = \int_0^1 \frac{1}{1 + u}\,du$

$\displaystyle = \Big[\ln|1 + u|\Big]_0^1$

$= \ln 2$

$u = \tan\dfrac{\theta}{2}$

67. $\displaystyle\int \frac{\sin\theta}{3 - 2\cos\theta}\,d\theta = \frac{1}{2} \int \frac{2\sin\theta}{3 - 2\cos\theta}\,d\theta$

$\displaystyle = \frac{1}{2} \ln|u| + C$

$\displaystyle = \frac{1}{2} \ln(3 - 2\cos\theta) + C$

$u = 3 - 2\cos\theta,\, du = 2\sin\theta\,d\theta$

69. $\displaystyle\int \frac{\cos\sqrt{\theta}}{\sqrt{\theta}}\,d\theta = 2\int \cos\sqrt{\theta}\left(\frac{1}{2\sqrt{\theta}}\right)d\theta$

$= 2\sin\sqrt{\theta} + C$

$u = \sqrt{\theta},\, du = \dfrac{1}{2\sqrt{\theta}}\,d\theta$

71. $\displaystyle A = \int_0^8 \frac{x}{\sqrt{x + 1}}\,dx$

$\displaystyle = \left[\frac{-2(2 - x)}{3}\sqrt{x + 1}\right]_0^8$

$\displaystyle = 12 - \left(-\frac{4}{3}\right)$

$\displaystyle = \frac{40}{3} \approx 13.333 \text{ square units}$

73. Arctangent Formula, Formula 23,

$$\int \frac{1}{u^2 + 1}\,du, \, u = e^x$$

75. Substitution: $u = x^2,\, du = 2x\,dx$

Then Formula 81.

77. Cannot be integrated.

79. Answers will vary. For example,

$$\int (2x)e^{2x}\,dx$$

can be integrated by first letting $u = 2x$ and then using Formula 82.

81. $W = \int_0^5 2000xe^{-x}\, dx$

$\qquad = -2000 \int_0^5 -xe^{-x}\, dx$

$\qquad = 2000 \int_0^5 (-x)e^{-x}(-1)\, dx$

$\qquad = 2000\left[(-x)e^{-x} - e^{-x}\right]_0^5$

$\qquad = 2000\left(-\dfrac{6}{e^5} + 1\right)$

$\qquad \approx 1919.145 \text{ ft} \cdot \text{lbs}$

83. (a) $V = 20(2)\int_0^3 \dfrac{2}{\sqrt{1 + y^2}}\, dy$ $\qquad\qquad W = 148\left(80\ln\left(3 + \sqrt{10}\right)\right)$

$\qquad = \left[80\ln\left|y + \sqrt{1 + y^2}\right|\,\right]_0^3$ $\qquad\qquad = 11{,}840\ln\left(3 + \sqrt{10}\right)$

$\qquad = 80\ln\left(3 + \sqrt{10}\right)$ $\qquad\qquad\qquad \approx 21{,}530.4 \text{ lb}$

$\qquad \approx 145.5 \text{ cubic feet}$

(b) By symmetry, $\bar{x} = 0$.

$\qquad M = \rho(2)\int_0^3 \dfrac{2}{\sqrt{1 + y^2}}\, dy = \left[4\rho\ln\left|y + \sqrt{1 + y^2}\right|\,\right]_0^3 = 4\rho\ln\left(3 + \sqrt{10}\right)$

$\qquad M_x = 2\rho\int_0^3 \dfrac{2y}{\sqrt{1 + y^2}}\, dy = \left[4\rho\sqrt{1 + y^2}\,\right]_0^3 = 4\rho\left(\sqrt{10} - 1\right)$

$\qquad \bar{y} = \dfrac{M_x}{M} = \dfrac{4\rho\left(\sqrt{10} - 1\right)}{4\rho\ln\left(3 + \sqrt{10}\right)} \approx 1.19$

$\qquad$ Centroid: $(\bar{x}, \bar{y}) \approx (0, 1.19)$

85. (a) $\int_0^4 \dfrac{k}{2 + 3x}\, dx = 10$ $\qquad\qquad\qquad\qquad$ (b) $\int_0^4 \dfrac{15.417}{2 + 3x}\, dx$

$\qquad k = \dfrac{10}{\displaystyle\int_0^4 \dfrac{1}{2 + 3x}\, dx} \approx \dfrac{10}{0.6486}$

$\qquad\quad = 15.417\ \left(= \dfrac{30}{\ln 7}\right)$

87. False. You might need to convert your integral using substitution or algebra.

Section 7.7   Indeterminate Forms and L'Hôpital's Rule

1. $\displaystyle\lim_{x \to 0} \dfrac{\sin 5x}{\sin 2x} \approx 2.5\left(\text{exact: } \dfrac{5}{2}\right)$

x	-0.1	-0.01	-0.001	0.001	0.01	0.1
$f(x)$	2.4132	2.4991	2.500	2.500	2.4991	2.4132

3. $\lim\limits_{x\to\infty} x^5 e^{-x/100} \approx 0$

x	1	10	10^2	10^3	10^4	10^5
$f(x)$	0.9900	90,484	3.7×10^9	4.5×10^{10}	0	0

5. (a) $\lim\limits_{x\to3} \dfrac{2(x-3)}{x^2-9} = \lim\limits_{x\to3} \dfrac{2(x-3)}{(x+3)(x-3)} = \lim\limits_{x\to3} \dfrac{2}{x+3} = \dfrac{1}{3}$

　　(b) $\lim\limits_{x\to3} \dfrac{2(x-3)}{x^2-9} = \lim\limits_{x\to3} \dfrac{(d/dx)[2(x-3)]}{(d/dx)[x^2-9]} = \lim\limits_{x\to3} \dfrac{2}{2x} = \dfrac{2}{6} = \dfrac{1}{3}$

7. (a) $\lim\limits_{x\to3} \dfrac{\sqrt{x+1}-2}{x-3} = \lim\limits_{x\to3} \dfrac{\sqrt{x+1}-2}{x-3} \cdot \dfrac{\sqrt{x+1}+2}{\sqrt{x+1}+2} = \lim\limits_{x\to3} \dfrac{(x+1)-4}{(x-3)\left[\sqrt{x+1}+2\right]} = \lim\limits_{x\to3} \dfrac{1}{\sqrt{x+1}+2} = \dfrac{1}{4}$

　　(b) $\lim\limits_{x\to3} \dfrac{\sqrt{x+1}-2}{x-3} = \lim\limits_{x\to3} \dfrac{(d/dx)\left[\sqrt{x+1}-2\right]}{(d/dx)[x-3]} = \lim\limits_{x\to3} \dfrac{1/\left(2\sqrt{x+1}\right)}{1} = \dfrac{1}{4}$

9. (a) $\lim\limits_{x\to\infty} \dfrac{5x^2-3x+1}{3x^2-5} = \lim\limits_{x\to\infty} \dfrac{5-(3/x)+(1/x^2)}{3-(5/x^2)} = \dfrac{5}{3}$

　　(b) $\lim\limits_{x\to\infty} \dfrac{5x^2-3x+1}{3x^2-5} = \lim\limits_{x\to\infty} \dfrac{(d/dx)[5x^2-3x+1]}{(d/dx)[3x^2-5]} = \lim\limits_{x\to\infty} \dfrac{10x-3}{6x} = \lim\limits_{x\to\infty} \dfrac{(d/dx)[10x-3]}{(d/dx)[6x]} = \lim\limits_{x\to\infty} \dfrac{10}{6} = \dfrac{5}{3}$

11. $\lim\limits_{x\to2} \dfrac{x^2-x-2}{x-2} = \lim\limits_{x\to2} \dfrac{2x-1}{1} = 3$

13. $\lim\limits_{x\to0} \dfrac{\sqrt{4-x^2}-2}{x} = \lim\limits_{x\to0} \dfrac{-x/\sqrt{4-x^2}}{1} = 0$

15. $\lim\limits_{x\to0} \dfrac{e^x-(1-x)}{x} = \lim\limits_{x\to0} \dfrac{e^x+1}{1} = 2$

17. Case 1: $n=1$

$$\lim_{x\to0^+} \frac{e^x-(1+x)}{x} = \lim_{x\to0^+} \frac{e^x-1}{1} = 0$$

Case 2: $n=2$

$$\lim_{x\to0^+} \frac{e^x-(1+x)}{x^2} = \lim_{x\to0^+} \frac{e^x-1}{2x} = \lim_{x\to0^+} \frac{e^x}{2} = \frac{1}{2}$$

Case 3: $n\geq3$

$$\lim_{x\to0^+} \frac{e^x-(1+x)}{x^n} = \lim_{x\to0^+} \frac{e^x-1}{nx^{n-1}} = \lim_{x\to0^+} \frac{e^x}{n(n-1)x^{n-2}} = \infty$$

19. $\lim\limits_{x\to0} \dfrac{\sin 2x}{\sin 3x} = \lim\limits_{x\to0} \dfrac{2\cos 2x}{3\cos 3x} = \dfrac{2}{3}$

21. $\lim\limits_{x\to0} \dfrac{\arcsin x}{x} = \lim\limits_{x\to0} \dfrac{1/\sqrt{1-x^2}}{1} = 1$

23. $\lim\limits_{x\to\infty} \dfrac{3x^2-2x+1}{2x^2+3} = \lim\limits_{x\to\infty} \dfrac{6x-2}{4x}$

$$= \lim_{x\to\infty} \frac{6}{4} = \frac{3}{2}$$

25. $\lim\limits_{x\to\infty} \dfrac{x^2+2x+3}{x-1} = \lim\limits_{x\to\infty} \dfrac{2x+2}{1} = \infty$

27. $\lim\limits_{x\to\infty} \dfrac{x^3}{e^{x/2}} = \lim\limits_{x\to\infty} \dfrac{3x^2}{(1/2)e^{x/2}}$

$$= \lim_{x\to\infty} \frac{6x}{(1/4)e^{x/2}} = \lim_{x\to\infty} \frac{6}{(1/8)e^{x/2}} = 0$$

29. $\lim\limits_{x\to\infty} \dfrac{x}{\sqrt{x^2+1}} = \lim\limits_{x\to\infty} \dfrac{1}{\sqrt{1+(1/x^2)}} = 1$

Note: L'Hôpital's Rule does not work on this limit. See Exercise 79.

31. $\lim\limits_{x\to\infty} \dfrac{\cos x}{x} = 0$ by Squeeze Theorem

$$\left(\frac{\cos x}{x} \le \frac{1}{x}\right)$$

33. $\lim\limits_{x\to\infty} \dfrac{\ln x}{x^2} = \lim\limits_{x\to\infty} \dfrac{1/x}{2x} = \lim\limits_{x\to\infty} \dfrac{1}{2x^2} = 0$

35. $\lim\limits_{x\to\infty} \dfrac{e^x}{x^2} = \lim\limits_{x\to\infty} \dfrac{e^x}{2x} = \lim\limits_{x\to\infty} \dfrac{e^x}{2} = \infty$

37. (a) $\lim\limits_{x\to0^+} (-x\ln x) = (-0)(-\infty) = (0)(\infty)$

(b) $\lim\limits_{x\to0^+} (-x\ln x) = \lim\limits_{x\to0^+} \dfrac{\ln x}{-1/x}$

$\qquad = \lim\limits_{x\to0^+} \dfrac{1/x}{1/x^2}$

$\qquad = \lim\limits_{x\to0^+} x = 0$

(c)

39. (a) $\lim\limits_{x\to\infty} \left(x\sin\dfrac{1}{x}\right) = (\infty)(0)$

(b) $\lim\limits_{x\to\infty} x\sin\dfrac{1}{x} = \lim\limits_{x\to\infty} \dfrac{\sin(1/x)}{1/x}$

$\qquad = \lim\limits_{x\to\infty} \dfrac{(-1/x^2)\cos(1/x)}{-1/x^2}$

$\qquad = \lim\limits_{x\to\infty} \cos\left(\dfrac{1}{x}\right) = 1$

(c)

41. (a) $\lim\limits_{x\to0^+} x^{1/x} = 0^\infty = 0$, not indeterminant (See Exercise 95)

(b) Let $y = x^{1/x}$

$\ln y = \ln x^{1/x} = \dfrac{1}{x}\ln x$.

Since $x\to0^+$, $\dfrac{1}{x}\ln x \to (\infty)(-\infty) = -\infty$. Hence,

$\ln y \to -\infty \Rightarrow y\to0^+$.

Therefore, $\lim\limits_{x\to0^+} x^{1/x} = 0$.

(c)

43. (a) $\lim\limits_{x\to\infty} x^{1/x} = \infty^0$

(b) Let $y = \lim\limits_{x\to\infty} x^{1/x}$.

$\ln y = \lim\limits_{x\to\infty} \dfrac{\ln x}{x} = \lim\limits_{x\to\infty} \left(\dfrac{1/x}{1}\right) = 0$

Thus, $\ln y = 0 \Rightarrow y = e^0 = 1$. Therefore,

$\lim\limits_{x\to\infty} x^{1/x} = 1$.

(c)

45. (a) $\lim\limits_{x\to0^+} (1+x)^{1/x} = 1^\infty$

(b) Let $y = \lim\limits_{x\to0^+} (1+x)^{1/x}$.

$\ln y = \lim\limits_{x\to0^+} \dfrac{\ln(1+x)}{x}$

$\qquad = \lim\limits_{x\to0^+} \left(\dfrac{1/(1+x)}{1}\right) = 1$

Thus, $\ln y = 1 \Rightarrow y = e^1 = e$.

Therefore, $\lim\limits_{x\to0^+} (1+x)^{1/x} = e$.

(c)

47. (a) $\lim\limits_{x \to 0^+} [3(x)^{x/2}] = 0^0$

(b) Let $y = \lim\limits_{x \to 0^+} 3(x)^{x/2}$.

$$\ln y = \lim\limits_{x \to 0^+} \left[\ln 3 + \frac{x}{2} \ln x \right]$$

$$= \lim\limits_{x \to 0^+} \left[\ln 3 + \frac{\ln x}{2/x} \right]$$

$$= \lim\limits_{x \to 0^+} \ln 3 + \lim\limits_{x \to 0^+} \frac{1/x}{-2/x^2}$$

$$= \lim\limits_{x \to 0^+} \ln 3 - \lim\limits_{x \to 0^+} \frac{x}{2}$$

$$= \ln 3$$

Hence, $\lim\limits_{x \to 0^+} 3(x)^{x/2} = 3$.

(c)

49. (a) $\lim\limits_{x \to 1^+} (\ln x)^{x-1} = 0^0$

(b) Let $y = \lim\limits_{x \to 1^+} (\ln x)^{x-1}$.

$$= \lim\limits_{x \to 1^+} (x - 1)\ln x = 0$$

Hence, $\lim\limits_{x \to 1^+} (\ln x)^{x-1} = 1$

(c)

51. (a) $\lim\limits_{x \to 2^+} \left(\frac{8}{x^2 - 4} - \frac{x}{x - 2} \right) = \infty - \infty$

(b) $\lim\limits_{x \to 2^+} \left(\frac{8}{x^2 - 4} - \frac{x}{x - 2} \right) = \lim\limits_{x \to 2^+} \frac{8 - x(x + 2)}{x^2 - 4}$

$$= \lim\limits_{x \to 2^+} \frac{(2 - x)(4 + x)}{(x + 2)(x - 2)}$$

$$= \lim\limits_{x \to 2^+} \frac{-(x + 4)}{x + 2} = \frac{-3}{2}$$

(c)

53. (a) $\lim\limits_{x \to 1^+} \left(\frac{3}{\ln x} - \frac{2}{x - 1} \right) = \infty - \infty$

(b) $\lim\limits_{x \to 1^+} \left(\frac{3}{\ln x} - \frac{2}{x - 1} \right) = \lim\limits_{x \to 1^+} \frac{3x - 3 - 2\ln x}{(x - 1)\ln x}$

$$= \lim\limits_{x \to 1^+} \frac{3 - (2/x)}{[(x - 1)/x] + \ln x} = \infty$$

(c)

55. (a)

(b) $\lim\limits_{x \to 3} \dfrac{x - 3}{\ln(2x - 5)} = \lim\limits_{x \to 3} \dfrac{1}{2/(2x - 5)}$

$$= \lim\limits_{x \to 3} \frac{2x - 5}{2} = \frac{1}{2}$$

57. (a)

(b) $\lim\limits_{x \to \infty} \left(\sqrt{x^2 + 5x + 2} - x \right) = \lim\limits_{x \to \infty} \left(\sqrt{x^2 + 5x + 2} - x \right) \dfrac{\left(\sqrt{x^2 + 5x + 2} + x \right)}{\left(\sqrt{x^2 + 5x + 2} + x \right)}$

$$= \lim\limits_{x \to \infty} \frac{(x^2 + 5x + 2) - x^2}{\sqrt{x^2 + 5x + 2} + x}$$

$$= \lim\limits_{x \to \infty} \frac{5x + 2}{\sqrt{x^2 + 5x + 2} + x}$$

$$= \lim\limits_{x \to \infty} \frac{5 + (2/x)}{\sqrt{1 + (5/x) + (2/x^2)} + 1} = \frac{5}{2}$$

59. $\dfrac{0}{0}, \dfrac{\infty}{\infty}, 0 \cdot \infty, 1^{\infty}, 0^{0}, \infty - \infty, \infty^{0}$

61. (a) Let $f(x) = x^2 - 25$ and $g(x) = x - 5$.

(b) Let $f(x) = (x - 5)^2$ and $g(x) = x^2 - 25$.

(c) Let $f(x) = x^2 - 25$ and $g(x) = (x - 5)^3$.

63. $\lim\limits_{x \to \infty} \dfrac{x^2}{e^{5x}} = \lim\limits_{x \to \infty} \dfrac{2x}{5e^{5x}} = \lim\limits_{x \to \infty} \dfrac{2}{25e^{5x}} = 0$

65.
$$\lim_{x \to \infty} \frac{(\ln x)^3}{x} = \lim_{x \to \infty} \frac{3(\ln x)^2(1/x)}{1}$$
$$= \lim_{x \to \infty} \frac{3(\ln x)^2}{x}$$
$$= \lim_{x \to \infty} \frac{6(\ln x)(1/x)}{1}$$
$$= \lim_{x \to \infty} \frac{6(\ln x)}{x} = \lim_{x \to \infty} \frac{6}{x} = 0$$

67.
$$\lim_{x \to \infty} \frac{(\ln x)^n}{x^m} = \lim_{x \to \infty} \frac{n(\ln x)^{n-1}/x}{mx^{m-1}}$$
$$= \lim_{x \to \infty} \frac{n(\ln x)^{n-1}}{mx^m}$$
$$= \lim_{x \to \infty} \frac{n(n-1)(\ln x)^{n-2}}{m^2 x^m}$$
$$= \cdots = \lim_{x \to \infty} \frac{n!}{m^n x^m} = 0$$

69.

x	10	10^2	10^4	10^6	10^8	10^{10}
$\dfrac{(\ln x)^4}{x}$	2.811	4.498	0.720	0.036	0.001	0.000

71. $y = x^{1/x}, \; x > 0$

Horizontal asymptote: $y = 1$ (See Exercise 43)

$$\ln y = \frac{1}{x} \ln x$$
$$\frac{1}{y}\frac{dy}{dx} = \frac{1}{x}\left(\frac{1}{x}\right) + (\ln x)\left(-\frac{1}{x^2}\right)$$
$$\frac{dy}{dx} = x^{1/x}\left(\frac{1}{x^2}\right)(1 - \ln x) = x^{(1/x)-2}(1 - \ln x) = 0$$

Critical number: $x = e$

Intervals: $(0, e)$ (e, ∞)

Sign of dy/dx: $+$ $-$

$y = f(x)$: Increasing   Decreasing

Relative maximum: $(e, e^{1/e})$

73. $y = 2xe^{-x}$

$$\lim_{x \to \infty} \frac{2x}{e^x} = \lim_{x \to \infty} \frac{2}{e^x} = 0$$

Horizontal asymptote: $y = 0$

$$\frac{dy}{dx} = 2x(-e^{-x}) + 2e^{-x}$$
$$= 2e^{-x}(1 - x) = 0$$

Critical number: $x = 1$

Intervals: $(-\infty, 1)$ $(1, \infty)$

Sign of dy/dx: $+$ $-$

$y = f(x)$: Increasing   Decreasing

Relative maximum: $\left(1, \dfrac{2}{e}\right)$

75. $\lim\limits_{x \to 0} \dfrac{e^{2x} - 1}{e^x} = \dfrac{0}{1} = 0$

Limit is not of the form $0/0$ or ∞/∞.
L'Hôpital's Rule does not apply.

77. $\lim\limits_{x \to \infty} x \cos \dfrac{1}{x} = \infty(1) = \infty$

Limit is not of the form $0/0$ or ∞/∞.
L'Hôpital's Rule does not apply.

79. (a) $\displaystyle\lim_{x\to\infty}\frac{x}{\sqrt{x^2+1}}=\lim_{x\to\infty}\frac{x/x}{\sqrt{x^2+1}/x}$

$\displaystyle=\lim_{x\to\infty}\frac{1}{\sqrt{x^2+1}/\sqrt{x^2}}$

$\displaystyle=\lim_{x\to\infty}\frac{1}{\sqrt{1+(1/x^2)}}$

$\displaystyle=\frac{1}{\sqrt{1+0}}=1$

(b) $\displaystyle\lim_{x\to\infty}\frac{x}{\sqrt{x^2+1}}=\lim_{x\to\infty}\frac{1}{x/\sqrt{x^2+1}}$

$\displaystyle=\lim_{x\to\infty}\frac{\sqrt{x^2+1}}{x}=\lim_{x\to\infty}\frac{x/\sqrt{x^2+1}}{1}$

$\displaystyle=\lim_{x\to\infty}\frac{x}{\sqrt{x^2+1}}$

Applying L'Hôpital's rule twice results in the original limit, so L'Hôpital's rule fails.

(c)

81. $\displaystyle\lim_{k\to0}\frac{32\left(1-e^{-kt}+\dfrac{v_0ke^{-kt}}{32}\right)}{k}=\lim_{k\to0}\frac{32(1-e^{-kt})}{k}+\lim_{k\to0}\left(v_0e^{-kt}\right)$

$\displaystyle=\lim_{k\to0}\frac{32(0+te^{-kt})}{1}+\lim_{k\to0}\left(\frac{v_0}{e^{kt}}\right)=32t+v_0$

83. Area of triangle: $\dfrac{1}{2}(2x)(1-\cos x)=x-x\cos x$

Shaded area: Area of rectangle $-$ Area under curve

$$2x(1-\cos x)-2\int_0^x(1-\cos t)\,dt=2x(1-\cos x)-2\Big[t-\sin t\Big]_0^x$$

$$=2x(1-\cos x)-2(x-\sin x)=2\sin x-2x\cos x$$

Ratio: $\displaystyle\lim_{x\to0}\frac{x-x\cos x}{2\sin x-2x\cos x}=\lim_{x\to0}\frac{1+x\sin x-\cos x}{2\cos x+2x\sin x-2\cos x}$

$\displaystyle=\lim_{x\to0}\frac{1+x\sin x-\cos x}{2x\sin x}$

$\displaystyle=\lim_{x\to0}\frac{x\cos x+\sin x+\sin x}{2x\cos x+2\sin x}$

$\displaystyle=\lim_{x\to0}\frac{x\cos x+2\sin x}{2x\cos x+2\sin x}\cdot\frac{1/\cos x}{1/\cos x}$

$\displaystyle=\lim_{x\to0}\frac{x+2\tan x}{2x+2\tan x}$

$\displaystyle=\lim_{x\to0}\frac{1+2\sec^2 x}{2+2\sec^2 x}=\frac{3}{4}$

85. $f(x)=x^3,\ g(x)=x^2+1,\ [0,1]$

$\dfrac{f(b)-f(a)}{g(b)-g(a)}=\dfrac{f'(c)}{g'(c)}$

$\dfrac{f(1)-f(0)}{g(1)-g(0)}=\dfrac{3c^2}{2c}$

$\dfrac{1}{1}=\dfrac{3c}{2}$

$c=\dfrac{2}{3}$

87. $f(x)=\sin x,\ g(x)=\cos x,\ \left[0,\dfrac{\pi}{2}\right]$

$\dfrac{f(\pi/2)-f(0)}{g(\pi/2)-g(0)}=\dfrac{f'(c)}{g'(c)}$

$\dfrac{1}{-1}=\dfrac{\cos c}{-\sin c}$

$-1=-\cot c$

$c=\dfrac{\pi}{4}$

89. False. L'Hôpital's Rule does not apply since

$$\lim_{x \to 0} (x^2 + x + 1) \neq 0.$$

$$\lim_{x \to 0^+} \frac{x^2 + x + 1}{x} = \lim_{x \to 0^+} \left(x + 1 + \frac{1}{x} \right) = 1 + \infty = \infty$$

91. True

93. (a) $\sin \theta = BD$

$\cos \theta = DO \implies AD = 1 - \cos \theta$

Area $\triangle ABD = \frac{1}{2} bh = \frac{1}{2}(1 - \cos \theta) \sin \theta = \frac{1}{2} \sin \theta - \frac{1}{2} \sin \theta \cos \theta$

(b) Area of sector: $\frac{1}{2} \theta$

Shaded area: $\frac{1}{2} \theta - $ Area $\triangle OBD = \frac{1}{2} \theta - \frac{1}{2}(\cos \theta)(\sin \theta) = \frac{1}{2} \theta - \frac{1}{2} \sin \theta \cos \theta$

(c) $R = \dfrac{(1/2) \sin \theta - (1/2) \sin \theta \cos \theta}{(1/2) \theta - (1/2) \sin \theta \cos \theta} = \dfrac{\sin \theta - \sin \theta \cos \theta}{\theta - \sin \theta \cos \theta}$

(d) $\displaystyle\lim_{\theta \to 0} R = \lim_{\theta \to 0} \frac{\sin \theta - (1/2) \sin 2\theta}{\theta - (1/2) \sin 2\theta}$

$$= \lim_{\theta \to 0} \frac{\cos \theta - \cos 2\theta}{1 - \cos 2\theta} = \lim_{\theta \to 0} \frac{-\sin \theta + 2 \sin 2\theta}{2 \sin 2\theta} = \lim_{\theta \to 0} \frac{-\cos \theta + 4 \cos 2\theta}{4 \cos 2\theta} = \frac{3}{4}$$

95. $\displaystyle\lim_{x \to a} f(x)^{g(x)}$

$y = f(x)^{g(x)}$

$\ln y = g(x) \ln f(x)$

$\displaystyle\lim_{x \to a} g(x) \ln f(x) = (\infty)(-\infty) = -\infty$

As $x \to a$, $\ln y \implies -\infty$, and hence $y = 0$. Thus,

$\displaystyle\lim_{x \to a} f(x)^{g(x)} = 0.$

97. $f'(a)(b - a) - \displaystyle\int_a^b f''(t)(t - b)\, dt = f'(a)(b - a) - \left\{ \left[f'(t)(t - b) \right]_a^b - \int_a^b f'(t)\, dt \right\}$

$$= f'(a)(b - a) + f'(a)(a - b) + \left[f(t) \right]_a^b = f(b) - f(a)$$

$dv = f''(t)dt \implies v = f'(t)$

$u = t - b \implies du = dt$

Section 7.8   Improper Integrals

1. Infinite discontinuity at $x = 0$.

$$\int_0^4 \frac{1}{\sqrt{x}}\, dx = \lim_{b \to 0^+} \int_b^4 \frac{1}{\sqrt{x}}\, dx$$

$$= \lim_{b \to 0^+} \left[2\sqrt{x} \right]_b^4$$

$$= \lim_{b \to 0^+} \left(4 - 2\sqrt{b} \right) = 4$$

Converges

3. Infinite discontinuity at $x = 1$.

$$\int_0^2 \frac{1}{(x-1)^2} \, dx = \int_0^1 \frac{1}{(x-1)^2} \, dx + \int_1^2 \frac{1}{(x-1)^2} \, dx$$

$$= \lim_{b \to 1^-} \int_0^b \frac{1}{(x-1)^2} \, dx + \lim_{c \to 1^+} \int_c^2 \frac{1}{(x-1)^2} \, dx$$

$$= \lim_{b \to 1^-} \left[-\frac{1}{x-1} \right]_0^b + \lim_{c \to 1^+} \left[-\frac{1}{x-1} \right]_c^2 = (\infty - 1) + (-1 + \infty)$$

Diverges

5. Infinite limit of integration.

$$\int_0^\infty e^{-x} \, dx = \lim_{b \to \infty} \int_0^b e^{-x} \, dx$$

$$= \lim_{b \to \infty} \left[-e^{-x} \right]_0^b = 0 + 1 = 1$$

Converges

7. $\int_{-1}^1 \frac{1}{x^2} \, dx \ne -2$

because the integrand is not defined at $x = 0$.
Diverges

9. $\int_1^\infty \frac{1}{x^2} \, dx = \lim_{b \to \infty} \int_1^b \frac{1}{x^2} \, dx$

$$= \lim_{b \to \infty} \left[-\frac{1}{x} \right]_1^b = 1$$

11. $\int_1^\infty \frac{3}{\sqrt[3]{x}} \, dx = \lim_{b \to \infty} \int_1^b 3x^{-1/3} \, dx$

$$= \lim_{b \to \infty} \left[\frac{9}{2} x^{2/3} \right]_1^b = \infty$$

Diverges

13. $\int_{-\infty}^0 xe^{-2x} \, dx = \lim_{b \to -\infty} \int_b^0 xe^{-2x} \, dx = \lim_{b \to -\infty} \frac{1}{4} \left[(-2x - 1)e^{-2x} \right]_b^0 = \lim_{b \to -\infty} \frac{1}{4} [-1 + (2b+1)e^{-2b}] = -\infty$ (Integration by parts)

Diverges

15. $\int_0^\infty x^2 e^{-x} \, dx = \lim_{b \to \infty} \int_0^b x^2 e^{-x} \, dx = \lim_{b \to \infty} \left[-e^{-x}(x^2 + 2x + 2) \right]_0^b = \lim_{b \to \infty} \left(-\frac{b^2 + 2b + 2}{e^b} + 2 \right) = 2$

Since $\lim_{b \to \infty} \left(-\frac{b^2 + 2b + 2}{e^b} \right) = 0$ by L'Hôpital's Rule.

17. $\int_0^\infty e^{-x} \cos x \, dx = \lim_{b \to \infty} \frac{1}{2} \left[e^{-x}(-\cos x + \sin x) \right]_0^b$

$$= \frac{1}{2} [0 - (-1)] = \frac{1}{2}$$

19. $\int_4^\infty \frac{1}{x(\ln x)^3} \, dx = \lim_{b \to \infty} \int_4^b (\ln x)^{-3} \frac{1}{x} \, dx$

$$= \lim_{b \to \infty} \left[-\frac{1}{2}(\ln x)^{-2} \right]_4^b$$

$$= -\frac{1}{2}(\ln b)^{-2} + \frac{1}{2}(\ln 4)^{-2}$$

$$= \frac{1}{2} \frac{1}{(2 \ln 2)^2} = \frac{1}{8(\ln 2)^2}$$

21. $\int_{-\infty}^\infty \frac{2}{4 + x^2} \, dx = \int_{-\infty}^0 \frac{2}{4 + x^2} \, dx + \int_0^\infty \frac{2}{4 + x^2} \, dx$

$$= \lim_{b \to -\infty} \int_b^0 \frac{2}{4 + x^2} \, dx + \lim_{c \to \infty} \int_0^c \frac{2}{4 + x^2} \, dx$$

$$= \lim_{b \to -\infty} \left[\arctan\left(\frac{x}{2}\right) \right]_b^0 + \lim_{c \to \infty} \left[\arctan\left(\frac{x}{2}\right) \right]_0^c$$

$$= \left(0 - \left(-\frac{\pi}{2} \right) \right) + \left(\frac{\pi}{2} - 0 \right) = \pi$$

23. $\displaystyle\int_0^\infty \frac{1}{e^x + e^{-x}}\,dx = \lim_{b\to\infty}\int_0^b \frac{e^x}{1 + e^{2x}}\,dx$

$\qquad\qquad\qquad = \lim_{b\to\infty}\Big[\arctan(e^x)\Big]_0^b$

$\qquad\qquad\qquad = \dfrac{\pi}{2} - \dfrac{\pi}{4} = \dfrac{\pi}{4}$

25. $\displaystyle\int_0^\infty \cos \pi x\,dx = \lim_{b\to\infty}\Big[\frac{1}{\pi}\sin \pi x\Big]_0^b$

Diverges since $\sin \pi x$ does not approach a limit as $x\to\infty$.

27. $\displaystyle\int_0^1 \frac{1}{x^2}\,dx = \lim_{b\to 0^+}\Big[\frac{-1}{x}\Big]_b^1 = -1 + \infty$

Diverges

29. $\displaystyle\int_0^8 \frac{1}{\sqrt[3]{8-x}}\,dx = \lim_{b\to 8^-}\int_0^b \frac{1}{\sqrt[3]{8-x}}\,dx = \lim_{b\to 8^-}\Big[\frac{-3}{2}(8-x)^{2/3}\Big]_0^b = 6$

31. $\displaystyle\int_0^1 x \ln x\,dx = \lim_{b\to 0^+}\Big[\frac{x^2}{2}\ln|x| - \frac{x^2}{4}\Big]_b^1 = \lim_{b\to 0^+}\Big[\frac{-1}{4} - \frac{b^2 \ln b}{2} + \frac{b^2}{4}\Big] = \frac{-1}{4}$ since $\displaystyle\lim_{b\to 0^+}(b^2 \ln b) = 0$ by L'Hôpital's Rule.

33. $\displaystyle\int_0^{\pi/2} \tan \theta\,d\theta = \lim_{b\to(\pi/2)^-}\Big[\ln|\sec \theta|\Big]_0^b = \infty$

Diverges

35. $\displaystyle\int_2^4 \frac{2}{x\sqrt{x^2-4}}\,dx = \lim_{b\to 2^+}\int_b^4 \frac{2}{x\sqrt{x^2-4}}\,dx$

$\qquad\qquad\qquad = \lim_{b\to 2^+}\Big[\arcsec\Big|\frac{x}{2}\Big|\Big]_b^4$

$\qquad\qquad\qquad = \lim_{b\to 2^+}\Big(\arcsec 2 - \arcsec\Big(\frac{b}{2}\Big)\Big)$

$\qquad\qquad\qquad = \dfrac{\pi}{3} - 0 = \dfrac{\pi}{3}$

37. $\displaystyle\int_2^4 \frac{1}{\sqrt{x^2-4}} = \lim_{b\to 2^+}\Big[\ln\Big|x + \sqrt{x^2-4}\Big|\Big]_b^4$

$\qquad\qquad = \ln\big(4 + 2\sqrt{3}\big) - \ln 2$

$\qquad\qquad = \ln\big(2 + \sqrt{3}\big) \approx 1.317$

39. $\displaystyle\int_0^2 \frac{1}{\sqrt[3]{x-1}}\,dx = \int_0^1 \frac{1}{\sqrt[3]{x-1}}\,dx + \int_1^2 \frac{1}{\sqrt[3]{x-1}}\,dx$

$\qquad\qquad = \lim_{b\to 1^-}\Big[\frac{3}{2}(x-1)^{2/3}\Big]_0^b + \lim_{c\to 1^+}\Big[\frac{3}{2}(x-1)^{2/3}\Big]_c^2 = \frac{-3}{2} + \frac{3}{2} = 0$

41. $\displaystyle\int_0^\infty \frac{4}{\sqrt{x}(x+6)}\,dx = \int_0^1 \frac{4}{\sqrt{x}(x+6)}\,dx + \int_1^\infty \frac{4}{\sqrt{x}(x+6)}\,dx$

Let $u = \sqrt{x}, u^2 = x, 2u\,du = dx.$

$\displaystyle\int \frac{4}{\sqrt{x}(x+6)}\,dx = \int \frac{4(2u\,du)}{u(u^2+6)} = 8\int \frac{du}{u^2+6} = \frac{8}{\sqrt{6}}\arctan\Big(\frac{u}{\sqrt{6}}\Big) + C = \frac{8}{\sqrt{6}}\arctan\Big(\frac{\sqrt{x}}{\sqrt{6}}\Big) + C$

Thus, $\displaystyle\int_0^\infty \frac{4}{\sqrt{x}(x+6)}\,dx = \lim_{b\to 0^+}\Big[\frac{8}{\sqrt{6}}\arctan\Big(\frac{\sqrt{x}}{\sqrt{6}}\Big)\Big]_b^1 + \lim_{c\to\infty}\Big[\frac{8}{\sqrt{6}}\arctan\Big(\frac{\sqrt{x}}{\sqrt{6}}\Big)\Big]_1^c$

$\qquad\qquad = \Big(\frac{8}{\sqrt{6}}\arctan\Big(\frac{1}{\sqrt{6}}\Big) - \frac{8}{\sqrt{6}}0\Big) + \Big(\frac{8}{\sqrt{6}}\frac{\pi}{2} - \frac{8}{\sqrt{6}}\arctan\Big(\frac{1}{\sqrt{6}}\Big)\Big)$

$\qquad\qquad = \dfrac{8\pi}{2\sqrt{6}} = \dfrac{2\pi\sqrt{6}}{3}.$

43. If $p = 1$, $\displaystyle\int_1^\infty \frac{1}{x} \, dx = \lim_{b \to \infty} \int_1^b \frac{1}{x} \, dx = \lim_{b \to \infty} \ln x \Big]_1^b$.

Diverges. For $p \neq 1$,

$$\int_1^\infty \frac{1}{x^p} \, dx = \lim_{b \to \infty} \left[\frac{x^{1-p}}{1-p} \right]_1^b = \lim_{b \to \infty} \left[\frac{b^{1-p}}{1-p} - \frac{1}{1-p} \right].$$

This converges to $\dfrac{1}{p-1}$ if $1 - p < 0$ or $p > 1$.

45. For $n = 1$ we have

$$\int_0^\infty xe^{-x} \, dx = \lim_{b \to \infty} \int_0^b xe^{-x} \, dx$$

$$= \lim_{b \to \infty} \left[-e^{-x}x - e^{-x} \right]_0^b \quad \text{(Parts: } u = x, dv = e^{-x} \, dx\text{)}$$

$$= \lim_{b \to \infty} \left[-e^{-b}b - e^{-b} + 1 \right]$$

$$= \lim_{b \to \infty} \left[\frac{-b}{e^b} - \frac{1}{e^b} + 1 \right] = 1 \quad \text{(L'Hôpital's Rule)}$$

Assume that $\displaystyle\int_0^\infty x^n e^{-x} \, dx$ converges. Then for $n + 1$ we have

$$\int x^{n+1} e^{-x} \, dx = -x^{n+1} e^{-x} + (n+1) \int x^n e^{-x} \, dx$$

by parts $(u = x^{n+1}, du = (n+1)x^n \, dx, dv = e^{-x} \, dx, v = -e^{-x})$.

Thus,

$$\int_0^\infty x^{n+1} e^{-x} \, dx = \lim_{b \to \infty} \left[-x^{n+1} e^{-x} \right]_0^b + (n+1) \int_0^\infty x^n e^{-x} \, dx = 0 + (n+1) \int_0^\infty x^n e^{-x} \, dx, \text{ which converges.}$$

47. $\displaystyle\int_0^1 \frac{1}{x^3} \, dx$ diverges.

(See Exercise 44, $p = 3 \not< 1$.)

49. $\displaystyle\int_1^\infty \frac{1}{x^3} \, dx = \frac{1}{3-1} = \frac{1}{2}$ converges.

(See Exercise 43, $p = 3$.)

51. Since $\dfrac{1}{x^2 + 5} \leq \dfrac{1}{x^2}$ on $[1, \infty)$ and $\displaystyle\int_1^\infty \frac{1}{x^2} \, dx$ converges by Exercise 43, $\displaystyle\int_1^\infty \frac{1}{x^2 + 5} \, dx$ converges.

53. Since $\dfrac{1}{\sqrt[3]{x(x-1)}} \geq \dfrac{1}{\sqrt[3]{x^2}}$ on $[2, \infty)$ and $\displaystyle\int_2^\infty \frac{1}{\sqrt[3]{x^2}} \, dx$ diverges by Exercise 43, $\displaystyle\int_2^\infty \frac{1}{\sqrt[3]{x(x-1)}} \, dx$ diverges.

55. Since $e^{-x^2} \leq e^{-x}$ on $[1, \infty)$ and $\displaystyle\int_0^\infty e^{-x} \, dx$ converges (see Exercise 5), $\displaystyle\int_0^\infty e^{-x^2} \, dx$ converges.

57. Answers will vary. See pages 540, 543.

59. $\displaystyle\int_{-1}^1 \frac{1}{x^3} \, dx = \int_{-1}^0 \frac{1}{x^3} \, dx + \int_0^1 \frac{1}{x^3} \, dx$

These two integrals diverge by Exercise 44.

61. $f(t) = 1$

$$F(s) = \int_0^\infty e^{-st} \, dt = \lim_{b \to \infty} \left[-\frac{1}{s} e^{-st} \right]_0^b = \frac{1}{s}, s > 0$$

63. $f(t) = t^2$

$$F(s) = \int_0^\infty t^2 e^{-st} \, dt = \lim_{b \to \infty} \left[\frac{1}{s^3} (-s^2 t^2 - 2st - 2)e^{-st} \right]_0^b$$

$$= \frac{2}{s^3}, s > 0$$

65. $f(t) = \cos at$

$$F(s) = \int_0^\infty e^{-st} \cos at \, dt$$

$$= \lim_{b \to \infty} \left[\frac{e^{-st}}{s^2 + a^2} (-s \cos at + a \sin at) \right]_0^b$$

$$= 0 + \frac{s}{s^2 + a^2} = \frac{s}{s^2 + a^2}, \, s > 0$$

67. $f(t) = \cosh at$

$$F(s) = \int_0^\infty e^{-st} \cosh at \, dt = \int_0^\infty e^{-st} \left(\frac{e^{at} + e^{-at}}{2} \right) dt = \frac{1}{2} \int_0^\infty \left[e^{t(-s+a)} + e^{t(-s-a)} \right] dt$$

$$= \lim_{b \to \infty} \frac{1}{2} \left[\frac{1}{(-s+a)} e^{t(-s+a)} + \frac{1}{(-s-a)} e^{t(-s-a)} \right]_0^b = 0 - \frac{1}{2} \left[\frac{1}{(-s+a)} + \frac{1}{(-s-a)} \right]$$

$$= \frac{-1}{2} \left[\frac{1}{(-s+a)} + \frac{1}{(-s-a)} \right] = \frac{s}{s^2 - a^2}, \, s > |a|$$

69. (a) $A = \int_0^\infty e^{-x} \, dx$

$$= \lim_{b \to \infty} \left[-e^{-x} \right]_0^b = 0 - (-1) = 1$$

(b) **Disk:**

$$V = \pi \int_0^\infty (e^{-x})^2 \, dx$$

$$= \lim_{b \to \infty} \pi \left[-\frac{1}{2} e^{-2x} \right]_0^b = \frac{\pi}{2}$$

(c) **Shell:**

$$V = 2\pi \int_0^\infty x e^{-x} \, dx$$

$$= \lim_{b \to \infty} \left\{ 2\pi \left[-e^{-x}(x+1) \right]_0^b \right\} = 2\pi$$

71. $x^{2/3} + y^{2/3} = 4$

$$\frac{2}{3} x^{-1/3} + \frac{2}{3} y^{-1/3} y' = 0$$

$$y' = \frac{-y^{1/3}}{x^{1/3}}$$

$$\sqrt{1 + (y')^2} = \sqrt{1 + \frac{y^{2/3}}{x^{2/3}}} = \sqrt{\frac{x^{2/3} + y^{2/3}}{x^{2/3}}} = \sqrt{\frac{4}{x^{2/3}}} = \frac{2}{x^{1/3}}, \, (x > 0)$$

$$s = 4 \int_0^8 \frac{2}{x^{1/3}} \, dx = \lim_{b \to 0^+} \left[8 \cdot \frac{3}{2} x^{2/3} \right]_b^8 = 48$$

73. $\Gamma(n) = \displaystyle\int_0^\infty x^{n-1}e^{-x}\,dx$

(a) $\Gamma(1) = \displaystyle\int_0^\infty e^{-x}\,dx = \lim_{b\to\infty}\left[-e^{-x}\right]_0^b = 1$

$\Gamma(2) = \displaystyle\int_0^\infty xe^{-x}\,dx = \lim_{b\to\infty}\left[-e^{-x}(x+1)\right]_0^b = 1$

$\Gamma(3) = \displaystyle\int_0^\infty x^2e^{-x}\,dx = \lim_{b\to\infty}\left[-x^2e^{-x} - 2xe^{-x} - 2e^{-x}\right]_0^b = 2$

(b) $\Gamma(n+1) = \displaystyle\int_0^\infty x^n e^{-x}\,dx = \lim_{b\to\infty}\left[-x^n e^{-x}\right]_0^b + \lim_{b\to\infty} n\int_0^b x^{n-1}e^{-x}\,dx = 0 + n\Gamma(n)$ ($u = x^n,\ dv = e^{-x}\,dx$)

(c) $\Gamma(n) = (n-1)!$

75. (a) $\displaystyle\int_{-\infty}^\infty \frac{1}{7}e^{-t/7}\,dt = \int_0^\infty \frac{1}{7}e^{-t/7}\,dt = \lim_{b\to\infty}\left[-e^{-t/7}\right]_0^b = 1$ (b) $\displaystyle\int_0^4 \frac{1}{7}e^{-t/7}\,dt = \left[-e^{-t/7}\right]_0^4 = -e^{-4/7} + 1$

$\approx 0.4353 = 43.53\%$

(c) $\displaystyle\int_0^\infty t\left[\frac{1}{7}e^{-t/7}\right]dt = \lim_{b\to\infty}\left[-te^{-t/7} - 7e^{-t/7}\right]_0^b$

$= 0 + 7 = 7$

77. (a) $C = 650{,}000 + \displaystyle\int_0^5 25{,}000\,e^{-0.06t}\,dt = 650{,}000 - \left[\frac{25{,}000}{0.06}e^{-0.06t}\right]_0^5 \approx \$757{,}992.41$

(b) $C = 650{,}000 + \displaystyle\int_0^{10} 25{,}000e^{-0.06t}\,dt \approx \$837{,}995.15$

(c) $C = 650{,}000 + \displaystyle\int_0^\infty 25{,}000e^{-0.06t}\,dt = 650{,}000 - \lim_{b\to\infty}\left[\frac{25{,}000}{0.06}e^{-0.06t}\right]_0^b \approx \$1{,}066{,}666.67$

79. Let $x = a\tan\theta$, $dx = a\sec^2\theta\,d\theta$, $\sqrt{a^2 + x^2} = a\sec\theta$.

$\displaystyle\int \frac{1}{(a^2 + x^2)^{3/2}}\,dx = \int \frac{a\sec^2\theta\,d\theta}{a^3\sec^3\theta} = \frac{1}{a^2}\int \cos\theta\,d\theta$

$= \frac{1}{a^2}\sin\theta = \frac{1}{a^2}\frac{x}{\sqrt{a^2 + x^2}}$

Hence,

$P = k\displaystyle\int_1^\infty \frac{1}{(a^2 + x^2)^{3/2}}\,dx = \frac{k}{a^2}\lim_{b\to\infty}\left[\frac{x}{\sqrt{a^2 + x^2}}\right]_1^b$

$= \frac{k}{a^2}\left[1 - \frac{1}{\sqrt{a^2 + 1}}\right] = \frac{k\left(\sqrt{a^2 + 1} - 1\right)}{a^2\sqrt{a^2 + 1}}.$

81. $\dfrac{10}{x^2 - 2x} = \dfrac{10}{x(x-2)} \Rightarrow x = 0, 2.$

You must analyze three improper integrals, and each must converge in order for the original integral to converge.

$\displaystyle\int_0^3 f(x)\,dx = \int_0^1 f(x)\,dx + \int_1^2 f(x)\,dx + \int_2^3 f(x)\,dx$

83. For $n = 1$,

$$I_1 = \int_0^\infty \frac{x}{(x^2 + 1)^4}\, dx = \lim_{b \to \infty} \frac{1}{2} \int_0^b (x^2 + 1)^{-4}(2x\, dx) = \lim_{b \to \infty} \left[-\frac{1}{6} \frac{1}{(x^2 + 1)^3} \right]_0^b = \frac{1}{6}.$$

For $n > 1$,

$$I_n = \int_0^\infty \frac{x^{2n-1}}{(x^2 + 1)^{n+3}}\, dx = \lim_{b \to \infty} \left[\frac{-x^{2n-2}}{2(n + 2)(x^2 + 1)^{n+2}} \right]_0^b + \frac{n - 1}{n + 2} \int_0^\infty \frac{x^{2n-3}}{(x^2 + 1)^{n+2}}\, dx = 0 + \frac{n - 1}{n + 2}(I_{n-1})$$

$$u = x^{2n-2},\, du = (2n - 2)x^{2n-3}\, dx,\, dv = \frac{x}{(x^2 + 1)^{n+3}}\, dx,\, v = \frac{-1}{2(n + 2)(x^2 + 1)^{n+2}}$$

(a) $\int_0^\infty \frac{x}{(x^2 + 1)^4}\, dx = \lim_{b \to \infty} \left[-\frac{1}{6(x^2 + 1)^3} \right]_0^b = \frac{1}{6}$

(b) $\int_0^\infty \frac{x^3}{(x^2 + 1)^5}\, dx = \frac{1}{4} \int_0^\infty \frac{x}{(x^2 + 1)^4}\, dx = \frac{1}{4}\left(\frac{1}{6}\right) = \frac{1}{24}$

(c) $\int_0^\infty \frac{x^5}{(x^2 + 1)^6} = \frac{2}{5} \int_0^\infty \frac{x^3}{(x^2 + 1)^5}\, dx = \frac{2}{5}\left(\frac{1}{24}\right) = \frac{1}{60}$

85. False. $f(x) = 1/(x + 1)$ is continuous on $[0, \infty)$, $\lim_{x \to \infty} 1/(x + 1) = 0$, but $\int_0^\infty \frac{1}{x + 1}\, dx = \lim_{b \to \infty} \left[\ln|x + 1| \right]_0^b = \infty$.

Diverges

87. True

Review Exercises for Chapter 7

1. $\int x\sqrt{x^2 - 1}\, dx = \frac{1}{2}\int (x^2 - 1)^{1/2}(2x)\, dx$

$\qquad = \frac{1}{2} \frac{(x^2 - 1)^{3/2}}{3/2} + C$

$\qquad = \frac{1}{3}(x^2 - 1)^{3/2} + C$

3. $\int \frac{x}{x^2 - 1}\, dx = \frac{1}{2}\int \frac{2x}{x^2 - 1}\, dx$

$\qquad = \frac{1}{2} \ln|x^2 - 1| + C$

5. $\int \frac{\ln(2x)}{x}\, dx = \frac{(\ln 2x)^2}{2} + C$

7. $\int \frac{16}{\sqrt{16 - x^2}}\, dx = 16 \arcsin\left(\frac{x}{4}\right) + C$

9. $\int e^{2x} \sin 3x\, dx = -\frac{1}{3}e^{2x} \cos 3x + \frac{2}{3}\int e^{2x} \cos 3x\, dx$

$\qquad = -\frac{1}{3}e^{2x} \cos 3x + \frac{2}{3}\left(\frac{1}{3}e^{2x} \sin 3x - \frac{2}{3}\int e^{2x} \sin 3x\, dx\right)$

$\frac{13}{9} \int e^{2x} \sin 3x\, dx = -\frac{1}{3}e^{2x} \cos 3x + \frac{2}{9}e^{2x} \sin 3x$

$\int e^{2x} \sin 3x\, dx = \frac{e^{2x}}{13}(2 \sin 3x - 3 \cos 3x) + C$

(1) $dv = \sin 3x\, dx \implies v = -\frac{1}{3} \cos 3x$

$\qquad u = e^{2x} \qquad \implies du = 2e^{2x}\, dx$

(2) $dv = \cos 3x\, dx \implies v = \frac{1}{3} \sin 3x$

$\qquad u = e^{2x} \qquad \implies du = 2e^{2x}\, dx$

11. $u = x, du = dx, dv = (x - 5)^{1/2} dx, v = \frac{2}{3}(x - 5)^{3/2}$

$$\int x\sqrt{x - 5}\, dx = \frac{2}{3}x(x - 5)^{3/2} - \int \frac{2}{3}(x - 5)^{3/2}\, dx$$

$$= \frac{2}{3}x(x - 5)^{3/2} - \frac{4}{15}(x - 5)^{5/2} + C$$

$$= (x - 5)^{3/2}\left[\frac{2}{3}x - \frac{4}{15}(x - 5)\right] + C$$

$$= (x - 5)^{3/2}\left[\frac{6}{15}x + \frac{4}{3}\right] + C$$

$$= \frac{2}{15}(x - 5)^{3/2}[3x + 10] + C$$

13. $\int x^2 \sin 2x\, dx = -\frac{1}{2}x^2 \cos 2x + \int x \cos 2x\, dx$

$$= -\frac{1}{2}x^2 \cos 2x + \frac{1}{2}x \sin 2x - \frac{1}{2}\int \sin 2x\, dx$$

$$= -\frac{1}{2}x^2 \cos 2x + \frac{x}{2}\sin 2x + \frac{1}{4}\cos 2x + C$$

$(1)\ dv = \sin 2x\, dx \implies v = -\frac{1}{2}\cos 2x$

$\quad u = x^2 \qquad\qquad \implies du = 2x\, dx$

$(2)\ dv = \cos 2x\, dx \implies v = \frac{1}{2}\sin 2x$

$\quad u = x \qquad\qquad \implies du = dx$

15. $\int x \arcsin 2x\, dx = \frac{x^2}{2}\arcsin 2x - \int \frac{x^2}{\sqrt{1 - 4x^2}}\, dx$

$$= \frac{x^2}{2}\arcsin 2x - \frac{1}{8}\int \frac{2(2x)^2}{\sqrt{1 - (2x)^2}}\, dx$$

$$= \frac{x^2}{2}\arcsin 2x - \frac{1}{8}\left(\frac{1}{2}\right)\left[-(2x)\sqrt{1 - 4x^2} + \arcsin 2x\right] + C \ \text{(by Formula 43 of Integration Tables)}$$

$$= \frac{1}{16}\left[(8x^2 - 1)\arcsin 2x + 2x\sqrt{1 - 4x^2}\right] + C$$

$dv = x\, dx \quad \implies v = \frac{x^2}{2}$

$u = \arcsin 2x \implies du = \frac{2}{\sqrt{1 - 4x^2}}\, dx$

17. $\int \cos^3(\pi x - 1)\, dx = \int [1 - \sin^2(\pi x - 1)]\cos(\pi x - 1)\, dx$

$$= \frac{1}{\pi}\left[\sin(\pi x - 1) - \frac{1}{3}\sin^3(\pi x - 1)\right] + C$$

$$= \frac{1}{3\pi}\sin(\pi x - 1)[3 - \sin^2(\pi x - 1)] + C$$

$$= \frac{1}{3\pi}\sin(\pi x - 1)[3 - (1 - \cos^2(\pi x - 1))] + C$$

$$= \frac{1}{3\pi}\sin(\pi x - 1)[2 + \cos^2(\pi x - 1)] + C$$

19. $\int \sec^4\left(\frac{x}{2}\right) dx = \int \left[\tan^2\left(\frac{x}{2}\right) + 1\right]\sec^2\left(\frac{x}{2}\right) dx$

$$= \int \tan^2\left(\frac{x}{2}\right)\sec^2\left(\frac{x}{2}\right) dx + \int \sec^2\left(\frac{x}{2}\right) dx$$

$$= \frac{2}{3}\tan^3\left(\frac{x}{2}\right) + 2\tan\left(\frac{x}{2}\right) + C = \frac{2}{3}\left[\tan^3\left(\frac{x}{2}\right) + 3\tan\left(\frac{x}{2}\right)\right] + C$$

21. $\int \frac{1}{1 - \sin \theta}\, d\theta = \int \frac{1 + \sin \theta}{\cos^2 \theta}\, d\theta = \int (\sec^2 \theta + \sec \theta \tan \theta)\, d\theta = \tan \theta + \sec \theta + C$

23. $\displaystyle\int \frac{-12}{x^2\sqrt{4-x^2}}\,dx = \int \frac{-24\cos\theta\,d\theta}{(4\sin^2\theta)(2\cos\theta)}$

$\qquad\qquad = -3\displaystyle\int \csc^2\theta\,d\theta$

$\qquad\qquad = 3\cot\theta + C$

$\qquad\qquad = \dfrac{3\sqrt{4-x^2}}{x} + C$

$x = 2\sin\theta, dx = 2\cos\theta\,d\theta, \sqrt{4-x^2} = 2\cos\theta$

25. $\qquad x = 2\tan\theta$

$\qquad dx = 2\sec^2\theta\,d\theta$

$\quad 4 + x^2 = 4\sec^2\theta$

$\displaystyle\int \frac{x^3}{\sqrt{4+x^2}}\,dx = \int \frac{8\tan^3\theta}{2\sec\theta}\, 2\sec^2\theta\,d\theta$

$\qquad\qquad = 8\displaystyle\int \tan^3\theta\sec\theta\,d\theta$

$\qquad\qquad = 8\displaystyle\int (\sec^2\theta - 1)\tan\theta\sec\theta\,d\theta$

$\qquad\qquad = 8\left[\dfrac{\sec^3\theta}{3} - \sec\theta\right] + C$

$\qquad\qquad = 8\left[\dfrac{(x^2+4)^{3/2}}{24} - \dfrac{\sqrt{x^2+4}}{2}\right] + C$

$\qquad\qquad = \sqrt{x^2+4}\left[\dfrac{1}{3}(x^2+4) - 4\right] + C$

$\qquad\qquad = \dfrac{1}{3}x^2\sqrt{x^2+4} - \dfrac{8}{3}\sqrt{x^2+4} + C$

$\qquad\qquad = \dfrac{1}{3}(x^2+4)^{1/2}(x^2 - 8) + C$

27. $\displaystyle\int \sqrt{4-x^2}\,dx = \int (2\cos\theta)(2\cos\theta)\,d\theta$

$\qquad\qquad = 2\displaystyle\int (1 + \cos 2\theta)\,d\theta$

$\qquad\qquad = 2\left(\theta + \dfrac{1}{2}\sin 2\theta\right) + C$

$\qquad\qquad = 2(\theta + \sin\theta\cos\theta) + C$

$\qquad\qquad = 2\left[\arcsin\left(\dfrac{x}{2}\right) + \dfrac{x}{2}\left(\dfrac{\sqrt{4-x^2}}{2}\right)\right] + C$

$\qquad\qquad = \dfrac{1}{2}\left[4\arcsin\left(\dfrac{x}{2}\right) + x\sqrt{4-x^2}\right] + C$

$x = 2\sin\theta, dx = 2\cos\theta\,d\theta, \sqrt{4-x^2} = 2\cos\theta$

29. (a) $x = 2 \tan \theta$,

$$dx = 2 \sec^2 \theta \, d\theta$$

$$\sqrt{4 + x^2} = 2 \sec \theta$$

$$\int \frac{x^3}{\sqrt{4 + x^2}} \, dx = \int \frac{8 \tan^3 \theta}{2 \sec \theta} 2 \sec^2 \theta \, d\theta$$

$$= 8 \int \frac{\sin^3 \theta}{\cos^4 \theta} \, d\theta$$

$$= 8 \int (1 - \cos^2 \theta) \cos^{-4} \theta \sin \theta \, d\theta$$

$$= 8 \int (\cos^{-4} \theta - \cos^{-2} \theta) \sin \theta \, d\theta$$

$$= 8 \left[\frac{\cos^{-3} \theta}{3} + \frac{\cos^{-1} \theta}{-1} \right] + C$$

$$= \frac{8}{3} \sec \theta (\sec^2 \theta - 3) + C$$

$$= \frac{8}{3} \frac{\sqrt{4 + x^2}}{2} \left(\frac{4 + x^2}{4} - 3 \right) + C$$

$$= \frac{1}{3} \sqrt{4 + x^2} (x^2 - 8) + C$$

(b) $\displaystyle \int \frac{x^3}{\sqrt{4 + x^2}} \, dx = \int \frac{x^2}{\sqrt{4 + x^2}} x \, dx$

$$= \int \frac{(u^2 - 4)}{u} u \, du$$

$$= \int (u^2 - 4) \, du$$

$$= \frac{1}{3} u^3 - 4u + C$$

$$= \frac{u}{3} (u^2 - 12) + C$$

$$= \frac{\sqrt{4 + x^2}}{3} (x^2 - 8) + C$$

$$u^2 = 4 + x^2, 2u \, du = 2x \, dx$$

(c) $\displaystyle \int \frac{x^3}{\sqrt{4 + x^2}} \, dx = x^2 \sqrt{4 + x^2} - \int 2x \sqrt{4 + x^2} \, dx$

$$= x^2 \sqrt{4 + x^2} - \frac{2}{3} (4 + x^2)^{3/2} + C$$

$$= \frac{\sqrt{4 + x^2}}{3} (x^2 - 8) + C$$

$$dv = \frac{x}{\sqrt{4 + x^2}} \, dx \implies v = \sqrt{4 + x^2}$$

$$u = x^2 \qquad \implies du = 2x \, dx$$

31. $\displaystyle \frac{x - 28}{x^2 - x - 6} = \frac{A}{x - 3} + \frac{B}{x + 2}$

$$x - 28 = A(x + 2) + B(x - 3)$$

$$x = -2 \implies -30 = B(-5) \implies B = 6$$

$$x = 3 \implies -25 = A(5) \implies A = -5$$

$$\int \frac{x - 28}{x^2 - x - 6} \, dx = \int \left(\frac{-5}{x - 3} + \frac{6}{x + 2} \right) dx = -5 \ln|x - 3| + 6 \ln|x + 2| + C$$

33. $\displaystyle \frac{x^2 + 2x}{(x - 1)(x^2 + 1)} = \frac{A}{x - 1} + \frac{Bx + C}{x^2 + 1}$

$$x^2 + 2x = A(x^2 + 1) + (Bx + C)(x - 1)$$

Let $x = 1$: $3 = 2A \implies A = \dfrac{3}{2}$

Let $x = 0$: $0 = A - C \implies C = \dfrac{3}{2}$

Let $x = 2$: $8 = 5A + 2B + C \implies B = -\dfrac{1}{2}$

$$\int \frac{x^2 + 2x}{x^3 - x^2 + x - 1} \, dx = \frac{3}{2} \int \frac{1}{x - 1} \, dx - \frac{1}{2} \int \frac{x - 3}{x^2 + 1} \, dx$$

$$= \frac{3}{2} \int \frac{1}{x - 1} \, dx - \frac{1}{4} \int \frac{2x}{x^2 + 1} \, dx + \frac{3}{2} \int \frac{1}{x^2 + 1} \, dx$$

$$= \frac{3}{2} \ln|x - 1| - \frac{1}{4} \ln|x^2 + 1| + \frac{3}{2} \arctan x + C$$

$$= \frac{1}{4} [6 \ln|x - 1| - \ln(x^2 + 1) + 6 \arctan x] + C$$

35. $\dfrac{x^2}{x^2 + 2x - 15} = 1 + \dfrac{15 - 2x}{x^2 + 2x - 15}$

$\dfrac{15 - 2x}{(x - 3)(x + 5)} = \dfrac{A}{x - 3} + \dfrac{B}{x + 5}$

$15 - 2x = A(x + 5) + B(x - 3)$

Let $x = 3$: $9 = 8A \implies A = \dfrac{9}{8}$

Let $x = -5$: $25 = -8B \implies B = -\dfrac{25}{8}$

$\displaystyle\int \dfrac{x^2}{x^2 + 2x - 15}\, dx = \int dx + \dfrac{9}{8}\int \dfrac{1}{x - 3}\, dx - \dfrac{25}{8}\int \dfrac{1}{x + 5}\, dx$

$\qquad\qquad\qquad\qquad = x + \dfrac{9}{8}\ln|x - 3| - \dfrac{25}{8}\ln|x + 5| + C$

37. $\displaystyle\int \dfrac{x}{(2 + 3x)^2}\, dx = \dfrac{1}{9}\left[\dfrac{2}{2 + 3x} + \ln|2 + 3x|\right] + C$

(Formula 4)

39. $\displaystyle\int \dfrac{x}{1 + \sin x^2}\, dx = \dfrac{1}{2}\int \dfrac{1}{1 + \sin u}\, du \qquad (u = x^2)$

$\qquad\qquad\qquad = \dfrac{1}{2}[\tan u - \sec u] + C \quad$ (Formula 56)

$\qquad\qquad\qquad = \dfrac{1}{2}[\tan x^2 - \sec x^2] + C$

41. $\displaystyle\int \dfrac{x}{x^2 + 4x + 8}\, dx = \dfrac{1}{2}\left[\ln|x^2 + 4x + 8| - 4\int \dfrac{1}{x^2 + 4x + 8}\, dx\right] \qquad$ (Formula 15)

$\qquad\qquad\qquad = \dfrac{1}{2}[\ln|x^2 + 4x + 8|] - 2\left[\dfrac{2}{\sqrt{32 - 16}}\arctan\left(\dfrac{2x + 4}{\sqrt{32 - 16}}\right)\right] + C \quad$ (Formula 14)

$\qquad\qquad\qquad = \dfrac{1}{2}\ln|x^2 + 4x + 8| - \arctan\left(1 + \dfrac{x}{2}\right) + C$

43. $\displaystyle\int \dfrac{1}{\sin \pi x \cos \pi x}\, dx = \dfrac{1}{\pi}\int \dfrac{1}{\sin \pi x \cos \pi x}(\pi)\, dx \quad (u = \pi x)$

$\qquad\qquad\qquad = \dfrac{1}{\pi}\ln|\tan \pi x| + C \qquad$ (Formula 58)

45. $dv = dx \implies v = x$

$u = (\ln x)^n \implies du = n(\ln x)^{n-1}\dfrac{1}{x}dx$

$\displaystyle\int (\ln x)^n\, dx = x(\ln x)^n - n\int (\ln x)^{n-1}\, dx$

47. $\displaystyle\int \theta \sin \theta \cos \theta\, d\theta = \dfrac{1}{2}\int \theta \sin 2\theta\, d\theta$

$\qquad = -\dfrac{1}{4}\theta \cos 2\theta + \dfrac{1}{4}\int \cos 2\theta\, d\theta = -\dfrac{1}{4}\theta \cos 2\theta + \dfrac{1}{8}\sin 2\theta + C = \dfrac{1}{8}(\sin 2\theta - 2\theta \cos 2\theta) + C$

$dv = \sin 2\theta\, d\theta \implies v = -\dfrac{1}{2}\cos 2\theta$

$u = \theta \qquad \implies du = d\theta$

49. $\int \dfrac{x^{1/4}}{1 + x^{1/2}}\, dx = 4 \int \dfrac{u(u^3)}{1 + u^2}\, du$

$\qquad\qquad = 4 \int \left(u^2 - 1 + \dfrac{1}{u^2 + 1} \right) du$

$\qquad\qquad = 4 \left(\dfrac{1}{3} u^3 - u + \arctan u \right) + C$

$\qquad\qquad = \dfrac{4}{3} \left[x^{3/4} - 3x^{1/4} + 3 \arctan(x^{1/4}) \right] + C$

$u = \sqrt[4]{x}, x = u^4, dx = 4u^3\, du$

51. $\int \sqrt{1 + \cos x}\, dx = \int \dfrac{\sin x}{\sqrt{1 - \cos x}}\, dx$

$\qquad\qquad = \int (1 - \cos x)^{-1/2} (\sin x)\, dx$

$\qquad\qquad = 2\sqrt{1 - \cos x} + C$

$u = 1 - \cos x, du = \sin x\, dx$

53. $\int \cos x \ln(\sin x)\, dx = \sin x \ln(\sin x) - \int \cos x\, dx$

$\qquad\qquad\qquad = \sin x \ln(\sin x) - \sin x + C$

$dv = \cos x\, dx \implies v = \sin x$

$u = \ln(\sin x) \implies du = \dfrac{\cos x}{\sin x}\, dx$

55. $y = \int \dfrac{9}{x^2 - 9}\, dx = \dfrac{3}{2} \ln \left| \dfrac{x - 3}{x + 3} \right| + C$

(by Formula 24 of Integration Tables)

57. $y = \int \ln(x^2 + x)\, dx = x \ln|x^2 + x| - \int \dfrac{2x^2 + x}{x^2 + x}\, dx$

$\qquad\qquad = x \ln|x^2 + x| - \int \dfrac{2x + 1}{x + 1}\, dx$

$\qquad\qquad = x \ln|x^2 + x| - \int 2\, dx + \int \dfrac{1}{x + 1}\, dx$

$\qquad\qquad = x \ln|x^2 + x| - 2x + \ln|x + 1| + C$

$dv = dx \qquad \implies v = x$

$u = \ln(x^2 + x) \implies du = \dfrac{2x + 1}{x^2 + x}\, dx$

59. $\int_{2}^{\sqrt{5}} x(x^2 - 4)^{3/2}\, dx = \left[\dfrac{1}{5}(x^2 - 4)^{5/2} \right]_{2}^{\sqrt{5}} = \dfrac{1}{5}$

61. $\int_{1}^{4} \dfrac{\ln x}{x}\, dx = \left[\dfrac{1}{2}(\ln x)^2 \right]_{1}^{4} = \dfrac{1}{2}(\ln 4)^2 = 2(\ln 2)^2 \approx 0.961$

63. $\int_{0}^{\pi} x \sin x\, dx = \left[-x \cos x + \sin x \right]_{0}^{\pi} = \pi$

65. $A = \int_{0}^{4} x\sqrt{4 - x}\, dx = \int_{2}^{0} (4 - u^2) u(-2u)\, du$

$\qquad = \int_{2}^{0} 2(u^4 - 4u^2)\, du$

$\qquad = \left[2 \left(\dfrac{u^5}{5} - \dfrac{4u^3}{3} \right) \right]_{2}^{0} = \dfrac{128}{15}$

$u = \sqrt{4 - x}, x = 4 - u^2, dx = -2u\, du$

67. By symmetry, $\bar{x} = 0, A = \dfrac{1}{2}\pi$.

$\bar{y} = \dfrac{2}{\pi} \left(\dfrac{1}{2} \right) \int_{-1}^{1} \left(\sqrt{1 - x^2} \right)^2 dx = \dfrac{1}{\pi} \left[x - \dfrac{1}{3} x^3 \right]_{-1}^{1} = \dfrac{4}{3\pi}$

$(\bar{x}, \bar{y}) = \left(0, \dfrac{4}{3\pi} \right)$

69. $s = \int_{0}^{\pi} \sqrt{1 + \cos^2 x}\, dx \approx 3.82$

71. $\displaystyle\lim_{x \to 1} \left[\dfrac{(\ln x)^2}{x - 1} \right] = \lim_{x \to 1} \left[\dfrac{2(1/x)\ln x}{1} \right] = 0$

73. $\displaystyle\lim_{x \to \infty} \dfrac{e^{2x}}{x^2} = \lim_{x \to \infty} \dfrac{2e^{2x}}{2x} = \lim_{x \to \infty} \dfrac{4e^{2x}}{2} = \infty$

75. $y = \displaystyle\lim_{x \to \infty} (\ln x)^{2/x}$

$\ln y = \displaystyle\lim_{x \to \infty} \dfrac{2 \ln(\ln x)}{x} = \lim_{x \to \infty} \left[\dfrac{2/(x \ln x)}{1} \right] = 0$

Since $\ln y = 0, y = 1$.

77. $\displaystyle\lim_{n \to \infty} 1000\left(1 + \frac{0.09}{n}\right)^n = 1000 \lim_{n \to \infty} \left(1 + \frac{0.09}{n}\right)^n$

Let $y = \displaystyle\lim_{n \to \infty} \left(1 + \frac{0.09}{n}\right)^n$.

$\ln y = \displaystyle\lim_{n \to \infty} n \ln\left(1 + \frac{0.09}{n}\right) = \lim_{n \to \infty} \frac{\ln\left(1 + \frac{0.09}{n}\right)}{\frac{1}{n}} = \lim_{n \to \infty} \left(\frac{\frac{-0.09/n^2}{1 + (0.09/n)}}{-\frac{1}{n^2}}\right) = \lim_{n \to \infty} \frac{0.09}{1 + \left(\frac{0.09}{n}\right)} = 0.09$

Thus, $\ln y = 0.09 \implies y = e^{0.09}$ and $\displaystyle\lim_{n \to \infty} 1000\left(1 + \frac{0.09}{n}\right)^n = 1000e^{0.09} \approx 1094.17$.

79. $\displaystyle\int_0^{16} \frac{1}{\sqrt[4]{x}}\,dx = \lim_{b \to 0^+} \left[\frac{4}{3}x^{3/4}\right]_b^{16} = \frac{32}{3}$

Converges

81. $\displaystyle\int_1^\infty x^2 \ln x\,dx = \lim_{b \to \infty} \left[\frac{x^3}{9}(-1 + 3\ln x)\right]_1^b = \infty$

Diverges

83. $\displaystyle\int_0^{t_0} 500{,}000 e^{-0.05t}\,dt = \left[\frac{500{,}000}{-0.05}e^{-0.05t}\right]_0^{t_0}$

$\qquad = \dfrac{-500{,}000}{0.05}(e^{-0.05t_0} - 1)$

$\qquad = 10{,}000{,}000(1 - e^{-0.05t_0})$

(a) $t_0 = 20$: \$6,321,205.59

(b) $t_0 \to \infty$: \$10,000,000

85. (a) $P(13 \le x < \infty) = \dfrac{1}{0.95\sqrt{2\pi}} \displaystyle\int_{13}^\infty e^{-(x-12.9)^2/2(0.95)^2}\,dx \approx 0.4581$

(b) $P(15 \le x < \infty) = \dfrac{1}{0.95\sqrt{2\pi}} \displaystyle\int_{15}^\infty e^{-(x-12.9)^2/2(0.95)^2}\,dx \approx 0.0135$

Problem Solving for Chapter 7

1. (a) $\displaystyle\int_{-1}^1 (1 - x^2)\,dx = \left[x - \frac{x^3}{3}\right]_{-1}^1 = 2\left(1 - \frac{1}{3}\right) = \frac{4}{3}$

$\displaystyle\int_{-1}^1 (1 - x^2)^2\,dx = \int_{-1}^1 (1 - 2x^2 + x^4)\,dx = \left[x - \frac{2x^3}{3} + \frac{x^5}{5}\right]_{-1}^1 = 2\left(1 - \frac{2}{3} + \frac{1}{5}\right) = \frac{16}{15}$

(b) Let $x = \sin u$, $dx = \cos u\,du$, $1 - x^2 = 1 - \sin^2 u = \cos^2 u$.

$\displaystyle\int_{-1}^1 (1 - x^2)^n\,dx = \int_{-\pi/2}^{\pi/2} (\cos^2 u)^n \cos u\,du$

$\qquad = \displaystyle\int_{-\pi/2}^{\pi/2} \cos^{2n+1} u\,du$

$\qquad = 2\left[\dfrac{2}{3} \cdot \dfrac{4}{5} \cdot \dfrac{6}{7} \cdots \dfrac{(2n)}{(2n + 1)}\right]$ \qquad (Wallis's Formula)

$\qquad = 2\left[\dfrac{2^2 \cdot 4^2 \cdot 6^2 \cdots (2n)^2}{2 \cdot 3 \cdot 4 \cdot 5 \cdots (2n)(2n + 1)}\right]$

$\qquad = \dfrac{2(2^{2n})(n!)^2}{(2n + 1)!} = \dfrac{2^{2n+1}(n!)^2}{(2n + 1)!}$

3.
$$\lim_{x \to \infty} \left(\frac{x+c}{x-c}\right)^x = 9$$

$$\lim_{x \to \infty} x \ln\left(\frac{x+c}{x-c}\right) = \ln 9$$

$$\lim_{x \to \infty} \frac{\ln(x+c) - \ln(x-c)}{1/x} = \ln 9$$

$$\lim_{x \to \infty} \frac{\dfrac{1}{x+c} - \dfrac{1}{x-c}}{-\dfrac{1}{x^2}} = \ln 9$$

$$\lim_{x \to \infty} \frac{-2c}{(x+c)(x-c)}(-x^2) = \ln 9$$

$$\lim_{x \to \infty} \left(\frac{2cx^2}{x^2 - c^2}\right) = \ln 9$$

$$2c = \ln 9$$

$$2c = 2 \ln 3$$

$$c = \ln 3$$

5. $\sin\theta = \dfrac{PB}{OP} = PB$, $\cos\theta = OB$

$AQ = \widehat{AP} = \theta$

$BR = OR + OB = OR + \cos\theta$

The triangles $\triangle AQR$ and $\triangle BPR$ are similar:

$$\frac{AR}{AQ} = \frac{BR}{BP} \implies \frac{OR + 1}{\theta} = \frac{OR + \cos\theta}{\sin\theta}$$

$$\sin\theta(OR) + \sin\theta = (OR)\theta + \theta\cos\theta$$

$$OR = \frac{\theta\cos\theta - \sin\theta}{\sin\theta - \theta}$$

$$\lim_{\theta \to 0^+} OR = \lim_{\theta \to 0^+} \frac{\theta\cos\theta - \sin\theta}{\sin\theta - \theta}$$

$$= \lim_{\theta \to 0^+} \frac{-\theta\sin\theta + \cos\theta - \cos\theta}{\cos\theta - 1}$$

$$= \lim_{\theta \to 0^+} \frac{-\theta\sin\theta}{\cos\theta - 1}$$

$$= \lim_{\theta \to 0^+} \frac{-\sin\theta - \theta\cos\theta}{-\sin\theta}$$

$$= \lim_{\theta \to 0^+} \frac{\cos\theta + \cos\theta - \theta\sin\theta}{\cos\theta}$$

$$= 2$$

7. (a)

Area ≈ 0.2986

(b) Let $x = 3\tan\theta$, $dx = 3\sec^2\theta\,d\theta$, $x^2 + 9 = 9\sec^2\theta$.

$$\int \frac{x^2}{(x^2+9)^{3/2}}\,dx = \int \frac{9\tan^2\theta}{(9\sec^2\theta)^{3/2}}(3\sec^2\theta\,d\theta)$$

$$= \int \frac{\tan^2\theta}{\sec\theta}\,d\theta$$

$$= \int \frac{\sin^2\theta}{\cos\theta}\,d\theta$$

$$= \int \frac{1 - \cos^2\theta}{\cos\theta}\,d\theta$$

$$= \ln|\sec\theta + \tan\theta| - \sin\theta + C$$

$$\text{Area} = \int_0^4 \frac{x^2}{(x^2+9)^{3/2}}\,dx = \left[\ln|\sec\theta + \tan\theta| - \sin\theta\right]_0^{\tan^{-1}(4/3)}$$

$$= \left[\ln\left(\frac{\sqrt{x^2+9}}{3} + \frac{x}{3}\right) - \frac{x}{\sqrt{x^2+9}}\right]_0^4$$

$$= \ln\left(\frac{5}{3} + \frac{4}{3}\right) - \frac{4}{5} = \ln 3 - \frac{4}{5}$$

(c) $x = 3\sinh u$, $dx = 3\cosh u\,du$, $x^2 + 9 = 9\sinh^2 u + 9 = 9\cosh^2 u$

$$A = \int_0^4 \frac{x^2}{(x^2+9)^{3/2}}\,dx = \int_0^{\sinh^{-1}(4/3)} \frac{9\sinh^2 u}{(9\cosh^2 u)^{3/2}}(3\cosh u\,du)$$

$$= \int_0^{\sinh^{-1}(4/3)} \tanh^2 u\,du$$

$$= \int_0^{\sinh^{-1}(4/3)} (1 - \text{sech}^2 u)\,du$$

$$= \left[u - \tanh u\right]_0^{\sinh^{-1}(4/3)}$$

$$= \sinh^{-1}\left(\frac{4}{3}\right) - \tanh\left(\sinh^{-1}\left(\frac{4}{3}\right)\right)$$

$$= \ln\left(\frac{4}{3} + \sqrt{\frac{16}{9} + 1}\right) - \tanh\left[\ln\left(\frac{4}{3} + \sqrt{\frac{16}{9} + 1}\right)\right]$$

$$= \ln\left(\frac{4}{3} + \frac{5}{3}\right) - \tanh\left(\ln\left(\frac{4}{3} + \frac{5}{3}\right)\right)$$

$$= \ln 3 - \tanh(\ln 3)$$

$$= \ln 3 - \frac{3 - (1/3)}{3 + (1/3)}$$

$$= \ln 3 - \frac{4}{5}$$

9. $y = \ln(1 - x^2), y' = \dfrac{-2x}{1 - x^2}$

$$1 + (y')^2 = 1 + \frac{4x^2}{(1 - x^2)^2} = \frac{1 - 2x^2 + x^4 + 4x^2}{(1 - x^2)^2} = \left(\frac{1 + x^2}{1 - x^2}\right)^2$$

$$\text{Arc length} = \int_0^{1/2} \sqrt{1 + (y')^2}\, dx$$

$$= \int_0^{1/2} \left(\frac{1 + x^2}{1 - x^2}\right) dx$$

$$= \int_0^{1/2} \left(-1 + \frac{2}{1 - x^2}\right) dx$$

$$= \int_0^{1/2} \left(-1 + \frac{1}{x + 1} + \frac{1}{1 - x}\right) dx$$

$$= \left[-x + \ln(1 + x) - \ln(1 - x)\right]_0^{1/2}$$

$$= \left(-\frac{1}{2} + \ln\frac{3}{2} - \ln\frac{1}{2}\right)$$

$$= -\frac{1}{2} + \ln 3 - \ln 2 + \ln 2$$

$$= \ln 3 - \frac{1}{2} \approx 0.5986$$

11. Consider $\displaystyle\int \frac{1}{\ln x}\, dx.$

Let $u = \ln x, du = \dfrac{1}{x}\, dx, x = e^u.$ Then $\displaystyle\int \frac{1}{\ln x}\, dx = \int \frac{1}{u} e^u\, du = \int \frac{e^u}{u}\, du.$

If $\displaystyle\int \frac{1}{\ln x}\, dx$ were elementary, then $\displaystyle\int \frac{e^u}{u}\, du$ would be too, which is false.

Hence, $\displaystyle\int \frac{1}{\ln x}\, dx$ is not elementary.

13. $x^4 + 1 = (x^2 + ax + b)(x^2 + cx + d)$

$$= x^4 + (a + c)x^3 + (ac + b + d)x^2 + (ad + bc)x + bd$$

$a = -c, b = d = 1, a = \sqrt{2}$

$$x^4 + 1 = \left(x^2 + \sqrt{2}x + 1\right)\left(x^2 - \sqrt{2}x + 1\right)$$

$$\int_0^1 \frac{1}{x^4 + 1}\, dx = \int_0^1 \frac{Ax + B}{x^2 + \sqrt{2}x + 1}\, dx + \int_0^1 \frac{Cx + D}{x^2 - \sqrt{2}x + 1}\, dx$$

$$= \int_0^1 \frac{\dfrac{1}{2} + \dfrac{\sqrt{2}}{4}x}{x^2 + \sqrt{2}x + 1}\, dx - \int_0^1 \frac{-\dfrac{1}{2} + \dfrac{\sqrt{2}}{4}x}{x^2 + \sqrt{2}x + 1}\, dx$$

$$= \frac{\sqrt{2}}{4}\left[\arctan\left(\sqrt{2}x + 1\right) + \arctan\left(\sqrt{2}x - 1\right)\right]_0^1 + \frac{\sqrt{2}}{8}\left[\ln\left(x^2 + \sqrt{2}x + 1\right) - \ln\left(x^2 - \sqrt{2}x + 1\right)\right]_0^1$$

$$= \frac{\sqrt{2}}{4}\left[\arctan\left(\sqrt{2} + 1\right) + \arctan\left(\sqrt{2} - 1\right)\right] + \frac{\sqrt{2}}{8}\left[\ln\left(2 + \sqrt{2}\right) - \ln\left(2 - \sqrt{2}\right)\right] - \frac{\sqrt{2}}{4}\left[\frac{\pi}{4} - \frac{\pi}{4}\right] - \frac{\sqrt{2}}{8}[0]$$

$$\approx 0.5554 + 0.3116$$

$$\approx 0.8670$$

15. Using a graphing utility,

(a) $\lim\limits_{x \to 0^+} \left(\cot x + \dfrac{1}{x} \right) = \infty$

(b) $\lim\limits_{x \to 0^+} \left(\cot x - \dfrac{1}{x} \right) = 0$

(c) $\lim\limits_{x \to 0^+} \left(\cot x + \dfrac{1}{x} \right)\left(\cot x - \dfrac{1}{x} \right) \approx -\dfrac{2}{3}.$

Analytically,

(a) $\lim\limits_{x \to 0^+} \left(\cot x + \dfrac{1}{x} \right) = \infty + \infty = \infty$

(b) $\lim\limits_{x \to 0^+} \left(\cot x - \dfrac{1}{x} \right) = \lim\limits_{x \to 0^+} \dfrac{x \cot x - 1}{x} = \lim\limits_{x \to 0^+} \dfrac{x \cos x - \sin x}{x \sin x}$

$\qquad\qquad = \lim\limits_{x \to 0^+} \dfrac{\cos x - x \sin x - \cos x}{\sin x + x \cos x} = \lim\limits_{x \to 0^+} \dfrac{-x \sin x}{\sin x + x \cos x}$

$\qquad\qquad = \lim\limits_{x \to 0^+} \dfrac{-\sin x - x \cos x}{\cos x + \cos x - x \sin x} = 0.$

(c) $\left(\cot x + \dfrac{1}{x} \right)\left(\cot x - \dfrac{1}{x} \right) = \cot^2 x - \dfrac{1}{x^2}$

$\qquad\qquad\qquad = \dfrac{x^2 \cot^2 x - 1}{x^2}$

$\lim\limits_{x \to 0^+} \dfrac{x^2 \cot^2 x - 1}{x^2} = \lim\limits_{x \to 0^+} \dfrac{2x \cot^2 x - 2x^2 \cot x \csc^2 x}{2x}$

$\qquad\qquad = \lim\limits_{x \to 0^+} \dfrac{\cot^2 x - x \cot x \csc^2 x}{1}$

$\qquad\qquad = \lim\limits_{x \to 0^+} \dfrac{\cos^2 x \sin x - x \cos x}{\sin^3 x}$

$\qquad\qquad = \lim\limits_{x \to 0^+} \dfrac{(1 - \sin^2 x)\sin x - x \cos x}{\sin^3 x}$

$\qquad\qquad = \lim\limits_{x \to 0^+} \dfrac{\sin x - x \cos x}{\sin^3 x} - 1$

Now, $\lim\limits_{x \to 0^+} \dfrac{\sin x - x \cos x}{\sin^3 x} = \lim\limits_{x \to 0^+} \dfrac{\cos x - \cos x + x \sin x}{3 \sin^2 x \cos x}$

$\qquad\qquad = \lim\limits_{x \to 0^+} \dfrac{x}{3 \sin x \cdot \cos x}$

$\qquad\qquad = \lim\limits_{x \to 0^+} \left(\dfrac{x}{\sin x} \right)\dfrac{1}{3 \cos x} = \dfrac{1}{3}.$

Thus, $\lim\limits_{x \to 0^+} \left(\cot x + \dfrac{1}{x} \right)\left(\cot x - \dfrac{1}{x} \right) = \dfrac{1}{3} - 1 = -\dfrac{2}{3}.$

The form $0 \cdot \infty$ is indeterminant.

17. $\dfrac{x^3 - 3x^2 + 1}{x^4 - 13x^2 + 12x} = \dfrac{P_1}{x} + \dfrac{P_2}{x-1} + \dfrac{P_3}{x+4} + \dfrac{P_4}{x-3} \Rightarrow c_1 = 0,\ c_2 = 1,\ c_3 = -4,\ c_4 = 3$

$N(x) = x^3 - 3x^2 + 1$

$D'(x) = 4x^3 - 26x + 12$

$P_1 = \dfrac{N(0)}{D'(0)} = \dfrac{1}{12}$

$P_2 = \dfrac{N(1)}{D'(1)} = \dfrac{-1}{-10} = \dfrac{1}{10}$

$P_3 = \dfrac{N(-4)}{D'(-4)} = \dfrac{-111}{-140} = \dfrac{111}{140}$

$P_4 = \dfrac{N(3)}{D'(3)} = \dfrac{1}{42}$

Thus, $\dfrac{x^3 - 3x^2 + 1}{x^4 - 13x^2 + 12x} = \dfrac{1/12}{x} + \dfrac{1/10}{x-1} + \dfrac{111/140}{x+4} + \dfrac{1/42}{x-3}.$

19. By parts,

$$\int_a^b f(x)g''(x)\,dx = \left[f(x)g'(x) \right]_a^b - \int_a^b f'(x)g'(x)\,dx$$

$$= -\int_a^b f'(x)g'(x)\,dx \quad 8$$

$$= \left[-f'(x)g(x) \right]_a^b + \int_a^b g(x)f''(x)\,dx$$

$$= \int_a^b f''(x)g(x)\,dx.$$

C H A P T E R   8
Infinite Series

Section 8.1   Sequences **401**

Section 8.2   Series and Convergence **406**

Section 8.3   The Integral Test and p-Series **411**

Section 8.4   Comparisons of Series **415**

Section 8.5   Alternating Series **418**

Section 8.6   The Ratio and Root Tests **422**

Section 8.7   Taylor Polynomials and Approximations **427**

Section 8.8   Power Series **432**

Section 8.9   Representation of Functions by Power Series . . **437**

Section 8.10   Taylor and Maclaurin Series **440**

Review Exercises **447**

Problem Solving **452**

CHAPTER 8
Infinite Series

Section 8.1   Sequences

Solutions to Odd-Numbered Exercises

1. $a_n = 2^n$

$a_1 = 2^1 = 2$

$a_2 = 2^2 = 4$

$a_3 = 2^3 = 8$

$a_4 = 2^4 = 16$

$a_5 = 2^5 = 32$

3. $a_n = \left(-\frac{1}{2}\right)^n$

$a_1 = \left(-\frac{1}{2}\right)^1 = -\frac{1}{2}$

$a_2 = \left(-\frac{1}{2}\right)^2 = \frac{1}{4}$

$a_3 = \left(-\frac{1}{2}\right)^3 = -\frac{1}{8}$

$a_4 = \left(-\frac{1}{2}\right)^4 = \frac{1}{16}$

$a_5 = \left(-\frac{1}{2}\right)^5 = -\frac{1}{32}$

5. $a_n = \sin \frac{n\pi}{2}$

$a_1 = \sin \frac{\pi}{2} = 1$

$a_2 = \sin \pi = 0$

$a_3 = \sin \frac{3\pi}{2} = -1$

$a_4 = \sin 2\pi = 0$

$a_5 = \sin \frac{5\pi}{2} = 1$

7. $a_n = \frac{(-1)^{n(n+1)/2}}{n^2}$

$a_1 = \frac{(-1)^1}{1^2} = -1$

$a_2 = \frac{(-1)^3}{2^2} = -\frac{1}{4}$

$a_3 = \frac{(-1)^6}{3^2} = \frac{1}{9}$

$a_4 = \frac{(-1)^{10}}{4^2} = \frac{1}{16}$

$a_5 = \frac{(-1)^{15}}{5^2} = -\frac{1}{25}$

9. $a_n = 5 - \frac{1}{n} + \frac{1}{n^2}$

$a_1 = 5 - 1 + 1 = 5$

$a_2 = 5 - \frac{1}{2} + \frac{1}{4} = \frac{19}{4}$

$a_3 = 5 - \frac{1}{3} + \frac{1}{9} = \frac{43}{9}$

$a_4 = 5 - \frac{1}{4} + \frac{1}{16} = \frac{77}{16}$

$a_5 = 5 - \frac{1}{5} + \frac{1}{25} = \frac{121}{25}$

11. $a_n = \frac{3^n}{n!}$

$a_1 = \frac{3}{1!} = 3$

$a_2 = \frac{3^2}{2!} = \frac{9}{2}$

$a_3 = \frac{3^3}{3!} = \frac{27}{6}$

$a_4 = \frac{3^4}{4!} = \frac{81}{24}$

$a_5 = \frac{3^5}{5!} = \frac{243}{120}$

13. $a_1 = 3,\ a_{k+1} = 2(a_k - 1)$

$a_2 = 2(a_1 - 1)$

$\quad = 2(3 - 1) = 4$

$a_3 = 2(a_2 - 1)$

$\quad = 2(4 - 1) = 6$

$a_4 = 2(a_3 - 1)$

$\quad = 2(6 - 1) = 10$

$a_5 = 2(a_4 - 1)$

$\quad = 2(10 - 1) = 18$

15. $a_1 = 32,\ a_{k+1} = \frac{1}{2}a_k$

$a_2 = \frac{1}{2}a_1 = \frac{1}{2}(32) = 16$

$a_3 = \frac{1}{2}a_2 = \frac{1}{2}(16) = 8$

$a_4 = \frac{1}{2}a_3 = \frac{1}{2}(8) = 4$

$a_5 = \frac{1}{2}a_4 = \frac{1}{2}(4) = 2$

17. Because $a_1 = 8/(1 + 1) = 4$ and $a_2 = 8/(2 + 1) = \frac{8}{3}$, the sequence matches graph (d).

19. This sequence decreases and $a_1 = 4$, $a_2 = 4(0.5) = 2$. Matches (c).

21.

$$a_n = \frac{2}{3}n, \, n = 1, \ldots, 10$$

23.

$$a_n = 16(-0.5)^{n-1}, \, n = 1, \ldots, 10$$

25.

$$a_n = \frac{2n}{n + 1}, \, n = 1, 2, \ldots, 10$$

27. $a_n = 3n - 1$

$a_5 = 3(5) - 1 = 14$

$a_6 = 3(6) - 1 = 17$

Add 3 to preceeding term.

29. $a_n = \dfrac{3}{(-2)^{n-1}}$

$a_5 = \dfrac{3}{(-2)^4} = \dfrac{3}{16}$

$a_6 = \dfrac{3}{(-2)^5} = -\dfrac{3}{32}$

Multiply the preceeding term by $-\frac{1}{2}$.

31. $\dfrac{10!}{8!} = \dfrac{8!(9)(10)}{8!}$

$= (9)(10) = 90$

33. $\dfrac{(n + 1)!}{n!} = \dfrac{n!(n + 1)}{n!}$

$= n + 1$

35. $\dfrac{(2n - 1)!}{(2n + 1)!} = \dfrac{(2n - 1)!}{(2n - 1)!(2n)(2n + 1)}$

$= \dfrac{1}{2n(2n + 1)}$

37. $\displaystyle\lim_{n \to \infty} \frac{5n^2}{n^2 + 2} = 5$

39. $\displaystyle\lim_{n \to \infty} \frac{2n}{\sqrt{n^2 + 1}} = \lim_{n \to \infty} \frac{2}{\sqrt{1 + (1/n^2)}}$

$= \dfrac{2}{1} = 2$

41. $\displaystyle\lim_{n \to \infty} \sin\left(\frac{1}{n}\right) = 0$

43.

The graph seems to indicate that the sequence converges to 1. Analytically,

$$\lim_{n \to \infty} a_n = \lim_{n \to \infty} \frac{n + 1}{n} = \lim_{x \to \infty} \frac{x + 1}{x} = \lim_{x \to \infty} 1 = 1.$$

45.

The graph seems to indicate that the sequence diverges. Analytically, the sequence is

$$\{a_n\} = \{0, -1, 0, 1, 0, -1, \ldots\}.$$

Hence, $\displaystyle\lim_{n \to \infty} a_n$ does not exist.

47. $\displaystyle\lim_{n \to \infty} (-1)^n \left(\frac{n}{n + 1}\right)$

does not exist (oscillates between -1 and 1), diverges.

49. $\displaystyle\lim_{n \to \infty} \frac{3n^2 - n + 4}{2n^2 + 1} = \frac{3}{2}$, converges

51. $\displaystyle\lim_{n \to \infty} \frac{1 + (-1)^n}{n} = 0$, converges

53. $\displaystyle\lim_{n \to \infty} \frac{\ln(n^3)}{2n} = \lim_{n \to \infty} \frac{3}{2} \frac{\ln(n)}{n}$

$= \displaystyle\lim_{n \to \infty} \frac{3}{2}\left(\frac{1}{n}\right) = 0$, converges

(L'Hôpital's Rule)

55. $\lim\limits_{n\to\infty}\left(\dfrac{3}{4}\right)^n = 0$, converges

57. $\lim\limits_{n\to\infty}\dfrac{(n+1)!}{n!} = \lim\limits_{n\to\infty}(n+1) = \infty$, diverges

59. $\lim\limits_{n\to\infty}\left(\dfrac{n-1}{n}-\dfrac{n}{n-1}\right) = \lim\limits_{n\to\infty}\dfrac{(n-1)^2-n^2}{n(n-1)}$

$\qquad\qquad = \lim\limits_{n\to\infty}\dfrac{1-2n}{n^2-n} = 0$, converges

61. $\lim\limits_{n\to\infty}\dfrac{n^p}{e^n} = 0$, converges

$\quad(p > 0, n \geq 2)$

63. $a_n = \left(1+\dfrac{k}{n}\right)^n$

$\qquad \lim\limits_{n\to\infty}\left(1+\dfrac{k}{n}\right)^n = \lim\limits_{u\to 0}\left[(1+u)^{1/u}\right]^k = e^k$

$\quad$ where $u = \dfrac{k}{n}$, converges

65. $\lim\limits_{n\to\infty}\dfrac{\sin n}{n} = \lim\limits_{n\to\infty}(\sin n)\dfrac{1}{n} = 0$, converges

67. $a_n = 3n - 2$

69. $a_n = n^2 - 2$

71. $a_n = \dfrac{n+1}{n+2}$

73. $a_n = \dfrac{(-1)^{n-1}}{2^{n-2}}$

75. $a_n = 1 + \dfrac{1}{n} = \dfrac{n+1}{n}$

77. $a_n = \dfrac{n}{(n+1)(n+2)}$

79. $a_n = \dfrac{(-1)^{n-1}}{1\cdot 3\cdot 5\cdots(2n-1)} = \dfrac{(-1)^{n-1}2^n n!}{(2n)!}$

81. $a_n = 4 - \dfrac{1}{n} < 4 - \dfrac{1}{n+1} = a_{n+1}$,

$\quad$ monotonic; $|a_n| < 4$ bounded.

83. $\dfrac{n}{2^{n+2}} \overset{?}{\geq} \dfrac{n+1}{2^{(n+1)+2}}$

$\quad 2^{n+3}n \overset{?}{\geq} 2^{n+2}(n+1)$

$\quad\quad 2n \overset{?}{\geq} n+1$

$\quad\quad\quad n \geq 1$

$\quad$ Hence, $\quad n \geq 1$

$\quad\quad\quad 2n \geq n+1$

$\quad\quad 2^{n+3}n \geq 2^{n+2}(n+1)$

$\quad\quad\quad \dfrac{n}{2^{n+2}} \geq \dfrac{n+1}{2^{(n+1)+2}}$

$\quad\quad\quad a_n \geq a_{n+1}$

$\quad$ True; monotonic; $|a_n| \leq \frac{1}{8}$, bounded

85. $a_n = (-1)^n\left(\dfrac{1}{n}\right)$

$\quad a_1 = -1$

$\quad a_2 = \dfrac{1}{2}$

$\quad a_3 = -\dfrac{1}{3}$

$\quad$ Not monotonic; $|a_n| \leq 1$, bounded

87. $a_n = \left(\frac{2}{3}\right)^n > \left(\frac{2}{3}\right)^{n+1} = a_{n+1}$

$\quad$ Monotonic; $|a_n| \leq \frac{2}{3}$, bounded

89. $a_n = \sin\left(\dfrac{n\pi}{6}\right)$

$\quad a_1 = 0.500$

$\quad a_2 = 0.8660$

$\quad a_3 = 1.000$

$\quad a_4 = 0.8660$

$\quad$ Not monotonic; $|a_n| \leq 1$, bounded

91. (a) $a_n = 5 + \dfrac{1}{n}$

$\left| 5 + \dfrac{1}{n} \right| \le 6 \Rightarrow \{a_n\}$ bounded

$a_n = 5 + \dfrac{1}{n} > 5 + \dfrac{1}{n+1}$

$= a_{n+1} \Rightarrow \{a_n\}$ monotonic

Therefore, $\{a_n\}$ converges.

(b)

$\lim\limits_{n \to \infty} \left(5 + \dfrac{1}{n} \right) = 5$

93. (a) $a_n = \dfrac{1}{3}\left(1 - \dfrac{1}{3^n} \right)$

$\left| \dfrac{1}{3}\left(1 - \dfrac{1}{3^n} \right) \right| < \dfrac{1}{3} \Rightarrow \{a_n\}$ bounded

$a_n = \dfrac{1}{3}\left(1 - \dfrac{1}{3^n} \right) < \dfrac{1}{3}\left(1 - \dfrac{1}{3^{n+1}} \right)$

$= a_{n+1} \Rightarrow \{a_n\}$ monotonic

Therefore, $\{a_n\}$ converges.

(b)

$\lim\limits_{n \to \infty} \left[\dfrac{1}{3}\left(1 - \dfrac{1}{3^n} \right) \right] = \dfrac{1}{3}$

95. $A_n = P\left[1 + \dfrac{r}{12} \right]^n$

(a) $\lim\limits_{n \to \infty} A_n = \infty$, divergent. The amount will grow arbitrarily large over time.

(b) $A_n = 9000\left[1 + \dfrac{0.115}{12} \right]^n$

$A_1 = \$9086.25$	$A_6 = \$9530.06$
$A_2 = \$9173.33$	$A_7 = \$9621.39$
$A_3 = \$9261.24$	$A_8 = \$9713.59$
$A_4 = \$9349.99$	$A_9 = \$9806.68$
$A_5 = \$9439.60$	$A_{10} = \$9900.66$

97. (a) A sequence is a function whose domain is the set of positive integers.

(b) A sequence converges if it has a limit.

(c) A bounded monotonic sequence is a sequence that has nondecreasing or nonincreasing terms, and an upper and lower bound.

99. $a_n = 10 - \dfrac{1}{n}$

101. $a_n = \dfrac{3n}{4n+1}$

103. (a) $A_n = (0.8)^n (2.5)$ billion

(b) $A_1 = \$2$ billion

$A_2 = \$1.6$ billion

$A_3 = \$1.28$ billion

$A_4 = \$1.024$ billion

(c) $\lim\limits_{n \to \infty} (0.8)^n (2.5) = 0$

105. (a) $a_n = -3.7262n^2 + 75.9167n + 684.25$

(b) For 2004, $n = 14$ and $a_{14} \approx 1017$, or \$1017.

107. $a_n = \dfrac{10^n}{n!}$

    (a) $a_9 = a_{10} = \dfrac{10^9}{9!}$

$$= \dfrac{1,000,000,000}{362,880}$$

$$= \dfrac{1,562,500}{567}$$

    (b) Decreasing

    (c) Factorials increase more rapidly than exponentials.

109. $\{a_n\} = \left\{\sqrt[n]{n}\right\} = \left\{n^{1/n}\right\}$

$a_1 = 1^{1/1} = 1$

$a_2 = \sqrt{2} \approx 1.4142$

$a_3 = \sqrt[3]{3} \approx 1.4422$

$a_4 = \sqrt[4]{4} \approx 1.4142$

$a_5 = \sqrt[5]{5} \approx 1.3797$

$a_6 = \sqrt[6]{6} \approx 1.3480$

Let $y = \lim\limits_{n\to\infty} n^{1/n}$.

$$\ln y = \lim_{n\to\infty}\left(\frac{1}{n}\ln n\right)$$

$$= \lim_{n\to\infty}\frac{\ln n}{n} = \lim_{n\to\infty}\frac{1/n}{1} = 0$$

Since $\ln y = 0$, we have $y = e^0 = 1$. Therefore, $\lim\limits_{n\to\infty}\sqrt[n]{n} = 1$.

111. $a_{n+2} = a_n + a_{n+1}$

    (a)

$a_1 = 1$	$a_7 = 8 + 5 = 13$
$a_2 = 1$	$a_8 = 13 + 8 = 21$
$a_3 = 1 + 1 = 2$	$a_9 = 21 + 13 = 34$
$a_4 = 2 + 1 = 3$	$a_{10} = 34 + 21 = 55$
$a_5 = 3 + 2 = 5$	$a_{11} = 55 + 34 = 89$
$a_6 = 5 + 3 = 8$	$a_{12} = 89 + 55 = 144$

    (b) $b_n = \dfrac{a_{n+1}}{a_n}, n \geq 1$

$b_1 = \dfrac{1}{1} = 1$	$b_6 = \dfrac{13}{8}$
$b_2 = \dfrac{2}{1} = 2$	$b_7 = \dfrac{21}{13}$
$b_3 = \dfrac{3}{2}$	$b_8 = \dfrac{34}{21}$
$b_4 = \dfrac{5}{3}$	$b_9 = \dfrac{55}{34}$
$b_5 = \dfrac{8}{5}$	$b_{10} = \dfrac{89}{55}$

    (c) $1 + \dfrac{1}{b_{n-1}} = 1 + \dfrac{1}{a_n/a_{n-1}}$

$$= 1 + \frac{a_{n-1}}{a_n}$$

$$= \frac{a_n + a_{n-1}}{a_n} = \frac{a_{n+1}}{a_n} = b_n$$

    (d) If $\lim\limits_{n\to\infty} b_n = \rho$, then $\lim\limits_{n\to\infty}\left(1 + \dfrac{1}{b_{n-1}}\right) = \rho$.

Since $\lim\limits_{n\to\infty} b_n = \lim\limits_{n\to\infty} b_{n-1}$ we have,

$1 + (1/\rho) = \rho$.

$\rho + 1 = \rho^2$

$0 = \rho^2 - \rho - 1$

$$\rho = \frac{1 \pm \sqrt{1+4}}{2} = \frac{1 \pm \sqrt{5}}{2}$$

Since a_n, and thus b_n, is positive,

$$\rho = \left(1 + \sqrt{5}\right)/2 \approx 1.6180.$$

113. True

115. True

117. $a_1 = \sqrt{2} \approx 1.4142$

$a_2 = \sqrt{2 + \sqrt{2}} \approx 1.8478$

$a_3 = \sqrt{2 + \sqrt{2 + \sqrt{2}}} \approx 1.9616$

$a_4 = \sqrt{2 + \sqrt{2 + \sqrt{2 + \sqrt{2}}}} \approx 1.9904$

$a_5 = \sqrt{2 + \sqrt{2 + \sqrt{2 + \sqrt{2 + \sqrt{2}}}}} \approx 1.9976$

$\{a_n\}$ is increasing and bounded by 2, and hence converges to L. Letting $\lim\limits_{n\to\infty} a_n = L$ implies that $\sqrt{2 + L} = L \Rightarrow L = 2$. Hence, $\lim\limits_{n\to\infty} a_n = 2$.

Section 8.2   Series and Convergence

1. $S_1 = 1$

$S_2 = 1 + \frac{1}{4} = 1.2500$

$S_3 = 1 + \frac{1}{4} + \frac{1}{9} \approx 1.3611$

$S_4 = 1 + \frac{1}{4} + \frac{1}{9} + \frac{1}{16} \approx 1.4236$

$S_5 = 1 + \frac{1}{4} + \frac{1}{9} + \frac{1}{16} + \frac{1}{25} \approx 1.4636$

3. $S_1 = 3$

$S_2 = 3 - \frac{9}{2} = -1.5$

$S_3 = 3 - \frac{9}{2} + \frac{27}{4} = 5.25$

$S_4 = 3 - \frac{9}{2} + \frac{27}{4} - \frac{81}{8} = -4.875$

$S_5 = 3 - \frac{9}{2} + \frac{27}{4} - \frac{81}{8} + \frac{243}{16} = 10.3125$

5. $S_1 = 3$

$S_2 = 3 + \frac{3}{2} = 4.5$

$S_3 = 3 + \frac{3}{2} + \frac{3}{4} = 5.250$

$S_4 = 3 + \frac{3}{2} + \frac{3}{4} + \frac{3}{8} = 5.625$

$S_5 = 3 + \frac{3}{2} + \frac{3}{4} + \frac{3}{8} + \frac{3}{16} = 5.8125$

7. $\displaystyle\sum_{n=0}^{\infty} 3\left(\frac{3}{2}\right)^n$ Geometric series

$r = \frac{3}{2} > 1$

Diverges by Theorem 8.6

9. $\displaystyle\sum_{n=0}^{\infty} 1000(1.055)^n$ Geometric series

$r = 1.055 > 1$

Diverges by Theorem 8.6

11. $\displaystyle\sum_{n=1}^{\infty} \frac{n}{n+1}$

$\displaystyle\lim_{n\to\infty} \frac{n}{n+1} = 1 \neq 0$

Diverges by Theorem 8.9

13. $\displaystyle\sum_{n=1}^{\infty} \frac{n^2}{n^2+1}$

$\displaystyle\lim_{n\to\infty} \frac{n^2}{n^2+1} = 1 \neq 0$

Diverges by Theorem 8.9

15. $\displaystyle\sum_{n=1}^{\infty} \frac{2^n+1}{2^{n+1}}$

$\displaystyle\lim_{n\to\infty} \frac{2^n+1}{2^{n+1}} = \lim_{n\to\infty} \frac{1+2^{-n}}{2} = \frac{1}{2} \neq 0$

Diverges by Theorem 8.9

17. $\displaystyle\sum_{n=0}^{\infty} \frac{9}{4}\left(\frac{1}{4}\right)^n = \frac{9}{4}\left[1 + \frac{1}{4} + \frac{1}{16} + \cdots\right]$

$S_0 = \frac{9}{4}, S_1 = \frac{9}{4}\cdot\frac{5}{4} = \frac{45}{16}, S_2 = \frac{9}{4}\cdot\frac{21}{16} \approx 2.95, \ldots$

Matches graph (c).

Analytically, the series is geometric:

$$\sum_{n=0}^{\infty} \left(\frac{9}{4}\right)\left(\frac{1}{4}\right)^n = \frac{9/4}{1-1/4} = \frac{9/4}{3/4} = 3$$

19. $\displaystyle\sum_{n=0}^{\infty} \frac{15}{4}\left(-\frac{1}{4}\right)^n = \frac{15}{4}\left[1 - \frac{1}{4} + \frac{1}{16} - \cdots\right]$

$S_0 = \frac{15}{4}, S_1 = \frac{45}{16}, S_2 \approx 3.05, \ldots$

Matches graph (a).

Analytically, the series is geometric:

$$\sum_{n=0}^{\infty} \frac{15}{4}\left(-\frac{1}{4}\right)^n = \frac{15/4}{1-(-1/4)} = \frac{15/4}{5/4} = 3$$

21. $\displaystyle\sum_{n=1}^{\infty} \frac{1}{n(n+1)} = \sum_{n=1}^{\infty}\left(\frac{1}{n} - \frac{1}{n+1}\right) = \left(1 - \frac{1}{2}\right) + \left(\frac{1}{2} - \frac{1}{3}\right) + \left(\frac{1}{3} - \frac{1}{4}\right) + \left(\frac{1}{4} - \frac{1}{5}\right) + \cdots . \quad S_n = 1 - \frac{1}{n+1}$

$\displaystyle\sum_{n=1}^{\infty} \frac{1}{n(n+1)} = \lim_{n\to\infty} S_n = \lim_{n\to\infty}\left(1 - \frac{1}{n+1}\right) = 1$

23. $\displaystyle\sum_{n=0}^{\infty} 2\left(\frac{3}{4}\right)^n$

Geometric series with $r = \frac{3}{4} < 1$.

Converges by Theorem 8.6

25. $\displaystyle\sum_{n=0}^{\infty} (0.9)^n$

Geometric series with $r = 0.9 < 1$.

Converges by Theorem 8.6

27. (a) $\displaystyle\sum_{n=1}^{\infty} \frac{6}{n(n+3)} = 2\sum_{n=1}^{\infty}\left(\frac{1}{n} - \frac{1}{n+3}\right)$

$$= 2\left[\left(1 - \frac{1}{4}\right) + \left(\frac{1}{2} - \frac{1}{5}\right) + \left(\frac{1}{3} - \frac{1}{6}\right) + \left(\frac{1}{4} - \frac{1}{7}\right) + \cdots\right]$$

$$= 2\left[1 + \frac{1}{2} + \frac{1}{3}\right] = \frac{11}{3} \approx 3.667$$

(b)

n	5	10	20	50	100
S_n	2.7976	3.1643	3.3936	3.5513	3.6078

(c)

(d) The terms of the series decrease in magnitude slowly. Thus, the sequence of partial sums approaches the sum slowly.

29. (a) $\displaystyle\sum_{n=1}^{\infty} 2(0.9)^{n-1} = \sum_{n=0}^{\infty} 2(0.9)^n = \frac{2}{1 - 0.9} = 20$

(c)

(b)

n	5	10	20	50	100
S_n	8.1902	13.0264	17.5685	19.8969	19.9995

(d) The terms of the series decrease in magnitude slowly. Thus, the sequence of partial sums approaches the sum slowly.

31. (a) $\displaystyle\sum_{n=1}^{\infty} 10(0.25)^{n-1} = \frac{10}{1 - 0.25} = \frac{40}{3} \approx 13.3333$

(c)

(b)

n	5	10	20	50	100
S_n	13.3203	13.3333	13.3333	13.3333	13.3333

(d) The terms of the series decrease in magnitude rapidly. Thus, the sequence of partial sums approaches the sum rapidly.

33. $\displaystyle\sum_{n=2}^{\infty} \frac{1}{n^2 - 1} = \sum_{n=2}^{\infty}\left(\frac{1/2}{n-1} - \frac{1/2}{n+1}\right) = \frac{1}{2}\sum_{n=2}^{\infty}\left(\frac{1}{n-1} - \frac{1}{n+1}\right)$

$$= \frac{1}{2}\left[\left(1 - \frac{1}{3}\right) + \left(\frac{1}{2} - \frac{1}{4}\right) + \left(\frac{1}{3} - \frac{1}{5}\right) + \left(\frac{1}{4} - \frac{1}{6}\right) + \cdots\right]$$

$$= \frac{1}{2}\left(1 + \frac{1}{2}\right) = \frac{3}{4}$$

35. $\displaystyle\sum_{n=1}^{\infty} \frac{8}{(n+1)(n+2)} = 8\sum_{n=1}^{\infty}\left(\frac{1}{n+1} - \frac{1}{n+2}\right) = 8\left[\left(\frac{1}{2} - \frac{1}{3}\right) + \left(\frac{1}{3} - \frac{1}{4}\right) + \left(\frac{1}{4} - \frac{1}{5}\right) + \cdots\right] = 8\left(\frac{1}{2}\right) = 4$

37. $\displaystyle\sum_{n=0}^{\infty}\left(\frac{1}{2}\right)^n = \frac{1}{1 - (1/2)} = 2$

39. $\displaystyle\sum_{n=0}^{\infty}\left(-\frac{1}{2}\right)^n = \frac{1}{1 - (-1/2)} = \frac{2}{3}$

41. $\displaystyle\sum_{n=0}^{\infty}\left(\frac{1}{10}\right)^n = \frac{1}{1 - (1/10)} = \frac{10}{9}$

43. $\displaystyle\sum_{n=0}^{\infty} 3\left(-\frac{1}{3}\right)^n = \frac{3}{1 - (-1/3)} = \frac{9}{4}$

45. $\displaystyle\sum_{n=0}^{\infty}\left(\frac{1}{2^n}-\frac{1}{3^n}\right)=\sum_{n=0}^{\infty}\left(\frac{1}{2}\right)^n-\sum_{n=0}^{\infty}\left(\frac{1}{3}\right)^n$

$\displaystyle=\frac{1}{1-(1/2)}-\frac{1}{1-(1/3)}$

$\displaystyle=2-\frac{3}{2}=\frac{1}{2}$

47. $\displaystyle 0.\overline{4}=\sum_{n=0}^{\infty}\frac{4}{10}\left(\frac{1}{10}\right)^n$

Geometric series with $a=\frac{4}{10}$ and $r=\frac{1}{10}$

$\displaystyle S=\frac{a}{1-r}=\frac{4/10}{1-(1/10)}=\frac{4}{9}$

49. $\displaystyle 0.07\overline{575}=\sum_{n=0}^{\infty}\frac{3}{40}\left(\frac{1}{100}\right)^n$

Geometric series with $a=\frac{3}{40}$ and $r=\frac{1}{100}$

$\displaystyle S=\frac{a}{1-r}=\frac{3/40}{99/100}=\frac{5}{66}$

51. $\displaystyle\sum_{n=1}^{\infty}\frac{n+10}{10n+1}$

$\displaystyle\lim_{n\to\infty}\frac{n+10}{10n+1}=\frac{1}{10}\neq 0$

Diverges by Theorem 8.9

53. $\displaystyle\sum_{n=1}^{\infty}\left(\frac{1}{n}-\frac{1}{n+2}\right)=\left(1-\frac{1}{3}\right)+\left(\frac{1}{2}-\frac{1}{4}\right)+\left(\frac{1}{3}-\frac{1}{5}\right)+\left(\frac{1}{4}-\frac{1}{6}\right)+\cdots=1+\frac{1}{2}=\frac{3}{2}$, converges

55. $\displaystyle\sum_{n=1}^{\infty}\frac{3n-1}{2n+1}$

$\displaystyle\lim_{n\to\infty}\frac{3n-1}{2n+1}=\frac{3}{2}\neq 0$

Diverges by Theorem 8.9

57. $\displaystyle\sum_{n=0}^{\infty}\frac{4}{2^n}=4\sum_{n=0}^{\infty}\left(\frac{1}{2}\right)^n$

Geometric series with $r=\frac{1}{2}$

Converges by Theorem 8.6

59. $\displaystyle\sum_{n=0}^{\infty}(1.075)^n$

Geometric series with $r=1.075$

Diverges by Theorem 8.6

61. $\displaystyle\sum_{n=2}^{\infty}\frac{n}{\ln n}$

$\displaystyle\lim_{n\to\infty}\frac{n}{\ln n}=\lim_{n\to\infty}\frac{1}{1/n}=\infty$

(by L'Hôpital's Rule) Diverges by Theorem 8.9

63. See definition, page 567.

65. The series given by

$$\sum_{n=0}^{\infty}ar^n=a+ar+ar^2+\cdots+ar^n+\cdots,\, a\neq 0$$

is a geometric series with ratio r. When $0<|r|<1$, the series converges to $\dfrac{a}{1-r}$. The series diverges if $|r|\geq 1$.

67. (a) x is the common ratio.

(b) $\displaystyle 1+x+x^2+\cdots=\sum_{n=0}^{\infty}x^n=\frac{1}{1-x},\, |x|<1$

Geometric series: $a=1,\, r=x,\, |x|<1$

(c) $\displaystyle y_1=\frac{1}{1-x},\, |x|<1$

$y_2=s_2=1+x$

69. $\displaystyle f(x)=3\left[\frac{1-0.5^x}{1-0.5}\right]$

Horizontal asymptote: $y=6$

$\displaystyle\sum_{n=0}^{\infty}3\left(\frac{1}{2}\right)^n$

$\displaystyle S=\frac{3}{1-(1/2)}=6$

The horizontal asymptote is the sum of the series. $f(n)$ is the nth partial sum.

71. $\dfrac{1}{n(n + 1)} < 0.001$

$10{,}000 < n^2 + n$

$0 < n^2 + n - 10{,}000$

$n = \dfrac{-1 \pm \sqrt{1^2 - 4(1)(-10{,}000)}}{2}$

Choosing the positive value for n we have $n \approx 99.5012$. The first *term* that is less than 0.001 is $n = 100$.

$\left(\dfrac{1}{8}\right)^n < 0.001$

$10{,}000 < 8^n$

This inequality is true when $n = 5$. This series converges at a faster rate.

73. $\displaystyle\sum_{i=0}^{n-1} 8000(0.9)^i = \dfrac{8000[1 - (0.9)^{(n-1)+1}]}{1 - 0.9}$

$= 80{,}000(1 - 0.9^n),\ n > 0$

75. $\displaystyle\sum_{i=0}^{n-1} 100(0.75)^i = \dfrac{100[1 - 0.75^{(n-1)+1}]}{1 - 0.75}$

$= 400(1 - 0.75^n)$ million dollars.

Sum $= 400$ million dollars

77. $D_1 = 16$

$D_2 = \underbrace{0.81(16)}_{\text{up}} + \underbrace{0.81(16)}_{\text{down}} = 32(0.81)$

$D_3 = 16(0.81)^2 + 16(0.81)^2 = 32(0.81)^2$

$\vdots$

$D = 16 + 32(0.81) + 32(0.81)^2 + \cdots = -16 + \displaystyle\sum_{n=0}^{\infty} 32(0.81)^n = -16 + \dfrac{32}{1 - 0.81} \approx 152.42$ feet

79. $P(n) = \dfrac{1}{2}\left(\dfrac{1}{2}\right)^n$

$P(2) = \dfrac{1}{2}\left(\dfrac{1}{2}\right)^2 = \dfrac{1}{8}$

$\displaystyle\sum_{n=0}^{\infty} \dfrac{1}{2}\left(\dfrac{1}{2}\right)^n = \dfrac{1/2}{1 - (1/2)} = 1$

81. (a) $\displaystyle\sum_{n=1}^{\infty}\left(\dfrac{1}{2}\right)^n = \sum_{n=0}^{\infty}\dfrac{1}{2}\left(\dfrac{1}{2}\right)^n = \dfrac{1}{2}\dfrac{1}{(1 - (1/2))} = 1$

(b) No, the series is not geometric.

(c) $\displaystyle\sum_{n=1}^{\infty} n\left(\dfrac{1}{2}\right)^n = 2$

83. Assuming that the payments are made at the beginning of each year,

Present Value $= \displaystyle\sum_{n=0}^{19} 50{,}000\left(\dfrac{1}{1.06}\right)^n$

$= 50{,}000\left(\dfrac{1 - 1.06^{-20}}{1 - 1.06^{-1}}\right)$

$\approx \$607{,}905.82$

The present value is less than \$1,000,000. After accruing interest over 20 years, it attains its full value.

85. $w = \displaystyle\sum_{i=0}^{n-1} 0.01(2)^i = \dfrac{0.01(1 - 2^n)}{1 - 2} = 0.01(2^n - 1)$

(a) When $n = 29$: $w = \$5{,}368{,}709.11$

(b) When $n = 30$: $w = \$10{,}737{,}418.23$

(c) When $n = 31$: $w = \$21{,}474{,}836.47$

87. $P = 50, r = 0.03, t = 20$

(a) $A = 50\left(\dfrac{12}{0.03}\right)\left[\left(1 + \dfrac{0.03}{12}\right)^{12(20)} - 1\right] \approx \$16,415.10$

(b) $A = \dfrac{50 - (e^{0.03(20)} - 1)}{e^{0.03/12} - 1} \approx \$16,421.83$

89. $P = 100, r = 0.04, t = 40$

(a) $A = 100\left(\dfrac{12}{0.04}\right)\left[\left(1 + \dfrac{0.04}{12}\right)^{12(40)} - 1\right] \approx \$118,196.13$

(b) $A = \dfrac{100(e^{0.04(40)} - 1)}{e^{0.04/12} - 1} \approx \$118,393.43$

91. (a) $a_n = 6110.1832(1.0544)^x = 6110.1832e^{0.05297n}$

(b) Adding the 10 values $a_0, a_1, \ldots, a_9$, you obtain 78,530 or \$78,530,000,000

(c) Total $= \displaystyle\sum_{n=0}^{9} 6110.1832e^{-0.05297n} \approx 78,449$ or \$78,449,000,000

93. $x = 0.749999\ldots = 0.74 + \displaystyle\sum_{n=0}^{\infty} 0.009(0.1)^n$

$= 0.74 + \dfrac{0.009}{1 - 0.1}$

$= 0.74 + 0.01 = 0.75$

95. By letting $S_0 = 0$, we have $a_n = \displaystyle\sum_{k=1}^{n} a_k - \sum_{k=1}^{n-1} a_k = S_n - S_{n-1}$. Thus,

$$\sum_{n=1}^{\infty} a_n = \sum_{n=1}^{\infty} (S_n - S_{n-1}) = \sum_{n=1}^{\infty} (S_n - S_{n-1} + c - c) = \sum_{n=1}^{\infty} [(c - S_{n-1}) - (c - S_n)].$$

97. Let $\displaystyle\sum a_n = \sum_{n=0}^{\infty} 1$ and $\displaystyle\sum b_n = \sum_{n=0}^{\infty} (-1)$.

Both are divergent series.

$$\sum (a_n + b_n) = \sum_{n=0}^{\infty} [1 + (-1)] = \sum_{n=0}^{\infty} [1 - 1] = 0$$

99. False. $\displaystyle\lim_{n \to \infty} \frac{1}{n} = 0$, but $\displaystyle\sum_{n=1}^{\infty} \frac{1}{n}$ diverges.

101. False

$$\sum_{n=1}^{\infty} ar^n = \left(\frac{a}{1 - r}\right) - a$$

The formula requires that the geometric series begins with $n = 0$.

103. Let H represent the half-life of the drug. If a patient receives n equal doses of P units each of this drug, administered at equal time interval of length t, the total amount of the drug in the patient's system at the time the last dose is administered is given by

$$T_n = P + Pe^{kt} + Pe^{2kt} + \cdots + Pe^{(n-1)kt}$$

where $k = -(\ln 2)/H$. One time interval *after* the last dose is administered is given by

$$T_{n+1} = Pe^{kt} + Pe^{2kt} + Pe^{3kt} + \cdots + Pe^{nkt}.$$

Two time intervals *after* the last dose is administered is given by

$$T_{n+2} = Pe^{2kt} + Pe^{3kt} + Pe^{4kt} + \cdots + Pe^{(n+1)kt}$$

and so on. Since $k < 0$, $T_{n+s} \to 0$ as $s \to \infty$, where s is an integer.

Section 8.3   The Integral Test and *p*-Series

1. $\displaystyle\sum_{n=1}^{\infty} \frac{1}{n+1}$

Let $f(x) = \dfrac{1}{x+1}$.

f is positive, continuous and decreasing for $x \geq 1$.

$$\int_1^{\infty} \frac{1}{x+1}\, dx = \left[\ln(x+1)\right]_1^{\infty} = \infty$$

Diverges by Theorem 8.10

3. $\displaystyle\sum_{n=1}^{\infty} e^{-n}$

Let $f(x) = e^{-x}$.

f is positive, continuous, and decreasing for $x \geq 1$.

$$\int_1^{\infty} e^{-x}\, dx = \left[-e^{-x}\right]_1^{\infty} = \frac{1}{e}$$

Converges by Theorem 8.10

5. $\displaystyle\sum_{n=1}^{\infty} \frac{1}{n^2+1}$

Let $f(x) = \dfrac{1}{x^2+1}$.

f is positive, continuous, and decreasing for $x \geq 1$.

$$\int_1^{\infty} \frac{1}{x^2+1}\, dx = \left[\arctan x\right]_1^{\infty} = \frac{\pi}{4}$$

Converges by Theorem 8.10

7. $\displaystyle\sum_{n=1}^{\infty} \frac{\ln(n+1)}{n+1}$

Let $f(x) = \dfrac{\ln(x+1)}{x+1}$.

f is positive, continuous, and decreasing for $x \geq 2$ since

$$f'(x) = \frac{1 - \ln(x+1)}{(x+1)^2} < 0 \text{ for } x \geq 2.$$

$$\int_1^{\infty} \frac{\ln(x+1)}{x+1}\, dx = \left[\frac{\ln^2(x+1)}{2}\right]_1^{\infty} = \infty$$

Diverges by Theorem 8.10

9. $\displaystyle\sum_{n=1}^{\infty} \frac{n^{k-1}}{n^k+c}$

Let $f(x) = \dfrac{x^{k-1}}{x^k+c}$.

f is positive, continuous, and decreasing for $x > \sqrt[k]{c(k-1)}$ since

$$f'(x) = \frac{x^{k-2}[c(k-1) - x^k]}{(x^k+c)^2} < 0$$

for $x > \sqrt[k]{c(k-1)}$.

$$\int_1^{\infty} \frac{x^{k-1}}{x^k+c}\, dx = \left[\frac{1}{k}\ln(x^k+c)\right]_1^{\infty} = \infty$$

Diverges by Theorem 8.10

11. $\displaystyle\sum_{n=1}^{\infty} \frac{1}{n^3}$

Let $f(x) = \dfrac{1}{x^3}$.

f is positive, continuous, and decreasing for $x \geq 1$.

$$\int_1^{\infty} \frac{1}{x^3}\, dx = \left[-\frac{1}{2x^2}\right]_1^{\infty} = \frac{1}{2}$$

Converges by Theorem 8.10

13. $\displaystyle\sum_{n=1}^{\infty} \frac{1}{\sqrt[5]{n}} = \sum_{n=1}^{\infty} \frac{1}{n^{1/5}}$

Divergent p-series with $p = \frac{1}{5} < 1$

15. $\displaystyle\sum_{n=1}^{\infty} \frac{1}{n^{1/2}}$

Divergent p-series with $p = \frac{1}{2} < 1$

17. $\displaystyle\sum_{n=1}^{\infty} \frac{1}{n^{3/2}}$

Convergent p-series with $p = \frac{3}{2} > 1$

19. $\displaystyle\sum_{n=1}^{\infty} \frac{1}{n^{1.04}}$

Convergent p-series with $p = 1.04 > 1$

21. $\sum_{n=1}^{\infty} \frac{2}{\sqrt[4]{n^3}} = \frac{2}{1} + \frac{2}{2^{3/4}} + \frac{2}{3^{3/4}} + \cdots$

$S_1 = 2$

$S_2 \approx 3.189$

$S_3 \approx 4.067$

Matches (a)

Diverges—p-series with $p = \frac{3}{4} < 1$

23. $\sum_{n=1}^{\infty} \frac{2}{n\sqrt{n}} = 2 + 2/2^{3/2} + 2/3^{3/2} + \cdots$

$S_1 = 2$

$S_2 \approx 2.707$

$S_3 \approx 3.092$

Matches (b)

Converges—p-series with $p = 3/2 > 1$

25. No. Theorem 8.9 says that if the series converges, then the terms a_n tend to zero. Some of the series in Exercises 21-24 converge because the terms tend to 0 very rapidly.

27. $\sum_{n=1}^{N} \frac{1}{n} = 1 + \frac{1}{2} + \frac{1}{3} + \frac{1}{4} + \cdots + \frac{1}{N} > M$

(a)

M	2	4	6	8
N	4	31	227	1674

(b) No. Since the terms are decreasing (approaching zero), more and more terms are required to increase the partial sum by 2.

29. $\sum_{n=2}^{\infty} \frac{1}{n(\ln n)^p}$

If $p = 1$, then the series diverges by the Integral Test. If $p \neq 1$,

$$\int_{2}^{\infty} \frac{1}{x(\ln x)^p}\, dx = \int_{2}^{\infty} (\ln x)^{-p} \frac{1}{x}\, dx = \left[\frac{(\ln x)^{-p+1}}{-p+1} \right]_{2}^{\infty}.$$

Converges for $-p + 1 < 0$ or $p > 1$.

31. Let f be positive, continuous, and decreasing for $x \geq 1$ and $a_n = f(n)$. Then,

$$\sum_{n=1}^{\infty} a_n \text{ and } \int_{1}^{\infty} f(x)\, dx$$

either both converge or both diverge (Theorem 8.10). See Example 1, page 578.

33. Your friend is not correct. The series

$$\sum_{n=10,000}^{\infty} \frac{1}{n} = \frac{1}{10,000} + \frac{1}{10,001} + \cdots$$

is the harmonic series, starting with the $10,000^{\text{th}}$ term, and hence diverges.

35. Since f is positive, continuous, and decreasing for $x \geq 1$ and $a_n = f(n)$, we have,

$$R_N = S - S_N = \sum_{n=1}^{\infty} a_n - \sum_{n=1}^{N} a_n = \sum_{n=N+1}^{\infty} a_n > 0.$$

Also, $R_N = S - S_N = \sum_{n=N+1}^{\infty} a_n \leq a_{N+1} + \int_{N+1}^{\infty} f(x)\, dx \leq \int_{N}^{\infty} f(x)\, dx$. Thus,

$$0 \leq R_N \leq \int_{N}^{\infty} f(x)\, dx.$$

37. $S_6 = 1 + \frac{1}{2^4} + \frac{1}{3^4} + \frac{1}{4^4} + \frac{1}{5^4} + \frac{1}{6^4} \approx 1.0811$

$R_6 \leq \int_{6}^{\infty} \frac{1}{x^4}\, dx = \left[-\frac{1}{3x^3} \right]_{6}^{\infty} \approx 0.0015$

$1.0811 \leq \sum_{n=1}^{\infty} \frac{1}{n^4} \leq 1.0811 + 0.0015 = 1.0826$

39. $S_{10} = \dfrac{1}{2} + \dfrac{1}{5} + \dfrac{1}{10} + \dfrac{1}{17} + \dfrac{1}{26} + \dfrac{1}{37} + \dfrac{1}{50} + \dfrac{1}{65} + \dfrac{1}{82} + \dfrac{1}{101} \approx 0.9818$

$R_{10} \le \displaystyle\int_{10}^{\infty} \dfrac{1}{x^2 + 1}\, dx = \left[\arctan x\right]_{10}^{\infty} = \dfrac{\pi}{2} - \arctan 10 \approx 0.0997$

$0.9818 \le \displaystyle\sum_{n=1}^{\infty} \dfrac{1}{n^2 + 1} \le 0.9818 + 0.0997 = 1.0815$

41. $S_4 = \dfrac{1}{e} + \dfrac{2}{e^4} + \dfrac{3}{e^9} + \dfrac{4}{e^{16}} \approx 0.4049$

$R_4 \le \displaystyle\int_{4}^{\infty} xe^{-x^2}\, dx = \left[-\dfrac{1}{2}e^{-x^2}\right]_{4}^{\infty} = \dfrac{e^{-16}}{2} \approx 5.6 \times 10^{-8}$

$0.4049 \le \displaystyle\sum_{n=1}^{\infty} ne^{-n^2} \le 0.4049 + 5.6 \times 10^{-8}$

43. $0 \le R_N \le \displaystyle\int_{N}^{\infty} \dfrac{1}{x^4}\, dx = \left[-\dfrac{1}{3x^3}\right]_{N}^{\infty} = \dfrac{1}{3N^3} < 0.001$

$\dfrac{1}{N^3} < 0.003$

$N^3 > 333.33$

$N > 6.93$

$N \ge 7$

45. $R_N \le \displaystyle\int_{N}^{\infty} e^{-5x}\, dx = \left[-\dfrac{1}{5}e^{-5x}\right]_{N}^{\infty} = \dfrac{e^{-5N}}{5} < 0.001$

$\dfrac{1}{e^{5N}} < 0.005$

$e^{5N} > 200$

$5N > \ln 200$

$N > \dfrac{\ln 200}{5}$

$N > 1.0597$

$N \ge 2$

47. $R_N \le \displaystyle\int_{N}^{\infty} \dfrac{1}{x^2 + 1}\, dx = \left[\arctan x\right]_{N}^{\infty}$

$= \dfrac{\pi}{2} - \arctan N < 0.001$

$-\arctan N < -1.5698$

$\arctan N > 1.5698$

$N > \tan 1.5698$

$N \ge 1004$

49. (a) $\displaystyle\sum_{n=2}^{\infty} \dfrac{1}{n^{1.1}}$. This is a convergent p-series with $p = 1.1 > 1$.

$\displaystyle\sum_{n=2}^{\infty} \dfrac{1}{n \ln n}$ is a divergent series. Use the Integral Test.

$\displaystyle\int_{2}^{\infty} \dfrac{1}{x \ln x}\, dx = \left[\ln|\ln x|\right]_{2}^{\infty} = \infty$

(b) $\displaystyle\sum_{n=2}^{6} \dfrac{1}{n^{1.1}} = \dfrac{1}{2^{1.1}} + \dfrac{1}{3^{1.1}} + \dfrac{1}{4^{1.1}} + \dfrac{1}{5^{1.1}} + \dfrac{1}{6^{1.1}} \approx 0.4665 + 0.2987 + 0.2176 + 0.1703 + 0.1393$

$\displaystyle\sum_{n=2}^{6} \dfrac{1}{n \ln n} = \dfrac{1}{2 \ln 2} + \dfrac{1}{3 \ln 3} + \dfrac{1}{4 \ln 4} + \dfrac{1}{5 \ln 5} + \dfrac{1}{6 \ln 6} \approx 0.7213 + 0.3034 + 0.1803 + 0.1243 + 0.0930$

For $n \ge 4$ the terms of the convergent series **seem** to be larger than those of the divergent series!

(c) $\dfrac{1}{n^{1.1}} < \dfrac{1}{n \ln n}$

$n \ln n < n^{1.1}$

$\ln n < n^{0.1}$

This inequality holds when $n \ge 3.5 \times 10^{15}$. Or, $n > e^{40}$. Then $\ln e^{40} = 40 < (e^{40})^{0.1} = e^4 \approx 55$.

51. (a) Let $f(x) = 1/x$. f is positive, continuous, and decreasing on $[1, \infty)$.

$$S_n - 1 \le \int_1^n \frac{1}{x}\,dx$$

$$S_n - 1 \le \ln n$$

Hence, $S_n \le 1 + \ln n$. Similarly,

$$S_n \ge \int_1^{n+1} \frac{1}{x}\,dx = \ln(n + 1).$$

Thus, $\ln(n + 1) \le S_n \le 1 + \ln n$.

(b) Since $\ln(n + 1) \le S_n \le 1 + \ln n$, we have $\ln(n + 1) - \ln n \le S_n - \ln n \le 1$. Also, since $\ln x$ is an increasing function, $\ln(n + 1) - \ln n > 0$ for $n \ge 1$. Thus, $0 \le S_n - \ln n \le 1$ and the sequence $\{a_n\}$ is bounded.

(c) $a_n - a_{n+1} = [S_n - \ln n] - [S_{n+1} - \ln(n + 1)] = \int_n^{n+1} \frac{1}{x}\,dx - \frac{1}{n + 1} \ge 0$

Thus, $a_n \ge a_{n+1}$ and the sequence is decreasing.

(d) Since the sequence is bounded and monotonic, it converges to a limit, γ.

(e) $a_{100} = S_{100} - \ln 100 \approx 0.5822$ (Actually $\gamma \approx 0.577216$.)

53. $\displaystyle\sum_{n=1}^{\infty} \frac{1}{2n - 1}$

Let $f(x) = \dfrac{1}{2x - 1}$.

f is positive, continuous, and decreasing for $x \ge 1$.

$$\int_1^{\infty} \frac{1}{2x - 1}\,dx = \left[\ln \sqrt{2x - 1}\, \right]_1^{\infty} = \infty$$

Diverges by Theorem 8.10

55. $\displaystyle\sum_{n=1}^{\infty} \frac{1}{n\sqrt[4]{n}} = \sum_{n=1}^{\infty} \frac{1}{n^{5/4}}$

p-series with $p = \frac{5}{4}$

Converges by Theorem 8.11

57. $\displaystyle\sum_{n=0}^{\infty} \left(\frac{2}{3}\right)^n$

Geometric series with $r = \frac{2}{3}$

Converges by Theorem 8.6

59. $\displaystyle\sum_{n=1}^{\infty} \frac{n}{\sqrt{n^2 + 1}}$

$$\lim_{n\to\infty} \frac{n}{\sqrt{n^2 + 1}} = \lim_{n\to\infty} \frac{1}{\sqrt{1 + (1/n^2)}} = 1 \ne 0$$

Diverges by Theorem 8.9

61. $\displaystyle\sum_{n=1}^{\infty} \left(1 + \frac{1}{n}\right)^n$

$$\lim_{n\to\infty} \left(1 + \frac{1}{n}\right)^n = e \ne 0$$

Fails nth Term Test

Diverges by Theorem 8.9

63. $\displaystyle\sum_{n=2}^{\infty} \frac{1}{n(\ln n)^3}$

Let $f(x) = \dfrac{1}{x(\ln x)^3}$.

f is positive, continuous and decreasing for $x \ge 2$.

$$\int_2^{\infty} \frac{1}{x(\ln x)^3}\,dx = \int_2^{\infty} (\ln x)^{-3}\frac{1}{x}\,dx = \left[\frac{(\ln x)^{-2}}{-2} \right]_2^{\infty} = \left[-\frac{1}{2(\ln x)^2} \right]_2^{\infty} = \frac{1}{2(\ln 2)^2}$$

Converges by Theorem 8.10. See Exercise 29.

Section 8.4   Comparisons of Series

1. (a) $\sum_{n=1}^{\infty} \frac{6}{n^{3/2}} = \frac{6}{1} + \frac{6}{2^{3/2}} + \cdots \quad S_1 = 6$

$\sum_{n=1}^{\infty} \frac{6}{n^{3/2} + 3} = \frac{6}{4} + \frac{6}{2^{3/2} + 3} + \cdots \quad S_1 = \frac{3}{2}$

$\sum_{n=1}^{\infty} \frac{6}{n\sqrt{n^2 + 0.5}} = \frac{6}{1\sqrt{1.5}} + \frac{6}{2\sqrt{4.5}} + \cdots \quad S_1 = \frac{6}{\sqrt{1.5}} \approx 4.9$

(b) The first series is a *p*-series. It converges ($p = 3/2 > 1$).

(c) The magnitude of the terms of the other two series are less than the corresponding terms at the convergent *p*-series. Hence, the other two series converge.

(d) The smaller the magnitude of the terms, the smaller the magnitude of the terms of the sequence of partial sums.

3. $\frac{1}{n^2 + 1} < \frac{1}{n^2}$

Therefore,

$\sum_{n=1}^{\infty} \frac{1}{n^2 + 1}$

converges by comparison with the convergent *p*-series

$\sum_{n=1}^{\infty} \frac{1}{n^2}.$

5. $\frac{1}{n - 1} > \frac{1}{n}$ for $n \geq 2$

Therefore,

$\sum_{n=2}^{\infty} \frac{1}{n - 1}$

diverges by comparison with the divergent *p*-series

$\sum_{n=2}^{\infty} \frac{1}{n}.$

7. $\frac{1}{3^n + 1} < \frac{1}{3^n}$

Therefore,

$\sum_{n=0}^{\infty} \frac{1}{3^n + 1}$

converges by comparison with the convergent geometric series

$\sum_{n=0}^{\infty} \left(\frac{1}{3}\right)^n.$

9. For $n \geq 3$, $\frac{\ln n}{n + 1} > \frac{1}{n + 1}.$

Therefore,

$\sum_{n=1}^{\infty} \frac{\ln n}{n + 1}$

diverges by comparison with the divergent series

$\sum_{n=1}^{\infty} \frac{1}{n + 1}.$

Note: $\sum_{n=1}^{\infty} \frac{1}{n + 1}$ diverges by the integral test.

11. For $n > 3$, $\frac{1}{n^2} > \frac{1}{n!}.$

Therefore,

$\sum_{n=0}^{\infty} \frac{1}{n!}$

converges by comparison with the convergent *p*-series

$\sum_{n=1}^{\infty} \frac{1}{n^2}.$

13. $\frac{1}{e^{n^2}} \leq \frac{1}{e^n}$

Therefore,

$\sum_{n=0}^{\infty} \frac{1}{e^{n^2}}$

converges by comparison with the convergent geometric series

$\sum_{n=0}^{\infty} \left(\frac{1}{e}\right)^n.$

15. $\lim\limits_{n\to\infty} \dfrac{n/(n^2+1)}{1/n} = \lim\limits_{n\to\infty} \dfrac{n^2}{n^2+1} = 1$

Therefore,

$$\sum_{n=1}^{\infty} \frac{n}{n^2+1}$$

diverges by a limit comparison with the divergent *p*-series

$$\sum_{n=1}^{\infty} \frac{1}{n}.$$

17. $\lim\limits_{n\to\infty} \dfrac{1/\sqrt{n^2+1}}{1/n} = \lim\limits_{n\to\infty} \dfrac{n}{\sqrt{n^2+1}} = 1$

Therefore,

$$\sum_{n=0}^{\infty} \frac{1}{\sqrt{n^2+1}}$$

diverges by a limit comparison with the divergent *p*-series

$$\sum_{n=1}^{\infty} \frac{1}{n}.$$

19. $\lim\limits_{n\to\infty} \dfrac{\frac{2n^2-1}{3n^5+2n+1}}{1/n^3} = \lim\limits_{n\to\infty} \dfrac{2n^5-n^3}{3n^5+2n+1} = \dfrac{2}{3}$

Therefore,

$$\sum_{n=1}^{\infty} \frac{2n^2-1}{3n^5+2n+1}$$

converges by a limit comparison with the convergent *p*-series

$$\sum_{n=1}^{\infty} \frac{1}{n^3}.$$

21. $\lim\limits_{n\to\infty} \dfrac{\frac{n+3}{n(n+2)}}{1/n} = \lim\limits_{n\to\infty} \dfrac{n^2+3n}{n^2+2n} = 1$

Therefore,

$$\sum_{n=1}^{\infty} \frac{n+3}{n(n+2)}$$

diverges by a limit comparison with the divergent *p*-series

$$\sum_{n=1}^{\infty} \frac{1}{n}.$$

23. $\lim\limits_{n\to\infty} \dfrac{1/(n\sqrt{n^2+1})}{1/n^2} = \lim\limits_{n\to\infty} \dfrac{n^2}{n\sqrt{n^2+1}} = 1$

Therefore,

$$\sum_{n=1}^{\infty} \frac{1}{n\sqrt{n^2+1}}$$

converges by a limit comparison with the convergent *p*-series

$$\sum_{n=1}^{\infty} \frac{1}{n^2}.$$

25. $\lim\limits_{n\to\infty} \dfrac{(n^{k-1})/(n^k+1)}{1/n} = \lim\limits_{n\to\infty} \dfrac{n^k}{n^k+1} = 1$

Therefore,

$$\sum_{n=1}^{\infty} \frac{n^{k-1}}{n^k+1}$$

diverges by a limit comparison with the divergent *p*-series

$$\sum_{n=1}^{\infty} \frac{1}{n}.$$

27. $\lim\limits_{n\to\infty} \dfrac{\sin(1/n)}{1/n} = \lim\limits_{n\to\infty} \dfrac{(-1/n^2)\cos(1/n)}{-1/n^2}$

$$= \lim_{n\to\infty} \cos\left(\frac{1}{n}\right) = 1$$

Therefore,

$$\sum_{n=1}^{\infty} \sin\left(\frac{1}{n}\right)$$

diverges by a limit comparison with the divergent *p*-series

$$\sum_{n=1}^{\infty} \frac{1}{n}.$$

29. $\displaystyle\sum_{n=1}^{\infty} \frac{\sqrt{n}}{n} = \sum_{n=1}^{\infty} \frac{1}{\sqrt{n}}$

Diverges

p-series with $p = \frac{1}{2}$

31. $\displaystyle\sum_{n=1}^{\infty} \frac{1}{3^n+2}$

Converges

Direct comparison with $\displaystyle\sum_{n=1}^{\infty} \left(\frac{1}{3}\right)^n$

33. $\displaystyle\sum_{n=1}^{\infty} \frac{n}{2n+3}$

Diverges; *n*th Term Test

$$\lim_{n\to\infty} \frac{n}{2n+3} = \frac{1}{2} \neq 0$$

35. $\displaystyle\sum_{n=1}^{\infty} \frac{n}{(n^2+1)^2}$

Converges; integral test

37. $\lim\limits_{n\to\infty} \dfrac{a_n}{1/n} = \lim\limits_{n\to\infty} na_n$. By given conditions $\lim\limits_{n\to\infty} na_n$ is finite and nonzero.

Therefore,

$$\sum_{n=1}^{\infty} a_n$$

diverges by a limit comparison with the *p*-series

$$\sum_{n=1}^{\infty} \frac{1}{n}.$$

41. $\displaystyle\sum_{n=1}^{\infty} \frac{1}{n^3 + 1}$

converges since the degree of the numerator is three less than the degree of the denominator.

45. See Theorem 8.12, page 583. One example is $\displaystyle\sum_{n=1}^{\infty} \frac{1}{n^2 + 1}$ converges because

$$\frac{1}{n^2 + 1} < \frac{1}{n^2} \text{ and } \sum_{n=1}^{\infty} \frac{1}{n^2}$$

converges (*p*-series).

49. $\dfrac{1}{200} + \dfrac{1}{400} + \dfrac{1}{600} + \cdots = \displaystyle\sum_{n=1}^{\infty} \frac{1}{200n}$, diverges (harmonic)

53. Some series diverge or converge very slowly. You cannot decide convergence or divergence of a series by comparing the first few terms.

57. True

59. Since $\displaystyle\sum_{n=1}^{\infty} b_n$ converges, $\lim\limits_{n\to\infty} b_n = 0$. There exists N such that $b_n < 1$ for $n > N$. Thus,

$$a_n b_n < a_n \text{ for } n > N \text{ and } \sum_{n=1}^{\infty} a_n b_n$$

converges by comparison to the convergent series $\displaystyle\sum_{i=1}^{\infty} a_n$.

61. $\displaystyle\sum \frac{1}{n^2}$ and $\displaystyle\sum \frac{1}{n^3}$ both converge, and hence so does $\displaystyle\sum\left(\frac{1}{n^2}\right)\left(\frac{1}{n^3}\right) = \sum \frac{1}{n^5}$.

39. $\dfrac{1}{2} + \dfrac{2}{5} + \dfrac{3}{10} + \dfrac{4}{17} + \dfrac{5}{26} + \cdots = \displaystyle\sum_{n=1}^{\infty} \frac{n}{n^2 + 1}$,

which diverges since the degree of the numerator is only one less than the degree of the denominator.

43. $\lim\limits_{n\to\infty} n\left(\dfrac{n^3}{5n^4 + 3}\right) = \lim\limits_{n\to\infty} \dfrac{n^4}{5n^4 + 3} = \dfrac{1}{5} \neq 0$

Therefore,

$$\sum_{n=1}^{\infty} \frac{n^3}{5n^4 + 3} \text{ diverges.}$$

47.

For $0 < a_n < 1$, $0 < a_n^2 < a_n < 1$. Hence, the lower terms are those of $\sum a_n^2$.

51. $\dfrac{1}{201} + \dfrac{1}{204} + \dfrac{1}{209} + \dfrac{1}{216} = \displaystyle\sum_{n=1}^{\infty} \frac{1}{200 + n^2}$, converges

55. False. Let $a_n = 1/n^3$ and $b_n = 1/n^2$. $0 < a_n \le b_n$ and both

$$\sum_{n=1}^{\infty} \frac{1}{n^3} \text{ and } \sum_{n=1}^{\infty} \frac{1}{n^2}$$

converge.

63. (a) Suppose Σb_n converges and Σa_n diverges. Then there exists N such that $0 < b_n < a_n$ for $n \geq N$. This means that $1 < a_n/b_n$ for $n \geq N$. Therefore, $\lim_{n\to\infty} a_n/b_n \neq 0$. Thus, Σa_n must also converge.

(b) Suppose Σb_n diverges and Σa_n converges. Then there exists N such that $0 < a_n < b_n$ for $n \geq N$. This means that $0 < a_n/b_n < 1$ for $n \geq N$. Therefore, $\lim_{n\to\infty} a_n/b_n \neq \infty$. Thus, Σa_n must also diverge.

65. Start with one triangle whose sides have length 9. At the nth step, each side is replaced by four smaller line segments each having $\frac{1}{3}$ the length of the original side.

#Sides	Length of sides
3	9
$3 \cdot 4$	$9\left(\frac{1}{3}\right)$
$3 \cdot 4^2$	$9\left(\frac{1}{3}\right)^2$
$\vdots$	
$3 \cdot 4^n$	$9\left(\frac{1}{3}\right)^n$

At the nth step there are $3 \cdot 4^n$ sides, each of length $9\left(\frac{1}{3}\right)^n$. At the next step, there are $3 \cdot 4^n$ new triangles of side $9\left(\frac{1}{3}\right)^{n+1}$. The area of an equilateral triangle of side x is $\frac{1}{4}\sqrt{3}\,x^2$. Thus, the new triangles each have area

$$\frac{\sqrt{3}}{4}\left[9\left(\frac{1}{3^{n+1}}\right)\right]^2 = \frac{9\sqrt{3}}{4} \cdot \frac{1}{3^{2n}} = \frac{9}{4}\sqrt{3}\left(\frac{1}{9}\right)^n$$

The area of the $3 \cdot 4^n$ new triangles is

$$(3 \cdot 4^n)\left[\frac{9\sqrt{3}}{4}\left(\frac{1}{9}\right)^n\right] = \frac{27\sqrt{3}}{4}\left(\frac{4}{9}\right)^n$$

The total area is the infinite sum

$$\frac{81\sqrt{3}}{4} + \sum_{n=0}^{\infty} \frac{27\sqrt{3}}{4}\left(\frac{4}{9}\right)^n = \frac{81\sqrt{3}}{4} + \frac{27\sqrt{3}}{4}\left(\frac{1}{1-4/9}\right) = \frac{81\sqrt{3}}{4} + \frac{27\sqrt{3}}{4}\left(\frac{9}{5}\right) = \frac{162\sqrt{3}}{5}.$$

The perimeter is infinite, since at step n there are $3 \cdot 4^n$ sides of length $9\left(\frac{1}{3}\right)^n$. Thus, the perimeter at step n is $27\left(\frac{4}{3}\right)^n \to \infty$.

Section 8.5   Alternating Series

1. $\displaystyle\sum_{n=1}^{\infty} \frac{6}{n^2} = \frac{6}{1} + \frac{6}{4} + \frac{6}{9} + \cdots$

$S_1 = 6, S_2 = 7.5$

Matches (b)

3. $\displaystyle\sum_{n=1}^{\infty} \frac{10}{n2^n} = \frac{10}{2} + \frac{10}{8} + \cdots$

$S_1 = 5, S_2 = 6.25$

Matches (c)

5. $\displaystyle\sum_{n=1}^{\infty} \frac{(-1)^{n-1}}{2n-1} = \frac{\pi}{4} \approx 0.7854$

(a)

n	1	2	3	4	5	6	7	8	9	10
S_n	1	0.6667	0.8667	0.7238	0.8349	0.7440	0.8209	0.7543	0.8131	0.7605

(b)

(c) The points alternate sides of the horizontal line that represents the sum of the series. The distance between successive points and the line decreases.

(d) The distance in part (c) is always less than the magnitude of the next term of the series.

7. $\displaystyle\sum_{n=1}^{\infty} \frac{(-1)^{n-1}}{n^2} = \frac{\pi^2}{12} \approx 0.8225$

(a)

n	1	2	3	4	5	6	7	8	9	10
S_n	1	0.75	0.8611	0.7986	0.8386	0.8108	0.8312	0.8156	0.8280	0.8180

(b)

(c) The points alternate sides of the horizontal line that represents the sum of the series. The distance between successive points and the line decreases.

(d) The distance in part (c) is always less than the magnitude of the next term in the series.

9. $\displaystyle\sum_{n=1}^{\infty} \frac{(-1)^{n+1}}{n}$

$a_{n+1} = \dfrac{1}{n+1} < \dfrac{1}{n} = a_n$

$\displaystyle\lim_{n\to\infty} \frac{1}{n} = 0$

Converges by Theorem 8.14.

11. $\displaystyle\sum_{n=1}^{\infty} \frac{(-1)^{n+1}}{2n-1}$

$a_{n+1} = \dfrac{1}{2(n+1)-1} < \dfrac{1}{2n-1} = a_n$

$\displaystyle\lim_{n\to\infty} \frac{1}{2n-1} = 0$

Converges by Theorem 8.14

13. $\displaystyle\sum_{n=1}^{\infty} \frac{(-1)^n n^2}{n^2+1}$

$\displaystyle\lim_{n\to\infty} \frac{n^2}{n^2+1} = 1$

Diverges by the nth Term Test

15. $\displaystyle\sum_{n=1}^{\infty} \frac{(-1)^n}{\sqrt{n}}$

$a_{n+1} = \dfrac{1}{\sqrt{n+1}} < \dfrac{1}{\sqrt{n}} = a_n$

$\displaystyle\lim_{n\to\infty} \frac{1}{\sqrt{n}} = 0$

Converges by Theorem 8.14

17. $\displaystyle\sum_{n=1}^{\infty} \frac{(-1)^{n+1}(n+1)}{\ln(n+1)}$

$\displaystyle\lim_{n\to\infty} \frac{n+1}{\ln(n+1)} = \lim_{n\to\infty} \frac{1}{1/(n+1)} = \lim_{n\to\infty} (n+1) = \infty$

Diverges by the nth Term Test

19. $\displaystyle\sum_{n=1}^{\infty} \sin\left[\frac{(2n-1)\pi}{2}\right] = \sum_{n=1}^{\infty} (-1)^{n+1}$

Diverges by the nth Term Test

21. $\displaystyle\sum_{n=1}^{\infty} \cos n\pi = \sum_{n=1}^{\infty} (-1)^n$

Diverges by the nth Term Test

23. $\displaystyle\sum_{n=0}^{\infty} \frac{(-1)^n}{n!}$

$a_{n+1} = \dfrac{1}{(n+1)!} < \dfrac{1}{n!} = a_n$

$\displaystyle\lim_{n\to\infty} \frac{1}{n!} = 0$

Converges by Theorem 8.14

25. $\sum_{n=1}^{\infty} \frac{(-1)^{n+1} \sqrt{n}}{n+2}$

$a_{n+1} = \frac{\sqrt{n+1}}{(n+1)+2} < \frac{\sqrt{n}}{n+2}$ for $n \geq 2$

$\lim_{n \to \infty} \frac{\sqrt{n}}{n+2} = 0$

Converges by Theorem 8.14

27. $\sum_{n=1}^{\infty} \frac{(-1)^{n+1}(2)}{e^n - e^{-n}} = \sum_{n=1}^{\infty} \frac{(-1)^{n+1}(2e^n)}{e^{2n} - 1}$

Let $f(x) = \frac{2e^x}{e^{2x} - 1}$. Then

$$f'(x) = \frac{-2e^x(e^{2x} + 1)}{(e^{2x} - 1)^2} < 0.$$

Thus, $f(x)$ is decreasing. Therefore, $a_{n+1} < a_n$, and

$$\lim_{n \to \infty} \frac{2e^n}{e^{2n} - 1} = \lim_{n \to \infty} \frac{2e^n}{2e^{2n}} = \lim_{n \to \infty} \frac{1}{e^n} = 0.$$

The series converges by Theorem 8.14.

29. $S_6 = \sum_{n=1}^{6} \frac{3(-1)^{n+1}}{n^2} = 2.4325$

$|R_6| = |S - S_6| \leq a_7 = \frac{3}{49} \approx 0.0612;\ 2.3713 \leq S \leq 2.4937$

31. $S_6 = \sum_{n=0}^{5} \frac{2(-1)^n}{n!} \approx 0.7333$

$|R_6| = |S - S_6| \leq a_7 = \frac{2}{6!} = 0.002778;\ 0.7305 \leq S \leq 0.7361$

33. $\sum_{n=0}^{\infty} \frac{(-1)^n}{n!}$

(a) By Theorem 8.15,

$$|R_N| \leq a_{N+1} = \frac{1}{(N+1)!} < 0.001.$$

This inequality is valid when $N = 6$.

(b) We may approximate the series by

$$\sum_{n=0}^{6} \frac{(-1)^n}{n!} = 1 - 1 + \frac{1}{2} - \frac{1}{6} + \frac{1}{24} - \frac{1}{120} + \frac{1}{720}$$

$$\approx 0.368.$$

(7 terms. Note that the sum begins with $n = 0$.)

35. $\sum_{n=0}^{\infty} \frac{(-1)^n}{(2n+1)!}$

(a) By Theorem 8.15,

$$|R_N| \leq a_{N+1} = \frac{1}{[2(N+1)+1]!} < 0.001.$$

This inequality is valid when $N = 2$.

(b) We may approximate the series by

$$\sum_{n=0}^{2} \frac{(-1)^n}{(2n+1)!} = 1 - \frac{1}{6} + \frac{1}{120} \approx 0.842.$$

(3 terms. Note that the sum begins with $n = 0$.)

37. $\sum_{n=1}^{\infty} \frac{(-1)^{n+1}}{n}$

(a) By Theorem 8.15,

$$|R_N| \leq a_{N+1} = \frac{1}{N+1} < 0.001.$$

This inequality is valid when $N = 1000$.

(b) We may approximate the series by

$$\sum_{n=1}^{1000} \frac{(-1)^{n+1}}{n} = 1 - \frac{1}{2} + \frac{1}{3} - \frac{1}{4} + \cdots - \frac{1}{1000}$$

$$\approx 0.693.$$

(1000 terms)

39. $\sum_{n=1}^{\infty} \frac{(-1)^{n+1}}{2n^3 - 1}$

By Theorem 8.15,

$$|R_N| \leq a_{N+1} = \frac{1}{2(N+1)^3 - 1} < 0.001.$$

This inequality is valid when $N = 7$.

41. $\displaystyle\sum_{n=1}^{\infty} \frac{(-1)^{n+1}}{(n+1)^2}$

$\displaystyle\sum_{n=1}^{\infty} \frac{1}{(n+1)^2}$ converges by comparison to the p-series

$$\sum_{n=1}^{\infty} \frac{1}{n^2}.$$

Therefore, the given series converge absolutely.

45. $\displaystyle\sum_{n=1}^{\infty} \frac{(-1)^{n+1} n^2}{(n+1)^2}$

$\displaystyle\lim_{n\to\infty} \frac{n^2}{(n+1)^2} = 1$

Therefore, the series diverges by the nth Term Test.

49. $\displaystyle\sum_{n=2}^{\infty} \frac{(-1)^n n}{n^3 - 1}$

$\displaystyle\sum_{n=2}^{\infty} \frac{n}{n^3 - 1}$

converges by a limit comparison to the convergent
p-series

$$\sum_{n=2}^{\infty} \frac{1}{n^2}.$$

Therefore, the given series converges absolutely.

53. $\displaystyle\sum_{n=0}^{\infty} \frac{\cos n\pi}{n + 1} = \sum_{n=0}^{\infty} \frac{(-1)^n}{n + 1}$

The given series converges by the Alternating Series Test,
but

$$\sum_{n=0}^{\infty} \frac{|\cos n\pi|}{n + 1} = \sum_{n=0}^{\infty} \frac{1}{n + 1}$$

diverges by a limit comparison to the divergent harmonic
series,

$$\sum_{n=1}^{\infty} \frac{1}{n}.$$

$\displaystyle\lim_{n\to\infty} \frac{|\cos n\pi|/(n + 1)}{1/n} = 1$, therefore the series

converges conditionally.

43. $\displaystyle\sum_{n=1}^{\infty} \frac{(-1)^{n+1}}{\sqrt{n}}$

The given series converges by the Alternating Series Test,
but does not converge absolutely since

$$\sum_{n=1}^{\infty} \frac{1}{\sqrt{n}}$$

is a divergent p-series. Therefore, the series converges
conditionally.

47. $\displaystyle\sum_{n=2}^{\infty} \frac{(-1)^n}{\ln(n)}$

The given series converges by the Alternating Series Test,
but does not converge absolutely since the series

$$\sum_{n=2}^{\infty} \frac{1}{\ln n}$$

diverges by comparison to the harmonic series

$$\sum_{n=1}^{\infty} \frac{1}{n}.$$

Therefore, the series converges conditionally.

51. $\displaystyle\sum_{n=0}^{\infty} \frac{(-1)^n}{(2n + 1)!}$

$\displaystyle\sum_{n=0}^{\infty} \frac{1}{(2n + 1)!}$

is convergent by comparison to the convergent geometric
series

$$\sum_{n=0}^{\infty} \left(\frac{1}{2}\right)^n$$

since

$$\frac{1}{(2n + 1)!} < \frac{1}{2^n} \text{ for } n > 0.$$

Therefore, the given series converges absolutely.

55. $\displaystyle\sum_{n=1}^{\infty} \frac{\cos n\pi}{n^2} = \sum_{n=1}^{\infty} \frac{(-1)^n}{n^2}$

$\displaystyle\sum_{n=1}^{\infty} \frac{1}{n^2}$ is a convergent p-series. Therefore, the given

series converges absolutely.

57. An alternating series is a series whose terms alternate in sign. See Theorem 8.14.

59. $\sum a_n$ is absolutely convergent if $\sum |a_n|$ converges.

$\sum a_n$ is conditionally convergent if $\sum |a_n|$ diverges, but $\sum a_n$ converges.

61. (b). The partial sums alternate above and below the horizontal line representing the sum.

63. Since $\sum_{n=1}^{\infty} |a_n|$ converges we have

$$\lim_{n \to \infty} |a_n| = 0.$$

Thus, there must exist an $N > 0$ such that $|a_N| < 1$ for all $n > N$ and it follows that $a_n^2 \le |a_n|$ for all $n > N$. Hence, by the Comparison Test,

$$\sum_{n=1}^{\infty} a_n^2$$

converges. Let $a_n = 1/n$ to see that the converse is false.

65. $\sum_{n=1}^{\infty} \dfrac{1}{n^2}$ converges, hence so does $\sum_{n=1}^{\infty} \dfrac{1}{n^4}$.

67. False

Let $a_n = \dfrac{(-1)^n}{n}$.

69. $\sum_{n=1}^{\infty} \dfrac{10}{n^{3/2}} = 10 \sum_{n=1}^{\infty} \dfrac{1}{n^{3/2}}$ convergent p-series

71. Diverges by nth Term Test. $\lim_{n \to \infty} a_n = \infty$

73. Convergent Geometric Series $\left(r = \frac{7}{8} < 1\right)$

75. Convergent Geometric Series $\left(r = \dfrac{1}{\sqrt{e}}\right)$ or Integral Test

77. Converges (absolutely) by Alternating Series Test

79. The first term of the series is zero, not one. You cannot regroup series terms arbitrarily.

Section 8.6   The Ratio and Root Tests

1. $\dfrac{(n+1)!}{(n-2)!} = \dfrac{(n+1)(n)(n-1)(n-2)!}{(n-2)!}$

$\qquad = (n+1)(n)(n-1)$

3. Use the Principle of Mathematical Induction. When $k = 1$, the formula is valid since $1 = \dfrac{(2(1))!}{2^1 \cdot 1!}$. Assume that

$$1 \cdot 3 \cdot 5 \cdots (2n - 1) = \dfrac{(2n)!}{2^n n!}$$

and show that

$$1 \cdot 3 \cdot 5 \cdots (2n - 1)(2n + 1) = \dfrac{(2n + 2)!}{2^{n+1}(n + 1)!}.$$

—CONTINUED—

3. **—CONTINUED—**

To do this, note that:

$$1 \cdot 3 \cdot 5 \cdots (2n - 1)(2n + 1) = [1 \cdot 3 \cdot 5 \cdots (2n - 1)](2n + 1)$$

$$= \frac{(2n)!}{2^n n!} \cdot (2n + 1) \quad \text{(Induction hypothesis)}$$

$$= \frac{(2n)!(2n + 1)}{2^n n!} \cdot \frac{(2n + 2)}{2(n + 1)}$$

$$= \frac{(2n)!(2n + 1)(2n + 2)}{2^{n+1} n!(n + 1)}$$

$$= \frac{(2n + 2)!}{2^{n+1}(n + 1)!}$$

The formula is valid for all $n \geq 1$.

5. $\displaystyle\sum_{n=1}^{\infty} n\left(\frac{3}{4}\right)^n = 1\left(\frac{3}{4}\right) + 2\left(\frac{9}{16}\right) + \cdots$

$S_1 = \dfrac{3}{4}, S_2 \approx 1.875$

Matches (d)

7. $\displaystyle\sum_{n=1}^{\infty} \frac{(-3)^{n+1}}{n!} = 9 - \frac{3^3}{2} + \cdots$

$S_1 = 9$

Matches (f)

9. $\displaystyle\sum_{n=1}^{\infty} \left(\frac{4n}{5n - 3}\right)^n = \frac{4}{2} + \left(\frac{8}{7}\right)^2 + \cdots$

$S_1 = 2$

Matches (a)

11. (a) Ratio Test: $\displaystyle\lim_{n \to \infty} \left|\frac{a_{n+1}}{a_n}\right| = \lim_{n \to \infty} \frac{(n + 1)^2(5/8)^{n+1}}{n^2(5/8)^n} = \lim_{n \to \infty} \left(\frac{n + 1}{n}\right)^2 \frac{5}{8} = \frac{5}{8} < 1.$ Converges

(b)

n	5	10	15	20	25
S_n	9.2104	16.7598	18.8016	19.1878	19.2491

(c)

(d) The sum is approximately 19.26.

(e) The more rapidly the terms of the series approach 0, the more rapidly the sequence of the partial sums approaches the sum of the series.

13. $\displaystyle\sum_{n=0}^{\infty} \frac{n!}{3^n}$

$$\lim_{n \to \infty} \left|\frac{a_{n+1}}{a_n}\right| = \lim_{n \to \infty} \left|\frac{(n + 1)!}{3^{n+1}} \cdot \frac{3^n}{n!}\right|$$

$$= \lim_{n \to \infty} \frac{n + 1}{3} = \infty$$

Therefore, by the Ratio Test, the series diverges.

15. $\displaystyle\sum_{n=1}^{\infty} n\left(\frac{3}{4}\right)^n$

$$\lim_{n \to \infty} \left|\frac{a_{n+1}}{a_n}\right| = \lim_{n \to \infty} \left|\frac{(n + 1)(3/4)^{n+1}}{n(3/4)^n}\right|$$

$$= \lim_{n \to \infty} \left|\frac{3(n + 1)}{4n}\right| = \frac{3}{4}$$

Therefore, by the Ratio Test, the series converges.

17. $\displaystyle\sum_{n=1}^{\infty} \frac{n}{2^n}$

$$\lim_{n \to \infty} \left|\frac{a_{n+1}}{a_n}\right| = \lim_{n \to \infty} \left|\frac{n + 1}{2^{n+1}} \cdot \frac{2^n}{n}\right|$$

$$= \lim_{n \to \infty} \frac{n + 1}{2n} = \frac{1}{2}$$

Therefore, by the Ratio Test, the series converges.

19. $\displaystyle\sum_{n=1}^{\infty} \frac{2^n}{n^2}$

$$\lim_{n \to \infty} \left|\frac{a_{n+1}}{a_n}\right| = \lim_{n \to \infty} \left|\frac{2^{n+1}}{(n + 1)^2} \cdot \frac{n^2}{2^n}\right|$$

$$= \lim_{n \to \infty} \frac{2n^2}{(n + 1)^2} = 2$$

Therefore, by the Ratio Test, the series diverges.

21. $\displaystyle\sum_{n=0}^{\infty} \frac{(-1)^n 2^n}{n!}$

$$\lim_{n\to\infty}\left|\frac{a_{n+1}}{a_n}\right| = \lim_{n\to\infty}\left|\frac{2^{n+1}}{(n+1)!}\cdot\frac{n!}{2^n}\right|$$

$$= \lim_{n\to\infty}\frac{2}{n+1} = 0$$

Therefore, by the Ratio Test, the series converges.

23. $\displaystyle\sum_{n=1}^{\infty} \frac{n!}{n3^n}$

$$\lim_{n\to\infty}\left|\frac{a_{n+1}}{a_n}\right| = \lim_{n\to\infty}\left|\frac{(n+1)!}{(n+1)3^{n+1}}\cdot\frac{n3^n}{n!}\right|$$

$$= \lim_{n\to\infty}\frac{n}{3} = \infty$$

Therefore, by the Ratio Test, the series diverges.

25. $\displaystyle\sum_{n=0}^{\infty} \frac{4^n}{n!}$

$$\lim_{n\to\infty}\left|\frac{a_{n+1}}{a_n}\right| = \lim_{n\to\infty}\left|\frac{4^{n+1}}{(n+1)!}\cdot\frac{n!}{4^n}\right|$$

$$= \lim_{n\to\infty}\frac{4}{n+1} = 0$$

Therefore, by the Ratio Test, the series converges.

27. $\displaystyle\sum_{n=0}^{\infty} \frac{3^n}{(n+1)^n}$

$$\lim_{n\to\infty}\left|\frac{a_{n+1}}{a_n}\right| = \lim_{n\to\infty}\left|\frac{3^{n+1}}{(n+2)^{n+1}}\cdot\frac{(n+1)^n}{3^n}\right| = \lim_{n\to\infty}\frac{3(n+1)^n}{(n+2)^{n+1}} = \lim_{n\to\infty}\frac{3}{n+2}\left(\frac{n+1}{n+2}\right)^n = (0)\left(\frac{1}{e}\right) = 0$$

To find $\displaystyle\lim_{n\to\infty}\left(\frac{n+1}{n+2}\right)^n$, let $y = \displaystyle\lim_{n\to\infty}\left(\frac{n+1}{n+2}\right)^n$. Then,

$$\ln y = \lim_{n\to\infty} n\ln\left(\frac{n+1}{n+2}\right) = \lim_{n\to\infty}\frac{\ln[(n+1)/(n+2)]}{1/n} = \frac{0}{0}$$

$$\ln y = \lim_{n\to\infty}\frac{[(1)/(n+1)] - [(1)/(n+2)]}{-(1/n^2)} = -1 \text{ by L'Hôpital's Rule}$$

$$y = e^{-1} = \frac{1}{e}.$$

Therefore, by the Ratio Test, the series converges.

29. $\displaystyle\sum_{n=0}^{\infty} \frac{4^n}{3^n + 1}$

$$\lim_{n\to\infty}\left|\frac{a_{n+1}}{a_n}\right| = \lim_{n\to\infty}\left|\frac{4^{n+1}}{3^{n+1}+1}\cdot\frac{3^n+1}{4^n}\right| = \lim_{n\to\infty}\frac{4(3^n+1)}{3^{n+1}+1} = \lim_{n\to\infty}\frac{4(1+1/3^n)}{3+1/3^n} = \frac{4}{3}$$

Therefore, by the Ratio Test, the series diverges.

31. $\displaystyle\sum_{n=0}^{\infty} \frac{(-1)^{n+1}n!}{1\cdot3\cdot5\cdots(2n+1)}$

$$\lim_{n\to\infty}\left|\frac{a_{n+1}}{a_n}\right| = \lim_{n\to\infty}\left|\frac{(n+1)!}{1\cdot3\cdot5\cdots(2n+1)(2n+3)}\cdot\frac{1\cdot3\cdot5\cdots(2n+1)}{n!}\right| = \lim_{n\to\infty}\frac{n+1}{2n+3} = \frac{1}{2}$$

Therefore, by the Ratio Test, the series converges.

Note: The first few terms of this series are $-1 + \dfrac{1}{1\cdot3} - \dfrac{2!}{1\cdot3\cdot5} + \dfrac{3!}{1\cdot3\cdot5\cdot7} - \cdots$

33. (a) $\displaystyle\sum_{n=1}^{\infty} \frac{1}{n^{3/2}}$

$$\lim_{n\to\infty} \left|\frac{a_{n+1}}{a_n}\right| = \lim_{n\to\infty} \left|\frac{1}{(n+1)^{3/2}} \cdot \frac{n^{3/2}}{1}\right| = \lim_{n\to\infty} \left(\frac{n}{n+1}\right)^{3/2} = 1 \qquad \text{Ratio Test is inconclusive.}$$

(b) $\displaystyle\sum_{n=1}^{\infty} \frac{1}{n^{1/2}}$

$$\lim_{n\to\infty} \left|\frac{a_{n+1}}{a_n}\right| = \lim_{n\to\infty} \left|\frac{1}{(n+1)^{1/2}} \cdot \frac{n^{1/2}}{1}\right| = \lim_{n\to\infty} \left(\frac{n}{n+1}\right)^{1/2} = 1 \qquad \text{Ratio Test is inconclusive.}$$

35. $\displaystyle\sum_{n=1}^{\infty} \left(\frac{n}{2n+1}\right)^n$

$$\lim_{n\to\infty} \sqrt[n]{|a_n|} = \lim_{n\to\infty} \sqrt[n]{\left(\frac{n}{2n+1}\right)^n}$$
$$= \lim_{n\to\infty} \frac{n}{2n+1} = \frac{1}{2}$$

Therefore, by the Root Test, the series converges.

37. $\displaystyle\sum_{n=2}^{\infty} \frac{(-1)^n}{(\ln n)^n}$

$$\lim_{n\to\infty} \sqrt[n]{|a_n|} = \lim_{n\to\infty} \sqrt[n]{\left|\frac{(-1)^n}{(\ln n)^n}\right|}$$
$$= \lim_{n\to\infty} \frac{1}{|\ln n|} = 0$$

Therefore, by the Root Test, the series converges.

39. $\displaystyle\sum_{n=1}^{\infty} \left(2\sqrt[n]{n} + 1\right)^n$

$$\lim_{n\to\infty} \sqrt[n]{|a_n|} = \lim_{n\to\infty} \sqrt[n]{\left(2\sqrt[n]{n} + 1\right)^n} = \lim_{n\to\infty} \left(2\sqrt[n]{n} + 1\right)$$

To find $\displaystyle\lim_{n\to\infty} \sqrt[n]{n}$, let $y = \displaystyle\lim_{n\to\infty} \sqrt[n]{n}$. Then

$$\ln y = \lim_{n\to\infty} \left(\ln \sqrt[n]{n}\right) = \lim_{n\to\infty} \frac{1}{n} \ln n = \lim_{n\to\infty} \frac{\ln n}{n} = \lim_{n\to\infty} \frac{1/n}{1} = 0.$$

Thus, $\ln y = 0$, so $y = e^0 = 1$ and $\displaystyle\lim_{n\to\infty} \left(2\sqrt[n]{n} + 1\right) = 2(1) + 1 = 3$. Therefore, by the Root Test, the series diverges.

41. $\displaystyle\sum_{n=3}^{\infty} \frac{1}{(\ln n)^n}$

$$\lim_{n\to\infty} \sqrt[n]{|a_n|} = \lim_{n\to\infty} \sqrt[n]{\frac{1}{(\ln n)^n}} = \lim_{n\to\infty} \frac{1}{\ln n} = 0$$

Therefore, by the Root Test, the series converges.

43. $\displaystyle\sum_{n=1}^{\infty} \frac{(-1)^{n+1} 5}{n}$

$$a_{n+1} = \frac{5}{n+1} < \frac{5}{n} = a_n$$

$$\lim_{n\to\infty} \frac{5}{n} = 0$$

Therefore, by the Alternating Series Test, the series converges (conditional convergence).

45. $\displaystyle\sum_{n=1}^{\infty} \frac{3}{n\sqrt{n}} = 3\sum_{n=1}^{\infty} \frac{1}{n^{3/2}}$

This is convergent *p*-series.

47. $\displaystyle\sum_{n=1}^{\infty} \frac{2n}{n+1}$

$$\lim_{n\to\infty} \frac{2n}{n+1} = 2 \neq 0$$

This diverges by the *n*th Term Test for Divergence.

49. $\displaystyle\sum_{n=1}^{\infty} \frac{(-1)^n 3^{n-2}}{2^n} = \sum_{n=1}^{\infty} \frac{(-1)^n 3^n 3^{-2}}{2^n} = \sum_{n=1}^{\infty} \frac{1}{9}\left(-\frac{3}{2}\right)^n$

Since $|r| = \frac{3}{2} > 1$, this is a divergent geometric series.

51. $\displaystyle\sum_{n=1}^{\infty} \frac{10n+3}{n2^n}$

$$\lim_{n\to\infty} \frac{(10n+3)/n2^n}{1/2^n} = \lim_{n\to\infty} \frac{10n+3}{n} = 10$$

Therefore, the series converges by a limit comparison test with the geometric series

$$\sum_{n=0}^{\infty} \left(\frac{1}{2}\right)^n.$$

53. $\displaystyle\sum_{n=1}^{\infty} \frac{\cos(n)}{2^n}$

$$\left|\frac{\cos(n)}{2^n}\right| \le \frac{1}{2^n}$$

Therefore, the series

$$\sum_{n=1}^{\infty} \left|\frac{\cos(n)}{2^n}\right|$$

converges by comparison with the geometric series

$$\sum_{n=0}^{\infty} \left(\frac{1}{2}\right)^n.$$

55. $\displaystyle\sum_{n=1}^{\infty} \frac{n7^n}{n!}$

$$\lim_{n\to\infty}\left|\frac{a_{n+1}}{a_n}\right| = \lim_{n\to\infty}\left|\frac{(n+1)7^{n+1}}{(n+1)!} \cdot \frac{n!}{n7^n}\right| = \lim_{n\to\infty}\frac{7}{n} = 0$$

Therefore, by the Ratio Test, the series converges.

57. $\displaystyle\sum_{n=1}^{\infty} \frac{(-1)^n 3^{n-1}}{n!}$

$$\lim_{n\to\infty}\left|\frac{a_{n+1}}{a_n}\right| = \lim_{n\to\infty}\left|\frac{3^n}{(n+1)!} \cdot \frac{n!}{3^{n-1}}\right| = \lim_{n\to\infty}\frac{3}{n+1} = 0$$

Therefore, by the Ratio Test, the series converges.

59. $\displaystyle\sum_{n=1}^{\infty} \frac{(-3)^n}{3 \cdot 5 \cdot 7 \cdots (2n+1)}$

$$\lim_{n\to\infty}\left|\frac{a_{n+1}}{a_n}\right| = \lim_{n\to\infty}\left|\frac{(-3)^{n+1}}{3 \cdot 5 \cdot 7 \cdots (2n+1)(2n+3)} \cdot \frac{3 \cdot 5 \cdot 7 \cdots (2n+1)}{(-3)^n}\right| = \lim_{n\to\infty}\frac{3}{2n+3} = 0$$

Therefore, by the Ratio Test, the series converges.

61. (a) and (c)

$$\sum_{n=1}^{\infty} \frac{n5^n}{n!} = \sum_{n=0}^{\infty} \frac{(n+1)5^{n+1}}{(n+1)!}$$

$$= 5 + \frac{(2)(5)^2}{2!} + \frac{(3)(5)^3}{3!} + \frac{(4)(5)^4}{4!} + \cdots$$

63. (a) and (b) are the same.

65. Replace n with $n+1$.

$$\sum_{n=1}^{\infty} \frac{n}{4^n} = \sum_{n=0}^{\infty} \frac{n+1}{4^{n+1}}$$

67. Since

$$\frac{3^{10}}{2^{10}\,10!} \approx 1.59 \times 10^{-5},$$

use 9 terms.

$$\sum_{k=1}^{9} \frac{(-3)^k}{2^k\,k!} \approx -0.7769$$

69. See Theorem 8.17, page 597.

71. No. Let $a_n = \dfrac{1}{n + 10{,}000}$.

The series $\displaystyle\sum_{n=1}^{\infty} \frac{1}{n + 10{,}000}$ diverges.

73. The series converges absolutely. See Theorem 8.17.

75. First, let

$$\lim_{n \to \infty} \sqrt[n]{|a_n|} = r < 1$$

and choose R such that $0 \le r < R < 1$. There must exist some $N > 0$ such that $\sqrt[n]{|a_n|} < R$ for all $n > N$. Thus, for $n > N$, we $|a_n| < R^n$ and since the geometric series

$$\sum_{n=0}^{\infty} R^n$$

converges, we can apply the Comparison Test to conclude that

$$\sum_{n=1}^{\infty} |a_n|$$

converges which in turn implies that $\sum_{n=1}^{\infty} a_n$ converges.

Second, let

$$\lim_{n \to \infty} \sqrt[n]{|a_n|} = r > R > 1.$$

Then there must exist some $M > 0$ such that $\sqrt[n]{|a_n|} > R$ for infinitely many $n > M$. Thus, for infinitely many $n > M$, we have $|a_n| > R^n > 1$ which implies that $\lim_{n \to \infty} a_n \ne 0$ which in turn implies that

$$\sum_{n=1}^{\infty} a_n \text{ diverges.}$$

Section 8.7   Taylor Polynomials and Approximations

1. $y = -\frac{1}{2}x^2 + 1$

Parabola

Matches (d)

3. $y = e^{-1/2}[(x + 1) + 1]$

Linear

Matches (a)

5. $f(x) = \dfrac{4}{\sqrt{x}} = 4x^{-1/2} \qquad f(1) = 4$

$f'(x) = -2x^{-3/2} \qquad f'(1) = -2$

$P_1(x) = f(1) + f'(1)(x - 1)$

$\qquad = 4 + (-2)(x - 1)$

$P_1(x) = -2x + 6$

P_1 is called the first degree Taylor polynomial for f at c.

7. $f(x) = \sec x \qquad f\left(\dfrac{\pi}{4}\right) = \sqrt{2}$

$f'(x) = \sec x \tan x \qquad f'\left(\dfrac{\pi}{4}\right) = \sqrt{2}$

$P_1(x) = f\left(\dfrac{\pi}{4}\right) + f'\left(\dfrac{\pi}{4}\right)\left(x - \dfrac{\pi}{4}\right)$

$P_1(x) = \sqrt{2} + \sqrt{2}\left(x - \dfrac{\pi}{4}\right)$

P_1 is called the first degree Taylor polynomial for f at c.

9. $f(x) = \dfrac{4}{\sqrt{x}} = 4x^{-1/2} \qquad f(1) = 4$

$f'(x) = -2x^{-3/2} \qquad f'(1) = -2$

$f''(x) = 3x^{-5/2} \qquad f''(1) = 3$

$P_2 = f(1) + f'(1)(x - 1) + \dfrac{f''(1)}{2}(x - 1)^2$

$\qquad = 4 - 2(x - 1) + \dfrac{3}{2}(x - 1)^2$

x	0	0.8	0.9	1.0	1.1	1.2	2
$f(x)$	Error	4.4721	4.2164	4.0	3.8139	3.6515	2.8284
$P_2(x)$	7.5	4.46	4.215	4.0	3.815	3.66	3.5

11. $f(x) = \cos x$

$P_2(x) = 1 - \frac{1}{2}x^2$

$P_4(x) = 1 - \frac{1}{2}x^2 + \frac{1}{24}x^4$

$P_6(x) = 1 - \frac{1}{2}x^2 + \frac{1}{24}x^4 - \frac{1}{720}x^6$

(a)

(b) $f'(x) = -\sin x$ $\qquad P_2'(x) = -x$

$\quad f''(x) = -\cos x$ $\qquad P_2''(x) = -1$

$\quad f''(0) = P_2''(0) = -1$

$\quad f'''(x) = \sin x$ $\qquad P_4'''(x) = x$

$\quad f^{(4)}(x) = \cos x$ $\qquad P_4^{(4)}(x) = 1$

$\quad f^{(4)}(0) = 1 = P_4^{(4)}(0)$

$\quad f^{(5)}(x) = -\sin x$ $\qquad P_6^{(5)}(x) = -x$

$\quad f^{(6)}(x) = -\cos x$ $\qquad P^{(6)}(x) = -1$

$\quad f^{(6)}(0) = -1 = P_6^{(6)}(0)$

(c) In general, $f^{(n)}(0) = P_n^{(n)}(0)$ for all n.

13. $f(x) = e^{-x}$ $\qquad f(0) = 1$

$\quad f'(x) = -e^{-x}$ $\qquad f'(0) = -1$

$\quad f''(x) = e^{-x}$ $\qquad f''(0) = 1$

$\quad f'''(x) = -e^{-x}$ $\qquad f'''(0) = -1$

$\quad P_3(x) = f(0) + f'(0)x + \frac{f''(0)}{2!}x^2 + \frac{f'''(0)}{3!}x^3$

$\qquad = 1 - x + \frac{x^2}{2} - \frac{x^3}{6}$

15. $f(x) = e^{2x}$ $\qquad f(0) = 1$

$\quad f'(x) = 2e^{2x}$ $\qquad f'(0) = 2$

$\quad f''(x) = 4e^{2x}$ $\qquad f''(0) = 4$

$\quad f'''(x) = 8e^{2x}$ $\qquad f'''(0) = 8$

$\quad f^{(4)}(x) = 16^{2x}$ $\qquad f^{(4)}(0) = 16$

$\quad P_4(x) = 1 + 2x + \frac{4}{2!}x^2 + \frac{8}{3!}x^3 + \frac{16}{4!}x^4$

$\qquad = 1 + 2x + 2x^2 + \frac{4}{3}x^3 + \frac{2}{3}x^4$

17. $f(x) = \sin x$ $\qquad f(0) = 0$

$\quad f'(x) = \cos x$ $\qquad f'(0) = 1$

$\quad f''(x) = -\sin x$ $\qquad f''(0) = 0$

$\quad f'''(x) = -\cos x$ $\qquad f'''(0) = -1$

$\quad f^{(4)}(x) = \sin x$ $\qquad f^{(4)}(0) = 0$

$\quad f^{(5)}(x) = \cos x$ $\qquad f^{(5)}(0) = 1$

$\quad P_5(x) = 0 + (1)x + \frac{0}{2!}x^2 + \frac{-1}{3!}x^3 + \frac{0}{4!}x^4 + \frac{1}{5!}x^5$

$\qquad = x - \frac{1}{6}x^3 + \frac{1}{120}x^5$

19. $f(x) = xe^x$ $\qquad f(0) = 0$

$\quad f'(x) = xe^x + e^x$ $\qquad f'(0) = 1$

$\quad f''(x) = xe^x + 2e^x$ $\qquad f''(0) = 2$

$\quad f'''(x) = xe^x + 3e^x$ $\qquad f'''(0) = 3$

$\quad f^{(4)}(x) = xe^x + 4e^x$ $\qquad f^{(4)}(0) = 4$

$\quad P_4(x) = 0 + x + \frac{2}{2!}x^2 + \frac{3}{3!}x^3 + \frac{4}{4!}x^4$

$\qquad = x + x^2 + \frac{1}{2}x^3 + \frac{1}{6}x^4$

21. $f(x) = \dfrac{1}{x+1}$ $\qquad f(0) = 1$

$\quad f'(x) = -\dfrac{1}{(x+1)^2}$ $\qquad f'(0) = -1$

$\quad f''(x) = \dfrac{2}{(x+1)^3}$ $\qquad f''(0) = 2$

$\quad f'''(x) = \dfrac{-6}{(x+1)^4}$ $\qquad f'''(0) = -6$

$\quad f^{(4)}(x) = \dfrac{24}{(x+1)^5}$ $\qquad f^{(4)}(0) = 24$

$\quad P_4(x) = 1 - x + \frac{2}{2!}x^2 + \frac{-6}{3!}x^3 + \frac{24}{4!}x^4$

$\qquad = 1 - x + x^2 - x^3 + x^4$

23. $f(x) = \sec x$ $\qquad f(0) = 1$

$\quad f'(x) = \sec x \tan x$ $\qquad f'(0) = 0$

$\quad f''(x) = \sec^3 x + \sec x \tan^2 x$ $\qquad f''(0) = 1$

$\quad P_2(x) = 1 + 0x + \frac{1}{2!}x^2 = 1 + \frac{1}{2}x^2$

25. $f(x) = \dfrac{1}{x}$ $f(1) = 1$

$f'(x) = -\dfrac{1}{x^2}$ $f'(1) = -1$

$f''(x) = \dfrac{2}{x^3}$ $f''(1) = 2$

$f'''(x) = -\dfrac{6}{x^4}$ $f'''(1) = -6$

$f^{(4)}(x) = \dfrac{24}{x^5}$ $f^{(4)}(1) = 24$

$P_4(x) = 1 - (x - 1) + \dfrac{2}{2!}(x - 1)^2 + \dfrac{-6}{3!}(x - 1)^3 + \dfrac{24}{4!}(x - 1)^4$

$\qquad = 1 - (x - 1) + (x - 1)^2 - (x - 1)^3 + (x - 1)^4$

27. $f(x) = \sqrt{x}$ $f(1) = 1$

$f'(x) = \dfrac{1}{2\sqrt{x}}$ $f'(1) = \dfrac{1}{2}$

$f''(x) = -\dfrac{1}{4x\sqrt{x}}$ $f''(1) = -\dfrac{1}{4}$

$f'''(x) = \dfrac{3}{8x^2\sqrt{x}}$ $f'''(1) = \dfrac{3}{8}$

$f^{(4)}(x) = -\dfrac{15}{16x^3\sqrt{x}}$ $f^{(4)}(1) = -\dfrac{15}{16}$

$P_4(x) = 1 + \dfrac{1}{2}(x - 1) - \dfrac{1}{8}(x - 1)^2$

$\qquad + \dfrac{1}{16}(x - 1)^3 - \dfrac{5}{128}(x - 1)^4$

29. $f(x) = \ln x$ $f(1) = 0$

$f'(x) = \dfrac{1}{x}$ $f'(1) = 1$

$f''(x) = -\dfrac{1}{x^2}$ $f''(1) = -1$

$f'''(x) = \dfrac{2}{x^3}$ $f'''(1) = 2$

$f^{(4)}(x) = -\dfrac{6}{x^4}$ $f^{(4)}(1) = -6$

$P_4(x) = 0 + (x - 1) - \dfrac{1}{2}(x - 1)^2$

$\qquad + \dfrac{1}{3}(x - 1)^3 - \dfrac{1}{4}(x - 1)^4$

31. $f(x) = \tan x$

$f'(x) = \sec^2 x$

$f''(x) = 2\sec^2 x \tan x$

$f'''(x) = 4\sec^2 x \tan^2 x + 2\sec^4 x$

$f^{(4)}(x) = 8\sec^2 x \tan^3 x + 16\sec^4 x \tan x$

$f^{(5)}(x) = 16\sec^2 x \tan^4 x + 88\sec^4 x \tan^2 x + 16\sec^6 x$

(a) $n = 3, c = 0$

$P_3(x) = 0 + x + \dfrac{0}{2!}x^2 + \dfrac{2}{3!}x^3 = x + \dfrac{1}{3}x^3$

(b) $n = 5, c = 0$

$P_5(x) = 0 + x + \dfrac{0}{2!}x^2 + \dfrac{2}{3!}x^3 + \dfrac{0}{4!}x^4 + \dfrac{16}{5!}x^5$

$\qquad = x + \dfrac{1}{3}x^3 + \dfrac{2}{15}x^5$

(c) $n = 3, c = \dfrac{\pi}{4}$

$Q_3(x) = 1 + 2\left(x - \dfrac{\pi}{4}\right) + \dfrac{4}{2!}\left(x - \dfrac{\pi}{4}\right)^2 + \dfrac{16}{3!}\left(x - \dfrac{\pi}{4}\right)^3$

$\qquad = 1 + 2\left(x - \dfrac{\pi}{4}\right) + 2\left(x - \dfrac{\pi}{4}\right)^2 + \dfrac{8}{3}\left(x - \dfrac{\pi}{4}\right)^3$

33. $f(x) = \sin x$

$P_1(x) = x$

$P_3(x) = x - \frac{1}{6}x^3$

$P_5(x) = x - \frac{1}{6}x^3 + \frac{1}{120}x^5$

$P_7(x) = x - \frac{1}{6}x^3 + \frac{1}{120}x^5 - \frac{1}{5040}x^7$

(a)

x	0.00	0.25	0.50	0.75	1.00
$\sin x$	0.0000	0.2474	0.4794	0.6816	0.8415
$P_1(x)$	0.0000	0.2500	0.5000	0.7500	1.0000
$P_3(x)$	0.0000	0.2474	0.4792	0.6797	0.8333
$P_5(x)$	0.0000	0.2474	0.4794	0.6817	0.8417
$P_7(x)$	0.0000	0.2474	0.4794	0.6816	0.8415

(b)

(c) As the distance increases, the accuracy decreases

35. $f(x) = \arcsin x$

(a) $P_3(x) = x + \dfrac{x^3}{6}$

(b)

x	-0.75	-0.50	-0.25	0	0.25	0.50	0.75
$f(x)$	-0.848	-0.524	-0.253	0	0.253	0.524	0.848
$P_3(x)$	-0.820	-0.521	-0.253	0	0.253	0.521	0.820

(c)

37. $f(x) = \cos x$

39. $f(x) = \ln(x^2 + 1)$

41. $f(x) = e^{-x} \approx 1 - x + \dfrac{x^2}{2} - \dfrac{x^3}{6}$

$f\left(\dfrac{1}{2}\right) \approx 0.6042$

43. $f(x) = \ln x \approx (x - 1) - \frac{1}{2}(x - 1)^2 + \frac{1}{3}(x - 1)^3 - \frac{1}{4}(x - 1)^4$

$f(1.2) \approx 0.1823$

45. $f(x) = \cos x;\, f^{(5)}(x) = -\sin x \implies$ Max on $[0, 0.3]$ is 1.

$R_4(x) \le \dfrac{1}{5!}(0.3)^5 = 2.025 \times 10^{-5}$

Note: You could use $R_5(x)$: $f^{(6)}(x) = -\cos x$, max on $[0, 0.3]$ is 1. $R_5(x) \le \dfrac{1}{6!}(0.3)^6 = 1.0125 \times 10^{-6}$

Exact Error: $0.000001 = 1.0 \times 10^{-6}$

47. $f(x) = \arcsin x; f^{(4)}(x) = \dfrac{x(6x^2 + 9)}{(1 - x^2)^{7/2}} \Longrightarrow$ Max on $[0, 0.4]$ is $f^{(4)}(0.4) \approx 7.3340.$

$R_3(x) \leq \dfrac{7.3340}{4!}(0.4)^4 \approx 0.00782 = 7.82 \times 10^{-3}$

49. $g(x) = \sin x$

$g^{(n+1)}(x) \leq 1$ for all x

$R_n(x) \leq \dfrac{1}{(n + 1)!}(0.3)^{n+1} < 0.001$

By trial and error, $n = 3$.

51. $f(x) = \ln(x + 1)$

$f^{(n+1)}(x) = \dfrac{(-1)^n n!}{(x + 1)^{n+1}} \Longrightarrow$ Max on $[0, 0.5]$ is $n!.$

$R_n \leq \dfrac{n!}{(n + 1)!}(0.5)^{n+1} = \dfrac{(0.5)^{n+1}}{n + 1} < 0.0001$

By trial and error, $n = 9$. (See Example 9.) Using 9 terms, $\ln(1.5) \approx 0.4055.$

53. $\qquad f(x) = e^x \approx 1 + x + \dfrac{x^2}{2} + \dfrac{x^3}{6},\ x < 0$

$R_3(x) = \dfrac{e^z}{4!}x^4 < 0.001$

$e^z x^4 < 0.024$

$xe^{z/4} < 0.3936$

$\qquad x < \dfrac{0.3936}{e^{z/4}} < 0.3936,\ z < 0$

$-0.3936 < x < 0$

55. The graph of the approximating polynomial P and the elementary function f both pass through the point $(c, f(c))$ and the slopes of P and f agree at $(c, f(c))$. Depending on the degree of P, the nth derivatives of P and f agree at $(c, f(c))$.

57. See definition on page 607.

59. The accuracy increases as the degree increases (for values within the interval of convergence).

61. (a) $f(x) = e^x$

$P_4(x) = 1 + x + \dfrac{1}{2}x^2 + \dfrac{1}{6}x^3 + \dfrac{1}{24}x^4$

$g(x) = xe^x$

$Q_5(x) = x + x^2 + \dfrac{1}{2}x^3 + \dfrac{1}{6}x^4 + \dfrac{1}{24}x^5$

$Q_5(x) = xP_4(x)$

(b) $f(x) = \sin x$

$P_5(x) = x - \dfrac{x^3}{3!} + \dfrac{x^5}{5!}$

$g(x) = x\sin x$

$Q_6(x) = xP_5(x) = x^2 - \dfrac{x^4}{3!} + \dfrac{x^6}{5!}$

(c) $g(x) = \dfrac{\sin x}{x} = \dfrac{1}{x}P_5(x) = 1 - \dfrac{x^2}{3!} + \dfrac{x^4}{5!}$

63. (a) $Q_2(x) = -1 + \dfrac{\pi^2(x + 2)^2}{32}$

(b) $R_2(x) = -1 + \dfrac{\pi^2(x - 6)^2}{32}$

(c) No. The polynomial will be linear.
Translations are possible at $x = -2 + 8n$.

65. Let f be an even function and P_n be the nth Maclaurin polynomial for f. Since f is even, f' is odd, f'' is even, f''' is odd, etc. All of the odd derivatives of f are odd and thus, all of the odd powers of x will have coefficients of zero. P_n will only have terms with even powers of x.

67. As you move away from $x = c$, the Taylor Polynomial becomes less and less accurate.

Section 8.8 Power Series

1. Centered at 0

3. Centered at 2

5. $\displaystyle\sum_{n=0}^{\infty} (-1)^n \frac{x^n}{n+1}$

$$L = \lim_{n \to \infty} \left| \frac{u_{n+1}}{u_n} \right| = \lim_{n \to \infty} \left| \frac{(-1)^{n+1} x^{n+1}}{n+2} \cdot \frac{n+1}{(-1)^n x^n} \right|$$

$$= \lim_{n \to \infty} \left| \frac{n+1}{n+2} \right| |x| = |x|$$

$$|x| < 1 \Rightarrow R = 1$$

7. $\displaystyle\sum_{n=1}^{\infty} \frac{(2x)^n}{n^2}$

$$L = \lim_{n \to \infty} \left| \frac{u_{n+1}}{u_n} \right| = \lim_{n \to \infty} \left| \frac{(2x)^{n+1}}{(n+1)^2} \cdot \frac{n^2}{(2x)^n} \right|$$

$$= \lim_{n \to \infty} \left| \frac{2n^2 x}{(n+1)^2} \right| = 2|x|$$

$$2|x| < 1 \Rightarrow R = \frac{1}{2}$$

9. $\displaystyle\sum_{n=0}^{\infty} \frac{(2x)^{2n}}{(2n)!}$

$$L = \lim_{n \to \infty} \left| \frac{u_{n+1}}{u_n} \right| = \lim_{n \to \infty} \left| \frac{(2x)^{2n+2}/(2n+2)!}{(2x)^{2n}/(2n)!} \right|$$

$$= \lim_{n \to \infty} \left| \frac{(2x)^2}{(2n+2)(2n+1)} \right| = 0$$

Thus, the series converges for all x. $R = \infty$.

11. $\displaystyle\sum_{n=0}^{\infty} \left(\frac{x}{2} \right)^n$

Since the series is geometric, it converges only if $|x/2| < 1$ or $-2 < x < 2$.

13. $\displaystyle\sum_{n=1}^{\infty} \frac{(-1)^n x^n}{n}$

$$\lim_{n \to \infty} \left| \frac{u_{n+1}}{u_n} \right| = \lim_{n \to \infty} \left| \frac{(-1)^{n+1} x^{n+1}}{n+1} \cdot \frac{n}{(-1)^n x^n} \right|$$

$$= \lim_{n \to \infty} \left| \frac{nx}{n+1} \right| = |x|$$

Interval: $-1 < x < 1$

When $x = 1$, the alternating series $\displaystyle\sum_{n=1}^{\infty} \frac{(-1)^n}{n}$ converges.

When $x = -1$, the p-series $\displaystyle\sum_{n=1}^{\infty} \frac{1}{n}$ diverges.

Therefore, the interval of convergence is $-1 < x \leq 1$.

15. $\displaystyle\sum_{n=0}^{\infty} \frac{x^n}{n!}$

$$\lim_{n \to \infty} \left| \frac{u_{n+1}}{u_n} \right| = \lim_{n \to \infty} \left| \frac{x^{n+1}}{(n+1)!} \cdot \frac{n!}{x^n} \right|$$

$$= \lim_{n \to \infty} \left| \frac{x}{n+1} \right| = 0$$

The series converges for all x. Therefore, the interval of convergence is $-\infty < x < \infty$.

17. $\displaystyle\sum_{n=0}^{\infty} (2n)! \left(\frac{x}{2} \right)^n$

$$\lim_{n \to \infty} \left| \frac{u_{n+1}}{u_n} \right| = \lim_{n \to \infty} \left| \frac{(2n+2)! x^{n+1}}{2^{n+1}} \cdot \frac{2^n}{(2n)! x^n} \right| = \lim_{n \to \infty} \left| \frac{(2n+2)(2n+1) x}{2} \right| = \infty$$

Therefore, the series converges only for $x = 0$.

19. $\displaystyle\sum_{n=1}^{\infty} \frac{(-1)^{n+1} x^n}{4^n}$

Since the series is geometric, it converges only if $|x/4| < 1$ or $-4 < x < 4$.

21. $\sum_{n=1}^{\infty} \dfrac{(-1)^{n+1}(x-5)^n}{n5^n}$

$\displaystyle\lim_{n\to\infty}\left|\dfrac{u_{n+1}}{u_n}\right| = \lim_{n\to\infty}\left|\dfrac{(-1)^{n+2}(x-5)^{n+1}}{(n+1)5^{n+1}}\cdot\dfrac{n5^n}{(-1)^{n+1}(x-5)^n}\right| = \lim_{n\to\infty}\left|\dfrac{n(x-5)}{5(n+1)}\right| = \dfrac{1}{5}|x-5|$

$R = 5$

Center: $x = 5$

Interval: $-5 < x - 5 < 5$ or $0 < x < 10$

When $x = 0$, the p-series $\sum_{n=1}^{\infty}\dfrac{-1}{n}$ diverges.

When $x = 10$, the alternating series $\sum_{n=1}^{\infty}\dfrac{(-1)^{n+1}}{n}$ converges.

Therefore, the interval of convergence is $0 < x \le 10$.

23. $\sum_{n=0}^{\infty} \dfrac{(-1)^{n+1}(x-1)^{n+1}}{n+1}$

$\displaystyle\lim_{n\to\infty}\left|\dfrac{u_{n+1}}{u_n}\right| = \lim_{n\to\infty}\left|\dfrac{(-1)^{n+2}(x-1)^{n+2}}{n+2}\cdot\dfrac{n+1}{(-1)^{n+1}(x-1)^{n+1}}\right| = \lim_{n\to\infty}\left|\dfrac{(n+1)(x-1)}{n+2}\right| = |x-1|$

$R = 1$

Center: $x = 1$

Interval: $-1 < x - 1 < 1$ or $0 < x < 2$

When $x = 0$, the series $\sum_{n=0}^{\infty}\dfrac{1}{n+1}$ diverges by the integral test.

When $x = 2$, the alternating series $\sum_{n=0}^{\infty}\dfrac{(-1)^{n+1}}{n+1}$ converges.

Therefore, the interval of convergence is $0 < x \le 2$.

25. $\sum_{n=1}^{\infty} \dfrac{(x-c)^{n-1}}{c^{n-1}}$

$\displaystyle\lim_{n\to\infty}\left|\dfrac{u_{n+1}}{u_n}\right| = \lim_{n\to\infty}\left|\dfrac{(x-c)^n}{c^n}\cdot\dfrac{c^{n-1}}{(x-c)^{n-1}}\right| = \dfrac{1}{c}|x-c|$

$R = c$

Center: $x = c$

Interval: $-c < x - c < c$ or $0 < x < 2c$

When $x = 0$, the series $\sum_{n=1}^{\infty}(-1)^{n-1}$ diverges.

When $x = 2c$, the series $\sum_{n=1}^{\infty}1$ diverges.

Therefore, the interval of convergence is $0 < x < 2c$.

27. $\sum_{n=1}^{\infty} \dfrac{n}{n+1}(-2x)^{n-1}$

$\displaystyle\lim_{n\to\infty}\left|\dfrac{u_{n+1}}{u_n}\right| = \lim_{n\to\infty}\left|\dfrac{(n+1)(-2x)^n}{n+2}\cdot\dfrac{n+1}{n(-2x)^{n-1}}\right|$

$\qquad\qquad\quad = \lim_{n\to\infty}\left|\dfrac{(-2x)(n+1)^2}{n(n+2)}\right| = 2|x|$

$R = \dfrac{1}{2}$

Interval: $-\dfrac{1}{2} < x < \dfrac{1}{2}$

When $x = -\dfrac{1}{2}$, the series $\sum_{n=1}^{\infty}\dfrac{n}{n+1}$ diverges by the nth Term Test.

When $x = \dfrac{1}{2}$, the alternating series $\sum_{n=1}^{\infty}\dfrac{(-1)^{n-1}n}{n+1}$ diverges.

Therefore, the interval of convergence is $-\dfrac{1}{2} < x < \dfrac{1}{2}$.

29. $\displaystyle\sum_{n=0}^{\infty} \frac{x^{2n+1}}{(2n+1)!}$

$$\lim_{n\to\infty}\left|\frac{u_{n+1}}{u_n}\right| = \lim_{n\to\infty}\left|\frac{x^{2n+3}}{(2n+3)!}\cdot\frac{(2n+1)!}{x^{2n+1}}\right|$$

$$= \lim_{n\to\infty}\left|\frac{x^2}{(2n+2)(2n+3)}\right| = 0$$

Therefore, the interval of convergence is $-\infty < x < \infty$.

31. $\displaystyle\sum_{n=1}^{\infty} \frac{k(k+1)\cdots(k+n-1)x^n}{n!}$

$$\lim_{n\to\infty}\left|\frac{u_{n+1}}{u_n}\right| = \lim_{n\to\infty}\left|\frac{k(k+1)\cdots(k+n-1)(k+n)x^{n+1}}{(n+1)!}\cdot\frac{n!}{k(k+1)\cdots(k+n-1)x^n}\right| = \lim_{n\to\infty}\left|\frac{(k+n)x}{n+1}\right| = |x|$$

$R = 1$

When $x = \pm 1$, the series diverges and the interval of convergence is $-1 < x < 1$.

$$\left[\frac{k(k+1)\cdots(k+n-1)}{1\cdot 2\cdots n} \geq 1\right]$$

33. $\displaystyle\sum_{n=1}^{\infty} \frac{(-1)^{n+1}3\cdot 7\cdot 11\cdots(4n-1)(x-3)^n}{4^n}$

$$\lim_{n\to\infty}\left|\frac{u_{n+1}}{u_n}\right| = \lim_{n\to\infty}\left|\frac{(-1)^{n+2}\cdot 3\cdot 7\cdot 11\cdots(4n-1)(4n+3)(x-3)^{n+1}}{4^{n+1}}\cdot\frac{4^n}{(-1)^{n+1}\cdot 3\cdot 7\cdot 11\cdots(4n-1)(x-3)^n}\right|$$

$$= \lim_{n\to\infty}\left|\frac{(4n+3)(x-3)}{4}\right| = \infty$$

$R = 0$

Center: $x = 3$

Therefore, the series converges only for $x = 3$.

35. (a) $\displaystyle f(x) = \sum_{n=0}^{\infty}\left(\frac{x}{2}\right)^n, \ -2 < x < 2$ (Geometric)

(b) $\displaystyle f'(x) = \sum_{n=1}^{\infty}\left(\frac{n}{2}\right)\left(\frac{x}{2}\right)^{n-1}, \ -2 < x < 2$

(c) $\displaystyle f''(x) = \sum_{n=2}^{\infty}\left(\frac{n}{2}\right)\left(\frac{n-1}{2}\right)\left(\frac{x}{2}\right)^{n-2}, \ -2 < x < 2$

(d) $\displaystyle \int f(x)\,dx = \sum_{n=0}^{\infty}\frac{2}{n+1}\left(\frac{x}{2}\right)^{n+1}, \ -2 \leq x < 2$

37. (a) $\displaystyle f(x) = \sum_{n=0}^{\infty}\frac{(-1)^{n+1}(x-1)^{n+1}}{n+1}, \ 0 < x \leq 2$

(b) $\displaystyle f'(x) = \sum_{n=0}^{\infty}(-1)^{n+1}(x-1)^n, \ 0 < x < 2$

(c) $\displaystyle f''(x) = \sum_{n=1}^{\infty}(-1)^{n+1}n(x-1)^{n-1}, \ 0 < x < 2$

(d) $\displaystyle \int f(x)\,dx = \sum_{n=1}^{\infty}\frac{(-1)^{n+1}(x-1)^{n+2}}{(n+1)(n+2)}, \ 0 \leq x \leq 2$

39. $\displaystyle g(1) = \sum_{n=0}^{\infty}\left(\frac{1}{3}\right)^n = 1 + \frac{1}{3} + \frac{1}{9} + \cdots$

$S_1 = 1, \ S_2 = 1.33$. Matches (c)

41. $\displaystyle g(3.1) = \sum_{n=0}^{\infty}\left(\frac{3.1}{3}\right)^n$ diverges. Matches (b)

43. A series of the form

$$\sum_{n=0}^{\infty} a_n(x - c)^n$$

is called a power series centered at c.

45. A single point, an interval, or the entire real line.

47. (a) $f(x) = \displaystyle\sum_{n=0}^{\infty} \frac{(-1)^n x^{2n+1}}{(2n+1)!}, \; -\infty < x < \infty$ (See Exercise 29.)

$$g(x) = \sum_{n=0}^{\infty} \frac{(-1)^n x^{2n}}{(2n)!}, \; -\infty < x < \infty$$

(b) $f'(x) = \displaystyle\sum_{n=0}^{\infty} \frac{(-1)^n x^{2n}}{(2n)!} = g(x)$

(c) $g'(x) = \displaystyle\sum_{n=1}^{\infty} \frac{(-1)^n x^{2n-1}}{(2n-1)!} = \sum_{n=0}^{\infty} \frac{(-1)^{n+1} x^{2n+1}}{(2n+1)!} = -\sum_{n=0}^{\infty} \frac{(-1)^n x^{2n+1}}{(2n+1)!} = -f(x)$

(d) $f(x) = \sin x$ and $g(x) = \cos x$

49.
$$y = \sum_{n=0}^{\infty} \frac{x^{2n}}{2^n n!}$$

$$y' = \sum_{n=1}^{\infty} \frac{2n x^{2n-1}}{2^n n!}$$

$$y'' = \sum_{n=1}^{\infty} \frac{2n(2n-1) x^{2n-2}}{2^n n!}$$

$$y'' - xy' - y = \sum_{n=1}^{\infty} \frac{2n(2n-1) x^{2n-2}}{2^n n!} - \sum_{n=1}^{\infty} \frac{2n x^{2n}}{2^n n!} - \sum_{n=0}^{\infty} \frac{x^{2n}}{2^n n!}$$

$$= \sum_{n=1}^{\infty} \frac{2n(2n-1) x^{2n-2}}{2^n n!} - \sum_{n=0}^{\infty} \frac{(2n+1) x^{2n}}{2^n n!}$$

$$= \sum_{n=0}^{\infty} \left[\frac{(2n+2)(2n+1) x^{2n}}{2^{n+1}(n+1)!} - \frac{(2n+1) x^{2n}}{2^n n!} \cdot \frac{2(n+1)}{2(n+1)} \right]$$

$$= \sum_{n=0}^{\infty} \frac{2(n+1) x^{2n} [(2n+1) - (2n+1)]}{2^{n+1}(n+1)!} = 0$$

51. $J_0(x) = \displaystyle\sum_{k=0}^{\infty} \frac{(-1)^k x^{2k}}{2^{2k}(k!)^2}$

(a) $\displaystyle\lim_{k \to \infty} \left| \frac{u_{k+1}}{u_k} \right| = \lim_{k \to \infty} \left| \frac{(-1)^{k+1} x^{2k+2}}{2^{2k+2}[(k+1)!]^2} \cdot \frac{2^{2k}(k!)^2}{(-1)^k x^{2k}} \right| = \lim_{k \to \infty} \left| \frac{(-1)x^2}{2^2(k+1)^2} \right| = 0$

Therefore, the interval of convergence is $-\infty < x < \infty$.

(b)
$$J_0 = \sum_{k=0}^{\infty} (-1)^k \frac{x^{2k}}{4^k (k!)^2}$$

$$J_0' = \sum_{k=1}^{\infty} (-1)^k \frac{2k x^{2k-1}}{4^k (k!)^2} = \sum_{k=0}^{\infty} (-1)^{k+1} \frac{(2k+2) x^{2k+1}}{4^{k+1}[(k+1)!]^2}$$

$$J_0'' = \sum_{k=1}^{\infty} (-1)^k \frac{2k(2k-1) x^{2k-2}}{4^k (k!)^2} = \sum_{k=0}^{\infty} (-1)^{k+1} \frac{(2k+2)(2k+1) x^{2k}}{4^{k+1}[(k+1)!]^2}$$

$$x^2 J_0'' + x J_0' + x^2 J_0 = \sum_{k=0}^{\infty} (-1)^{k+1} \frac{2(2k+1) x^{2k+2}}{4^{k+1}(k+1)! k!} + \sum_{k=0}^{\infty} (-1)^{k+1} \frac{2 x^{2k+2}}{4^{k+1}(k+1)! k!} + \sum_{k=0}^{\infty} (-1)^k \frac{x^{2k+2}}{4^k (k!)^2}$$

$$= \sum_{k=0}^{\infty} \frac{(-1)^k x^{2k+2}}{4^k (k!)^2} \left[(-1)\frac{2(2k+1)}{4(k+1)} + (-1)\frac{2}{4(k+1)} + 1 \right]$$

$$= \sum_{k=0}^{\infty} \frac{(-1)^k x^{2k+2}}{4^k (k!)^2} \left[\frac{-4k-2}{4k+4} - \frac{2}{4k+4} + \frac{4k+4}{4k+4} \right] = 0$$

—CONTINUED—

51. —CONTINUED—

(c) $P_6(x) = 1 - \dfrac{x^2}{4} + \dfrac{x^4}{64} - \dfrac{x^6}{2304}$

(d) $\displaystyle\int_0^1 J_0\,dx = \int_0^1 \sum_{k=0}^{\infty} \frac{(-1)^k x^{2k}}{4^k\,(k!)^2}\,dx$

$$= \left[\sum_{k=0}^{\infty} \frac{(-1)^k x^{2k+1}}{4^k(k!)^2(2k+1)} \right]_0^1$$

$$= \sum_{k=0}^{\infty} \frac{(-1)^k}{4^k(k!)^2(2k+1)}$$

$$= 1 - \frac{1}{12} + \frac{1}{320} \approx 0.92$$

(integral is approximately 0.9197304101)

53. $f(x) = \displaystyle\sum_{n=0}^{\infty} (-1)^n \frac{x^{2n}}{(2n)!} = \cos x$

(See Exercise 47.)

55. $f(x) = \displaystyle\sum_{n=0}^{\infty} (-1)^n x^n = \sum_{n=0}^{\infty} (-x)^n$

$$= \frac{1}{1-(-x)} = \frac{1}{1+x} \text{ for } -1 < x < 1$$

57. $\displaystyle\sum_{n=0}^{\infty} \left(\frac{x}{2}\right)^n$

(a) $\displaystyle\sum_{n=0}^{\infty} \left(\frac{3/4}{2}\right)^n = \sum_{n=0}^{\infty} \left(\frac{3}{8}\right)^n$

$$= \frac{1}{1-(3/8)} = \frac{8}{5} = 1.6$$

(b) $\displaystyle\sum_{n=0}^{\infty} \left(\frac{-3/4}{2}\right)^n = \sum_{n=0}^{\infty} \left(-\frac{3}{8}\right)^n$

$$= \frac{1}{1-(-3/8)} = \frac{8}{11} \approx 0.7272$$

(c) The alternating series converges more rapidly. The partial sums of the series of positive terms approach the sum from below. The partial sums of the alternating series alternate sides of the horizontal line representing the sum.

(d) $\displaystyle\sum_{n=0}^{N} \left(\frac{3}{2}\right)^n > M$

M	10	100	1000	10,000
N	4	9	15	21

59. False;

$$\sum_{n=0}^{\infty} \frac{(-1)^n x^n}{n2^n}$$

converges for $x = 2$ but diverges for $x = -2$.

61. True; the radius of convergence is $R = 1$ for both series.

Section 8.9   Representation of Functions by Power Series

1. (a) $\dfrac{1}{2-x} = \dfrac{1/2}{1-(x/2)} = \dfrac{a}{1-r}$

$\qquad = \displaystyle\sum_{n=0}^{\infty} \frac{1}{2}\left(\frac{x}{2}\right)^n = \sum_{n=0}^{\infty} \frac{x^n}{2^{n+1}}$

This series converges on $(-2, 2)$.

(b)
$$\frac{1}{2} + \frac{x}{4} + \frac{x^2}{8} + \frac{x^3}{16} + \cdots$$

$2-x\,\overline{)\,1}$

$\qquad \dfrac{1 - \dfrac{x}{2}}{}$

$\qquad\qquad \dfrac{x}{2}$

$\qquad\qquad \dfrac{x}{2} - \dfrac{x^2}{4}$

$\qquad\qquad\qquad \dfrac{x^2}{4}$

$\qquad\qquad\qquad \dfrac{x^2}{4} - \dfrac{x^3}{8}$

$\qquad\qquad\qquad\qquad \dfrac{x^3}{8}$

$\qquad\qquad\qquad\qquad \dfrac{x^3}{8} - \dfrac{x^4}{16}$

$\qquad\qquad\qquad\qquad\qquad \vdots$

3. (a) $\dfrac{1}{2+x} = \dfrac{1/2}{1-(-x/2)} = \dfrac{a}{1-r}$

$\qquad = \displaystyle\sum_{n=0}^{\infty} \frac{1}{2}\left(-\frac{x}{2}\right)^n = \sum_{n=0}^{\infty} \frac{(-1)^n x^n}{2^{n+1}}$

This series converges on $(-2, 2)$.

(b)
$$\frac{1}{2} - \frac{x}{4} + \frac{x^2}{8} - \frac{x^3}{16} + \cdots$$

$2+x\,\overline{)\,1}$

$\qquad \dfrac{1 + \dfrac{x}{2}}{}$

$\qquad\qquad -\dfrac{x}{2}$

$\qquad\qquad -\dfrac{x}{2} - \dfrac{x^2}{4}$

$\qquad\qquad\qquad \dfrac{x^2}{4}$

$\qquad\qquad\qquad \dfrac{x^2}{4} + \dfrac{x^3}{8}$

$\qquad\qquad\qquad\qquad -\dfrac{x^3}{8}$

$\qquad\qquad\qquad\qquad -\dfrac{x^3}{8} - \dfrac{x^4}{16}$

$\qquad\qquad\qquad\qquad\qquad \vdots$

5. Writing $f(x)$ in the form $a/(1-r)$, we have

$$\frac{1}{2-x} = \frac{1}{-3-(x-5)} = \frac{-1/3}{1+(1/3)(x-5)}$$

which implies that $a = -1/3$ and $r = (-1/3)(x-5)$.

Therefore, the power series for $f(x)$ is given by

$$\frac{1}{2-x} = \sum_{n=0}^{\infty} ar^n = \sum_{n=0}^{\infty} -\frac{1}{3}\left[-\frac{1}{3}(x-5)\right]^n$$

$$= \sum_{n=0}^{\infty} \frac{(x-5)^n}{(-3)^{n+1}}, \; |x-5| < 3 \text{ or } 2 < x < 8.$$

7. Writing $f(x)$ in the form $a/(1-r)$, we have

$$\frac{3}{2x-1} = \frac{-3}{1-2x} = \frac{a}{1-r}$$

which implies that $a = -3$ and $r = 2x$.

Therefore, the power series for $f(x)$ is given by

$$\frac{3}{2x-1} = \sum_{n=0}^{\infty} ar^n = \sum_{n=0}^{\infty} (-3)(2x)^n$$

$$= -3\sum_{n=0}^{\infty} (2x)^n, \; |2x| < 1 \text{ or } -\frac{1}{2} < x < \frac{1}{2}.$$

9. Writing $f(x)$ in the form $a/(1-r)$, we have

$$\frac{1}{2x-5} = \frac{-1}{11-2(x+3)}$$

$$= \frac{-1/11}{1-(2/11)(x+3)} = \frac{a}{1-r}$$

which implies that $a = -1/11$ and $r = (2/11)(x+3)$.
Therefore, the power series for $f(x)$ is given by

$$\frac{1}{2x-5} = \sum_{n=0}^{\infty} ar^n = \sum_{n=0}^{\infty} \left(-\frac{1}{11}\right)\left[\frac{2}{11}(x+3)\right]^n$$

$$= -\sum_{n=0}^{\infty} \frac{2^n(x+3)^n}{11^{n+1}},$$

$$|x+3| < \frac{11}{2} \text{ or } -\frac{17}{2} < x < \frac{5}{2}.$$

11. Writing $f(x)$ in the form $a/(1-r)$, we have

$$\frac{3}{x+2} = \frac{3}{2+x} = \frac{3/2}{1+(1/2)x} = \frac{a}{1-r}$$

which implies that $a = 3/2$ and $r = (-1/2)x$. Therefore, the power series for $f(x)$ is given by

$$\frac{3}{x+2} = \sum_{n=0}^{\infty} ar^n = \sum_{n=0}^{\infty} \frac{3}{2}\left(-\frac{1}{2}x\right)^n$$

$$= 3\sum_{n=0}^{\infty} \frac{(-1)^n x^n}{2^{n+1}} = \frac{3}{2}\sum_{n=0}^{\infty}\left(-\frac{x}{2}\right)^n,$$

$$|x| < 2 \text{ or } -2 < x < 2.$$

13. $\dfrac{3x}{x^2 + x - 2} = \dfrac{2}{x+2} + \dfrac{1}{x-1} = \dfrac{2}{2+x} + \dfrac{1}{-1+x} = \dfrac{1}{1+(1/2)x} + \dfrac{-1}{1-x}$

Writing $f(x)$ as a sum of two geometric series, we have

$$\frac{3x}{x^2 + x - 2} = \sum_{n=0}^{\infty}\left(-\frac{1}{2}x\right)^n + \sum_{n=0}^{\infty}(-1)(x)^n = \sum_{n=0}^{\infty}\left[\frac{1}{(-2)^n} - 1\right]x^n.$$

The interval of convergence is $-1 < x < 1$ since

$$\lim_{n\to\infty}\left|\frac{u_{n+1}}{u_n}\right| = \lim_{n\to\infty}\left|\frac{(1-(-2)^{n+1})x^{n+1}}{(-2)^{n+1}} \cdot \frac{(-2)^n}{(1-(-2)^n)x^n}\right| = \lim_{n\to\infty}\left|\frac{(1-(-2)^{n+1})x}{-2-(-2)^{n+1}}\right| = |x|.$$

15. $\dfrac{2}{1-x^2} = \dfrac{1}{1-x} + \dfrac{1}{1+x}$

Writing $f(x)$ as a sum of two geometric series, we have

$$\frac{2}{1-x^2} = \sum_{n=0}^{\infty} x^n + \sum_{n=0}^{\infty}(-x)^n = \sum_{n=0}^{\infty}(1+(-1)^n)x^n = \sum_{n=0}^{\infty} 2x^{2n}.$$

The interval of convergence is $|x^2| < 1$ or $-1 < x < 1$ since $\displaystyle\lim_{n\to\infty}\left|\frac{u_{n+1}}{u_n}\right| = \lim_{n\to\infty}\left|\frac{2x^{2n+2}}{2x^{2n}}\right| = |x^2|.$

17. $\dfrac{1}{1+x} = \displaystyle\sum_{n=0}^{\infty}(-1)^n x^n$

$$\frac{1}{1-x} = \sum_{n=0}^{\infty}(-1)^n(-x)^n = \sum_{n=0}^{\infty}(-1)^{2n}x^n = \sum_{n=0}^{\infty} x^n$$

$$h(x) = \frac{-2}{x^2-1} = \frac{1}{1+x} + \frac{1}{1-x} = \sum_{n=0}^{\infty}(-1)^n x^n + \sum_{n=0}^{\infty} x^n = \sum_{n=0}^{\infty}[(-1)^n + 1]x^n$$

$$= 2 + 0x + 2x^2 + 0x^3 + 2x^4 + 0x^5 + 2x^6 + \cdots = \sum_{n=0}^{\infty} 2x^{2n}, \; -1 < x < 1 \; \text{(See Exercise 15.)}$$

19. By taking the first derivative, we have $\dfrac{d}{dx}\left[\dfrac{1}{x+1}\right] = \dfrac{-1}{(x+1)^2}.$ Therefore,

$$\frac{-1}{(x+1)^2} = \frac{d}{dx}\left[\sum_{n=0}^{\infty}(-1)^n x^n\right] = \sum_{n=1}^{\infty}(-1)^n nx^{n-1}$$

$$= \sum_{n=0}^{\infty}(-1)^{n+1}(n+1)x^n, \; -1 < x < 1.$$

21. By integrating, we have $\displaystyle\int \frac{1}{x+1}\,dx = \ln(x+1).$ Therefore,

$$\ln(x+1) = \int\left[\sum_{n=0}^{\infty}(-1)^n x^n\right]dx = C + \sum_{n=0}^{\infty}\frac{(-1)^n x^{n+1}}{n+1}, \; -1 < x \le 1.$$

To solve for C, let $x = 0$ and conclude that $C = 0$. Therefore,

$$\ln(x+1) = \sum_{n=0}^{\infty}\frac{(-1)^n x^{n+1}}{n+1}, \; -1 < x \le 1.$$

23. $\dfrac{1}{x^2+1} = \displaystyle\sum_{n=0}^{\infty}(-1)^n(x^2)^n = \sum_{n=0}^{\infty}(-1)^n x^{2n}, \; -1 < x < 1$

25. Since, $\dfrac{1}{x+1} = \displaystyle\sum_{n=0}^{\infty}(-1)^n x^n$, we have $\dfrac{1}{4x^2+1} = \sum_{n=0}^{\infty}(-1)^n(4x^2)^n = \sum_{n=0}^{\infty}(-1)^n\,4^n x^{2n} = \sum_{n=0}^{\infty}(-1)^n(2x)^{2n}, \; -\dfrac{1}{2} < x < \dfrac{1}{2}.$

27. $x - \dfrac{x^2}{2} \le \ln(x + 1) \le x - \dfrac{x^2}{2} + \dfrac{x^3}{3}$

x	0.0	0.2	0.4	0.6	0.8	1.0
$x - \dfrac{x^2}{2}$	0.000	0.180	0.320	0.420	0.480	0.500
$\ln(x + 1)$	0.000	0.182	0.336	0.470	0.588	0.693
$x - \dfrac{x^2}{2} + \dfrac{x^3}{3}$	0.000	0.183	0.341	0.492	0.651	0.833

29. $g(x) = x$, line, Matches (c)   **31.** $g(x) = x - \dfrac{x^3}{3} + \dfrac{x^5}{5}$, Matches (a)   **33.** $f(x) = \arctan x$ is an odd function (symmetric to the origin)

In Exercises 35 and 37, arctan $x = \displaystyle\sum_{n=0}^{\infty} (-1)^n \dfrac{x^{2n+1}}{2n + 1}$.

35. $\arctan \dfrac{1}{4} = \displaystyle\sum_{n=0}^{\infty} (-1)^n \dfrac{(1/4)^{2n+1}}{2n + 1} = \sum_{n=0}^{\infty} \dfrac{(-1)^n}{(2n + 1)4^{2n+1}} = \dfrac{1}{4} - \dfrac{1}{192} + \dfrac{1}{5120} + \cdots$

Since $\dfrac{1}{5120} < 0.001$, we can approximate the series by its first two terms: $\arctan \dfrac{1}{4} \approx \dfrac{1}{4} - \dfrac{1}{192} \approx 0.245$.

37. $\dfrac{\arctan x^2}{x} = \dfrac{1}{x} \displaystyle\sum_{n=0}^{\infty} (-1)^n \dfrac{(x^2)^{2n+1}}{2n + 1} = \sum_{n=0}^{\infty} (-1)^n \dfrac{x^{4n+1}}{2n + 1}$

$\displaystyle\int \dfrac{\arctan x^2}{x}\, dx = \sum_{n=0}^{\infty} (-1)^n \dfrac{x^{4n+2}}{(4n + 2)(2n + 1)} + C$ (Note: $C = 0$)

$\displaystyle\int_0^{1/2} \dfrac{\arctan x^2}{x}\, dx = \sum_{n=0}^{\infty} (-1)^n \dfrac{1}{(4n + 2)(2n + 1)2^{4n+2}} = \dfrac{1}{8} - \dfrac{1}{1152} + \cdots$

Since $\dfrac{1}{1152} < 0.001$, we can approximate the series by its first term: $\displaystyle\int_0^{1/2} \dfrac{\arctan x^2}{x}\, dx \approx 0.125$

In Exercises 39 and 41, use $\dfrac{1}{1 - x} = \displaystyle\sum_{n=0}^{\infty} x^n, \; |x| < 1$.

39. (a) $\dfrac{1}{(1 - x)^2} = \dfrac{d}{dx}\left[\dfrac{1}{1 - x}\right] = \dfrac{d}{dx}\left[\displaystyle\sum_{n=0}^{\infty} x^n\right] = \sum_{n=1}^{\infty} nx^{n-1}, \; |x| < 1$

(b) $\dfrac{x}{(1 - x)^2} = x\displaystyle\sum_{n=1}^{\infty} nx^{n-1} = \sum_{n=1}^{\infty} nx^n, \; |x| < 1$

(c) $\dfrac{1 + x}{(1 - x)^2} = \dfrac{1}{(1 - x)^2} + \dfrac{x}{(1 - x)^2} = \displaystyle\sum_{n=1}^{\infty} n(x^{n-1} + x^n), \; |x| < 1$

$\qquad\qquad\qquad = \displaystyle\sum_{n=0}^{\infty} (2n + 1)x^n, \; |x| < 1$

(d) $\dfrac{x(1 + x)}{(1 - x)^2} = x\displaystyle\sum_{n=0}^{\infty} (2n + 1)x^n = \sum_{n=0}^{\infty} (2n + 1)x^{n+1}, \; |x| < 1$

41. $P(n) = \left(\dfrac{1}{2}\right)^n$

$E(n) = \displaystyle\sum_{n=1}^{\infty} nP(n) = \sum_{n=1}^{\infty} n\left(\dfrac{1}{2}\right)^n = \dfrac{1}{2}\sum_{n=1}^{\infty} n\left(\dfrac{1}{2}\right)^{n-1}$

$\quad = \dfrac{1}{2}\dfrac{1}{[1 - (1/2)]^2} = 2$

Since the probability of obtaining a head on a single toss is $\dfrac{1}{2}$, it is expected that, on average, a head will be obtained in two tosses.

43. Replace x with $(-x)$.

45. Replace x with $(-x)$ and multiply the series by 5.

47. Let $\arctan x + \arctan y = \theta$. Then,

$$\tan(\arctan x + \arctan y) = \tan \theta$$

$$\frac{\tan(\arctan x) + \tan(\arctan y)}{1 - \tan(\arctan x)\tan(\arctan y)} = \tan \theta$$

$$\frac{x + y}{1 - xy} = \tan \theta$$

$$\arctan\left(\frac{x + y}{1 - xy}\right) = \theta. \text{ Therefore, } \arctan x + \arctan y = \arctan\left(\frac{x + y}{1 - xy}\right) \text{ for } xy \neq 1.$$

49. (a) $2 \arctan \dfrac{1}{2} = \arctan \dfrac{1}{2} + \arctan \dfrac{1}{2} = \arctan\left[\dfrac{(1/2) + (1/2)}{1 - (1/2)^2}\right] = \arctan \dfrac{4}{3}$

$2 \arctan \dfrac{1}{2} - \arctan \dfrac{1}{7} = \arctan \dfrac{4}{3} + \arctan\left(-\dfrac{1}{7}\right) = \arctan\left[\dfrac{(4/3) - (1/7)}{1 + (4/3)(1/7)}\right] = \arctan \dfrac{25}{25} = \arctan 1 = \dfrac{\pi}{4}$

(b) $\pi = 8 \arctan \dfrac{1}{2} - 4 \arctan \dfrac{1}{7} \approx 8\left[\dfrac{1}{2} - \dfrac{(0.5)^3}{3} + \dfrac{(0.5)^5}{5} - \dfrac{(0.5)^7}{7}\right] - 4\left[\dfrac{1}{7} - \dfrac{(1/7)^3}{3} + \dfrac{(1/7)^5}{5} - \dfrac{(1/7)^7}{7}\right] \approx 3.14$

51. From Exercise 21, we have

$$\ln(x + 1) = \sum_{n=0}^{\infty} \frac{(-1)^n x^{n+1}}{n + 1} = \sum_{n=1}^{\infty} \frac{(-1)^{n-1} x^n}{n}$$

$$= \sum_{n=1}^{\infty} \frac{(-1)^{n+1} x^n}{n}.$$

Thus, $\displaystyle\sum_{n=1}^{\infty} (-1)^{n+1} \frac{1}{2^n n} = \sum_{n=1}^{\infty} \frac{(-1)^{n+1}(1/2)^n}{n}$

$$= \ln\left(\frac{1}{2} + 1\right) = \ln \frac{3}{2} \approx 0.4055$$

53. From Exercise 51, we have

$$\sum_{n=1}^{\infty} (-1)^{n+1} \frac{2^n}{5^n n} = \sum_{n=1}^{\infty} \frac{(-1)^{n+1}(2/5)^n}{n}$$

$$= \ln\left(\frac{2}{5} + 1\right) = \ln \frac{7}{5} \approx 0.3365.$$

55. From Exercise 54, we have

$$\sum_{n=0}^{\infty} (-1)^n \frac{1}{2^{2n+1}(2n + 1)} = \sum_{n=0}^{\infty} (-1)^n \frac{(1/2)^{2n+1}}{2n + 1} = \arctan \frac{1}{2} \approx 0.4636.$$

57. The series in Exercise 54 converges to its sum at a slower rate because its terms approach 0 at a much slower rate.

59. $f(x) = \displaystyle\sum_{n=1}^{\infty} (-1)^{n+1} \frac{(x - 1)^n}{n}, \quad 0 < x \leq 2$

$f(0.5) = \displaystyle\sum_{n=1}^{\infty} (-1)^{n+1} \frac{(-0.5)^n}{n} = \sum_{n=1}^{\infty} - \frac{(1/2)^n}{n}$

$\displaystyle\sum_{n=1}^{50} - \frac{(1/2)^n}{n} \approx -0.693147$

$\ln(0.5) \approx -0.693147$

Section 8.10   Taylor and Maclaurin Series

1. For $c = 0$, we have:

$$f(x) = e^{2x}$$

$$f^{(n)}(x) = 2^n e^{2x} \implies f^{(n)}(0) = 2^n$$

$$e^{2x} = 1 + 2x + \frac{4x^2}{2!} + \frac{8x^3}{3!} + \frac{16x^4}{4!} + \cdots = \sum_{n=0}^{\infty} \frac{(2x)^n}{n!}$$

3. For $c = \pi/4$, we have:

$$f(x) = \cos(x) \qquad f\left(\frac{\pi}{4}\right) = \frac{\sqrt{2}}{2}$$

$$f'(x) = -\sin(x) \qquad f'\left(\frac{\pi}{4}\right) = -\frac{\sqrt{2}}{2}$$

$$f''(x) = -\cos(x) \qquad f''\left(\frac{\pi}{4}\right) = -\frac{\sqrt{2}}{2}$$

$$f'''(x) = \sin(x) \qquad f'''\left(\frac{\pi}{4}\right) = \frac{\sqrt{2}}{2}$$

$$f^{(4)}(x) = \cos(x) \qquad f^{(4)}\left(\frac{\pi}{4}\right) = \frac{\sqrt{2}}{2}$$

and so on. Therefore, we have:

$$\cos x = \sum_{n=0}^{\infty} \frac{f^{(n)}(\pi/4)[x - (\pi/4)]^n}{n!}$$

$$= \frac{\sqrt{2}}{2}\left[1 - \left(x - \frac{\pi}{4}\right) - \frac{[x - (\pi/4)]^2}{2!} + \frac{[x - (\pi/4)]^3}{3!} + \frac{[x - (\pi/4)]^4}{4!} - \cdots\right]$$

$$= \frac{\sqrt{2}}{2}\sum_{n=0}^{\infty} \frac{(-1)^{n(n+1)/2}[x - (\pi/4)]^n}{n!}.$$

[**Note:** $(-1)^{n(n+1)/2} = 1, -1, -1, 1, 1, -1, -1, 1, \ldots$]

5. For $c = 1$, we have,

$$f(x) = \ln x \qquad f(1) = 0$$

$$f'(x) = \frac{1}{x} \qquad f'(1) = 1$$

$$f''(x) = -\frac{1}{x^2} \qquad f''(1) = -1$$

$$f'''(x) = \frac{2}{x^3} \qquad f'''(1) = 2$$

$$f^{(4)}(x) = -\frac{6}{x^4} \qquad f^{(4)}(1) = -6$$

$$f^{(5)}(x) = \frac{24}{x^5} \qquad f^{(5)}(1) = 24$$

and so on. Therefore, we have:

$$\ln x = \sum_{n=0}^{\infty} \frac{f^{(n)}(1)(x - 1)^n}{n!}$$

$$= 0 + (x - 1) - \frac{(x - 1)^2}{2!} + \frac{2(x - 1)^3}{3!} - \frac{6(x - 1)^4}{4!} + \frac{24(x - 1)^5}{5!} - \cdots$$

$$= (x - 1) - \frac{(x - 1)^2}{2} + \frac{(x - 1)^3}{3} - \frac{(x - 1)^4}{4} + \frac{(x - 1)^5}{5} - \cdots$$

$$= \sum_{n=0}^{\infty} (-1)^n \frac{(x - 1)^{n+1}}{n + 1}$$

7. For $c = 0$, we have:

$$f(x) = \sin 2x \qquad\qquad f(0) = 0$$
$$f'(x) = 2\cos 2x \qquad\qquad f'(0) = 2$$
$$f''(x) = -4\sin 2x \qquad\qquad f''(0) = 0$$
$$f'''(x) = -8\cos 2x \qquad\qquad f'''(0) = -8$$
$$f^{(4)}(x) = 16\sin 2x \qquad\qquad f^{(4)}(0) = 0$$
$$f^{(5)}(x) = 32\cos 2x \qquad\qquad f^{(5)}(0) = 32$$
$$f^{(6)}(x) = -64\sin 2x \qquad\qquad f^{(6)}(0) = 0$$
$$f^{(7)}(x) = -128\cos 2x \qquad\qquad f^{(7)}(0) = -128$$

and so on. Therefore, we have:

$$\sin 2x = \sum_{n=0}^{\infty} \frac{f^{(n)}(0)x^n}{n!} = 0 + 2x + \frac{0x^2}{2!} - \frac{8x^3}{3!} + \frac{0x^4}{4!} + \frac{32x^5}{5!} + \frac{0x^6}{6!} - \frac{128x^7}{7!} + \cdots$$

$$= 2x - \frac{8x^3}{3!} + \frac{32x^5}{5!} - \frac{128x^7}{7!} + \cdots = \sum_{n=0}^{\infty} \frac{(-1)^n(2x)^{2n+1}}{(2n+1)!}$$

9. For $c = 0$, we have:

$$f(x) = \sec(x) \qquad\qquad\qquad f(0) = 1$$
$$f'(x) = \sec(x)\tan(x) \qquad\qquad\qquad f'(0) = 0$$
$$f''(x) = \sec^3(x) + \sec(x)\tan^2(x) \qquad\qquad\qquad f''(0) = 1$$
$$f'''(x) = 5\sec^3(x)\tan(x) + \sec(x)\tan^3(x) \qquad\qquad\qquad f'''(0) = 0$$
$$f^{(4)}(x) = 5\sec^5(x) + 18\sec^3(x)\tan^2(x) + \sec(x)\tan^4(x) \qquad f^{(4)}(0) = 5$$

$$\sec(x) = \sum_{n=0}^{\infty} \frac{f^{(n)}(0)x^n}{n!} = 1 + \frac{x^2}{2!} + \frac{5x^4}{4!} + \cdots$$

11. The Maclaurin series for $f(x) = \cos x$ is $\sum_{n=0}^{\infty} \frac{(-1)x^{2n}}{(2n)!}$.

Because $f^{(n+1)}(x) = \pm\sin x$ or $\pm\cos x$, we have $|f^{(n+1)}(z)| \le 1$ for all z. Hence by Taylor's Theorem,

$$0 \le |R_n(x)| = \left|\frac{f^{(n+1)}(z)}{(n+1)!}x^{n+1}\right| \le \frac{|x|^{n+1}}{(n+1)!}.$$

Since $\lim_{n\to\infty} \frac{|x|^{n+1}}{(n+1)!} = 0$, it follows that $R_n(x) \to 0$ as $n \to \infty$. Hence, the Maclaurin series for $\cos x$ converges to $\cos x$ for all x.

13. Since $(1+x)^{-k} = 1 - kx + \frac{k(k+1)x^2}{2!} - \frac{k(k+1)(k+2)x^3}{3!} + \cdots$, we have

$$(1+x)^{-2} = 1 - 2x + \frac{2(3)x^2}{2!} - \frac{2(3)(4)x^3}{3!} + \frac{2(3)(4)(5)x^4}{5!} - \cdots = 1 - 2x + 3x^2 - 4x^3 + 5x^4 - \cdots$$

$$= \sum_{n=0}^{\infty} (-1)^n(n+1)x^n.$$

15. $\dfrac{1}{\sqrt{4+x^2}} = \left(\dfrac{1}{2}\right)\left[1 + \left(\dfrac{x}{2}\right)^2\right]^{-1/2}$ and since $(1+x)^{-1/2} = 1 + \displaystyle\sum_{n=1}^{\infty} \dfrac{(-1)^n 1 \cdot 3 \cdot 5 \cdots (2n-1)x^n}{2^n n!}$, we have

$$\dfrac{1}{\sqrt{4+x^2}} = \dfrac{1}{2}\left[1 + \sum_{n=1}^{\infty} \dfrac{(-1)^n 1 \cdot 3 \cdot 5 \cdots (2n-1)(x/2)^{2n}}{2^n n!}\right] = \dfrac{1}{2} + \sum_{n=1}^{\infty} \dfrac{(-1)^n 1 \cdot 3 \cdot 5 \cdots (2n-1)x^{2n}}{2^{3n+1}n!}.$$

17. Since $(1+x)^{1/2} = 1 + \dfrac{x}{2} + \displaystyle\sum_{n=2}^{\infty} \dfrac{(-1)^{n+1} 1 \cdot 3 \cdot 5 \cdots (2n-3)x^n}{2^n n!}$ (Exercise 14)

we have $(1+x^2)^{1/2} = 1 + \dfrac{x^2}{2} + \displaystyle\sum_{n=2}^{\infty} \dfrac{(-1)^{n+1} 1 \cdot 3 \cdot 5 \cdots (2n-3)x^{2n}}{2^n n!}.$

19. $e^x = \displaystyle\sum_{n=0}^{\infty} \dfrac{x^n}{n!} = 1 + x + \dfrac{x^2}{2!} + \dfrac{x^3}{3!} + \dfrac{x^4}{4!} + \dfrac{x^5}{5!} + \cdots$

$e^{x^2/2} = \displaystyle\sum_{n=0}^{\infty} \dfrac{(x^2/2)^n}{n!} = \displaystyle\sum_{n=0}^{\infty} \dfrac{x^{2n}}{2^n n!} = 1 + \dfrac{x^2}{2} + \dfrac{x^4}{2^2 2!} + \dfrac{x^6}{2^3 3!} + \dfrac{x^8}{2^4 4!} + \cdots$

21. $\sin x = \displaystyle\sum_{n=0}^{\infty} \dfrac{(-1)^n x^{2n+1}}{(2n+1)!} = x - \dfrac{x^3}{3!} + \dfrac{x^5}{5!} - \dfrac{x^7}{7!} + \cdots$

$\sin 2x = \displaystyle\sum_{n=0}^{\infty} \dfrac{(-1)^n (2x)^{2n+1}}{(2n+1)!} = \displaystyle\sum_{n=0}^{\infty} \dfrac{(-1)^n 2^{2n+1} x^{2n+1}}{(2n+1)!} = 2x - \dfrac{8x^3}{3!} + \dfrac{32x^5}{5!} - \dfrac{128x^7}{7!} + \cdots$

23. $\cos x = \displaystyle\sum_{n=0}^{\infty} \dfrac{(-1)^n x^{2n}}{(2n)!} = 1 - \dfrac{x^2}{2!} + \dfrac{x^4}{4!} - \cdots$

$\cos x^{3/2} = \displaystyle\sum_{n=0}^{\infty} \dfrac{(-1)^n (x^{3/2})^{2n}}{(2n)!} = \displaystyle\sum_{n=0}^{\infty} \dfrac{(-1)^n x^{3n}}{(2n)!} = 1 - \dfrac{x^3}{2!} + \dfrac{x^6}{4!} - \cdots$

25. $e^x = 1 + x + \dfrac{x^2}{2!} + \dfrac{x^3}{3!} + \dfrac{x^4}{4!} + \dfrac{x^5}{5!} + \cdots$

$e^{-x} = 1 - x + \dfrac{x^2}{2!} - \dfrac{x^3}{3!} + \dfrac{x^4}{4!} - \dfrac{x^5}{5!} + \cdots$

$e^x - e^{-x} = 2x + \dfrac{2x^3}{3!} + \dfrac{2x^5}{5!} + \dfrac{2x^7}{7!} + \cdots$

$\sinh(x) = \dfrac{1}{2}(e^x - e^{-x}) = x + \dfrac{x^3}{3!} + \dfrac{x^5}{5!} + \dfrac{x^7}{7!} + \cdots = \displaystyle\sum_{n=0}^{\infty} \dfrac{x^{2n+1}}{(2n+1)!}$

27. $\cos^2(x) = \dfrac{1}{2}[1 + \cos(2x)]$

$= \dfrac{1}{2}\left[1 + 1 - \dfrac{(2x)^2}{2!} + \dfrac{(2x)^4}{4!} - \dfrac{(2x)^6}{6!} - \cdots\right]$

$= \dfrac{1}{2}\left[1 + \displaystyle\sum_{n=0}^{\infty} \dfrac{(-1)^n (2x)^{2n}}{(2n)!}\right]$

29. $x \sin x = x\left(x - \dfrac{x^3}{3!} + \dfrac{x^5}{5!} - \cdots\right)$

$= x^2 - \dfrac{x^4}{3!} + \dfrac{x^6}{5!} - \cdots$

$= \displaystyle\sum_{n=0}^{\infty} \dfrac{(-1)^n x^{2n+2}}{(2n+1)!}$

31. $\dfrac{\sin x}{x} = \dfrac{x - (x^3/3!) + (x^5/5!) - \cdots}{x}$

$= 1 - \dfrac{x^2}{2!} + \dfrac{x^4}{4!} - \cdots$

$= \displaystyle\sum_{n=0}^{\infty} \dfrac{(-1)^n x^{2n}}{(2n+1)!}, x \neq 0$

33.
$$e^{ix} = 1 + ix + \frac{(ix)^2}{2!} + \frac{(ix)^3}{3!} + \frac{(ix)^4}{4!} + \cdots = 1 + ix - \frac{x^2}{2!} - \frac{ix^3}{3!} + \frac{x^4}{4!} + \frac{ix^5}{5!} - \frac{x^6}{6!} - \cdots$$

$$e^{-ix} = 1 - ix + \frac{(-ix)^2}{2!} + \frac{(-ix)^3}{3!} + \frac{(-ix)^4}{4!} + \cdots = 1 - ix - \frac{x^2}{2!} + \frac{ix^3}{3!} + \frac{x^4}{4!} - \frac{ix^5}{5!} - \frac{x^6}{6!} + \cdots$$

$$e^{ix} - e^{-ix} = 2ix - \frac{2ix^3}{3!} + \frac{2ix^5}{5!} - \frac{2ix^7}{7!} + \cdots$$

$$\frac{e^{ix} - e^{-ix}}{2i} = x - \frac{x^3}{3!} + \frac{x^5}{5!} - \frac{x^7}{7!} + \cdots = \sum_{n=0}^{\infty} \frac{(-1)^n x^{2n+1}}{(2n+1)!} = \sin(x)$$

35. $f(x) = e^x \sin x$

$$= \left(1 + x + \frac{x^2}{2} + \frac{x^3}{6} + \frac{x^4}{24} + \cdots\right)\left(x - \frac{x^3}{6} + \frac{x^5}{120} - \cdots\right)$$

$$= x + x^2 + \left(\frac{x^3}{2} - \frac{x^3}{6}\right) + \left(\frac{x^4}{6} - \frac{x^4}{6}\right) + \left(\frac{x^5}{120} - \frac{x^5}{12} + \frac{x^5}{24}\right) + \cdots$$

$$= x + x^2 + \frac{x^3}{3} - \frac{x^5}{30} + \cdots$$

37. $h(x) = \cos x \ln(1 + x)$

$$= \left(1 - \frac{x^2}{2} + \frac{x^4}{24} - \cdots\right)\left(x - \frac{x^2}{2} + \frac{x^3}{3} - \frac{x^4}{4} + \frac{x^5}{5} - \cdots\right)$$

$$= x - \frac{x^2}{2} + \left(\frac{x^3}{3} - \frac{x^3}{2}\right) + \left(\frac{x^4}{4} - \frac{x^4}{4}\right) + \left(\frac{x^5}{5} - \frac{x^5}{6} + \frac{x^5}{24}\right) + \cdots$$

$$= x - \frac{x^2}{2} - \frac{x^3}{6} + \frac{3x^5}{40} + \cdots$$

39. $g(x) = \dfrac{\sin x}{1 + x}$. Divide the series for $\sin x$ by $(1 + x)$. $g(x) = x - x^2 + \dfrac{5x^3}{6} - \dfrac{5x^4}{6} + \cdots$

$$
\begin{array}{r}
x - x^2 + \frac{5x^2}{6} - \frac{5x^4}{6} + \\[4pt]
1 + x \overline{\smash{\big)}\, x + 0x^2 - \frac{x^3}{6} + 0x^4 + \frac{x^5}{120} + \cdots} \\
\underline{x + x^2} \\
-x^2 - \frac{x^3}{6} \\
\underline{-x^2 - \frac{x^3}{6}} \\
\frac{5x^3}{6} + 0x^4 \\
\underline{\frac{5x^3}{6} + \frac{5x^4}{6}} \\
-\frac{5x^4}{6} + \frac{x^5}{120} \\
\underline{-\frac{5x^4}{6} - \frac{5x^5}{6}} \\
\vdots
\end{array}
$$

41. $y = x^2 - \dfrac{x^4}{3!} = x\left(x - \dfrac{x^3}{3!}\right) \approx x \sin x$.

Matches (a)

43. $y = x + x^2 + \dfrac{x^3}{2!} = x\left(1 + x + \dfrac{x^2}{2!}\right) \approx xe^x$.

Matches (c)

45. $\int_0^x (e^{-t^2} - 1)\, dt = \int_0^x \left[\left(\sum_{n=0}^{\infty} \frac{(-1)^n t^{2n}}{n!} \right) - 1 \right] dt$

$= \int_0^x \left[\sum_{n=0}^{\infty} \frac{(-1)^{n+1} t^{2n+2}}{(n+1)!} \right] dt = \left[\sum_{n=0}^{\infty} \frac{(-1)^{n+1} t^{2n+3}}{(2n+3)(n+1)!} \right]_0^x = \sum_{n=0}^{\infty} \frac{(-1)^{n+1} x^{2n+3}}{(2n+3)(n+1)!}$

47. Since $\ln x = \sum_{n=0}^{\infty} \frac{(-1)^n (x-1)^{n+1}}{n+1} = (x-1) - \frac{(x-1)^2}{2} + \frac{(x-1)^3}{3} - \frac{(x-1)^4}{4} + \cdots \quad (0 < x \le 2)$

we have $\ln 2 = 1 - \frac{1}{2} + \frac{1}{3} - \frac{1}{4} + \cdots = \sum_{n=1}^{\infty} (-1)^{n+1} \frac{1}{n} \approx 0.6931.$ (10,001 terms)

49. Since $e^x = \sum_{n=0}^{\infty} \frac{x^n}{n!} = 1 + x + \frac{x^2}{2!} + \frac{x^3}{3!} + \cdots,$

we have $e^2 = 1 + 2 + \frac{2^2}{2!} + \frac{2^3}{3!} + \cdots = \sum_{n=0}^{\infty} \frac{2^n}{n!} \approx 7.3891.$ (12 terms)

51. Since

$$\cos x = \sum_{n=0}^{\infty} \frac{(-1)^n x^{2n}}{(2n)!} = 1 - \frac{x^2}{2!} + \frac{x^4}{4!} - \frac{x^6}{6!} + \frac{x^8}{8!} - \cdots$$

$$1 - \cos x = \frac{x^2}{2!} - \frac{x^4}{4!} + \frac{x^6}{6!} - \frac{x^8}{8!} + \cdots = \sum_{n=0}^{\infty} \frac{(-1)^n x^{2n+2}}{(2n+2)!}$$

$$\frac{1 - \cos}{x} = \frac{x}{2!} - \frac{x^3}{4!} + \frac{x^5}{6!} - \frac{x^7}{8!} + \cdots = \sum_{n=0}^{\infty} \frac{(-1)^n x^{2n+1}}{(2n+2)!}$$

we have $\lim_{x \to 0} \frac{1 - \cos x}{x} = \lim_{x \to 0} \sum_{n=0}^{\infty} \frac{(-1) x^{2n+1}}{(2n+2)!} = 0.$

53. $\int_0^1 \frac{\sin x}{x}\, dx = \int_0^1 \left[\sum_{n=0}^{\infty} \frac{(-1)^n x^{2n}}{(2n+1)!} \right] dx = \left[\sum_{n=0}^{\infty} \frac{(-1)^n x^{2n+1}}{(2n+1)(2n+1)!} \right]_0^1 = \sum_{n=0}^{\infty} \frac{(-1)^n}{(2n+1)(2n+1)!}$

Since $1/(7 \cdot 7!) < 0.0001$, we need three terms:

$$\int_0^1 \frac{\sin x}{x}\, dx = 1 - \frac{1}{3 \cdot 3!} + \frac{1}{5 \cdot 5!} - \cdots \approx 0.9461.$$ (Using three non-zero terms)

Note: We are using $\lim_{x \to 0^+} \frac{\sin x}{x} = 1.$

55. $\int_0^{\pi/2} \sqrt{x} \cos x\, dx = \int_0^{\pi/2} \left[\sum_{n=0}^{\infty} \frac{(-1)^n x^{(4n+1)/2}}{(2n)!} \right] dx = \left[\sum_{n=0}^{\infty} \frac{(-1)^n x^{(4n+3)/2}}{\left(\frac{4n+3}{2} \right)(2n)!} \right]_0^{\pi/2} = \left[\sum_{n=0}^{\infty} \frac{(-1)^n 2 x^{(4n+3)/2}}{(4n+3)(2n)!} \right]_0^{\pi/2}$

Since $2(\pi/2)^{23/2}/23.10! < 0.0001$, we need five terms

$$\int_0^1 \sqrt{x} \cos x\, dx = 2 \left[\frac{(\pi/2)^{3/2}}{3} - \frac{(\pi/2)^{7/2}}{14} + \frac{(\pi/2)^{11/2}}{264} - \frac{(\pi/2)^{15/2}}{10,800} + \frac{(\pi/2)^{19/2}}{766,080} \right] \approx 0.7040.$$

57. $\int_{0.1}^{0.3} \sqrt{1 + x^3}\, dx = \int_{0.1}^{0.3} \left(1 + \frac{x^3}{2} - \frac{x^6}{8} + \frac{x^9}{16} - \frac{5x^{12}}{128} + \cdots \right) dx = \left[x + \frac{x^4}{8} - \frac{x^7}{56} + \frac{x^{10}}{160} - \frac{5x^{13}}{1664} + \cdots \right]_{0.1}^{0.3}$

Since $\frac{1}{56}(0.3^7 - 0.1^7) < 0.0001$, we need two terms

$$\int_{0.1}^{0.3} \sqrt{1 + x^3}\, dx = \left[(0.3 - 0.1) + \frac{1}{8}(0.3^4 - 0.1^4) \right] \approx 0.201.$$

59. From Exercise 19, we have

$$\frac{1}{\sqrt{2\pi}}\int_0^1 e^{-x^2/2}\,dx = \frac{1}{\sqrt{2\pi}}\int_0^1 \sum_{n=0}^{\infty}\frac{(-1)^n x^{2n}}{2^n n!}\,dx = \frac{1}{\sqrt{2\pi}}\left[\sum_{n=0}^{\infty}\frac{(-1)^n x^{2n+1}}{2^n n!(2n+1)}\right]_0^1 = \frac{1}{\sqrt{2\pi}}\sum_{n=0}^{\infty}\frac{(-1)^n}{2^n n!(2n+1)}$$

$$\approx \frac{1}{\sqrt{2\pi}}\left[1 - \frac{1}{2\cdot 1\cdot 3} + \frac{1}{2^2\cdot 2!\cdot 5} - \frac{1}{2^3\cdot 3!\cdot 7}\right] \approx 0.3414.$$

61. $f(x) = x\cos 2x = \displaystyle\sum_{n=0}^{\infty}\frac{(-1)^n 4^n x^{2n+1}}{(2n)!}$

$$P_5(x) = x - 2x^3 + \frac{2x^5}{3}$$

The polynomial is a reasonable approximation on the interval $\left[-\frac{3}{4}, \frac{3}{4}\right]$.

63. $f(x) = \sqrt{x}\ln x,\ c = 1$

$$P_5(x) = (x-1) - \frac{(x-1)^3}{24} + \frac{(x-1)^4}{24} - \frac{71(x-1)^5}{1920}$$

The polynomial is a reasonable approximation on the interval $\left[\frac{1}{4}, 2\right]$.

65. See Guidelines, page 636.

67. (a) Replace x with $(-x)$.

    (c) Multiply series by x.

    (b) Replace x with $3x$.

    (d) Replace x with $2x$, then replace x with $-2x$, and add the two together.

69. $y = \left(\tan\theta - \dfrac{g}{kv_0\cos\theta}\right)x - \dfrac{g}{k^2}\ln\left(1 - \dfrac{kx}{v_0\cos\theta}\right)$

$$= (\tan\theta)x - \frac{gx}{kv_0\cos\theta} - \frac{g}{k^2}\left[-\frac{kx}{v_0\cos\theta} - \frac{1}{2}\left(\frac{kx}{v_0\cos\theta}\right)^2 - \frac{1}{3}\left(\frac{kx}{v_0\cos\theta}\right)^3 - \frac{1}{4}\left(\frac{kx}{v_o\cos\theta}\right)^4 - \cdots\right]$$

$$= (\tan\theta)x - \frac{gx}{kv_0\cos\theta} + \frac{gx}{kv_0\cos\theta} + \frac{gx^2}{2v_0^2\cos^2\theta} + \frac{gkx^3}{3v_0^3\cos^3\theta} + \frac{gk^2x^4}{4v_0^4\cos^4\theta} + \cdots\Bigg]$$

$$= (\tan\theta)x + \frac{gx^2}{2v_0^2\cos^2\theta} + \frac{kgx^3}{3v_0^3\cos^3\theta} + \frac{k^2gx^4}{4v_0^4\cos^4\theta} + \cdots$$

71. $f(x) = \begin{cases} e^{-1/x^2}, & x \neq 0 \\ 0, & x = 0 \end{cases}$

(a)

(b) $f'(0) = \displaystyle\lim_{x\to 0}\frac{f(x)-f(0)}{x-0} = \lim_{x\to 0}\frac{e^{-1/x^2}-0}{x}$

Let $y = \displaystyle\lim_{x\to 0}\frac{e^{-1/x^2}}{x}$. Then

$$\ln y = \lim_{x\to 0}\ln\left(\frac{e^{-1/x^2}}{x}\right) = \lim_{x\to 0^+}\left[-\frac{1}{x^2} - \ln x\right] = \lim_{x\to 0^+}\left[\frac{-1 - x^2\ln x}{x^2}\right] = -\infty.$$

Thus, $y = e^{-\infty} = 0$ and we have $f'(0) = 0$.

(c) $\displaystyle\sum_{n=0}^{\infty}\frac{f^{(n)}(0)}{n!}x^n = f(0) + \frac{f'(0)x}{1!} + \frac{f''(0)x^2}{2!} + \cdots = 0 \neq f(x)$

This series converges to f at $x = 0$ only.

73. By the Ratio Test: $\displaystyle\lim_{n\to\infty}\left|\frac{x^{n+1}}{(n+1)!}\cdot\frac{n!}{x^n}\right| = \lim_{n\to\infty}\frac{|x|}{n+1} = 0$ which shows that $\displaystyle\sum_{n=0}^{\infty}\frac{x^n}{n!}$ converges for all x.

Review Exercises for Chapter 8

1. $a_n = \dfrac{1}{n!}$

3. $a_n = 4 + \dfrac{2}{n}$: $6, 5, 4.67, \ldots$

Matches (a)

5. $a_n = 10(0.3)^{n-1}$: $10, 3, \ldots$

Matches (d)

7. $a_n = \dfrac{5n + 2}{n}$

The sequence seems to converge to 5.

$$\lim_{n\to\infty} a_n = \lim_{n\to\infty} \frac{5n + 2}{n}$$

$$= \lim_{n\to\infty} \left(5 + \frac{2}{n}\right) = 5$$

9. $\displaystyle\lim_{n\to\infty} \frac{n + 1}{n^2} = 0$

Converges

11. $\displaystyle\lim_{n\to\infty} \frac{n^3}{n^2 + 1} = \infty$

13. $\displaystyle\lim_{n\to\infty} \left(\sqrt{n+1} - \sqrt{n}\right) = \lim_{n\to\infty} \left(\sqrt{n+1} - \sqrt{n}\right)\frac{\sqrt{n+1} + \sqrt{n}}{\sqrt{n+1} + \sqrt{n}} = \lim_{n\to\infty} \frac{1}{\sqrt{n+1} + \sqrt{n}} = 0$ Converges

15. $\displaystyle\lim_{n\to\infty} \frac{\sin\sqrt{n}}{\sqrt{n}} = 0$

Converges

17. $A_n = 5000\left(1 + \dfrac{0.05}{4}\right)^n = 5000(1.0125)^n$

$n = 1, 2, 3$

(a) $A_1 = 5062.50$ $A_5 \approx 5320.41$

$A_2 \approx 5125.78$ $A_6 \approx 5386.92$

$A_3 \approx 5189.85$ $A_7 \approx 5454.25$

$A_4 \approx 5254.73$ $A_8 \approx 5522.43$

(b) $A_{40} \approx 8218.10$

19. (a)

k	5	10	15	20	25
S_k	13.2	113.3	873.8	6448.5	50,500.3

The series diverges $\left(\text{geometric } r = \dfrac{3}{2} > 1\right)$

(b)

21. (a)

k	5	10	15	20	25
S_k	0.4597	0.4597	0.4597	0.4597	0.4597

The series converges by the Alternating Series Test.

(b)

23. Converges. Geometric series, $r = 0.82$, $|r| < 1$.

25. Diverges. nth Term Test. $\displaystyle\lim_{n\to\infty} a_n \neq 0$.

27. $\displaystyle\sum_{n=0}^{\infty}\left(\frac{2}{3}\right)^n$

Geometric series with $a = 1$ and $r = \frac{2}{3}$.

$$S = \frac{a}{1-r} = \frac{1}{1-(2/3)} = \frac{1}{1/3} = 3$$

29. $\displaystyle\sum_{n=0}^{\infty}\left(\frac{1}{2^n} - \frac{1}{3^n}\right) = \sum_{n=0}^{\infty}\left(\frac{1}{2}\right)^n - \sum_{n=0}^{\infty}\left(\frac{1}{3}\right)^n$

$$= \frac{1}{1-(1/2)} - \frac{1}{1-(1/3)} = 2 - \frac{3}{2} = \frac{1}{2}$$

31. $0.\overline{09} = 0.09 + 0.0009 + 0.000009 + \cdots = 0.09(1 + 0.01 + 0.0001 + \cdots) = \displaystyle\sum_{n=0}^{\infty}(0.09)(0.01)^n = \frac{0.09}{1-0.01} = \frac{1}{11}$

33. $D_1 = 8$

$D_2 = 0.7(8) + 0.7(8) = 16(0.7)$

$\vdots$

$D = 8 + 16(0.7) + 16(0.7)^2 + \cdots + 16(0.7)^n + \cdots$

$\quad = -8 + \displaystyle\sum_{n=0}^{\infty}16(0.7)^n = -8 + \frac{16}{1-0.7} = 45\frac{1}{3}$ meters

35. See Exercise 86 in Section 8.2.

$$A = \frac{P(e^{rt}-1)}{e^{r/12}-1}$$

$$= \frac{200(e^{(0.06)(2)}-1)}{e^{0.06/12}-1}$$

$$\approx \$5087.14$$

37. $\displaystyle\int_1^{\infty}x^{-4}\ln(x)\,dx = \lim_{b\to\infty}\left[-\frac{\ln x}{3x^3} - \frac{1}{9x^3}\right]_1^b$

$$= 0 + \frac{1}{9} = \frac{1}{9}$$

By the Integral Test, the series converges.

39. $\displaystyle\sum_{n=1}^{\infty}\left(\frac{1}{n^2} - \frac{1}{n}\right) = \sum_{n=1}^{\infty}\frac{1}{n^2} - \sum_{n=1}^{\infty}\frac{1}{n}$

Since the second series is a divergent p-series while the first series is a convergent p-series, the difference diverges.

41. $\displaystyle\sum_{n=1}^{\infty}\frac{1}{\sqrt{n^3+2n}}$

$$\lim_{n\to\infty}\frac{1/\sqrt{n^3+2n}}{1/(n^{3/2})} = \lim_{n\to\infty}\frac{n^{3/2}}{\sqrt{n^3+2n}} = 1$$

By a limit comparison test with the convergent p-series

$\displaystyle\sum_{n=1}^{\infty}\frac{1}{n^{3/2}}$, the series converges.

43. $\displaystyle\sum_{n=1}^{\infty}\frac{1\cdot3\cdot5\cdots(2n-1)}{2\cdot4\cdot6\cdots(2n)}$

$$a_n = \frac{1\cdot3\cdot5\cdots(2n-1)}{2\cdot4\cdot6\cdots(2n)}$$

$$= \left(\frac{3}{2}\cdot\frac{5}{4}\cdots\frac{2n-1}{2n-2}\right)\frac{1}{2n} > \frac{1}{2n}$$

Since $\displaystyle\sum_{n=1}^{\infty}\frac{1}{2n} = \frac{1}{2}\sum_{n=1}^{\infty}\frac{1}{n}$ diverges (harmonic series), so does the original series.

45. Converges by the Alternating Series Test (Conditional convergence)

47. Diverges by the nth Term Test

49. $\displaystyle\sum_{n=1}^{\infty}\frac{n}{e^{n^2}}$

$$\lim_{n\to\infty}\left|\frac{a_{n+1}}{a_n}\right| = \lim_{n\to\infty}\left|\frac{n+1}{e^{(n+1)^2}}\cdot\frac{e^{n^2}}{n}\right|$$

$$= \lim_{n\to\infty}\left|\frac{e^{n^2}(n+1)}{e^{n^2+2n+1}n}\right|$$

$$= \lim_{n\to\infty}\left(\frac{1}{e^{2n+1}}\right)\left(\frac{n+1}{n}\right)$$

$$= (0)(1) = 0 < 1$$

By the Ratio Test, the series converges.

51. $\displaystyle\sum_{n=1}^{\infty}\frac{2^n}{n^3}$

$$\lim_{n\to\infty}\left|\frac{a_{n+1}}{a_n}\right| = \lim_{n\to\infty}\left|\frac{2^{n+1}}{(n+1)^3}\cdot\frac{n^3}{2^n}\right|$$

$$= \lim_{n\to\infty}\frac{2n^3}{(n+1)^3} = 2$$

Therefore, by the Ratio Test, the series diverges.

53. (a) Ratio Test: $\lim\limits_{n\to\infty}\left|\dfrac{a_{n+1}}{a_n}\right| = \lim\limits_{n\to\infty}\dfrac{(n+1)(3/5)^{n+1}}{n(3/5)^n}$

$$= \lim_{n\to\infty}\left(\frac{n+1}{n}\right)\left(\frac{3}{5}\right) = \frac{3}{5} < 1$$

Converges

(b)

x	5	10	15	20	25
S_n	2.8752	3.6366	3.7377	3.7488	3.7499

(c)

(d) The sum is approximately 3.75.

55. (a) $\displaystyle\int_N^\infty \frac{1}{x^2}\,dx = \left[-\frac{1}{x}\right]_N^\infty = \frac{1}{N}$

N	5	10	20	30	40
$\displaystyle\sum_{n=1}^{N}\frac{1}{n^2}$	1.4636	1.5498	1.5962	1.6122	1.6202
$\displaystyle\int_N^\infty \frac{1}{x^2}\,dx$	0.2000	0.1000	0.0500	0.0333	0.0250

(b) $\displaystyle\int_N^\infty \frac{1}{x^5}\,dx = \left[-\frac{1}{4x^4}\right]_N^\infty = \frac{1}{4N^4}$

N	5	10	20	30	40
$\displaystyle\sum_{n=1}^{N}\frac{1}{n^5}$	1.0367	1.0369	1.0369	1.0369	1.0369
$\displaystyle\int_N^\infty \frac{1}{x^5}\,dx$	0.0004	0.0000	0.0000	0.0000	0.0000

The series in part (b) converges more rapidly. The integral values represent the remainders of the partial sums.

57. $f(x) = e^{-x/2}$ $\qquad f(0) = 1$

$f'(x) = -\dfrac{1}{2}e^{-x/2}$ $\qquad f'(0) = -\dfrac{1}{2}$

$f''(x) = \dfrac{1}{4}e^{-x/2}$ $\qquad f''(0) = \dfrac{1}{4}$

$f'''(x) = -\dfrac{1}{8}e^{-x/2}$ $\qquad f'''(0) = -\dfrac{1}{8}$

$P_3(x) = f(0) + f'(0)x + f''(0)\dfrac{x^2}{2!} + f'''(0)\dfrac{x^3}{3!}$

$\qquad = 1 - \dfrac{1}{2}x + \dfrac{1}{4}\dfrac{x^2}{2!} - \dfrac{1}{8}\dfrac{x^3}{3!}$

$\qquad = 1 - \dfrac{1}{2}x + \dfrac{1}{8}x^2 - \dfrac{1}{48}x^3$

59. Since $\dfrac{(95\pi)^9}{180^9\cdot 9!} < 0.001$, use four terms $\sin(95°) = \sin\left(\dfrac{95\pi}{180}\right) \approx \dfrac{95\pi}{180} - \dfrac{(95\pi)^3}{180^3 3!} + \dfrac{(95\pi)^5}{180^5 5!} - \dfrac{(95\pi)^7}{180^7 7!} \approx 0.99594$

61. $\ln(1.75) \approx (0.75) - \dfrac{(0.75)^2}{2} + \dfrac{(0.75)^3}{3} - \dfrac{(0.75)^4}{4} + \dfrac{(0.75)^5}{5} - \dfrac{(0.75)^6}{6} + \cdots - \dfrac{(0.75)^{14}}{14} \approx 0.559062$

63. $f(x) = \cos x, \ c = 0$

$$R_n(x) = \frac{f^{(n+1)}(z)}{(n+1)!} x^{n+1}$$

$$|f^{(n+1)}(z)| \le 1 \implies R_n(x) \le \frac{x^{n+1}}{(n+1)!}$$

(a) $R_n(x) \le \dfrac{(0.5)^{n+1}}{(n+1)!} < 0.001$

This inequality is true for $n = 4$.

(c) $R_n(x) \le \dfrac{(0.5)^{n+1}}{(n+1)!} < 0.0001$

This inequality is true for $n = 5$.

(b) $R_n(x) \le \dfrac{(1)^{n+1}}{(n+1)!} < 0.001$

This inequality is true for $n = 6$.

(d) $R_n(x) \le \dfrac{2^{n+1}}{(n+1)!} < 0.0001$

This inequality is true for $n = 10$.

65. $\displaystyle\sum_{n=0}^{\infty} \left(\frac{x}{10}\right)^n$

Geometric series which converges only if $|x/10| < 1$ or $-10 < x < 10$.

67. $\displaystyle\sum_{n=0}^{\infty} \frac{(-1)^n (x-2)^n}{(n+1)^2}$

$$\lim_{n\to\infty} \left| \frac{u_{n+1}}{u_n} \right| = \lim_{n\to\infty} \left| \frac{(-1)^{n+1}(x-2)^{n+1}}{(n+2)^2} \cdot \frac{(n+1)^2}{(-1)^n(x-2)^n} \right|$$

$$= |x-2|$$

$R = 1$

Center: 2

Since the series converges when $x = 1$ and when $x = 3$, the interval of convergence is $1 \le x \le 3$.

69. $\displaystyle\sum_{n=0}^{\infty} n!(x-2)^n$

$$\lim_{n\to\infty} \left| \frac{u_{n+1}}{u_n} \right| = \lim_{n\to\infty} \left| \frac{(n+1)!(x-2)^{n+1}}{n!(x-2)^n} \right| = \infty$$

which implies that the series converges only at the center $x = 2$.

71.

$$y = \sum_{n=0}^{\infty} (-1)^n \frac{x^{2n}}{4^n (n!)^2}$$

$$y' = \sum_{n=1}^{\infty} \frac{(-1)^n (2n) x^{2n-1}}{4^n (n!)^2} = \sum_{n=0}^{\infty} \frac{(-1)^{n+1}(2n+2)x^{2n+1}}{4^{n+1}[(n+1)!]^2}$$

$$y'' = \sum_{n=0}^{\infty} \frac{(-1)^{n+1}(2n+2)(2n+1)x^{2n}}{4^{n+1}[(n+1)!]^2}$$

$$x^2 y'' + xy' + x^2 y = \sum_{n=0}^{\infty} \frac{(-1)^{n+1}(2n+2)(2n+1)x^{2n+2}}{4^{n+1}[(n+1)!]^2} + \sum_{n=0}^{\infty} \frac{(-1)^{n+1}(2n+2)x^{2n+2}}{4^{n+1}[(n+1)!]^2} + \sum_{n=0}^{\infty} (-1)^n \frac{x^{2n+2}}{4^n (n!)^2}$$

$$= \sum_{n=0}^{\infty} \left[(-1)^{n+1} \frac{(2n+2)(2n+1)}{4^{n+1}[(n+1)!]^2} + \frac{(-1)^{n+1}(2n+2)}{4^{n+1}[(n+1)!]^2} + \frac{(-1)^n}{4^n (n!)^2} \right] x^{2n+2}$$

$$= \sum_{n=0}^{\infty} \left[\frac{(-1)^{n+1}(2n+2)(2n+1+1)}{4^{n+1}[(n+1)!]^2} + (-1)^n \frac{1}{4^n (n!)^2} \right] x^{2n+2}$$

$$= \sum_{n=0}^{\infty} \left[\frac{(-1)^{n+1} 4(n+1)^2}{4^{n+1}[(n+1)!]^2} + (-1)^n \frac{1}{4^n (n!)^2} \right] x^{2n+2}$$

$$= \sum_{n=0}^{\infty} \left[\frac{(-1)^{n+1} 1}{4^n (n!)^2} + (-1)^n \frac{1}{4^n (n!)^2} \right] x^{2n+2} = 0$$

73. $\dfrac{2}{3-x} = \dfrac{2/3}{1-(x/3)} = \dfrac{a}{1-r}$

$$\sum_{n=0}^{\infty} \frac{2}{3} \left(\frac{x}{3}\right)^n = \sum_{n=0}^{\infty} \frac{2x^n}{3^{n+1}}$$

75. Derivative: $\displaystyle\sum_{n=1}^{\infty} \frac{2n x^{n-1}}{3^{n+1}}$

77. $1 + \frac{2}{3}x + \frac{4}{9}x^2 + \frac{8}{27}x^3 + \cdots = \sum_{n=0}^{\infty} \left(\frac{2x}{3}\right)^n = \frac{1}{1-(2x/3)} = \frac{3}{3-2x}, \quad -\frac{3}{2} < x < \frac{3}{2}$

79. $f(x) = \sin(x)$

$f'(x) = \cos(x)$

$f''(x) = -\sin(x)$

$f'''(x) = -\cos(x), \cdots$

$\sin(x) = \sum_{n=0}^{\infty} \frac{f^{(n)}(x)[x-(3\pi/4)]^n}{n!}$

$\quad = \frac{\sqrt{2}}{2} - \frac{\sqrt{2}}{2}\left(x - \frac{3\pi}{4}\right) - \frac{\sqrt{2}}{2\cdot 2!}\left(x - \frac{3\pi}{4}\right)^2 + \cdots = \frac{\sqrt{2}}{2}\sum_{n=0}^{\infty} \frac{(-1)^{n(n+1)/2}[x-(3\pi/4)]^n}{n!}$

81. $3^x = (e^{\ln(3)})^x = e^{x\ln(3)}$ and since $e^x = \sum_{n=0}^{\infty} \frac{x^n}{n!}$, we have

$\quad 3^x = \sum_{n=0}^{\infty} \frac{(x\ln 3)^n}{n!}$

$\quad\quad = 1 + x\ln 3 + \frac{x^2 \ln^2 3}{2!} + \frac{x^3 \ln^3 3}{3!} + \frac{x^4 \ln^4 3}{4!} + \cdots.$

83. $f(x) = \frac{1}{x}$

$f'(x) = -\frac{1}{x^2}$

$f''(x) = \frac{2}{x^3}$

$f'''(x) = -\frac{6}{x^4}, \cdots$

$\frac{1}{x} = \sum_{n=0}^{\infty} \frac{f^{(n)}(-1)(x+1)^n}{n!}$

$\quad = \sum_{n=0}^{\infty} \frac{-n!(x+1)^n}{n!} = -\sum_{n=0}^{\infty}(x+1)^n, \ -2 < x < 0$

85. $(1+x)^k = 1 + kx + \frac{k(k-1)x^2}{2!} + \frac{k(k-1)(k-2)x^3}{3!} + \cdots$

$\quad (1+x)^{1/5} = 1 + \frac{x}{5} + \frac{(1/5)(-4/5)x^2}{2!} + \frac{1/5(-4/5)(-9/5)x^3}{3!} + \cdots$

$\quad\quad = 1 + \frac{1}{5}x - \frac{1\cdot 4x^2}{5^2 2!} + \frac{1\cdot 4\cdot 9x^3}{5^3 3!} - \cdots$

$\quad\quad = 1 + \frac{x}{5} + \sum_{n=2}^{\infty} \frac{(-1)^{n+1}4\cdot 9\cdot 14\cdots(5n-6)x^n}{5^n n!}$

$\quad\quad = 1 + \frac{x}{5} - \frac{2}{25}x^2 + \frac{6}{125}x^3 - \cdots$

87. $\ln x = \sum_{n=1}^{\infty} (-1)^{n+1}\frac{(x-1)^n}{n}, \quad 0 < x \le 2$

$\quad \ln\left(\frac{5}{4}\right) = \sum_{n=1}^{\infty} (-1)^{n+1}\left(\frac{(5/4)-1}{n}\right)^n$

$\quad\quad = \sum_{n=1}^{\infty} (-1)^{n+1}\frac{1}{4^n n} \approx 0.2231$

89. $e^x = \sum_{n=0}^{\infty} \frac{x^n}{n!}, \quad -\infty < x < \infty$

$\quad e^{1/2} = \sum_{n=0}^{\infty} \frac{(1/2)^n}{n!} = \sum_{n=0}^{\infty} \frac{1}{2^n n!} \approx 1.6487$

91. $\cos x = \sum\limits_{n=0}^{\infty} (-1)^n \dfrac{x^{2n}}{(2n)!}, \quad -\infty < x < \infty$

$\cos\left(\dfrac{2}{3}\right) = \sum\limits_{n=0}^{\infty} (-1)^n \dfrac{2^{2n}}{3^{2n}(2n)!} \approx 0.7859$

93. The series for Exercise 41 converges very slowly because the terms approach 0 at a slow rate.

95. (a) $f(x) = e^{2x} \qquad f(0) = 1$

$\qquad f'(x) = 2e^{2x} \qquad f'(0) = 2$

$\qquad f''(x) = 4e^{2x} \qquad f''(0) = 4$

$\qquad f'''(x) = 8e^{2x} \qquad f'''(0) = 8$

$\qquad e^{2x} = 1 + 2x + \dfrac{4x^2}{2!} + \dfrac{8x^3}{3!} + \cdots$

$\qquad\qquad = 1 + 2x + 2x^2 + \dfrac{4}{3}x^3 + \cdots$

(b) $e^x = \sum\limits_{n=0}^{\infty} \dfrac{x^n}{n!}$

$\qquad e^{2x} = \sum\limits_{n=0}^{\infty} \dfrac{(2x)^n}{n!} = 1 + 2x + \dfrac{4x^2}{2!} + \dfrac{8x^3}{3!} + \cdots$

$\qquad\qquad = 1 + 2x + 2x^2 + \dfrac{4}{3}x^3 + \cdots$

(c) $e^{2x} = e^x \cdot e^x = \left(1 + x + \dfrac{x^2}{2} + \dfrac{x^3}{6} + \cdots\right)\left(1 + x + \dfrac{x^2}{2} + \dfrac{x^3}{6} + \cdots\right)$

$\qquad = 1 + (x + x) + \left(x^2 + \dfrac{x^2}{2} + \dfrac{x^2}{2}\right) + \left(\dfrac{x^3}{6} + \dfrac{x^3}{6} + \dfrac{x^3}{2} + \dfrac{x^3}{2}\right) + \cdots = 1 + 2x + 2x^2 + \dfrac{4}{3}x^3 + \cdots$

97. $\sin t = \sum\limits_{n=0}^{\infty} \dfrac{(-1)^n t^{2n+1}}{(2n+1)!}$

$\qquad \dfrac{\sin t}{t} = \sum\limits_{n=0}^{\infty} \dfrac{(-1)^n t^{2n}}{(2n+1)!}$

$\qquad \displaystyle\int_0^x \dfrac{\sin t}{t}\, dt = \left[\sum\limits_{n=0}^{\infty} \dfrac{(-1)^n t^{2n+1}}{(2n+1)(2n+1)!}\right]_0^x$

$\qquad\qquad = \sum\limits_{n=0}^{\infty} \dfrac{(-1)^n x^{2n+1}}{(2n+1)(2n+1)!}$

99. $\dfrac{1}{1+t} = \sum\limits_{n=0}^{\infty} (-1)^n t^n$

$\qquad \ln(1+t) = \displaystyle\int \dfrac{1}{1+t}\, dt = \sum\limits_{n=0}^{\infty} \dfrac{(-1)^n t^{n+1}}{n+1}$

$\qquad \dfrac{\ln(t+1)}{t} = \sum\limits_{n=0}^{\infty} \dfrac{(-1)^n t^n}{n+1}$

$\qquad \displaystyle\int_0^x \dfrac{\ln(t+1)}{t}\, dt = \left[\sum\limits_{n=0}^{\infty} \dfrac{(-1)^n t^{n+1}}{(n+1)^2}\right]_0^x = \sum\limits_{n=0}^{\infty} \dfrac{(-1)^n x^{n+1}}{(n+1)^2}$

101. $\arctan x = x - \dfrac{x^3}{3} + \dfrac{x^5}{5} - \dfrac{x^7}{7} + \dfrac{x^9}{9} - \cdots$

$\qquad \dfrac{\arctan x}{\sqrt{x}} = \sqrt{x} - \dfrac{x^{5/2}}{3} + \dfrac{x^{9/2}}{5} - \dfrac{x^{13/2}}{7} + \dfrac{x^{17/2}}{9} - \cdots$

$\lim\limits_{x \to 0^+} \dfrac{\arctan x}{\sqrt{x}} = 0$

By L'Hôpital's Rule, $\lim\limits_{x \to 0^+} \dfrac{\arctan x}{\sqrt{x}} = \lim\limits_{x \to 0^+} \dfrac{\left(\dfrac{1}{1+x^2}\right)}{\left(\dfrac{1}{2\sqrt{x}}\right)} = \lim\limits_{x \to 0^+} \dfrac{2\sqrt{x}}{1+x^2} = 0.$

Problem Solving for Chapter 8

1. (a) $1\left(\dfrac{1}{3}\right) + 2\left(\dfrac{1}{9}\right) + 4\left(\dfrac{1}{27}\right) + \cdots = \sum\limits_{n=0}^{\infty} \dfrac{1}{3}\left(\dfrac{2}{3}\right)^n = \dfrac{1/3}{1-(2/3)} = 1$

(b) $0, \dfrac{1}{3}, \dfrac{2}{3}, 1$, etc.

(c) $\lim\limits_{n \to \infty} C_n = 1 - \sum\limits_{n=0}^{\infty} \dfrac{1}{3}\left(\dfrac{2}{3}\right)^n = 1 - 1 = 0$

3. If there are n rows, then $a_n = \dfrac{n(n+1)}{2}$.

For one circle,

$$a_1 = 1 \text{ and } r_1 = \frac{1}{3}\left(\frac{\sqrt{3}}{2}\right) = \frac{\sqrt{3}}{6} = \frac{1}{2\sqrt{3}}$$

For three circles,

$$a_2 = 3 \text{ and } 1 = 2\sqrt{3}r_2 + 2r_2$$

$$r_2 = \frac{1}{2 + 2\sqrt{3}}$$

For six circles,

$$a_3 = 6 \text{ and } 1 = 2\sqrt{3}r_3 + 4r_3$$

$$r_3 = \frac{1}{2\sqrt{3} + 4}$$

Continuing this pattern, $r_n = \dfrac{1}{2\sqrt{3} + 2(n-1)}$.

Total Area $= (\pi r_n^2)a_n = \pi\left(\dfrac{1}{2\sqrt{3} + 2(n-1)}\right)^2 \dfrac{n(n+1)}{2}$

$$A_n = \frac{\pi}{2} \frac{n(n+1)}{\left[2\sqrt{3} + 2(n+1)\right]^2}$$

$$\lim_{n\to\infty} A_n = \frac{\pi}{2} \cdot \frac{1}{4} = \frac{\pi}{8}$$

5. (a) $\sum a_n x^n = 1 + 2x + 3x^2 + x^3 + 2x^4 + 3x^5 + \cdots$

$$= (1 + x^3 + x^6 + \cdots) + 2(x + x^4 + x^7 + \cdots) + 3(x^2 + x^5 + x^8 + \cdots)$$

$$= (1 + x^3 + x^6 + \cdots)[1 + 2x + 3x^2]$$

$$= (1 + 2x + 3x^2)\frac{1}{1 - x^3}$$

$R = 1$ because each series in the second line has $R = 1$.

(b) $\sum a_n x^n = (a_0 + a_1 x + \cdots + a_{p-1}x^{p-1}) + (a_0 x^p + a_1 x^{p+1} + \cdots) + \cdots$

$$= a_0(1 + x^p + \cdots) + a_1 x(1 + x^p + \cdots) + \cdots + a_{p-1}x^{p-1}(1 + x^p + \cdots)$$

$$= (a_0 + a_1 x + \cdots + a_{p-1}x^{p-1})(1 + x^p + \cdots)$$

$$= (a_0 + a_1 x + \cdots + a_{p-1}x^{p-1})\frac{1}{1 - x^p}.$$

$R = 1$

(Assume all $a_n > 0$.)

7. $e^x = 1 + x + \dfrac{x^2}{2!} + \cdots$

$$xe^x = x + x^2 + \dfrac{x^3}{2!} + \cdots = \sum_{n=0}^{\infty} \dfrac{x^{n+1}}{n!}$$

$$\int xe^x\, dx = xe^x - e^x + C = \sum_{n=0}^{\infty} \dfrac{x^{n+2}}{(n+2)n!}$$

Letting $x = 0$, $C = 1$. Letting $x = 1$,

$$1 = \sum_{n=0}^{\infty} \dfrac{1}{(n+2)n!} = \dfrac{1}{2} + \sum_{n=1}^{\infty} \dfrac{1}{(n+2)n!}.$$

Thus, $\displaystyle\sum_{n=1}^{\infty} \dfrac{1}{(n+2)n!} = \dfrac{1}{2}$.

9. Let $a_1 = \displaystyle\int_0^{\pi} \dfrac{\sin x}{x}\, dx$, $a_2 = -\displaystyle\int_{\pi}^{2\pi} \dfrac{\sin x}{x}\, dx$, $a_3 = \displaystyle\int_{2\pi}^{3\pi} \dfrac{\sin x}{x}\, dx$, etc.

Then,

$$\int_0^{\infty} \dfrac{\sin x}{x}\, dx = a_1 - a_2 + a_3 - a_4 + \cdots.$$

Since $\displaystyle\lim_{n\to\infty} a_n = 0$ and $a_{n+1} < a_n$, this series converges.

11. (a) $a_1 = 3.0$

$a_2 \approx 1.73205$

$a_3 \approx 2.17533$

$a_4 \approx 2.27493$

$a_5 \approx 2.29672$

$a_6 \approx 2.30146$

$\displaystyle\lim_{n\to\infty} a_n = \dfrac{1 + \sqrt{13}}{2}$ [See part (b) for proof.]

(b) Use mathematical induction to show the sequence is increasing. Clearly, $a_2 = \sqrt{a + a_1} = \sqrt{a\sqrt{a}} > \sqrt{a} = a_1$.

Now assume $a_n > a_{n-1}$. Then

$$a_n + a > a_{n-1} + a$$

$$\sqrt{a_n + a} > \sqrt{a_{n-1} + a}$$

$$a_{n+1} > a_n.$$

Use mathematical induction to show that the sequence is bounded above by a. Clearly, $a_1 = \sqrt{a} < a$.

Now assume $a_n < a$. Then $a > a_n$ and $a - 1 > 1$ implies

$$a(a - 1) > a_n(1)$$

$$a^2 - a > a_n$$

$$a^2 > a_n + a$$

$$a > \sqrt{a_n + a} = a_{n+1}.$$

Hence, the sequence converges to some number L. To find L, assume $a_{n+1} \approx a_n \approx L$:

$$L = \sqrt{a + L} \implies L^2 = a + L \implies L^2 - L - a = 0$$

$$L = \dfrac{1 \pm \sqrt{1 + 4a}}{2}.$$

Hence, $L = \dfrac{1 + \sqrt{1 + 4a}}{2}$.

13. (a) $\displaystyle\sum_{n=1}^{\infty} \frac{1}{2^{n+(-1)^n}} = \frac{1}{2^{1-1}} + \frac{1}{2^{2+1}} + \frac{1}{2^{3-1}} + \frac{1}{2^{4+1}} + \frac{1}{2^{5-1}} + \cdots$

$$S_1 = \frac{1}{2^0} = 1$$

$$S_1 = 1 + \frac{1}{8} = \frac{9}{8}$$

$$S_3 = \frac{9}{8} + \frac{1}{4} = \frac{11}{8}$$

$$S_4 = \frac{11}{8} + \frac{1}{32} = \frac{45}{32}$$

$$S_5 = \frac{45}{32} + \frac{1}{16} = \frac{47}{32}$$

(b) $\displaystyle\frac{a_{n+1}}{a_n} = \frac{2^{n+(-1)^n}}{2^{(n+1)+(-1)^{n+1}}} = \frac{2^{(-1)^n}}{2^{1+(-1)^{n+1}}}$

This sequence is $\frac{1}{8}, 2, \frac{1}{8}, 2, \ldots$ which diverges.

(c) $\displaystyle\sqrt[n]{\frac{1}{2^{n+(-1)^n}}} = \left(\frac{1}{2^n \cdot 2^{(-1)^n}}\right)^{1/n}$

$$= \frac{1}{2 \cdot \sqrt[n]{2^{(-1)^n}}} \to \frac{1}{2} < 1 \text{ converges because } \{2^{(-1)^n}\} = \frac{1}{2}, 2, \frac{1}{2}, 2, \ldots \text{ and } \sqrt[n]{1/2} \to 1 \text{ and } \sqrt[n]{2} \to 1.$$

15. $S_6 = 130 + 70 + 40 = 240$

$S_7 = 240 + 130 + 70 = 440$

$S_8 = 440 + 240 + 130 = 810$

$S_9 = 810 + 440 + 240 = 1490$

$S_{10} = 1490 + 810 + 440 = 2740$

C H A P T E R 9
Conics, Parametric Equations, and Polar Coordinates

Section 9.1   Conics and Calculus **457**

Section 9.2   Plane Curves and Parametric Equations **468**

Section 9.3   Parametric Equations and Calculus **472**

Section 9.4   Polar Coordinates and Polar Graphs **478**

Section 9.5   Area and Arc Length in Polar Coordinates **485**

Section 9.6   Polar Equations of Conics and Kepler's Laws **490**

Review Exercises **494**

Problem Solving **502**

CHAPTER 9
Conics, Parametric Equations, and Polar Coordinates

Section 9.1   Conics and Calculus

Solutions to Odd-Numbered Exercises

1. $y^2 = 4x$

Vertex: $(0, 0)$

$p = 1 > 0$

Opens to the right
Matches graph (h).

3. $(x + 3)^2 = -2(y - 2)$

Vertex: $(-3, 2)$

$p = -\frac{1}{2} < 0$

Opens downward
Matches graph (e).

5. $\dfrac{x^2}{9} + \dfrac{y^2}{4} = 1$

Center: $(0, 0)$
Ellipse
Matches (f)

7. $\dfrac{y^2}{16} - \dfrac{x^2}{1} = 1$

Hyperbola
Center: $(0, 0)$

Vertical transverse axis.
Matches (c)

9. $y^2 = -6x = 4\left(-\frac{3}{2}\right)x$

Vertex: $(0, 0)$

Focus: $\left(-\frac{3}{2}, 0\right)$

Directrix: $x = \frac{3}{2}$

11. $(x + 3) + (y - 2)^2 = 0$

$$(y - 2)^2 = 4\left(-\frac{1}{4}\right)(x + 3)$$

Vertex: $(-3, 2)$

Focus: $(-3.25, 2)$

Directrix: $x = -2.75$

13. $y^2 - 4y - 4x = 0$

$y^2 - 4y + 4 = 4x + 4$

$(y - 2)^2 = 4(1)(x + 1)$

Vertex: $(-1, 2)$

Focus: $(0, 2)$

Directrix: $x = -2$

15. $x^2 + 4x + 4y - 4 = 0$

$x^2 + 4x + 4 = -4y + 4 + 4$

$(x + 2)^2 = 4(-1)(y - 2)$

Vertex: $(-2, 2)$

Focus: $(-2, 1)$

Directrix: $y = 3$

17. $y^2 + x + y = 0$

$y^2 + y + \frac{1}{4} = -x + \frac{1}{4}$

$\left(y + \frac{1}{2}\right)^2 = 4\left(-\frac{1}{4}\right)\left(x - \frac{1}{4}\right)$

Vertex: $\left(\frac{1}{4}, -\frac{1}{2}\right)$

Focus: $\left(0, -\frac{1}{2}\right)$

Directrix: $x = \frac{1}{2}$

19. $y^2 - 4x - 4 = 0$

$y^2 = 4x + 4$

$= 4(1)(x + 1)$

Vertex: $(-1, 0)$

Focus: $(0, 0)$

Directrix: $x = -2$

21. $(y - 2)^2 = 4(-2)(x - 3)$

$y^2 - 4y + 8x - 20 = 0$

23. $(x - h)^2 = 4p(y - k)$

$x^2 = 4(6)(y - 4)$

$x^2 - 24y + 96 = 0$

25. $y = 4 - x^2$

$x^2 + y - 4 = 0$

27. Since the axis of the parabola is vertical, the form of the equation is $y = ax^2 + bx + c$. Now, substituting the values of the given coordinates into this equation, we obtain

$3 = c,\ 4 = 9a + 3b + c,\ 11 = 16a + 4b + c.$

Solving this system, we have $a = \frac{5}{3}, b = -\frac{14}{3}, c = 3$. Therefore,

$y = \frac{5}{3}x^2 - \frac{14}{3}x + 3$ or $5x^2 - 14x - 3y + 9 = 0.$

29. $x^2 + 4y^2 = 4$

$\frac{x^2}{4} + \frac{y^2}{1} = 1$

$a^2 = 4, b^2 = 1, c^2 = 3$

Center: $(0, 0)$

Foci: $\left(\pm\sqrt{3}, 0\right)$

Vertices: $(\pm 2, 0)$

$e = \frac{\sqrt{3}}{2}$

31. $\frac{(x - 1)^2}{9} + \frac{(y - 5)^2}{25} = 1$

$a^2 = 25, b^2 = 9, c^2 = 16$

Center: $(1, 5)$

Foci: $(1, 9), (1, 1)$

Vertices: $(1, 10), (1, 0)$

$e = \frac{4}{5}$

33. $9x^2 + 4y^2 + 36x - 24y + 36 = 0$

$9(x^2 + 4x + 4) + 4(y^2 - 6y + 9) = -36 + 36 + 36$

$= 36$

$\frac{(x + 2)^2}{4} + \frac{(y - 3)^2}{9} = 1$

$a^2 = 9, b^2 = 4, c^2 = 5$

Center: $(-2, 3)$

Foci: $\left(-2, 3 \pm \sqrt{5}\right)$

Vertices: $(-2, 6), (-2, 0)$

$e = \frac{\sqrt{5}}{3}$

35. $12x^2 + 20y^2 - 12x + 40y - 37 = 0$

$12\left(x^2 - x + \dfrac{1}{4}\right) + 20(y^2 + 2y + 1) = 37 + 3 + 20$

$$= 60$$

$$\dfrac{[x - (1/2)]^2}{5} + \dfrac{(y + 1)^2}{3} = 1$$

$a^2 = 5,\ b^2 = 3,\ c^2 = 2$

Center: $\left(\dfrac{1}{2}, -1\right)$

Foci: $\left(\dfrac{1}{2} \pm \sqrt{2}, -1\right)$

Vertices: $\left(\dfrac{1}{2} \pm \sqrt{5}, -1\right)$

Solve for y:

$20(y^2 + 2y + 1) = -12x^2 + 12x + 37 + 20$

$$(y + 1)^2 = \dfrac{57 + 12x - 12x^2}{20}$$

$$y = -1 \pm \sqrt{\dfrac{57 + 12x - 12x^2}{20}}$$

(Graph each of these separately.)

37. $x^2 + 2y^2 - 3x + 4y + 0.25 = 0$

$\left(x^2 - 3x + \dfrac{9}{4}\right) + 2(y^2 + 2y + 1) = -\dfrac{1}{4} + \dfrac{9}{4} + 2 = 4$

$$\dfrac{[x - (3/2)]^2}{4} + \dfrac{(y + 1)^2}{2} = 1$$

$a^2 = 4,\ b^2 = 2,\ c^2 = 2$

Center: $\left(\dfrac{3}{2}, -1\right)$

Foci: $\left(\dfrac{3}{2} \pm \sqrt{2}, -1\right)$

Vertices: $\left(-\dfrac{1}{2}, -1\right), \left(\dfrac{7}{2}, -1\right)$

Solve for y: $2(y^2 + 2y + 1) = -x^2 + 3x - \dfrac{1}{4} + 2$

$$(y + 1)^2 = \dfrac{1}{2}\left(\dfrac{7}{4} + 3x - x^2\right)$$

$$y = -1 \pm \sqrt{\dfrac{7 + 12x - 4x^2}{8}}$$

(Graph each of these separately.)

39. Center: $(0, 0)$
Focus: $(2, 0)$
Vertex: $(3, 0)$
Horizontal major axis

$a = 3,\ c = 2 \Rightarrow b = \sqrt{5}$

$$\dfrac{x^2}{9} + \dfrac{y^2}{5} = 1$$

41. Vertices: $(3, 1), (3, 9)$
Minor axis length: 6
Vertical major axis
Center: $(3, 5)$

$a = 4,\ b = 3$

$$\dfrac{(x - 3)^2}{9} + \dfrac{(y - 5)^2}{16} = 1$$

43. Center: $(0, 0)$
Horizontal major axis
Points on ellipse: $(3, 1), (4, 0)$

Since the major axis is horizontal,

$$\left(\dfrac{x^2}{a^2}\right) + \left(\dfrac{y^2}{b^2}\right) = 1.$$

Substituting the values of the coordinates of the given points into this equation, we have

$$\left(\dfrac{9}{a^2}\right) + \left(\dfrac{1}{b^2}\right) = 1, \text{ and } \dfrac{16}{a^2} = 1.$$

The solution to this system is $a^2 = 16,\ b^2 = 16/7$.

Therefore,

$$\dfrac{x^2}{16} + \dfrac{y^2}{16/7} = 1,\ \dfrac{x^2}{16} + \dfrac{7y^2}{16} = 1.$$

45. $\dfrac{y^2}{1} - \dfrac{x^2}{4} = 1$

$a = 1, b = 2, c = \sqrt{5}$

Center: $(0, 0)$

Vertices: $(0, \pm 1)$

Foci: $\left(0, \pm\sqrt{5}\right)$

Asymptotes: $y = \pm\dfrac{1}{2}x$

47. $\dfrac{(x-1)^2}{4} - \dfrac{(y+2)^2}{1} = 1$

$a = 2, b = 1, c = \sqrt{5}$

Center: $(1, -2)$

Vertices: $(-1, -2), (3, -2)$

Foci: $\left(1 \pm \sqrt{5}, -2\right)$

Asymptotes: $y = -2 \pm\dfrac{1}{2}(x - 1)$

49. $\quad 9x^2 - y^2 - 36x - 6y + 18 = 0$

$9(x^2 - 4x + 4) - (y^2 + 6y + 9) = -18 + 36 - 9$

$$\dfrac{(x-2)^2}{1} - \dfrac{(y+3)^2}{9} = 1$$

$a = 1, b = 3, c = \sqrt{10}$

Center: $(2, -3)$

Vertices: $(1, -3), (3, -3)$

Foci: $\left(2 \pm \sqrt{10}, -3\right)$

Asymptotes: $y = -3 \pm 3(x - 2)$

51. $\quad x^2 - 9y^2 + 2x - 54y - 80 = 0$

$(x^2 + 2x + 1) - 9(y^2 + 6y + 9) = 80 + 1 - 81 = 0$

$(x + 1)^2 - 9(y + 3)^2 = 0$

$$y + 3 = \pm\dfrac{1}{3}(x + 1)$$

Degenerate hyperbola is two lines intersecting at $(-1, -3)$.

53. $\quad 9y^2 - x^2 + 2x + 54y + 62 = 0$

$9(y^2 + 6y + 9) - (x^2 - 2x + 1) = -62 - 1 + 81 = 18$

$$\dfrac{(y+3)^2}{2} - \dfrac{(x-1)^2}{18} = 1$$

$a = \sqrt{2}, b = 3\sqrt{2}, c = 2\sqrt{5}$

Center: $(1, -3)$

Vertices: $\left(1, -3 \pm \sqrt{2}\right)$

Foci: $\left(1, -3 \pm 2\sqrt{5}\right)$

Solve for y:

$9(y^2 + 6y + 9) = x^2 - 2x - 62 + 81$

$(y + 3)^2 = \dfrac{x^2 - 2x + 19}{9}$

$y = -3 \pm \dfrac{1}{3}\sqrt{x^2 - 2x + 19}$

(Graph each curve separately.)

55. $\quad 3x^2 - 2y^2 - 6x - 12y - 27 = 0$

$3(x^2 - 2x + 1) - 2(y^2 + 6y + 9) = 27 + 3 - 18 = 12$

$$\dfrac{(x-1)^2}{4} - \dfrac{(y+3)^2}{6} = 1$$

$a = 2, b = \sqrt{6}, c = \sqrt{10}$

Center: $(1, -3)$

Vertices: $(-1, -3), (3, -3)$

Foci: $\left(1 \pm \sqrt{10}, -3\right)$

Solve for y:

$2(y^2 + 6y + 9) = 3x^2 - 6x - 27 + 18$

$(y + 3)^2 = \dfrac{3x^2 - 6x - 9}{2}$

$y = -3 \pm \sqrt{\dfrac{3(x^2 - 2x - 3)}{2}}$

(Graph each curve separately.)

57. Vertices: $(\pm 1, 0)$

Asymptotes: $y = \pm 3x$

Horizontal transverse axis

Center: $(0, 0)$

$$a = 1, \pm\frac{b}{a} = \pm\frac{b}{1} = \pm 3 \implies b = 3$$

Therefore, $\dfrac{x^2}{1} - \dfrac{y^2}{9} = 1$.

59. Vertices: $(2, \pm 3)$

Point on graph: $(0, 5)$

Vertical transverse axis

Center: $(2, 0)$

$a = 3$

Therefore, the equation is of the form

$$\frac{y^2}{9} - \frac{(x - 2)^2}{b^2} = 1.$$

Substituting the coordinates of the point $(0, 5)$, we have

$$\frac{25}{9} - \frac{4}{b^2} = 1 \quad \text{or} \quad b^2 = \frac{9}{4}.$$

Therefore, the equation is $\dfrac{y^2}{9} - \dfrac{(x - 2)^2}{9/4} = 1$.

61. Center: $(0, 0)$

Vertex: $(0, 2)$

Focus: $(0, 4)$

Vertical transverse axis

$a = 2, c = 4, b^2 = c^2 - a^2 = 12$

Therefore, $\dfrac{y^2}{4} - \dfrac{x^2}{12} = 1$.

63. Vertices: $(0, 2), (6, 2)$

Asymptotes: $y = \dfrac{2}{3}x, y = 4 - \dfrac{2}{3}x$

Horizontal transverse axis

Center: $(3, 2)$

$a = 3$

Slopes of asymptotes: $\pm\dfrac{b}{a} = \pm\dfrac{2}{3}$

Thus, $b = 2$. Therefore,

$$\frac{(x - 3)^2}{9} - \frac{(y - 2)^2}{4} = 1.$$

65. (a) $\dfrac{x^2}{9} - y^2 = 1, \dfrac{2x}{9} - 2yy' = 0, \dfrac{x}{9y} = y'$

At $x = 6$: $y = \pm\sqrt{3}, y' = \dfrac{\pm 6}{9\sqrt{3}} = \dfrac{\pm 2\sqrt{3}}{9}$

At $\left(6, \sqrt{3}\right)$: $y - \sqrt{3} = \dfrac{2\sqrt{3}}{9}(x - 6)$

$\quad$ or $2x - 3\sqrt{3}y - 3 = 0$

At $\left(6, -\sqrt{3}\right)$: $y + \sqrt{3} = \dfrac{-2\sqrt{3}}{9}(x - 6)$

$\quad$ or $2x + 3\sqrt{3}y - 3 = 0$

(b) From part (a) we know that the slopes of the normal lines must be $\mp 9/\left(2\sqrt{3}\right)$.

At $\left(6, \sqrt{3}\right)$: $y - \sqrt{3} = -\dfrac{9}{2\sqrt{3}}(x - 6)$

$\quad$ or $9x + 2\sqrt{3}y - 60 = 0$

At $\left(6, -\sqrt{3}\right)$: $y + \sqrt{3} = \dfrac{9}{2\sqrt{3}}(x - 6)$

$\quad$ or $9x - 2\sqrt{3}y - 60 = 0$

67.

$$x^2 + 4y^2 - 6x + 16y + 21 = 0$$
$$x^2 - 6x + 4(y^2 + 4y) = -21$$
$$x^2 - 6x + 9 + 4(y^2 + 4y + 4) = -21 + 9 + 16$$
$$(x - 3)^2 + 4(y + 2)^2 = 4$$
$$\frac{(x - 3)^2}{4} + \frac{(y + 2)^2}{1} = 1$$

Ellipse

69.

$$y^2 - 4y - 4x = 0$$
$$y^2 - 4y + 4 - 4x = 4$$
$$(y - 2)^2 - 4x = 4$$
$$(y - 2)^2 = 4x + 4$$
$$(y - 2)^2 = 4(x + 1)$$

Parabola

71.

$$4x^2 + 4y^2 - 16y + 15 = 0$$
$$4x^2 + 4(y^2 - 4y) = -15$$
$$x^2 + (y^2 - 4y) = -\frac{15}{4}$$
$$x^2 + y^2 - 4y + 4 = -\frac{15}{4} + 4$$
$$x^2 + (y - 2)^2 = \frac{1}{4}$$

Circle

73.
$$9x^2 + 9y^2 - 36x + 6y + 34 = 0$$
$$9x^2 - 36x + 9y^2 + 6y = -34$$
$$9(x^2 - 4x + 4) + 9\left(y^2 + \frac{2}{3}y + \frac{1}{9}\right) = -34 + 36 + 1$$
$$9(x - 2)^2 = 9\left(y = \frac{1}{3}\right)^2 = 3$$
$$(x - 2)^2 + \left(y + \frac{1}{3}\right)^2 = \frac{1}{3}$$

Circle

75.
$$3(x - 1)^2 = 6 + 2(y + 1)^2$$
$$3(x - 1)^2 - (y + 1)^2 = 6$$
$$\frac{(x - 1)^2}{2} - \frac{(y + 1)^2}{3} = 1$$

Hyperbola

77. (a) A parabola is the set of all points (x, y) that are equidistant from a fixed line (directrix) and a fixed point (focus) not on the line.

(b) $(x - h)^2 = 4p(y - k)$ or $(y - k)^2 = 4p(x - h)$

(c) See Theorem 9.2.

79. (a) A hyperbola is the set of all points (x, y) for which the absolute value of the difference between the distances from two distance fixed points (foci) is constant.

(b) $\dfrac{(x - h)^2}{a^2} - \dfrac{(y - k)^2}{b^2} = 1$ or $\dfrac{(y - k)^2}{a^2} - \dfrac{(x - h)^2}{b^2} = 1$

(c) $y = k \pm \dfrac{b}{a}(x - h)$ or $y = k \pm \dfrac{a}{b}(x - h)$

81. Assume that the vertex is at the origin.
$$x^2 = 4py$$
$$(3)^2 = 4p(1)$$
$$\frac{9}{4} = p$$

The pipe is located $\frac{9}{4}$ meters from the vertex.

83. $y = ax^2$

$y' = 2ax$

The equation of the tangent line is
$$y - ax_0^2 = 2ax_0(x - x_0)$$
$$\text{or } y = 2ax_0 x - ax_0^2.$$

Let $y = 0$. Then:
$$-ax_0^2 = 2ax_0 x - 2ax_0^2$$
$$ax_0^2 = 2ax_0 x$$
$$x = \frac{x_0}{2}$$

Therefore, $\left(\dfrac{x_0}{2}, 0\right)$ is the x-intercept.

85. (a) Consider the parabola $x^2 = 4py$. Let m_0 be the slope of the one tangent line at (x_1, y_1) and therefore, $-1/m_0$ is the slope of the second at (x_2, y_2). Differentiating, $2x = 4py'$ or $y' = \dfrac{x}{2p}$, and we have:

$$m_0 = \frac{1}{2p}x_1 \text{ or } x_1 = 2pm_0$$

$$\frac{-1}{m_0} = \frac{1}{2p}x_2 \text{ or } x_2 = \frac{-2p}{m_0}$$

Substituting these values of x into the equation $x^2 = 4py$, we have the coordinates of the points of tangency $(2pm_0, pm_0^2)$ and $(-2p/m_0, p/m_0^2)$ and the equations of the tangent lines are

$$(y - pm_0^2) = m_0(x - 2pm_0) \quad \text{and} \quad \left(y - \frac{p}{m_0^2}\right) = \frac{-1}{m_0}\left(x + \frac{2p}{m_0}\right).$$

The point of intersection of these lines is

$$\left(\frac{p(m_0^2 - 1)}{m_0}, -p\right)$$ and is on the directrix, $y = -p$.

—CONTINUED—

85. —CONTINUED—

(b) $x^2 - 4x - 4y + 8 = 0$

$\qquad (x - 2)^2 = 4(y - 1)$. Vertex $(2, 1)$

$\qquad 2x - 4 - 4\dfrac{dy}{dx} = 0$

$\qquad\qquad \dfrac{dy}{dx} = \dfrac{1}{2}x - 1$

At $(-2, 5)$, $dy/dx = -2$. At $\left(3, \frac{5}{4}\right)$, $dy/dx = \frac{1}{2}$.

Tangent line at $(-2, 5)$: $y - 5 = -2(x + 2) \Rightarrow 2x + y - 1 = 0$.

Tangent line at $\left(3, \frac{5}{4}\right)$: $y - \frac{5}{4} = \frac{1}{2}(x - 3) \Rightarrow 2x - 4y - 1 = 0$.

Since $m_1 m_2 = (-2)\left(\frac{1}{2}\right) = -1$, the lines are perpendicular.

Point of intersection: $-2x + 1 = \dfrac{1}{2}x - \dfrac{1}{4}$

$$-\dfrac{5}{2}x = -\dfrac{5}{4}$$

$$x = \dfrac{1}{2}$$

$$y = 0$$

Directrix: $y = 0$ and the point of intersection $\left(\frac{1}{2}, 0\right)$ lies on this line.

87. $y = x - x^2$

$\dfrac{dy}{dx} = 1 - 2x$

At (x_1, y_1) on the mountain, $m = 1 - 2x_1$. Also, $m = \dfrac{y_1 - 1}{x_1 + 1}$.

$\qquad \dfrac{y_1 - 1}{x_1 + 1} = 1 - 2x_1$

$\qquad (x_1 - x_1{}^2) - 1 = (1 - 2x_1)(x_1 + 1)$

$\qquad -x_1{}^2 + x_1 - 1 = -2x_1{}^2 - x_1 + 1$

$\qquad x_1{}^2 + 2x_1 - 2 = 0$

$\qquad\qquad x_1 = \dfrac{-2 \pm \sqrt{2^2 - 4(1)(-2)}}{2(1)} = \dfrac{-2 \pm 2\sqrt{3}}{2} = -1 \pm \sqrt{3}$

Choosing the positive value for x_1, we have $x_1 = -1 + \sqrt{3}$.

$\qquad m = 1 - 2\left(-1 + \sqrt{3}\right) = 3 - 2\sqrt{3}$

$\qquad m = \dfrac{0 - 1}{x_0 + 1} = -\dfrac{1}{x_0 + 1}$

Thus, $-\dfrac{1}{x_0 + 1} = 3 - 2\sqrt{3}$

$\qquad \dfrac{-1}{3 - 2\sqrt{3}} = x_0 + 1$

$\qquad \dfrac{3 + 2\sqrt{3}}{3} - 1 = x_0$

$\qquad\qquad \dfrac{2\sqrt{3}}{3} = x_0$.

The closest the receiver can be to the hill is $\left(2\sqrt{3}/3\right) - 1 \approx 0.155$.

89. Parabola

Vertex: $(0, 4)$

$$x^2 = 4p(y - 4)$$

$$4^2 = 4p(0 - 4)$$

$$p = -1$$

$$x^2 = -4(y - 4)$$

$$y = 4 - \frac{x^2}{4}$$

Circle

Center: $(0, k)$

Radius: 8

$$x^2 + (y - k)^2 = 64$$

$$4^2 + (0 - k)^2 = 64$$

$$k^2 = 48$$

$$k = -4\sqrt{3} \quad \text{(Center is on the negative } y\text{-axis.)}$$

$$x^2 + \left(y + 4\sqrt{3}\right)^2 = 64$$

$$y = -4\sqrt{3} \pm \sqrt{64 - x^2}$$

Since the y-value is positive when $x = 0$, we have $y = -4\sqrt{3} + \sqrt{64 - x^2}$.

$$A = 2\int_0^4 \left[\left(4 - \frac{x^2}{4}\right) - \left(-4\sqrt{3} + \sqrt{64 - x^2}\right)\right] dx$$

$$= 2\left[4x - \frac{x^3}{12} + 4\sqrt{3}x - \frac{1}{2}\left(x\sqrt{64 - x^2} + 64 \arcsin\frac{x}{8}\right)\right]_0^4$$

$$= 2\left[16 - \frac{64}{12} + 16\sqrt{3} - 2\sqrt{48} - 32 \arcsin\frac{1}{2}\right]$$

$$= \frac{16\left(4 + 3\sqrt{3} - 2\pi\right)}{3} \approx 15.536 \text{ square feet}$$

91. (a) Assume that $y = ax^2$.

$$20 = a(60)^2 \implies a = \frac{2}{360} = \frac{1}{180} \implies y = \frac{1}{180}x^2$$

(b) $f(x) = \frac{1}{180}x^2, f'(x) = \frac{1}{90}x$

$$S = 2\int_0^{60} \sqrt{1 + \left(\frac{1}{90}x\right)^2}\, dx = \frac{2}{90}\int_0^{60} \sqrt{90^2 + x^2}\, dx$$

$$= \frac{2}{90} \frac{1}{2}\left[x\sqrt{90^2 + x^2} + 90^2 \ln\left|x + \sqrt{90^2 + x^2}\right|\right]_0^{60} \quad \text{(formula 26)}$$

$$= \frac{1}{90}\left[60\sqrt{11,700} + 90^2 \ln\left(60 + \sqrt{11,700}\right) - 90^2 \ln 90\right]$$

$$= \frac{1}{90}\left[1800\sqrt{13} + 90^2 \ln\left(60 + 30\sqrt{13}\right) - 90^2 \ln 90\right]$$

$$= 20\sqrt{13} + 90 \ln\left(\frac{60 + 30\sqrt{13}}{90}\right)$$

$$= 10\left[2\sqrt{13} + 9 \ln\left(\frac{2 + \sqrt{13}}{3}\right)\right] \approx 128.4 \text{ m}$$

93. $x^2 = 4py, p = \frac{1}{4}, \frac{1}{2}, 1, \frac{3}{2}, 2$

As p increases, the graph becomes wider.

95.

97. $a = \dfrac{5}{2}, b = 2, c = \sqrt{\left(\dfrac{5}{2}\right)^2 - (2)^2} = \dfrac{3}{2}$

The tacks should be placed 1.5 feet from the center. The string should be $2a = 5$ feet long.

99. $e = \dfrac{c}{a}$

$A + P = 2a$

$\quad a = \dfrac{A + P}{2}$

$\quad c = a - P = \dfrac{A + P}{2} - P = \dfrac{A - P}{2}$

$\quad e = \dfrac{c}{a} = \dfrac{(A - P)/2}{(A + P)/2} = \dfrac{A - P}{A + P}$

101. $e = \dfrac{A - P}{A + P} = \dfrac{35.34au - 0.59au}{35.34au + 0.59au} \approx 0.9672$

103. $\dfrac{x^2}{10^2} + \dfrac{y^2}{5^2} = 1$

$\dfrac{2x}{10^2} + \dfrac{2yy'}{5^2} = 0$

$\qquad y' = \dfrac{-5^2 x}{10^2 y} = \dfrac{-x}{4y}$

At $(-8, 3)$: $y' = \dfrac{8}{12} = \dfrac{2}{3}$

The equation of the tangent line is $y - 3 = \dfrac{2}{3}(x + 8)$. It will cross the y-axis when $x = 0$ and $y = \dfrac{2}{3}(8) + 3 = \dfrac{25}{3}$.

105. $16x^2 + 9y^2 + 96x + 36y + 36 = 0$

$32x + 18yy' + 96 + 36y' = 0$

$\qquad y'(18y + 36) = -(32x + 96)$

$\qquad\qquad y' = \dfrac{-(32x + 96)}{18y + 36}$

$y' = 0$ when $x = -3$. y' is undefined when $y = -2$.

At $x = -3, y = 2$ or -6.

Endpoints of major axis: $(-3, 2), (-3, -6)$

At $y = -2, x = 0$ or -6.

Endpoints of minor axis: $(0, -2), (-6, -2)$

Note: Equation of ellipse is $\dfrac{(x + 3)^2}{9} + \dfrac{(y + 2)^2}{16} = 1$

107. (a) $A = 4\displaystyle\int_0^2 \dfrac{1}{2}\sqrt{4 - x^2}\,dx = \left[x\sqrt{4 - x^2} + 4\arcsin\left(\dfrac{x}{2}\right)\right]_0^2 = 2\pi$ $\;[\text{or, } A = \pi ab = \pi(2)(1) = 2\pi]$

(b) Disk: $\qquad V = 2\pi\displaystyle\int_0^2 \dfrac{1}{4}(4 - x^2)\,dx = \dfrac{1}{2}\pi\left[4x - \dfrac{1}{3}x^3\right]_0^2 = \dfrac{8\pi}{3}$

$\qquad\qquad\qquad y = \dfrac{1}{2}\sqrt{4 - x^2}$

$\qquad\qquad\qquad y' = \dfrac{-x}{2\sqrt{4 - x^2}}$

$\qquad \sqrt{1 + (y')^2} = \sqrt{1 + \dfrac{x^2}{16 - 4x^2}} = \sqrt{\dfrac{16 - 3x^2}{4y}}$

$S = 2(2\pi)\displaystyle\int_0^2 y\left(\dfrac{\sqrt{16 - 3x^2}}{4y}\right)dx = \dfrac{\pi}{2\sqrt{3}}\left[\sqrt{3}x\sqrt{16 - 3x^2} + 16\arcsin\left(\dfrac{\sqrt{3}x}{4}\right)\right]_0^2 = \dfrac{2\pi}{9}\left(9 + 4\sqrt{3}\pi\right) \approx 21.48$

—CONTINUED—

107. —CONTINUED—

(c) **Shell:** $V = 2\pi \int_0^2 x\sqrt{4 - x^2}\,dx = -\pi \int_0^2 -2x(4 - x^2)^{1/2}\,dx = -\frac{2\pi}{3}\Big[(4 - x^2)^{3/2}\Big]_0^2 = \frac{16\pi}{3}$

$$x = 2\sqrt{1 - y^2}$$

$$x' = \frac{-2y}{\sqrt{1 - y^2}}$$

$$\sqrt{1 + (x')^2} = \sqrt{1 + \frac{4y^2}{1 - y^2}} = \frac{\sqrt{1 + 3y^2}}{\sqrt{1 - y^2}}$$

$$S = 2(2\pi)\int_0^1 2\sqrt{1 - y^2}\,\frac{\sqrt{1 + 3y^2}}{\sqrt{1 - y^2}}\,dy = 8\pi\int_0^1 \sqrt{1 + 3y^2}\,dy$$

$$= \frac{8\pi}{2\sqrt{3}}\Big[\sqrt{3}y\sqrt{1 + 3y^2} + \ln\big|\sqrt{3}y + \sqrt{1 + 3y^2}\big|\Big]_0^1 = \frac{4\pi}{3}\big|6 + \sqrt{3}\ln(2 + \sqrt{3})\big| \approx 34.69$$

109. From Example 5,

$$C = 4a\int_0^{\pi/2} \sqrt{1 - e^2 \sin^2\theta}\,d\theta$$

For $\dfrac{x^2}{25} + \dfrac{y^2}{49} = 1$, we have

$a = 7, b = 5, c = \sqrt{49 - 25} = 2\sqrt{6}, e = \dfrac{c}{a} = \dfrac{2\sqrt{6}}{7}.$

$$C = 4(7)\int_0^{\pi/2} \sqrt{1 - \frac{24}{49}\sin^2\theta}\,d\theta$$

$$\approx 28(1.3558) \approx 37.9614$$

111. Area circle $= \pi r^2 = 100\pi$

Area ellipse $= \pi a b = \pi a(10)$

$2(100\pi) = 10\pi a \implies a = 20$

Hence, the length of the major axis is $2a = 40$.

113. The transverse axis is horizontal since $(2, 2)$ and $(10, 2)$ are the foci (see definition of hyperbola).

Center: $(6, 2)$

$c = 4, 2a = 6, b^2 = c^2 - a^2 = 7$

Therefore, the equation is

$$\frac{(x - 6)^2}{9} - \frac{(y - 2)^2}{7} = 1.$$

115. $2a = 10 \implies a = 5$

$c = 6 \implies b = \sqrt{11}$

117. Time for sound of bullet hitting target to reach (x, y): $\dfrac{2c}{v_m} + \dfrac{\sqrt{(x - c)^2 + y^2}}{v_s}$

Time for sound of rifle to reach (x, y): $\dfrac{\sqrt{(x + c)^2 + y^2}}{v_s}$

Since the times are the same, we have: $\dfrac{2c}{v_m} + \dfrac{\sqrt{(x - c)^2 + y^2}}{v_s} = \dfrac{\sqrt{(x + c)^2 + y^2}}{v_s}$

$$\frac{4c^2}{v_m^2} + \frac{4c}{v_m v_s}\sqrt{(x - c)^2 + y^2} + \frac{(x - c)^2 + y^2}{v_s^2} = \frac{(x + c)^2 + y^2}{v_s^2}$$

$$\sqrt{(x - c)^2 + y^2} = \frac{v_m^2 x - v_s^2 c}{v_s v_m}$$

$$\Big(1 - \frac{v_m^2}{v_s^2}\Big)x^2 + y^2 = \Big(\frac{v_s^2}{v_m^2} - 1\Big)c^2$$

$$\frac{x^2}{c^2 v_s^2/v_m^2} - \frac{y^2}{c^2(v_m^2 - v_s^2)/v_m^2} = 1$$

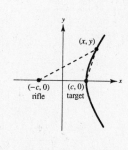

119. The point (x, y) lies on the line between $(0, 10)$ and $(10, 0)$. Thus, $y = 10 - x$. The point also lies on the hyperbola $(x^2/36) - (y^2/64) = 1$. Using substitution, we have:

$$\frac{x^2}{36} - \frac{(10 - x)^2}{64} = 1$$

$$16x^2 - 9(10 - x)^2 = 576$$

$$7x^2 + 180x - 1476 = 0$$

$$x = \frac{-180 \pm \sqrt{180^2 - 4(7)(-1476)}}{2(7)} = \frac{-180 \pm 192\sqrt{2}}{14} = \frac{-90 \pm 96\sqrt{2}}{7}$$

Choosing the positive value for x we have:

$$x = \frac{-90 + 96\sqrt{2}}{7} \approx 6.538 \text{ and } y = \frac{160 - 96\sqrt{2}}{7} \approx 3.462$$

121.
$$\frac{x^2}{a^2} + \frac{2y^2}{b^2} = 1 \implies \frac{2y^2}{b^2} = 1 - \frac{x^2}{a^2}, \; c^2 = a^2 - b^2$$

$$\frac{x^2}{a^2 - b^2} - \frac{2y^2}{b^2} = 1 \implies \frac{2y^2}{b^2} = \frac{x^2}{a^2 - b^2} - 1$$

$$1 - \frac{x^2}{a^2} = \frac{x^2}{a^2 - b^2} - 1 \implies 2 = x^2\left(\frac{1}{a^2} + \frac{1}{a^2 - b^2}\right)$$

$$x^2 = \frac{2a^2(a^2 - b^2)}{2a^2 - b^2} \implies x = \pm\frac{\sqrt{2}a\sqrt{a^2 - b^2}}{\sqrt{2a^2 - b^2}} = \pm\frac{\sqrt{2}ac}{\sqrt{2a^2 - b^2}}$$

$$\frac{2y^2}{b^2} = 1 - \frac{1}{a^2}\left(\frac{2a^2c^2}{2a^2 - b^2}\right) \implies \frac{2y^2}{b^2} = \frac{b^2}{2a^2 - b^2}$$

$$y^2 = \frac{b^4}{2(2a^2 - b^2)} \implies y = \pm\frac{b^2}{\sqrt{2}\sqrt{2a^2 - b^2}}$$

There are four points of intersection: $\left(\dfrac{\sqrt{2}ac}{\sqrt{2a^2 - b^2}}, \pm\dfrac{b^2}{\sqrt{2}\sqrt{2a^2 - b^2}}\right), \left(-\dfrac{\sqrt{2}ac}{\sqrt{2a^2 - b^2}}, \pm\dfrac{b^2}{\sqrt{2}\sqrt{2a^2 - b^2}}\right)$

$$\frac{x^2}{a^2} + \frac{2y^2}{b^2} = 1 \implies \frac{2x}{a^2} + \frac{4yy'}{b^2} = 0 \implies y'_e = -\frac{b^2x}{2a^2y}$$

$$\frac{x^2}{a^2 - b^2} - \frac{2y^2}{b^2} = 1 \implies \frac{2x}{c^2} - \frac{4yy'}{b^2} = 0 \implies y'_h = \frac{b^2x}{2c^2y}$$

At $\left(\dfrac{\sqrt{2}ac}{\sqrt{2a^2 - b^2}}, \dfrac{b^2}{\sqrt{2}\sqrt{2a^2 - b^2}}\right)$, the slopes of the tangent lines are:

$$y'_e = \frac{-b^2\left(\dfrac{\sqrt{2}ac}{\sqrt{2a^2 - b^2}}\right)}{2a^2\left(\dfrac{b^2}{\sqrt{2}\sqrt{2a^2 - b^2}}\right)} = -\frac{c}{a} \quad \text{and} \quad y'_h = \frac{b^2\left(\dfrac{\sqrt{2}ac}{\sqrt{2a^2 - b^2}}\right)}{2c^2\left(\dfrac{b^2}{\sqrt{2}\sqrt{2a^2 - b^2}}\right)} = \frac{a}{c}$$

Since the slopes are negative reciprocals, the tangent lines are perpendicular. Similarly, the curves are perpendicular at the other three points of intersection.

123. False. See the definition of a parabola.

125. True

127. False. $y^2 - x^2 + 2x + 2y = 0$ yields two intersecting lines.

129. True

Section 9.2   Plane Curves and Parametric Equations

1. $x = \sqrt{t},\ y = 1 - t$

(a)

t	0	1	2	3	4
x	0	1	$\sqrt{2}$	$\sqrt{3}$	2
y	1	0	-1	-2	-3

(b)

(c)

(d) $x^2 = t$

$y = 1 - x^2,\ x \geq 0$

---

3. $x = 3t - 1$

$y = 2t + 1$

$y = 2\left(\dfrac{x+1}{3}\right) + 1$

$2x - 3y + 5 = 0$

5. $x = t + 1$

$y = t^2$

$y = (x - 1)^2$

7. $x = t^3$

$y = \frac{1}{2}t^2$

$x = t^3$ implies $t = x^{1/3}$

$y = \frac{1}{2}x^{2/3}$

---

9. $x = \sqrt{t},\ t \geq 0$

$y = t - 2$

$y = x^2 - 2,\ x \geq 0$

11. $x = t - 1$

$y = \dfrac{t}{t - 1}$

$y = \dfrac{x + 1}{x}$

13. $x = 2t$

$y = |t - 2|$

$y = \left|\dfrac{x}{2} - 2\right| = \dfrac{|x - 4|}{2}$

---

15. $x = e^t,\ x > 0$

$y = e^{3t} + 1$

$y = x^3 + 1,\ x > 0$

17. $x = \sec \theta$

$y = \cos \theta$

$0 \le \theta < \dfrac{\pi}{2}, \dfrac{\pi}{2} < \theta \le \pi$

$xy = 1$

$y = \dfrac{1}{x}$

$|x| \ge 1, \ |y| \le 1$

19. $x = 3 \cos \theta, \ y = 3 \sin \theta$

Squaring both equations and adding, we have

$x^2 + y^2 = 9.$

21. $x = 4 \sin 2\theta$

$y = 2 \cos 2\theta$

$\dfrac{x^2}{16} = \sin^2 2\theta$

$\dfrac{y^2}{4} = \cos^2 2\theta$

$\dfrac{x^2}{16} + \dfrac{y^2}{4} = 1$

23. $x = 4 + 2 \cos \theta$

$y = -1 + \sin \theta$

$\dfrac{(x-4)^2}{4} = \cos^2 \theta$

$\dfrac{(y+1)^2}{1} = \sin^2 \theta$

$\dfrac{(x-4)^2}{4} + \dfrac{(y+1)^2}{1} = 1$

25. $x = 4 + 2 \cos \theta$

$y = -1 + 4 \sin \theta$

$\dfrac{(x-4)^2}{4} = \cos^2 \theta$

$\dfrac{(y+1)^2}{16} = \sin^2 \theta$

$\dfrac{(x-4)^2}{4} + \dfrac{(y+1)^2}{16} = 1$

27. $x = 4 \sec \theta$

$y = 3 \tan \theta$

$\dfrac{x^2}{16} = \sec^2 \theta$

$\dfrac{y^2}{9} = \tan^2 \theta$

$\dfrac{x^2}{16} - \dfrac{y^2}{9} = 1$

29. $x = t^3$

$y = 3 \ln t$

$y = 3 \ln \sqrt[3]{x} = \ln x$

31. $x = e^{-t}$

$y = e^{3t}$

$e^t = \dfrac{1}{x}$

$e^t = \sqrt[3]{y}$

$\sqrt[3]{y} = \dfrac{1}{x}$

$y = \dfrac{1}{x^3}$

$x > 0$

$y > 0$

33. By eliminating the parameters in (a) – (d), we get $y = 2x + 1$. They differ from each other in orientation and in restricted domains. These curves are all smooth except for (b).

(a) $x = t$, $y = 2t + 1$

(b) $x = \cos \theta$ $y = 2 \cos \theta + 1$

     $-1 \leq x \leq 1$ $-1 \leq y \leq 3$

     $\dfrac{dx}{d\theta} = \dfrac{dy}{d\theta} = 0$ when $\theta = 0, \pm \pi, \pm 2\pi, \ldots$

(c) $x = e^{-t}$ $y = 2e^{-t} + 1$

     $x > 0$ $y > 1$

(d) $x = e^{t}$ $y = 2e^{t} + 1$

     $x > 0$ $y > 1$

35. The curves are identical on $0 < \theta < \pi$. They are both smooth. Represent $y = 2(1 - x^2)$

37. (a)

(b) The orientation of the second curve is reversed.

(c) The orientation will be reversed.

(d) Many answers possible. For example, $x = 1 + t$, $y = 1 + 2t$, and $x = 1 - t$, $x = 1 - 2t$.

39.
$$x = x_1 + t(x_2 - x_1)$$
$$y = y_1 + t(y_2 - y_1)$$
$$\frac{x - x_1}{x_2 - x_1} = t$$
$$y = y_1 + \left(\frac{x - x_1}{x_2 - x_1}\right)(y_2 - y_1)$$
$$y - y_1 = \frac{y_2 - y_1}{x_2 - x_1}(x - x_1)$$
$$y - y_1 = m(x - x_1)$$

41.
$$x = h + a \cos \theta$$
$$y = k + b \sin \theta$$
$$\frac{x - h}{a} = \cos \theta$$
$$\frac{y - k}{b} = \sin \theta$$
$$\frac{(x - h)^2}{a^2} + \frac{(y - k)^2}{b^2} = 1$$

43. From Exercise 39 we have
$$x = 5t$$
$$y = -2t.$$
Solution not unique

45. From Exercise 40 we have
$$x = 2 + 4 \cos \theta$$
$$y = 1 + 4 \sin \theta.$$
Solution not unique

47. From Exercise 41 we have
$$a = 5, c = 4 \implies b = 3$$
$$x = 5 \cos \theta$$
$$y = 3 \sin \theta.$$
Center: $(0, 0)$
Solution not unique

49. From Exercise 42 we have

$a = 4, c = 5 \implies b = 3$

$x = 4 \sec \theta$

$y = 3 \tan \theta.$

Center: $(0, 0)$

Solution not unique

51. $y = 3x - 2$

Example

$x = t, \qquad y = 3t - 2$

$x = t - 3, \quad y = 3t - 11$

53. $y = x^3$

Example

$x = t, \qquad y = t^3$

$x = \sqrt[3]{t}, \qquad y = t$

$x = \tan t, \quad y = \tan^3 t$

55. $x = 2(\theta - \sin \theta)$

$y = 2(1 - \cos \theta)$

Not smooth at $\theta = 2n\pi$

57. $x = \theta - \frac{3}{2} \sin \theta$

$y = 1 - \frac{3}{2} \cos \theta$

59. $x = 3 \cos^3 \theta$

$y = 3 \sin^3 \theta$

Not smooth at $(x, y) = (\pm 3, 0)$ and $(0, \pm 3)$, or $\theta = \frac{1}{2} n\pi$.

61. $x = 2 \cot \theta$

$y = 2 \sin^2 \theta$

Smooth everywhere

63. See definition on page 665.

65. A plane curve C, represented by $x = f(t)$, $y = g(t)$, is smooth if f' and g' are continuous and not simultaneously 0. See page 670.

67. $x = 4 \cos \theta$

$y = 2 \sin 2\theta$

Matches (d)

69. $x = \cos \theta + \theta \sin \theta$

$y = \sin \theta - \theta \cos \theta$

Matches (b)

71. When the circle has rolled θ radians, we know that the center is at $(a\theta, a)$.

$$\sin \theta = \sin(180° - \theta) = \frac{|AC|}{b} = \frac{|BD|}{b} \quad \text{or} \quad |BD| = b \sin \theta$$

$$\cos \theta = -\cos(180° - \theta) = \frac{|AP|}{-b} \quad \text{or} \quad |AP| = -b \cos \theta$$

Therefore, $x = a\theta - b \sin \theta$ and $y = a - b \cos \theta$

73. False

$x = t^2 \implies x \geq 0$

$x = t^2 \implies y \geq 0$

The graph of the parametric equations is only a portion of the line $y = x$.

75. (a) $100 \text{ mi/hr} = \dfrac{(100)(5280)}{3600} = \dfrac{440}{3} \text{ ft/sec}$

$$x = (v_0 \cos \theta)t = \left(\frac{440}{3} \cos \theta\right)t$$

$$y = h + (v_0 \sin \theta)t - 16t^2$$

$$= 3 + \left(\frac{440}{3} \sin \theta\right)t - 16t^2$$

(b)

It is not a home run—when $x = 400$, $y < 10$.

(c)

Yes, it's a home run when $x = 400$, $y > 10$.

(d) We need to find the angle θ (and time t) such that

$$x = \left(\frac{440}{3} \cos \theta\right)t = 400$$

$$y = 3 + \left(\frac{440}{3} \sin \theta\right)t - 16t^2 = 10.$$

From the first equation $t = 1200/440 \cos \theta$. Substituting into the second equation,

$$10 = 3 + \left(\frac{440}{3} \sin \theta\right)\left(\frac{1200}{440 \cos \theta}\right) - 16\left(\frac{1200}{440 \cos \theta}\right)^2$$

$$7 = 400 \tan \theta - 16\left(\frac{120}{44}\right)^2 \sec^2 \theta$$

$$= 400 \tan \theta - 16\left(\frac{120}{44}\right)^2 (\tan^2 \theta + 1).$$

We now solve the quadratic for $\tan \theta$:

$$16\left(\frac{120}{44}\right)^2 \tan^2 \theta - 400 \tan \theta + 7 + 16\left(\frac{120}{44}\right)^2 = 0$$

$$\tan \theta \approx 0.35185 \implies \theta \approx 19.4°$$

Section 9.3   Parametric Equations and Calculus

1. $\dfrac{dy}{dx} = \dfrac{dy/dt}{dx/dt} = \dfrac{-4}{2t} = \dfrac{-2}{t}$

3. $\dfrac{dy}{dx} = \dfrac{dy/d\theta}{dx/d\theta} = \dfrac{-2 \cos \theta \sin \theta}{2 \sin \theta \cos \theta} = -1$

$$\left[\text{Note: } x + y = 1 \implies y = 1 - x \text{ and } \frac{dy}{d\theta} = -1\right]$$

5. $x = 2t$, $y = 3t - 1$

$$\frac{dy}{dx} = \frac{dy/dt}{dx/dt} = \frac{3}{2}$$

$$\frac{d^2y}{dx^2} = 0 \text{ Line}$$

7. $x = t + 1$, $y = t^2 + 3t$

$$\frac{dy}{dx} = \frac{2t + 3}{1} = 1 \text{ when } t = -1.$$

$$\frac{d^2y}{dx^2} = 2 \text{ concave upwards}$$

9. $x = 2 \cos \theta$, $y = 2 \sin \theta$

$$\frac{dy}{dx} = \frac{2 \cos \theta}{-2 \sin \theta} = -\cot \theta = -1 \text{ when } \theta = \frac{\pi}{4}.$$

$$\frac{d^2y}{dx^2} = \frac{\csc^2 \theta}{-2 \sin \theta} = \frac{-\csc^3 \theta}{2} = -\sqrt{2} \text{ when } \theta = \frac{\pi}{4}.$$

concave downward

11. $x = 2 + \sec \theta$, $y = 1 + 2 \tan \theta$

$$\frac{dy}{dx} = \frac{2 \sec^2 \theta}{\sec \theta \tan \theta}$$

$$= \frac{2 \sec \theta}{\tan \theta} = 2 \csc \theta = 4 \text{ when } \theta = \frac{\pi}{6}.$$

$$\frac{d^2y}{dx^2} = \frac{-2 \csc \theta \cot \theta}{\sec \theta \tan \theta}$$

$$= -2 \cot^3 \theta = -6\sqrt{3} \text{ when } \theta = \frac{\pi}{6}.$$

concave downward

13. $x = \cos^3 \theta, \ y = \sin^3 \theta$

$$\frac{dy}{dx} = \frac{3 \sin^2 \theta \cos \theta}{-3 \cos^2 \theta \sin \theta}$$

$$= -\tan \theta = -1 \text{ when } \theta = \frac{\pi}{4}.$$

$$\frac{d^2y}{dx^2} = \frac{-\sec^2 \theta}{-3 \cos^2 \theta \sin \theta} = \frac{1}{3 \cos^4 \theta \sin \theta}$$

$$= \frac{\sec^4 \theta \csc \theta}{3} = \frac{4\sqrt{2}}{3} \text{ when } \theta = \frac{\pi}{4}.$$

concave upward

15. $x = 2 \cot \theta, \ y = 2 \sin^2 \theta$

$$\frac{dy}{dx} = \frac{4 \sin \theta \cos \theta}{-2 \csc^2 \theta} = -2 \sin^3 \theta \cos \theta$$

At $\left(-\dfrac{2}{\sqrt{3}}, \dfrac{3}{2}\right)$, $\theta = \dfrac{2\pi}{3}$, and $\dfrac{dy}{dx} = \dfrac{3\sqrt{3}}{8}$.

Tangent line: $\quad\quad\quad y - \dfrac{3}{2} = \dfrac{3\sqrt{3}}{8}\left(x + \dfrac{2}{\sqrt{3}}\right)$

$$3\sqrt{3}x - 8y + 18 = 0$$

At $(0, 2)$, $\theta = \dfrac{\pi}{2}$, and $\dfrac{dy}{dx} = 0$.

Tangent line: $y - 2 = 0$

At $\left(2\sqrt{3}, \dfrac{1}{2}\right)$, $\theta = \dfrac{\pi}{6}$, and $\dfrac{dy}{dx} = -\dfrac{\sqrt{3}}{8}$.

Tangent line: $\quad\quad\quad y - \dfrac{1}{2} = -\dfrac{\sqrt{3}}{8}(x - 2\sqrt{3})$

$$\sqrt{3}x + 8y - 10 = 0$$

17. $x = 2t, \ y = t^2 - 1, \ t = 2$

(a)

(b) At $t = 2$, $(x, y) = (4, 3)$, and

$$\frac{dx}{dt} = 2, \ \frac{dy}{dt} = 4, \ \frac{dy}{dx} = 2$$

(c) $\dfrac{dy}{dx} = 2$. At $(4, 3)$, $y - 3 = 2(x - 4)$

$$y = 2x - 5$$

(d)

19. $x = t^2 - t + 2, \ y = t^3 - 3t, \ t = -1$

(a)

(b) At $t = -1$, $(x, y) = (4, 2)$, and

$$\frac{dx}{dt} = -3, \ \frac{dy}{dt} = 0, \ \frac{dy}{dx} = 0$$

(c) $\dfrac{dy}{dx} = 0$. At $(4, 2)$, $y - 2 = 0(x - 4)$

$$y = 2$$

(d)

21. $x = 2 \sin 2t, \ y = 3 \sin t$ crosses itself at the origin, $(x, y) = (0, 0)$.

At this point, $t = 0$ or $t = \pi$.

$$\frac{dy}{dx} = \frac{3 \cos t}{4 \cos 2t}$$

At $t = 0$: $\dfrac{dy}{dx} = \dfrac{3}{4}$ and $y = \dfrac{3}{4}x$. Tangent Line

At $t = \pi$, $\dfrac{dy}{dx} = -\dfrac{3}{4}$ and $y = \dfrac{-3}{4}x$ Tangent Line

23. $x = \cos\theta + \theta\sin\theta$, $y = \sin\theta - \theta\cos\theta$

Horizontal tangents: $\dfrac{dy}{d\theta} = \theta\sin\theta = 0$ when $\theta = \pm\pi,\ \pm 2\pi,\ \pm 3\pi,\ \ldots$.

Points: $(-1, [2n-1]\pi)$, $(1, 2n\pi)$ where n is an integer.

Points shown: $(1, 0)$, $(-1, \pi)$, $(1, -2\pi)$

Vertical tangents: $\dfrac{dx}{d\theta} = \theta\cos\theta = 0$ when $\theta = \pm\dfrac{\pi}{2},\ \pm\dfrac{3\pi}{2},\ \pm\dfrac{5\pi}{2},\ \ldots$.

Note: $\theta = 0$ corresponds to the cusp at $(x, y) = (1, 0)$.

$\dfrac{dy}{dx} = \dfrac{\theta\sin\theta}{\theta\cos\theta} = \tan\theta = 0$ at $\theta = 0$.

Points: $\left(\dfrac{(-1)^{n+1}(2n-1)\pi}{2}, (-1)^{n+1}\right)$ Points shown: $\left(\dfrac{\pi}{2}, 1\right)$, $\left(-\dfrac{3\pi}{2}, -1\right)$, $\left(\dfrac{5\pi}{2}, 1\right)$

25. $x = 1 - t$, $y = t^2$

Horizontal tangents: $\dfrac{dy}{dt} = 2t = 0$ when $t = 0$.

Point: $(1, 0)$

Vertical tangents: $\dfrac{dx}{dt} = -1 \neq 0$; none

27. $x = 1 - t$, $y = t^3 - 3t$

Horizontal tangents: $\dfrac{dy}{dt} = 3t^2 - 3 = 0$ when $t = \pm 1$.

Points: $(0, -2)$, $(2, 2)$

Vertical tangents: $\dfrac{dx}{dt} = -1 \neq 0$; none

29. $x = 3\cos\theta$, $y = 3\sin\theta$

Horizontal tangents: $\dfrac{dy}{d\theta} = 3\cos\theta = 0$ when $\theta = \dfrac{\pi}{2}, \dfrac{3\pi}{2}$.

Points: $(0, 3)$, $(0, -3)$

Vertical tangents: $\dfrac{dx}{d\theta} = -3\sin\theta = 0$ when $\theta = 0, \pi$.

Points: $(3, 0)$, $(-3, 0)$

31. $x = 4 + 2\cos\theta$, $y = -1 + \sin\theta$

Horizontal tangents: $\dfrac{dy}{d\theta} = \cos\theta = 0$ when $\theta = \dfrac{\pi}{2}, \dfrac{3\pi}{2}$.

Points: $(4, 0)$, $(4, -2)$

Vertical tangents: $\dfrac{dx}{d\theta} = -2\sin\theta = 0$ when $\theta = 0, \pi$.

Points: $(6, -1)$, $(2, -1)$

33. $x = \sec\theta$, $y = \tan\theta$

Horizontal tangents: $\dfrac{dy}{d\theta} = \sec^2\theta \neq 0$; none

Vertical tangents: $\dfrac{dx}{d\theta} = \sec\theta\tan\theta = 0$ when $x = 0, \pi$.

Points: $(1, 0)$, $(-1, 0)$

35. $x = t^2$, $y = 2t$, $0 \leq t \leq 2$

$\dfrac{dx}{dt} = 2t$, $\dfrac{dy}{dt} = 2$, $\left(\dfrac{dx}{dt}\right)^2 + \left(\dfrac{dy}{dt}\right)^2 = 4t^2 + 4 = 4(t^2 + 1)$

$s = 2\displaystyle\int_0^2 \sqrt{t^2 + 1}\, dt$

$= \left[t\sqrt{t^2 + 1} + \ln\left| t + \sqrt{t^2 + 1} \right| \right]_0^2$

$= 2\sqrt{5} + \ln\left(2 + \sqrt{5}\right) \approx 5.916$

37. $x = e^{-t}\cos t, y = e^{-t}\sin t, 0 \le t \le \dfrac{\pi}{2}$

$\dfrac{dx}{dt} = -e^{-t}(\sin t + \cos t), \dfrac{dy}{dt} = e^{-t}(\cos t - \sin t)$

$s = \displaystyle\int_0^{\pi/2} \sqrt{\left(\dfrac{dx}{dt}\right)^2 + \left(\dfrac{dy}{dt}\right)^2}\, dt$

$= \displaystyle\int_0^{\pi/2} \sqrt{2e^{-2t}}\, dt = -\sqrt{2}\int_0^{\pi/2} e^{-t}(-1)\, dt$

$= \left[-\sqrt{2}e^{-t}\right]_0^{\pi/2} = \sqrt{2}(1 - e^{-\pi/2}) \approx 1.12$

39. $x = \sqrt{t}, y = 3t - 1, \dfrac{dx}{dt} = \dfrac{1}{2\sqrt{t}}, \dfrac{dy}{dt} = 3$

$s = \displaystyle\int_0^1 \sqrt{\dfrac{1}{4t} + 9}\, dt = \dfrac{1}{2}\int_0^1 \dfrac{\sqrt{1 + 36t}}{\sqrt{t}}\, dt$

$= \dfrac{1}{6}\displaystyle\int_0^6 \sqrt{1 + u^2}\, du$

$= \dfrac{1}{12}\left[\ln\left(\sqrt{1 + u^2} + u\right) + u\sqrt{1 + u^2}\right]_0^6$

$= \dfrac{1}{12}\left[\ln\left(\sqrt{37} + 6\right) + 6\sqrt{37}\right] \approx 3.249$

$u = 6\sqrt{t}, \ du = \dfrac{3}{\sqrt{t}}\, dt$

41. $x = a\cos^3 \theta, y = a\sin^3 \theta, \dfrac{dx}{d\theta} = -3a\cos^2 \theta \sin \theta,$

$\dfrac{dy}{d\theta} = 3a\sin^2 \theta \cos \theta$

$s = 4\displaystyle\int_0^{\pi/2} \sqrt{9a^2\cos^4 \theta \sin^2 + 9a^2\sin^4 \theta \cos^2 \theta}\, d\theta$

$= 12a\displaystyle\int_0^{\pi/2} \sin \theta \cos \theta\sqrt{\cos^2 \theta + \sin^2 \theta}\, d\theta$

$= 6a\displaystyle\int_0^{\pi/2} \sin 2\theta\, d\theta = \left[-3a\cos 2\theta\right]_0^{\pi/2} = 6a$

43. $x = a(\theta - \sin \theta), y = a(1 - \cos \theta),$

$\dfrac{dx}{d\theta} = a(1 - \cos \theta), \dfrac{dy}{d\theta} = a\sin \theta$

$s = 2\displaystyle\int_0^\pi \sqrt{a^2(1 - \cos \theta)^2 + a^2\sin^2 \theta}\, d\theta$

$= 2\sqrt{2}a\displaystyle\int_0^\pi \sqrt{1 - \cos \theta}\, d\theta$

$= 2\sqrt{2}a\displaystyle\int_0^\pi \dfrac{\sin \theta}{\sqrt{1 + \cos \theta}}\, d\theta$

$= \left[-4\sqrt{2}a\sqrt{1 + \cos \theta}\right]_0^\pi = 8a$

45. $x = (90\cos 30°)t, y = (90\sin 30°)t - 16t^2$

(a)

(b) Range: $\left(t = \dfrac{45}{16}\right)$

(c) $\dfrac{dx}{dt} = 90\cos 30°, \dfrac{dy}{dt} = 90\sin 30° - 32t.$

$y = 0$ for $t = \dfrac{45}{16}.$

$s = \displaystyle\int_0^{45/16} \sqrt{(90\cos 30°)^2 + (90\sin 30° - 32t)^2}\, dt$

≈ 230.8 ft

(d) $y = 0 \Rightarrow (90\sin \theta)t = 16t^2 \Rightarrow t = 0, \dfrac{90}{16}\sin \theta$

$x = (90\cos \theta)t = (90\cos \theta)\dfrac{90}{16}\sin \theta$

$= \dfrac{90^2}{16}\sin \theta \cos \theta = \dfrac{90^2}{32}\sin 2\theta$

$x'(\theta) = \dfrac{90^2}{32}2\cos 2\theta = 0 \Rightarrow \theta = \dfrac{\pi}{4}$

By the First Derivative Test, $\theta = \dfrac{\pi}{4}(45°)$ maximizes the range ($x = 253.125$ feet).

To maximize the arc length, we have

$\dfrac{dx}{dt} = 90\cos \theta, \dfrac{dy}{dt} = 90\sin \theta - 32t$

$s = \displaystyle\int_0^{(90/16)\sin \theta} \sqrt{(90\cos \theta)^2 + (90\sin \theta - 32t)^2}\, dt$

$= \dfrac{2025}{8}\sin \theta + \dfrac{2025}{16}\cos^2 \theta \ln\left[\dfrac{1 + \sin \theta}{1 - \sin \theta}\right]$

Using a graphing utility, we see that s is a maximum of approximately 303.67 feet at $\theta \approx 0.9855(56.5°)$.

47. (a) $x = t - \sin t$ $x = 2t - \sin(2t)$

$y = 1 - \cos t$ $y = 1 - \cos(2t)$

$0 \le t \le 2\pi$ $0 \le t \le \pi$

(b) The average speed of the particle on the second path is twice the average speed of a particle on the first path.

(c) $x = \frac{1}{2}t - \sin(\frac{1}{2}t)$

$y = 1 - \cos(\frac{1}{2}t)$

The time required for the particle to traverse the same path is $t = 4\pi$.

49. $x = t,\ y = 2t,\ \dfrac{dx}{dt} = 1,\ \dfrac{dy}{dt} = 2$

(a) $S = 2\pi \displaystyle\int_0^4 2t\sqrt{1 + 4}\,dt = 4\sqrt{5}\pi \int_0^4 t\,dt$

$= \left[2\sqrt{5}\pi t^2 \right]_0^4 = 32\pi\sqrt{5}$

(b) $S = 2\pi \displaystyle\int_0^4 t\sqrt{1 + 4}\,dt = 2\sqrt{5}\pi \int_0^4 t\,dt$

$= \left[\sqrt{5}\pi t^2 \right]_0^4 = 16\pi\sqrt{5}$

51. $x = 4\cos\theta,\ y = 4\sin\theta,\ \dfrac{dx}{d\theta} = -4\sin\theta,\ \dfrac{dy}{d\theta} = 4\cos\theta$

$S = 2\pi \displaystyle\int_0^{\pi/2} 4\cos\theta\sqrt{(-4\sin\theta)^2 + (4\cos\theta)^2}\,d\theta$

$= 32\pi \displaystyle\int_0^{\pi/2} \cos\theta\,d\theta = \left[32\pi\sin\theta \right]_0^{\pi/2} = 32\pi$

53. $x = a\cos^3\theta,\ y = a\sin^3\theta,\ \dfrac{dx}{d\theta} = -3a\cos^2\theta\sin\theta,\ \dfrac{dy}{d\theta} = 3a\sin^2\theta\cos\theta$

$S = 4\pi \displaystyle\int_0^{\pi/2} a\sin^3\theta\sqrt{9a^2\cos^4\theta\sin^2\theta + 9a^2\sin^4\theta\cos^2\theta}\,d\theta = 12a^2\pi \int_0^{\pi/2} \sin^4\theta\cos\theta\,d\theta = \frac{12\pi a^2}{5}\left[\sin^5\theta \right]_0^{\pi/2} = \frac{12}{5}\pi a^2$

55. $\dfrac{dy}{dx} = \dfrac{dy/dt}{dx/dt}$

See Theorem 9.7, page 675.

57. One possible answer is the graph given by

$x = t,\ y = -t.$

59. $s = \displaystyle\int_a^b \sqrt{\left(\frac{dx}{dt}\right)^2 + \left(\frac{dy}{dt}\right)^2}\,dt$

See Theorem 9.8, page 678.

61. $x = r\cos\phi,\ y = r\sin\phi$

$S = 2\pi \displaystyle\int_0^\theta r\sin\phi\sqrt{r^2\sin^2\phi + r^2\cos^2\phi}\,d\phi$

$= 2\pi r^2 \displaystyle\int_0^\theta \sin\phi\,d\phi$

$= \left[-2\pi r^2\cos\phi \right]_0^\theta$

$= 2\pi r^2(1 - \cos\theta)$

63. $x = \sqrt{t}$, $y = 4 - t$, $0 \leq t \leq 4$

$$A = \int_0^4 (4 - t)\frac{1}{2\sqrt{t}}\,dt = \frac{1}{2}\int_0^4 (4t^{-1/2} - t^{1/2})\,dt = \left[\frac{1}{2}\left(8\sqrt{t} - \frac{2}{3}t\sqrt{t}\right)\right]_0^4 = \frac{16}{3}$$

$$\bar{x} = \frac{3}{16}\int_0^4 (4 - t)\sqrt{t}\left(\frac{1}{2\sqrt{t}}\right)dt = \frac{3}{32}\int_0^4 (4 - t)\,dt = \left[\frac{3}{32}\left(4t - \frac{t^2}{2}\right)\right]_0^4 = \frac{3}{4}$$

$$\bar{y} = \frac{3}{32}\int_0^4 (4 - t)^2\,\frac{1}{2\sqrt{t}}\,dt = \frac{3}{64}\int_0^4 [16t^{-1/2} - 8t^{1/2} + t^{3/2}]\,dt = \frac{3}{64}\left[32\sqrt{t} - \frac{16}{3}t\sqrt{t} + \frac{2}{5}t^2\sqrt{t}\right]_0^4 = \frac{8}{5}$$

$$(\bar{x}, \bar{y}) = \left(\frac{3}{4}, \frac{8}{5}\right)$$

65. $x = 3\cos\theta$, $y = 3\sin\theta$, $\dfrac{dx}{d\theta} = -3\sin\theta$

$$V = 2\pi\int_{\pi/2}^0 (3\sin\theta)^2(-3\sin\theta)\,d\theta$$

$$= -54\pi\int_{\pi/2}^0 \sin^3\theta\,d\theta$$

$$= -54\pi\int_{\pi/2}^0 (1 - \cos^2\theta)\sin\theta\,d\theta$$

$$= -54\pi\left[-\cos\theta + \frac{\cos^3\theta}{3}\right]_{\pi/2}^0 = 36\pi$$

67. $x = 2\sin^2\theta$

$y = 2\sin^2\theta\tan\theta$

$\dfrac{dx}{d\theta} = 4\sin\theta\cos\theta$

$$A = \int_0^{\pi/2} 2\sin^2\theta\tan\theta(4\sin\theta\cos\theta)\,d\theta = 8\int_0^{\pi/2}\sin^4\theta\,d\theta$$

$$= 8\left[\frac{-\sin^3\theta\cos\theta}{4} - \frac{3}{8}\sin\theta\cos\theta + \frac{3}{8}\theta\right]_0^{\pi/2} = \frac{3\pi}{2}$$

69. πab is area of ellipse (d).   **71.** $6\pi a^2$ is area of cardioid (f).   **73.** $\frac{8}{3}ab$ is area of hourglass (a).

75. (a) $x = \dfrac{1 - t^2}{1 + t^2}$, $y = \dfrac{2t}{1 + t^2}$, $-20 \leq t \leq 20$

The graph (for $-\infty < t < \infty$) is the circle $x^2 + y^2 = 1$, except the point $(-1, 0)$.

Verify: $x^2 + y^2 = \left(\dfrac{1 - t^2}{1 + t^2}\right)^2 + \left(\dfrac{2t}{1 + t^2}\right)^2 = \dfrac{1 - 2t^2 + t^4 + 4t^2}{(1 + t^2)^2} = \dfrac{(1 + t^2)^2}{(1 + t^2)^2} = 1$

(b) As t increases from -20 to 0, the speed increases, and as t increases from 0 to 20, the speed decreases.

77. False

$$\frac{d^2y}{dx^2} = \frac{\dfrac{d}{dt}\left[\dfrac{g'(t)}{f'(t)}\right]}{f'(t)} = \frac{f'(t)g''(t) - g'(t)f''(t)}{[f'(t)]^3}$$

Section 9.4   Polar Coordinates and Polar Graphs

1. $\left(4, \dfrac{\pi}{2}\right)$

$x = 4\cos\left(\dfrac{\pi}{2}\right) = 0$

$y = 4\sin\left(\dfrac{\pi}{2}\right) = 4$

$(x, y) = (0, 4)$

3. $\left(-4, -\dfrac{\pi}{3}\right)$

$x = -4\cos\left(-\dfrac{\pi}{3}\right) = -2$

$y = -4\sin\left(-\dfrac{\pi}{3}\right) = 2\sqrt{3}$

$(x, y) = \left(-2, 2\sqrt{3}\right)$

5. $\left(\sqrt{2}, 2.36\right)$

$x = \sqrt{2}\cos(2.36) \approx -1.004$

$y = \sqrt{2}\sin(2.36) \approx 0.996$

$(x, y) = (-1.004, 0.996)$

7. $(r, \theta) = \left(5, \dfrac{3\pi}{4}\right)$

$(x, y) = (-3.5355, 3.5355)$

9. $(r, \theta) = (-3.5, 2.5)$

$(x, y) = (2.804, -2.095)$

11. $(x, y) = (1, 1)$

$r = \pm\sqrt{2}$

$\tan\theta = 1$

$\theta = \dfrac{\pi}{4}, \dfrac{5\pi}{4}, \left(\sqrt{2}, \dfrac{\pi}{4}\right), \left(-\sqrt{2}, \dfrac{5\pi}{4}\right)$

13. $(x, y) = (-3, 4)$

$r = \pm\sqrt{9 + 16} = \pm 5$

$\tan\theta = -\dfrac{4}{3}$

$\theta \approx 2.214, 5.356, (5, 2.214), (-5, 5.356)$

15. $(x, y) = (3, -2)$

$(r, \theta) = (3.606, -0.588)$

17. $(x, y) = \left(\dfrac{5}{2}, \dfrac{4}{3}\right)$

$(r, \theta) = (2.833, 0.490)$

19. (a) $(x, y) = (4, 3.5)$

(b) $(r, \theta) = (4, 3.5)$

21. $x^2 + y^2 = a^2$

$r = a$

23. $y = 4$

$r \sin \theta = 4$

$r = 4 \csc \theta$

25. $3x - y + 2 = 0$

$3r \cos \theta - r \sin \theta + 2 = 0$

$r(3 \cos \theta - \sin \theta) = -2$

$$r = \frac{-2}{3 \cos \theta - \sin \theta}$$

27. $y^2 = 9x$

$r^2 \sin^2 \theta = 9r \cos \theta$

$$r = \frac{9 \cos \theta}{\sin^2 \theta}$$

$r = 9 \csc^2 \theta \cos \theta$

29. $r = 3$

$r^2 = 9$

$x^2 + y^2 = 9$

31. $r = \sin \theta$

$r^2 = r \sin \theta$

$x^2 + y^2 = y$

$$x^2 + \left(y - \frac{1}{2}\right)^2 = \frac{1}{4}$$

$x^2 + y^2 - y = 0$

33. $r = \theta$

$\tan r = \tan \theta$

$$\tan \sqrt{x^2 + y^2} = \frac{y}{x}$$

$$\sqrt{x^2 + y^2} = \arctan \frac{y}{x}$$

35. $r = 3 \sec \theta$

$r \cos \theta = 3$

$x = 3$

$x - 3 = 0$

37. $r = 3 - 4 \cos \theta$

$0 \le \theta < 2\pi$

39. $r = 2 + \sin \theta$

$0 \le \theta < 2\pi$

41. $r = \dfrac{2}{1 + \cos \theta}$

Traced out once on
$-\pi < \theta < \pi$

43. $r = 2 \cos\left(\dfrac{3\theta}{2}\right)$

$0 \le \theta < 4\pi$

45. $r^2 = 4 \sin 2\theta$

$0 \le \theta < \dfrac{\pi}{2}$

47.
$$r = 2(h \cos \theta + k \sin \theta)$$
$$r^2 = 2r(h \cos \theta + k \sin \theta)$$
$$r^2 = 2[h(r \cos \theta) + k(r \sin \theta)]$$
$$x^2 + y^2 = 2(hx + ky)$$
$$x^2 + y^2 - 2hx - 2ky = 0$$
$$(x^2 - 2hx + h^2) + (y^2 - 2ky + k^2) = 0 + h^2 + k^2$$
$$(x - h)^2 + (y - k)^2 = h^2 + k^2$$

Radius: $\sqrt{h^2 + k^2}$

Center: (h, k)

49. $\left(4, \dfrac{2\pi}{3}\right), \left(2, \dfrac{\pi}{6}\right)$

$$d = \sqrt{4^2 + 2^2 - 2(4)(2) \cos\left(\dfrac{2\pi}{3} - \dfrac{\pi}{6}\right)}$$

$$= \sqrt{20 - 16 \cos \dfrac{\pi}{2}} = 2\sqrt{5} \approx 4.5$$

51. $(2, 0.5), (7, 1.2)$

$$d = \sqrt{2^2 + 7^2 - 2(2)(7) \cos(0.5 - 1.2)}$$

$$= \sqrt{53 - 28 \cos(-0.7)} \approx 5.6$$

53. $r = 2 + 3 \sin \theta$

$$\dfrac{dy}{dx} = \dfrac{3 \cos \theta \sin \theta + \cos \theta(2 + 3 \sin \theta)}{3 \cos \theta \cos \theta - \sin \theta(2 + 3 \sin \theta)}$$

$$= \dfrac{2 \cos \theta(3 \sin \theta + 1)}{3 \cos 2\theta - 2 \sin \theta} = \dfrac{2 \cos \theta(3 \sin \theta + 1)}{6 \cos^2 \theta - 2 \sin \theta - 3}$$

At $\left(5, \dfrac{\pi}{2}\right), \dfrac{dy}{dx} = 0.$

At $(2, \pi), \dfrac{dy}{dx} = -\dfrac{2}{3}.$

At $\left(-1, \dfrac{3\pi}{2}\right), \dfrac{dy}{dx} = 0.$

55. (a), (b) $r = 3(1 - \cos \theta)$

$(r, \theta) = \left(3, \dfrac{\pi}{2}\right) \Rightarrow (x, y) = (0, 3)$

Tangent line: $y - 3 = -1(x - 0)$

$$y = -x + 3$$

(c) At $\theta = \dfrac{\pi}{2}, \dfrac{dy}{dx} = -1.0.$

57. (a), (b) $r = 3 \sin \theta$

$(r, \theta) = \left(\dfrac{3\sqrt{3}}{2}, \dfrac{\pi}{3}\right) \Rightarrow (x, y) = \left(\dfrac{3\sqrt{3}}{4}, \dfrac{9}{4}\right)$

Tangent line: $y - \dfrac{9}{4} = -\sqrt{3}\left(x - \dfrac{3\sqrt{3}}{4}\right)$

$$y = -\sqrt{3}x + \dfrac{9}{2}$$

(c) At $\theta = \dfrac{\pi}{3}, \dfrac{dy}{dx} = -\sqrt{3} \approx -1.732.$

59. $r = 1 - \sin\theta$

$$\frac{dy}{d\theta} = (1 - \sin\theta)\cos\theta - \cos\theta\sin\theta$$

$$= \cos\theta(1 - 2\sin\theta) = 0$$

$$\cos\theta = 0, \ \sin\theta = \frac{1}{2} \implies \theta = \frac{\pi}{2}, \frac{3\pi}{2}, \frac{\pi}{6}, \frac{5\pi}{6}$$

Horizontal tangents: $\left(2, \frac{3\pi}{2}\right), \left(\frac{1}{2}, \frac{\pi}{6}\right), \left(\frac{1}{2}, \frac{5\pi}{6}\right)$

$$\frac{dx}{d\theta} = (-1 + \sin\theta)\sin\theta - \cos\theta\cos\theta$$

$$= -\sin\theta + \sin^2\theta + \sin^2\theta - 1$$

$$= 2\sin^2\theta - \sin\theta - 1$$

$$= (2\sin\theta + 1)(\sin\theta - 1) = 0$$

$$\sin\theta = 1, \ \sin\theta = -\frac{1}{2} \implies \theta = \frac{\pi}{2}, \frac{7\pi}{6}, \frac{11\pi}{6}$$

Vertical tangents: $\left(\frac{3}{2}, \frac{7\pi}{6}\right), \left(\frac{3}{2}, \frac{11\pi}{6}\right)$

61. $r = 2\csc\theta + 3$

$$\frac{dy}{d\theta} = (2\csc\theta + 3)\cos\theta + (-2\csc\theta\cot\theta)\sin\theta$$

$$= 3\cos\theta = 0$$

$$\theta = \frac{\pi}{2}, \frac{3\pi}{2}$$

Horizontal: $\left(5, \frac{\pi}{2}\right), \left(1, \frac{3\pi}{2}\right)$

63. $r = 4\sin\theta\cos^2\theta$

Horizontal tangents:

   $(0, 0), (1.4142, 0.7854), (1.4142, 2.3562)$

65. $r = 2\csc\theta + 5$

Horizontal tangents: $\left(7, \frac{\pi}{2}\right), \left(3, \frac{3\pi}{2}\right)$

67.
$$r = 3\sin\theta$$
$$r^2 = 3r\sin\theta$$
$$x^2 + y^2 = 3y$$
$$x^2 + \left(y - \frac{3}{2}\right)^2 = \frac{9}{4}$$

Circle $r = \frac{3}{2}$

Center: $\left(0, \frac{3}{2}\right)$

Tangent at the pole: $\theta = 0$

69. $r = 2(1 - \sin\theta)$

Cardioid

Symmetric to y-axis, $\theta = \frac{\pi}{2}$

71. $r = 2\cos(3\theta)$

Rose curve with three petals

Symmetric to the polar axis

Relative extrema: $(2, 0), \left(-2, \frac{\pi}{3}\right), \left(2, \frac{2\pi}{3}\right)$

θ	0	$\frac{\pi}{6}$	$\frac{\pi}{4}$	$\frac{\pi}{3}$	$\frac{\pi}{2}$	$\frac{2\pi}{3}$	$\frac{5\pi}{6}$	π
r	2	0	$-\sqrt{2}$	-2	0	2	0	-2

Tangents at the pole: $\theta = \frac{\pi}{6}, \frac{\pi}{2}, \frac{5\pi}{6}$

73. $r = 3 \sin 2\theta$

Rose curve with four petals

Symmetric to the polar axis, $\theta = \dfrac{\pi}{2}$, and pole

Relative extrema: $\left(\pm 3, \dfrac{\pi}{4}\right), \left(\pm 3, \dfrac{5\pi}{4}\right)$

Tangents at the pole: $\theta = 0, \dfrac{\pi}{2}$

($\theta = \pi, \ 3\pi/2$ give the same tangents.)

75. $r = 5$

Circle radius: 5

$x^2 + y^2 = 25$

77. $r = 4(1 + \cos\theta)$

Cardioid

79. $r = 3 - 2\cos\theta$

Limaçon

Symmetric to polar axis

θ	0	$\dfrac{\pi}{3}$	$\dfrac{\pi}{2}$	$\dfrac{2\pi}{3}$	π
r	1	2	3	4	5

81. $r = 3 \csc\theta$

$r \sin\theta = 3$

     $y = 3$

Horizontal line

83. $r = 2\theta$

Spiral of Archimedes

Symmetric to $\theta = \dfrac{\pi}{2}$

θ	0	$\dfrac{\pi}{4}$	$\dfrac{\pi}{2}$	$\dfrac{3\pi}{4}$	π	$\dfrac{5\pi}{4}$	$\dfrac{3\pi}{2}$
r	0	$\dfrac{\pi}{2}$	π	$\dfrac{3\pi}{2}$	2π	$\dfrac{5\pi}{2}$	3π

Tangent at the pole: $\theta = 0$

85. $r^2 = 4\cos(2\theta)$

Lemniscate

Symmetric to the polar axis, $\theta = \dfrac{\pi}{2}$, and pole

Relative extrema: $(\pm 2, 0)$

θ	0	$\dfrac{\pi}{6}$	$\dfrac{\pi}{4}$
r	± 2	$\pm\sqrt{2}$	0

Tangents at the pole: $\theta = \dfrac{\pi}{4}, \dfrac{3\pi}{4}$

87. Since

$$r = 2 - \sec \theta = 2 - \frac{1}{\cos \theta},$$

the graph has polar axis symmetry and the lengths at the pole are

$$\theta = \frac{\pi}{3}, \frac{-\pi}{3}.$$

Furthermore,

$$r \Rightarrow -\infty \text{ as } \theta \Rightarrow \frac{\pi}{2}^{-}$$

$$r \Rightarrow \infty \text{ as } \theta \Rightarrow -\frac{\pi}{2}^{+}.$$

Also, $r = 2 - \dfrac{1}{\cos \theta} = 2 - \dfrac{r}{r \cos \theta} = 2 - \dfrac{r}{x}$

$$rx = 2x - r$$

$$r = \frac{2x}{1 + x}.$$

Thus, $r \Rightarrow \pm\infty$ as $x \Rightarrow -1$.

89. $r = \dfrac{2}{\theta}$

Hyperbolic spiral

$r \Rightarrow \infty$ as $\theta \Rightarrow 0$

$$r = \frac{2}{\theta} \Rightarrow \theta = \frac{2}{r} = \frac{2 \sin \theta}{r \sin \theta} = \frac{2 \sin \theta}{y}$$

$$y = \frac{2 \sin \theta}{\theta}$$

$$\lim_{\theta \to 0} \frac{2 \sin \theta}{\theta} = \lim_{\theta \to 0} \frac{2 \cos \theta}{1} = 2$$

91. The rectangular coordinate system consists of all points of the form (x, y) where x is the directed distance from the y-axis to the point, and y is the directed distance from the x-axis to the point. Every point has a unique representation.

The polar coordinate system uses (r, θ) to designate the location of a point.

r is the directed distance to the origin and θ is the angle the point makes with the positive x-axis, measured clockwise.

Point do not have a unique polar representation.

93. $r = a$ circle

$\theta = b$ line

95. $r = 2 \sin \theta$ circle

Matches (c)

97. $r = 3(1 + \cos \theta)$

Cardioid

Matches (a)

99. $r = 4 \sin \theta$

(a) $0 \le \theta \le \dfrac{\pi}{2}$

(b) $\dfrac{\pi}{2} \le \theta \le \pi$

(c) $-\dfrac{\pi}{2} \le \theta \le \dfrac{\pi}{2}$

101. Let the curve $r = f(\theta)$ be rotated by ϕ to form the curve $r = g(\theta)$. If (r_1, θ_1) is a point on $r = f(\theta)$, then $(r_1, \theta_1 + \phi)$ is on $r = g(\theta)$. That is,

$$g(\theta_1 + \phi) = r_1 = f(\theta_1).$$

Letting $\theta = \theta_1 + \phi$, or $\theta_1 = \theta - \phi$, we see that

$$g(\theta) = g(\theta_1 + \phi) = f(\theta_1) = f(\theta - \phi).$$

103. $r = 2 - \sin \theta$

(a) $r = 2 - \sin\left(\theta - \dfrac{\pi}{4}\right) = 2 - \dfrac{\sqrt{2}}{2}(\sin \theta - \cos \theta)$

(b) $r = 2 - \sin\left(\theta - \dfrac{\pi}{2}\right) = 2 - (-\cos \theta) = 2 + \cos \theta$

(c) $r = 2 - \sin(\theta - \pi) = 2 - (-\sin \theta) = 2 + \sin \theta$

(d) $r = 2 - \sin\left(\theta - \dfrac{3\pi}{2}\right) = 2 - \cos \theta$

105. (a) $r = 1 - \sin \theta$

(b) $r = 1 - \sin\left(\theta - \dfrac{\pi}{4}\right)$

Rotate the graph of

$$r = 1 - \sin \theta$$

through the angle $\pi/4$.

107. $\tan \psi = \dfrac{r}{dr/d\theta} = \dfrac{2(1 - \cos \theta)}{2 \sin \theta}$

At $\theta = \pi$, $\tan \psi$ is undefined $\implies \psi = \dfrac{\pi}{2}$.

109. $\tan \psi = \dfrac{r}{dr/d\theta} = \dfrac{2 \cos 3\theta}{-6 \sin 3\theta}$

At $\theta = \dfrac{\pi}{4}$, $\tan \psi = \dfrac{1}{3} \implies \psi \approx 18.4°$

111. $r = \dfrac{6}{1 - \cos\theta} = 6(1 - \cos\theta)^{-1} \Rightarrow \dfrac{dr}{d\theta} = \dfrac{6\sin\theta}{(1 - \cos\theta)^2}$

$$\tan\psi = \frac{r}{\dfrac{dr}{d\theta}} = \frac{\dfrac{6}{(1 - \cos\theta)}}{\dfrac{6\sin\theta}{(1 - \cos\theta)^2}} = \frac{1 - \cos\theta}{\sin\theta}$$

At $\theta = \dfrac{2\pi}{3}$, $\tan\psi = \dfrac{1 - \left(-\dfrac{1}{2}\right)}{\dfrac{\sqrt{3}}{2}} = \sqrt{3}$.

$\psi = \dfrac{\pi}{3}$, (60°)

113. True

115. True

Section 9.5   Area and Arc Length in Polar Coordinates

1. (a) $r = 8\sin\theta$

$A = \pi(4)^2 = 16\pi$

(b) $A = 2\left(\dfrac{1}{2}\right)\displaystyle\int_0^{\pi/2}\left[8\sin\theta\right]^2 d\theta$

$\qquad = 64\displaystyle\int_0^{\pi/2}\sin^2\theta\, d\theta$

$\qquad = 32\displaystyle\int_0^{\pi/2}(1 - \cos 2\theta)\, d\theta$

$\qquad = 32\left[\theta - \dfrac{\sin 2\theta}{2}\right]_0^{\pi/2} = 16\pi$

3. $A = 2\left[\dfrac{1}{2}\displaystyle\int_0^{\pi/6}(2\cos 3\theta)^2\, d\theta\right] = 2\left[\theta + \dfrac{1}{6}\sin 6\theta\right]_0^{\pi/6} = \dfrac{\pi}{3}$

5. $A = 2\left[\dfrac{1}{2}\displaystyle\int_0^{\pi/4}(\cos 2\theta)^2\, d\theta\right]$

$\qquad = \dfrac{1}{2}\left[\theta + \dfrac{1}{4}\sin 4\theta\right]_0^{\pi/4} = \dfrac{\pi}{8}$

7. $A = 2\left[\dfrac{1}{2}\displaystyle\int_{-\pi/2}^{\pi/2}(1 - \sin\theta)^2\, d\theta\right]$

$\qquad = \left[\dfrac{3}{2}\theta + 2\cos\theta - \dfrac{1}{4}\sin 2\theta\right]_{-\pi/2}^{\pi/2} = \dfrac{3\pi}{2}$

9. $A = 2\left[\dfrac{1}{2}\displaystyle\int_{2\pi/3}^{\pi}(1 + 2\cos\theta)^2\, d\theta\right]$

$\qquad = \left[3\theta + 4\sin\theta + \sin 2\theta\right]_{2\pi/3}^{\pi} = \dfrac{2\pi - 3\sqrt{3}}{2}$

11. The area inside the outer loop is

$$2\left[\dfrac{1}{2}\displaystyle\int_0^{2\pi/3}(1 + 2\cos\theta)^2\, d\theta\right] = \left[3\theta + 4\sin\theta + \sin 2\theta\right]_0^{2\pi/3} = \dfrac{4\pi + 3\sqrt{3}}{2}.$$

From the result of Exercise 9, the area between the loops is

$$A = \left(\dfrac{4\pi + 3\sqrt{3}}{2}\right) - \left(\dfrac{2\pi - 3\sqrt{3}}{2}\right) = \pi + 3\sqrt{3}.$$

13. $r = 1 + \cos\theta$

$r = 1 - \cos\theta$

Solving simultaneously,

$$1 + \cos\theta = 1 - \cos\theta$$

$$2\cos\theta = 0$$

$$\theta = \frac{\pi}{2}, \frac{3\pi}{2}.$$

Replacing r by $-r$ and θ by $\theta + \pi$ in the first equation and solving, $-1 + \cos\theta = 1 - \cos\theta$, $\cos\theta = 1$, $\theta = 0$. Both curves pass through the pole, $(0, \pi)$, and $(0, 0)$, respectively.

Points of intersection: $\left(1, \frac{\pi}{2}\right), \left(1, \frac{3\pi}{2}\right), (0, 0)$

15. $r = 1 + \cos\theta$

$r = 1 - \sin\theta$

Solving simultaneously,

$$1 + \cos\theta = 1 - \sin\theta$$

$$\cos\theta = -\sin\theta$$

$$\tan\theta = -1$$

$$\theta = \frac{3\pi}{4}, \frac{7\pi}{4}.$$

Replacing r by $-r$ and θ by $\theta + \pi$ in the first equation and solving, $-1 + \cos\theta = 1 - \sin\theta$, $\sin\theta + \cos\theta = 2$, which has no solution. Both curves pass through the pole, $(0, \pi)$, and $(0, \pi/2)$, respectively.

Points of intersection: $\left(\frac{2 - \sqrt{2}}{2}, \frac{3\pi}{4}\right), \left(\frac{2 + \sqrt{2}}{2}, \frac{7\pi}{4}\right), (0, 0)$

17. $r = 4 - 5\sin\theta$

$r = 3\sin\theta$

Solving simultaneously,

$$4 - 5\sin\theta = 3\sin\theta$$

$$\sin\theta = \frac{1}{2}$$

$$\theta = \frac{\pi}{6}, \frac{5\pi}{6}.$$

Both curves pass through the pole, $(0, \arcsin 4/5)$, and $(0, 0)$, respectively.

Points of intersection: $\left(\frac{3}{2}, \frac{\pi}{6}\right), \left(\frac{3}{2}, \frac{5\pi}{6}\right), (0, 0)$

19. $r = \frac{\theta}{2}$

$r = 2$

Solving simultaneously, we have

$$\theta/2 = 2, \theta = 4.$$

Points of intersection:

$$(2, 4), (-2, -4)$$

21. $r = 4\sin 2\theta$

$r = 2$

$r = 4\sin 2\theta$ is the equation of a rose curve with four petals and is symmetric to the polar axis, $\theta = \pi/2$, and the pole. Also, $r = 2$ is the equation of a circle of radius 2 centered at the pole. Solving simultaneously,

$$4\sin 2\theta = 2$$

$$2\theta = \frac{\pi}{6}, \frac{5\pi}{6}$$

$$\theta = \frac{\pi}{12}, \frac{5\pi}{12}.$$

Therefore, the points of intersection for one petal are $(2, \pi/12)$ and $(2, 5\pi/12)$. By symmetry, the other points of intersection are $(2, 7\pi/12)$, $(2, 11\pi/12)$, $(2, 13\pi/12)$, $(2, 17\pi/12)$, $(2, 19\pi/12)$, and $(2, 23\pi/12)$.

23. $r = 2 + 3 \cos \theta$

$r = \dfrac{\sec \theta}{2}$

The graph of $r = 2 + 3 \cos \theta$ is a limaçon with an inner loop $(b > a)$ and is symmetric to the polar axis. The graph of $r = (\sec \theta)/2$ is the vertical line $x = 1/2$. Therefore, there are four points of intersection. Solving simultaneously,

$$2 + 3 \cos \theta = \frac{\sec \theta}{2}$$

$$6 \cos^2 \theta + 4 \cos \theta - 1 = 0$$

$$\cos \theta = \frac{-2 \pm \sqrt{10}}{6}$$

$$\theta = \arccos\left(\frac{-2 + \sqrt{10}}{6}\right) \approx 1.376$$

$$\theta = \arccos\left(\frac{-2 - \sqrt{10}}{6}\right) \approx 2.6068.$$

Points of intersection: $(-0.581, \pm 2.607), (2.581, \pm 1.376)$

25. $r = \cos \theta$

$r = 2 - 3 \sin \theta$

Points of intersection:

$$(0, 0), (0.935, 0.363), (0.535, -1.006)$$

The graphs reach the pole at different times (θ values).

27. From Exercise 21, the points of intersection for one petal are $(2, \pi/12)$ and $(2, 5\pi/12)$. The area within one petal is

$$A = \frac{1}{2} \int_0^{\pi/12} (4 \sin 2\theta)^2 \, d\theta + \frac{1}{2} \int_{\pi/12}^{5\pi/12} (2)^2 \, d\theta + \frac{1}{2} \int_{5\pi/12}^{\pi/2} (4 \sin 2\theta)^2 \, d\theta$$

$$= 16 \int_0^{\pi/12} \sin^2(2\theta) \, d\theta + 2 \int_{\pi/12}^{5\pi/12} d\theta \text{ (by symmetry of the petal)}$$

$$= 8\left[\theta - \frac{1}{4} \sin 4\theta\right]_0^{\pi/12} + \left[2\theta\right]_{\pi/12}^{5\pi/12} = \frac{4\pi}{3} - \sqrt{3}.$$

Total area $= 4\left(\dfrac{4\pi}{3} - \sqrt{3}\right) = \dfrac{16\pi}{3} - 4\sqrt{3} = \dfrac{4}{3}\left(4\pi - 3\sqrt{3}\right)$

29. $A = 4\left[\dfrac{1}{2} \displaystyle\int_0^{\pi/2} (3 - 2 \sin \theta)^2 \, d\theta\right]$

$$= 2\left[11\theta + 12 \cos \theta - \sin(2\theta)\right]_0^{\pi/2} = 11\pi - 24$$

31. $A = 2\left[\dfrac{1}{2} \displaystyle\int_0^{\pi/6} (4 \sin \theta)^2 \, d\theta + \dfrac{1}{2} \displaystyle\int_{\pi/6}^{\pi/2} (2)^2 \, d\theta\right]$

$$= 16\left[\frac{1}{2} \theta - \frac{1}{4} \sin(2\theta)\right]_0^{\pi/6} + \left[4\theta\right]_{\pi/6}^{\pi/2}$$

$$= \frac{8\pi}{3} - 2\sqrt{3} = \frac{2}{3}\left(4\pi - 3\sqrt{3}\right)$$

33. $A = 2\left[\dfrac{1}{2} \displaystyle\int_0^{\pi} [a(1 + \cos \theta)]^2 \, d\theta\right] - \dfrac{a^2 \pi}{4}$

$$= a^2\left[\frac{3}{2}\theta + 2 \sin \theta + \frac{\sin 2\theta}{4}\right]_0^{\pi} - \frac{a^2 \pi}{4}$$

$$= \frac{3a^2 \pi}{2} - \frac{a^2 \pi}{4} = \frac{5a^2 \pi}{4}$$

35. $A = \dfrac{\pi a^2}{8} + \dfrac{1}{2} \displaystyle\int_{\pi/2}^{\pi} [a(1 + \cos\theta)]^2 \, d\theta$

$= \dfrac{\pi a^2}{8} + \dfrac{a^2}{2} \displaystyle\int_{\pi/2}^{\pi} \left(\dfrac{3}{2} + 2\cos\theta + \dfrac{\cos 2\theta}{2} \right) d\theta$

$= \dfrac{\pi a^2}{8} + \dfrac{a^2}{2} \left[\dfrac{3}{2}\theta + 2\sin\theta + \dfrac{\sin 2\theta}{4} \right]_{\pi/2}^{\pi}$

$= \dfrac{\pi a^2}{8} + \dfrac{a^2}{2} \left[\dfrac{3\pi}{2} - \dfrac{3\pi}{4} - 2 \right] = \dfrac{a^2}{2} [\pi - 2]$

37. (a) $r = a\cos^2\theta$

$r^3 = ar^2 \cos^2\theta$

$(x^2 + y^2)^{3/2} = ax^2$

(b)

(c) $A = 4\left(\dfrac{1}{2}\right) \displaystyle\int_0^{\pi/2} [(6\cos^2\theta)^2 - (4\cos^2\theta)^2] \, d\theta = 40 \displaystyle\int_0^{\pi/2} \cos^4\theta \, d\theta = 10 \displaystyle\int_0^{\pi/2} (1 + \cos 2\theta)^2 \, d\theta$

$= 10 \displaystyle\int_0^{\pi/2} \left(1 + 2\cos 2\theta + \dfrac{1 - \cos 4\theta}{2} \right) d\theta = 10 \left[\dfrac{3}{2}\theta + \sin 2\theta + \dfrac{1}{8}\sin 4\theta \right]_0^{\pi/2} = \dfrac{15\pi}{2}$

39. $r = a\cos(n\theta)$

For $n = 1$:

$r = a\cos\theta$

$A = \pi\left(\dfrac{a}{2}\right)^2 = \dfrac{\pi a^2}{4}$

For $n = 2$:

$r = a\cos 2\theta$

$A = 8\left(\dfrac{1}{2}\right) \displaystyle\int_0^{\pi/4} (a\cos 2\theta)^2 \, d\theta = \dfrac{\pi a^2}{2}$

For $n = 3$:

$r = a\cos 3\theta$

$A = 6\left(\dfrac{1}{2}\right) \displaystyle\int_0^{\pi/6} (a\cos 3\theta)^2 \, d\theta = \dfrac{\pi a^2}{4}$

For $n = 4$:

$r = a\cos 4\theta$

$A = 16\left(\dfrac{1}{2}\right) \displaystyle\int_0^{\pi/8} (a\cos 4\theta)^2 \, d\theta = \dfrac{\pi a^2}{2}$

In general, the area of the region enclosed by $r = a\cos(n\theta)$ for $n = 1, 2, 3, \ldots$ is $(\pi a^2)/4$ if n is odd and is $(\pi a^2)/2$ if n is even.

41. $r = a$

$r' = 0$

$s = \int_0^{2\pi} \sqrt{a^2 + 0^2}\, d\theta = \left[a\theta \right]_0^{2\pi} = 2\pi a$

(circumference of circle of radius a)

43. $r = 1 + \sin\theta$

$r' = \cos\theta$

$s = 2 \int_{\pi/2}^{3\pi/2} \sqrt{(1 + \sin\theta)^2 + (\cos\theta)^2}\, d\theta$

$= 2\sqrt{2} \int_{\pi/2}^{3\pi/2} \sqrt{1 + \sin\theta}\, d\theta$

$= 2\sqrt{2} \int_{\pi/2}^{3\pi/2} \dfrac{-\cos\theta}{\sqrt{1 - \sin\theta}}\, d\theta$

$= \left[4\sqrt{2}\, \sqrt{1 - \sin\theta} \right]_{\pi/2}^{3\pi/2}$

$= 4\sqrt{2}\left(\sqrt{2} - 0 \right) = 8$

45. $r = 2\theta,\ 0 \le \theta \le \dfrac{\pi}{2}$

Length ≈ 4.16

47. $r = \dfrac{1}{\theta},\ \pi \le \theta \le 2\pi$

Length ≈ 0.71

49. $r = \sin(3\cos\theta),\ 0 \le \theta \le \pi$

Length ≈ 4.39

51. $r = 6\cos\theta$

$r' = -6\sin\theta$

$S = 2\pi \int_0^{\pi/2} 6\cos\theta\sin\theta\sqrt{36\cos^2\theta + 36\sin^2\theta}\, d\theta$

$= 72\pi \int_0^{\pi/2} \sin\theta\cos\theta\, d\theta$

$= \left[36\pi \sin^2\theta \right]_0^{\pi/2}$

$= 36\pi$

53. $r = e^{a\theta}$

$r' = ae^{a\theta}$

$S = 2\pi \int_0^{\pi/2} e^{a\theta}\cos\theta\sqrt{(e^{a\theta})^2 + (ae^{a\theta})^2}\, d\theta$

$= 2\pi\sqrt{1 + a^2} \int_0^{\pi/2} e^{2a\theta}\cos\theta\, d\theta$

$= 2\pi\sqrt{1 + a^2}\left[\dfrac{e^{2a\theta}}{4a^2 + 1}(2a\cos\theta + \sin\theta) \right]_0^{\pi/2}$

$= \dfrac{2\pi\sqrt{1 + a^2}}{4a^2 + 1}\left(e^{\pi a} - 2a \right)$

55. $r = 4\cos 2\theta$

$r' = -8\sin 2\theta$

$S = 2\pi \int_0^{\pi/4} 4\cos 2\theta\sin\theta\sqrt{16\cos^2 2\theta + 64\sin^2 2\theta}\, d\theta$

$= 32\pi \int_0^{\pi/4} \cos 2\theta\sin\theta\sqrt{\cos^2 2\theta + 4\sin^2 2\theta}\, d\theta \approx 21.87$

57.   Area $= \dfrac{1}{2}\int_\alpha^\beta [f(\theta)]^2\, d\theta = \dfrac{1}{2}\int_\alpha^\beta r^2\, d\theta$

Arc length $= \int_\alpha^\beta \sqrt{f(\theta)^2 + f'(\theta)^2}\, d\theta = \int_\alpha^\beta \sqrt{r^2 + \left(\dfrac{dr}{d\theta}\right)^2}\, d\theta$

59. (a) is correct: $s \approx 33.124$.

61. Revolve $r = a$ about the line $r = b \sec \theta$ where $b > a > 0$.

$$f(\theta) = a$$

$$f'(\theta) = 0$$

$$S = 2\pi \int_0^{2\pi} [b - a \cos \theta] \sqrt{a^2 + 0^2} \, d\theta$$

$$= 2\pi a \left[b\theta - a \sin \theta \right]_0^{2\pi}$$

$$= 2\pi a (2\pi b) = 4\pi^2 ab$$

63. False. $f(\theta) = 1$ and $g(\theta) = -1$ have the same graphs.

65. In parametric form,

$$s = \int_a^b \sqrt{\left(\frac{dx}{dt}\right)^2 + \left(\frac{dy}{dt}\right)^2} \, dt.$$

Using θ instead of t, we have $x = r \cos \theta = f(\theta) \cos \theta$ and $y = r \sin \theta = f(\theta) \sin \theta$. Thus,

$$\frac{dx}{d\theta} = f'(\theta) \cos \theta - f(\theta) \sin \theta \text{ and } \frac{dy}{d\theta} = f'(\theta) \sin \theta + f(\theta) \cos \theta.$$

It follows that

$$\left(\frac{dx}{d\theta}\right)^2 + \left(\frac{dy}{d\theta}\right)^2 = [f(\theta)]^2 + [f'(\theta)]^2.$$

Therefore, $s = \displaystyle\int_\alpha^\beta \sqrt{[f(\theta)]^2 + [f'(\theta)]^2} \, d\theta$

Section 9.6   Polar Equations of Conics and Kepler's Laws

1. $r = \dfrac{2e}{1 + e \cos \theta}$

(a) $e = 1, r = \dfrac{2}{1 + \cos \theta}$, parabola

(b) $e = 0.5, r = \dfrac{1}{1 + 0.5 \cos \theta} = \dfrac{2}{2 + \cos \theta}$, ellipse

(c) $e = 1.5, r = \dfrac{3}{1 + 1.5 \cos \theta} = \dfrac{6}{2 + 3 \cos \theta}$, hyperbola

3. $r = \dfrac{2e}{1 - e \sin \theta}$

(a) $e = 1, r = \dfrac{2}{1 - \sin \theta}$, parabola

(b) $e = 0.5, r = \dfrac{1}{1 - 0.5 \sin \theta} = \dfrac{2}{2 - \sin \theta}$, ellipse

(c) $e = 1.5, r = \dfrac{3}{1 - 1.5 \sin \theta} = \dfrac{6}{2 - 3 \sin \theta}$, hyperbola

5. $r = \dfrac{4}{1 + e \sin \theta}$

(a)

(b)

The conic is an ellipse. As $e \to 1^-$, the ellipse becomes more elliptical, and as $e \to 0^+$, it becomes more circular.

The conic is a parabola.

(c)

The conic is a hyperbola. As $e \to 1^+$, the hyperbolas opens more slowly, and as $e \to \infty$, they open more rapidly.

7. Parabola; Matches (c)

9. Hyperbola; Matches (a)

11. Ellipse; Matches (b)

13. $r = \dfrac{-1}{1 - \sin \theta}$

Parabola since $e = 1$

Vertex: $\left(-\dfrac{1}{2}, \dfrac{3\pi}{2}\right)$

15. $r = \dfrac{6}{2 + \cos \theta}$

$\quad = \dfrac{3}{1 + (1/2) \cos \theta}$

Ellipse since $e = \dfrac{1}{2} < 1$

Vertices: $(2, 0), (6, \pi)$

17. $r(2 + \sin \theta) = 4$

$r = \dfrac{4}{2 + \sin \theta}$

$\quad = \dfrac{2}{1 + (1/2) \sin \theta}$

Ellipse since $e = \dfrac{1}{2} < 1$

Vertices: $\left(\dfrac{4}{3}, \dfrac{\pi}{2}\right), \left(4, \dfrac{3\pi}{2}\right)$

19. $r = \dfrac{5}{-1 + 2 \cos \theta} = \dfrac{-5}{1 - 2 \cos \theta}$

Hyperbola since $e = 2 > 1$

Vertices: $(5, 0), \left(-\dfrac{5}{3}, \pi\right)$

21. $r = \dfrac{3}{2 + 6 \sin \theta} = \dfrac{3/2}{1 + 3 \sin \theta}$

Hyperbola since $e = 3 > 1$

Vertices: $\left(\dfrac{3}{8}, \dfrac{\pi}{2}\right), \left(-\dfrac{3}{4}, \dfrac{3\pi}{2}\right)$

23.   Ellipse

25.   Parabola

27. $r = \dfrac{-1}{1 - \sin\left(\theta - \dfrac{\pi}{4}\right)}$

Rotate the graph of

$$r = \dfrac{-1}{1 - \sin\theta}$$

counterclockwise through the angle $\dfrac{\pi}{4}$.

29. $r = \dfrac{6}{2 + \cos\left(\theta + \dfrac{\pi}{6}\right)}$

Rotate the graph of

$$r = \dfrac{6}{2 + \cos\theta}$$

clockwise through the angle $\dfrac{\pi}{6}$.

31. Change θ to $\theta + \dfrac{\pi}{4}$: $r = \dfrac{5}{5 + 3\cos\left(\theta + \dfrac{\pi}{4}\right)}$.

33. Parabola

$e = 1, x = -1, d = 1$

$$r = \dfrac{ed}{1 - e\cos\theta} = \dfrac{1}{1 - \cos\theta}$$

35. Ellipse

$e = \dfrac{1}{2}, y = 1, d = 1$

$$r = \dfrac{ed}{1 + e\sin\theta}$$

$$= \dfrac{1/2}{1 + (1/2)\sin\theta}$$

$$= \dfrac{1}{2 + \sin\theta}$$

37. Hyperbola

$e = 2, x = 1, d = 1$

$$r = \dfrac{ed}{1 + e\cos\theta} = \dfrac{2}{1 + 2\cos\theta}$$

39. Parabola

Vertex: $\left(1, -\dfrac{\pi}{2}\right)$

$e = 1, d = 2, r = \dfrac{2}{1 - \sin\theta}$

41. Ellipse

Vertices: $(2, 0), (8, \pi)$

$e = \dfrac{3}{5}, d = \dfrac{16}{3}$

$$r = \dfrac{ed}{1 + e\cos\theta}$$

$$= \dfrac{16/5}{1 + (3/5)\cos\theta}$$

$$= \dfrac{16}{5 + 3\cos\theta}$$

43. Hyperbola

Vertices: $\left(1, \dfrac{3\pi}{2}\right), \left(9, \dfrac{3\pi}{2}\right)$

$e = \dfrac{5}{4}, d = \dfrac{9}{5}$

$$r = \dfrac{ed}{1 - e\sin\theta}$$

$$= \dfrac{9/4}{1 - (5/4)\sin\theta}$$

$$= \dfrac{9}{4 - 5\sin\theta}$$

45. Ellipse if $0 < e < 1$, parabola if $e = 1$, hyperbola if $e > 1$.

47. (a) Hyperbola ($e = 2 > 1$)

    (b) Ellipse $\left(e = \dfrac{1}{2} < 1\right)$

    (c) Parabola ($e = 1$)

    (d) Rotated hyperbola ($e = 3$)

49. $a = 5, c = 4, e = \dfrac{4}{5}, b = 3$

$$r^2 = \frac{9}{1 - (16/25)\cos^2\theta}$$

51. $a = 3, b = 4, c = 5, e = \dfrac{5}{3}$

$$r^2 = \frac{-16}{1 - (25/9)\cos^2\theta}$$

53. $A = 2\left[\dfrac{1}{2}\displaystyle\int_0^\pi \left(\dfrac{3}{2 - \cos\theta}\right)^2 d\theta\right]$

$$= 9\int_0^\pi \frac{1}{(2 - \cos\theta)^2}\, d\theta \approx 10.88$$

55. Vertices: $(126{,}000, 0), (4119, \pi)$

$$a = \frac{126{,}000 + 4119}{2} = 65{,}059.5, c = 65{,}059.5 - 4119 = 60{,}940.5, e = \frac{c}{a} = \frac{40{,}627}{43{,}373}, d = 4119\left(\frac{84{,}000}{40{,}627}\right)$$

$$r = \frac{ed}{1 - e\cos\theta} = \frac{4119(84{,}000/43{,}373)}{1 - (40{,}627/43{,}373)\cos\theta} = \frac{345{,}996{,}000}{43{,}373 - 40{,}627\cos\theta}$$

When $\theta = 60°, r = \dfrac{345{,}996{,}000}{23{,}059.5} \approx 15{,}004.49.$

Distance between the surface of the earth and the satellite is $r - 4000 = 11{,}004.49$ miles.

57. $a = 92.957 \times 10^6$ mi, $e = 0.0167$

$$r = \frac{(1 - e^2)a}{1 - e\cos\theta} = \frac{92{,}931{,}075.2223}{1 - 0.0167\cos\theta}$$

Perihelion distance: $a(1 - e) \approx 91{,}404{,}618$ mi

Aphelion distance: $a(1 + e) \approx 94{,}509{,}382$ mi

59. $a = 5.900 \times 10^9$ km, $e = 0.2481$

$$r = \frac{(1 - e^2)a}{1 - e\cos\theta} \approx \frac{5.537 \times 10^9}{1 - 0.2481\cos\theta}$$

Perihelion distance: $a(1 - e) = 4.436 \times 10^9$ km

Aphelion distance: $a(1 + e) = 7.364 \times 10^9$ km

61. $r = \dfrac{5.537 \times 10^9}{1 - 0.2481\cos\theta}$

(a) $A = \dfrac{1}{2}\displaystyle\int_0^{\pi/9}\left[\dfrac{5.537 \times 10^9}{1 - 0.2481\cos\theta}\right]^2 d\theta \approx 9.341 \times 10^{18}$ km^2

$$248\left[\frac{\dfrac{1}{2}\displaystyle\int_0^{\pi/9}\left[\dfrac{5.537 \times 10^9}{1 - 0.2481\cos\theta}\right]^2 d\theta}{\dfrac{1}{2}\displaystyle\int_0^{2\pi}\left[\dfrac{5.537 \times 10^9}{1 - 0.2481\cos\theta}\right]^2 d\theta}\right] \approx 21.867 \text{ yr}$$

(b) $\dfrac{1}{2}\displaystyle\int_\pi^{\alpha - \pi}\left[\dfrac{5.537 \times 10^9}{1 - 0.2481\cos\theta}\right]^2 d\theta = 9.341 \times 10^{18}$

$\alpha \approx \pi + 0.8995$ rad

In part (a) the ray swept through a smaller angle to generate the same area since the length of the ray is longer than in part (b).

(c) $r' = \dfrac{(-5.537 \times 10^9)(0.2481\sin\theta)}{(1 - 0.2481\cos\theta)^2}$

$$s = \int_0^{\pi/9}\sqrt{\left(\frac{5.537 \times 10^9}{1 - 0.2481\cos\theta}\right)^2 + \left[\frac{-1.3737297 \times 10^9 \sin\theta}{(1 - 0.2481\cos\theta)^2}\right]^2}\, d\theta \approx 2.559 \times 10^9 \text{ km}$$

$\dfrac{2.559 \times 10^9 \text{ km}}{21.867 \text{ yr}} \approx 1.17 \times 10^8$ km/yr

$$s = \int_\pi^{\pi + 0.899}\sqrt{\left(\frac{5.537 \times 10^9}{1 - 0.2481\cos\theta}\right)^2 + \left[\frac{-1.3737297 \times 10^9 \sin\theta}{(1 - 0.2481\cos\theta)^2}\right]^2}\, d\theta \approx 4.119 \times 10^9 \text{ km}$$

$\dfrac{4.119 \times 10^9 \text{ km}}{21.867 \text{ yr}} \approx 1.88 \times 10^8$ km/yr

63. $r_1 = \dfrac{ed}{1 + \sin\theta}$ and $r_2 = \dfrac{ed}{1 - \sin\theta}$

Points of intersection: $(ed, 0), (ed, \pi)$

$r_1:\ \dfrac{dy}{dx} = \dfrac{\left(\dfrac{ed}{1 + \sin\theta}\right)(\cos\theta) + \left(\dfrac{-ed\cos\theta}{(1 + \sin\theta)^2}\right)(\sin\theta)}{\left(\dfrac{-ed}{1 + \sin\theta}\right)(\sin\theta) + \left(\dfrac{-ed\cos\theta}{(1 + \sin\theta)^2}\right)(\cos\theta)}$

At $(ed, 0)$, $\dfrac{dy}{dx} = -1$. At (ed, π), $\dfrac{dy}{dx} = 1$.

$r_2:\ \dfrac{dy}{dx} = \dfrac{\left(\dfrac{ed}{1 - \sin\theta}\right)(\cos\theta) + \left(\dfrac{ed\cos\theta}{(1 - \sin\theta)^2}\right)(\sin\theta)}{\left(\dfrac{-ed}{1 - \sin\theta}\right)(\sin\theta) + \left(\dfrac{ed\cos\theta}{(1 - \sin\theta)^2}\right)(\cos\theta)}$

At $(ed, 0)$, $\dfrac{dy}{dx} = 1$. At (ed, π), $\dfrac{dy}{dx} = -1$.

Therefore, at $(ed, 0)$ we have $m_1m_2 = (-1)(1) = -1$, and at (ed, π) we have $m_1m_2 = 1(-1) = -1$. The curves intersect at right angles.

Review Exercises for Chapter 9

1. Matches (d) - ellipse

3. Matches (a) - parabola

5. $16x^2 + 16y^2 - 16x + 24y - 3 = 0$

$\left(x^2 - x + \dfrac{1}{4}\right) + \left(y^2 + \dfrac{3}{2}y + \dfrac{9}{16}\right) = \dfrac{3}{16} + \dfrac{1}{4} + \dfrac{9}{16}$

$\left(x - \dfrac{1}{2}\right)^2 + \left(y + \dfrac{3}{4}\right)^2 = 1$

Circle

Center: $\left(\dfrac{1}{2}, -\dfrac{3}{4}\right)$

Radius: 1

7. $3x^2 - 2y^2 + 24x + 12y + 24 = 0$

$3(x^2 + 8x + 16) - 2(y^2 - 6y + 9) = -24 + 48 - 18$

$\dfrac{(x + 4)^2}{2} - \dfrac{(y - 3)^2}{3} = 1$

Hyperbola

Center: $(-4, 3)$

Vertices: $\left(-4 \pm \sqrt{2}, 3\right)$

Asymptotes: $y = 3 \pm \sqrt{\dfrac{3}{2}}(x + 4)$

9. $3x^2 + 2y^2 - 12x + 12y + 29 = 0$

$3(x^2 - 4x + 4) + 2(y^2 + 6y + 9) = -29 + 12 + 18$

$\dfrac{(x - 2)^2}{1/3} + \dfrac{(y + 3)^2}{1/2} = 1$

Ellipse

Center: $(2, -3)$

Vertices: $\left(2, -3 \pm \dfrac{\sqrt{2}}{2}\right)$

11. Vertex: $(0, 2)$

Directrix: $x = -3$

Parabola opens to the right

$p = 3$

$(y - 2)^2 = 4(3)(x - 0)$

$y^2 - 4y - 12x + 4 = 0$

13. Vertices: $(-3, 0), (7, 0)$

Foci: $(0, 0), (4, 0)$

Horizontal major axis

Center: $(2, 0)$

$a = 5, c = 2, b = \sqrt{21}$

$$\frac{(x - 2)^2}{25} + \frac{y^2}{21} = 1$$

15. Vertices: $(\pm 4, 0)$

Foci: $(\pm 6, 0)$

Center: $(0, 0)$

Horizontal transverse axis

$a = 4, c = 6, b = \sqrt{36 - 16} = 2\sqrt{5}$

$$\frac{x^2}{16} - \frac{y^2}{20} = 1$$

17. $\dfrac{x^2}{9} + \dfrac{y^2}{4} = 1, a = 3, b = 2, c = \sqrt{5}, e = \dfrac{\sqrt{5}}{3}$

By Example 5 of Section 9.1,

$$C = 12 \int_0^{\pi/2} \sqrt{1 - \left(\frac{5}{9}\right) \sin^2 \theta} \, d\theta \approx 15.87.$$

19. $y = x - 2$ has a slope of 1. The perpendicular slope is -1.

$$y = x^2 - 2x + 2$$

$$\frac{dy}{dx} = 2x - 2 = -1 \text{ when } x = \frac{1}{2} \text{ and } y = \frac{5}{4}.$$

Perpendicular line: $\quad y - \dfrac{5}{4} = -1\left(x - \dfrac{1}{2}\right)$

$$4x + 4y - 7 = 0$$

21. (a) $V = (\pi a b)(\text{Length}) = 12\pi(16) = 192\pi \text{ ft}^3$

(b) $F = 2(62.4) \displaystyle\int_{-3}^{3} (3 - y)\frac{4}{3}\sqrt{9 - y^2} \, dy = \frac{8}{3}(62.4)\left[3 \int_{-3}^{3} \sqrt{9 - y^2} \, dy - \int_{-3}^{3} y\sqrt{9 - y^2} \, dy\right]$

$= \dfrac{8}{3}(62.4)\left[\dfrac{3}{2}\left(y\sqrt{9 - y^2} + 9\arcsin\dfrac{y}{3}\right) + \dfrac{1}{3}(9 - y^2)^{3/2}\right]_{-3}^{3}$

$= \dfrac{8}{3}(62.4)\left[\dfrac{3}{2}\left(\dfrac{9\pi}{2}\right) - \dfrac{3}{2}\left(-\dfrac{9\pi}{2}\right)\right] = \dfrac{8}{3}(62.4)\left(\dfrac{27\pi}{2}\right) \approx 7057.274$

(c) You want $\frac{3}{4}$ of the total area of 12π covered. Find h so that

$$2\int_0^h \frac{4}{3}\sqrt{9 - y^2} \, dy = 3\pi$$

$$\int_0^h \sqrt{9 - y^2} \, dy = \frac{9\pi}{8}$$

$$\frac{1}{2}\left[y\sqrt{9 - y^2} + 9\arcsin\left(\frac{y}{3}\right)\right]_0^h = \frac{9\pi}{8}$$

$$h\sqrt{9 - h^2} + 9\arcsin\left(\frac{h}{3}\right) = \frac{9\pi}{4}.$$

$x = \frac{4}{3}\sqrt{9 - y^2}$

Area of filled tank above x-axis is 3π.

Area of filled tank below x-axis is 6π.

By Newton's Method, $h \approx 1.212$. Therefore, the total height of the water is $1.212 + 3 = 4.212$ ft.

(d) Area of ends $= 2(12\pi) = 24\pi$

Area of sides $= (\text{Perimeter})(\text{Length})$

$= 16\displaystyle\int_0^{\pi/2}\left(\sqrt{1 - \left(\frac{7}{16}\right)\sin^2\theta}\right)d\theta(16) \quad$ [from Example 5 of Section 9.1]

$\approx 256\left(\dfrac{\pi/2}{12}\right)\left[\sqrt{1 - \left(\dfrac{7}{16}\right)\sin^2(0)} + 4\sqrt{1 - \left(\dfrac{7}{16}\right)\sin^2\left(\dfrac{\pi}{8}\right)} + 2\sqrt{1 - \left(\dfrac{7}{16}\right)\sin^2\left(\dfrac{\pi}{4}\right)}\right.$

$\left. + 4\sqrt{1 - \left(\dfrac{7}{16}\right)\sin^2\left(\dfrac{3\pi}{8}\right)} + \sqrt{1 - \left(\dfrac{7}{16}\right)\sin^2\left(\dfrac{\pi}{2}\right)}\right] \approx 353.65$

Total area $= 24\pi + 353.65 \approx 429.05$

23. $x = 1 + 4t, y = 2 - 3t$

$t = \dfrac{x - 1}{4} \Rightarrow y = 2 - 3\left(\dfrac{x - 1}{4}\right)$

$y = -\dfrac{3}{4}x + \dfrac{11}{4}$

$4y + 3x - 11 = 0$

Line

25. $x = 6 \cos \theta, y = 6 \sin \theta$

$\left(\dfrac{x}{6}\right)^2 + \left(\dfrac{y}{6}\right)^2 = 1$

$x^2 + y^2 = 36$

Circle

27. $x = 2 + \sec \theta, y = 3 + \tan \theta$

$(x - 2)^2 = \sec^2 \theta = 1 + \tan^2 \theta = 1 + (y - 3)^2$

$(x - 2)^2 - (y - 3)^2 = 1$

Hyperbola

29. $x = 3 + (3 - (-2))t = 3 + 5t$

$y = 2 + (2 - 6)t = 2 - 4t$

(other answers possible)

31. $\dfrac{(x + 3)^2}{16} + \dfrac{(y - 4)^2}{9} = 1$

Let $\dfrac{(x + 3)^2}{16} = \cos^2 \theta$ and $\dfrac{(y - 4)^2}{9} = \sin^2 \theta.$

Then $x = -3 + 4 \cos \theta$ and $y = 4 + 3 \sin \theta.$

33. $x = \cos 3\theta + 5 \cos \theta$

$y = \sin 3\theta + 5 \sin \theta$

35. (a) $x = 2 \cot \theta, y = 4 \sin \theta \cos \theta, 0 < \theta < \pi$

(b) $(4 + x^2)y = (4 + 4 \cot^2 \theta)4 \sin \theta \cos \theta$

$= 16 \csc^2 \theta \cdot \sin \theta \cdot \cos \theta$

$= 16 \dfrac{\cos \theta}{\sin \theta}$

$= 8(2 \cot \theta)$

$= 8x$

37. $x = 1 + 4t$

$y = 2 - 3t$

(a) $\dfrac{dy}{dx} = -\dfrac{3}{4}$

No horizontal tangents

(b) $t = \dfrac{x - 1}{4}$

$y = 2 - \dfrac{3}{4}(x - 1) = \dfrac{-3x + 11}{4}$

(c)

39. $x = \dfrac{1}{t}$

$y = 2t + 3$

(a) $\dfrac{dy}{dx} = \dfrac{2}{-1/t^2} = -2t^2$

No horizontal tangents
$(t \neq 0)$

(b) $t = \dfrac{1}{x}$

$y = \dfrac{2}{x} + 3$

(c)

41. $x = \dfrac{1}{2t + 1}$

$y = \dfrac{1}{t^2 - 2t}$

(a) $\dfrac{dy}{dx} = \dfrac{\dfrac{-(2t-2)}{(t^2 - 2t)^2}}{\dfrac{-2}{(2t+1)^2}} = \dfrac{(t-1)(2t+1)^2}{t^2(t-2)^2} = 0$ when $t = 1$.

Point of horizontal tangency: $\left(\frac{1}{3}, -1\right)$

(b) $2t + 1 = \dfrac{1}{x} \Rightarrow t = \dfrac{1}{2}\left(\dfrac{1}{x} - 1\right)$

$$y = \dfrac{1}{\dfrac{1}{2}\left(\dfrac{1-x}{x}\right)\left[\dfrac{1}{2}\left(\dfrac{1-x}{x}\right) - 2\right]}$$

$$= \dfrac{4x^2}{(1-x)^2 - 4x(1-x)} = \dfrac{4x^2}{(5x-1)(x-1)}, x \neq 0$$

(c)

43. $x = 3 + 2\cos\theta$

$y = 2 + 5\sin\theta$

(a) $\dfrac{dy}{dx} = \dfrac{5\cos\theta}{-2\sin\theta} = -2.5\cot\theta = 0$ when $\theta = \dfrac{\pi}{2}, \dfrac{3\pi}{2}$.

Points of horizontal tangency: $(3, 7), (3, -3)$

(b) $\dfrac{(x-3)^2}{4} + \dfrac{(y-2)^2}{25} = 1$

(c)

45. $x = \cos^3\theta$

$y = 4\sin^3\theta$

(a) $\dfrac{dy}{dx} = \dfrac{12\sin^2\theta\cos\theta}{3\cos^2\theta(-\sin\theta)} = \dfrac{-4\sin\theta}{\cos\theta} = -4\tan\theta = 0$ when $\theta = 0, \pi$.

But, $\dfrac{dy}{dt} = \dfrac{dx}{dt} = 0$ at $\theta = 0, \pi$. Hence no points of horizontal tangency.

(b) $x^{2/3} + \left(\dfrac{y}{4}\right)^{2/3} = 1$

(c)

47. $x = \cot\theta$

$y = \sin 2\theta = 2\sin\theta\cos\theta$

(a), (c)

(b) At $\theta = \dfrac{\pi}{6}, \dfrac{dx}{d\theta} = -4, \dfrac{dy}{d\theta} = 1$, and $\dfrac{dy}{dx} = -\dfrac{1}{4}$

49. $x = r(\cos\theta + \theta\sin\theta)$

$y = r(\sin\theta - \theta\cos\theta)$

$\dfrac{dx}{d\theta} = r\theta\cos\theta$

$\dfrac{dy}{d\theta} = r\theta\sin\theta$

$s = r\displaystyle\int_0^\pi \sqrt{\theta^2\cos^2\theta + \theta^2\sin^2\theta}\,d\theta$

$= r\displaystyle\int_0^\pi \theta\,d\theta = \dfrac{r}{2}\Big[\theta^2\Big]_0^\pi = \dfrac{1}{2}\pi^2 r$

51. $(x, y) = (4, -4)$

$r = \sqrt{4^2 + (-4)^2} = 4\sqrt{2}$

$\theta = 7\dfrac{\pi}{4}$

$(r, \theta) = \left(4\sqrt{2}, \dfrac{7\pi}{4}\right), \left(-4\sqrt{2}, \dfrac{3\pi}{4}\right)$

53.

$r = 3\cos\theta$

$r^2 = 3r\cos\theta$

$x^2 + y^2 = 3x$

$x^2 + y^2 - 3x = 0$

55.

$r = -2(1 + \cos\theta)$

$r^2 = -2r(1 + \cos\theta)$

$x^2 + y^2 = -2\big(\pm\sqrt{x^2 + y^2}\big) - 2x$

$(x^2 + y^2 + 2x)^2 = 4(x^2 + y^2)$

57.

$r^2 = \cos 2\theta = \cos^2\theta - \sin^2\theta$

$r^4 = r^2\cos^2\theta - r^2\sin^2\theta$

$(x^2 + y^2)^2 = x^2 - y^2$

59.

$r = 4\cos 2\theta\sec\theta$

$= 4(2\cos^2\theta - 1)\left(\dfrac{1}{\cos\theta}\right)$

$r\cos\theta = 8\cos^2\theta - 4$

$x = 8\left(\dfrac{x^2}{x^2 + y^2}\right) - 4$

$x^3 + xy^2 = 4x^2 - 4y^2$

$y^2 = x^2\left(\dfrac{4 - x}{4 + x}\right)$

61. $(x^2 + y^2)^2 = ax^2 y$

$r^4 = a(r^2\cos^2\theta)(r\sin\theta)$

$r = a\cos^2\theta\sin\theta$

63. $x^2 + y^2 = a^2\left(\arctan\dfrac{y}{x}\right)^2$

$r^2 = a^2\theta^2$

65. $r = 4$

Circle of radius 4

Centered at the pole

Symmetric to polar axis,

$\theta = \pi/2$, and pole

67. $r = -\sec\theta = \dfrac{-1}{\cos\theta}$

$r\cos\theta = -1, x = -1$

Vertical line

69. $r = -2(1 + \cos\theta)$

Cardioid

Symmetric to polar axis

θ	0	$\dfrac{\pi}{3}$	$\dfrac{\pi}{2}$	$\dfrac{2\pi}{3}$	π
r	-4	-3	-2	-1	0

71. $r = 4 - 3\cos\theta$

Limaçon

Symmetric to polar axis

θ	0	$\dfrac{\pi}{3}$	$\dfrac{\pi}{2}$	$\dfrac{2\pi}{3}$	π
r	1	$\dfrac{5}{2}$	4	$\dfrac{11}{2}$	7

73. $r = -3\cos(2\theta)$

Rose curve with four petals

Symmetric to polar axis, $\theta = \dfrac{\pi}{2}$, and pole

Relative extrema: $(-3, 0), \left(3, \dfrac{\pi}{2}\right), (-3, \pi), \left(3, \dfrac{3\pi}{2}\right)$

Tangents at the pole: $\theta = \dfrac{\pi}{4}, \dfrac{3\pi}{4}$

75. $r^2 = 4\sin^2(2\theta)$

$r = \pm 2\sin(2\theta)$

Rose curve with four petals

Symmetric to the polar axis, $\theta = \dfrac{\pi}{2}$, and pole

Relative extrema: $\left(\pm 2, \dfrac{\pi}{4}\right), \left(\pm 2, \dfrac{3\pi}{4}\right)$

Tangents at the pole: $\theta = 0, \dfrac{\pi}{2}$

77. $r = \dfrac{3}{\cos[\theta - (\pi/4)]}$

Graph of $r = 3\sec\theta$ rotated through an angle of $\pi/4$

79. $r = 4\cos 2\theta \sec\theta$

Strophoid

Symmetric to the polar axis

$r \Rightarrow -\infty$ as $\theta \Rightarrow \dfrac{\pi^-}{2}$

$r \Rightarrow -\infty$ as $\theta \Rightarrow \dfrac{-\pi^+}{2}$

81. $r = 1 - 2\cos\theta$

(a) The graph has polar symmetry and the tangents at the pole are

$$\theta = \frac{\pi}{3}, -\frac{\pi}{3}.$$

(b) $\dfrac{dy}{dx} = \dfrac{2\sin^2\theta + (1 - 2\cos\theta)\cos\theta}{2\sin\theta\cos\theta - (1 - 2\cos\theta)\sin\theta}$

Horizontal tangents: $-4\cos^2\theta + \cos\theta + 2 = 0$, $\cos\theta = \dfrac{-1 \pm \sqrt{1 + 32}}{-8} = \dfrac{1 \pm \sqrt{33}}{8}$

When $\cos\theta = \dfrac{1 \pm \sqrt{33}}{8}$, $r = 1 - 2\left(\dfrac{1 + \sqrt{33}}{8}\right) = \dfrac{3 \mp \sqrt{33}}{4}$,

$$\left[\dfrac{3 - \sqrt{33}}{4}, \arccos\left(\dfrac{1 + \sqrt{33}}{8}\right)\right] \approx (-0.686, 0.568)$$

$$\left[\dfrac{3 - \sqrt{33}}{4}, -\arccos\left(\dfrac{1 + \sqrt{33}}{8}\right)\right] \approx (-0.686, -0.568)$$

$$\left[\dfrac{3 + \sqrt{33}}{4}, \arccos\left(\dfrac{1 - \sqrt{33}}{8}\right)\right] \approx (2.186, 2.206)$$

$$\left[\dfrac{3 + \sqrt{33}}{4}, -\arccos\left(\dfrac{1 - \sqrt{33}}{8}\right)\right] \approx (2.186, -2.206).$$

Vertical tangents:

$$\sin\theta(4\cos\theta - 1) = 0, \sin\theta = 0, \cos\theta = \frac{1}{4},$$

$$\theta = 0, \pi, \theta = \pm\arccos\left(\frac{1}{4}\right), (-1, 0), (3, \pi)$$

$$\left(\frac{1}{2}, \pm\arccos\frac{1}{4}\right) \approx (0.5, \pm1.318)$$

(c)

83. Circle: $r = 3\sin\theta$

$$\frac{dy}{dx} = \frac{3\cos\theta\sin\theta + 3\sin\theta\cos\theta}{3\cos\theta\cos\theta - 3\sin\theta\sin\theta} = \frac{\sin 2\theta}{\cos^2\theta - \sin^2\theta} = \tan 2\theta \text{ at } \theta = \frac{\pi}{6}, \frac{dy}{dx} = \sqrt{3}$$

Limaçon: $r = 4 - 5\sin\theta$

$$\frac{dy}{dx} = \frac{-5\cos\theta\sin\theta + (4 - 5\sin\theta)\cos\theta}{-5\cos\theta\cos\theta - (4 - 5\sin\theta)\sin\theta} \text{ at } \theta = \frac{\pi}{6}, \frac{dy}{dx} = \frac{\sqrt{3}}{9}$$

Let α be the angle between the curves:

$$\tan\alpha = \frac{\sqrt{3} - (\sqrt{3}/9)}{1 + (1/3)} = \frac{2\sqrt{3}}{3}.$$

Therefore, $\alpha = \arctan\left(\dfrac{2\sqrt{3}}{3}\right) \approx 49.1°.$

85. $r = 1 + \cos\theta, r = 1 - \cos\theta$

The points $(1, \pi/2)$ and $(1, 3\pi/2)$ are the two points of intersection (other than the pole). The slope of the graph of $r = 1 + \cos\theta$ is

$$m_1 = \frac{dy}{dx} = \frac{r'\sin\theta + r\cos\theta}{r'\cos\theta - r\sin\theta} = \frac{-\sin^2\theta + \cos\theta(1 + \cos\theta)}{-\sin\theta\cos\theta - \sin\theta(1 + \cos\theta)}.$$

At $(1, \pi/2)$, $m_1 = -1/-1 = 1$ and at $(1, 3\pi/2)$, $m_1 = -1/1 = -1$. The slope of the graph of $r = 1 - \cos\theta$ is

$$m_2 = \frac{dy}{dx} = \frac{\sin^2\theta + \cos\theta(1 - \cos\theta)}{\sin\theta\cos\theta - \sin\theta(1 - \cos\theta)}.$$

At $(1, \pi/2)$, $m_2 = 1/-1 = -1$ and at $(1, 3\pi/2)$, $m_2 = 1/1 = 1$. In both cases, $m_1 = -1/m_2$ and we conclude that the graphs are orthogonal at $(1, \pi/2)$ and $(1, 3\pi/2)$.

87. $r = 2 + \cos\theta$

$$A = 2\left[\frac{1}{2}\int_0^\pi (2 + \cos\theta)^2\, d\theta\right] \approx 14.14 \quad \left(\frac{9\pi}{2}\right)$$

89. $r = \sin\theta \cdot \cos^2\theta$

$$A = 2\left[\frac{1}{2}\int_0^{\pi/2} (\sin\theta\cos^2\theta)^2\, d\theta\right]$$

$$\approx 0.10 \quad \left(\frac{\pi}{32}\right)$$

91. $r^2 = 4\sin 2\theta$

$$A = 2\left[\frac{1}{2}\int_0^{\pi/2} 4\sin 2\theta\, d\theta\right] = 4$$

93. $r = 4\cos\theta, r = 2$

$$A = 2\left[\frac{1}{2}\int_0^{\pi/3} 4\, d\theta + \frac{1}{2}\int_{\pi/3}^{\pi/2} (4\cos\theta)^2\, d\theta\right] \approx 4.91$$

95. $s = 2\int_0^\pi \sqrt{a^2(1 - \cos\theta)^2 + a^2\sin^2\theta}\, d\theta$

$$= 2\sqrt{2}\, a\int_0^\pi \sqrt{1 - \cos\theta}\, d\theta = 2\sqrt{2}\, a\int_0^\pi \frac{\sin\theta}{\sqrt{1 + \cos\theta}}\, d\theta = \left[-4\sqrt{2}\, a(1 + \cos\theta)^{1/2}\right]_0^\pi = 8a$$

97. $r = \dfrac{2}{1 - \sin\theta},\ e = 1$

Parabola

99. $r = \dfrac{6}{3 + 2\cos\theta} = \dfrac{2}{1 + (2/3)\cos\theta},\ e = \dfrac{2}{3}$

Ellipse

101. $r = \dfrac{4}{2 - 3 \sin \theta} = \dfrac{2}{1 - (3/2)\sin \theta}, e = \dfrac{3}{2}$

Hyperbola

103. Circle

Center: $\left(5, \dfrac{\pi}{2}\right) = (0, 5)$ in rectangular coordinates

Solution point: $(0, 0)$

$$x^2 + (y - 5)^5 = 25$$

$$x^2 + y^2 - 10y = 0$$

$$r^2 - 10r \sin \theta = 0$$

$$r = 10 \sin \theta$$

105. Parabola

Vertex: $(2, \pi)$

Focus: $(0, 0)$

$e = 1, d = 4$

$r = \dfrac{4}{1 - \cos \theta}$

107. Ellipse

Vertices: $(5, 0), (1, \pi)$

Focus: $(0, 0)$

$a = 3, c = 2, e = \dfrac{2}{3}, d = \dfrac{5}{2}$

$$r = \dfrac{\left(\dfrac{2}{3}\right)\left(\dfrac{5}{2}\right)}{1 - \left(\dfrac{2}{3}\right)\cos \theta} = \dfrac{5}{3 - 2 \cos \theta}$$

Problem Solving for Chapter 9

1. (a)

(b) $x^2 = 4y$

$2x = 4y'$

$y' = \dfrac{1}{2}x$

$y - 4 = 2(x - 4) \implies y = 2x - 4$ Tangent line at $(4, 4)$

$y - \dfrac{1}{4} = -\dfrac{1}{2}(x + 1) \implies y = -\dfrac{1}{2}x - \dfrac{1}{4}$ Tangent line at $\left(-1, \dfrac{1}{4}\right)$

Tangent lines have slopes of 2 and $-1/2 \implies$ perpendicular.

(c) Intersection:

$2x - 4 = -\dfrac{1}{2}x - \dfrac{1}{4}$

$8x - 16 = -2x - 1$

$10x = 15$

$x = \dfrac{3}{2} \implies \left(\dfrac{3}{2}, -1\right)$

Point of intersection, $(3/2, -1)$, is on directrix $y = -1$.

3. Consider $x^2 = 4py$ with focus $F = (0, p)$.

Let $P = (a, b)$ be point on parabola.

$zx = 4py' \implies y' = \dfrac{x}{2p}$

$y - b = \dfrac{a}{2p}(x - a)$ Tangent line

For $x = 0$, $y = b + \dfrac{a}{2p}(-a) = b - \dfrac{a^2}{2p} = b - \dfrac{4pb}{2p} = -b.$

Thus, $Q = (0, -b)$.

$\triangle FQP$ is isosceles because

$|FQ| = p + b$

$$|FP| = \sqrt{(a - 0)^2 + (b - p)^2} = \sqrt{a^2 + b^2 - 2bp + p^2}$$
$$= \sqrt{4pb + b^2 - 2bp + p^2}$$
$$= \sqrt{(b + p)^2}$$
$$= b + p.$$

Thus, $\measuredangle FQP = \measuredangle BPA = \measuredangle FPQ$.

5. (a) In $\triangle OCB$, $\cos \theta = \dfrac{2a}{OB} \implies OB = 2a \cdot \sec \theta.$

In $\triangle OAC$, $\cos \theta = \dfrac{OA}{2a} \implies OA = 2a \cdot \cos \theta.$

$r = OP = AB = OB - OA = 2a(\sec \theta - \cos \theta)$

$$= 2a\left(\dfrac{1}{\cos \theta} - \cos \theta\right)$$

$$= 2a \cdot \dfrac{\sin^2 \theta}{\cos \theta}$$

$$= 2a \cdot \tan \theta \sin \theta$$

(c) $r = 2a \tan \theta \sin \theta$

$r \cos \theta = 2a \sin^2 \theta$

$r^3 \cos \theta = 2a\, r^2 \sin^2 \theta$

$(x^2 + y^2)x = 2ay^2$

$y^2 = \dfrac{x^3}{(2a - x)}$

(b) $x = r \cos \theta = (2a \tan \theta \sin \theta)\cos \theta = 2a \sin^2 \theta$

$y = r \sin \theta = (2a \tan \theta \sin \theta)\sin \theta = 2a \tan \theta \cdot \sin^2 \theta,\ -\dfrac{\pi}{2} < \theta < \dfrac{\pi}{2}$

Let $t = \tan \theta$, $-\infty < t < \infty$.

Then $\sin^2 \theta = \dfrac{t^2}{1 + t^2}$ and $x = 2a\dfrac{t^2}{1 + t^2}, y = 2a\dfrac{t^3}{1 + t^2}.$

7. $y = a(1 - \cos \theta) \implies \cos \theta = \dfrac{a - y}{a}$

$\theta = \arccos\left(\dfrac{a - y}{a}\right)$

$x = a(\theta - \sin \theta)$

$$= a\left(\arccos\left(\dfrac{a - y}{a}\right) - \sin\left(\arccos\left(\dfrac{a - y}{a}\right)\right)\right)$$

$$= a\left(\arccos\left(\dfrac{a - y}{a}\right) - \dfrac{\sqrt{2ay - y^2}}{a}\right)$$

$x = a \cdot \arccos\left(\dfrac{a - y}{a}\right) - \sqrt{2ay - y^2}, 0 \le y \le 2a$

9. For $t = \dfrac{\pi}{2}, \dfrac{3\pi}{2}, \dfrac{5\pi}{2}, \dfrac{7\pi}{2}, \ldots$

$$y = \frac{2}{\pi}, \frac{-2}{3\pi}, \frac{2}{5\pi}, \frac{-2}{7\pi}, \ldots$$

Hence, the curve has length greater that

$$S = \frac{2}{\pi} + \frac{2}{3\pi} + \frac{2}{5\pi} + \frac{2}{7\pi} + \cdots$$

$$= \frac{2}{\pi}\left(1 + \frac{1}{3} + \frac{1}{5} + \frac{1}{7} + \cdots\right)$$

$$> \frac{2}{\pi}\left(\frac{1}{2} + \frac{1}{4} + \frac{1}{6} + \frac{1}{8} + \cdots\right)$$

$$= \infty.$$

11. (a) Area $= \displaystyle\int_0^{\alpha} \frac{1}{2}r^2 \, d\theta$

$$= \frac{1}{2}\int_0^{\alpha} \sec^2 \theta \, d\theta$$

$x = 1$
$r = \sec\theta$

(b) $\tan \alpha = \dfrac{h}{1} \implies$ Area $= \dfrac{1}{2}(1)\tan \alpha$

$$\implies \tan \alpha = \int_0^{\alpha} \sec^2 \theta \, d\theta$$

(c) Differentiating, $\dfrac{d}{d\alpha}(\tan \alpha) = \sec^2 \alpha.$

13. If a dog is located at (r, θ) in the first quadrant, then its neighbor is at $\left(r, \theta + \dfrac{\pi}{2}\right)$:

$$(x, y) = (r\cos\theta, r\sin\theta) \text{ and } (x, y) = (-r\sin\theta, r\cos\theta).$$

The slope joining these points is

$$\frac{r\cos\theta - r\sin\theta}{-r\sin\theta - r\cos\theta} = \frac{\sin\theta - \cos\theta}{\sin\theta + \cos\theta} = \text{ slope of tangent line at } (r, \theta).$$

$$\frac{\dfrac{dr}{d\theta}\sin\theta + r\cos\theta}{\dfrac{dr}{d\theta}\cos\theta - r\sin\theta} = \frac{\sin\theta - \cos\theta}{\sin\theta + \cos\theta}$$

$$\implies \frac{dr}{d\theta} = -r$$

$$\frac{dr}{r} = -d\theta$$

$$\ln r = -\theta + C_1$$

$$r = e^{-\theta + C_1}$$

$$r = Ce^{-\theta}$$

$$r\left(\frac{\pi}{4}\right) = \frac{d}{\sqrt{2}} \implies r = Ce^{-\pi/4} = \frac{d}{\sqrt{2}} \implies C = \frac{d}{\sqrt{2}}e^{\pi/4}$$

Finally, $r = \dfrac{d}{\sqrt{2}}e^{((\pi/4) - \theta)}, \theta \geq \dfrac{\pi}{4}.$

15. (a) The first plane makes an angle of $70°$ with the positive x-axis, and is 150 miles from P:

$$x_1 = \cos 70°(150 - 375t)$$

$$y_1 = \sin 70°(150 - 375t)$$

Similarly for the second plane,

$$x_2 = \cos 135°(190 - 450t)$$

$$= \cos 45°(-190 + 450t)$$

$$y_2 = \sin 135°(190 - 450t)$$

$$= \sin 45°(190 - 450t)$$

(b) $d = \sqrt{(x_2 - x_1)^2 + (y_2 - y_1)^2}$

$$= [[\cos 45(-190 + 450t) - \cos 70(150 - 375t)]^2 + [\sin 45(190 - 450t) - \sin 70(150 - 375t)]^2]^{1/2}$$

(c)

280
0
0   1

The minimum distance is 7.59 miles when $t = 0.4145$.

17.

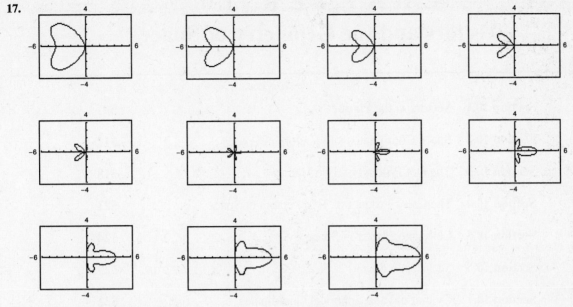

$n = 1, 2, 3, 4, 5$ produce "bells"; $n = -1, -2, -3, -4, -5$ produce "hearts".

C H A P T E R 1 0
Vectors and the Geometry of Space

Section 10.1 Vectors in the Plane 507

Section 10.2 Space Coordinates and Vectors in Space 512

Section 10.3 The Dot Product of Two Vectors 518

Section 10.4 The Cross Product of Two Vectors in Space 521

Section 10.5 Lines and Planes in Space 524

Section 10.6 Surfaces in Space 529

Section 10.7 Cylindrical and Spherical Coordinates 532

Review Exercises 536

Problem Solving 541

CHAPTER 10
Vectors and the Geometry of Space

Section 10.1   Vectors in the Plane
Solutions to Odd-Numbered Exercises

1. (a) $\mathbf{v} = \langle 5 - 1, 3 - 1 \rangle = \langle 4, 2 \rangle$

(b)

3. (a) $\mathbf{v} = \langle -4 - 3, -2 - (-2) \rangle = \langle -7, 0 \rangle$

(b)

5. $\mathbf{u} = \langle 5 - 3, 6 - 2 \rangle = \langle 2, 4 \rangle$

$\mathbf{v} = \langle 1 - (-1), 8 - 4 \rangle = \langle 2, 4 \rangle$

$\mathbf{u} = \mathbf{v}$

7. $\mathbf{u} = \langle 6 - 0, -2 - 3 \rangle = \langle 6, -5 \rangle$

$\mathbf{v} = \langle 9 - 3, 5 - 10 \rangle = \langle 6, -5 \rangle$

$\mathbf{u} = \mathbf{v}$

9. (b) $\mathbf{v} = \langle 5 - 1, 5 - 2 \rangle = \langle 4, 3 \rangle$

(a) and (c).

11. (b) $\mathbf{v} = \langle 6 - 10, -1 - 2 \rangle = \langle -4, -3 \rangle$

(a) and (c).

13. (b) $\mathbf{v} = \langle 6 - 6, 6 - 2 \rangle = \langle 0, 4 \rangle$

(a) and (c).

15. (b) $\mathbf{v} = \left\langle \frac{1}{2} - \frac{3}{2}, 3 - \frac{4}{3} \right\rangle = \left\langle -1, \frac{5}{3} \right\rangle$

(a) and (c).

17. (a) $2\mathbf{v} = \langle 4, 6 \rangle$

(b) $-3\mathbf{v} = \langle -6, -9 \rangle$

—CONTINUED—

507

17. —CONTINUED—

(c) $\frac{7}{2}\mathbf{v} = \left\langle 7, \frac{21}{2} \right\rangle$

(d) $\frac{2}{3}\mathbf{v} = \left\langle \frac{4}{3}, 2 \right\rangle$

19.

21.

23. (a) $\frac{2}{3}\mathbf{u} = \frac{2}{3}\langle 4, 9 \rangle = \left\langle \frac{8}{3}, 6 \right\rangle$

(b) $\mathbf{v} - \mathbf{u} = \langle 2, -5 \rangle - \langle 4, 9 \rangle = \langle -2, -14 \rangle$

(c) $2\mathbf{u} + 5\mathbf{v} = 2\langle 4, 9 \rangle + 5\langle 2, -5 \rangle = \langle 18, -7 \rangle$

25. $\mathbf{v} = \frac{3}{2}(2\mathbf{i} - \mathbf{j}) = 3\mathbf{i} - \frac{3}{2}\mathbf{j}$

$= \left\langle 3, -\frac{3}{2} \right\rangle$

27. $\mathbf{v} = (2\mathbf{i} - \mathbf{j}) + 2(\mathbf{i} + 2\mathbf{j})$

$= 4\mathbf{i} + 3\mathbf{j} = \langle 4, 3 \rangle$

29. $u_1 - 4 = -1$

$u_2 - 2 = 3$

$u_1 = 3$

$u_2 = 5$

$Q = (3, 5)$

31. $\|\mathbf{v}\| = \sqrt{16 + 9} = 5$

33. $\|\mathbf{v}\| = \sqrt{36 + 25} = \sqrt{61}$

35. $\|\mathbf{v}\| = \sqrt{0 + 16} = 4$

37. $\|\mathbf{u}\| = \sqrt{3^2 + 12^2} = \sqrt{153}$

$\mathbf{v} = \dfrac{\mathbf{u}}{\|\mathbf{u}\|} = \dfrac{\langle 3, 12 \rangle}{\sqrt{153}} = \left\langle \dfrac{3}{\sqrt{153}}, \dfrac{12}{\sqrt{153}} \right\rangle$

$= \left\langle \dfrac{\sqrt{17}}{17}, \dfrac{4\sqrt{17}}{17} \right\rangle$ unit vector

39. $\|\mathbf{u}\| = \sqrt{\left(\dfrac{3}{2}\right)^2 + \left(\dfrac{5}{2}\right)^2} = \dfrac{\sqrt{34}}{2}$

$\mathbf{v} = \dfrac{\mathbf{u}}{\|\mathbf{u}\|} = \dfrac{\langle (3/2), (5/2) \rangle}{\sqrt{34}/2} = \left\langle \dfrac{3}{\sqrt{34}}, \dfrac{5}{\sqrt{34}} \right\rangle$

$= \left\langle \dfrac{3\sqrt{34}}{34}, \dfrac{5\sqrt{34}}{34} \right\rangle$ unit vector

41. $\|\mathbf{u}\| = \langle 1, -1 \rangle, \mathbf{v} = \langle -1, 2 \rangle$

  (a) $\|\mathbf{u}\| = \sqrt{1 + 1} = \sqrt{2}$

  (b) $\|\mathbf{v}\| = \sqrt{1 + 4} = \sqrt{5}$

  (c) $\mathbf{u} + \mathbf{v} = \langle 0, 1 \rangle$

     $\|\mathbf{u} + \mathbf{v}\| = \sqrt{0 + 1} = 1$

  (d) $\dfrac{\mathbf{u}}{\|\mathbf{u}\|} = \dfrac{1}{\sqrt{2}} \langle 1, -1 \rangle$

     $\left\| \dfrac{\mathbf{u}}{\|\mathbf{u}\|} \right\| = 1$

  (e) $\dfrac{\mathbf{v}}{\|\mathbf{v}\|} = \dfrac{1}{\sqrt{5}} \langle -1, 2 \rangle$

     $\left\| \dfrac{\mathbf{v}}{\|\mathbf{v}\|} \right\| = 1$

  (f) $\dfrac{\mathbf{u} + \mathbf{v}}{\|\mathbf{u} + \mathbf{v}\|} = \langle 0, 1 \rangle$

     $\left\| \dfrac{\mathbf{u} + \mathbf{v}}{\|\mathbf{u} + \mathbf{v}\|} \right\| = 1$

43. $\mathbf{u} = \left\langle 1, \dfrac{1}{2} \right\rangle, \mathbf{v} = \langle 2, 3 \rangle$

  (a) $\|\mathbf{u}\| = \sqrt{1 + \dfrac{1}{4}} = \dfrac{\sqrt{5}}{2}$

  (b) $\|\mathbf{v}\| = \sqrt{4 + 9} = \sqrt{13}$

  (c) $\mathbf{u} + \mathbf{v} = \left\langle 3, \dfrac{7}{2} \right\rangle$

     $\|\mathbf{u} + \mathbf{v}\| = \sqrt{9 + \dfrac{49}{4}} = \dfrac{\sqrt{85}}{2}$

  (d) $\dfrac{\mathbf{u}}{\|\mathbf{u}\|} = \dfrac{2}{\sqrt{5}} \left\langle 1, \dfrac{1}{2} \right\rangle$

     $\left\| \dfrac{\mathbf{u}}{\|\mathbf{u}\|} \right\| = 1$

  (e) $\dfrac{\mathbf{v}}{\|\mathbf{v}\|} = \dfrac{1}{\sqrt{13}} \langle 2, 3 \rangle$

     $\left\| \dfrac{\mathbf{v}}{\|\mathbf{v}\|} \right\| = 1$

  (f) $\dfrac{\mathbf{u} + \mathbf{v}}{\|\mathbf{u} + \mathbf{v}\|} = \dfrac{2}{\sqrt{85}} \left\langle 3, \dfrac{7}{2} \right\rangle$

     $\left\| \dfrac{\mathbf{u} + \mathbf{v}}{\|\mathbf{u} + \mathbf{v}\|} \right\| = 1$

45. $\mathbf{u} = \langle 2, 1 \rangle$

    $\|\mathbf{u}\| = \sqrt{5} \approx 2.236$

    $\mathbf{v} = \langle 5, 4 \rangle$

    $\|\mathbf{v}\| = \sqrt{41} \approx 6.403$

  $\mathbf{u} + \mathbf{v} = \langle 7, 5 \rangle$

  $\|\mathbf{u} + \mathbf{v}\| = \sqrt{74} \approx 8.602$

  $\|\mathbf{u} + \mathbf{v}\| \leq \|\mathbf{u}\| + \|\mathbf{v}\|$

47. $\dfrac{\mathbf{u}}{\|\mathbf{u}\|} = \dfrac{1}{\sqrt{2}} \langle 1, 1 \rangle$

  $4 \left(\dfrac{\mathbf{u}}{\|\mathbf{u}\|} \right) = 2\sqrt{2} \langle 1, 1 \rangle$

  $\mathbf{v} = \langle 2\sqrt{2}, 2\sqrt{2} \rangle$

49. $\dfrac{\mathbf{u}}{\|\mathbf{u}\|} = \dfrac{1}{2\sqrt{3}} \langle \sqrt{3}, 3 \rangle$

  $2 \left(\dfrac{\mathbf{u}}{\|\mathbf{u}\|} \right) = \dfrac{1}{\sqrt{3}} \langle \sqrt{3}, 3 \rangle$

    $\mathbf{v} = \langle 1, \sqrt{3} \rangle$

51. $\mathbf{v} = 3[(\cos 0°)\mathbf{i} + (\sin 0°)\mathbf{j}] = 3\mathbf{i} = \langle 3, 0 \rangle$

53. $\mathbf{v} = 2[(\cos 150°)\mathbf{i} + (\sin 150°)\mathbf{j}]$

    $= -\sqrt{3}\mathbf{i} + \mathbf{j} = \langle -\sqrt{3}, 1 \rangle$

55. $\mathbf{u} = \mathbf{i}$

    $\mathbf{v} = \dfrac{3\sqrt{2}}{2}\mathbf{i} + \dfrac{3\sqrt{2}}{2}\mathbf{j}$

  $\mathbf{u} + \mathbf{v} = \left(\dfrac{2 + 3\sqrt{2}}{2} \right)\mathbf{i} + \dfrac{3\sqrt{2}}{2}\mathbf{j}$

57. $\mathbf{u} = 2(\cos 4)\mathbf{i} + 2(\sin 4)\mathbf{j}$

$\mathbf{v} = (\cos 2)\mathbf{i} + (\sin 2)\mathbf{j}$

$\mathbf{u} + \mathbf{v} = (2\cos 4 + \cos 2)\mathbf{i} + (2\sin 4 + \sin 2)\mathbf{j}$

59. A scalar is a real number. A vector is represented by a directed line segment. A vector has both length and direction.

61. To normalize $\mathbf{v} \neq 0$, you find a unit vector $\mathbf{u}$ in the direction of $\mathbf{v}$:

$$\mathbf{u} = \frac{\mathbf{v}}{\|\mathbf{v}\|}.$$

For Exercises 63–67, $a\mathbf{u} + b\mathbf{w} = a(\mathbf{i} + 2\mathbf{j}) + b(\mathbf{i} - \mathbf{j}) = (a + b)\mathbf{i} + (2a - b)\mathbf{j}.$

63. $\mathbf{v} = 2\mathbf{i} + \mathbf{j}$. Therefore, $a + b = 2, 2a - b = 1$. Solving simultaneously, we have $a = 1, b = 1$.

65. $\mathbf{v} = 3\mathbf{i}$. Therefore, $a + b = 3, 2a - b = 0$. Solving simultaneously, we have $a = 1, b = 2$.

67. $\mathbf{v} = \mathbf{i} + \mathbf{j}$. Therefore, $a + b = 1, 2a - b = 1$. Solving simultaneously, we have $a = \frac{2}{3}, b = \frac{1}{3}$.

69. $y = x^3, y' = 3x^2 = 3$ at $x = 1$.

(a) $m = 3$. Let $\mathbf{w} = \langle 1, 3 \rangle$, then

$$\frac{\mathbf{w}}{\|\mathbf{w}\|} = \pm\frac{1}{\sqrt{10}} \langle 1, 3 \rangle.$$

(b) $m = -\frac{1}{3}$. Let $\mathbf{w} = \langle 3, -1 \rangle$, then

$$\frac{\mathbf{w}}{\|\mathbf{w}\|} = \pm\frac{1}{\sqrt{10}} \langle 3, -1 \rangle.$$

71. $f(x) = \sqrt{25 - x^2}$

$$f'(x) = \frac{-x}{\sqrt{25 - x^2}} = \frac{-3}{4} \text{ at } x = 3.$$

(a) $m = -\frac{3}{4}$. Let $\mathbf{w} = \langle -4, 3 \rangle$, then

$$\frac{\mathbf{w}}{\|\mathbf{w}\|} = \pm\frac{1}{5} \langle -4, 3 \rangle.$$

(b) $m = \frac{4}{3}$. Let $\mathbf{w} = \langle 3, 4 \rangle$, then

$$\frac{\mathbf{w}}{\|\mathbf{w}\|} = \pm\frac{1}{5} \langle 3, 4 \rangle$$

73. $\mathbf{u} = \frac{\sqrt{2}}{2}\mathbf{i} + \frac{\sqrt{2}}{2}\mathbf{j}$

$\mathbf{u} + \mathbf{v} = \sqrt{2}\mathbf{j}$

$\mathbf{v} = (\mathbf{u} + \mathbf{v}) - \mathbf{u} = -\frac{\sqrt{2}}{2}\mathbf{i} + \frac{\sqrt{2}}{2}\mathbf{j}$

75. Programs will vary.

77. $\|\mathbf{F}_1\| = 2, \theta_{\mathbf{F}_1} = 33°$

$\|\mathbf{F}_2\| = 3, \theta_{\mathbf{F}_2} = -125°$

$\|\mathbf{F}_3\| = 2.5, \theta_{\mathbf{F}_3} = 110°$

$\|\mathbf{R}\| = \|\mathbf{F}_1 + \mathbf{F}_2 + \mathbf{F}_3\| \approx 1.33$

$\theta_{\mathbf{R}} = \theta_{\mathbf{F}_1 + \mathbf{F}_2 + \mathbf{F}_3} \approx 132.5°$

79. (a) $180(\cos 30°\mathbf{i} + \sin 30°\mathbf{j}) + 275\mathbf{i} \approx 430.88\mathbf{i} + 90\mathbf{j}$

Direction: $\alpha \approx \arctan\left(\dfrac{90}{430.88}\right) \approx 0.206 (\approx 11.8°)$

Magnitude: $\sqrt{430.88^2 + 90^2} \approx 440.18$ newtons

(b) $M = \sqrt{(275 + 180\cos\theta)^2 + (180\sin\theta)^2}$

$\alpha = \arctan\left[\dfrac{180\sin\theta}{275 + 180\cos\theta}\right]$

—CONTINUED—

79. —CONTINUED—

(c)

θ	0°	30°	60°	90°	120°	150°	180°
M	455	440.2	396.9	328.7	241.9	149.3	95
α	0°	11.8°	23.1°	33.2°	40.1°	37.1°	0

(d)

(e) M decreases because the forces change from acting in the same direction to acting in the opposite direction as θ increases from 0° to 180°.

81. $\mathbf{F}_1 + \mathbf{F}_2 + \mathbf{F}_3 = (75 \cos 30°\mathbf{i} + 75 \sin 30°\mathbf{j}) + (100 \cos 45°\mathbf{i} + 100 \sin 45°\mathbf{j}) + (125 \cos 120°\mathbf{i} + 125 \sin 120°\mathbf{j})$

$$= \left(\frac{75}{2}\sqrt{3} + 50\sqrt{2} - \frac{125}{2}\right)\mathbf{i} + \left(\frac{75}{2} + 50\sqrt{2} + \frac{125}{2}\sqrt{3}\right)\mathbf{j}$$

$$\|\mathbf{R}\| = \|\mathbf{F}_1 + \mathbf{F}_2 + \mathbf{F}_3\| \approx 228.5 \text{ lb}$$

$$\theta_\mathbf{R} = \theta_{\mathbf{F}_1 + \mathbf{F}_2 + \mathbf{F}_3} \approx 71.3°$$

83. (a) The forces act along the same direction. $\theta = 0°$.

(b) The forces cancel out each other. $\theta = 180°$.

(c) No, the magnitude of the resultant can not be greater than the sum.

85. $(-4, -1), (6, 5), (10, 3)$

87. $\mathbf{u} = \overrightarrow{CB} = \|\mathbf{u}\|(\cos 30° \, \mathbf{i} + \sin 30° \, \mathbf{j})$

$\mathbf{v} = \overrightarrow{CA} = \|\mathbf{v}\|(\cos 130° \, \mathbf{i} + \sin 130° \, \mathbf{j})$

Vertical components: $\|\mathbf{u}\| \sin 30° + \|\mathbf{v}\| \sin 130° = 2000$

Horizontal components: $\|\mathbf{u}\| \cos 30° + \|\mathbf{v}\| \cos 130° = 0$

Solving this system, you obtain

$\|\mathbf{u}\| \approx 1305.5$ pounds and $\|\mathbf{v}\| \approx 1758.8$ pounds.

89. Horizontal component $= \|\mathbf{v}\| \cos \theta = 1200 \cos 6° \approx 1193.43 \text{ ft/sec}$

Vertical component $= \|\mathbf{v}\| \sin \theta = 1200 \sin 6° \approx 125.43 \text{ ft/sec}$

91. $\mathbf{u} = 900[\cos 148° \, \mathbf{i} + \sin 148° \, \mathbf{j}]$

$\mathbf{v} = 100[\cos 45° \, \mathbf{i} + \sin 45° \, \mathbf{j}]$

$\mathbf{u} + \mathbf{v} = [900 \cos 148° + 100 \cos 45°]\mathbf{i} + [900 \sin 148° + 100 \sin 45°]\mathbf{j}$

$\approx -692.53 \, \mathbf{i} + 547.64 \, \mathbf{j}$

$\theta \approx \arctan\left(\dfrac{547.64}{-692.53}\right) \approx -38.34°.$ 38.34° North of West.

$\|\mathbf{u} + \mathbf{v}\| \approx \sqrt{(-692.53)^2 + (547.64)^2} \approx 882.9 \text{ km/hr.}$

93. $\mathbf{F}_1 + \mathbf{F}_2 + \mathbf{F}_3 = \mathbf{0}$

$-3600\mathbf{j} + T_2(\cos 35°\mathbf{i} - \sin 35° \, \mathbf{j}) + T_3(\cos 92°\mathbf{i} + \sin 92°\mathbf{j}) = 0$

$T_2 \cos 35° + T_3 \cos 92° = 0$

$-T_2 \sin 35° + T_3 \sin 92° = 3600$

$T_2 = \dfrac{-T_3 \cos 92°}{\cos 35°} \implies \dfrac{T_3 \cos 92°}{\cos 35°} \sin 35° + T_3 \sin 92° = 3600$ and $T_3(0.97495) \approx 3600 \implies T_3 \approx 3692.48$

Finally, $T_2 \approx 157.32$

95. Let the triangle have vertices at $(0, 0)$, $(a, 0)$, and (b, c). Let $\mathbf{u}$ be the vector joining $(0, 0)$ and (b, c), as indicated in the figure. Then $\mathbf{v}$, the vector joining the midpoints, is

$\mathbf{v} = \left(\dfrac{a + b}{2} - \dfrac{a}{2}\right)\mathbf{i} + \dfrac{c}{2}\mathbf{j}$

$= \dfrac{b}{2}\mathbf{i} + \dfrac{c}{2}\mathbf{j} = \dfrac{1}{2}(b\mathbf{i} + c\mathbf{j}) = \dfrac{1}{2}\mathbf{u}$

97. $\mathbf{w} = \|\mathbf{u}\|\mathbf{v} + \|\mathbf{v}\|\mathbf{u}$

$= \|\mathbf{u}\|[\|\mathbf{v}\| \cos \theta_\mathbf{v}\mathbf{i} + \|\mathbf{v}\| \sin \theta_\mathbf{v}\mathbf{j}] + \|\mathbf{v}\|[\|\mathbf{u}\| \cos \theta_\mathbf{u}\mathbf{i} + \|\mathbf{u}\| \sin \theta_\mathbf{u}\mathbf{j}] = \|\mathbf{u}\| \|\mathbf{v}\|[(\cos \theta_\mathbf{u} + \cos \theta_\mathbf{v})\mathbf{i} + (\sin \theta_\mathbf{u} + \sin \theta_\mathbf{v})\mathbf{j}]$

$= 2\|\mathbf{u}\| \|\mathbf{v}\|\left[\cos\left(\dfrac{\theta_\mathbf{u} + \theta_\mathbf{v}}{2}\right)\cos\left(\dfrac{\theta_\mathbf{u} - \theta_\mathbf{v}}{2}\right)\mathbf{i} + \sin\left(\dfrac{\theta_\mathbf{u} + \theta_\mathbf{v}}{2}\right)\cos\left(\dfrac{\theta_\mathbf{u} - \theta_\mathbf{v}}{2}\right)\mathbf{j}\right]$

$\tan \theta_\mathbf{w} = \dfrac{\sin\left(\dfrac{\theta_\mathbf{u} + \theta_\mathbf{v}}{2}\right)\cos\left(\dfrac{\theta_\mathbf{u} - \theta_\mathbf{v}}{2}\right)}{\cos\left(\dfrac{\theta_\mathbf{u} + \theta_\mathbf{v}}{2}\right)\cos\left(\dfrac{\theta_\mathbf{u} - \theta_\mathbf{v}}{2}\right)} = \tan\left(\dfrac{\theta_\mathbf{u} + \theta_\mathbf{v}}{2}\right)$

Thus, $\theta_\mathbf{w} = (\theta_\mathbf{u} + \theta_\mathbf{v})/2$ and $\mathbf{w}$ bisects the angle between $\mathbf{u}$ and $\mathbf{v}$.

99. True

101. True

103. False

$\|a\mathbf{i} + b\mathbf{j}\| = \sqrt{2}\,|a|$

Section 10.2   Space Coordinates and Vectors in Space

1.

3.

5. $A(2, 3, 4)$

    $B(-1, -2, 2)$

7. $x = -3, y = 4, z = 5$: $(-3, 4, 5)$

9. $y = z = 0, x = 10$: $(10, 0, 0)$

11. The z-coordinate is 0.

13. The point is 6 units above the xy-plane.

15. The point is on the plane parallel to the yz-plane that passes through $x = 4$.

17. The point is to the left of the xz-plane.

19. The point is on or between the planes $y = 3$ and $y = -3$.

21. The point (x, y, z) is 3 units below the xy-plane, and below either quadrant I or III.

23. The point could be above the xy-plane and thus above quadrants II or IV, or below the xy-plane, and thus below quadrants I or III.

25. $d = \sqrt{(5 - 0)^2 + (2 - 0)^2 + (6 - 0)^2}$

    $= \sqrt{25 + 4 + 36} = \sqrt{65}$

27. $d = \sqrt{(6 - 1)^2 + (-2 - (-2))^2 + (-2 - 4)^2}$

    $= \sqrt{25 + 0 + 36} = \sqrt{61}$

29. $A(0, 0, 0), B(2, 2, 1), C(2, -4, 4)$

    $|AB| = \sqrt{4 + 4 + 1} = 3$

    $|AC| = \sqrt{4 + 16 + 16} = 6$

    $|BC| = \sqrt{0 + 36 + 9} = 3\sqrt{5}$

    $|BC|^2 = |AB|^2 + |AC|^2$

    Right triangle

31. $A(1, -3, -2), B(5, -1, 2), C(-1, 1, 2)$

    $|AB| = \sqrt{16 + 4 + 16} = 6$

    $|AC| = \sqrt{4 + 16 + 16} = 6$

    $|BC| = \sqrt{36 + 4 + 0} = 2\sqrt{10}$

    Since $|AB| = |AC|$, the triangle is isosceles.

33. The z-coordinate is changed by 5 units:

    $(0, 0, 5), (2, 2, 6), (2, -4, 9)$

35. $\left(\dfrac{5 + (-2)}{2}, \dfrac{-9 + 3}{2}, \dfrac{7 + 3}{2} \right) = \left(\dfrac{3}{2}, -3, 5 \right)$

37. Center: $(0, 2, 5)$

    Radius: 2

    $(x - 0)^2 + (y - 2)^2 + (z - 5)^2 = 4$

    $x^2 + y^2 + z^2 - 4y - 10z + 25 = 0$

39. Center: $\dfrac{(2, 0, 0) + (0, 6, 0)}{2} = (1, 3, 0)$

    Radius: $\sqrt{10}$

    $(x - 1)^2 + (y - 3)^2 + (z - 0)^2 = 10$

    $x^2 + y^2 + z^2 - 2x - 6y = 0$

41. $x^2 + y^2 + z^2 - 2x + 6y + 8z + 1 = 0$

    $(x^2 - 2x + 1) + (y^2 + 6y + 9) + (z^2 + 8z + 16) = -1 + 1 + 9 + 16$

    $(x - 1)^2 + (y + 3)^2 + (z + 4)^2 = 25$

    Center: $(1, -3, -4)$

    Radius: 5

43. $9x^2 + 9y^2 + 9z^2 - 6x + 18y + 1 = 0$

$$x^2 + y^2 + z^2 - \frac{2}{3}x + 2y + \frac{1}{9} = 0$$

$$\left(x^2 - \frac{2}{3}x + \frac{1}{9}\right) + (y^2 + 2y + 1) + z^2 = -\frac{1}{9} + \frac{1}{9} + 1$$

$$\left(x - \frac{1}{3}\right)^2 + (y + 1)^2 + (z - 0)^2 = 1$$

Center: $\left(\frac{1}{3}, -1, 0\right)$

Radius: 1

45. $x^2 + y^2 + z^2 \le 36$

Solid ball of radius 6 centered at origin.

47. (a) $\mathbf{v} = (2 - 4)\mathbf{i} + (4 - 2)\mathbf{j} + (3 - 1)\mathbf{k}$

   $= -2\mathbf{i} + 2\mathbf{j} + 2\mathbf{k} = \langle -2, 2, 2 \rangle$

(b)

49. (a) $\mathbf{v} = (0 - 3)\mathbf{i} + (3 - 3)\mathbf{j} + (3 - 0)\mathbf{k}$

   $= -3\mathbf{i} + 3\mathbf{k} = \langle -3, 0, 3 \rangle$

(b)

51. $\langle 4 - 3, 1 - 2, 6 - 0 \rangle = \langle 1, -1, 6 \rangle$

$\|\langle 1, -1, 6 \rangle\| = \sqrt{1 + 1 + 36} = \sqrt{38}$

Unit vector: $\dfrac{\langle 1, -1, 6 \rangle}{\sqrt{38}} = \left\langle \dfrac{1}{\sqrt{38}}, \dfrac{-1}{\sqrt{38}}, \dfrac{6}{\sqrt{38}} \right\rangle$

53. $\langle -5 - (-4), 3 - 3, 0 - 1 \rangle = \langle -1, 0, -1 \rangle$

$\|\langle -1, 0, -1 \rangle\| = \sqrt{1 + 1} = \sqrt{2}$

Unit vector: $\left\langle \dfrac{-1}{\sqrt{2}}, 0, \dfrac{-1}{\sqrt{2}} \right\rangle$

55. (b) $\mathbf{v} = (3 + 1)\mathbf{i} + (3 - 2)\mathbf{j} + (4 - 3)\mathbf{k}$

   $= 4\mathbf{i} + \mathbf{j} + \mathbf{k} = \langle 4, 1, 1 \rangle$

(a) and (c).

57. $(q_1, q_2, q_3) - (0, 6, 2) = (3, -5, 6)$

$Q = (3, 1, 8)$

59. (a) $2\mathbf{v} = \langle 2, 4, 4 \rangle$

(b) $-\mathbf{v} = \langle -1, -2, -2 \rangle$

—CONTINUED—

59. —CONTINUED—

(c) $\frac{3}{2}\mathbf{v} = \left\langle \frac{3}{2}, 3, 3 \right\rangle$

(d) $0\mathbf{v} = \langle 0, 0, 0 \rangle$

61. $\mathbf{z} = \mathbf{u} - \mathbf{v} = \langle 1, 2, 3 \rangle - \langle 2, 2, -1 \rangle = \langle -1, 0, 4 \rangle$

63. $\mathbf{z} = 2\mathbf{u} + 4\mathbf{v} - \mathbf{w} = \langle 2, 4, 6 \rangle + \langle 8, 8, -4 \rangle - \langle 4, 0, -4 \rangle = \langle 6, 12, 6 \rangle$

65. $2\mathbf{z} - 3\mathbf{u} = 2\langle z_1, z_2, z_3 \rangle - 3\langle 1, 2, 3 \rangle = \langle 4, 0, -4 \rangle$

$2z_1 - 3 = 4 \implies z_1 = \frac{7}{2}$

$2z_2 - 6 = 0 \implies z_2 = 3$

$2z_3 - 9 = -4 \implies z_3 = \frac{5}{2}$

$\mathbf{z} = \left\langle \frac{7}{2}, 3, \frac{5}{2} \right\rangle$

67. (a) and (b) are parallel since $\langle -6, -4, 10 \rangle = -2\langle 3, 2, -5 \rangle$ and $\left\langle 2, \frac{4}{3}, -\frac{10}{3} \right\rangle = \frac{2}{3}\langle 3, 2, -5 \rangle$.

69. $\mathbf{z} = -3\mathbf{i} + 4\mathbf{j} + 2\mathbf{k}$

(a) is parallel since $-6\mathbf{i} + 8\mathbf{j} + 4\mathbf{k} = 2\mathbf{z}$.

71. $P(0, -2, -5)$, $Q(3, 4, 4)$, $R(2, 2, 1)$

$\overrightarrow{PQ} = \langle 3, 6, 9 \rangle$

$\overrightarrow{PR} = \langle 2, 4, 6 \rangle$

$\langle 3, 6, 9 \rangle = \frac{3}{2}\langle 2, 4, 6 \rangle$

Therefore, $\overrightarrow{PQ}$ and $\overrightarrow{PR}$ are parallel. The points are collinear.

73. $P(1, 2, 4)$, $Q(2, 5, 0)$, $R(0, 1, 5)$

$\overrightarrow{PQ} = \langle 1, 3, -4 \rangle$

$\overrightarrow{PR} = \langle -1, -1, 1 \rangle$

Since $\overrightarrow{PQ}$ and $\overrightarrow{PR}$ are not parallel, the points are not collinear.

75. $A(2, 9, 1)$, $B(3, 11, 4)$, $C(0, 10, 2)$, $D(1, 12, 5)$

$\overrightarrow{AB} = \langle 1, 2, 3 \rangle$

$\overrightarrow{CD} = \langle 1, 2, 3 \rangle$

$\overrightarrow{AC} = \langle -2, 1, 1 \rangle$

$\overrightarrow{BD} = \langle -2, 1, 1 \rangle$

Since $\overrightarrow{AB} = \overrightarrow{CD}$ and $\overrightarrow{AC} = \overrightarrow{BD}$, the given points form the vertices of a parallelogram.

77. $\|\mathbf{v}\| = 0$

79. $\mathbf{v} = \langle 1, -2, -3 \rangle$

$\|\mathbf{v}\| = \sqrt{1 + 4 + 9} = \sqrt{14}$

81. $\mathbf{v} = \langle 0, 3, -5 \rangle$

$\|\mathbf{v}\| = \sqrt{0 + 9 + 25} = \sqrt{34}$

83. $\mathbf{u} = \langle 2, -1, 2 \rangle$

$\|\mathbf{u}\| = \sqrt{4 + 1 + 4} = 3$

(a) $\dfrac{\mathbf{u}}{\|\mathbf{u}\|} = \dfrac{1}{3}\langle 2, -1, 2 \rangle$

(b) $-\dfrac{\mathbf{u}}{\|\mathbf{u}\|} = -\dfrac{1}{3}\langle 2, -1, 2 \rangle$

85. $\mathbf{u} = \langle 3, 2, -5 \rangle$

$\|\mathbf{u}\| = \sqrt{9 + 4 + 25} = \sqrt{38}$

(a) $\dfrac{\mathbf{u}}{\|\mathbf{u}\|} = \dfrac{1}{\sqrt{38}}\langle 3, 2, -5 \rangle$

(b) $-\dfrac{\mathbf{u}}{\|\mathbf{u}\|} = -\dfrac{1}{\sqrt{38}}\langle 3, 2, -5 \rangle$

87. Programs will vary.

89. $c\mathbf{v} = \langle 2c, 2c, -c \rangle$

$\|c\mathbf{v}\| = \sqrt{4c^2 + 4c^2 + c^2} = 5$

$9c^2 = 25$

$c = \pm\dfrac{5}{3}$

91. $\mathbf{v} = 10\dfrac{\mathbf{u}}{\|\mathbf{u}\|} = 10\left\langle 0, \dfrac{1}{\sqrt{2}}, \dfrac{1}{\sqrt{2}}\right\rangle$

$= \left\langle 0, \dfrac{10}{\sqrt{2}}, \dfrac{10}{\sqrt{2}}\right\rangle$

93. $\mathbf{v} = \dfrac{3}{2}\dfrac{\mathbf{u}}{\|\mathbf{u}\|} = \dfrac{3}{2}\left\langle \dfrac{2}{3}, \dfrac{-2}{3}, \dfrac{1}{3}\right\rangle = \left\langle 1, -1, \dfrac{1}{2}\right\rangle$

95. $\mathbf{v} = 2[\cos(\pm 30°)\mathbf{j} + \sin(\pm 30°)\mathbf{k}]$

$= \sqrt{3}\mathbf{j} \pm \mathbf{k} = \langle 0, \sqrt{3}, \pm 1 \rangle$

97.

$\mathbf{v} = \langle -3, -6, 3 \rangle$

$\dfrac{2}{3}\mathbf{v} = \langle -2, -4, 2 \rangle$

$(4, 3, 0) + (-2, -4, 2) = (2, -1, 2)$

99. (a)

(b) $\mathbf{w} = a\mathbf{u} + b\mathbf{v} = a\mathbf{i} + (a + b)\mathbf{j} + b\mathbf{k} = \mathbf{0}$

$a = 0, a + b = 0, b = 0$

Thus, a and b are both zero.

(c) $a\mathbf{i} + (a + b)\mathbf{j} + b\mathbf{k} = \mathbf{i} + 2\mathbf{j} + \mathbf{k}$

$a = 1, b = 1$

$\mathbf{w} = \mathbf{u} + \mathbf{v}$

(d) $a\mathbf{i} + (a + b)\mathbf{j} + b\mathbf{k} = \mathbf{i} + 2\mathbf{j} + 3\mathbf{k}$

$a = 1, a + b = 2, b = 3$

Not possible

101. $d = \sqrt{(x_2 - x_1)^2 + (y_2 - y_1)^2 + (z_2 - z_1)^2}$

103. Two nonzero vectors $\mathbf{u}$ and $\mathbf{v}$ are parallel if $\mathbf{u} = c\mathbf{v}$ for some scalar c.

105. (a) The height of the right triangle is $h = \sqrt{L^2 - 18^2}$. The vector $\overrightarrow{PQ}$ is given by

$\overrightarrow{PQ} = \langle 0, -18, h \rangle$.

The tension vector $\mathbf{T}$ in each wire is

$\mathbf{T} = c\langle 0, -18, h \rangle$ where $ch = \dfrac{24}{3} = 8$.

Hence, $\mathbf{T} = \dfrac{8}{h}\langle 0, -18, h \rangle$ and

$T = \|\mathbf{T}\| = \dfrac{8}{h}\sqrt{18^2 + h^2} = \dfrac{8}{\sqrt{L^2 - 18^2}}\sqrt{18^2 + (L^2 - 18^2)} = \dfrac{8L}{\sqrt{L^2 - 18^2}}$

(b)

L	20	25	30	35	40	45	50
T	18.4	11.5	10	9.3	9.0	8.7	8.6

—CONTINUED—

105. —CONTINUED—

(c)

$x = 18$ is a vertical asymptote and $y = 8$ is a horizontal asymptote.

(d) $\displaystyle\lim_{L \to 18^+} \frac{8L}{\sqrt{L^2 - 18^2}} = \infty$

$\displaystyle\lim_{L \to \infty} \frac{8L}{\sqrt{L^2 - 18^2}} = \lim_{L \to \infty} \frac{8}{\sqrt{1 - (18/L)^2}} = 8$

(e) From the table, $T = 10$ implies $L = 30$ inches.

107. Let α be the angle between $\mathbf{v}$ and the coordinate axes.

$$\mathbf{v} = (\cos \alpha)\mathbf{i} + (\cos \alpha)\mathbf{j} + (\cos \alpha)\mathbf{k}$$

$$\|\mathbf{v}\| = \sqrt{3} \cos \alpha = 1$$

$$\cos \alpha = \frac{1}{\sqrt{3}} = \frac{\sqrt{3}}{3}$$

$$\mathbf{v} = \frac{\sqrt{3}}{3}(\mathbf{i} + \mathbf{j} + \mathbf{k}) = \frac{\sqrt{3}}{3}\langle 1, 1, 1 \rangle$$

109. $\overrightarrow{AB} = \langle 0, 70, 115 \rangle$, $\mathbf{F}_1 = C_1\langle 0, 70, 115 \rangle$

$\overrightarrow{AC} = \langle -60, 0, 115 \rangle$, $\mathbf{F}_2 = C_2\langle -60, 0, 115 \rangle$

$\overrightarrow{AD} = \langle 45, -65, 115 \rangle$, $\mathbf{F}_3 = C_3\langle 45, -65, 115 \rangle$

$\mathbf{F} = \mathbf{F}_1 + \mathbf{F}_2 + \mathbf{F}_3 = \langle 0, 0, 500 \rangle$

Thus:

$$-60C_2 + 45C_3 = 0$$
$$70C_1 \qquad - 65C_3 = 0$$
$$115(C_1 + C_2 + C_3) = 500$$

Solving this system yields $C_1 = \frac{104}{69}$, $C_2 = \frac{28}{23}$, and $C_3 = \frac{112}{69}$. Thus:

$$\|\mathbf{F}_1\| \approx 202.919N$$

$$\|\mathbf{F}_2\| \approx 157.909N$$

$$\|\mathbf{F}_3\| \approx 226.521N$$

111. $d(AP) = 2d(BP)$

$$\sqrt{x^2 + (y + 1)^2 + (z - 1)^2} = 2\sqrt{(x - 1)^2 + (y - 2)^2 + z^2}$$

$$x^2 + y^2 + z^2 + 2y - 2z + 2 = 4(x^2 + y^2 + z^2 - 2x - 4y + 5)$$

$$0 = 3x^2 + 3y^2 + 3z^2 - 8x - 18y + 2z + 18$$

$$-6 + \frac{16}{9} + 9 + \frac{1}{9} = \left(x^2 - \frac{8}{3}x + \frac{16}{9}\right) + (y^2 - 6y + 9) + \left(z^2 + \frac{2}{3}z + \frac{1}{9}\right)$$

$$\frac{44}{9} = \left(x - \frac{4}{3}\right)^2 + (y - 3)^2 + \left(z + \frac{1}{3}\right)^2$$

Sphere; center: $\left(\dfrac{4}{3}, 3, -\dfrac{1}{3}\right)$, radius: $\dfrac{2\sqrt{11}}{3}$

Section 10.3   The Dot Product of Two Vectors

1. $\mathbf{u} = \langle 3, 4 \rangle, \mathbf{v} = \langle 2, -3 \rangle$

(a) $\mathbf{u} \cdot \mathbf{v} = 3(2) + 4(-3) = -6$

(b) $\mathbf{u} \cdot \mathbf{u} = 3(3) + 4(4) = 25$

(c) $\|\mathbf{u}\|^2 = 25$

(d) $(\mathbf{u} \cdot \mathbf{v})\mathbf{v} = -6\langle 2, -3 \rangle = \langle -12, 18 \rangle$

(e) $\mathbf{u} \cdot (2\mathbf{v}) = 2(\mathbf{u} \cdot \mathbf{v}) = 2(-6) = -12$

3. $\mathbf{u} = \langle 2, -3, 4 \rangle, \mathbf{v} = \langle 0, 6, 5 \rangle$

(a) $\mathbf{u} \cdot \mathbf{v} = 2(0) + (-3)(6) + (4)(5) = 2$

(b) $\mathbf{u} \cdot \mathbf{u} = 2(2) + (-3)(-3) + 4(4) = 29$

(c) $\|\mathbf{u}\|^2 = 29$

(d) $(\mathbf{u} \cdot \mathbf{v})\mathbf{v} = 2\langle 0, 6, 5 \rangle = \langle 0, 12, 10 \rangle$

(e) $\mathbf{u} \cdot (2\mathbf{v}) = 2(\mathbf{u} \cdot \mathbf{v}) = 2(2) = 4$

5. $\mathbf{u} = 2\mathbf{i} - \mathbf{j} + \mathbf{k}, \mathbf{v} = \mathbf{i} - \mathbf{k}$

(a) $\mathbf{u} \cdot \mathbf{v} = 2(1) + (-1)(0) + 1(-1) = 1$

(b) $\mathbf{u} \cdot \mathbf{u} = 2(2) + (-1)(-1) + (1)(1) = 6$

(c) $\|\mathbf{u}\|^2 = 6$

(d) $(\mathbf{u} \cdot \mathbf{v})\mathbf{v} = \mathbf{v} = \mathbf{i} - \mathbf{k}$

(e) $\mathbf{u} \cdot (2\mathbf{v}) = 2(\mathbf{u} \cdot \mathbf{v}) = 2$

7. $\mathbf{u} = \langle 3240, 1450, 2235 \rangle$

$\mathbf{v} = \langle 2.22, 1.85, 3.25 \rangle$

$\mathbf{u} \cdot \mathbf{v} = \$17{,}139.05$

This gives the total amount that the person earned on his products.

9. $\dfrac{\mathbf{u} \cdot \mathbf{v}}{\|\mathbf{u}\|\,\|\mathbf{v}\|} = \cos \theta$

$\mathbf{u} \cdot \mathbf{v} = (8)(5) \cos \dfrac{\pi}{3} = 20$

11. $\mathbf{u} = \langle 1, 1 \rangle, \mathbf{v} = \langle 2, -2 \rangle$

$\cos \theta = \dfrac{\mathbf{u} \cdot \mathbf{v}}{\|\mathbf{u}\|\,\|\mathbf{v}\|} = \dfrac{0}{\sqrt{2}\sqrt{8}} = 0$

$\theta = \dfrac{\pi}{2}$

13. $\mathbf{u} = 3\mathbf{i} + \mathbf{j}, \mathbf{v} = -2\mathbf{i} + 4\mathbf{j}$

$\cos \theta = \dfrac{\mathbf{u} \cdot \mathbf{v}}{\|\mathbf{u}\|\,\|\mathbf{v}\|} = \dfrac{-2}{\sqrt{10}\sqrt{20}} = \dfrac{-1}{5\sqrt{2}}$

$\theta = \arccos\left(-\dfrac{1}{5\sqrt{2}}\right) \approx 98.1°$

15. $\mathbf{u} = \langle 1, 1, 1 \rangle, \mathbf{v} = \langle 2, 1, -1 \rangle$

$\cos \theta = \dfrac{\mathbf{u} \cdot \mathbf{v}}{\|\mathbf{u}\|\,\|\mathbf{v}\|} = \dfrac{2}{\sqrt{3}\sqrt{6}} = \dfrac{\sqrt{2}}{3}$

$\theta = \arccos \dfrac{\sqrt{2}}{3} \approx 61.9°$

17. $\mathbf{u} = 3\mathbf{i} + 4\mathbf{j}, \mathbf{v} = -2\mathbf{j} + 3\mathbf{k}$

$\cos \theta = \dfrac{\mathbf{u} \cdot \mathbf{v}}{\|\mathbf{u}\|\,\|\mathbf{v}\|} = \dfrac{-8}{5\sqrt{13}} = \dfrac{-8\sqrt{13}}{65}$

$\theta = \arccos\left(-\dfrac{8\sqrt{13}}{65}\right) \approx 116.3°$

19. $\mathbf{u} = \langle 4, 0 \rangle, \mathbf{v} = \langle 1, 1 \rangle$

$\mathbf{u} \neq c\mathbf{v} \implies$ not parallel

$\mathbf{u} \cdot \mathbf{v} = 4 \neq 0 \implies$ not orthogonal

Neither

21. $\mathbf{u} = \langle 4, 3 \rangle, \mathbf{v} = \left\langle \dfrac{1}{2}, -\dfrac{2}{3} \right\rangle$

$\mathbf{u} \neq c\mathbf{v} \implies$ not parallel

$\mathbf{u} \cdot \mathbf{v} = 0 \implies$ orthogonal

23. $\mathbf{u} = \mathbf{j} + 6\mathbf{k}, \mathbf{v} = \mathbf{i} - 2\mathbf{j} - \mathbf{k}$

$\mathbf{u} \neq c\mathbf{v} \implies$ not parallel

$\mathbf{u} \cdot \mathbf{v} = -8 \neq 0 \implies$ not orthogonal

Neither

25. $\mathbf{u} = \langle 2, -3, 1 \rangle, \mathbf{v} = \langle -1, -1, -1 \rangle$

$\mathbf{u} \neq c\mathbf{v} \implies$ not parallel

$\mathbf{u} \cdot \mathbf{v} = 0 \implies$ orthogonal

27. $\mathbf{u} = \mathbf{i} + 2\mathbf{j} + 2\mathbf{k}, \|\mathbf{u}\| = 3$

$\cos \alpha = \dfrac{1}{3}$

$\cos \beta = \dfrac{2}{3}$

$\cos \gamma = \dfrac{2}{3}$

$\cos^2 \alpha + \cos^2 \beta + \cos^2 \gamma = \dfrac{1}{9} + \dfrac{4}{9} + \dfrac{4}{9} = 1$

29. $\mathbf{u} = \langle 0, 6, -4 \rangle, \|\mathbf{u}\| = \sqrt{52} = 2\sqrt{13}$

$\cos \alpha = 0$

$\cos \beta = \dfrac{3}{\sqrt{13}}$

$\cos \gamma = -\dfrac{2}{\sqrt{13}}$

$\cos^2 \alpha + \cos^2 \beta + \cos^2 \gamma = 0 + \dfrac{9}{13} + \dfrac{4}{13} = 1$

31. $\mathbf{u} = \langle 3, 2, -2 \rangle \quad \|\mathbf{u}\| = \sqrt{17}$

$\cos \alpha = \dfrac{3}{\sqrt{17}} \Rightarrow \alpha \approx 0.7560 \text{ or } 43.3°$

$\cos \beta = \dfrac{2}{\sqrt{17}} \Rightarrow \beta \approx 1.0644 \text{ or } 61.0°$

$\cos \gamma = \dfrac{-2}{\sqrt{17}} \Rightarrow \gamma \approx 2.0772 \text{ or } 119.0°$

33. $\mathbf{u} = \langle -1, 5, 2 \rangle \quad \|\mathbf{u}\| = \sqrt{30}$

$\cos \alpha = \dfrac{-1}{\sqrt{30}} \Rightarrow \alpha \approx 1.7544 \text{ or } 100.5°$

$\cos \beta = \dfrac{5}{\sqrt{30}} \Rightarrow \beta \approx 0.4205 \text{ or } 24.1°$

$\cos \gamma = \dfrac{2}{\sqrt{30}} \Rightarrow \gamma \approx 1.1970 \text{ or } 68.6°$

35. $\mathbf{F}_1: C_1 = \dfrac{50}{\|\mathbf{F}_1\|} \approx 4.3193$

$\mathbf{F}_2: C_2 = \dfrac{80}{\|\mathbf{F}_2\|} \approx 5.4183$

$\mathbf{F} = \mathbf{F}_1 + \mathbf{F}_2$

$\approx 4.3193 \langle 10, 5, 3 \rangle + 5.4183 \langle 12, 7, -5 \rangle$

$= \langle 108.2126, 59.5246, -14.1336 \rangle$

$\|\mathbf{F}\| \approx 124.310 \text{ lb}$

$\cos \alpha \approx \dfrac{108.2126}{\|\mathbf{F}\|} \Rightarrow \alpha \approx 29.48°$

$\cos \beta \approx \dfrac{59.5246}{\|\mathbf{F}\|} \Rightarrow \beta \approx 61.39°$

$\cos \gamma \approx \dfrac{-14.1336}{\|\mathbf{F}\|} \Rightarrow \gamma \approx 96.53°$

37. Let s = length of a side.

$\mathbf{v} = \langle s, s, s \rangle$

$\|\mathbf{v}\| = s\sqrt{3}$

$\cos \alpha = \cos \beta = \cos \gamma = \dfrac{s}{s\sqrt{3}} = \dfrac{1}{\sqrt{3}}$

$\alpha = \beta = \gamma = \arccos\left(\dfrac{1}{\sqrt{3}}\right) \approx 54.7°$

39. $\overrightarrow{OA} = \langle 0, 10, 10 \rangle$

$\cos \alpha = \dfrac{0}{\sqrt{0^2 + 10^2 + 10^2}} = 0 \Rightarrow \alpha = 90°$

$\cos \beta = \cos \gamma = \dfrac{10}{\sqrt{0^2 + 10^2 + 10^2}}$

$= \dfrac{1}{\sqrt{2}} \Rightarrow \beta = \gamma = 45°$

41. $\mathbf{w}_2 = \mathbf{u} - \mathbf{w}_1 = \langle 6, 7 \rangle - \langle 2, 8 \rangle = \langle 4, -1 \rangle$

43. $\mathbf{w}_2 = \mathbf{u} - \mathbf{w}_1 = \langle 0, 3, 3 \rangle - \langle -2, 2, 2 \rangle = \langle 2, 1, 1 \rangle$

45. $\mathbf{u} = \langle 2, 3 \rangle, \mathbf{v} = \langle 5, 1 \rangle$

(a) $\mathbf{w}_1 = \left(\dfrac{\mathbf{u} \cdot \mathbf{v}}{\|\mathbf{v}\|^2}\right)\mathbf{v} = \dfrac{13}{26}\langle 5, 1 \rangle = \left\langle \dfrac{5}{2}, \dfrac{1}{2} \right\rangle$

(b) $\mathbf{w}_2 = \mathbf{u} - \mathbf{w}_1 = \left\langle -\dfrac{1}{2}, \dfrac{5}{2} \right\rangle$

47. $\mathbf{u} = \langle 2, 1, 2 \rangle$, $\mathbf{v} = \langle 0, 3, 4 \rangle$

    (a) $\mathbf{w}_1 = \left(\dfrac{\mathbf{u} \cdot \mathbf{v}}{\|\mathbf{v}\|^2} \right) \mathbf{v}$

        $= \dfrac{11}{25} \langle 0, 3, 4 \rangle = \left\langle 0, \dfrac{33}{25}, \dfrac{44}{25} \right\rangle$

    (b) $\mathbf{w}_2 = \mathbf{u} - \mathbf{w}_1 = \left\langle 2, -\dfrac{8}{25}, \dfrac{6}{25} \right\rangle$

49. $\mathbf{u} \cdot \mathbf{v} = \langle u_1, u_2, u_3 \rangle \cdot \langle v_1, v_2, v_3 \rangle = u_1 v_1 + u_2 v_2 + u_3 v_3$

51. (a) Orthogonal, $\theta = \dfrac{\pi}{2}$ (b) Acute, $0 < \theta < \dfrac{\pi}{2}$ (c) Obtuse, $\dfrac{\pi}{2} < \theta < \pi$

53. See page 738. Direction cosines of $\mathbf{v} = \langle v_1, v_2, v_3 \rangle$ are

    $\cos \alpha = \dfrac{v_1}{\|\mathbf{v}\|}$, $\cos \beta = \dfrac{v_2}{\|\mathbf{v}\|}$, $\cos \gamma = \dfrac{v_3}{\|\mathbf{v}\|}$.

    α, β, and γ are the direction angles. See Figure 10.26.

55. (a) $\left(\dfrac{\mathbf{u} \cdot \mathbf{v}}{\|\mathbf{v}\|^2} \right) \mathbf{v} = \mathbf{u} \implies \mathbf{u} = c\mathbf{v} \implies \mathbf{u}$ and $\mathbf{v}$ are parallel.

    (b) $\left(\dfrac{\mathbf{u} \cdot \mathbf{v}}{\|\mathbf{v}\|^2} \right) \mathbf{v} = \mathbf{0} \implies \mathbf{u} \cdot \mathbf{v} = 0 \implies \mathbf{u}$ and $\mathbf{v}$ are orthogonal.

57. Programs will vary.

59. Programs will vary.

61. Because $\mathbf{u}$ appears to be perpendicular to $\mathbf{v}$, the projection of $\mathbf{u}$ onto $\mathbf{v}$ is $\mathbf{0}$. Analytically,

    $\text{proj}_{\mathbf{v}} \mathbf{u} = \dfrac{\mathbf{u} \cdot \mathbf{v}}{\|\mathbf{v}\|^2} \mathbf{v} = \dfrac{\langle 2, -3 \rangle \cdot \langle 6, 4 \rangle}{\|\langle 6, 4 \rangle\|^2} \langle 6, 4 \rangle = 0\langle 6, 4 \rangle = \mathbf{0}$.

63. $\mathbf{u} = \dfrac{1}{2}\mathbf{i} - \dfrac{2}{3}\mathbf{j}$. Want $\mathbf{u} \cdot \mathbf{v} = 0$.

    $\mathbf{v} = 8\mathbf{i} + 6\mathbf{j}$ and $-\mathbf{v} = -8\mathbf{i} - 6\mathbf{j}$ are orthogonal to $\mathbf{u}$.

65. $\mathbf{u} = \langle 3, 1, -2 \rangle$. Want $\mathbf{u} \cdot \mathbf{v} = 0$.

    $\mathbf{v} = \langle 0, 2, 1 \rangle$ and $-\mathbf{v} = \langle 0, -2, -1 \rangle$ are orthogonal to $\mathbf{u}$.

67. (a) Gravitational Force $\mathbf{F} = -48{,}000\,\mathbf{j}$

    $\mathbf{v} = \cos 10° \,\mathbf{i} + \sin 10° \,\mathbf{j}$

    $\mathbf{w}_1 = \dfrac{\mathbf{F} \cdot \mathbf{v}}{\|\mathbf{v}\|^2} \mathbf{v} = (\mathbf{F} \cdot \mathbf{v})\mathbf{v} = (-48{,}000)(\sin 10°)\mathbf{v}$

                      $\approx -8335.1(\cos 10° \,\mathbf{i} + \sin 10° \,\mathbf{j})$

    $\|\mathbf{w}_1\| \approx 8335.1$ lb

    (b) $\mathbf{w}_2 = \mathbf{F} \cdot \mathbf{w}_1 = -48{,}000\,\mathbf{j} + 8335.1(\cos 10° \,\mathbf{i} + \sin 10° \,\mathbf{j})$

        $= 8208.5\,\mathbf{i} - 46{,}552.6\,\mathbf{j}$

    $\|\mathbf{w}_2\| \approx 47{,}270.8$ lb

69. $\mathbf{F} = 85 \left(\dfrac{1}{2}\mathbf{i} + \dfrac{\sqrt{3}}{2}\mathbf{j} \right)$

    $\mathbf{v} = 10\mathbf{i}$

    $W = \mathbf{F} \cdot \mathbf{v} = 425$ ft $\cdot$ lb

71. $\overrightarrow{PQ} = \langle 4, 7, 5 \rangle$

    $\mathbf{v} = \langle 1, 4, 8 \rangle$

    $W = \overrightarrow{PQ} \cdot \mathbf{v} = 72$

73. False. Let $\mathbf{u} = \langle 2, 4 \rangle$, $\mathbf{v} = \langle 1, 7 \rangle$ and $\mathbf{w} = \langle 5, 5 \rangle$. Then $\mathbf{u} \cdot \mathbf{v} = 2 + 28 = 30$ and $\mathbf{u} \cdot \mathbf{w} = 10 + 20 = 30$.

75. In a rhombus, $\|\mathbf{u}\| = \|\mathbf{v}\|$. The diagonals are $\mathbf{u} + \mathbf{v}$ and $\mathbf{u} - \mathbf{v}$.

    $(\mathbf{u} + \mathbf{v}) \cdot (\mathbf{u} - \mathbf{v}) = (\mathbf{u} + \mathbf{v}) \cdot \mathbf{u} - (\mathbf{u} + \mathbf{v}) \cdot \mathbf{v}$

                                  $= \mathbf{u} \cdot \mathbf{u} + \mathbf{v} \cdot \mathbf{u} - \mathbf{u} \cdot \mathbf{v} - \mathbf{v} \cdot \mathbf{v}$

                                  $= \|\mathbf{u}\|^2 - \|\mathbf{v}\|^2 = 0$

Therefore, the diagonals are orthogonal.

77. $\mathbf{u} = \langle \cos\alpha, \sin\alpha, 0 \rangle, \mathbf{v} = \langle \cos\beta, \sin\beta, 0 \rangle$

The angle between $\mathbf{u}$ and $\mathbf{v}$ is $\alpha - \beta$. (Assuming that $\alpha > \beta$). Also,

$$\cos(\alpha - \beta) = \frac{\mathbf{u} \cdot \mathbf{v}}{\|\mathbf{u}\| \, \|\mathbf{v}\|} = \frac{\cos\alpha\cos\beta + \sin\alpha\sin\beta}{(1)(1)} = \cos\alpha\cos\beta + \sin\alpha\sin\beta.$$

79. $\|\mathbf{u} - \mathbf{v}\|^2 = (\mathbf{u} - \mathbf{v}) \cdot (\mathbf{u} - \mathbf{v})$

$\qquad = (\mathbf{u} - \mathbf{v}) \cdot \mathbf{u} - (\mathbf{u} - \mathbf{v}) \cdot \mathbf{v}$

$\qquad = \mathbf{u} \cdot \mathbf{u} - \mathbf{v} \cdot \mathbf{u} - \mathbf{u} \cdot \mathbf{v} + \mathbf{v} \cdot \mathbf{v}$

$\qquad = \|\mathbf{u}\|^2 - \mathbf{u} \cdot \mathbf{v} - \mathbf{u} \cdot \mathbf{v} + \|\mathbf{v}\|^2$

$\qquad = \|\mathbf{u}\|^2 + \|\mathbf{v}\|^2 - 2\mathbf{u} \cdot \mathbf{v}$

81. $\|\mathbf{u} + \mathbf{v}\|^2 = (\mathbf{u} + \mathbf{v}) \cdot (\mathbf{u} + \mathbf{v})$

$\qquad = (\mathbf{u} + \mathbf{v}) \cdot \mathbf{u} + (\mathbf{u} + \mathbf{v}) \cdot \mathbf{v}$

$\qquad = \mathbf{u} \cdot \mathbf{u} + \mathbf{v} \cdot \mathbf{u} + \mathbf{u} \cdot \mathbf{v} + \mathbf{v} \cdot \mathbf{v}$

$\qquad = \|\mathbf{u}\|^2 + 2\mathbf{u} \cdot \mathbf{v} + \|\mathbf{v}\|^2$

$\qquad \le \|\mathbf{u}\|^2 + 2\|\mathbf{u}\| \, \|\mathbf{v}\| + \|\mathbf{v}\|^2$ from Exercise 80

$\qquad \le (\|\mathbf{u}\| + \|\mathbf{v}\|)^2$

Therefore, $\|\mathbf{u} + \mathbf{v}\| \le \|\mathbf{u}\| + \|\mathbf{v}\|$.

Section 10.4   The Cross Product of Two Vectors in Space

1. $\mathbf{j} \times \mathbf{i} = \begin{vmatrix} \mathbf{i} & \mathbf{j} & \mathbf{k} \\ 0 & 1 & 0 \\ 1 & 0 & 0 \end{vmatrix} = -\mathbf{k}$

3. $\mathbf{j} \times \mathbf{k} = \begin{vmatrix} \mathbf{i} & \mathbf{j} & \mathbf{k} \\ 0 & 1 & 0 \\ 0 & 0 & 1 \end{vmatrix} = \mathbf{i}$

5. $\mathbf{i} \times \mathbf{k} = \begin{vmatrix} \mathbf{i} & \mathbf{j} & \mathbf{k} \\ 1 & 0 & 0 \\ 0 & 0 & 1 \end{vmatrix} = -\mathbf{j}$

7. (a) $\mathbf{u} \times \mathbf{v} = \begin{vmatrix} \mathbf{i} & \mathbf{j} & \mathbf{k} \\ -2 & 3 & 4 \\ 3 & 7 & 2 \end{vmatrix} = \langle -22, 16, -23 \rangle$

(b) $\mathbf{v} \times \mathbf{u} = -(\mathbf{u} \times \mathbf{v}) = \langle 22, -16, 23 \rangle$

(c) $\mathbf{v} \times \mathbf{v} = \begin{vmatrix} \mathbf{i} & \mathbf{j} & \mathbf{k} \\ 3 & 7 & 2 \\ 3 & 7 & 2 \end{vmatrix} = \mathbf{0}$

9. (a) $\mathbf{u} \times \mathbf{v} = \begin{vmatrix} \mathbf{i} & \mathbf{j} & \mathbf{k} \\ 7 & 3 & 2 \\ 1 & -1 & 5 \end{vmatrix} = \langle 17, -33, -10 \rangle$

(b) $\mathbf{v} \times \mathbf{u} = -(\mathbf{u} \times \mathbf{v}) = \langle -17, 33, 10 \rangle$

(c) $\mathbf{v} \times \mathbf{v} = \mathbf{0}$

11. $\mathbf{u} = \langle 2, -3, 1 \rangle, \mathbf{v} = \langle 1, -2, 1 \rangle$

$\mathbf{u} \times \mathbf{v} = \begin{vmatrix} \mathbf{i} & \mathbf{j} & \mathbf{k} \\ 2 & -3 & 1 \\ 1 & -2 & 1 \end{vmatrix} = -\mathbf{i} - \mathbf{j} - \mathbf{k} = \langle -1, -1, -1 \rangle$

$\mathbf{u} \cdot (\mathbf{u} \times \mathbf{v}) = 2(-1) + (-3)(-1) + (1)(-1) = 0 \implies \mathbf{u} \perp \mathbf{u} \times \mathbf{v}$

$\mathbf{v} \cdot (\mathbf{u} \times \mathbf{v}) = 1(-1) + (-2)(-1) + (1)(-1) = 0 \implies \mathbf{v} \perp \mathbf{u} \times \mathbf{v}$

13. $\mathbf{u} = \langle 12, -3, 0 \rangle$, $\mathbf{v} = \langle -2, 5, 0 \rangle$

$$\mathbf{u} \times \mathbf{v} = \begin{vmatrix} \mathbf{i} & \mathbf{j} & \mathbf{k} \\ 12 & -3 & 0 \\ -2 & 5 & 0 \end{vmatrix} = 54\mathbf{k} = \langle 0, 0, 54 \rangle$$

$$\mathbf{u} \cdot (\mathbf{u} \times \mathbf{v}) = 12(0) + (-3)(0) + 0(54)$$

$$= 0 \Rightarrow \mathbf{u} \perp \mathbf{u} \times \mathbf{v}$$

$$\mathbf{v} \cdot (\mathbf{u} \times \mathbf{v}) = -2(0) + 5(0) + 0(54)$$

$$= 0 \Rightarrow \mathbf{v} \perp \mathbf{u} \times \mathbf{v}$$

15. $\mathbf{u} = \mathbf{i} + \mathbf{j} + \mathbf{k}$, $\mathbf{v} = 2\mathbf{i} + \mathbf{j} - \mathbf{k}$

$$\mathbf{u} \times \mathbf{v} = \begin{vmatrix} \mathbf{i} & \mathbf{j} & \mathbf{k} \\ 1 & 1 & 1 \\ 2 & 1 & -1 \end{vmatrix} = -2\mathbf{i} + 3\mathbf{j} - \mathbf{k} = \langle -2, 3, -1 \rangle$$

$$\mathbf{u} \cdot (\mathbf{u} \times \mathbf{v}) = 1(-2) + 1(3) + 1(-1)$$

$$= 0 \Rightarrow \mathbf{u} \perp \mathbf{u} \times \mathbf{v}$$

$$\mathbf{v} \cdot (\mathbf{u} \times \mathbf{v}) = 2(-2) + 1(3) + (-1)(-1)$$

$$= 0 \Rightarrow \mathbf{v} \perp \mathbf{u} \times \mathbf{v}$$

17.

19. $(-\mathbf{v}) \, x\mathbf{u} = -(\mathbf{v} \, x\mathbf{u}) = \mathbf{u} \times \mathbf{v}$

21. $\mathbf{u} = \langle 4, -3.5, 7 \rangle$

$\mathbf{v} = \langle -1, 8, 4 \rangle$

$\mathbf{u} \times \mathbf{v} = \left\langle -70, -23, \dfrac{57}{2} \right\rangle$

$\dfrac{\mathbf{u} \times \mathbf{v}}{\|\mathbf{u} \times \mathbf{v}\|} = \left\langle \dfrac{-140}{\sqrt{24,965}}, \dfrac{-46}{\sqrt{24,965}}, \dfrac{57}{\sqrt{24,965}} \right\rangle$

23. $\mathbf{u} = -3\mathbf{i} + 2\mathbf{j} - 5\mathbf{k}$

$\mathbf{v} = \dfrac{1}{2}\mathbf{i} - \dfrac{3}{4}\mathbf{j} + \dfrac{1}{10}\mathbf{k}$

$\mathbf{u} \times \mathbf{v} = \left\langle -\dfrac{71}{20}, -\dfrac{11}{5}, \dfrac{5}{4} \right\rangle$

$\dfrac{\mathbf{u} \times \mathbf{v}}{\|\mathbf{u} \times \mathbf{v}\|} = \dfrac{20}{\sqrt{7602}} \left\langle -\dfrac{71}{20}, -\dfrac{11}{5}, \dfrac{5}{4} \right\rangle$

$= \left\langle -\dfrac{71}{\sqrt{7602}}, -\dfrac{44}{\sqrt{7602}}, \dfrac{25}{\sqrt{7602}} \right\rangle$

25. Programs will vary.

27. $\mathbf{u} = \mathbf{j}$

$\mathbf{v} = \mathbf{j} + \mathbf{k}$

$$\mathbf{u} \times \mathbf{v} = \begin{vmatrix} \mathbf{i} & \mathbf{j} & \mathbf{k} \\ 0 & 1 & 0 \\ 0 & 1 & 1 \end{vmatrix} = \mathbf{i}$$

$A = \|\mathbf{u} \times \mathbf{v}\| = \|\mathbf{i}\| = 1$

29. $\mathbf{u} = \langle 3, 2, -1 \rangle$

$\mathbf{v} = \langle 1, 2, 3 \rangle$

$$\mathbf{u} \times \mathbf{v} = \begin{vmatrix} \mathbf{i} & \mathbf{j} & \mathbf{k} \\ 3 & 2 & -1 \\ 1 & 2 & 3 \end{vmatrix} = \langle 8, -10, 4 \rangle$$

$A = \|\mathbf{u} \times \mathbf{v}\| = \|\langle 8, -10, 4 \rangle\| = \sqrt{180} = 6\sqrt{5}$

31. $A(1, 1, 1,), B(2, 3, 4), C(6, 5, 2), D(7, 7, 5)$

$\overrightarrow{AB} = \langle 1, 2, 3 \rangle$, $\overrightarrow{AC} = \langle 5, 4, 1 \rangle$, $\overrightarrow{CD} = \langle 1, 2, 3 \rangle$, $\overrightarrow{BD} = \langle 5, 4, 1 \rangle$

Since $\overrightarrow{AB} = \overrightarrow{CD}$ and $\overrightarrow{AC} = \overrightarrow{BD}$, the figure is a parallelogram. $\overrightarrow{AB}$ and $\overrightarrow{AC}$ are adjacent sides and

$$\overrightarrow{AB} \times \overrightarrow{AC} = \begin{vmatrix} \mathbf{i} & \mathbf{j} & \mathbf{k} \\ 1 & 2 & 3 \\ 5 & 4 & 1 \end{vmatrix} = -10\mathbf{i} + 14\mathbf{j} - 6\mathbf{k}.$$

$A = \|\overrightarrow{AB} \times \overrightarrow{AC}\| = \sqrt{332} = 2\sqrt{83}$

33. $A(0, 0, 0), B(1, 2, 3), C(-3, 0, 0)$

$\overrightarrow{AB} = \langle 1, 2, 3 \rangle$, $\overrightarrow{AC} = \langle -3, 0, 0 \rangle$

$$\overrightarrow{AB} \times \overrightarrow{AC} = \begin{vmatrix} \mathbf{i} & \mathbf{j} & \mathbf{k} \\ 1 & 2 & 3 \\ -3 & 0 & 0 \end{vmatrix} = -9\mathbf{j} + 6\mathbf{k}$$

$A = \dfrac{1}{2} \|\overrightarrow{AB} \times \overrightarrow{AC}\| = \dfrac{1}{2}\sqrt{117} = \dfrac{3}{2}\sqrt{13}$

35. $A(2, -7, 3), B(-1, 5, 8), C(4, 6, -1)$

$\overrightarrow{AB} = \langle -3, 12, 5 \rangle, \overrightarrow{AC} = \langle 2, 13, -4 \rangle$

$$\overrightarrow{AB} \times \overrightarrow{AC} = \begin{vmatrix} \mathbf{i} & \mathbf{j} & \mathbf{k} \\ -3 & 12 & 5 \\ 2 & 13 & -4 \end{vmatrix} = \langle -113, -2, -63 \rangle$$

Area $= \frac{1}{2} \| \overrightarrow{AB} \times \overrightarrow{AC} \| = \frac{1}{2}\sqrt{16{,}742}$

37. $\mathbf{F} = -20\mathbf{k}$

$\overrightarrow{PQ} = \dfrac{1}{2}(\cos 40°\mathbf{j} + \sin 40°\mathbf{k})$

$$\overrightarrow{PQ} \times \mathbf{F} = \begin{vmatrix} \mathbf{i} & \mathbf{j} & \mathbf{k} \\ 0 & \cos 40°/2 & \sin 40°/2 \\ 0 & 0 & -20 \end{vmatrix} = -10\cos 40°\mathbf{i}$$

$\| \overrightarrow{PQ} \times \mathbf{F} \| = 10 \cos 40° \approx 7.66 \text{ ft} \cdot \text{lb}$

39. (a) $\overrightarrow{OA} = \dfrac{3}{2}\mathbf{k}$

$\mathbf{F} = -60(\sin\theta\mathbf{j} + \cos\theta\mathbf{k})$

$$\overrightarrow{OA} \times \mathbf{F} = \begin{vmatrix} \mathbf{i} & \mathbf{j} & \mathbf{k} \\ 0 & 0 & 3/2 \\ 0 & -60\sin\theta & -60\cos\theta \end{vmatrix} = 90\sin\theta\mathbf{i}$$

$\| \overrightarrow{OA} \times \mathbf{F} \| = 90\sin\theta$

(b) When $\theta = 45°$: $\| \overrightarrow{OA} \times \mathbf{F} \| = 90\left(\dfrac{\sqrt{2}}{2}\right) = 45\sqrt{2} \approx 63.64$.

(c) Let $T = 90\sin\theta$.

$\dfrac{dT}{d\theta} = 90\cos\theta = 0$ when $\theta = 90°$.

This is what we expected. When $\theta = 90°$ the pipe wrench is horizontal.

41. $\mathbf{u} \cdot (\mathbf{v} \times \mathbf{w}) = \begin{vmatrix} 1 & 0 & 0 \\ 0 & 1 & 0 \\ 0 & 0 & 1 \end{vmatrix} = 1$

43. $\mathbf{u} \cdot (\mathbf{v} \times \mathbf{w}) = \begin{vmatrix} 2 & 0 & 1 \\ 0 & 3 & 0 \\ 0 & 0 & 1 \end{vmatrix} = 6$

45. $\mathbf{u} \cdot (\mathbf{v} \times \mathbf{w}) = \begin{vmatrix} 1 & 1 & 0 \\ 0 & 1 & 1 \\ 1 & 0 & 1 \end{vmatrix} = 2$

$V = |\mathbf{u} \cdot (\mathbf{v} \times \mathbf{w})| = 2$

47. $\mathbf{u} = \langle 3, 0, 0 \rangle$

$\mathbf{v} = \langle 0, 5, 1 \rangle$

$\mathbf{w} = \langle 2, 0, 5 \rangle$

$\mathbf{u} \cdot (\mathbf{v} \times \mathbf{w}) = \begin{vmatrix} 3 & 0 & 0 \\ 0 & 5 & 1 \\ 2 & 0 & 5 \end{vmatrix} = 75$

$V = |\mathbf{u} \cdot (\mathbf{v} \times \mathbf{w})| = 75$

49. $\mathbf{u} \times \mathbf{v} = \langle u_1, u_2, u_3 \rangle \cdot \langle v_1, v_2, v_3 \rangle = (u_2 v_3 - u_3 v_2)\mathbf{i} - (u_1 v_3 - u_3 v_1)\mathbf{j} + (u_1 v_2 - u_2 v_1)\mathbf{k}$

51. The magnitude of the cross product will increase by a factor of 4.

53. If the vectors are ordered pairs, then the cross product does not exist. False.

55. True

57. $\mathbf{u} = \langle u_1, u_2, u_3 \rangle$, $\mathbf{v} = \langle v_1, v_2, v_3 \rangle$, $\mathbf{w} = \langle w_1, w_2, w_3 \rangle$

$$\mathbf{u} \times (\mathbf{v} + \mathbf{w}) = \begin{vmatrix} \mathbf{i} & \mathbf{j} & \mathbf{k} \\ u_1 & u_2 & u_3 \\ v_1 + w_1 & v_2 + w_2 & v_3 + w_3 \end{vmatrix}$$

$$= [u_2(v_3 + w_3) - u_3(v_2 + w_2)]\mathbf{i} - [u_1(v_3 + w_3) - u_3(v_1 + w_1)]\mathbf{j} + [u_1(v_2 + w_2) - u_2(v_1 + w_1)]\mathbf{k}$$

$$= (u_2v_3 - u_3v_2)\mathbf{i} - (u_1v_3 - u_3v_1)\mathbf{j} + (u_1v_2 - u_2v_1)\mathbf{k} + (u_2w_3 - u_3w_2)\mathbf{i} -$$

$$(u_1w_3 - u_3w_1)\mathbf{j} + (u_1w_2 - u_2w_1)\mathbf{k}$$

$$= (\mathbf{u} \times \mathbf{v}) + (\mathbf{u} \times \mathbf{w})$$

59. $\mathbf{u} = \langle u_1, u_2, u_3 \rangle$

$$\mathbf{u} \times \mathbf{u} = \begin{vmatrix} \mathbf{i} & \mathbf{j} & \mathbf{k} \\ u_1 & u_2 & u_3 \\ u_1 & u_2 & u_3 \end{vmatrix} = (u_2u_3 - u_3u_2)\mathbf{i} - (u_1u_3 - u_3u_1)\mathbf{j} + (u_1u_2 - u_2u_1)\mathbf{k} = \mathbf{0}$$

61. $\qquad \mathbf{u} \times \mathbf{v} = (u_2v_3 - u_3v_2)\mathbf{i} - (u_1v_3 - u_3v_1)\mathbf{j} + (u_1v_2 - u_2v_1)\mathbf{k}$

$(\mathbf{u} \times \mathbf{v}) \cdot \mathbf{u} = (u_2v_3 - u_3v_2)u_1 + (u_3v_1 - u_1v_3)u_2 + (u_1v_2 - u_2v_1)u_3 = 0$

$(\mathbf{u} \times \mathbf{v}) \cdot \mathbf{v} = (u_2v_3 - u_3v_2)v_1 + (u_3v_1 - u_1v_3)v_2 + (u_1v_2 - u_2v_1)v_3 = 0$

Thus, $\mathbf{u} \times \mathbf{v} \perp \mathbf{u}$ and $\mathbf{u} \times \mathbf{v} \perp \mathbf{v}$.

63. $\|\mathbf{u} \times \mathbf{v}\| = \|\mathbf{u}\|\,\|\mathbf{v}\| \sin \theta$

If $\mathbf{u}$ and $\mathbf{v}$ are orthogonal, $\theta = \pi/2$ and $\sin \theta = 1$. Therefore, $\|\mathbf{u} \times \mathbf{v}\| = \|\mathbf{u}\|\,\|\mathbf{v}\|$.

Section 10.5   Lines and Planes in Space

1. $x = 1 + 3t$, $y = 2 - t$, $z = 2 + 5t$

(a)

(b) When $t = 0$ we have $P = (1, 2, 2)$. When $t = 3$ we have $Q = (10, -1, 17)$.

$$\overrightarrow{PQ} = \langle 9, -3, 15 \rangle$$

The components of the vector and the coefficients of t are proportional since the line is parallel to $\overrightarrow{PQ}$.

(c) $y = 0$ when $t = 2$. Thus, $x = 7$ and $z = 12$.
Point: $(7, 0, 12)$

$x = 0$ when $t = -\dfrac{1}{3}$. Point: $\left(0, \dfrac{7}{3}, \dfrac{1}{3}\right)$

$z = 0$ when $t = -\dfrac{2}{5}$. Point: $\left(-\dfrac{1}{5}, \dfrac{12}{5}, 0\right)$

3. Point: $(0, 0, 0)$

Direction vector: $\mathbf{v} = \langle 1, 2, 3 \rangle$

Direction numbers: $1, 2, 3$

(a) Parametric: $x = t$, $y = 2t$, $z = 3t$

(b) Symmetric: $x = \dfrac{y}{2} = \dfrac{z}{3}$

5. Point: $(-2, 0, 3)$

Direction vector: $\mathbf{v} = \langle 2, 4, -2 \rangle$

Direction numbers: $2, 4, -2$

(a) Parametric: $x = -2 + 2t$, $y = 4t$, $z = 3 - 2t$

(b) Symmetric: $\dfrac{x + 2}{2} = \dfrac{y}{4} = \dfrac{z - 3}{-2}$

7. Point: $(1, 0, 1)$

Direction vector: $\mathbf{v} = 3\mathbf{i} - 2\mathbf{j} + \mathbf{k}$

Direction numbers: $3, -2, 1$

(a) Parametric: $x = 1 + 3t, y = -2t, z = 1 + t$

(b) Symmetric: $\dfrac{x - 1}{3} = \dfrac{y}{-2} = \dfrac{z - 1}{1}$

9. Points: $(5, -3, -2), \left(\dfrac{-2}{3}, \dfrac{2}{3}, 1\right)$

Direction vector: $\mathbf{v} = \dfrac{17}{3}\mathbf{i} - \dfrac{11}{3}\mathbf{j} - 3\mathbf{k}$

Direction numbers: $17, -11, -9$

(a) Parametric: $x = 5 + 17t, y = -3 - 11t, z = -2 - 9t$

(b) Symmetric: $\dfrac{x - 5}{17} = \dfrac{y + 3}{-11} = \dfrac{z + 2}{-9}$

11. Points: $(2, 3, 0), (10, 8, 12)$

Direction vector: $\langle 8, 5, 12 \rangle$

Direction numbers: $8, 5, 12$

(a) Parametric: $x = 2 + 8t, y = 3 + 5t, z = 12t$

(b) Symmetric: $\dfrac{x - 2}{8} = \dfrac{y - 3}{5} = \dfrac{z}{12}$

13. Point: $(2, 3, 4)$

Direction vector: $\mathbf{v} = \mathbf{k}$

Direction numbers: $0, 0, 1$

Parametric: $x = 2, y = 3, z = 4 + t$

15. Point: $(-2, 3, 1)$

Direction vector: $\mathbf{v} = 4\mathbf{i} - \mathbf{k}$

Direction numbers: $4, 0, -1$

Parametric: $x = -2 + 4t, y = 3, z = 1 - t$

Symmetric: $\dfrac{x + 2}{4} = \dfrac{z - 1}{-1}, y = 3$

(a) On line $(t = 1)$

(b) On line $(t = -1)$

(c) Not on line $(y \neq 3)$

(d) Not on line $\left(\dfrac{6 + 2}{4} \neq \dfrac{-2 - 1}{-1}\right)$

17. L_1: $\mathbf{v} = \langle -3, 2, 4 \rangle$ $(6, -2, 5)$ on line

L_2: $\mathbf{v} = \langle 6, -4, -8 \rangle$ $(6, -2, 5)$ on line

L_3: $\mathbf{v} = \langle -6, 4, 8 \rangle$ $(6, -2, 5)$ not on line

L_4: $\mathbf{v} = \langle 6, 4, -6 \rangle$ not parallel to L_1, L_2, nor L_3

Hence, L_1 and L_2 are identical.

$L_1 = L_2$ and L_3 are parallel.

19. At the point of intersection, the coordinates for one line equal the corresponding coordinates for the other line. Thus,

(i) $4t + 2 = 2s + 2$, (ii) $3 = 2s + 3$, and (iii) $-t + 1 = s + 1$.

From (ii), we find that $s = 0$ and consequently, from (iii), $t = 0$. Letting $s = t = 0$, we see that equation (i) is satisfied and therefore the two lines intersect. Substituting zero for s or for t, we obtain the point $(2, 3, 1)$.

$\mathbf{u} = 4\mathbf{i} - \mathbf{k}$ (First line)

$\mathbf{v} = 2\mathbf{i} + 2\mathbf{j} + \mathbf{k}$ (Second line)

$\cos \theta = \dfrac{|\mathbf{u} \cdot \mathbf{v}|}{\|\mathbf{u}\| \|\mathbf{v}\|} = \dfrac{8 - 1}{\sqrt{17}\sqrt{9}} = \dfrac{7}{3\sqrt{17}} = \dfrac{7\sqrt{17}}{51}$

21. Writing the equations of the lines in parametric form we have

$x = 3t$ $y = 2 - t$ $z = -1 + t$

$x = 1 + 4s$ $y = -2 + s$ $z = -3 - 3s.$

For the coordinates to be equal, $3t = 1 + 4s$ and $2 - t = -2 + s$. Solving this system yields $t = \frac{17}{7}$ and $s = \frac{11}{7}$. When using these values for s and t, the z coordinates are not equal. The lines do not intersect.

23. $x = 2t + 3$ $x = -2s + 7$

$y = 5t - 2$ $y = s + 8$

$z = -t + 1$ $z = 2s - 1$

Point of intersection: $(7, 8, -1)$

Note: $t = 2$ and $s = 0$

25. $4x - 3y - 6z = 6$

(a) $P = (0, 0, -1), Q = (0, -2, 0), R = (3, 4, -1)$

$\overrightarrow{PQ} = \langle 0, -2, 1 \rangle, \overrightarrow{PR} = \langle 3, 4, 0 \rangle$

(b) $\overrightarrow{PQ} \times \overrightarrow{PR} = \begin{vmatrix} \mathbf{i} & \mathbf{j} & \mathbf{k} \\ 0 & -2 & 1 \\ 3 & 4 & 0 \end{vmatrix} = \langle -4, 3, 6 \rangle$

The components of the cross product are proportional to the coefficients of the variables in the equation. The cross product is parallel to the normal vector.

27. Point: $(2, 1, 2)$

$\mathbf{n} = \mathbf{i} = \langle 1, 0, 0 \rangle$

$1(x - 2) + 0(y - 1) + 0(z - 2) = 0$

$x - 2 = 0$

29. Point: $(3, 2, 2)$

Normal vector: $\mathbf{n} = 2\mathbf{i} + 3\mathbf{j} - \mathbf{k}$

$2(x - 3) + 3(y - 2) - 1(z - 2) = 0$

$2x + 3y - z = 10$

31. Point: $(0, 0, 6)$

Normal vector: $\mathbf{n} = -\mathbf{i} + \mathbf{j} - 2\mathbf{k}$

$-1(x - 0) + 1(y - 0) - 2(z - 6) = 0$

$-x + y - 2z + 12 = 0$

$x - y + 2z = 12$

33. Let $\mathbf{u}$ be the vector from $(0, 0, 0)$ to $(1, 2, 3)$:

$\mathbf{u} = \mathbf{i} + 2\mathbf{j} + 3\mathbf{k}$

Let $\mathbf{v}$ be the vector from $(0, 0, 0)$ to $(-2, 3, 3)$:

$\mathbf{v} = -2\mathbf{i} + 3\mathbf{j} + 3\mathbf{k}$

Normal vector: $\mathbf{u} \times \mathbf{v} = \begin{vmatrix} \mathbf{i} & \mathbf{j} & \mathbf{k} \\ 1 & 2 & 3 \\ -2 & 3 & 3 \end{vmatrix}$

$= -3\mathbf{i} + (-9)\mathbf{j} + 7\mathbf{k}$

$-3(x - 0) - 9(y - 0) + 7(z - 0) = 0$

$3x + 9y - 7z = 0$

35. Let $\mathbf{u}$ be the vector from $(1, 2, 3)$ to $(3, 2, 1)$: $\mathbf{u} = 2\mathbf{i} - 2\mathbf{k}$

Let $\mathbf{v}$ be the vector from $(1, 2, 3)$ to $(-1, -2, 2)$: $\mathbf{v} = -2\mathbf{i} - 4\mathbf{j} - \mathbf{k}$

Normal vector: $\left(\frac{1}{2}\mathbf{u}\right) \times (-\mathbf{v}) = \begin{vmatrix} \mathbf{i} & \mathbf{j} & \mathbf{k} \\ 1 & 0 & -1 \\ 2 & 4 & 1 \end{vmatrix} = 4\mathbf{i} - 3\mathbf{j} + 4\mathbf{k}$

$4(x - 1) - 3(y - 2) + 4(z - 3) = 0$

$4x - 3y + 4z = 10$

37. $(1, 2, 3)$, Normal vector: $\mathbf{v} = \mathbf{k}, 1(z - 3) = 0, z = 3$

39. The direction vectors for the lines are $\mathbf{u} = -2\mathbf{i} + \mathbf{j} + \mathbf{k}$, $\mathbf{v} = -3\mathbf{i} + 4\mathbf{j} - \mathbf{k}$.

Normal vector: $\mathbf{u} \times \mathbf{v} = \begin{vmatrix} \mathbf{i} & \mathbf{j} & \mathbf{k} \\ -2 & 1 & 1 \\ -3 & 4 & -1 \end{vmatrix} = -5(\mathbf{i} + \mathbf{j} + \mathbf{k})$

Point of intersection of the lines: $(-1, 5, 1)$

$(x + 1) + (y - 5) + (z - 1) = 0$

$x + y + z = 5$

41. Let $\mathbf{v}$ be the vector from $(-1, 1, -1)$ to $(2, 2, 1)$: $\mathbf{v} = 3\mathbf{i} + \mathbf{j} + 2\mathbf{k}$

Let $\mathbf{n}$ be a vector normal to the plane $2x - 3y + z = 3$: $\mathbf{n} = 2\mathbf{i} - 3\mathbf{j} + \mathbf{k}$

Since $\mathbf{v}$ and $\mathbf{n}$ both lie in the plane P, the normal vector to P is

$\mathbf{v} \times \mathbf{n} = \begin{vmatrix} \mathbf{i} & \mathbf{j} & \mathbf{k} \\ 3 & 1 & 2 \\ 2 & -3 & 1 \end{vmatrix} = 7\mathbf{i} + \mathbf{j} - 11\mathbf{k}$

$7(x - 2) + 1(y - 2) - 11(z - 1) = 0$

$7x + y - 11z = 5$

43. Let $\mathbf{u} = \mathbf{i}$ and let $\mathbf{v}$ be the vector from $(1, -2, -1)$ to $(2, 5, 6)$: $\mathbf{v} = \mathbf{i} + 7\mathbf{j} + 7\mathbf{k}$

Since $\mathbf{u}$ and $\mathbf{v}$ both lie in the plane P, the normal vector to P is:

$$\mathbf{u} \times \mathbf{v} = \begin{vmatrix} \mathbf{i} & \mathbf{j} & \mathbf{k} \\ 1 & 0 & 0 \\ 1 & 7 & 7 \end{vmatrix} = -7\mathbf{j} + 7\mathbf{k} = -7(\mathbf{j} - \mathbf{k})$$

$$[y - (-2)] - [z - (-1)] = 0$$

$$y - z = -1$$

45. The normal vectors to the planes are

$$\mathbf{n}_1 = \langle 5, -3, 1 \rangle, \mathbf{n}_2 = \langle 1, 4, 7 \rangle, \cos\theta = \frac{|\mathbf{n}_1 \cdot \mathbf{n}_2|}{\|\mathbf{n}_1\| \|\mathbf{n}_2\|} = 0.$$

Thus, $\theta = \pi/2$ and the planes are orthogonal.

47. The normal vectors to the planes are

$$\mathbf{n}_1 = \mathbf{i} - 3\mathbf{j} + 6\mathbf{k}, \ \mathbf{n}_2 = 5\mathbf{i} + \mathbf{j} - \mathbf{k},$$

$$\cos\theta = \frac{|\mathbf{n}_1 \cdot \mathbf{n}_2|}{\|\mathbf{n}_1\| \|\mathbf{n}_2\|} = \frac{|5 - 3 - 6|}{\sqrt{46}\sqrt{27}} = \frac{4\sqrt{138}}{414}.$$

Therefore, $\theta = \arccos\left(\frac{4\sqrt{138}}{414}\right) \approx 83.5°$.

49. The normal vectors to the planes are $\mathbf{n}_1 = \langle 1, -5, -1 \rangle$ and $\mathbf{n}_2 = \langle 5, -25, -5 \rangle$. Since $\mathbf{n}_2 = 5\mathbf{n}_1$, the planes are parallel, but not equal.

51. $4x + 2y + 6z = 12$

53. $2x - y + 3z = 4$

55. $y + z = 5$

57. $x = 5$

59. $2x + y - z = 6$

Generated by Maple

61. $-5x + 4y - 6z + 8 = 0$

Generated by Maple

63. P_1: $\mathbf{n} = \langle 3, -2, 5 \rangle$ $(1, -1, 1)$ on plane

P_2: $\mathbf{n} = \langle -6, 4, -10 \rangle$ $(1, -1, 1)$ not on plane

P_3: $\mathbf{n} = \langle -3, 2, 5 \rangle$

P_4: $\mathbf{n} = \langle 75, -50, 125 \rangle$ $(1, -1, 1)$ on plane

P_1 and P_4 are identical.

$P_1 = P_4$ is parallel to P_2.

65. Each plane passes through the points

$(c, 0, 0)$, $(0, c, 0)$, and $(0, 0, c)$.

67. The normals to the planes are $\mathbf{n}_1 = 3\mathbf{i} + 2\mathbf{j} - \mathbf{k}$ and $\mathbf{n}_2 = \mathbf{i} - 4\mathbf{j} + 2\mathbf{k}$. The direction vector for the line is

$$\mathbf{n}_2 \times \mathbf{n}_1 = \begin{vmatrix} \mathbf{i} & \mathbf{j} & \mathbf{k} \\ 1 & -4 & 2 \\ 3 & 2 & -1 \end{vmatrix} = 7(\mathbf{j} + 2\mathbf{k}).$$

Now find a point of intersection of the planes.

$$\begin{aligned} 6x + 4y - 2z &= 14 \\ x - 4y + 2z &= 0 \\ 7x &= 14 \\ x &= 2 \end{aligned}$$

Substituting 2 for x in the second equation, we have $-4y + 2z = -2$ or $z = 2y - 1$. Letting $y = 1$, a point of intersection is $(2, 1, 1)$.

$$x = 2, y = 1 + t, z = 1 + 2t$$

69. Writing the equation of the line in parametric form and substituting into the equation of the plane we have:

$$x = \frac{1}{2} + t, \ y = \frac{-3}{2} - t, \ z = -1 + 2t$$

$$2\left(\frac{1}{2} + t\right) - 2\left(\frac{-3}{2} - t\right) + (-1 + 2t) = 12, \ t = \frac{3}{2}$$

Substituting $t = 3/2$ into the parametric equations for the line we have the point of intersection $(2, -3, 2)$. The line does not lie in the plane.

71. Writing the equation of the line in parametric form and substituting into the equation of the plane we have:

$$x = 1 + 3t, \ y = -1 - 2t, \ z = 3 + t$$

$$2(1 + 3t) + 3(-1 - 2t) = 10, -1 = 10, \text{ contradiction}$$

Therefore, the line does not intersect the plane.

73. Point: $Q(0, 0, 0)$

Plane: $2x + 3y + z - 12 = 0$

Normal to plane: $\mathbf{n} = \langle 2, 3, 1 \rangle$

Point in plane: $P(6, 0, 0)$

Vector $\overrightarrow{PQ} = \langle -6, 0\ 0 \rangle$

$$D = \frac{|\overrightarrow{PQ} \cdot \mathbf{n}|}{\|\mathbf{n}\|} = \frac{|-12|}{\sqrt{14}} = \frac{6\sqrt{14}}{7}$$

75. Point: $Q(2, 8, 4)$

Plane: $2x + y + z = 5$

Normal to plane: $\mathbf{n} = \langle 2, 1, 1 \rangle$

Point in plane: $P\langle 0, 0, 5 \rangle$

Vector: $\overrightarrow{PQ} = \langle 2, 8, -1 \rangle$

$$D = \frac{|\overrightarrow{PQ} \cdot \mathbf{n}|}{\|\mathbf{n}\|} = \frac{11}{\sqrt{6}} = \frac{11\sqrt{6}}{6}$$

77. The normal vectors to the planes are $\mathbf{n}_1 = \langle 1, -3, 4 \rangle$ and $\mathbf{n}_2 = \langle 1, -3, 4 \rangle$. Since $\mathbf{n}_1 = \mathbf{n}_2$, the planes are parallel. Choose a point in each plane.

$P = (10, 0, 0)$ is a point in $x - 3y + 4z = 10$.
$Q = (6, 0, 0)$ is a point in $x - 3y + 4z = 6$.

$$\overrightarrow{PQ} = \langle -4, 0, 0 \rangle, D = \frac{|\overrightarrow{PQ} \cdot \mathbf{n}_1|}{\|\mathbf{n}_1\|} = \frac{4}{\sqrt{26}} = \frac{2\sqrt{26}}{13}$$

79. The normal vectors to the planes are $\mathbf{n}_1 = \langle -3, 6, 7 \rangle$ and $\mathbf{n}_2 = \langle 6, -12, -14 \rangle$. Since $\mathbf{n}_2 = -2\mathbf{n}_1$, the planes are parallel. Choose a point in each plane.

$P = (0, -1, 1)$ is a point in $-3x + 6y + 7z = 1$.

$Q = \left(\frac{25}{6}, 0, 0\right)$ is a point in $6x - 12y - 14z = 25$.

$$\overrightarrow{PQ} = \left\langle \frac{25}{6}, 1, -1 \right\rangle$$

$$D = \frac{|\overrightarrow{PQ} \cdot \mathbf{n}_1|}{\|\mathbf{n}_1\|} = \frac{|-27/2|}{\sqrt{94}} = \frac{27}{2\sqrt{94}} = \frac{27\sqrt{94}}{188}$$

81. $\mathbf{u} = \langle 4, 0, -1 \rangle$ is the direction vector for the line. $Q(1, 5, -2)$ is the given point, and $P(-2, 3, 1)$ is on the line. Hence, $\overrightarrow{PQ} = \langle 3, 2, -3 \rangle$ and

$$\overrightarrow{PQ} \times \mathbf{u} = \begin{vmatrix} \mathbf{i} & \mathbf{j} & \mathbf{k} \\ 3 & 2 & -3 \\ 4 & 0 & -1 \end{vmatrix} = \langle -2, -9, -8 \rangle$$

$$D = \frac{\|\overrightarrow{PQ} \times \mathbf{u}\|}{\|\mathbf{u}\|} = \frac{\sqrt{149}}{\sqrt{17}} = \frac{\sqrt{2533}}{17}$$

83. The parametric equations of a line L parallel to $\mathbf{v} = \langle a, b, c, \rangle$ and passing through the point $P(x_1, y_1, z_1)$ are

$$x = x_1 + at, y = y_1 + bt, z = z_1 + ct.$$

The symmetric equations are

$$\frac{x - x_1}{a} = \frac{y - y_1}{b} = \frac{z - z_1}{c}.$$

85. Solve the two linear equations representing the planes to find two points of intersection. Then find the line determined by the two points.

87. (a) Sphere

$(x - 3)^2 + (y + 2)^2 + (z - 5)^2 = 16$

$x^2 + y^2 + z^2 - 6x + 4y - 10z + 22 = 0$

(b) Parallel planes

$4x - 3y + z = 10 \pm 4\|\mathbf{n}\| = 10 \pm 4\sqrt{26}$

89. (a) $z = 28.7 - 1.83x - 1.09y$

Year	1980	1985	1990	1994	1995	1996	1997
z (approx.)	16.16	14.23	9.81	8.60	8.42	8.27	8.23

(b) An increase in x or y will cause a decrease in z. In fact, any increase in two variables will cause a decrease in the third.

(c)

91. True

Section 10.6   Surfaces in Space

1. Ellipsoid

Matches graph (c)

3. Hyperboloid of one sheet

Matches graph (f)

5. Elliptic paraboloid

Matches graph (d)

7. $z = 3$

Plane parallel to the xy-coordinate plane

9. $y^2 + z^2 = 9$

The x-coordinate is missing so we have a cylindrical surface with rulings parallel to the x-axis. The generating curve is a circle.

11. $y = x^2$

The z-coordinate is missing so we have a cylindrical surface with rulings parallel to the z-axis. The generating curve is a parabola.

13. $4x^2 + y^2 = 4$

$$\frac{x^2}{1} + \frac{y^2}{4} = 1$$

The z-coordinate is missing so we have a cylindrical surface with rulings parallel to the z-axis. The generating curve is an ellipse.

15. $z = \sin y$

The x-coordinate is missing so we have a cylindrical surface with rulings parallel to the x-axis. The generating curve is the sine curve.

17. $z = x^2 + y^2$

(a) You are viewing the paraboloid from the x-axis: $(20, 0, 0)$

(b) You are viewing the paraboloid from above, but not on the z-axis: $(10, 10, 20)$

(c) You are viewing the paraboloid from the z-axis: $(0, 0, 20)$

(d) You are viewing the paraboloid from the y-axis: $(0, 20, 0)$

19. $\dfrac{x^2}{1} + \dfrac{y^2}{4} + \dfrac{z^2}{1} = 1$

Ellipsoid

xy-trace: $\dfrac{x^2}{1} + \dfrac{y^2}{4} = 1$ ellipse

xz-trace: $x^2 + z^2 = 1$ circle

yz-trace: $\dfrac{y^2}{4} + \dfrac{z^2}{1} = 1$ ellipse

21. $16x^2 - y^2 + 16z^2 = 4$

$4x^2 - \dfrac{y^2}{4} + 4z^2 = 1$

Hyperboloid on one sheet

xy-trace: $4x^2 - \dfrac{y^2}{4} = 1$ hyperbola

xz-trace: $4(x^2 + z^2) = 1$ circle

yz-trace: $\dfrac{-y^2}{4} + 4z^2 = 1$ hyperbola

23. $x^2 - y + z^2 = 0$

Elliptic paraboloid

xy-trace: $y = x^2$

xz-trace: $x^2 + z^2 = 0$,
    point $(0, 0, 0)$

yz-trace: $y = z^2$

$y = 1$: $x^2 + z^2 = 1$

25. $x^2 - y^2 + z = 0$

Hyperbolic paraboloid

xy-trace: $y = \pm x$

xz-trace: $z = -x^2$

yz-trace: $z = y^2$

$y = \pm 1$: $z = 1 - x^2$

27. $z^2 = x^2 + \dfrac{y^2}{4}$

Elliptic Cone

xy-trace: point $(0, 0, 0)$

xz-trace: $z = \pm x$

yz-trace: $z = \dfrac{\pm 1}{2}y$

$z = \pm 1$: $x^2 + \dfrac{y^2}{4} = 1$

29. $16x^2 + 9y^2 + 16z^2 - 32x - 36y + 36 = 0$

$16(x^2 - 2x + 1) + 9(y^2 - 4y + 4) + 16z^2 = -36 + 16 + 36$

$16(x - 1)^2 + 9(y - 2)^2 + 16z^2 = 16$

$\dfrac{(x - 1)^2}{1} + \dfrac{(y - 2)^2}{16/9} + \dfrac{z^2}{1} = 1$

Ellipsoid with center $(1, 2, 0)$.

31. $z = 2 \sin x$

33. $z^2 = x^2 + 4y^2$

$z = \pm \sqrt{x^2 + 4y^2}$

35. $x^2 + y^2 = \left(\dfrac{2}{z}\right)^2$

$y = \pm \sqrt{\dfrac{4}{z^2} - x^2}$

37. $z = 4 - \sqrt{|xy|}$

39. $4x^2 - y^2 + 4z^2 = -16$

$z = \pm \sqrt{\dfrac{y^2}{4} - x^2 - 4}$

41. $z = 2\sqrt{x^2 + y^2}$

$z = 2$

$2\sqrt{x^2 + y^2} = 2$

$x^2 + y^2 = 1$

43. $x^2 + y^2 = 1$

$x + z = 2$

$z = 0$

45. $x^2 + z^2 = [r(y)]^2$ and $z = r(y) = \pm 2\sqrt{y}$; therefore,

$x^2 + z^2 = 4y$.

47. $x^2 + y^2 = [r(z)]^2$ and $y = r(z) = \dfrac{z}{2}$; therefore,

$x^2 + y^2 = \dfrac{z^2}{4}, \ 4x^2 + 4y^2 = z^2$.

49. $y^2 + z^2 = [r(x)]^2$ and $y = r(x) = \dfrac{2}{x}$; therefore,

$y^2 + z^2 = \left(\dfrac{2}{x}\right)^2, \ y^2 + z^2 = \dfrac{4}{x^2}$.

51. $x^2 + y^2 - 2z = 0$

$x^2 + y^2 = \left(\sqrt{2z}\right)^2$

Equation of generating curve: $y = \sqrt{2z}$ or $x = \sqrt{2z}$

53. Let C be a curve in a plane and let L be a line not in a parallel plane. The set of all lines parallel to L and intersecting C is called a cylinder.

55. See pages 765 and 766.

57. $V = 2\pi \displaystyle\int_0^4 x(4x - x^2)\,dx$

$= 2\pi \left[\dfrac{4x^3}{3} - \dfrac{x^4}{4}\right]_0^4 = \dfrac{128\pi}{3}$

59. $z = \dfrac{x^2}{2} + \dfrac{y^2}{4}$

(a) When $z = 2$ we have $2 = \dfrac{x^2}{2} + \dfrac{y^2}{4}$, or $1 = \dfrac{x^2}{4} + \dfrac{y^2}{8}$

Major axis: $2\sqrt{8} = 4\sqrt{2}$

Minor axis: $2\sqrt{4} = 4$

$c^2 = a^2 - b^2, c^2 = 4, c = 2$

Foci: $(0, \pm 2, 2)$

(b) When $z = 8$ we have $8 = \dfrac{x^2}{2} + \dfrac{y^2}{4}$, or $1 = \dfrac{x^2}{16} + \dfrac{y^2}{32}$.

Major axis: $2\sqrt{32} = 8\sqrt{2}$

Minor axis: $2\sqrt{16} = 8$

$c^2 = 32 - 16 = 16, c = 4$

Foci: $(0, \pm 4, 8)$

61. If (x, y, z) is on the surface, then

$$(y + 2)^2 = x^2 + (y - 2)^2 + z^2$$

$$y^2 + 4y + 4 = x^2 + y^2 - 4y + 4 + z^2$$

$$x^2 + z^2 = 8y$$

Elliptic paraboloid

Traces parallel to xz-plane are circles.

63. $\dfrac{x^2}{3963^2} + \dfrac{y^2}{3963^2} + \dfrac{z^2}{3942^2} = 1$

65. $z = \dfrac{y^2}{b^2} - \dfrac{x^2}{a^2}, z = bx + ay$

$$bx + ay = \frac{y^2}{b^2} - \frac{x^2}{a^2}$$

$$\frac{1}{a^2}\left(x^2 + a^2bx + \frac{a^4b^2}{4}\right) = \frac{1}{b^2}\left(y^2 - ab^2y + \frac{a^2b^4}{4}\right)$$

$$\frac{\left(x + \dfrac{a^2b}{2}\right)^2}{a^2} = \frac{\left(y - \dfrac{ab^2}{2}\right)^2}{b^2}$$

$$y = \pm\frac{b}{a}\left(x + \frac{a^2b}{2}\right) + \frac{ab^2}{2}$$

Letting $x = at$, you obtain the two intersecting lines
$x = at, y = -bt, z = 0$ and $x = at, y = bt + ab^2$
$z = 2abt + a^2b^2$.

67. The Klein bottle *does not* have both an "inside" and an "outside." It is formed by inserting the small open end through the side of the bottle and making it contiguous with the top of the bottle.

Section 10.7 Cylindrical and Spherical Coordinates

1. $(5, 0, 2)$, cylindrical
$x = 5 \cos 0 = 5$
$y = 5 \sin 0 = 0$
$z = 2$
$(5, 0, 2)$, rectangular

3. $\left(2, \dfrac{\pi}{3}, 2\right)$, cylindrical
$x = 2 \cos \dfrac{\pi}{3} = 1$
$y = 2 \sin \dfrac{\pi}{3} = \sqrt{3}$
$z = 2$
$\left(1, \sqrt{3}, 2\right)$, rectangular

5. $\left(4, \dfrac{7\pi}{6}, 3\right)$, cylindrical
$x = 4 \cos \dfrac{7\pi}{6} = -2\sqrt{3}$
$y = 4 \sin \dfrac{7\pi}{6} = -2$
$z = 3$
$\left(-2\sqrt{3}, -2, 3\right)$, rectangular

7. (0, 5, 1), rectangular

$r = \sqrt{(0)^2 + (5)^2} = 5$

$\theta = \arctan\dfrac{5}{0} = \dfrac{\pi}{2}$

$z = 1$

$\left(5, \dfrac{\pi}{2}, 1\right)$, cylindrical

9. $\left(1, \sqrt{3}, 4\right)$, rectangular

$r = \sqrt{1^2 + \left(\sqrt{3}\right)^2} = 2$

$\theta = \arctan\sqrt{3} = \dfrac{\pi}{3}$

$z = 4$

$\left(2, \dfrac{\pi}{3}, 4\right)$, cylindrical

11. (2, −2, −4), rectangular

$r = \sqrt{2^2 + (-2)^2} = 2\sqrt{2}$

$\theta = \arctan(-1) = -\dfrac{\pi}{4}$

$z = -4$

$\left(2\sqrt{2}, \dfrac{-\pi}{4}, -4\right)$, cylindrical

13. $x^2 + y^2 + z^2 = 10$ rectangular equation

$r^2 + z^2 = 10$ cylindrical equation

15. $y = x^2$ rectangular equation

$r \sin\theta = (r\cos\theta)^2$

$\sin\theta = r\cos^2\theta$

$r = \sec\theta \cdot \tan\theta$ cylindrical equation

17. $r = 2$

$\sqrt{x^2 + y^2} = 2$

$x^2 + y^2 = 4$

19. $\theta = \dfrac{\pi}{6}$

$\tan\dfrac{\pi}{6} = \dfrac{y}{x}$

$\dfrac{1}{\sqrt{3}} = \dfrac{y}{x}$

$x = \sqrt{3}y$

$x - \sqrt{3}y = 0$

21. $r = 2\sin\theta$

$r^2 = 2r\sin\theta$

$x^2 + y^2 = 2y$

$x^2 + y^2 - 2y = 0$

$x^2 + (y - 1)^2 = 1$

23. $r^2 + z^2 = 4$

$x^2 + y^2 + z^2 = 4$

25. (4, 0, 0), rectangular

$\rho = \sqrt{4^2 + 0^2 + 0^2} = 4$

$\theta = \arctan 0 = 0$

$\phi = \arccos 0 = \dfrac{\pi}{2}$

$\left(4, 0, \dfrac{\pi}{2}\right)$, spherical

27. $\left(-2, 2\sqrt{3}, 4\right)$, rectangular

$\rho = \sqrt{(-2)^2 + \left(2\sqrt{3}\right)^2 + 4^2} = 4\sqrt{2}$

$\theta = \arctan\left(-\sqrt{3}\right) = \dfrac{2\pi}{3}$

$\phi = \arccos\dfrac{1}{\sqrt{2}} = \dfrac{\pi}{4}$

$\left(4\sqrt{2}, \dfrac{2\pi}{3}, \dfrac{\pi}{4}\right)$, spherical

29. $\left(\sqrt{3}, 1, 2\sqrt{3}\right)$, rectangular

$\rho = \sqrt{3 + 1 + 12} = 4$

$\theta = \arctan\dfrac{1}{\sqrt{3}} = \dfrac{\pi}{6}$

$\phi = \arccos\dfrac{\sqrt{3}}{2} = \dfrac{\pi}{6}$

$\left(4, \dfrac{\pi}{6}, \dfrac{\pi}{6}\right)$, spherical

31. $\left(4, \dfrac{\pi}{6}, \dfrac{\pi}{4}\right)$, spherical

$x = 4 \sin \dfrac{\pi}{4} \cos \dfrac{\pi}{6} = \sqrt{6}$

$y = 4 \sin \dfrac{\pi}{4} \sin \dfrac{\pi}{6} = \sqrt{2}$

$z = 4 \cos \dfrac{\pi}{4} = 2\sqrt{2}$

$\left(\sqrt{6}, \sqrt{2}, 2\sqrt{2}\right)$, rectangular

33. $\left(12, \dfrac{-\pi}{4}, 0\right)$, spherical

$x = 12 \sin 0 \cos\left(\dfrac{-\pi}{4}\right) = 0$

$y = 12 \sin 0 \sin\left(\dfrac{-\pi}{4}\right) = 0$

$z = 12 \cos 0 = 12$

$(0, 0, 12)$, rectangular

35. $\left(5, \dfrac{\pi}{4}, \dfrac{3\pi}{4}\right)$, spherical

$x = 5 \sin \dfrac{3\pi}{4} \cos \dfrac{\pi}{4} = \dfrac{5}{2}$

$y = 5 \sin \dfrac{3\pi}{4} \sin \dfrac{\pi}{4} = \dfrac{5}{2}$

$z = 5 \cos \dfrac{3\pi}{4} = -\dfrac{5\sqrt{2}}{2}$

$\left(\dfrac{5}{2}, \dfrac{5}{2}, -\dfrac{5\sqrt{2}}{2}\right)$, rectangular

37. (a) Programs will vary.

(b) $(x, y, z) = (3, -4, 2)$

$(\rho, \theta, \phi) \approx (5.385, -0.927, 1.190)$

39. $x^2 + y^2 + z^2 = 36$ rectangular equation

$\rho^2 = 36$ spherical equation

41. $x^2 + y^2 = 9$ rectangular equation

$\rho^2 \sin^2 \phi \cos^2 \theta + \rho^2 \sin^2 \phi \sin^2 \theta = 9$

$\rho^2 \sin^2 \phi = 9$

$\rho \sin \phi = 3$

$\rho = 3 \csc \phi$ spherical equation

43. $\rho = 2$

$x^2 + y^2 + z^2 = 4$

45. $\phi = \dfrac{\pi}{6}$

$\cos \phi = \dfrac{z}{\sqrt{x^2 + y^2 + z^2}}$

$\dfrac{\sqrt{3}}{2} = \dfrac{z}{\sqrt{x^2 + y^2 + z^2}}$

$\dfrac{3}{4} = \dfrac{z^2}{x^2 + y^2 + z^2}$

$3x^2 + 3y^2 - z^2 = 0, \ z \geq 0$

47. $\rho = 4 \cos \phi$

$\sqrt{x^2 + y^2 + z^2} = \dfrac{4z}{\sqrt{x^2 + y^2 + z^2}}$

$x^2 + y^2 + z^2 - 4z = 0$

$x^2 + y^2 + (z - 2)^2 = 4, \ z \geq 0$

49. $\rho = \csc \phi$

$\rho \sin \phi = 1$

$\sqrt{x^2 + y^2} = 1$

$x^2 + y^2 = 1$

51. $\left(4, \dfrac{\pi}{4}, 0\right)$, cylindrical

$\rho = \sqrt{4^2 + 0^2} = 4$

$\theta = \dfrac{\pi}{4}$

$\phi = \arccos 0 = \dfrac{\pi}{2}$

$\left(4, \dfrac{\pi}{4}, \dfrac{\pi}{2}\right)$, spherical

53. $\left(4, \dfrac{\pi}{2}, 4\right)$, cylindrical

$\rho = \sqrt{4^2 + 4^2} = 4\sqrt{2}$

$\theta = \dfrac{\pi}{2}$

$\phi = \arccos\left(\dfrac{4}{4\sqrt{2}}\right) = \dfrac{\pi}{4}$

$\left(4\sqrt{2}, \dfrac{\pi}{2}, \dfrac{\pi}{4}\right)$, spherical

55. $\left(4, \dfrac{-\pi}{6}, 6\right)$, cylindrical

$\rho = \sqrt{4^2 + 6^2} = 2\sqrt{13}$

$\theta = \dfrac{-\pi}{6}$

$\phi = \arccos \dfrac{3}{\sqrt{13}}$

$\left(2\sqrt{13}, \dfrac{-\pi}{6}, \arccos \dfrac{3}{\sqrt{13}}\right),$
spherical

57. $(12, \pi, 5)$, cylindrical

$\rho = \sqrt{12^2 + 5^2} = 13$

$\theta = \pi$

$\phi = \arccos \dfrac{5}{13}$

$\left(13, \pi, \arccos \dfrac{5}{13}\right)$, spherical

59. $\left(10, \dfrac{\pi}{6}, \dfrac{\pi}{2}\right)$, spherical

$r = 10 \sin \dfrac{\pi}{2} = 10$

$\theta = \dfrac{\pi}{6}$

$z = 10 \cos \dfrac{\pi}{2} = 0$

$\left(10, \dfrac{\pi}{6}, 0\right)$, cylindrical

61. $\left(36, \pi, \dfrac{\pi}{2}\right)$, spherical

$r = \rho \sin \phi = 36 \sin \dfrac{\pi}{2} = 36$

$\theta = \pi$

$z = \rho \cos \phi = 36 \cos \dfrac{\pi}{2} = 0$

$(36, \pi, 0)$, cylindrical

63. $\left(6, -\dfrac{\pi}{6}, \dfrac{\pi}{3}\right)$, spherical

$r = 6 \sin \dfrac{\pi}{3} = 3\sqrt{3}$

$\theta = -\dfrac{\pi}{6}$

$z = 6 \cos \dfrac{\pi}{3} = 3$

$\left(3\sqrt{3}, -\dfrac{\pi}{6}, 3\right)$, cylindrical

65. $\left(8, \dfrac{7\pi}{6}, \dfrac{\pi}{6}\right)$, spherical

$r = 8 \sin \dfrac{\pi}{6} = 4$

$\theta = \dfrac{7\pi}{6}$

$z = 8 \cos \dfrac{\pi}{6} = \dfrac{8\sqrt{3}}{2}$

$\left(4, \dfrac{7\pi}{6}, 4\sqrt{3}\right)$, cylindrical

Rectangular	*Cylindrical*	*Spherical*
67. $(4, 6, 3)$	$(7.211, 0.983, 3)$	$(7.810, 0.983, 1.177)$
69. $(4.698, 1.710, 8)$	$\left(5, \dfrac{\pi}{9}, 8\right)$	$(9.434, 0.349, 0.559)$
71. $(-7.071, 12.247, 14.142)$	$(14.142, 2.094, 14.142)$	$\left(20, \dfrac{2\pi}{3}, \dfrac{\pi}{4}\right)$
73. $(3, -2, 2)$	$(3.606, -0.588, 2)$	$(4.123, -0.588, 1.064)$
75. $\left(\dfrac{5}{2}, \dfrac{4}{3}, \dfrac{-3}{2}\right)$	$(2.833, 0.490, -1.5)$	$(3.206, 0.490, 2.058)$
77. $(-3.536, 3.536, -5)$	$\left(5, \dfrac{3\pi}{4}, -5\right)$	$(7.071, 2.356, 2.356)$
79. $(2.804, -2.095, 6)$	$(-3.5, 2.5, 6)$	$(6.946, 5.642, 0.528)$

[Note: Use the cylindrical coordinates $(3.5, 5.642, 6)$]

81. $r = 5$
Cylinder
Matches graph (d)

83. $\rho = 5$
Sphere
Matches graph (c)

85. $r^2 = z, x^2 + y^2 = z$
Paraboloid
Matches graph (f)

87. Rectangular to cylindrical: $r^2 = x^2 + y^2$

$\tan \theta = \dfrac{y}{x}$

$z = z$

Cylindrical to rectangular: $x = r \cos \theta$

$y = r \sin \theta$

$z = z$

89. Rectangular to spherical: $\rho^2 = x^2 + y^2 + z^2$

$\tan \theta = \dfrac{y}{x}$

$\phi = \arccos \left(\dfrac{z}{\sqrt{x^2 + y^2 + z^2}}\right)$

Spherical to rectangular: $x = \rho \sin \phi \cos \theta$

$y = \rho \sin \phi \sin \theta$

$z = \rho \cos \phi$

91. $x^2 + y^2 + z^2 = 16$

(a) $r^2 + z^2 = 16$

(b) $\rho^2 = 16, \rho = 4$

93. $x^2 + y^2 + z^2 - 2z = 0$

(a) $r^2 + z^2 - 2z = 0, r^2 + (z - 1)^2 = 1$

(b) $\rho^2 - 2\rho \cos \phi = 0, \rho(\rho - 2 \cos \phi) = 0,$

$\rho = 2 \cos \phi$

95. $x^2 + y^2 = 4y$

(a) $r^2 = 4r \sin \theta, r = 4 \sin \theta$

(b) $\rho^2 \sin^2 \phi = 4\rho \sin \phi \sin \theta,$

$\rho \sin \phi(\rho \sin \phi - 4 \sin \theta) = 0,$

$\rho = \dfrac{4 \sin \theta}{\sin \phi}, \rho = 4 \sin \theta \csc \phi$

97. $x^2 - y^2 = 9$

(a) $r^2 \cos^2 \theta - r^2 \sin^2 \theta = 9,$

$r^2 = \dfrac{9}{\cos^2 \theta - \sin^2 \theta}$

(b) $\rho^2 \sin^2 \phi \cos^2 \theta - \rho^2 \sin^2 \phi \sin^2 \theta = 9,$

$\rho^2 \sin^2 \phi = \dfrac{9}{\cos^2 \theta - \sin^2 \theta},$

$\rho^2 = \dfrac{9 \csc^2 \phi}{\cos^2 \theta - \sin^2 \theta}$

99. $0 \leq \theta \leq \dfrac{\pi}{2}$

$0 \leq r \leq 2$

$0 \leq z \leq 4$

101. $0 \leq \theta \leq 2\pi$

$0 \leq r \leq a$

$r \leq z \leq a$

103. $0 \leq \theta \leq 2\pi$

$0 \leq \phi \leq \dfrac{\pi}{6}$

$0 \leq \rho \leq a \sec \phi$

105. Rectangular

$0 \leq x \leq 10$

$0 \leq y \leq 10$

$0 \leq z \leq 10$

107. Spherical

$4 \leq \rho \leq 6$

109. $z = \sin \theta, r = 1$

$z = \dfrac{y}{r} = \dfrac{y}{1} = y$

The curve of intersection is the ellipse formed by the intersection of the plane $z = y$ and the cylinder $r = 1$.

Review Exercises for Chapter 10

1. $P = (1, 2), Q = (4, 1), R = (5, 4)$

(a) $\mathbf{u} = \overrightarrow{PQ} = \langle 3, -1 \rangle = 3\mathbf{i} - \mathbf{j},$

$\mathbf{v} = \overrightarrow{PR} = \langle 4, 2 \rangle = 4\mathbf{i} + 2\mathbf{j}$

(b) $\|\mathbf{v}\| = \sqrt{4^2 + 2^2} = 2\sqrt{5}$

(c) $2\mathbf{u} + \mathbf{v} = \langle 6, -2 \rangle + \langle 4, 2 \rangle = \langle 10, 0 \rangle = 10\mathbf{i}$

3. $\mathbf{v} = \|\mathbf{v}\| \cos \theta \, \mathbf{i} + \|\mathbf{v}\| \sin \theta \, \mathbf{j} = 8 \cos 120° \, \mathbf{i} + 8 \sin 120° \, \mathbf{j}$

$= -4\mathbf{i} + 4\sqrt{3}\mathbf{j}$

5. $120 \cos \theta = 100$

$$\theta = \arccos\left(\frac{5}{6}\right)$$

$$\tan \theta = \frac{2}{y} \implies y = \frac{2}{\tan \theta}$$

$$y = \frac{2}{\tan[\arccos(5/6)]} = \frac{2}{\sqrt{11}/5} = \frac{10}{\sqrt{11}} \approx 3.015 \text{ ft}$$

7. $z = 0,\ y = 4,\ x = -5$: $(-5, 4, 0)$

9. Looking down from the positive x-axis towards the yz-plane, the point is either in the first quadrant ($y > 0,\ z > 0$) or in the third quadrant ($y < 0,\ z < 0$). The x-coordinate can be any number.

11. $(x - 3)^2 + (y + 2)^2 + (z - 6)^2 = \left(\frac{15}{2}\right)^2$

13. $(x^2 - 4x + 4) + (y^2 - 6y + 9) + z^2 = -4 + 4 + 9$

$(x - 2)^2 + (y - 3)^2 + z^2 = 9$

Center: $(2, 3, 0)$

Radius: 3

15. $\mathbf{v} = \langle 4 - 2, 4 + 1, -7 - 3 \rangle = \langle 2, 5, -10 \rangle$

17. $\mathbf{v} = \langle -1 - 3, 6 - 4, 9 + 1 \rangle = \langle -4, 2, 10 \rangle$

$\mathbf{w} = \langle 5 - 3, 3 - 4, -6 + 1 \rangle = \langle 2, -1, -5 \rangle$

Since $-2\mathbf{w} = \mathbf{v}$, the points lie in a straight line.

19. Unit vector: $\dfrac{\mathbf{u}}{\|\mathbf{u}\|} = \dfrac{\langle 2, 3, 5 \rangle}{\sqrt{38}} = \left\langle \dfrac{2}{\sqrt{38}}, \dfrac{3}{\sqrt{38}}, \dfrac{5}{\sqrt{38}} \right\rangle$

21. $P = (5, 0, 0),\ Q = (4, 4, 0),\ R = (2, 0, 6)$

(a) $\mathbf{u} = \overrightarrow{PQ} = \langle -1, 4, 0 \rangle = -\mathbf{i} + 4\mathbf{j}$,

$\mathbf{v} = \overrightarrow{PR} = \langle -3, 0, 6 \rangle = -3\mathbf{i} + 6\mathbf{k}$

(b) $\mathbf{u} \cdot \mathbf{v} = (-1)(-3) + 4(0) + 0(6) = 3$

(c) $\mathbf{v} \cdot \mathbf{v} = 9 + 36 = 45$

23. $\mathbf{u} = \langle 7, -2, 3 \rangle,\ \mathbf{v} = \langle -1, 4, 5 \rangle$

Since $\mathbf{u} \cdot \mathbf{v} = 0$, the vectors are orthogonal.

25. $\mathbf{u} = 5\left(\cos\frac{3\pi}{4}\mathbf{i} + \sin\frac{3\pi}{4}\mathbf{j}\right) = \frac{5\sqrt{2}}{2}[-\mathbf{i} + \mathbf{j}]$

$\mathbf{v} = 2\left(\cos\frac{2\pi}{3}\mathbf{i} + \sin\frac{2\pi}{3}\mathbf{j}\right) = -\mathbf{i} + \sqrt{3}\mathbf{j}$

$\mathbf{u} \cdot \mathbf{v} = \frac{5\sqrt{2}}{2}\left(1 + \sqrt{3}\right)$

$\|\mathbf{u}\| = 5$

$\|\mathbf{v}\| = 2$

$\cos \theta = \frac{|\mathbf{u} \cdot \mathbf{v}|}{\|\mathbf{u}\|\,\|\mathbf{v}\|} = \frac{\left(5\sqrt{2}/2\right)\left(1 + \sqrt{3}\right)}{5(2)} = \frac{\sqrt{2} + \sqrt{6}}{4}$

$\theta = \arccos\frac{\sqrt{2} + \sqrt{6}}{4} = 15°$ or $\frac{3\pi}{4} - \frac{2\pi}{3} = \frac{\pi}{12}$ or $15°$

27. $\mathbf{u} = \langle 10, -5, 15 \rangle,\ \mathbf{v} = \langle -2, 1, -3 \rangle$

$\mathbf{u} = -5\mathbf{v} \implies \mathbf{u}$ is parallel to $\mathbf{v}$ and in the opposite direction.

$\theta = \pi$

29. There are many correct answers. For example: $\mathbf{v} = \pm\langle 6, -5, 0\rangle$.

In Exercises 31–39, $\mathbf{u} = \langle 3, -2, 1\rangle$, $\mathbf{v} = \langle 2, -4, -3\rangle$, $\mathbf{w} = \langle -1, 2, 2\rangle$.

31. $\mathbf{u} \cdot \mathbf{u} = 3(3) + (-2)(-2) + (1)(1)$

$\qquad = 14 = \left(\sqrt{14}\right)^2 = \|\mathbf{u}\|^2$

33. $\text{proj}_{\mathbf{u}}\mathbf{w} = \left(\dfrac{\mathbf{u} \cdot \mathbf{w}}{\|\mathbf{u}\|^2}\right)\mathbf{u}$

$\qquad = -\dfrac{5}{14}\langle 3, -2, 1\rangle$

$\qquad = \left\langle -\dfrac{15}{14}, \dfrac{10}{14}, -\dfrac{5}{14}\right\rangle$

$\qquad = \left\langle -\dfrac{15}{14}, \dfrac{5}{7}, -\dfrac{5}{14}\right\rangle$

35. $\mathbf{n} = \mathbf{v} \times \mathbf{w} = \begin{vmatrix} \mathbf{i} & \mathbf{j} & \mathbf{k} \\ 2 & -4 & -3 \\ -1 & 2 & 2 \end{vmatrix} = -2\mathbf{i} - \mathbf{j}$

$\|\mathbf{n}\| = \sqrt{5}$

$\dfrac{\mathbf{n}}{\|\mathbf{n}\|} = \dfrac{1}{\sqrt{5}}(-2\mathbf{i} - \mathbf{j})$, unit vector or $\dfrac{1}{\sqrt{5}}(2\mathbf{i} + \mathbf{j})$

37. $V = |\mathbf{u} \cdot (\mathbf{v} \times \mathbf{w})|$

$\qquad = |\langle 3, -2, 1\rangle \cdot \langle -2, -1, 0\rangle| = |-4| = 4$

39. Area parallelogram $= \|\mathbf{u} \times \mathbf{v}\| = \|\langle 10, 11, -8\rangle\| = \sqrt{10^2 + 11^2 + (-8)^2}$ (See Exercises 36, 38)

$\qquad\qquad\qquad = \sqrt{285}$

41. $\mathbf{F} = c(\cos 20°\mathbf{j} + \sin 20°\mathbf{k})$

$\overrightarrow{PQ} = 2\mathbf{k}$

$\overrightarrow{PQ} \times \mathbf{F} = \begin{vmatrix} \mathbf{i} & \mathbf{j} & \mathbf{k} \\ 0 & 0 & 2 \\ 0 & c\cos 20° & c\sin 20° \end{vmatrix} = -2c\cos 20°\mathbf{i}$

$200 = \|\overrightarrow{PQ} \times \mathbf{F}\| = 2c\cos 20°$

$c = \dfrac{100}{\cos 20°}$

$\mathbf{F} = \dfrac{100}{\cos 20°}(\cos 20°\mathbf{j} + \sin 20°\mathbf{k}) = 100(\mathbf{j} + \tan 20°\mathbf{k})$

$\|\mathbf{F}\| = 100\sqrt{1 + \tan^2 20°} = 100\sec 20° \approx 106.4 \text{ lb}$

43. $\mathbf{v} = \mathbf{j}$

(a) $x = 1,\ y = 2 + t,\ z = 3$

(b) None

45. $3x - 3y - 7z = -4,\ x - y + 2z = 3$

Solving simultaneously, we have $z = 1$. Substituting $z = 1$ into the second equation we have $y = x - 1$. Substituting for x in this equation we obtain two points on the line of intersection, $(0, -1, 1)$, $(1, 0, 1)$. The direction vector of the line of intersection is $\mathbf{v} = \mathbf{i} + \mathbf{j}$.

(a) $x = t,\ y = -1 + t,\ z = 1$

(b) $x = y + 1,\ z = 1$

47. The two lines are parallel as they have the same direction numbers, $-2, 1, 1$. Therefore, a vector parallel to the plane is $\mathbf{v} = -2\mathbf{i} + \mathbf{j} + \mathbf{k}$. A point on the first line is $(1, 0, -1)$ and a point on the second line is $(-1, 1, 2)$. The vector $\mathbf{u} = 2\mathbf{i} - \mathbf{j} - 3\mathbf{k}$ connecting these two points is also parallel to the plane. Therefore, a normal to the plane is

$$\mathbf{v} \times \mathbf{u} = \begin{vmatrix} \mathbf{i} & \mathbf{j} & \mathbf{k} \\ -2 & 1 & 1 \\ 2 & -1 & -3 \end{vmatrix}$$

$$= -2\mathbf{i} - 4\mathbf{j} = -2(\mathbf{i} + 2\mathbf{j}).$$

Equation of the plane: $(x - 1) + 2y = 0$

$$x + 2y = 1$$

49. $Q(1, 0, 2)$ point

$2x - 3y + 6z = 6$

A point P on the plane is $(3, 0, 0)$.

$$\overrightarrow{PQ} = \langle -2, 0, 2 \rangle$$

$$\mathbf{n} = \langle 2, -3, 6 \rangle \quad \text{normal to plane}$$

$$D = \frac{|\overrightarrow{PQ} \cdot \mathbf{n}|}{\|\mathbf{n}\|} = \frac{8}{7}$$

51. $Q(3, -2, 4)$ point

$P(5, 0, 0)$ point on plane

$\mathbf{n} = \langle 2, -5, 1 \rangle$ normal to plane

$$\overrightarrow{PQ} = \langle -2, -2, 4 \rangle$$

$$D = \frac{|\overrightarrow{PQ} \cdot \mathbf{n}|}{\|\mathbf{n}\|} = \frac{10}{\sqrt{30}} = \frac{\sqrt{30}}{3}$$

53. $x + 2y + 3z = 6$

Plane

Intercepts: $(6, 0, 0)$, $(0, 3, 0)$, $(0, 0, 2)$

55. $y = \frac{1}{2}z$

Plane with rulings parallel to the x-axis

57. $\dfrac{x^2}{16} + \dfrac{y^2}{9} + z^2 = 1$

Ellipsoid

xy-trace: $\dfrac{x^2}{16} + \dfrac{y^2}{9} = 1$

xz-trace: $\dfrac{x^2}{16} + z^2 = 1$

yz-trace: $\dfrac{y^2}{9} + z^2 = 1$

59. $\dfrac{x^2}{16} - \dfrac{y^2}{9} + z^2 = -1$

$\dfrac{y^2}{9} - \dfrac{x^2}{16} - z^2 = 1$

Hyperboloid of two sheets

xy-trace: $\dfrac{y^2}{9} - \dfrac{x^2}{16} = 1$

xz-trace: None

yz-trace: $\dfrac{y^2}{9} - z^2 = 1$

61. (a)
$$x^2 + y^2 = [r(z)]^2$$
$$= \left[\sqrt{2(z-1)}\right]^2$$
$$x^2 + y^2 - 2z + 2 = 0$$

(b) $V = 2\pi \int_0^2 x\left[3 - \left(\frac{1}{2}x^2 + 1\right)\right] dx$

$\quad = 2\pi \int_0^2 \left(2x - \frac{1}{2}x^3\right) dx$

$\quad = 2\pi\left[x^2 - \frac{x^4}{8}\right]_0^2$

$\quad = 4\pi \approx 12.6 \text{ cm}^3$

(c) $V = 2\pi \int_{1/2}^2 x\left[3 - \left(\frac{1}{2}x^2 + 1\right)\right] dx$

$\quad = 2\pi \int_{1/2}^2 \left(2x - \frac{1}{2}x^3\right) dx$

$\quad = 2\pi\left[x^2 - \frac{x^4}{8}\right]_{1/2}^2$

$\quad = 4\pi - \frac{31\pi}{64} = \frac{225\pi}{64} \approx 11.04 \text{ cm}^3$

63. $\left(-2\sqrt{2}, 2\sqrt{2}, 2\right)$, rectangular

(a) $r = \sqrt{\left(-2\sqrt{2}\right)^2 + \left(2\sqrt{2}\right)^2} = 4, \ \theta = \arctan(-1) = \frac{3\pi}{4}, \ z = 2, \ \left(4, \frac{3\pi}{4}, 2\right),$ cylindrical

(b) $\rho = \sqrt{\left(-2\sqrt{2}\right)^2 + \left(2\sqrt{2}\right)^2 + (2)^2} = 2\sqrt{5}, \ \theta = \frac{3\pi}{4}, \ \phi = \arccos\frac{2}{2\sqrt{5}} = \arccos\frac{1}{\sqrt{5}}, \ \left(2\sqrt{5}, \frac{3\pi}{4}, \arccos\frac{\sqrt{5}}{5}\right),$ spherical

65. $\left(100, -\frac{\pi}{6}, 50\right)$, cylindrical

$\rho = \sqrt{100^2 + 50^2} = 50\sqrt{5}$

$\theta = -\frac{\pi}{6}$

$\phi = \arccos\left(\frac{50}{50\sqrt{5}}\right) = \arccos\left(\frac{1}{\sqrt{5}}\right) \approx 63.4° \text{ or } 1.107$

$\left(50\sqrt{5}, -\frac{\pi}{6}, 63.4°\right),$ spherical

67. $\left(25, -\frac{\pi}{4}, \frac{3\pi}{4}\right)$, spherical

$r^2 = \left(25 \sin\left(\frac{3\pi}{4}\right)\right)^2 \Rightarrow r = 25\frac{\sqrt{2}}{2}$

$\theta = -\frac{\pi}{4}$

$z = \rho \cos\phi = 25 \cos\frac{3\pi}{4} = -25\frac{\sqrt{2}}{2}$

$\left(25\frac{\sqrt{2}}{2}, -\frac{\pi}{4}, -\frac{25\sqrt{2}}{2}\right),$ cylindrical

69. $x^2 - y^2 = 2z$

(a) Cylindrical: $r^2 \cos^2\theta - r^2 \sin^2\theta = 2z, \ r^2 \cos 2\theta = 2z$

(b) Spherical: $\rho^2 \sin^2\phi \cos^2\theta - \rho^2 \sin^2\phi \sin^2\theta = 2\rho \cos\phi, \ \rho \sin^2\phi \cos 2\theta - 2\cos\phi = 0, \ \rho = 2 \sec 2\theta \cos\phi \csc^2\phi$

Problem Solving for Chapter 10

1.
$$\mathbf{a} + \mathbf{b} + \mathbf{c} = 0$$
$$\mathbf{b} \times (\mathbf{a} + \mathbf{b} + \mathbf{c}) = 0$$
$$(\mathbf{b} \times \mathbf{a}) + (\mathbf{b} \times \mathbf{c}) = 0$$
$$\|\mathbf{a} \times \mathbf{b}\| = \|\mathbf{b} \times \mathbf{c}\|$$
$$\|\mathbf{b} \times \mathbf{c}\| = \|\mathbf{b}\| \, \|\mathbf{c}\| \sin A$$
$$\|\mathbf{a} \times \mathbf{b}\| = \|\mathbf{a}\| \, \|\mathbf{b}\| \sin C$$

Then,
$$\frac{\sin A}{\|\mathbf{a}\|} = \frac{\|\mathbf{b} \times \mathbf{c}\|}{\|\mathbf{a}\| \, \|\mathbf{b}\| \, \|\mathbf{c}\|}$$
$$= \frac{\|\mathbf{a} \times \mathbf{b}\|}{\|\mathbf{a}\| \, \|\mathbf{b}\| \, \|\mathbf{c}\|}$$
$$= \frac{\sin C}{\|\mathbf{c}\|}.$$

The other case, $\dfrac{\sin A}{\|\mathbf{a}\|} = \dfrac{\sin B}{\|\mathbf{b}\|}$ is similar.

3. Label the figure as indicated.

From the figure, you see that
$$\overrightarrow{SP} = \frac{1}{2}\mathbf{a} - \frac{1}{2}\mathbf{b} = \overrightarrow{RQ} \text{ and}$$
$$\overrightarrow{SR} = \frac{1}{2}\mathbf{a} + \frac{1}{2}\mathbf{b} = \overrightarrow{PQ}.$$

Since $\overrightarrow{SP} = \overrightarrow{RQ}$ and $\overrightarrow{SR} = \overrightarrow{PQ}$, *PSRQ* is a parallelogram.

5. (a) $\mathbf{u} = \langle 0, 1, 1 \rangle$ direction vector of line determined by P_1 and P_2.

$$D = \frac{\|\overrightarrow{P_1Q} \times \mathbf{u}\|}{\|\mathbf{u}\|}$$
$$= \frac{\|\langle 2, 0, -1 \rangle \times \langle 0, 1, 1 \rangle\|}{\sqrt{2}}$$
$$= \frac{\|\langle 1, -2, 2 \rangle\|}{\sqrt{2}} = \frac{3}{\sqrt{2}} = \frac{3\sqrt{2}}{2}$$

(b) The shortest distance to the line **segment** is $\|P_1Q\| = \|\langle 2, 0, -1 \rangle\| = \sqrt{5}$.

7. (a) $V = \pi \displaystyle\int_0^1 \left(\sqrt{z}\right)^2 dz = \left[\pi \dfrac{z^2}{2}\right]_0^1 = \dfrac{1}{2}\pi$

Note: $\dfrac{1}{2}$(base)(altitude) $= \dfrac{1}{2}\pi(1) = \dfrac{1}{2}\pi$

(b) $\dfrac{x^2}{a^2} + \dfrac{y^2}{b^2} = z$: (slice at $z = c$)

$$\frac{x^2}{\left(\sqrt{c}a\right)^2} + \frac{y^2}{\left(\sqrt{c}b\right)^2} = 1$$

At $z = c$, figure is ellipse of area

$$\pi\left(\sqrt{c}a\right)\left(\sqrt{c}b\right) = \pi abc.$$

$$V = \int_0^k \pi abc \cdot dc = \left[\frac{\pi abc^2}{2}\right]_0^k = \frac{\pi abk^2}{2}$$

(c) $V = \dfrac{1}{2}(\pi abk)k = \dfrac{1}{2}$(base)(height)

9. (a) $\rho = 2 \sin \phi$

Torus

(b) $\rho = 2 \cos \phi$

Sphere

11. From Exercise 64, Section 10.4, $(\mathbf{u} \times \mathbf{v}) \times (\mathbf{w} \times \mathbf{z}) = [(\mathbf{u} \times \mathbf{v}) \cdot \mathbf{z}]\mathbf{w} - [(\mathbf{u} \times \mathbf{v}) \cdot \mathbf{w}]\mathbf{z}.$

13. (a) $\mathbf{u} = \|\mathbf{u}\|(\cos 0\,\mathbf{i} + \sin 0\,\mathbf{j}) = \|\mathbf{u}\|\mathbf{i}$

Downward force $\mathbf{w} = -\mathbf{j}$

$\mathbf{T} = \|\mathbf{T}\|(\cos(90° + \theta)\mathbf{i} + \sin(90° + \theta)\mathbf{j})$

$\quad = \|\mathbf{T}\|(-\sin\theta\,\mathbf{i} + \cos\theta\,\mathbf{j})$

$\mathbf{0} = \mathbf{u} + \mathbf{w} + \mathbf{T} = \|\mathbf{u}\|\mathbf{i} - \mathbf{j} + \|\mathbf{T}\|(-\sin\theta\,\mathbf{i} + \cos\theta\,\mathbf{j})$

$\|\mathbf{u}\| = \sin\theta\,\|\mathbf{T}\|$

$1 = \cos\theta\,\|\mathbf{T}\|$

If $\theta = 30°$, $\|\mathbf{u}\| = (1/2)\|\mathbf{T}\|$ and $1 = \left(\sqrt{3}/2\right)\|\mathbf{T}\|$

$\Rightarrow \|\mathbf{T}\| = \dfrac{2}{\sqrt{3}} \approx 1.1547 \text{ lb}$

and

$\|\mathbf{u}\| = \dfrac{1}{2}\left(\dfrac{2}{\sqrt{3}}\right) \approx 0.5774 \text{ lb}$

(b) From part (a), $\|\mathbf{u}\| = \tan\theta$ and $\|\mathbf{T}\| = \sec\theta$.

Domain: $0 \le \theta \le 90°$

(c)

θ	0°	10°	20°	30°	40°	50°	60°
T	1	1.0154	1.0642	1.1547	1.3054	1.5557	2
$\|\mathbf{u}\|$	0	0.1763	0.3640	0.5774	0.8391	1.1918	1.7321

(d)

(e) Both are increasing functions.

(f) $\displaystyle \lim_{\theta \to \pi/2^-} T = \infty$ and $\displaystyle \lim_{\theta \to \pi/2^-} \|\mathbf{u}\| = \infty$.

15. Let $\theta = \alpha - \beta$, the angle between $\mathbf{u}$ and $\mathbf{v}$. Then

$\sin(\alpha - \beta) = \dfrac{\|\mathbf{u} \times \mathbf{v}\|}{\|\mathbf{u}\|\,\|\mathbf{v}\|} = \dfrac{\|\mathbf{v} \times \mathbf{u}\|}{\|\mathbf{u}\|\,\|\mathbf{v}\|}.$

For $\mathbf{u} = \langle \cos\alpha, \sin\alpha, 0 \rangle$ and $\mathbf{v} = \langle \cos\beta, \sin\beta, 0 \rangle$, $\|\mathbf{u}\| = \|\mathbf{v}\| = 1$ and

$\mathbf{v} \times \mathbf{u} = \begin{vmatrix} \mathbf{i} & \mathbf{j} & \mathbf{k} \\ \cos\beta & \sin\beta & 0 \\ \cos\alpha & \sin\alpha & 0 \end{vmatrix} = (\sin\alpha\cos\beta - \cos\alpha\sin\beta)\mathbf{k}.$

Thus, $\sin(\alpha - \beta) = \|\mathbf{v} \times \mathbf{u}\| = \sin\alpha\cos\beta - \cos\alpha\sin\beta$.

17. From Theorem 10.13 and Theorem 10.7 (6) we have

$D = \dfrac{|\overrightarrow{PQ} \cdot \mathbf{n}|}{\|\mathbf{n}\|}$

$\quad = \dfrac{|\mathbf{w} \cdot (\mathbf{u} \times \mathbf{v})|}{\|\mathbf{u} \times \mathbf{v}\|} = \dfrac{|(\mathbf{u} \times \mathbf{v}) \cdot \mathbf{w}|}{\|\mathbf{u} \times \mathbf{v}\|} = \dfrac{|\mathbf{u} \cdot (\mathbf{v} \times \mathbf{w})|}{\|\mathbf{u} \times \mathbf{v}\|}.$

19. a_1, b_1, c_1, and a_2, b_2, c_2 are two sets of direction numbers for the same line. The line is parallel to both $\mathbf{u} = a_1\mathbf{i} + b_1\mathbf{j} + c_1\mathbf{k}$ and $\mathbf{v} = a_2\mathbf{i} + b_2\mathbf{j} + c_2\mathbf{k}$. Therefore, $\mathbf{u}$ and $\mathbf{v}$ are parallel, and there exists a scalar d such that $\mathbf{u} = d\mathbf{v}$, $a_1\mathbf{i} + b_1\mathbf{j} + c_1\mathbf{k} = d(a_2\mathbf{i} + b_2\mathbf{j} + c_2\mathbf{k})$, $a_1 = a_2 d$, $b_1 = b_2 d$, $c_1 = c_2 d$.